INORGANIC CHEMISTRY

INORGANIC CHEMISTRY

A MODERN INTRODUCTION

THERALD MOELLER
Arizona State University

A Wiley-Interscience Publication

JOHN WILEY & SONS

New York   Chichester   Brisbane   Toronto   Singapore

Library of Congress Cataloging in Publication Data:

Moeller, Therald.
   Inorganic chemistry, a modern introduction.

   "A Wiley-Interscience publication."
   Includes indexes.
   1. Chemistry, Inorganic.   I. Title.

QD151.2.M63   546   81-16455
ISBN 0-471-61230-8   AACR2

Printed in the United States of America

10 9 8 7 6 5 4 3 2

PREFACE

So much has been said and written about the "Renaissance of Inorganic Chemistry" during the past 35 to 40 years that it is no longer essential to dwell on this subject. Much less time and space have been devoted to the problems engendered in instruction in inorganic chemistry by the very magnitude and diversity of the subject; by the exponential increase in publication in every area of the discipline; by the inevitable specialization of teachers and practitioners in relatively narrow areas; and at least in the United States of America, by the belief among many non-inorganic chemists that both one semester of undergraduate instruction in the area is adequate and, that unlike organic and physical chemistry, instruction in inorganic chemistry beyond the freshman level should be elective.

What has been presented in so many junior-level, senior-level, and beginning graduate-level courses has lacked uniformity and has very often reflected to a large degree the instructor's own training and research interests. Coincident with this approach has been a strong emphasis on concept and theory, often with insufficient attention to factual aspects. A case in point involves the d-transition metal complexes where crystal-field, molecular-orbital, and ligand-field interpretations have provided such fascinating and useful interrelationships among properties, bonding, and molecular structure that an uninitiated student looking at almost any current journal in inorganic chemistry could be convinced that only this area is important. What a tragedy this could be in the sense of the overall technical and societal significance of, for example, nonmetal chemistry and the many voids existing in our knowledge of this area.

Fortunately, there is increasing evidence of concern and a developing belief that instruction dealing with breadth in inorganic chemistry is essential. The discipline is so broad and so varied that total coverage in a single volume is impossible. Textbooks have grown in size, and this one is no exception, and have either attempted encyclopedic treatments or have elected to cover only specialized subjects.

Since 1940, the author has been privileged to have witnessed the modern development of inorganic chemistry; to have known personally on an international basis many of the persons whose contributions to inorganic chemistry have been significant; to have taught inorganic chemistry to innumerable advanced undergraduate and graduate students; to have been deeply involved with graduate research involving both his own students and those of his distinguished colleagues; and to have had direct associations with the center of research and instruction at the University of Illinois at Urbana-Champaign, Inorganic Syntheses, and the Division of Inorganic Chemistry of the American Chemical Society. For all of these he is deeply grateful.

This textbook is directed primarily to the advanced undergraduate and the beginning graduate student. It is hoped that the more advanced graduate student, the instructor, and the person in research and industry may also find it useful.

The approach used is based on a definition and an implementation of that definition involving synthesis and evaluation of the product in terms of structure, bonding, and reactions, as outlined in Chapter 1. Following this introduction, atomic structure, the periodic classification, the chemical bond, molecular structure and overall stereochemistry, coordination chemistry in both the classical and current senses, and chemical reactions are presented. The treatment is modern, with hundreds of citations of the original and review literature, but only moderately mathematical. Throughout, familiarization with the literature of inorganic chemistry and recognition of contributions to the subject are stressed.

Prefaces commonly close with recognitions of contributions by specific individuals and requests for uninhibited comments by users and readers. To list all the persons to whom I have become indebted through the years for their help, counsel, and instruction is impossible. Nevertheless, my undying gratitude goes to former colleagues, to innumerable friends and acquaintances—both alive and deceased—and to the host of outstanding students with whom I have been associated. To all of these, thank you! Early suggestions by Mary L. Good relative to organization and atomic structure were appreciated. Final manuscript copy could never have been prepared without the efforts of two superb typists, Verna M. Campbell and Betty Landon—the best in my experience. Special recognition and gratitude are due my wife Ellyn, whose encouragement and faith have been and continue to be inspirations to me. These she has always offered in spite of the many demands occasioned by my profession. Finally, forgive me for saying that even though circumstance during the Great Depression channeled me into academic inorganic chemistry, no more totally rewarding and interesting a profession could possibly have come my way!

THERALD MOELLER

Tempe, Arizona
February 1982

CONTENTS

INORGANIC CHEMISTRY

CHAPTER ONE

AN INTRODUCTION TO MODERN INORGANIC CHEMISTRY

In 1952, the author introduced students to inorganic chemistry with the statement: "Inorganic chemistry is not general chemistry."[1] Although, as a consequence of the rather general practice of restricting discussion of inorganic topics to freshman courses in chemistry, such a statement was wholly appropriate at that time, it is no longer necessary. During the past three decades, interest and research in inorganic areas have increased so rapidly and diversified so extensively that inorganic chemistry has changed from a primarily factual branch of the science to a full-fledged and highly pervasive component that is a blend of fact with bonding and structure theory. Modern inorganic chemistry relies strongly on physical chemistry, but it relates increasingly to all other areas of the science from geochemistry and cosmochemistry through biological chemistry.

Modern inorganic chemistry has become so all inclusive that it is very difficult to define succinctly. Rather than avoiding this difficulty by merely saying, as some have done, that "inorganic chemistry is what inorganic chemists do," it seems better to offer a positive definition, such as: *Inorganic chemistry is the experimental investigation and theoretical interpretation of the properties and reactions of all the elements and of all their compounds, except the hydrocarbons and most of their derivatives.*

In a sense, this definition is a reversion to the broadly classical approach of considering all chemistry to be either inorganic or organic. No arbitrary line of demarcation can be established, however, since there is now, and fortunately for the progress of the entire science, necessary and desirable overlap through bioinorganic chemistry, organometallic chemistry, environmental chemistry, and so on. Analytical chemistry and physical chemistry, including of course theoretical chemistry, provide the tools, the methods, and the concepts and theories that allow for the evaluation and interpretation of properties of all chemical substances, irrespective of their compositions. Specific aspects of inorganic chemistry are found in geochemistry, cosmochemistry, solid-state chemistry, industrial chemistry, and biochemistry, and thus directly or indirectly in seemingly unrelated areas such as microbiology, molecular biology, mining and metallurgy, ceramic and glass technology, polymer production and utilization, energy sources, waste disposal, and a host of others.

Not every aspect of the subject can, or should be, treated in a general textbook on modern inorganic chemistry. The scope of inorganic chemistry which provides the framework for the treatment in this book is outlined in Table 1-1. Our primary task is to integrate, within this framework, both experimentally determined *fact* and theoretically developed *postulation* or *interpretation*, each in proper perspective. A complete and balanced understanding of modern inorganic chemistry is possible only when (1) factual information can be correlated, interpreted, and/or extended by application of appropriate concept or theory, and (2) conceptual or theoretical interpretation can be adequately tested and/or proved by experimental measurement. To neglect or to ignore either facet is to obtain an arbitrarily restricted and distorted view of modern inorganic chemistry.

It will become increasingly apparent as the subject is developed that there are many gaps in factual knowledge and many factual observations that are not in accord with or are incapable of being accounted for by existing theoretical approaches. The former problem can be solved by appropriate experimentation; the latter only by the continuing development of more sophisticated theoretical approaches. The ideal goal of complete agreement between fact and theory is not likely to be reached in inorganic chemistry because among all the elements and their innumerable compounds there exist major differences in properties that reflect differences in bonding and crystal and molecular structures. Even though the total number of isolable organic compounds exceeds the total

Table 1-1 The Scope of Modern Inorganic Chemistry

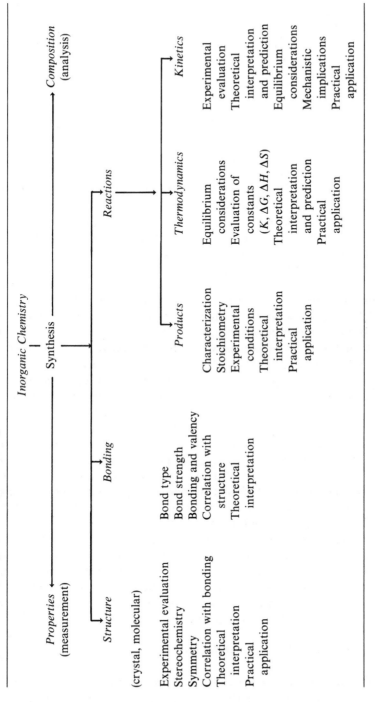

Inorganic Chemistry

Properties (measurement) ← Synthesis → *Composition* (analysis)

Synthesis → *Reactions*

Properties → *Structure*, *Bonding*

Reactions → *Products*, *Thermodynamics*, *Kinetics*

Structure
(crystal, molecular)

Experimental evaluation
Stereochemistry
Symmetry
Correlation with bonding
Theoretical
  interpretation
Practical
  application

Bonding

Bond type
Bond strength
Bonding and valency
Correlation with
  structure
Theoretical
  interpretation

Products

Characterization
Stoichiometry
Experimental
  conditions
Theoretical
  interpretation
Practical
  application

Thermodynamics

Equilibrium
  considerations
Evaluation of
  constants
  (K, ΔG, ΔH, ΔS)
Theoretical
  interpretation
  and prediction
Practical
  application

Kinetics

Experimental
  evaluation
Theoretical
  interpretation
  and prediction
Equilibrium
  considerations
Mechanistic
  implications
Practical
  application

3

number of isolable inorganic compounds, the limited number of bond types and possible structural relationships among the hydrocarbons and most of their derivatives make at least a close approach to this goal much more likely in organic chemistry.

1. THE DEVELOPMENT OF MODERN INORGANIC CHEMISTRY

The origins of inorganic chemistry are lost in antiquity. Although a few of the metallic elements were obtained from natural sources by the ancients and used by them and a variety of inorganic compounds were prepared and investigated by the alchemists, it was not until the nineteenth century that the foundations for the science as it exists today were laid. Investigations during the late 1700s and early 1800s resulted in the discovery and characterization of a number of the elements, the analyses of many minerals, the syntheses of a variety of inorganic compounds, and the first real attempts to describe matter on an atomic basis.[2, 3] The investigators involved were essentially generalists, who concerned themselves with all aspects of chemistry, but increasingly their efforts indicated distinct differences between the chemistry of carbon-containing, or organic, compounds and noncarbon, or inorganic, compounds.

After about 1860,[3] investigations were definitely oriented toward either the organic area or the inorganic area. Developments in the former were spectacular; those in the latter important but somewhat less than spectacular. Organic chemistry, for the reasons already mentioned above, lent itself readily to systematization and prediction; inorganic chemistry developed quite empirically. Between about 1900 and World War II, with only a few exceptions, inorganic chemistry remained essentially static, with a minimum of research effort, the presence of a limited number of centers emphasizing training in the discipline, and the development of a wide spread belief that all of inorganic chemistry could be presented in freshman courses. The atomic energy projects of World War II catalyzed the development of inorganic chemistry as we now know it and led to what has been called the "Renaissance of Inorganic Chemistry." The application of structure and reaction mechanism techniques has given to it the dimensions of understanding and prediction that are rapidly making it a cohesive discipline and that have resulted in exponential increases in research, training, and publication. It is instructive and important to our presentation to mention briefly a few of the significant contributions to the development of modern inorganic chemistry.[4, 5]

1-1. The Atomic Nature of Matter

The concept that matter is constructed of tiny, fundamental particles undoubtedly originated with the early Greek philosophers. Indeed, the term *atom* derives from the Greek *átomos* (indivisible). Modern atomic theory is generally traced to John Dalton, who in 1802 summarized in quantitative terms ideas developed by Isaac Newton, Bryan Higgins, William Higgins, and others.[6] The discovery of the more fundamental particles (electron, proton, neutron) and the evaluation of their properties culminated (1) by way of the Bohr theory of the hydrogen atom (1913), in the ultimate quantum-mechanical elucidation of the electronic configurations of atoms and ions that determine chemical characteristics of inorganic species, and (2) by way of nuclear transformations (natural and artificial), in elucidation of the cosmic origins of the elements and in the syntheses of new elements. The atomic nature of matter and its implications are discussed in detail in Chapter 2.

1-2. The Systematic Classification of the Elements

Early observations of the similarities existing among certain elements (e.g., the alkali metals Li, Na, K and the halogens Cl, Br, I) and their compounds culminated in elucidation of the *periodic law* and the independent publication by Lothar Meyer[7] and Dimitri Mendeleev[8] of the *periodic table*.* By bringing together elements of closely similar properties in a suitable geometrical arrangement, the periodic table did, and still does, more to systematize chemistry, to allow reasonable predictions of properties, and to stimulate research than any other single generalization. Current periodic tables, by bringing together elements the atoms of which have the same fundamental types of electronic configurations, show clearly the importance of these configurations in determining the properties of those elements. The periodic table, its origins, its significance, and its applications are discussed in Chapter 3.

1-3. The Noble Gas Elements

The first periodic tables listed no elements lying between the halogens and the alkali metals. Yet when in 1894–1895 Lord Rayleigh, William Ramsay, and others, identified neon, argon, krypton, and xenon as components of the atmosphere and Ramsay showed that the gas liberated when the uranium mineral cleveite is heated is identical with the helium identified earlier in the solar spectrum,[2,9,10] it was possible to accommodate this family of chemically inert elements within the periodic table merely by the insertion of another group (group O, or more modernly, group VIIIA). Later radon, discovered as a decay product in the uranium series, was added as the heaviest member of the family.

When the significance of the outer electronic configuration of an atom in determining its chemical behavior became apparent, the eight-electron (or two-electron for the He atom) outermost arrangement characteristic of a noble gas atom was considered to be an arrangement of chemical stability and used as a reference point in developing theories of the chemical bond· (Ch. 4).[11,12] The

---

* The periodic law stated originally that the chemical and physical properties of the elements are periodic functions of their *atomic weights*. In the current statement, *atomic numbers* is substituted for *atomic weights*.

lack of absolute inertness was first demonstrated only relatively recently by the syntheses of the xenon compounds $Xe(PtF_6)_n (1 > n < 2)$ and XeF_4 by oxidation of the element with platinum(VI) fluoride[13] and elemental fluorine,[14] respectively.

1-4. Werner's Coordination Theory

As the nineteenth century progressed, investigators continued to isolate compounds, such as hydrates, ammonates, and double salts, which could be regarded as complex species formed by the combination in stoichiometric quantities of independently existing species but differing from those latter species in properties. A case in point was the isolation by various reactions of aqueous cobalt(II) salt solutions, aqueous ammonia, and oxidizing agents of the compounds $CoCl_3 \cdot 6\,NH_3$ (yellow—*luteo* cobalt chloride), $CoCl_3 \cdot 5\,NH_3 \cdot H_2O$ (pink—*roseo* cobalt chloride), $CoCl_3 \cdot 5\,NH_3$ (violet—*purpureo* cobalt chloride), $CoCl_3 \cdot 4\,NH_3$ (violet—*violeo* cobalt chloride), and $CoCl_3 \cdot 4\,NH_3$ (green—*praseo* cobalt chloride). These compounds differed not only in color, but also in chemical behavior and often in numbers of ions formed in aqueous solutions (Chs. 7, 8).

Ultimate explanation of the formation and properties of these substances and systematization were provided in 1893 by Alfred Werner,[15] who suggested (1) that metals simultaneously possess primary (ionizable) and secondary (nonionizable) valence, (2) that the secondary valence ($=coordination\ number$) is fixed and must be satisfied by either neutral groups or anions, and (3) that groups held by secondary valence are arranged in a definite geometry (e.g., six groups octahedrally) about the metal atom. Transfer of these ideas to an electronic basis by N. V. Sidgwick[16] and T. M. Lowry,[17] subsequent development of orbital theories of bonding (Ch. 8), the elucidation of molecular structures by physical measurements (Chs. 6–8), the correlation between properties and bonding (in particular of complexes of the d-transition metals), and the applications of a number of these compounds in catalysis and other processes have produced the largest single research effort in the history of inorganic chemistry. Indeed, cursory examination of current research publications and monographs can easily convince an uninitiated individual that modern inorganic chemistry is solely the chemistry of complexes, or coordination compounds. Of course, that is not really true.

1-5. Nonexistent Compounds

Quite apart from compounds of the noble gas elements, there are a number of compounds of more reactive elements that, by comparison with known compounds of elements in the same periodic family, should form but the isolation of which either defied experimenters for a long period of time or still defies them.[18] Examples include arsenic(V) chloride, selenium(VI) oxide, and perbromates, where the analogous compounds of the lighter and heavier elements within the same periodic family (i.e., PCl_5 and $SbCl_5$, SO_3 and TeO_3, and ClO_4^- and IO_4^-) were synthesized readily many years ago.

In spite of varied theoretical justifications for their lack of existence, selenium(VI) oxide and perbromic acid and its salts have been obtained in high states of purity, and arsenic(V) chloride has been identified in various adducts. Selenium(VI) oxide was obtained initially as a white crystalline solid by either heating potassium selenate (K_2SeO_4) with excess sulfur(VI) oxide and distilling away the sulfur(VI) oxide from the SeO_3–SO_3 layer that separates[19] or dehydrating selenic acid with phosphorus(V) oxide and recovering the compound by vacuum sublimation.[20] Perbromate ion was obtained in ponderable quantity by oxidation of the bromate ion electrolytically or with xenon(II) fluoride.[21] A more practical synthesis involves oxidation of bromate ion with elemental fluorine under alkaline conditions.[22] Interestingly, at least $6\,M$ solutions of perbromic acid can be stored without decomposition, and crystalline potassium perbromate undergoes thermal decomposition only above 280°C. Arsenic(V) chloride has not been isolated as such, but oxidation of arsenic(III) chloride with elemental chlorine yields complexes such as $[AsCl_4]^+[MCl_{n+1}]^-$ ($MCl_n = AlCl_3$, $GaCl_3$, PCl_5, $SbCl_5$)[23] and $AsCl_5 \cdot OP(CH_3)_3$,[24] where reduction of arsenic(V) by the chloride ion is apparently inhibited.

Many other inorganic compounds that were earlier believed to be incapable of existence have been obtained by new techniques and the use of new reagents.

1-6. Lanthanide Chemistry

Lanthanide chemistry dates to the isolation of two supposedly pure "earths" (oxides): *yttria* by J. Gadolin in 1794 and *ceria* by M. H. Klaproth, J. J. Berzelius, and W. Hisinger in 1803. The application of tedious methods of fractionation extending into 1907 resolved the original yttria into pure compounds of yttrium, gadolinium, terbium, dysprosium, holmium, erbium, thulium, ytterbium, and lutetium and the original ceria into compounds of lanthanum, cerium, praseodymium, neodymium, samarium, europium, and gadolinium.[25] The remaining lanthanide, promethium, was first identified positively in products from the neutron-induced fission of uranium-235.[26] It was as a consequence of the need to isolate lanthanide species from other substances produced by nuclear fission that the relatively rapid, quite selective, and highly automated ion-exchange and solvent-extraction processes of separation, which ultimately made all of the lanthanide elements available in commercial quantities, were developed.

The almost unique radiant energy absorption-emission, magnetic, catalytic, and metallic properties associated with electrons in nonvalence $4f$ energy levels have led to a number of commercial developments. Furthermore, the development of the chemistry of the actinide ($5f$) elements has been markedly simplified and substantially accelerated by knowledge of the chemistry of the lanthanide elements.

1-7. Expansion of the Periodic Table

By 1920, it was generally assumed that uranium is the heaviest element, and the only spaces prior to uranium in the periodic table that remained unfilled were at atomic numbers 43, 61, 72, 75, 85, and 87.

In terms of extensions of the Bohr theory, it became apparent that the lanthanide series ended at atomic number $(Z) = 71$ and that element 72 should be a congener of zirconium. A careful examination of zirconium ores led D. Coster and G. von Hevesy (later de Hevesy) to announce the discovery of hafnium in 1923.[27] Separation of hafnium from zirconium by large-scale solvent extraction techniques came only after it was discovered that whereas both of these metals resist corrosion in nuclear reactors, the isotopes of hafnium absorb neutrons sufficiently well so as to reduce substantially the neutron flux in such reactors when natural zirconium is used.

In 1925, a long search for the then-unknown congeners of manganese culminated in the identification of element 75 (rhenium) in gadolinite.[28] Parallel work by W. I. Noddack and I. Noddack (neé Tacke) reporting the discovery of element 43 could not be repeated, but in 1936, C. Perrier and E. Segrè separated small quantities of this element from molybdenum that had been bombarded with deuterons.[29] Technetium is now recovered in substantial quantities as a fission product.

Element 87 (francium) was first identified positively in 1939 as a product of the branched-chain decay of the nuclide ^{227}Ac.[30] Element 85 (astatine) was first identified positively in 1940 among the products obtained from the bombardment of bismuth with deuterons.[31] The isolation of element 61 from fission products has already been mentioned.

The syntheses of transuranium elements by nuclear bombardment reactions began with identification of element 93 (neptunium)[32] in 1940, closely followed by that of element 94 (plutonium).[33] Subsequently, elements 95 to 103 were synthesized as members of the 5 f-transition series, as well as at least elements 104, 105, and 106, all by nuclear reactions. Theoretical studies of the systematics of nuclear stability which suggested potentially enhanced nuclear stability at $Z = 114$ (eka-lead) and prompted prediction of electronic configurations and properties, encouraged attempts to bridge a gap of instability by bombardment with massive nuclear particles and stimulated searches in natural sources for various "superheavy" elements.[34, 35] These researches have added very substantially to inorganic chemistry in terms of refinement of techniques of separation and identification and development of methods for handling and studying substances in submicrogram quantities.

1-8. Reactions in Nonaqueous Media

The long-believed uniqueness of water as a medium for ionic reactions of inorganic species was questioned by the pioneering researches of H. P. Cady,[36] and later E. C. Franklin and C. A. Kraus,[37] on liquid ammonia as an electrolytic solvent and by P. Walden[38] on liquid sulfur(IV) oxide, also as an electrolytic solvent. In subsequent years, many anhydrous substances [e.g., HF, HCN, CH_3COOH, $H_2NC_2H_4NH_2$, C_5H_5N, H_2SO_4, HSO_3F, HNO_3, $SeOCl_2$, $COCl_2$, BrF_3, N_2O_4, CH_3CN, $POCl_3$, IF_5, $PO\{N(CH_3)_2\}_3$] have been investigated as media in which metathetical and/or redox reactions can occur. Around reactions of self-dissociation, definitions of acids and bases have been developed (Ch. 9). Syntheses that are impossible to carry out in water have been effected. Reactions involving particularly strong reducing agents (e.g., alkali metals) have been achieved in strongly basic solvents (such as liquid ammonia or ethylenediamine), and those involving strong oxidizing agents (e.g., $S_2O_6F_2$) have been carried out in strongly acidic solvents (such as HSO_3F or HF). An extension of such studies to reactions in molten salts has added yet another dimension to modern inorganic chemistry.[39,40]

2. THE STUDY OF MODERN INORGANIC CHEMISTRY

To blend all these facets of inorganic chemistry, and many others that have not yet been mentioned, into a coordinated presentation that proceeds logically to topics that are increasingly comprehensive is no small task. This presentation begins with atomic structure and electronic configuration as an ultimate basis; proceeds through the periodic table to bonding, molecular structure, and stereochemistry; and culminates in discussions of various aspects of chemical reactivity and chemical reactions. This background of principles, liberally extended by many examples, thus provides a basis for detailed study of the individual elements and their compounds and the reactions that both the elements and their compounds undergo.

3. THE LITERATURE OF MODERN INORGANIC CHEMISTRY

The pattern, already introduced in the preceding sections of this chapter, of citing direct sources of information or referring to more extensive discussions of particular topics, is continued in succeeding chapters. The value of consulting these literature sources as one studies the discussions that follow lies in the enrichment that can result. Such aspects as the development of scientific thought and practice, the refinement of experimentation, the development of in-strumental approaches, the stimulation of thinking and the genesis of new ideas, and the overall expansion of knowledge depend upon the more detailed treatments that the literature provides. No one person can ever hope to know or understand everything that qualifies as inorganic chemistry. He or she can, however, by diligent use of the literature, learn where to locate detailed information about almost every topic that he or she may desire to pursue in depth.

In Appendix I, the literature that is most useful to the inorganic chemist is summarized under the general headings of comprehensive reference works, advanced-level textbooks, inorganic laboratory practice, summary and review volumes, and journals. Brief comments relating to the scope and limitations of each item are included.

By a wide margin, the most comprehensive source of published information is *L. Gmelin's Handbuch der Anorganischen Chemie*, compiled by the Gmelin-Institut für Anorganische Chemie und Grenzgebiete der Max-Planck-Gesellschaft zur Förderung der Wissenschaften, Frankfurt (Main), currently under the direction of H. C. Ekkehard Fluck.[41,42] Each volume in this series is an authoritative and critical survey of published information, replete with pertinent literature citations. Details of coverage and usage are given in Appendix I.

4. THE NOMENCLATURE OF INORGANIC CHEMISTRY

Nomenclature in inorganic chemistry developed with much less system than that in organic chemistry, and often was little more than a hodgepodge combining names based upon properties with those attempting to reflect compositions. Typical cases that illustrate this point are: corrosive sublimate, bichloride of mercury, and mercuric chloride for the compound $HgCl_2$; and copperas, green vitriol, and hydrated ferrous sulfate for the compound $FeSO_4 \cdot 7H_2O$.

In the 1920 edition of his classic textbook,[43] A. Werner summarized a system of nomenclature for coordination entities (see Sec. 7.1-3 for details) that, with only a limited number of modifications, is still in use. Systematic approaches to the naming of other inorganic substances, however, did not appear until 1957, when a commission of the International Union of Pure and Applied Chemistry (IUPAC) compiled the first set of systematic rules of nomenclature. This system has since been expanded and extended to cover all classes of inorganic compounds.[44-46] The essential details are given in Appendix II. It is neither necessary nor desirable that these rules be committed to memory as such. Rather they should be referred to constantly and used wherever necessary. In subsequent chapters, every attempt is made to use the IUPAC rules of nomenclature. Correctly used, these rules not only indicate without ambiguity the composition of the species in question but also specify the oxidation state of at least one element and/or ionic charge if necessary.

REFERENCES

1. T. Moeller, *Inorganic Chemistry: An Advanced Textbook*, John Wiley & Sons, New York (1952).

2. M. E. Weeks and H. M. Leicester, *Discovery of the Elements*, 7th ed., Journal of Chemical Education Circulation Services, Easton, Pa. (1968). A comprehensive and completely docu-

mented account of the discovery of all the elements, including much biographical information on the discoverers.

3. A. J. Ihde, *The Development of Modern Chemistry*, Harper & Row, New York (1964). An authoritative and highly informative account of the whole of chemistry. Chapters 14 and 22 are devoted specifically to inorganic chemistry.

4. L. F. Audrieth, *Ind. Eng. Chem.*, **43**, 269 (1951). A general review of academic and industrial inorganic chemistry in the United States subsequent to the founding of the American Chemical Society in 1876.

5. J. A. Zubieta and J. J. Zuckerman, *Chem. Eng. News*, Apr. 6, 1976, pp. 64–79. A review of 100 years of development in inorganic chemistry in commemoration of the 100th birthday of the American Chemical Society.

6. R. Siegfried, *J. Chem. Educ.*, **33**, 263 (1956).

7. L. Meyer, *Ann. Chem.*, Suppl. **7**, 354 (1870).

8. D. Mendeleev, *J. Russ. Phys.-Chem. Soc.*, **1**, 60 (1869); *Z. Chem.*, **5**, 405 (1869); *Ann. Chem.*, Suppl. **8**, 133 (1871).

9. F. P. Gross, *J. Chem. Educ.*, **18**, 533 (1941).

10. J. H. Wolfenden, *J. Chem. Educ.*, **46**, 569 (1969).

11. W. Kossel, *Ann. Phys.* (*Leipzig*), [4], **49**, 229 (1916).

12. G. N. Lewis, *J. Am. Chem. Soc.*, **38**, 762 (1916).

13. N. Bartlett, *Proc. Chem. Soc.*, **1962**, 218.

14. H. H. Claassen, H. Selig, and J. G. Malm, *J. Am. Chem. Soc.*, **84**, 3593 (1962).

15. A. Werner, *Z. Anorg. Chem.*, **3**, 267 (1893).

16. N. V. Sidgwick, *J. Chem. Soc.*, **123**, 725 (1923).

17. T. M. Lowry, *J. Soc. Chem. Ind.*, **42**, 316 (1923).

18. W. E. Dasent, *Non-Existent Compounds*, Marcel Dekker, New York (1965).

19. H. A. Lehrmann and G. Krüger, *Z. Anorg. Allg. Chem.*, **267**, 315 (1952).

20. B. Toul and K. Dostál, *Chem. Listy*, **46**, 132 (1952).

21. E. H. Appelman, *J. Am. Chem. Soc.*, **90**, 1900 (1968); *Acc. Chem. Res.*, **6**, 113 (1973).

22. E. H. Appleman, *Inorg. Chem.*, **8**, 223 (1969).

23. L. Kolditz and W. Schmidt, *Z. Anorg. Allg. Chem.*, **296**, 189 (1958).

24. I. Lindqvist and G. Olofsson, *Acta Chem. Scand.*, **13**, 1753 (1959).

25. T. Moeller, *The Chemistry of the Lanthanides*, Pergamon Press, Elmsford, N.Y. (1975).

26. J. A. Marinsky, L. E. Glendenin, and C. D. Coryell, *J. Am. Chem. Soc.*, **69**, 2781 (1947).

27. D. Coster and G. von Hevesy, *Naturwissenschaften*, **11**, 133 (1923). See G. von Hevesy, *Chem. Rev.*, **2**, 1 (1926).

28. W. I. Noddack, I. Tacke, and O. Berg, *Naturwissenschaften*, **13**, 567 (1925).

29. C. Perrier and E. Segrè, *J. Chem. Phys.*, **5**, 712 (1937); **7**, 155 (1939).

30. M. Perey, *C. R. Acad. Sci*, **208**, 97 (1939). See also M. Perey, *J. chim. Phys.*, **43**, 155, 262, 269 (1946).

31. D. R. Corson, K. R. MacKenzie, and E. Segrè, *Phys. Rev.*, **57**, 459, 1087 (1940); **58**, 672 (1940).

32. E. M. McMillan and P. H. Abelson, *Phys. Rev.*, **57**, 1185 (1940).

33. G. T. Seaborg, E. M. McMillan, J. W. Kennedy, and A. C. Wahl, *Phys. Rev.* **69**, 366, 367 (1946). (Received in 1941.)

34. G. T. Seaborg, *J. Chem. Educ.*, **46**, 626 (1969); *Am. Sci.*, **68**, 279 (1980).

35. B. Fricke, *Struct. Bonding*, **21**, 89 (1975).

36. H. P. Cady, *J. Phys. Chem.*, **1**, 707 (1897).

37. E. C. Franklin and C. A. Kraus, *Am. Chem. J.*, **21**, 8 (1899).

38. P. Walden, *Chem. Ber.*, **32**, 2862 (1899).

39. H. Bloom and J. W. Hastie, in *Non-Aqueous Solvent Systems*, T. C. Waddington (Ed.), Ch. 9, Academic Press, New York (1965).

40. J. H. R. Clarke and G. J. Hills, *Chem. Br.*, **9**, 12 (1973).

41. W. Lippert, *Chem. Br.*, **8**, 483 (1972).

42. M. Becke-Goehring, *J. Chem. Educ.*, **50**, 406 (1973).

43. A. Werner, *Neuere Anschauung auf dem Gebiete der anorganischen Chemie*, T. Viewig u. Sohn, Braunschweig (1920).

44. Commission on Nomenclature of Inorganic Chemistry, *Pure Appl. Chem.*, **28**(1), 1 (1971). Also *Nomenclature of Inorganic Chemistry, Definitive Rules 1970*, Butterworth, London (1971). (The "Red Book.")

45. Commission on Nomenclature of Inorganic Chemistry, *How to Name an Inorganic Substance*, Pergamon Press, Oxford (1977). A guide to the use of the *Definitive Rules*, with a revised and enlarged section on names for ions and radicals.

46. W. C. Fernelius, K. Loening, and R. M. Adams, *J. Chem. Educ.*, **49**, 488 (1972).

SUGGESTED SUPPLEMENTARY REFERENCES

Anonymous, "Report of the Inorganic Chemistry Subcommittee of the Curriculum Committee (James E. Huheey, Chairman)," *J. Chem. Educ.*, **49**, 327 (1972).

L. F. Audrieth, *Ind. Eng. Chem.*, **43**, 269 (1951). General review of academic and industrial inorganic chemistry in the United States subsequent to 1876.

M. Gorman, "Some Reflections on the Inorganic Course," *J. Chem. Educ.*, **50** 772 (1973).

Symposium, "Inorganic Chemistry in the Curriculum: What Should Be left In and What Should Be Left Out," Fall 1980 Meeting, American Chemical Society, (Las Vegas), *J. Chem. Educ.*, **57**, 761–769 (1980). Papers by F. Basolo, R. A. Laudise, T. J. Meyer, H. B. Gray, M. J. Sienko, R. J. Angelici, F. A. Cotton, plus panel summary.

Symposium, "Instruction in Inorganic Chemistry," Fall 1958 Meeting, American Chemical Society (Chicago), *J. Chem. Educ.*, **36**, 441–457, 502–518 (1959). Papers by L. F. Audrieth, P. M. Gross, Jr., L. Brewer, H. Taube, S. Y. Tyree, Jr., R. W. Parry, R. E. Heath, R. T. Sanderson, S. Kirschner, W. L. Jolly, H. H. Sisler.

H. M. Woodburn, "Retrieval and Use of the Literature of Inorganic Chemistry," *J. Chem. Educ.*, **49**, 689 (1972).

J. A. Zubieta and J. J. Zuckerman, *Chem. Eng. News*, Apr. 6, 1976, pp. 64–79. Review of 100 years of development in inorganic chemistry.

CHAPTER TWO

FACETS OF ATOMIC STRUCTURE AND SOME PROPERTIES RELATED THERETO

Implementation of our definition of inorganic chemistry (p. 2) requires that we have not only knowledge of the physical and chemical properties of the elements but also understanding, insofar as is possible, of the fundamental bases for the observed properties of a pure substance and differences among the properties of various substances. It is, of course, obvious in terms of the existing level of scientific sophistication that observed properties and differences among them are atomic in origin and that our explanations must be basically atomic in nature. The purposes of this chapter are to review the significant developments that led to atomic structure, both nuclear and extranuclear, as we now conceive it to be; to treat modern concepts in some detail; and ultimately to apply these concepts. The basic combination of fact and theory so resulting is then extended in subsequent chapters to chemical periodicity, bonding, molecular structure and symmetry, coordination chemistry, and chemical reactions of various types and under various conditions.

1. DEVELOPMENT OF MODERN ATOMIC THEORY — BRIEF SUMMARY[1]

It is generally agreed that modern atomic theory can be traced to the summation of ideas propounded by John Dalton in 1802 (Sec. 1.1-1). This summation stated that (1) matter is composed of tiny real particles (atoms); (2) atoms of a pure substance can be neither subdivided nor changed one into another; (3) atoms are incapable of being either created or destroyed; (4) all of the atoms of a specific pure substance are identical in weight, size, and other properties, whereas atoms of one pure substance differ from those of another pure substance in weight and other characteristics; and (5) chemical combination results from the union of atoms in definite numerical proportions.

Although certain of these statements are now known to be incorrect, the fundamental importance of Dalton's theory in influencing chemical thought and in providing bases for current concepts of atomic structure and behavior is evident.

Subsequent developments were concerned largely with expansion of this atomic theory and increasingly more comprehensive and more nearly quantitative explanations of its tenets. Prominent among these developments were the hypothesis that the hydrogen atom is the fundamental building block (W. Prout, 1816); demonstration of the quantitative equivalence between electrical energy and the amount of chemical oxidation or reduction (M. Faraday, 1833); initial nearly precise determinations of atomic weights (J. S. Stas, 1860); elucidation of the properties of cathode rays (W. Crookes, 1879); empirical mathematical interpretation of lines in the Balmer series (Sec. 2.2-3) of the emission spectrum of hydrogen (J. R. Rydberg, 1890); quantitative evaluations of the charge-to-mass ratio (J. J. Thomson, 1897) and the charge of the electron (R. A. Millikan, 1910–1913); discovery of x-rays (W. K. Röntgen, 1895); initial observations on natural radioactivity (A.-H. Becquerel, 1896) and subsequent discovery of polonium and radium (Marie and Pierre Curie, 1898); elucidation of the fundamentals of quantum theory (M. Planck, 1900); development of the concept of the nuclear atom (E. Rutherford, 1911); elucidation of a workable theory of the structure of the hydrogen atom (N. Bohr, 1913); assignment of atomic numbers (H. G. J. Moseley, 1913–1914); applications of quantum theory to the electronic structures of atoms (W. Heisenberg, 1926; E. Schrödinger, 1926); discoveries of the neutron (J. Chadwick, 1932) and deuterium (H. C. Urey, 1932); discoveries of artificial nuclear transmutation (E. Rutherford, 1919) and artificial radioactivity (Irène Curie and F. Joliot, 1934); elucidation of nuclear fission (O. Hahn and F. Strassmann, 1939); syntheses of transuranium elements (G. T. Seaborg et al., 1940–1961); and realization of explosive nuclear fusion (E. Teller, 1952).

1-1. Fundamental Subatomic Particles

Although the bewildering array of reported subatomic particles continues to increase in number as experimental techniques of bombardment of atoms or ions with excited subatomic particles are carried out at higher and higher energies,* the properties of the elements and their compounds that are of direct concern to the inorganic chemist are adequately treated in terms of only neutrons, protons, and electrons. The properties of these particles are summarized in Table 2-1.

---

* For an interesting summary, see M. Waldrop, *Chem. Eng. News*, Jan. 21, 1980, pp. 44–52.

Table 2-1   Characteristics of Fundamental Subatomic Particles

Property	Electron	Proton	Neutron
Designation	e^-, $_{-1}^{0}e$, β^-	p^+, $_1^1p$, P, $_1^1H^+$	n, $_0^1n$, N
Mass, relative to rest mass of electron[a]	1.000	1836.12	1838.65
Mass (amu)[b]	0.0005486	1.00757	1.00893
Charge (esu $\times 10^{10}$)	-4.8029	$+4.8029$	0
Charge, relative to charge of electron $= -1$	-1	$+1$	0
Spin ($h/2\pi$ units)[c]	$\frac{1}{2}$	$\frac{1}{2}$	$\frac{1}{2}$
Half-life (min)	Stable	Stable	12.8

[a] Because of variation of mass with velocity, the mass at rest or very low velocity (i.e., the rest mass) is used for comparisons.

[b] 1 amu $= 1.6606 \times 10^{-24}$ g.

[c] Expressed as integral or half-integral multiples of $h/2\pi$ (p. 23).

The term *electron* (Greek *élektron*, amber) was first applied in 1874 by G. J. Stoney, but to the unit charge on a univalent anion. As a tiny, energetic, negative particle, the electron was identified and characterized by Thomson[2] and Millikan.[3] Electrons are characteristic of all matter and of sufficiently small mass to exhibit wave, as well as particulate, properties (p. 36).[4] As a chemical entity, the electron is a significant component of chemical reactions[5] and is so considered in later chapters.

The *proton* (Greek *protos*, first) was initially identified as the unipositive particle of largest charge-to-mass ratio identifiable in beams of positive rays (Kanalstrahlen) found behind holes in the cathodes of cathode-ray tubes.[6] Identity between these particles and positive hydrogen ions is evident. It is concluded that all atomic species contain protons.

Although the existence of the *neutron* was predicted in 1920 by W. D. Harkins, O. Masson, and E. Rutherford, the difficulties of experimentally detecting an atomic particle that could neither ionize other substances except by direct collision nor be deflected by magnetic or electrostatic fields were such that discovery was delayed until 1932, when the penetrating radiation generated, for example, by bombarding beryllium atoms with alpha particles was so identified.[7] Outside an atomic nucleus, a neutron decays spontaneously to a proton, an electron, and a neutrino, with half-life 12.8 min.

1-2. The Nuclear Atom

Based fundamentally upon (1) Rutherford's interpretations of alpha-particle scattering by thin metal foils as resulting from repulsions by heavy, positively charged atomic nuclei;[8] (2) Moseley's interpretation of regular displacements of the frequencies of certain comparable lines in x-ray spectra with increasing atomic weight of the target element in terms of atomic numbers (i.e., numbers of electrons in target atoms);[9] and (3) the discovery of the neutron,[7] we now conceive of an atom as consisting of a tiny (radius ca. 10^{-13} cm., or 10^{-6} nm), compact, heavy, positively charged nucleus of protons and neutrons (except for the $_1^1H$ nucleus) surrounded by a much larger (radius ca. 10^{-8} cm., or 10^{-1} nm), lighter atmosphere of electrons. Each nuclear species (or *nuclide*) is described uniformly by its nuclear charge Z (=number of protons, P, or proton number = number of electrons), its neutron number N (=number of neutrons), and its mass number A ($=P+N=Z+N$). Values of Z range from 0 (i.e., the neutron) through at least 106; values of A range from 1 through at least 260. In accordance with internationally accepted practice, each nuclide, its ionic charge if any, and its quantity if necessary are symbolized as

$$_Z^A E_y^{x\pm}$$

where E is the symbol of the element in question, $x^\pm$ the ionic charge, and y the number of atoms or ions of E in the molecular aggregate in question.

Relationships to Inorganic Chemistry. The properties of atomic nuclei are important to the primordial and continuing syntheses of the elements and thus determine both the number of the elements that can exist and the abundances of these elements. The extranuclear structures of the atoms largely determine the chemical properties and many of the physical properties of the elements. Thus they control chemical reactions, account for similarities and differences among the elements and their compounds, and provide the ultimate basis for the periodic classification and much that can be learned from it (Chs. 3–9).

2. NUCLEAR STABILITY AND THE ORIGINS AND ABUNDANCES OF THE ELEMENTS

Atomic nuclei convert, either spontaneously (*via* radioactive decay) or as a consequence of appropriate bombardment at the atomic level (e.g., by neutrons, protons, other atomic nuclei) into different atomic nuclei. Nuclear stability is measured by either its half-life (if the change is spontaneous) or its ease of conversion (if the change is by bombardment). A nucleus of half-life equal to or longer than ca. 10^{10} years is considered to be stable.

Spontaneous nuclear conversion may involve any of the processes summarized in Table 2-2. Although it is not pertinent to this discussion to delineate all the details of each of these processes or to explore current thoughts as why they all occur, it is important to our continuing discussion to note briefly the influences of both neutron/proton ratio and sheer mass on nuclear stability. For more detailed treatments, standard works on nuclear and/or radiochemistry can be consulted.[10-13]

In Fig. 2-1, the neutron number (N) is plotted against the proton number (P) for stable, naturally occurring nuclides. An average curve drawn through the points reflects reasonably the neutron/proton ratios for maximum nuclear

Table 2-2 Spontaneous Nuclear Decay Processes

Process	Designation	Change in Z	Change in A	Examples[a,b]
Beta emission	β^-	+1	0	$^{14}_{6}C \xrightarrow{5570\ y} {}^{14}_{7}N + {}^{0}_{-1}e + \bar{\nu}_e$ $^{234}_{90}UX_1 \xrightarrow{24.1\ d} {}^{234}_{91}UX_2 + {}^{0}_{-1}e + \bar{\nu}_e$
Positron emission	β^+	−1	0	$^{19}_{10}Ne \xrightarrow{18.2\ s} {}^{19}_{9}F + {}^{0}_{+1}e + \nu_e$ $^{23}_{12}Mg \xrightarrow{10.7\ s} {}^{23}_{11}Na + {}^{0}_{+1}e + \nu_e$
K-Electron capture	E.C.	−1	0	$^{55}_{26}Fe + {}^{0}_{-1}e \xrightarrow{2.94\ y} {}^{55}_{25}Mn + \nu_e$ $^{73}_{33}As + {}^{0}_{-1}e \xrightarrow{76\ d} {}^{73}_{32}Ge + \nu_e$
Isomeric transition	I.T.	0	0	$^{52m}_{25}Mn \xrightarrow{21\ m} {}^{52}_{25}Mn$ $^{60m}_{27}Co \xrightarrow{10.5\ m} {}^{60}_{27}Co^c$
Alpha emission	α	−2	−4	$^{238}_{92}UI \xrightarrow{4.5\times10^9\ y} {}^{234}_{90}Th + {}^{4}_{2}He$ $^{226}_{88}Ra \xrightarrow{1622\ y} {}^{222}_{86}Rn + {}^{4}_{2}He$
Gamma emission	γ	0	0	Energy release accompanying especially IT, β^-, or β^+ processes

[a] Half-lives indicated (s = second, m = minute, d = day, y = year).

[b] ν_e = neutrino, $\bar{\nu}_e$ = antineutrino.

[c] m = isomeric short-lived ($<10^{-3}$ s) state in radioactive equilibrium with parent activity.

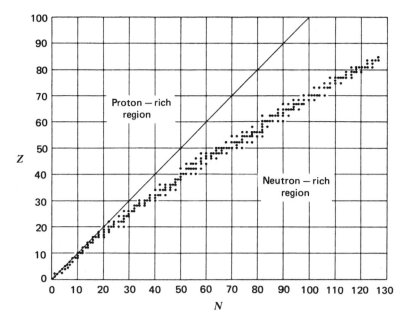

Figure 2-1 Relationships between neutron and proton numbers for naturally occurring stable nuclei. [G. Friedlander, J. W. Kennedy, and J. M. Miller, *Nuclear and Radiochemistry*, 2nd ed., John Wiley & Sons, New York (1964).]

stabilities.* This curve coincides with the line for the condition $N/P = 1$ for only the lighter nuclei and deviates more and more in the direction $N/P > 1$ as the mass number increases, possibly because of necessary compensation by neutron–proton attractions to overcome increasingly larger proton–proton repulsions.

For a given nuclear charge, only a limited number of nuclides are "stable," suggesting that for stability N/P has a limited range of values. A reasonable conclusion is that for a given nuclear charge, either an "excess" of neutrons (neutron-rich, large N/P) or an "excess" of protons (proton-rich, small N/P) results in nuclear instability. These regions of potential instability are indicated in Fig. 2-1. A neutron-rich nucleus gains stability by decreasing N through β^- emission, which has the effect of converting a neutron into a proton,

$$_0^1 n \rightarrow {}_1^1 p + {}_{-1}^0 e \quad (\text{or } \beta^-)$$

This type of radioactive decay is particularly common among naturally occurring nuclides and will be shown to be important in the cosmic syntheses of the heavier elements. Neutron emission, a self-evident means of reducing N, is observed for only a limited number of nuclides (e.g., excited $_4^9$Be, $^{17}_8$O, $^{87}_{36}$Kr,

---

* Figure 3-7 incorporates nuclear stability as a third coordinate in a three-dimensional representation.

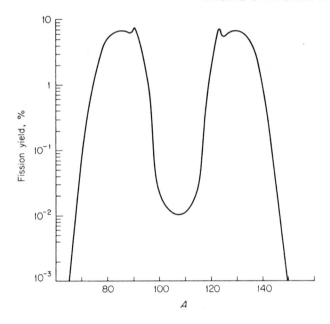

Figure 2-2 Yields in the nuclear fission of ^{235}U. [G. Friedlander, J. W. Kennedy, and J. M. Miller, *Nuclear and Radiochemistry*, 2nd ed., John Wiley & Sons (1964).]

$^{137}_{54}$Xe). A proton-rich nucleus gains stability by increasing N through β^+ emission or electron capture, either of which has the effect of converting a proton into a neutron,

$$^1_1p \rightarrow {}^1_0n + {}^{\,\,0}_{+1}e(\beta^+)$$

$$^1_1p + {}^{\,\,0}_{-1}e \rightarrow {}^1_0n$$

Inasmuch as there are few naturally occurring proton-rich nuclides, reactions of this type are essentially limited to artificially radioactive species.*

Nuclear instability associated with excess mass is most commonly relieved by emission of alpha particles (Table 2-2). Many of the naturally occurring heavy radionuclides decay in this way (e.g., $^{238}_{92}$U, $^{235}_{92}$U, $^{234}_{92}$U, $^{232}_{90}$Th, $^{226}_{88}$Ra). The absorption of neutrons by certain heavy nuclides (e.g., $^{235}_{92}$U, $^{239}_{94}$Pu) leads to fission, that is, to complete nuclear disruption, with the formation of nuclides of much smaller mass numbers.[14-16,†] Thus the fission yield of $^{235}_{92}$U as shown graphically in Fig. 2-2, indicates maxima at $A = 95$, $Z = 40$ to 42 and $A = 139$, $Z = 56$ to 58 and a minimum at $A = 117$. Spontaneous fission is characteristic of certain still heavier nuclides [e.g., $^{252}_{96}$Cm (ca. 3%), $^{256}_{100}$Fm].

---

*   The "effects" of interconversion of nuclear protons and neutrons require a means of energy transfer, which is apparently dependent upon mesons.

†   For an interesting historical account of "detours" leading to the discovery of nuclear fission, see K. Starke, *J. Chem. Educ.*, **56**, 771 (1979).

2-1. Mass–Energy Equivalence and Nuclear Binding Energy

For a given nuclide, the measured isotopic mass is always less than the summed up masses of the nucleons present. Thus for the nuclide $^{10}_{5}B$, the isotopic mass of which is 10.0129 amu, the calculated mass is 10.0825 amu ($= 5 \times 1.00757$ amu $+ 5 \times 1.00893$ amu). The "mass loss" of 0.0696 amu in the presumed formation of this nuclide from five neutrons and five protons is equivalent, in terms of the Einstein relationship,* to the release of 1.54×10^{11} kcal, 6.44×10^{11} kJ, or 64.8 MeV mole^{-1}. The energy equivalent to a "mass loss," or "mass defect," of this type is a measure of the strength of nuclear binding and is referred to as the *nuclear binding energy*. For comparative purposes, a nuclear binding energy is commonly expressed per nucleon, that is, as the ratio of total binding energy to mass number.

A plot of nuclear binding energy vs. mass number (Fig. 2-3) shows a maximum in the vicinity of ^{56}Fe. The lighter nuclei, with smaller nuclear binding energies, are those most readily transmuted by bombardment; and the heaviest nuclei, again with smaller nuclear binding energies, are the ones that decay

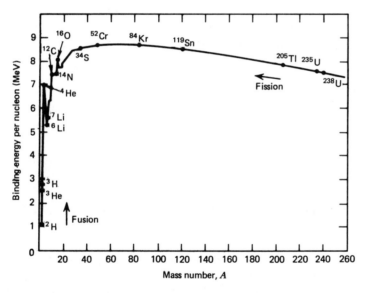

Figure 2-3 Nuclear binding energy as a function of mass number for naturally occuring nuclides (maximum values: ^{52}Cr, ^{55}Mn, ^{56}Fe, ^{60}Ni, ^{65}Cu—all 8.7 MeV). [T. Moeller, J. C. Bailar, Jr., J. Kleinberg, C. O. Guss, M. E. Castellion, and C. Metz, *Chemistry with Inorganic Qualitative Analysis*, Fig. 8.3, p. 201, Academic Press, New York (1980).]

---

* That is, $E = mc^2$, where E is energy, m is mass, and c is the velocity of light. Useful equivalences are: 1 eV = 1.6×10^{-12} erg, 1 MeV = 1.6×10^{-6} erg, 1 kcal = ca. 4.2×10^{-10} MeV = 4.184 kJ, 1 amu = 931 MeV (from the Einstein relationship).

spontaneously. Every nuclear transformation is accompanied by a change in nuclear binding energy that reflects a change in mass. These mass losses are not large, but the equivalent energy quantities released are very large. Thus the fission of the $^{235}_{92}U$ nuclide, which produces about 2.5 additional neutrons per atom, releases about 200 MeV, or 4.8×10^{11} kcal mole^{-1}. Nuclear fusion of $^{1}_{1}H$, which may be formulated via the summation of a number of steps, as

$$4^{1}_{1}H \rightarrow ^{4}_{2}He + 2^{0}_{1}e \quad (\text{or } \beta^{+})$$

releases about 5.9×10^6 kcal mole^{-1} of $^{4}_{2}He$.* (For comparison, the complete combustion of acetylene releases 3.2×10^2 kcal mole^{-1}.)

2-2. Nuclear Energy Levels

Examination of naturally occurring stable nuclei reveals the empirical facts given in Table 2-3. These data suggest that stability is associated with the pairing of nucleons, which pairing could easily be associated with the pairing of nuclear spins and the existence of nuclear energy levels and nuclear shells, somewhat comparable with extranuclear levels and shells involving electrons (Sec. 3.2-3).[17] A detailed examination of this concept reveals maximum nuclear stability at neutron numbers 2, 8, 20, 50, 82, and 126 and at proton numbers 2, 8, 20, 50, and 82. Among the observed facts that correlate with these "magic numbers" are abundance data (Table 2-4) and at least the following:

1. Elements with even Z have larger numbers of isotopes than those of odd Z.
2. For $N = 50$, six stable nuclides exist; for $N = 82$, seven stable nuclides exist.
3. The nuclides $^{87}_{36}Kr$ (51 neutrons) and $^{137}_{54}Xe$ (83 neutrons) are among the few neutron emitters.
4. Nuclei with fewer neutrons than the magic numbers have larger cross sections for neutron capture than nuclei that have magic numbers of neutrons.
5. Discontinuities in nuclear binding energies are observed at $N = 126$ and $Z = 82$.

Both nuclear binding energy and nuclear energy levels, as measures of nuclear stability, are related to the interactions of nuclear particles. The "liquid-drop" concept† provides a useful model in that maximum stability could be associated

---

* For an extensive discussion of nuclear fusion and its potentialities for the production of power, see D. A. Dingee, *Chem. Eng. News*, Apr. 2, 1979, pp. 32–47.

† The liquid-drop model, as proposed by N. Bohr and J. A. Wheeler [*Phys. Rev.*, **56**, 426 (1939)], suggests that each nucleus must be characterized by a critical energy that must be exceeded before elongation and ultimate subdivision can occur. A useful discussion of this approach, as either an alternative or as a supplement to the concept of nuclear energy levels, is given by S. Glasstone, *Sourcebook on Atomic Energy*, 2nd ed., pp. 371–384, D. Van Nostrand, Princeton, N.J. (1958).

Table 2-3 Types of Stable, Naturally Occurring Nuclei

Composition	Designation	Number of Nuclei
Z even, N even (A even)	Even–even	164
Z even, N odd (A odd)	Even–odd	55
Z odd, N even (A odd)	Odd–even	50
Z odd, N odd (A even)	Odd–odd	4[a]

[a] Specifically ^2_1H, ^6_3Li, $^{10}_5B$, $^{14}_7N$.

with a circumstance where nuclear interactions of attraction result in a spherical drop, whereas instability could result as the drop deforms and ultimately fragments via alpha emission or fission. The magic numbers apparently represent combinations of complete pairing where spherical nuclei result. In the absence of pairing, as in the odd situations noted in Table 2-3, distortions from the spherical drop can result in instability.

2-3. Nuclear Spin

Both the neutron and the proton have intrinsic angular momenta* of magnitude $\frac{1}{2}(h/2\pi)$, which is often noted as merely $\frac{1}{2}$. In a nucleus, these momenta combine vectorially to give a net *nuclear spin*, which amounts to an *even* integral number of units of $\frac{1}{2}$ if an even number of nuclear particles is present and to an *odd* integral number of units of $\frac{1}{2}$ if an odd number of nuclear particles is present. The intrinsic angular momentum of a nucleus (I) is *zero* or *integral* for nuclei of even A (e.g., 1 for ^2_1H, 0 for $^{12}_6C$, 0 for $^{16}_8O$, 4 for $^{40}_{19}K$) and *half-integral* for nuclei of odd A (e.g., $\frac{1}{2}$ for ^1_1H, $\frac{3}{2}$ for $^{11}_5B$, $\frac{1}{2}$ for $^{13}_6C$, $\frac{1}{2}$ for $^{15}_7N$, $\frac{1}{2}$ for $^{19}_9F$, $\frac{1}{2}$ for $^{31}_{15}P$). All nuclei of even A and even Z have zero spin in their normal states.

As noted above, the pairing of nuclear spins is associated with enhanced nuclear stability. Furthermore, nuclear spin results in a nuclear magnetic moment.† This is the property upon which nuclear magnetic resonance spectroscopy (Sec. 6.5-4), a technique that has proved to be particularly useful for establishing atomic environments in and the molecular structures of many compounds containing species such as ^1H, ^2H, ^{11}B, ^{13}C, ^{15}N, ^{19}F, ^{31}P, and ^{129}Xe, nuclear quadrupole resonance, and Mössbauer spectroscopy are based.

2-4. The Elemental Composition of the Cosmos (and the Total Earth)

Based upon data from the spectra of heavenly bodies, analyses of meteorites and other extraterrestrial samples, information from space probes, and analyses of

---

* Angular momenta of electrons and units thereof are discussed in Sec. 2.3-3.

† Magnetic moments associated with the motion of electrons are discussed in Sec. 2.4-4.

Table 2-4 Abundances of the Elements

Atomic Number	Symbol	Igneous Rocks of Crust of Earth[a] Grams/Metric Ton (ppm)	Percent by Weight	Atoms/10^6 Atoms Si	Cosmos[a,b] Atomic Abundance (Si = 1 × 10^6)
1	H	0.003	3×10^{-7}	7.6×10^{-2}	3.18×10^{10}
2	He				2.21×10^{9}
3	Li	65	6.5×10^{-3}	9.1×10^{2}	49.5
4	Be	6	6×10^{-4}	67	0.81
5	B	3–	3×10^{-4}	28	350
6	C	320–	3.2×10^{-2}	2.7×10^{3}	1.18×10^{7}
7	N	46.3	4.6×10^{-3}	3.3×10^{2}	3.74×10^{6}
8	O	466,000	46.6	2.96×10^{6}	2.15×10^{7}
9	F	600–900	$(6{-}9) \times 10^{-2}$	$(3.2{-}4.8) \times 10^{3}$	2.45×10^{3}
10	Ne	0.00007	7×10^{-9}	3.5×10^{-4}	3.44×10^{6}
11	Na	28,300	2.83	1.24×10^{5}	6.0×10^{4}
12	Mg	20,900	2.09	8.76×10^{4}	1.061×10^{6}
13	Al	81,300	8.13	3.05×10^{5}	8.5×10^{4}
14	Si	277,200	27.72	1.00×10^{6}	1.00×10^{6}
15	P	1,180	0.118	3.8×10^{3}	9.60×10^{3}
16	S	520	5.2×10^{-2}	1.6×10^{3}	5.0×10^{5}
17	Cl	314	3.14×10^{-2}	9×10^{2}	5.70×10^{3}
18	Ar	0.04	4×10^{-6}	0.1	1.172×10^{5}
19	K	25,900	2.59	4.42×10^{4}	4.20×10^{3}
20	Ca	36,300	3.63	9.17×10^{4}	7.21×10^{4}
21	Sc	5	5×10^{-4}	11	35
22	Ti	4,400	0.44	9.2×10^{3}	2.775×10^{3}
23	V	150	1.5×10^{-2}	3.0×10^{2}	262

24	Cr	200	2.0×10^{-2}	3.9×10^{2}	1.27×10^{4}
25	Mn	1,000	0.10	1.8×10^{3}	9.30×10^{3}
26	Fe	50,000	5.0	9.13×10^{4}	8.3×10^{5}
27	Co	23	2.3×10^{-3}	40	2.21×10^{3}
28	Ni	80	8.0×10^{-3}	1.4×10^{2}	4.80×10^{4}
29	Cu	70	7.0×10^{-3}	1.1×10^{2}	5.40×10^{2}
30	Zn	132	1.32×10^{-2}	2.0×10^{2}	1.244×10^{3}
31	Ga	15	1.5×10^{-3}	22	48
32	Ge	7	7×10^{-4}	9.5	1.15×10^{2}
33	As	5	5×10^{-4}	6.7	6.6
34	Se	0.09	9×10^{-6}	0.12	67.2
35	Br	1.62	1.6×10^{-4}	2.0	13.5
36	Kr	—	—	—	46.8
37	Rb	310	3.1×10^{-2}	3.6×10^{2}	5.88
38	Sr	300	3.0×10^{-2}	3.5×10^{2}	26.9
39	Y	28.1	2.81×10^{-3}	30.7	4.8
40	Zr	220	2.2×10^{-2}	2.6×10^{2}	28
41	Nb	24	2.4×10^{-3}	26	1.4
42	Mo	2.5–15	$(2.5–15) \times 10^{-4}$	3–16	4.0
43	Tc	Present			
44	Ru	0.001	1×10^{-7}	1×10^{-3}	1.9
45	Rh	0.010	1.0×10^{-6}	9×10^{-3}	0.4
46	Pd	0.10	1.0×10^{-5}	9×10^{-2}	1.3
47	Ag	0.15	1.5×10^{-5}	0.13	0.45
48	Cd	0.1	1×10^{-5}	7×10^{-2}	1.48
49	In				0.189
50	Sn	40	4.0×10^{-3}	34.3	3.6
51	Sb	1	1×10^{-4}	0.83	0.316
52	Te	0.0018 (?)	1.8×10^{-7} (?)	—	6.42
53	I	0.3	3×10^{-5}	0.24	1.09
54	Xe	—	—	—	5.38

Table 2-4 (*Continued*)

Atomic Number	Symbol	Igneous Rocks of Crust of Earth[a]: Grams/Metric Ton (ppm)	Igneous Rocks of Crust of Earth[a]: Percent by Weight	Igneous Rocks of Crust of Earth[a]: Atoms/10^6 Atoms Si	Cosmos[a,b] Atomic Abundance ($Si = 1 \times 10^6$)
55	Cs	7	7×10^{-4}	5.3	0.387
56	Ba	250	2.5×10^{-2}	1.8×10^{2}	4.8
57	La	18.3	1.83×10^{-3}	12.8	0.445
58	Ce	46.1	4.61×10^{-3}	32.1	1.18
59	Pr	5.53	5.53×10^{-4}	3.89	0.149
60	Nd	23.9	2.39×10^{-3}	16.2	0.78
61	Pm				
62	Sm	6.47	6.47×10^{-4}	4.19	0.226
63	Eu	1.06	1.06×10^{-4}	0.68	0.085
64	Gd	6.36	6.36×10^{-4}	3.94	0.297
65	Tb	0.91	9.1×10^{-5}	0.56	0.055
66	Dy	4.47	4.47×10^{-4}	2.69	0.36
67	Ho	1.15	1.15×10^{-4}	0.68	0.079
68	Er	2.47	2.47×10^{-4}	1.44	0.225
69	Tm	0.20	2.0×10^{-5}	0.115	0.034
70	Yb	2.66	2.66×10^{-4}	1.49	0.216
71	Lu	0.75	7.5×10^{-5}	0.37	0.036
72	Hf	4.5	4.5×10^{-4}	3.0	0.21
73	Ta	2.1	2.1×10^{-4}	1.2	0.021
74	W	1.5–69	$(1.5–69) \times 10^{-4}$	0.82—38	0.16
75	Re	0.001	1×10^{-7}	5.4×10^{-4}	0.053
76	Os	Present	—	—	0.75
77	Ir	0.001	1×10^{-7}	5×10^{-4}	0.717

78	Pt	0.005	5×10^{-7}	2.7×10^{-3}	1.4
79	Au	0.005	5×10^{-7}	2.6×10^{-3}	0.202
80	Hg	0.077–0.5	$(0.8–50) \times 10^{-6}$	0039–0.25	0.4
81	Tl	0.3–3	$(3–30) \times 10^{-5}$	0.15–1.5	0.192
82	Pb	16	1.6×10^{-3}	8.0	4
83	Bi	0.2	2×10^{-5}	9×10^{-2}	0.143
84	Po	3×10^{-10}	3×10^{-14}	1.4×10^{-10}	
85	At	Present	—	—	
86	Rn	Present	—	—	
87	Fr	Present	—	—	
88	Ra	1.3×10^{-6}	1.3×10^{-10}	5.8×10^{-7}	
89	Ac	3×10^{-10}	3×10^{-14}	1.3×10^{-10}	
90	Th	11.5	1.15×10^{-3}	5.0	0.058
91	Pa	8×10^{-7}	8×10^{-11}	3.5×10^{-7}	
92	U	4	4×10^{-4}	1.6	0.0262
93	Np	Probably present	—	—	
94	Pu	Present	—	—	

[a] 10^6 atoms of Si is a useful point of reference.

[b] A. G. W. Cameron, *Space Sci. Rev.*, **15**, 121 (1973).

selected terrestrial samples,[18-20] abundance data for the elements have been summarized for both the crust of the earth (outermost 36 km) and the cosmos (Table 2-4). The total earth is assumed to be a representative fragment of the solar system and the solar system a representative fragment of the cosmos.[21-25] Relative cosmic abundances are plotted as a function of atomic number in Fig. 2-4.

A detailed examination of cosmic abundance data indicates a rapid exponential decrease (by a factor of ca. 10^{10}) from elements of small atomic number to about $Z = 42$ (Mo), after which elemental abundances remain reasonably constant. It is apparent also that the abundances of elements with $Z > 28$ (i.e., beyond Ni) vary much less than those with $Z < 28$, suggesting different modes of formation of nuclides in these two regions. Elements with even Z are more abundant than their immediate neighbors of odd Z (the so-called Oddo–Harkins rule).[26] Only 11 elements (i.e., H, He, C, N, O, Ne, Mg, Si, S, Fe, Ni—all with $Z < 29$) are really abundant. That cosmic abundance data and these generalizations are related to nuclear stability is indicated by the following:

1.   The very large abundances of hydrogen and helium. These elements may thus be regarded as reasonable primordial nuclear fuels.

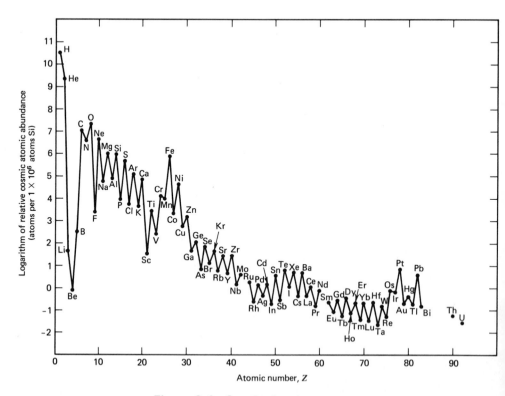

Figure 2-4   Cosmic abundance curve.

2. The comparatively small abundances of lithium, beryllium, and boron. Nuclides of these elements are known to be easily transmuted by nuclear bombardment.

3. The larger abundances of "alpha-particle" nuclides (i.e., those the mass numbers of which are multiples of 4, e.g., ^{16}O, ^{20}Ne, ^{24}Mg, ^{40}Ca, ^{48}Ti) as compared with neighboring nuclides. The nuclear stability of the two proton–two neutron combination is indicated.

4. The remarkably large abundance of the nuclide ^{56}Fe (by a factor of 10^4). This abundant nuclide lies at the maximum in the nuclear binding energy curve (Fig. 2-3).

5. Maximum abundances in pairs among the heavier elements (e.g., at $A = 80$, 90; 130, 138; 196, 208). The influence of magic numbers of nucleons is evident.

6. Lack of abundance of proton-rich nuclei. A reasonable conclusion is primordial elemental syntheses involving neutron absorption.

These, and other, data suggest very strongly[27-34] that the cosmological syntheses of the elements involved (and continue to involve) successive processes of *thermonuclear fusion* (or *thermal cooking*), *neutron absorption-β$^-$ emission*, and *spontaneous fission*, beginning with hydrogen as the ultimate fuel. These processes are summarized in Table 2-5 and depicted graphically, as a time–temperature relationship, in Fig. 2-5. It is assumed (1) that as hydrogen condensed (or now condenses) in stars, conversion of gravitational energy into thermal energy so increased the temperature that initially exothermic fusion to helium occurred (or occurs) and that the necessary neutrons were (or are) provided by natural nuclear reactions, such as fission or a sequence such as*

$$^{20}\text{Ne}(p, \gamma)^{21}\text{Na}(—, \beta^+)^{21}\text{Ne}(\alpha, n)^{24}\text{Mg}$$

The schematic summary in Table 2-5 requires some clarification. Hydrogen burning does not involve an unlikely direct four-body conversion to helium and may proceed by more reasonable pathways: for example,

$$^1\text{H}(p, \beta^+)^2\text{D}(p, \gamma)^3\text{He}(^3\text{He}, 2\,p)^4\text{He}$$

Furthermore, ^{4}He may also result as the net product from a sequence such as

$$^{12}\text{C}(p, \gamma)^{13}\text{N}(—, \beta^+)^{13}\text{C}(p, \gamma)^{14}\text{N}(p, \gamma)^{15}\text{O}(—, \beta^+)^{15}\text{N}(p, {}^4\text{He})^{12}\text{C}$$

that is, a carbon–nitrogen cycle, which also yields non-alpha-particle nuclides of the light elements. Furthermore, rapid proton capture (*p* process) by heavier nuclides probably accounts for proton-rich nuclei. Spallation reactions†

---

* The notation ^{20}Ne$(p, \gamma)^{21}$Na, for example, indicates that the nuclide ^{20}Ne is bombarded with protons and thereby yields γ radiation plus the nuclide ^{21}Na.

† In a spallation reaction, a given nuclide emits several nucleons and gives rise to various lighter nuclides.

Table 2-5 Cosmological Nuclear Syntheses of the Elements

Nuclear Process	Schematic Nuclear Reaction	Approximate Temperature (°K)	Approximate Ultimate Nuclear Binding Energy MeV/nucleon
Thermal cooking			
H burning	$^1H \to {}^4He$ + heat	10^7	7.1
He burning	$^4He \to {}^{12}C$, ^{16}O + heat	10^8	8.0
C burning	$^{12}C \to {}^{20}Ne$, ^{23}Na, ^{23}Mg, $^{24}Mg + \alpha, p, n$ + heat	10^9	8.6
O burning	$^{16}O \to {}^{28}Si$, ^{31}P, ^{31}S, $^{32}S + \alpha, p, n$ + heat	10^9	8.6
α absorption (Si burning, e process)	$^{28}Si + \alpha, p, n \to {}^{32}S \ldots {}^{56}Fe$	10^9	8.8
Neutron absorption			
β^- emission[a] (s and r processes)	$^{56}Fe + n \to$ nuclides of larger mass and charge $+ \beta^-$	10^9	<8.8
Spontaneous fission	large $A_{E \to \text{fission products}}$ + heat	$>10^{10}$	<8.8

[a] Successive process, such as $^{56}Fe(n, \beta^-){}^{57}Co(n, \beta^-){}^{58}Ni$.

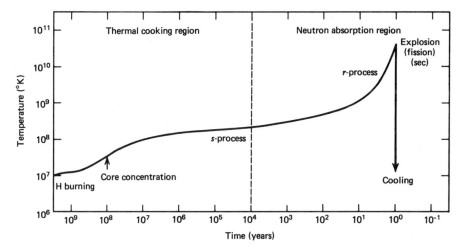

Figure 2-5 Time–temperature relationships in nuclear syntheses.

(x process) involving carbon and oxygen nuclides can account for the formation of lithium, beryllium, and boron.

Neutron absorption–β^- emission reactions account for the formation of some lighter nuclides but are most significant in the stepwise syntheses of nuclides beyond iron in mass and charge. Neutron capture by a nucleus increases A by one unit and increases the N/P ratio, commonly sufficiently to cause β^- emission and thus increase Z by one unit. Two processes may be distinguished: the s process, in which neutron capture is *slow* compared with β^- emission, and the r process, in which neutron capture is *rapid* compared with β^- emission. As shown in Fig. 2-6, the s process characterizes nuclides up to ^{56}Fe, where the r process can also occur. The s process leads ultimately to stable ^{209}Bi, beyond which α decay is spontaneous. The r process forms other nuclides up to ^{254}Cf and ends in spontaneous fission (at $A = 260$), which feeds materials back in the $A = $ ca. 108 and 146 regions. Approaches to nuclear stability are evident in the magic number species ($N = 50, 82, 126$).

Terrestrial Abundances of the Elements. The data in Table 2-4 indicate substantial differences between cosmic and crustal abundances. In the formation of the earth as we now know it, transition from the fluid state to the solid state resulted in broad fractionation into a zonal structure in terms of density changes and crystallization.[19, 35] Broadly, these zones, and their depths from the surface, are the continental crust (to ca. 36 km), the mantle (36 to 2900 km), and the core (2900 km to center). Of these, currently only the crust is available to us as a source of the elements and their compounds.* In the formation of the earth, according to Goldschmidt,[19] a primary differentiation based upon both den-

* Plus, of course, the atmosphere and bodies of water.

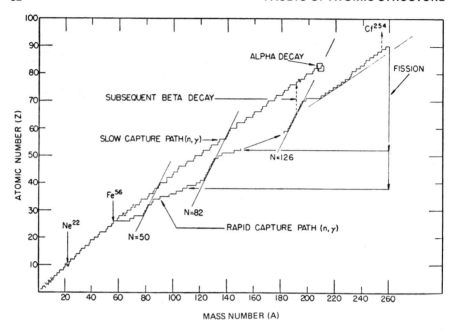

Figure 2-6 Nuclear synetheses by neutron capture—β^- emission reactions. [E. M. and G. R. Burbidge, W. Fowler, and F. Hoyle, *Rev. Mod. Phys.*, **29**, 547 (1957).]

sities and affinities for possible chemical phases resulted in distribution of the elements among these zones. Primarily, the crust contains silicates and oxides, the mantle combinations of olivine [i.e., $(Mg, Fe)_2SiO_4$] and iron (II) sulfide, and the core elemental iron and nickel. Secondary redistributions in the crust have resulted from volcanic activity, erosion, and other processes.

Based upon their relative affinities for metallic, sulfidic, silicate, and gaseous phases, Goldschmidt has classified the elements into four groups as indicated in Table 2-6. The siderophilic ("iron loving") elements include the relatively chemically inactive metals. The chalcophilic ("copper loving," but associated with sulfur since copper is commonly found as sulfidic minerals) elements are those of ionization energies larger than 6 to 7 eV, which form covalently bonded sulfides. The lithophilic ("stone loving," from their occurrence as silicates) elements are those of ionization energies less than 6 to 7 eV, which bond strongly and ionically to oxygen. The atmophilic ("vapor loving") elements are those that are gaseous at ambient temperatures. Cooccurrence of siderophilic elements in nature is often determined by isomorphism (e.g., GaAs as a minor material in ZnS). Cooccurrence of lithophilic elements is commonly determined by similarities in cationic radii (e.g., Ga^{3+}, $r = 0.62$ Å, with Al^{3+}, $r = 0.51$ Å, in silicate and oxide minerals). In Table 2-7, ions listed in each cationic radius bracket are often found together in silicate systems, even though they may be chemically unrelated.

Table 2-6 Geochemical Classification of the Elements[19]

Siderophilic	Chalcophilic	Lithophilic	Atmophilic
Fe,[a] Co,[a] Ni[a]	Cu,[d] Ag	Li, Na, K, Rb, Cs	H,[d] N, O[d]
Ru, Rh, Pd	Zn, Cd, Hg	Be, Mg, Ca, Sr, Ba	He, Ne, Ar, Kr, Xe
Os, Ir, Pt	Ga, In, Tl	B, Al, Sc, Y, La-Lu	
Au, Re,[b] Mo[b]	Ge,[d] Sn,[d] Pb	Si, Ti, Zr, Hf, Th	
Ge,[a] Sn,[a] W[b]	As,[d] Sb,[d] Bi	P, V, Nb, Ta	
C,[c] Cu,[a] Ga[a]	S, Se, Te	O, Cr, U	
Ge,[a] As,[b] Sb[b]	Fe,[d] Mo, Os[d]	H, F, Cl, Br, I	
	Ru,[d] Rh,[d] Pd[d]	Fe,[d] Mn, Zn,[d] Ga[d]	

[a] Chalcophilic and lithophilic in crust of earth.

[b] Chalcophilic in crust of earth.

[c] Lithophilic in crust of earth.

[d] Position of secondary affinity.

Cosmo- and Geochemistry and Inorganic Chemistry. The origin and overall abundances of the elements and the abundances and distributions of the elements in the earth, which are the primary concerns of *cosmochemistry* and *geochemistry*, respectively, are important as bases for implementation of our definition of inorganic chemistry (p. 2), both in terms of providing direct application of our concepts of atomic structure and in terms of indicating availability and sources of inorganic substances. It is significant to recall again that modern inorganic chemistry had its origins in the analyses of minerals. Further useful information on geochemistry can be obtained from several excellent sources.[19, 37-39]

Table 2-7 Ionic Species Occurring Together in Minerals[36]

Ionic, or Crystal, Radius (Å)	Ionic Species[a]
0.1–0.3	B^{3+}, C^{4+}, N^{5+}, S^{6+}
0.3–0.5	Be^{2+}, Si^{4+}, P^{5+}, V^{5+}
0.5–0.7	Li^+, Mg^{2+}, Al^{3+}, Ga^{3+}, Fe^{3+}, Cr^{3+}, V^{3+}, Ti^{4+}, Ge^{4+}, Mo^{6+}, W^{6+}
0.7–0.9	Ni^{2+}, Co^{2+}, Fe^{2+}, Zn^{2+}, Sc^{3+}, In^{3+}, Zr^{4+}, Hf^{4+}, Sn^{4+}, Nb^{5+}, Ta^{5+}
0.9–1.1	Na^+, Ca^{2+}, Cd^{2+}, Y^{3+}, $Gd^{3+} - Lu^{3+}$, Ce^{4+}, Th^{4+}, U^{4+}
1.1–1.4	K^+, Sr^{2+}, $La^{3+} - Eu^{3+}$
1.4–1.7	Rb^+, Cs^+, Tl^+, Ba^{2+}, Ra^{2+}

[a] Several species are formally considered as ions but have no real existence as ions (e.g., N^{5+}, S^{6+}, B^{3+}).

3. THE EXTRANUCLEAR STRUCTURES OF THE ATOMS

Under conditions of high-temperature excitation, gaseous atoms emit radiant energy, which by means of suitable instruments, can be resolved into sharply defined emission spectra characteristic of the specific elements in question. Thus the emission spectrum of atomic hydrogen, the simplest of the atomic species, is recorded as series of lines of precisely measurable wavelengths that are describable as the Lyman series (ultraviolet), Balmer series (visible), Paschen series (near infrared), Brackett series (far infrared), and Pfund series (even farther infrared). Lines in the first three series were characterized empirically in the late 1800s by the relationship

$$\frac{1}{\lambda} = \bar{v} = R\left(\frac{1}{n_1^2} - \frac{1}{n_2^2}\right) \tag{2-1}$$

where λ is the wavelength, $\bar{v}$ the wave number, n_1 and n_2 are small integers with $n_1 < n_2$, and R is the so-called Rydberg constant $(= 109{,}677.58 \text{ cm}^{-1}$ for hydrogen).* Values of n_1 for the Lyman, Balmer, and Paschen series are, respectively, 1, 2, and 3; for the more recently discovered Brackett and Pfund series, 4 and 5, respectively. It is significant that energy, recorded as spectral lines, is emitted discontinuously not only for hydrogen but for all atomic species.

In 1900, M. Planck,[40] using observations of this type, concluded that, contrary to then-existing theory, energy can neither be emitted nor absorbed continuously but rather only in definite quantities that are multiples of its frequency. Electromagnetic radiation is thus considered to be *quantized*, that is, restricted to certain definite values based upon units called *quanta*. Mathematically, permitted energy values and frequency were related as

$$E = hv \tag{2-2}$$

where h is a universal action constant (Planck's constant) with the magnitude 6.624×10^{-27} erg sec. Planck's conclusion provided the basis for all modern quantum theory: theory that attempts to account for the properties and behavior of all particles of atomic and subatomic dimensions.

3-1. Bohr Theory of the Hydrogen Atom — Principles and Limitations

In 1913, N. Bohr[41] proposed a quantum mechanical–classical hybrid interpretation of the emission spectrum of gaseous hydrogen in terms of an atomic structure, an interpretation that provided a basis for subsequent

---

* Electromagnetic radiation consists of oscillating and mutually perpendicular electric (E) and magnetic (H) components propagated as sine-wave patterns at a rate, *in vacuo*, of 2.9978×10^{10} cm sec^{-1} (c). The length of a given sine cycle [i.e., the wavelength (λ)] is expressed in (cm, Å, or nm) cycle^{-1}. The frequency (v), in cycles sec^{-1}, is given by the ratio c/λ. The wave number ($\bar{v}$), which is a measure of the energy of a photon, is the reciprocal of λ, with units (cm, Å, or nm)$^{-1}$.

structural postulations. Bohr's approach was based upon two fundamental postulates: (1) the single electron in the hydrogen atom revolves about the nucleus in any one of many circular orbits (stationary states), for each of which the energy of the electron has a fixed and constant value, and (2) energy is absorbed (in a definite quantity) when the electron is raised from one stationary state to another of higher energy (*excitation*) and emitted in the same quantity when the reverse process occurs (*emission*, giving spectral lines). Energy is thus quantized in terms of the equation

$$E_2 - E_1 = hv \tag{2-3}$$

where $E_2 > E_1$. A necessary assumption was that the orbital angular momentum of the electron ($=mvr$, where m is the mass, v the velocity, and r the radius of orbit) can have only simple multiple values, that is,

$$mvr = \frac{nh}{2\pi} \tag{2-4}$$

where $n = 1, 2, 3, \ldots$ and is a so-called *quantum number*. Since r increases with n in Eq. 2-4, values of n indicate the order of orbits of increasing size as the distance from the nucleus increases.

To satisfy his condition that the electron revolves in an orbit of fixed r, Bohr equated the centrifugal force on the electron ($=mv^2/r$) to the Coulombic force of attraction by the nucleus ($=Ze^2/r^2$), where e is the charge on the electron. By ultimate solution, he obtained

$$E = -\frac{2\pi^2 me^4 Z^2}{h^2 n^2} \tag{2-5}$$

for the energy associated with the electron in any quantum state. For the transition from one energy state E_2 to a lower state E_1 and incorporation of Eq. 2-3, he then obtained for emitted energy

$$E_2 - E_1 = hv = \frac{2\pi^2 me^4 Z^2}{h^2} \left(\frac{1}{n_1^2} - \frac{1}{n_2^2} \right) \tag{2-6}$$

which, in terms of wave number for an observed emission line can be rewritten

$$\bar{v} = \frac{2\pi^2 me^4 Z^2}{h^3 c} \left(\frac{1}{n_1^2} - \frac{1}{n_2^2} \right) \tag{2-7}$$

Substitution of numerical values for constants gives

$$\frac{2\pi^2 me^4 Z^2}{h^3 c} = 109,737 \text{ cm}^{-1} \tag{2-8}$$

in excellent agreement with the Rydberg value.*

---

* More nearly exactly, the electron revolves about the center of gravity of the system, and nuclear motion requires that the reduced mass (μ) [i.e., $\mu = mM/(m + M)$, where M = nuclear mass] be substituted for M. When this is done, modified Eq. 2-8 gives 109,678 cm^{-1}.

Suitable modifications of the mathematics involved allow one to calculate a Bohr radius, which for $n = 1$ is 0.529 Å or 0.529×10^{-8} cm ($=a_0$ in the literature), a value of fundamental importance (p. 43).

By means of Eq. 2-7, it is possible to calculate the wave numbers and energies of lines in the emission spectrum of atomic hydrogen with excellent accuracy, thus giving one confidence in the Bohr model of the hydrogen atom. Unfortunately, comparable models do not give accurate results for species containing more than a single electron. Furthermore, at high resolution and in the presence of magnetic fields, emission lines that cannot be accounted for on the basis of the Bohr theory are noted. A fundamental problem is the inconsistency between the Bohr theory and W. Heisenberg's principle of uncertainty,[42] which notes the impossibility of determining simultaneously both the position and the orbital angular momentum of the electron. Yet correct values of E_n result. Any satisfactory model that ignores probability cannot be constructed.

3-2. Development of a Wave-Mechanical Model for Atomic Systems

The failure of the Bohr model to predict the properties of any atomic species containing more than one electron led to several models based upon noncircular orbits, interpenetration of orbits, and so on, all designed to account for atomic spectra and electronic distributions in more complex atoms.[43-45] Although it was thus possible to arrive at the correct distributions of electrons among major energy levels and, at least to some degree, among sublevels, it was apparent that only a one-electron system could be treated rigorously.

Significant to continuing development was L. de Broglie's extension to the electron of the dualistic view that light has both wave and particle properties.[46] The momentum ($mc = p$) of a photon is calculated by equating the Einstein (Sec. 2.2-1) and Planck (Sec. 2-3) relationships, converting frequency (v) to wavelength (λ) and solving to give

$$p = mc = \frac{h}{\lambda} \tag{2-9}$$

Similarly, substitution of the velocity of an electron (v) for that of light (c) gives for the electron

$$p = mv = \frac{h}{\lambda} \tag{2-10}$$

or

$$\lambda = \frac{h}{mv} \tag{2-11}$$

where λ then becomes a "de Broglie wavelength." The wavelike property of electrons was confirmed in 1927 by the observed diffraction of electrons impinging on the surface of a crystal of nickel.[47]

A standard, or stationary, wave results when a particle acquires periodic motion and returns to its initial path after a certain number of wavelengths. This condition obtains only when the frequency of the wave, or the energy of the particle, is such that an integral number of waves just completes the path. If this requirement is not fulfilled, interference and destruction of the waves result. Such a situation is shown diagrammatically in Fig. 2-7 for a circular orbit (of radius r) of an electron about a nucleus, the solid line indicating reinforcement at the correct frequency and the dashed lines destruction by interference, and thus theoretically impossible quantum states. Mathematically, the condition for reinforced wave motion is described as

$$n\lambda = 2\pi r \qquad (2\text{-}12)$$

where n is an integer (1, 2, 3, ...). It is significant that combination of this equation with Eq. 2-11 and simplification gives Eq. 2-4, thereby indicating consistency with a fundamental Bohr postulation in terms of a wave characteristic of the electron.

Quantum-mechanical approaches to electronic configurations of atoms were developed independently in 1926 by W. Heisenberg,[42] who used matrix mechanics as a basis, and by E. Schrödinger,[48] who used wave mechanics. The two

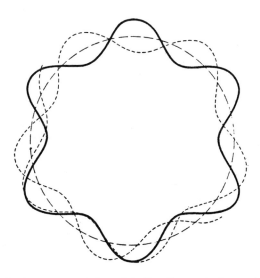

Figure 2-7 Qualitative representation of de Broglie waves for an electron in a circular orbit. De Broglie waves for the circular orbits of an electron about the nucleus of an atom (qualitative). Solid line represents a stationary state (standing wave); dashed line, a quantum theoretically impossible state (waves destroyed by interference). [G. Herzberg, *Atomic Spectra and Atomic Structure*, Fig. 15, p. 31, Dover Publications, New York (1944).]

methods are mathematically equivalent, but only the wave-mechanical approach is treated in subsequent sections. It is of interest that Schrödinger's treatment predated experimental verification[47] of de Broglie's hypothesis of duality.

3-3. The Schrödinger Wave-Mechanical Model of the Hydrogen Atom

The basic concept of the Schrödinger approach is that a particle in motion can be represented by a wave function, ψ, which specifies the coordinates and the oscillatory behavior of that particle, that is, all that one can know about the properties of the system when it is in a steady state that is not changing as time changes. For such a one-dimensional system, involving a particle of mass m in a potential V, ψ is then the solution to the equation

$$\frac{d^2\psi}{dx^2} + \frac{2m}{h^2}(E - V)\psi = 0 \tag{2-13}$$

where E is energy and x a Cartesian coordinate. Every particle in a collection of particles is described by a wave function, which depends on the coordinates of that particle. It has no simple physical meaning. The product of the wave function ψ and its complex coordinate ψ^* is proportional to the *probability density* associated with the particle. The probability that the particle is located in a small element of volume $dx\,dy\,dz$ is given by the product $(\psi\psi^*)$ (volume).

Quantum-mechanical quantities are represented by operators.† Each observable quantity has a corresponding operator. Expression of Eq. 2-13 in terms of operators involves a classical relationship of energy in terms of momenta and coordinates,

$$H = T + V \tag{2-14}$$

where H is a function known as a Hamiltonian and T is a kinetic energy. For a particle of mass m moving in the x direction at a potential dependent solely on x, then

$$H = \frac{p_x^2}{2m} + V \tag{2-15}$$

for which the corresponding quantum-mechanical Hamiltonian operator $\mathscr{H}$ is

$$\mathscr{H} = -\frac{h^2}{2m}\frac{d^2}{dx^2} + V \tag{2-16}$$

In this case, the function on which $\mathscr{H}$ operates is the wave function ψ, giving

$$\mathscr{H}\psi = E\psi \tag{2-17}$$

---

† An operator is a description of a mathematical operation that is to be applied to a given function to give a new function.

Substitution of Eqs. 2-16 and 2-17 then gives the quantum-mechanical equivalent of Eq. 2-13, or

$$-\frac{h^2}{2m}\frac{d^2\psi}{dx^2} + V\psi = E\psi$$

or

$$\frac{d^2\psi}{dx^2} + \frac{2m}{h^2}(E - V)\psi = 0 \tag{2-18}$$

Then for a particle moving in three dimensions,

$$\frac{\partial^2\psi}{\partial x^2} + \frac{\partial^2\psi}{\partial y^2} + \frac{\partial^2\psi}{\partial z^2} + \frac{2m}{h^2}(E - V)\psi = 0 \tag{2-19}$$

or using the Laplacian operator ∇^2, where

$$\nabla^2 = \frac{\partial^2}{\partial x^2} + \frac{\partial^2}{\partial y^2} + \frac{\partial^2}{\partial z^2} \tag{2-20}$$

we have

$$\nabla^2\psi + \frac{2m}{h^2}(E - V)\psi = 0 \tag{2-21}$$

Either Eq. 2-19 or 2-21 is then the Schrödinger equation for a single particle moving in three directions as described. Wave functions that are single-valued, continuous, and have finite values of the integral of the square of the function exist only for certain specific values of E. These values are called *eigenvalues*, and the corresponding wave functions are termed eigenfunctions (from the German *eigenfunktion*, characteristic function). The eigenvalues are the stationary energy states of the system. The primary functions of quantum mechanics are to determine appropriate eigenfunctions and corresponding eigenvalues.

Extension to atomic systems requires some additional considerations. The simplest atomic system is one of nuclear mass M and charge $+e$, with a single electron of mass m_e and charge $-e$. If it is assumed that the electron moves at a distance r around the stationary nucleus, the nucleus can be taken as the origin in a coordinate system. The energy of the electron consists of a kinetic term

$$T = \frac{m_e v_e^2}{2} \tag{2-22}$$

and a potential term

$$V = -\frac{e^2}{r} \tag{2-23}$$

The total Hamiltonian for the energy of the electron then becomes

$$H = \frac{m_e v^2}{2} - \frac{e^2}{r} \tag{2-24}$$

which converts, in terms of earlier considerations, into

$$\mathcal{H} = \frac{1}{2m_e}\left(-\frac{h^2\nabla^2}{4\pi^2}\right) - \frac{e^2}{r}$$

$$= -\frac{h^2}{8\pi m_e}\nabla^2 - \frac{e^2}{r} \qquad (2\text{-}25)$$

Then substitution into Eq. 2-17 yields

$$\left(-\frac{h^2}{8\pi^2 m_e}\nabla^2 - \frac{e^2}{r}\right)\psi = E\psi \qquad (2\text{-}26)$$

which rearranges to

$$\nabla^2\psi + \frac{8\pi^2 m_e}{h^2}\left(E + \frac{e^2}{r}\right)\psi = \nabla^2\psi + \frac{8\pi^2 m_e}{h^2}(E - V)\psi = 0 \qquad (2\text{-}27)$$

Equation 2-27, written either in this way or in the form of partials, is the Schrödinger wave equation for the hydrogen atom. Related equations can be written for two-electron, three-electron, and so on, species, but solutions can be obtained only for one-electron species.

The wave function ψ in Eq. 2-27 depends upon the coordinates x, y, and z. For a specified set of values of x, y, and z (i.e., at a given point in space relative to the atomic nucleus), only one value of ψ can be meaningful. Thus ψ is a probability function (p. 42). Although the resulting concept of probability of concentration of charge density for a single electron is somewhat nebulous, it is apparent that over a period of time the motion of the electron will result in a cloud of charge density about the nucleus. Furthermore, this cloud will be bounded by a surface which has a high probability of including the electron at all times.

Quantum Numbers and Orbitals. Equation 2-27 is most readily solved by first transforming from Cartesian coordinates to polar coordinates.* The relationships between the two coordinate systems are indicated in Fig. 2-8. The trigonometric equivalences are

$$\frac{y}{x} = \tan\theta$$

$$x = r\sin\theta\cos\phi$$

$$y = r\sin\theta\sin\phi$$

$$z = r\cos\theta$$

and, since $r^2 = x^2 + y^2 + z^2$,

$$\cos\theta = \frac{z}{(x^2 + y^2 + z^2)^{\frac{1}{2}}}$$

---

* For details, one may consult treatises on quantum mechanics or physical chemistry.

Equation 2-27 then transforms to

$$\frac{1}{r^2}\frac{\partial}{\partial r}\left(r^2\frac{\partial\psi}{\partial r}\right) + \frac{1}{r^2\sin\theta}\frac{\partial}{\partial\theta}\left(\sin\theta\frac{\partial\psi}{\partial\theta}\right) + \frac{1}{r^2\sin\theta}\frac{\partial^2\psi}{\partial\phi^2}$$

$$+ \frac{8\pi^2 m_e}{h^2}\left(E + \frac{e^2}{r}\right) = 0 \tag{2-28}$$

This equation, in turn, indicates that ψ is itself a function of r, θ, and ϕ as

$$\psi(r,\,\theta,\,\phi) = R(r)\Theta(\theta)\Phi(\phi) \tag{2-29}$$

where R is a radial, or spherical, function expressing spatial distribution of electronic charge density relative to the nucleus, and Θ and Φ are angular functions expressing orientation of this spatial distribution.

By appropriate mathematical manipulations, separate solvable differential equations for the R, Θ, and Φ functions can be developed. Each of these equations leads to the introduction of a separate *quantum number*, each of which describes a degree of freedom for the electron in the hydrogen atom. These quantum numbers are, respectively:

1. The *principal quantum number* (n), which measures the radial distribution electronic charge density and has the same modular significance as the n in the Bohr theory (p. 35). Permissible solutions of the Schrödinger equation correspond to $n = 1, 2, 3, ..., \infty$. Eigenvalues for the hydrogen atom are given by Eq. 2-5, as derived by Bohr but consistent with the quantum-mechanical treatment.

2. The *azimuthal*, or *angular momentum*, *quantum number* (l), which measures the angular momentum of the electron in terms of the relationship

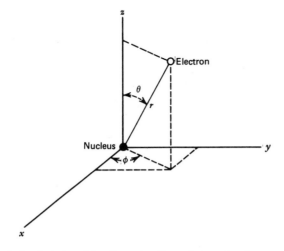

Figure 2-8 Relationships between Cartesian and polar coordinates.

$(h/2\pi)\sqrt{l(l+1)}$. Permitted values of l are 0, 1, 2, ..., $n-1$. The l values are symbolized: $l = 0$, or s; $l = 1$, or p; $l = 2$, or d; $l = 3$, or f; $l = 4$, or g; and so on.

3. The *magnetic quantum number* (m_l), which measures the removal of degeneracy (i.e., equal energy) upon application of a magnetic field.* Permitted values of m_l are 0, ± 1, ± 2, ..., $\pm l$ (i.e., a total $2l + 1$ values). The range is $-l$, $-(l-1)$, ..., -1, 0, $+1$, ..., $(l-1)$, l.

The wave function ψ for a given combination of n, l, and m_l values is called an *orbital*. Bohr orbits and orbitals are not synonymous. An orbit represents a specific pathway followed by an electron as it revolves about a nucleus. An orbital is a probability function that can be interpreted in a physical sense as a region in space relative to the nucleus where the probability of finding an electron of particular energy characteristics has a specified value (e.g., 90%, 95%, etc.). Orbitals are designated in terms of the characteristic n, l, and m_l values, and are referred to in terms of the general notation

$$nl^x_{m_l}$$

where n is given as a number, l is given as the letter equivalent of its numerical value (i.e., s, p, d, f), m_l is designated in terms of probabilities relative to Cartesian coordinates (e.g., x, y, z, z^2; see p. 46), and x represents the number of electrons in the specific orbital or in a set of orbitals of stated n and l values. Typical notations are $1s^1$, $3s^2$, $2p^6$, $2p_x^1$, $3d^5$, $4d_{z^2}^1$, $5d_{xy}^2$.

A fourth quantum number is essential to an understanding of orbital capacities, although it does not emerge from the preceding discussion and is not essential to the treatment of the hydrogen atom. An electron may have two states of angular momentum, either $+\frac{1}{2}(h/2\pi)$ or $-\frac{1}{2}(h/2\pi)$; energies of these two states are unaffected by the potential energy functions of the atomic or molecular species but are altered by an external magnetic field. Although in this sense electron spin as such is not involved, the term "*spin*" is applied, and the situation is described by the *spin quantum number*, m_s, the values of which are $+\frac{1}{2}$ and $-\frac{1}{2}$ in units of $h/2\pi$. More important, this circumstance means that for each m_l value two electrons differing only in "spin" can be present.

The relationships among the quantum numbers and the various types of orbitals are summarized in Table 2-8. For the one-electron hydrogen atom, or any other one-electron species of nuclear charge Ze (e.g., He^+, Li^{2+}, Be^{3+}), Table 2-8 merely summarizes very broadly the energy states to which that electron is restricted. Extensions to multielectron species are discussed later (pp. 53–57).

Hydrogenic Wave Functions. Solutions of equations for the radial (R) and angular (Θ, Φ) wave functions for the hydrogen atom and hydrogenlike species

---

* The magnetic quantum number was introduced originally to account for the observed splitting of certain spectral lines upon application of a magnetic field (Zeeman effect). However, it appears naturally in the solution of the Φ equation when the values permissible for m_l are specified.

Table 2-8 Orbitals and Distributions of Electrons

			Possible Number of Electrons	
n	l	m_l	By Orbital	By Quantum Level
1	$0(s)$	0	2	2
2	$0(s)$	0	2	8
	$1(p)$	$-1, 0, +1$	6	
3	$0(s)$	0	2	18
	$1(p)$	$-1, 0, +1$	6	
	$2(d)$	$-2, -1, 0, +1, +2$	10	
4	$0(s)$	0	2	32
	$1(p)$	$-1, 0, +1$	6	
	$2(d)$	$-2, -1, 0, +1, +2$	10	
	$3(f)$	$-3, -2, -1, 0, +1, +2, +3$	14	
5	$0(s)$	0	2	50
	$1(p)$	$-1, 0, +1$	6	
	$2(d)$	$-2, -1, 0, +1, +2$	10	
	$3(f)$	$-3, -2, -1, 0, +1, +2, +3$	14	
	$4(g)$	$-4, -3, -2, -1, 0, +1, +2, +3, +4$	18	

are summarized in Tables 2-9 to 2-11. The hydrogenic wave functions, ψ, are summarized in Table 2-12.

It is instructive to represent orbitals graphically, either in terms of direct plots of wave functions or by means of contour surfaces of constant probability of electronic charge density. It is not uncommon to find the two types intermingled or not distinguished from each other. In either type of representation, the radial and angular functions are conveniently handled separately.

Radial functions, $R_{n,l}(r)$, for the hydrogen atom are plotted directly against r/a_0, a representation of distance from the nucleus relative to the first Bohr orbit, in Fig. 2-9(a). It is of interest that (1) curves for s orbitals suggest maximum probability of electronic charge density, and (2) the presence of $n-l$ nodes (i.e., points where the function changes sign) indicates maxima of electronic charge density and regions of low probability within the electronic atmosphere of the atom. The former reflects only a radial contribution. The latter is evident also in the probability plots in Fig. 2-9(b) and may be reflected in bonding (p. 194). The radial function R is converted, by squaring and multiplying by r^2 into a probability function which indicates the likelihood of finding electronic charge density in a region between r and $r + dr$. Comparable radial plots indicating probability are given in Fig. 2-9(b).

Table 2-9 Radial Function $R(r)$ for Hydrogenic Species

n	l	Orbital	R^a Symbol	R in Terms of r and $a_0{}^b$
1	0	$1s$	$R_{1,0}$	$2\left(\dfrac{Z}{a_0}\right)^{3/2} e^{-Zr/a_0}$
2	0	$2s$	$R_{2,0}$	$\left(\dfrac{1}{2\sqrt{2}}\right)\left(\dfrac{Z}{a_0}\right)^{3/2}\left(1-\dfrac{Zr}{2a_0}\right)e^{-Zr/2a_0}$
	1	$2p$	$R_{2,1}$	$\left(\dfrac{1}{2\sqrt{6}}\right)\left(\dfrac{Z}{a_0}\right)^{3/2} r\, e^{-Zr/2a_0}$
3	0	$3s$	$R_{3,0}$	$\left(\dfrac{2}{3\sqrt{3}}\right)\left(\dfrac{Z}{a_0}\right)^{3/2}\left(1-\dfrac{2Zr}{3a_0}+\dfrac{2Z^2r^2}{27a_0^2}\right)e^{-Zr/3a_0}$
	1	$3p$	$R_{3,1}$	$\left(\dfrac{8}{27\sqrt{6}}\right)\left(\dfrac{Z}{a_0}\right)^{3/2}\left(\dfrac{Zr}{a_0}-\dfrac{Z^2r^2}{6a_0^2}\right)e^{-Zr/3a_0}$
	2	$3d$	$R_{3,2}$	$\left(\dfrac{4}{81\sqrt{30}}\right)\left(\dfrac{Z}{a_0}\right)^{7/2} r^2 e^{-Zr/3a_0}$

[a] Subscripts are numerical values of n and l, respectively.
[b] a_0 = first Bohr radius (p. 36); e = base for natural logarithms.

Table 2-10 Angular Function Θ for Hydrogenic Species

l	m_l	Θ Symbol	Θ in Terms of θ
0	0	$\Theta_{0,0}$	$\sqrt{2}/2$
1	0	$\Theta_{1,0}$	$\left(\dfrac{\sqrt{6}}{2}\right)(\cos\theta)$
	± 1	$\Theta_{1,\pm 1}$	$\left(\dfrac{\sqrt{3}}{2}\right)(\sin\theta)$
2	0	$\Theta_{2,0}$	$\left(\dfrac{\sqrt{10}}{4}\right)(3\cos^2\theta-1)$
	± 1	$\Theta_{2,\pm 1}$	$\left(\dfrac{\sqrt{15}}{2}\right)(\sin\theta\cos\theta)$
	± 2	$\Theta_{2,\pm 2}$	$\left(\dfrac{\sqrt{15}}{4}\right)(\sin^2\theta)$

[a] Subscripts are numerical values of l and m_l, respectively.

Comparable plots of the angular function (Θ, Φ) are shown for s and p orbitals in Fig. 2-10 and d orbitals in Fig. 2-11. The positive and negative signs included in the lobes in these two figures and subsequent graphical representations of orbitals are the signs of the appropriate wave functions in the directions indicated. When the sign of the wave function does not change on moving from a given point to an equivalent point opposite the center (e.g., for s, d_{z^2}, $d_{x^2-y^2}$), the orbital is said to have *gerade* (*German*, even), or g, symmetry. When the wave function changes sign in this operation (e.g., p orbitals), the orbital has *ungerade* (i.e., uneven), or u, symmetry. The significances of this type, and other types, of orbital symmetry will become apparent in later discussions.

Neither Fig. 2-10 nor Fig. 2-11 has any meaning in a modular sense. Each such diagram represents the solution to a mathematical equation, not a two-dimensional model. Contour maps representing constant probabilities of electronic charge density are deducible from appropriate hydrogenic probability functions. Maps of this type are indicative of the spatial extensions of orbitals, for example, as shown in Fig. 2-12.[49] Three-dimensional representations, although more difficult to construct, are more effective. Representations of probability by densities of dots (Figs. 2-13, 2-14) that can be observed with a stereoviewer have been used effectively.[50] More commonly, three-dimensional representations based upon boundary surfaces of constant value of the angular function $\Theta(\theta)\Phi(\phi)$ are used. This type of representation is shown in Fig. 2-15 for $1s$, $2p$, and $3d$ orbitals. Nodal planes, where $\psi\psi^* = 0$, are apparent for p and d orbitals. Yet another type of representation, corresponding to a cut through the nucleus as a basal plane and electron density plotted perpendicular to that plane, is shown in Fig. 2-16.[51]

Table 2-11 Angular Function Φ for Hydrogenic Species

m_l	Φ Symbol	Φ in Terms of ϕ^a
0	Φ_0	$\dfrac{1}{\sqrt{2\pi}}$
1	Φ_1	$\left(\dfrac{1}{\sqrt{2\pi}}\right)e^{i\phi}$ or $\left(\dfrac{1}{\sqrt{\pi}}\right)(\cos \phi)$
−1	Φ_{-1}	$\left(\dfrac{1}{\sqrt{2\pi}}\right)e^{-i\phi}$ or $\left(\dfrac{1}{\sqrt{\pi}}\right)(\sin \phi)$
2	Φ_2	$\left(\dfrac{1}{\sqrt{2\pi}}\right)e^{2i\phi}$ or $\left(\dfrac{1}{\sqrt{\pi}}\right)(\cos 2\phi)$
−2	Φ_{-2}	$\left(\dfrac{1}{\sqrt{2\pi}}\right)e^{-2i\phi}$ or $\left(\dfrac{1}{\sqrt{\pi}}\right)(\sin 2\phi)$

^{a} $i = \sqrt{-1}$.

Table 2-12 Hydrogenic Wave Functions

n	l	m_l	Orbital	Wave Function
1	0	0	$1s$	$\psi_{1s} = \left(\dfrac{1}{\sqrt{\pi}}\right)\left(\dfrac{Z}{a_0}\right)^{3/2} e^{-Zr/a_0}$
2	0	0	$2s$	$\psi_{2s} = \left(\dfrac{1}{4\sqrt{2\pi}}\right)\left(\dfrac{Z}{a_0}\right)^{3/2}\left(2 - \dfrac{Zr}{a_0}\right)e^{-Zr/2a_0}$
	1	0	$2p_z$	$\psi_{2p_z} = \left(\dfrac{1}{4\sqrt{2\pi}}\right)\left(\dfrac{Z}{a_0}\right)^{5/2} re^{-Zr/2a_0}(\cos\theta)$
		+1	$2p_x$	$\psi_{2p_x} = \left(\dfrac{1}{4\sqrt{2\pi}}\right)\left(\dfrac{Z}{a_0}\right)^{5/2} re^{-Zr/2a_0}(\sin\theta\,\cos\phi)$
		−1	$2p_y$	$\psi_{2p_y} = \left(\dfrac{1}{4\sqrt{2\pi}}\right)\left(\dfrac{Z}{a_0}\right)^{5/2} re^{-Zr/2a_0}(\sin\theta\,\sin\phi)$
3	0	0	$3s$	$\psi_{3s} = \left(\dfrac{1}{81\sqrt{3\pi}}\right)\left(\dfrac{Z}{a_0}\right)^{3/2}\left(27 - \dfrac{18Zr}{a_0} + \dfrac{2Z^2r^2}{a_0^2}\right)e^{-Zr/3a_0}$
	1	0	$3p_z$	$\psi_{3p_z} = \left(\dfrac{\sqrt{2}}{81\sqrt{\pi}}\right)\left(\dfrac{Z}{a_0}\right)^{5/2}\left(6 - \dfrac{Zr}{a_0}\right)re^{-Zr/3a_0}(\cos\theta)$
		+1	$3p_x$	$\psi_{3p_x} = \left(\dfrac{\sqrt{2}}{81\sqrt{\pi}}\right)\left(\dfrac{Z}{a_0}\right)^{5/2}\left(6 - \dfrac{Zr}{a_0}\right)re^{-Zr/3a_0}(\sin\theta\,\cos\phi)$
		−1	$3p_y$	$\psi_{3p_y} = \left(\dfrac{\sqrt{2}}{81\sqrt{\pi}}\right)\left(\dfrac{Z}{a_0}\right)^{5/2}\left(6 - \dfrac{Zr}{a_0}\right)re^{-Zr/3a_0}(\sin\theta\,\sin\phi)$

2	0	$3d_{z^2}$	$\psi_{3d_{z^2}} = \left(\dfrac{1}{81\sqrt{6\pi}}\right)\left(\dfrac{Z}{a_0}\right)^{7/2} r^2 e^{-Zr/3a_0}(3\cos^2\theta - 1)$
	$+1$	$3d_{xz}$	$\psi_{3d_{xz}} = \left(\dfrac{\sqrt{2}}{81\sqrt{\pi}}\right)\left(\dfrac{Z}{a_0}\right)^{7/2} r^2 e^{-Zr/3a_0}(\sin\theta\cos\theta\cos\phi)$
	-1	$3d_{yz}$	$\psi_{3d_{yz}} = \left(\dfrac{\sqrt{2}}{81\sqrt{\pi}}\right)\left(\dfrac{Z}{a_0}\right)^{7/2} r^2 e^{-Zr/3a_0}(\sin\theta\cos\theta\sin\phi)$
	$+2$	$3d_{x^2-y^2}$	$\psi_{3d_{x^2-y^2}} = \left(\dfrac{1}{81\sqrt{2\pi}}\right)\left(\dfrac{Z}{a_0}\right)^{7/2} r^2 e^{-Zr/3a_0}(\sin^2\theta\cos 2\phi)$
	-1	$3d_{xy}$	$\psi_{3d_{xy}} = \left(\dfrac{1}{81\sqrt{2\pi}}\right)\left(\dfrac{Z}{a_0}\right)^{7/2} r^2 e^{-Zr/3a_0}(\sin^2\theta\sin 2\phi)$

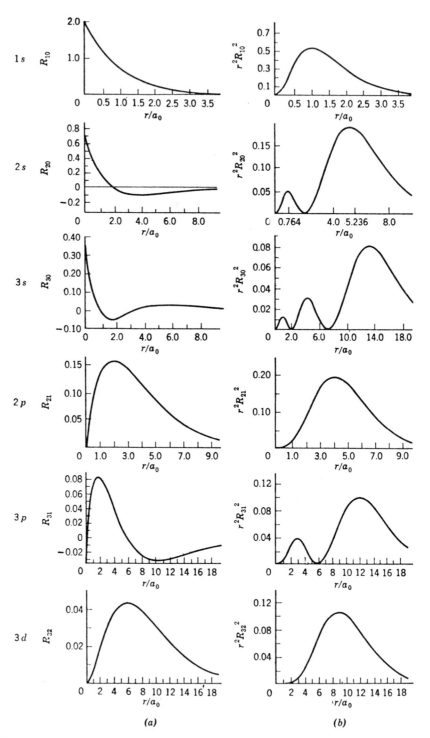

(a) (b)

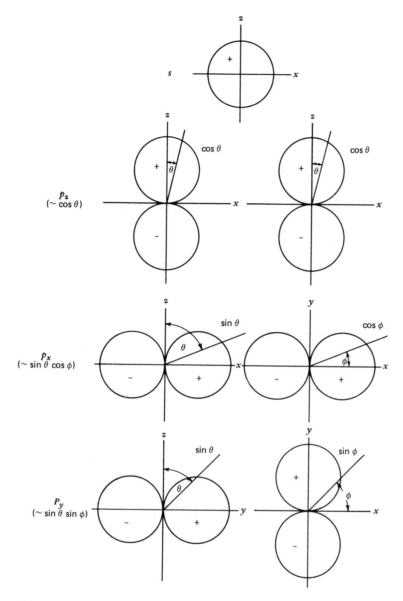

Figure 2-10 Representations of angular wave function for $l=0, 1$ and angles θ and ϕ (s and p orbitals).

Figure 2-9 Representations of radial wave functions for the hydrogen atom. (a) Radial functions alone; (b) probability densities. [M. Karplus and R. N. Porter, *Atoms and Molecules*, W. A. Benjamin, Menlo Park, Calif. (1970), as reproduced in F. Daniels and R. A. Alberty, *Physical Chemistry*, 4th ed., p. 414, John Wiley & Sons, New York (1975).]

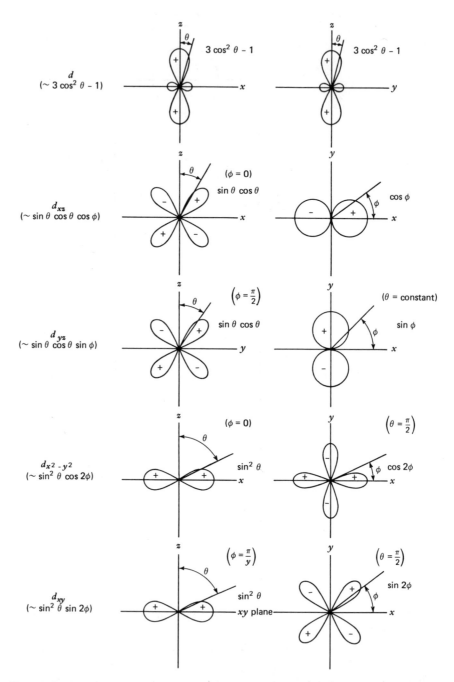

Figure 2-11 Representations of angular wave function for $l=2$ and θ and ϕ (*d* orbitals).

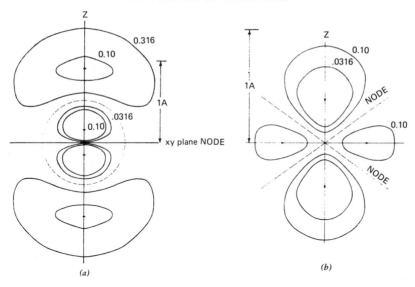

Figure 2-12 Electron-density contour maps for (a) $3p_z$ (Cl) and (b) $3d_{z^2}$ [Ti(III)] orbitals. Contours of constant ψ^2 at values of maxima indicated. [E. A. Ogryzlo and G. B. Porter, *J. Chem. Educ.*, **40**, 256 (1963), Figs. 4, 6.]

Although not included in previous discussions, hydrogenic wave functions for *f* orbitals have been determined.[52-55] Figures 2-17 and 2-18 are representations of cubic harmonic functions for the seven 4*f* orbitals. But little attention is commonly paid to this aspect of the 4*f* orbitals because of the limited, or nonexistent, participation of these orbitals in bonding (pp. 112–113).

It is apparent from all of these graphical representations that the hydrogenic *s* orbitals, irrespective of *n* values, represent spherically symmetrical distributions

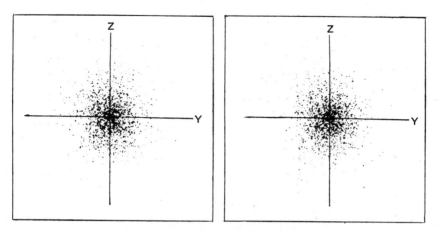

Figure 2-13 Electron-density map for hydrogenic 1*s* orbital. [D. T. Cromer, *J. Chem. Educ.*, **45**, 626 (1968), Fig. 1.]

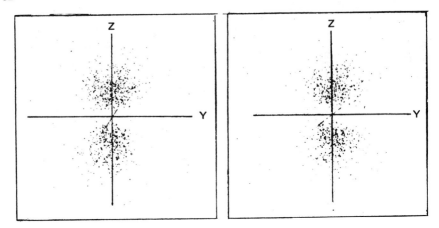

Figure 2-14   Electron-density map for hydrogenic $2p$ orbital. [D. T. Cromer, *J. Chem. Educ.*, **45**, 626 (1968), Fig. 5.]

of electronic charge density about the nucleus. The p orbitals of given n value are oriented at 90° to each other and are conveniently aligned along the three Cartesian coordinates. Five mutually independent $3d$ orbitals can be distinguished. The designations used are the common ones, but other choices could be made. The $d_{x^2-y^2}$ and d_{z^2} orbitals represent concentrations of electronic charge density along the Cartesian coordinates. By choice, major concentrations along the x and y axes and along the z axis are represented, respectively, by the $d_{x^2-y^2}$ and d_{z^2} orbitals, but the d_{z^2} orbital has a lesser concentration of electronic charge density in the xy plane, represented graphically as a torus. Electronic charge density in any one of the d_{xy}, d_{xz}, and d_{yz} orbitals is concentrated between the Cartesian axes and, respectively, in the xy, xz, and yz planes.

The various projections of hydrogenic orbitals are extremely useful in describing the atomic structures of one-electron species and, with substantial modifications and less accuracy, of multielectron species. However, they must not be regarded as physical models of electronic charge distribution and must not be so used. Only the proper manipulation of eigenfunctions can lead to quantitative descriptions of atomic and molecular systems.

3-4. Wave-Mechanical Models of Multielectron Atoms

The success of the Schrödinger approach to the structure of the hydrogen atom and other one-electron species suggests its extension to multielectron species. However, it proves to be impossible to solve the Schrödinger relationship exactly for any species containing more than one electron because of the necessity of considering also all electron–electron repulsions and electron–electron distances. Only approximations can be developed.

One of the better approaches was suggested by D. R. Hartree[56] and extended by V. Fock.[57] It is unnecessary here to do more than outline the method. As a

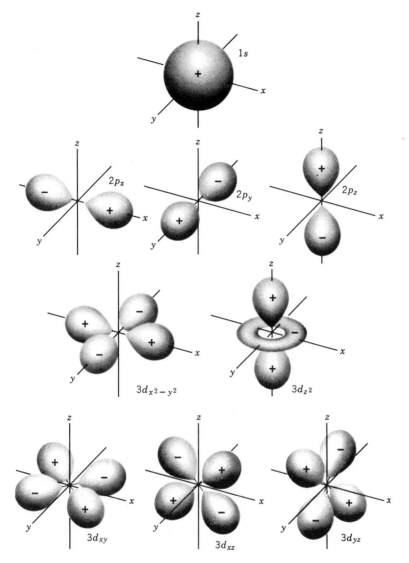

Figure 2-15 Three-dimensional representations showing angular dependence of $1s$, $2p$, and $3d$ orbitals at fixed r. [F. Daniels and R. A. Alberty, *Physical Chemistry*, 4th ed., p. 416, John Wiley & Sons, Inc., New York (1975).]

first approximation, the interaction of a particular electron with all other electrons is estimated, the other electrons all being assumed to obey hydrogen-like wave functions (perhaps best by using effective nuclear charges).[58] A wave function for the electron in question is then solved, and the same procedure is followed for every electron present. The new set of wave functions is used to repeat the process, and the entire operation is continued until no changes in

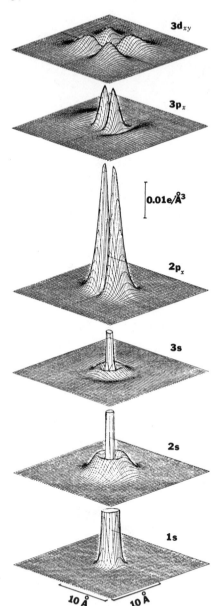

Figure 2-16 Electron-density maps for hydrogenic $1s$, $2s$, $3s$, $2p_x$, $3p_x$, and $3d_{xy}$ orbitals. [J. R. Van Wazer and I. Absar, Electron Densities in Molecules and Molecular Orbitals, Fig. 1.1, p. 2, Academic Press, New York (1975).]

wave functions result. The potential energies of the electrons are then said to be *self-consistent*. Resulting radial distribution curves, such as those given in Fig. 2-19 for the outermost s electrons of the Li, Na, K, and Rb atoms, indicate clearly that the electrons lie in energy levels comparable with those of hydrogenic species and suggest that the quantum numbers and orbital descriptions developed for these one-electron species are at least reasonable for multielectron species.

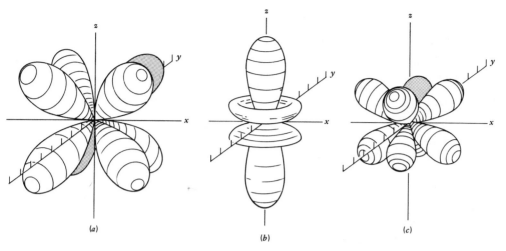

Figure 2-17 Representations of 4f orbitals. (a) $f_{x,y,z}$; (b) $f_{z(5z^2-3r^2)}$; (c) $f_{z(x^2-3y^2)}$ [J. T. Waber and J. E. Hockett, in *Proceedings of the Fourth Conference on Rare Earth Research*, L. Eyring (Ed.), Fig. 1, p. 285, Gordon and Breach, New York (1965).].

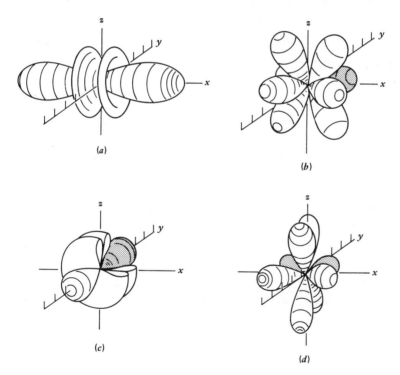

Figure 2-18 Representations of 4f orbitals. (a) $f_{x(5x^2-3r^2)}$; (b) $f_{x(z^2-y^2)}$; (c) $f_{y(5y^2-3r^2)}$; (d) $f_{y(z^2-x^2)}$. [J. T. Waber and J. E. Hockett, in *Proceedings of the Fourth Conference on Rare Earth Research*, L. Eyring (Ed.), Fig. 3, p. 287, Gordon and Breach, New York (1965).]

55

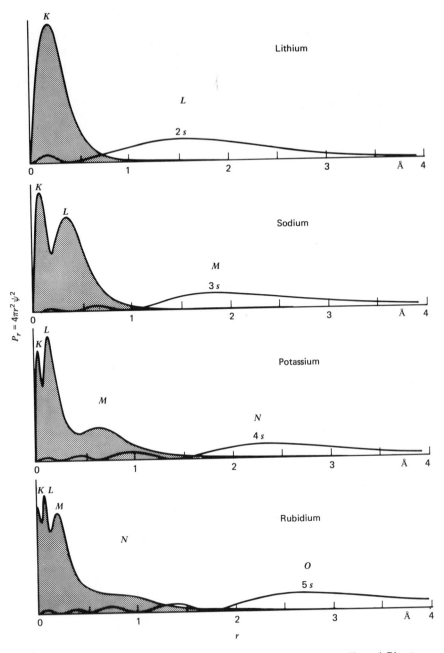

Figure 2-19 Radial distribution curves for electrons in Li, Na, K, and Rb atoms.

The procedure that is followed then is to develop the ground-state electronic configurations of atoms of the various elements in terms of the hydrogenic orbitals already described and the appropriate quantum numbers.

3-5. Ground-State Electronic Configurations of Gaseous Atoms

Development of a table of ground-state electronic configurations is dependent upon several principles:

1. **The Aufbau (German, building up) principle.** States that atoms are built up by successive additions of protons and neutrons to nuclei and additions of sufficient electrons to balance nuclear charges, each added electron occupying the available atomic orbital of lowest energy. Atomic orbitals are thus occupied in the same sequence as developed for the hydrogen atom, except that variations in this sequence develop since, as Z increases, electrons in the

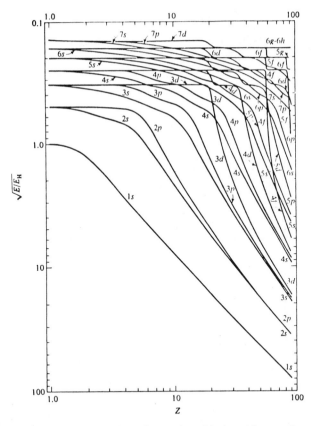

Figure 2-20 Variations in energies of atomic orbitals with atomic number. [As redrawn by M. Kasha from R. Latter, *Phys. Rev.*, **99**, 510 (1955).]

various possible orbitals are differently shielded from nuclear attraction by electrons in orbitals of lower energy. This situation is illustrated in Fig. 2-20, where energies of atomic orbitals are related to atomic number (nuclear charge). A similar relationship between principal quantum number and atomic number is shown in Fig. 2-21. Crossovers involving d and f orbitals are particularly significant in determining when these inner orbitals are occupied in preference to the orbitals of larger principal quantum numbers. For simplicity, the sequence illustrated in Fig. 2-22 applies with only a few exceptions.

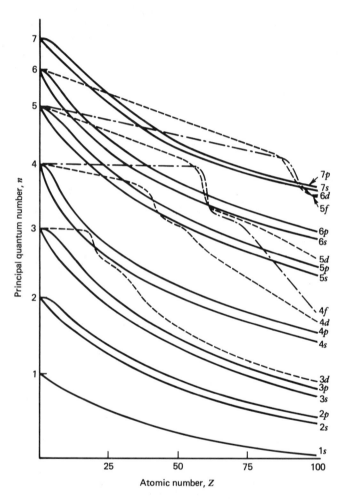

Figure 2-21   Variation in energies of atomic orbitals with atomic number in terms of principal quantum numbers. [K. M. Mackay and R. A. Mackay, *Introduction to Modern Inorganic Chemistry*, Fig. 7.4, p. 95, International Textbook Company, Glasgow (1968).]

2. **Spin quantum number.** As noted earlier, cancellation of the magnetic moment of electrons by the pairing of "spins" allows a maximum capacity of two electrons for each orbital characterized by specific n, l, and m_l values.

3. **Pauli exclusion principle.** States that in a given atomic or molecular species no two electrons can be characterized by the same set of *four* quantum numbers.

4. **Hund principle of maximum multiplicity.** States that the ground-state electronic configuration is that of maximum spin multiplicity (S = resultant of all m_s values) and, within this restriction, of maximum angular multiplicity (L = resultant of all m_l values). In other words, for orbitals of same n and l values, those of different m_l values are all occupied singly before pairing occurs. As examples, we note the ground-state configurations

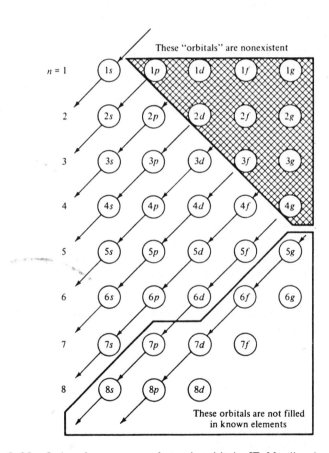

Figure 2-22 Order of occupancy of atomic orbitals. [T. Moeller, *Inorganic Chemistry*, Fig. 3.5, p. 97, John Wiley & Sons, New York (1952), modified as in J. Huheey, *Inorganic Chemistry*, 2nd ed., p. 27, Harper & Row, New York (1978).]

N: $\quad Z = 7;\quad 1s^2\ 2s^2\ 2p_x^1\ 2p_y^1\ 2p_z^1$

Mn: $\quad Z = 25;\quad 1s^2\ 2s^2\ 2p^6\ 3s^2\ 3p^6\ 3d_{z^2}^1\ 3d_{x^2-y^2}^1\ 3d_{xy}^1\ 3d_{xz}^1\ 3d_{yz}^1\ 4s^2$

Based upon these principles, and reflecting the interpretations of emission spectra, atomic beam resonance studies, and other configuration-dependent techniques, tables of electronic configurations of gaseous atoms in their ground, or lowest energy, states can be constructed. Inasmuch as the electronic configurations of atoms of the noble gas elements are characterized by closed energy shells (i.e., completely occupied inner orbital arrangements) and thus prove to be configurations of substantial chemical stability, they may be regarded as *core configurations*. They are so indicated in Table 2-13 and then used as points of reference in Table 2-14.

The absence of energy crossovers among orbitals in regions of small nuclear charge (Figs. 2-20, 2-21) allows exact establishment of configurations for atoms of the lighter elements. For atoms of the heavy elements, however, such crossovers and smaller differences among the energies of available orbitals make the assignments of exact configurations much more difficult. Those given in Table 2-14 appear to be the best choices currently available to us.

Although the configurations given in Table 2-14 most commonly involve the regular filling of orbitals of a given $nl - m_l$ type, there are numerous ones that, in this sense, appear to be "irregular." These irregularities usually reflect either the extra stability of half-occupied orbitals (e.g., $3d^5$ at Mn, $4d^5$ at Mo, $4f^7$ at Eu, $5f^7$ at Am) or of completely occupied orbitals (e.g., $3d^{10}$ at Cu, $4d^{10}$ at Pd and Ag, $4f^{14}$ at Yb, $5f^{14}$ at No) or the closeness in energy of two types of orbitals (e.g., $4f$ and $5d$ at Pr-Sm and Tb-Tm, and of the $5f$ and $6d$ in the region $Z = 90$ to 103).

The periodic recurrence of particular types of ground-state electronic configurations is, of course, a fundamental basis for the modern periodic table (Ch. 3). Based upon types of ground-state configurations, three general types of elements are distinguishable: representative or main-group elements (ns^1 through ns^2np^6 "outermost arrangements"), d-transition or subgroup elements

Table 2-13   Ground-State Electronic Configurations of Atoms of Noble Gas Elements

Symbol	Z	Complete Configuration	[Core] Notation
He	2	$1s^2$	[He]
Ne	10	$1s^22s^22p^6$	[Ne]
Ar	18	$1s^22s^22p^63s^23p^6$	[Ar]
Kr	36	$1s^22s^22p^63s^23p^63d^{10}4s^24p^6$	[Kr]
Xe	54	$1s^22s^22p^63s^23p^63d^{10}4s^24p^64d^{10}5s^25p^6$	[Xe]
Rn	86	$1s^22s^22p^63s^23p^63d^{10}4s^24p^64d^{10}4f^{14}5s^25p^65d^{10}6s^26p^6$	[Rn]

$[(n-1)d^1ns^2$ through $(n-1)d^{10}ns^2$, where $n = 4, 5, ...]$, and f-transition or inner transition elements $[(n-2)f^1(n-1)d^{0-1}ns^2$ through $(n-2)f^{14}(n-1)d^1ns^2$, where $n = 6, 7, ...]$. The general characteristics and placement of these elements in the periodic table are discussed in Ch. 3 (Sec. 3.2-1).

3-6. Atomic Energy States and Atomic Term Symbols

The energy states of atoms are specified in spectroscopic terms by the angular momentum designations S, L, and J, where S represents the net spin quantum number, L the net orbital angular momentum, and J the vector sum of spin and orbital angular momenta. Since both S and L are zero for closed shells, only the outermost electrons need be considered in this description of atomic energy states. As pointed out in previous sections, each outermost electron has a spin of $\pm\frac{1}{2}$ and is described by a particular value of l. Combinations of spin and angular momenta give the net S and L values, respectively, for the atom as a whole. The possible J values are then given by (so-called Russell–Saunders coupling)

$$J = (L - S), (L - S + 1), ..., (L + S) \tag{2-30}$$

For each outer electron, there are $(2l + 1)$ possible values for the magnetic quantum number m_l. The sum of these possible combinations of m_l values gives an M_L value. Each L state suggests M_L values of $0, \pm 1, ..., L$. In writing a term designation, the same approach is used as for l values, except that capital letters are used (i.e., S, P, D, F, G, ... for $L = 0, 1, 2, 3, 4, ...$). Similarly, a spin of $\pm\frac{1}{2}$ for each outer electron gives values of M_S. Each S state suggests M_S values of $(-S)$, $(-S+1), ..., S$, or $2S + 1$ total values. This M_S value is given as an upper left superscript with the letter designation of L (e.g., ^1S, ^3P, ^1D, etc.). The value of J is given as a lower right subscript, as $^2P_{3/2}$, $^1S_{1/2}$, and so on. Vectorial combination of individual s and l values gives an array of M_S and M_L values, from which a set of S and L values that will produce the same array is deduced.

Thus for the hydrogen atom (ground state $1s^1$), $l = 0$ and $m_l = 0$ and $m_s = \pm\frac{1}{2}$. Thus M_L and M_S are both zero, $l = 0$, $S = \frac{1}{2}$, and $J = \frac{1}{2}$; and the term designation, or symbol, is $^2S_{1/2}$. For a helium atom $(1s^2)$, $L = 0$, $S = 0$, and $J = 0$; so the term symbol is $^1S_0$. For a carbon atom $(1s^2 2s^2 2p^2)$, only the $2p$ arrangement is considered, for which the possibilities are $S = 1, 0$; $M_S = -1, 0, +1$; $L = 2, 1, 0$; $M_L = -2, -1, 0, +1, +2$; and $J = 3, 2, 1, 0$. All possible combinations of M_S and M_L are given in Table 2-15, and those forbidden by the Pauli exclusion principle are crossed out. In terms of the $(2S + 1)$ and $(2L + 1)$ limitations, the possible terms for this $2p^2$ (or np^2) configuration become ^1S, ^3P, and ^1D. Designations for various ground-state configurations are summarized in Table 2-16.*

---

* For details, the interested reader should consult treatments such as G. Herzberg, *Atomic Spectra and Atomic Structure*, Ch. 3, Dover Publications, New York (1944); J. C. Davis, Jr., *Advanced Physical Chemistry: Molecules, Structure, and Spectra*, Ronald Press, New York (1965); R. M. Hochstrasser, *Behavior of Electrons in Atoms*, Ch. 5, W. A. Benjamin, New York (1964).

Table 2-14 Ground-State Electronic Configurations of Gaseous Atoms[a]

Z	Symbol	Configuration as [Core] plus "Outermost" Orbitals	Z	Symbol	Configuration as [Core] plus "Outermost" Orbitals
1	H	$1s^1$	55	Cs	$[Xe]6s^1$
2	He	$1s^2$, or [He]	56	Ba	$[Xe]6s^2$
3	Li	$[He]2s^1$	57	La	$[Xe]5d^16s^2$
4	Be	$[He]2s^2$	58	Ce	$[Xe]4f^15d^16s^2$
5	B	$[He]2s^22p^1$	59	Pr	$[Xe]4f^36s^2$
6	C	$[He]2s^22p^2$	60	Nd	$[Xe]4f^46s^2$
7	N	$[He]2s^22p^3$	61	Pm	$[Xe]4f^56s^2$
8	O	$[He]2s^22p^4$	62	Sm	$[Xe]4f^66s^2$
9	F	$[He]2s^22p^5$	63	Eu	$[Xe]4f^76s^2$
10	Ne	$[He]2s^22p^6$, or [Ne]	64	Gd	$[Xe]4f^75d^16s^2$
11	Na	$[Ne]3s^1$	65	Tb	$[Xe]4f^96s^2$
12	Mg	$[Ne]3s^2$	66	Dy	$[Xe]4f^{10}6s^2$
13	Al	$[Ne]3s^23p^1$	67	Ho	$[Xe]4f^{11}6s^2$
14	Si	$[Ne]3s^23p^2$	68	Er	$[Xe]4f^{12}6s^2$
15	P	$[Ne]3s^23p^3$	69	Tm	$[Xe]4f^{13}6s^2$
16	S	$[Ne]3s^23p^4$	70	Yb	$[Xe]4f^{14}6s^2$
17	Cl	$[Ne]3s^23p^5$	71	Lu	$[Xe]4f^{14}5d^16s^2$
18	Ar	$[Ne]3s^23p^6$, or [Ar]	72	Hf	$[Xe]4f^{14}5d^26s^2$
19	K	$[Ar]4s^1$	73	Ta	$[Xe]4f^{14}5d^36s^2$
20	Ca	$[Ar]4s^2$	74	W	$[Xe]4f^{14}5d^46s^2$
21	Sc	$[Ar]3d^14s^2$	75	Re	$[Xe]4f^{14}5d^56s^2$
22	Ti	$[Ar]3d^24s^2$	76	Os	$[Xe]4f^{14}5d^66s^2$
23	V	$[Ar]3d^34s^2$	77	Ir	$[Xe]4f^{14}5d^76s^2$
24	Cr	$[Ar]3d^54s^1$	78	Pt	$[Xe]4f^{14}5d^96s^1$
25	Mn	$[Ar]3d^54s^2$	79	Au	$[Xe]4f^{14}5d^{10}6s^1$
26	Fe	$[Ar]3d^64s^2$	80	Hg	$[Xe]4f^{14}5d^{10}6s^2$

27	Co	$[\text{Ar}]3d^7 4s^2$
28	Ni	$[\text{Ar}]3d^8 4s^2$
29	Cu	$[\text{Ar}]3d^{10} 4s^1$
30	Zn	$[\text{Ar}]3d^{10} 4s^2$
31	Ga	$[\text{Ar}]3d^{10} 4s^2 4p^1$
32	Ge	$[\text{Ar}]3d^{10} 4s^2 4p^2$
33	As	$[\text{Ar}]3d^{10} 4s^2 4p^3$
34	Se	$[\text{Ar}]3d^{10} 4s^2 4p^4$
35	Br	$[\text{Ar}]3d^{10} 4s^2 4p^5$
36	Kr	$[\text{Ar}]3d^{10} 4s^2 4p^6$, or $[\text{Kr}]$
37	Rb	$[\text{Kr}]5s^1$
38	Sr	$[\text{Kr}]5s^2$
39	Y	$[\text{Kr}]4d^1 5s^2$
40	Zr	$[\text{Kr}]4d^2 5s^2$
41	Nb	$[\text{Kr}]4d^3 5s^2$
42	Mo	$[\text{Kr}]4d^5 5s^1$
43	Tc	$[\text{Kr}]4d^5 5s^2$
44	Ru	$[\text{Kr}]4d^7 5s^1$
45	Rh	$[\text{Kr}]4d^8 5s^1$
46	Pd	$[\text{Kr}]4d^{10}$
47	Ag	$[\text{Kr}]4d^{10} 5s^1$
48	Cd	$[\text{Kr}]4d^{10} 5s^2$
49	In	$[\text{Kr}]4d^{10} 5s^2 5p^1$
50	Sn	$[\text{Kr}]4d^{10} 5s^2 5p^2$
51	Sb	$[\text{Kr}]4d^{10} 5s^2 5p^3$
52	Te	$[\text{Kr}]4d^{10} 5s^2 5p^4$
53	I	$[\text{Kr}]4d^{10} 5s^2 5p^5$
54	Xe	$[\text{Kr}]4d^{10} 5s^2 5p^6$, or $[\text{Xe}]$
81	Tl	$[\text{Xe}]4f^{14} 5d^{10} 6s^2 6p^1$
82	Pb	$[\text{Xe}]4f^{14} 5d^{10} 6s^2 6p^2$
83	Bi	$[\text{Xe}]4f^{14} 5d^{10} 6s^2 6p^3$
84	Po	$[\text{Xe}]4f^{14} 5d^{10} 6s^2 6p^4$
85	At	$[\text{Xe}]4f^{14} 5d^{10} 6s^2 6p^5$
86	Rn	$[\text{Xe}]4f^{14} 5d^{10} 6s^2 6p^6$, or $[\text{Rn}]$
87	Fr	$[\text{Rn}]7s^1$
88	Ra	$[\text{Rn}]7s^2$
89	Ac	$[\text{Rn}]6d^1 7s^2$
90	Th	$[\text{Rn}]6d^2 7s^2$
91	Pa	$[\text{Rn}]5f^2 6d^1 7s^2$
92	U	$[\text{Rn}]5f^3 6d^1 7s^2$
93	Np	$[\text{Rn}]5f^4 6d^1 7s^2$
94	Pu	$[\text{Rn}]5f^6 7s^2$
95	Am	$[\text{Rn}]5f^7 7s^2$
96	Cm	$[\text{Rn}]5f^7 6d^1 7s^2$
97	Bk	$[\text{Rn}]5f^9 7s^2$
98	Cf	$[\text{Rn}]5f^{10} 7s^2$
99	Es	$[\text{Rn}]5f^{11} 7s^2$
100	Fm	$[\text{Rn}]5f^{12} 7s^2$
101	Md	$[\text{Rn}]5f^{13} 7s^2$
102	No	$[\text{Rn}]5f^{14} 7s^2$
103	Lr	$[\text{Rn}]5f^{14} 6d^1 7s^2$
104	?	$[\text{Rn}]5f^{14} 6d^2 7s^2$ (?)
105	?	$[\text{Rn}]5f^{14} 6d^3 7s^2$ (?)
106	?	$[\text{Rn}]5f^{14} 6d^4 7s^2$ (?)

[a] Data from C. E. Moore, *Ionization Potentials and Ionization Limits Derived from the Analyses of Optical Spectra*, NSRDS-NBS 34, National Bureau of Standards, Washington, D.C. (1970); for actinides from G. T. Seaborg, *Annu. Rev. Nucl. Sci.*, **18**, 53 (1968).

Table 2-15 M_S and M_L **Values for p^2 Configuration in the Carbon Atom**

	M_S^a		
M_L	+1	0	−1
−2	$(-1^+, -1^+)$	$(-1^+, -1^-)$	$(-1^-, -1^-)$
−1	$(0^+, -1^+)$	$(0^+, -1^-)$ $(0^-, -1^+)$	$(0^-, -1^-)$
0	$(0^+, 0^+)$ $(1^+, -1^+)$	$(0^+, 0^-)$ $(1^+, -1^-)$ $(-1^+, 1^-)$	$(0^-, 0^-)$ $(1^-, -1^-)$
+1	$(0^+, 1^+)$	$(0^+, 1^-)$ $(0^-, 1^+)$	$(0^-, 1^-)$
+2	$(1^+, 1^+)$	$(1^+, 1^-)$	$(1^-, 1^-)$

[a] Superscripts, + and −, represent signs of m_s quantum numbers.

A term symbol specifies an energy state resulting from spin-orbit coupling in a specific atomic configuration. The Hund principle of maximum multiplicity (Sec. 2.3-5) indicates that the ground-state term for the carbon atom (above) will be ^3P rather than either of the other possibilities. Of two states of the same multiplicity, the one with the larger value of L will have the lower energy (i.e., ^1D rather than ^1S for the $2p^2$ configuration). Furthermore, for subshells that are less than half filled, states with smaller J values are lower in energy, whereas for those that are more than half filled, the reverse is true. Thus all predictions lead to $^3P_2$ as the term symbol for the carbon atom. Relatively simple rules for writing ground-state Russell–Saunders terms have been given by M. Gorman, together with useful examples.[59] This information will be of value in later discussions of bonding (Sec. 5.1-1).

4. SOME FUNDAMENTAL PROPERTIES RELATED TO EXTRANUCLEAR ATOMIC STRUCTURES

The chemical characteristics of the elements, and many of their physical properties as well, are determined quite generally by the extranuclear structures of their atoms and most specifically by the electronic configurations in the highest quantum levels (i.e., the "outermost" electrons). It follows that particular broad sets of properties may be expected for elements the atoms of which have the same type of ground-state electronic configuration. Although such regularities are commonly observed, there exist significant specific differences and variations that cannot be explained so simply. It is a major goal of inorganic chemistry to provide reasonable explanations for all observed similarities, variations, differences, and trends in the properties of the elements and their

Table 2-16 Terms Arising from Various Ground States[a]

Ground State	Terms
	NONEQUIVALENT ELECTRONS
$s\ s$	$^1S,\ ^3S$
$s\ p$	$^1P,\ ^3P$
$s\ d$	$^1D\ ^3D$
$p\ p$	$^1S,\ ^1P,\ ^1D,\ ^3S,\ ^3P,\ ^3D$
$p\ d$	$^1P,\ ^1D,\ ^1F,\ ^3P,\ ^3D,\ ^3F$
$d\ d$	$^1S,\ ^1P,\ ^1D,\ ^1F,\ ^1G,\ ^3S,\ ^3P,\ ^3D,\ ^3F,\ ^3G$
$s\ s\ s$	$^2S,\ ^2S,\ ^4S$
$s\ s\ p$	$^2P,\ ^2P,\ ^4P$
$s\ s\ d$	$^2D,\ ^2D,\ ^4D$
$s\ p\ p$	$^2S,\ ^2P,\ ^2D,\ ^2S,\ ^2P,\ ^2D,\ ^4S,\ ^4P,\ ^4D$
$s\ p\ d$	$^2P,\ ^2D,\ ^2F,\ ^2P,\ ^2D,\ ^2F,\ ^4P,\ ^4D,\ ^4F$
$p\ p\ p$	$^2S(2),\ ^2P(6),\ ^2D(4),\ ^2F(2),\ ^4S(1),\ ^4P(3),\ ^4D(2),\ ^4F(1)$
$p\ p\ d$	$^2S(2),\ ^2P(4),\ ^2D(6),\ ^2F(4),\ ^2G(2),\ ^4S(1),\ ^4P(2),\ ^4D(3),\ ^4F(2),\ ^4G(1)$
$p\ d\ f$	$^2S(2),\ ^2P(4),\ ^2D(6),\ ^2F(6),\ ^2G(6),\ ^2H(4),\ ^2I(2),\ ^4S(1),\ ^4P(2),\ ^4D(3),$ $^4F(3),\ ^4G(3),\ ^4H(2),\ ^4I(1)$
	EQUIVALENT ELECTRONS
$s^2,\ p^6,\ d^{10}$	^1S
$p,\ p^5$	^2P
$p^2,\ p^4$	$^2S,\ ^1D,\ ^3P$
p^3	$^2P,\ ^2D,\ ^4S$
$d,\ d^9$	^2D
$d^2,\ d^8$	$^1S,\ ^1D,\ ^1G,\ ^3P,\ ^3F$
$d^3,\ d^7$	$^2P,\ ^2D(2),\ ^2F,\ ^2G,\ ^2H,\ ^4P,\ ^4F$
$d^4,\ d^6$	$^1S(2),\ ^1D(2),\ ^1F,\ ^1G(2),\ ^1I,\ ^3P(2),\ ^3D,\ ^3F(2),\ ^3G,\ ^3H,\ ^5D$
d^5	$^2S,\ ^2P,\ ^2D(3),\ ^2F(2),\ ^2G(2),\ ^2H,\ ^2I,\ ^4P,\ ^4D,\ ^4F,\ ^4G,\ ^6S$

[a] Adapted from G. Herzberg, *Atomic Spectra and Atomic Structure*, Dover Publications, New York (1944), p. 132.

compounds. Such a goal is currently unattainable, but it can be approached by presenting several properties that will be useful in subsequent discussions: namely, size relationships, factors affecting electron loss and gain, and magnetic behavior.

4-1. Shielding, or Screening, and Effective Nuclear Charge

Reference has been made to the shielding, or screening, effects of underlying electronic arrangements upon the attraction of the nucleus for the outermost electrons in an atom. If, however, the energies of the ns^1 electrons for the alkali metal atoms are calculated from Eq. 2-5 on the assumption that in each atom

Table 2-17   Comparison of n with n^* for Li and Cu Family Elements

Symbol	n^* (n in parentheses)			
	s	p	d	f
Li	1.59 (2)	1.96 (2)	3.00 (3)	4.00 (4)
Na	1.63 (3)	2.12 (3)	2.99 (3)	4.00 (4)
K	1.77 (4)	2.23 (4)	2.85 (3)	3.99 (4)
Rb	1.80 (5)	2.28 (5)	2.77 (4)	3.99 (4)
Cs	1.87 (6)	2.33 (6)	2.55 (5)	2.98 (4)
Cu	1.33 (4)	1.86 (4)	2.98 (4)	4.00 (4)
Ag	1.34 (5)	1.87 (5)	2.98 (5)	3.99 (4)
Au	1.21 (6)	1.72 (6)	2.98 (6)	

the electrons in inner filled orbitals shield that electron from all but a single positive charge, they prove to be substantially larger than the measured values. These differences are believed to be the consequences of differences in the degrees to which underlying orbitals are "penetrated" by electrons from other orbitals. The data for $3s$, $3p$, and $3d$ orbitals in Fig. 2-9(b) show this penetration, and as a consequence shielding, to decrease as $s > p > d$. The f orbitals are even less penetrating and less shielding than the d orbitals. To account in a more nearly quantitative way for this penetrating effect one may either replace n in Eq. 2-5 by n^*, an *effective principal quantum* number,* or replace Z by Z_{eff}, an effective nuclear charge definable as

$$Z_{eff} = Z - S \tag{2-31}$$

where S is a shielding or screening constant.

Values of n^* are compared with those of n for the elements of the Li and Cu families in Table 2-17.[60] It is apparent that the effect of the more deeply penetrating s electron is much larger than that for the less deeply penetrating p electron, and so on; that the effects are larger for heavier atoms (with more electron shells) than for smaller atoms; and that 18-electron shells are more effectively penetrated than 8-electron shells.

It is more common to evaluate screening constants and effective nuclear charges. For this purpose, one may use "Slater's rules," as derived for the calculation of improved atomic wave functions.[58, 61, 62] These rules for determining S are:

1.   The electrons are grouped by orbitals, each group having a different value of S, as: $(1s)$; $(2s, 2p)$; $(3s, 3p, 3d)$; $(4s, 4p)$; $(4d)$; $(4f)$; $(5s, 5p)$; and so on.

---

*   The effective quantum number n^* is the value that must be substituted for n in Eq. 2-5 to allow calculation of energy (E).

2. The shielding constant is the sum of the following contributions:
   (a) Zero from any shell outside (larger n) the one considered.
   (b) An amount 0.35 from each other electron in the group (same n) considered. For the $1s$ group, 0.30 is better than 0.35.
   (c) For an s, p shell, an amount 0.85 from each electron in the next inner shell $(n-1)$, and an amount 1.00 from each electron still further in ($<n-1$).
   (d) For an nd or nf shell, an amount 1.00 from each electron lying inside it ($n-1$, except where $n=4$).

These rules are illustrated by the following examples:

1. A $2p$ electron in a carbon atom ($Z=6$; $1s^2\ 2s^2\ 2p^2$). The grouping is $(1s)^2(2s\ 2p)^4$, whence for a $2p$ electron

$$S = (3 \times 0.35) + (2 \times 0.85) = 2.75$$

$$Z_{eff} = 6 - 2.75 = 3.25$$

2. A $4s$ electron in a gallium atom ($Z=31$; $1s^2\ 2s^2\ 2p^6\ 3s^2\ 3p^6\ 3d^{10}\ 4s^2\ 4p^1$). The grouping is $(1s)^2(2s\ 2p)^8(3s\ 3p\ 3d)^{18}(4s\ 4p)^3$, whence for a $4s$ electron

$$S = (2 \times 0.35) + (18 \times 0.85) + (10 \times 1.00) = 26.0$$

$$Z_{eff} = 31 - 26.0 = 5$$

3. A $3d$ electron in a gallium atom. Using the grouping shown above

$$S = (17 \times 1.00) + (10 \times 1.00) = 27.0$$

$$Z_{eff} = 31 - 27.0 = 4$$

Slater's rules do not allow the exact calculation of the energies of electrons, but they are important in rationalizing variations in radii, ionization energies, and electronegativities. Effective nuclear charges calculated in a more detailed way from a larger number of parameters for atoms from hydrogen through krypton are more nearly accurate.[63]

4-2. Atomic Radii

An inevitable consequence of the development of the atomic theory was the assignment of sizes to atomic and ionic particles. An early view that such particles can be considered as rigid spheres in contact with each other in solid substances led logically to the concept that each particle has a definite volume, diameter, and radius. On this basis, systematic crystal chemistry was developed, and numerous properties were related to differences in the sizes of atoms and ions.

Although this concept is now somewhat more pictorial than exact, it still has broad utility. The impossibility of representing an atom or ion in terms of a three-dimensional model and the necessity of dealing with probabilities of concentration of electronic charge density, as noted earlier in this chapter, preclude the assignment of radii that are applicable under all conditions. However, radial distribution curves (Figs. 2-9, 2-19) do suggest concentrations of charge density within rather well defined distances, as do the general orbital representations (Figs. 2-10 to 2-18), and thus give credence to the concept of electronic spheres of influence of rather well defined sizes. It is on this basis that the assignment of atomic, covalent, and ionic radii must be considered.

Covalent radii and *ionic radii* are readily understood in their relationships to covalent bonding and ionic bonding, respectively, and they are considered in detail in subsequent chapters on bonding (Chs. 4-7). The term *atomic radius*, however, has been used loosely and applied to a variety of situations. Because of the impossibility of isolating an individual atom and determining for it a radius that will correctly describe that atom in both its natural environment and in nonionic compounds, a radius of this type must be determined from measured internuclear distances and a correct apportionment of such distances. The apportionment must take into account the environment of the atom: that is, the number and arrangement nearest-neighboring atoms present, the nature of interactions between atoms, and the presence or absence of bonding. Details of the experimental measurement and apportionment of internuclear distances are given for ionic crystals and covalent molecules in later chapters. Here it is necessary only to distinguish among various types of "atomic" radii, to develop a tabulation of internally consistent radii, and to point out trends in the tabulated values.

Van der Waals Radii. In the solid state, the nonmetallic elements exist as aggregations of monoatomic (noble gas elements) or polyatomic (e.g., Cl_2, O_2, S_8, N_2, P_4, C_x) molecules. These molecules are not chemically bonded to each other but are held in their equilibrium positions by the interaction of fluctuating dipoles, the so-called London dispersion forces or van der Waals forces (Sec. 5.1-8). Half of a measured internuclear distance for a noble gas element or half of a measured internuclear distance between two atoms in two adjacent molecules is then a nonbonded atomic radius or a van der Waals radius. Thus for solid xenon the internuclear distance is 4.36 Å, giving a van der Waals radius of 2.18 Å, whereas for solid chlorine, consisting of Cl_2 molecules, the corresponding values are 3.60 and 1.80 Å.*

Although a van der Waals radius might be assumed to be invariant, it is dependent upon the extent to which an atom is compressed. Thus the radius of the xenon atom in solid xenon(IV) fluoride, XeF_4, is ca. 1.7 Å rather than 2.18 Å, a consequence of the lack of Xe–Xe contacts through interposition of F atoms. Furthermore, when one deals with gases, where molecules are in constant

---

* Radii are expressed in either angstrom units (Å), the older unit and the more common one in most of the literature, or picometers (pm), the newer SI unit. Conversion factors are 1 Å/10^{-8} cm, 1 pm/10^{-10} cm, 1 Å/100 pm.

unrestricted motion and collide with each other, still different values, called Lennard-Jones radii, are calculated by halving the distances of closest approach. These radii have been related by J. E. Huheey[64] to the crystallographic nonbonded radii.

Van der Waals radii are listed in Table 2-18. Early estimates that van der Waals and ionic or crystal radii (Sec. 4.3-2) are comparable[65] have not been completely justified by more recent data.[66]

Covalent Radii. In subsequent discussions (Sec. 5.1-1), the covalent bond is described in terms of interactions between the electron clouds of two atoms. Half the measured internuclear distance between two identical atoms in a single covalent bond then is the covalent radius of the atom in question. Thus, in the Cl_2 molecule the internuclear distance is 1.98 Å, giving a single-bonded covalent radius of 0.99 Å for the Cl atom. This radius is somewhat smaller than the van der Waals radius because the latter is not dependent on orbital interactions or overlap. Other types of covalent radii are discussed later in this section.

Metallic Radii. In crystals of most of the metallic elements, each atom has 12 nearest-neighbor atoms (Sec. 4.4-1). Taking this geometrical fact into account and assuming that orbital interactions between metallic atoms comparable with those between nonmetallic atoms occur, it is possible to derive a set of single-bonded metallic radii.[67]

Internally Consistent Sets of Atomic Radii. Single-bonded covalent radii for the nonmetallic elements and single-bonded metallic radii have been shown to be sufficiently closely related that they may be collected together into a single tabulation that is broadly useful, in particular as bases to account for observed trends among the elements and to a more limited degree for approximating bond lengths. Such a tabulation is included in Table 2-18.

Another set of internally consistent atomic radii, the Bragg–Slater radii,[58,68,69] is also included in Table 2-18. These radii can be added, irrespective of the bond type involved in the interaction between two atoms, to give rather good correlations with observed internuclear distances. Slater[58] noted that there is a close correlation between the radius at which radial electronic density reaches a maximum value

$$r_m = -\frac{(n^*)^2}{Z - S} \tag{2-32}$$

and the commonly accepted covalent and ionic radii. For example, an "ionic radius is roughly one at which the radial density becomes 10% of its maximum value." In general terms, the Bragg–Slater radii represent the radii of inner cores of electrons, which in turn determine how closely two atomic species can approach each other.

Table 2-18 Radii of Atoms

Z	Element Symbol	van der Waals $(r_W)^a$	Radius (Å) Covalent-metallic $(r_{CM})^b$	Bragg-Slater $(r_{BS})^c$
1	H	1.20	0.37	0.25
2	He	1.40	$(0.5)^d$	
3	Li		1.52 (metal); 1.34 (Li_2)	1.45
4	Be		1.11	1.05
5	B		0.80	0.85
6	C	1.70	0.77	0.70
7	N	1.55	0.74	0.65
8	O	1.52	0.74	0.60
9	F	1.47	0.71	0.50
10	Ne	1.54	$(0.65)^d$	
11	Na		1.86 (metal); 1.54 (Na_2)	1.80
12	Mg		1.60	1.50
13	Al		1.43	1.25
14	Si	2.10	1.18	1.10
15	P	1.80	1.10	1.00
16	S	1.80	1.03	1.00
17	Cl	1.75	0.99	1.00
18	Ar	1.88	$(0.95)^d$	
19	K		2.27 (metal); 1.96 (K_2)	2.20
20	Ca		1.97	1.80
21	Sc		1.61	1.40
22	Ti		1.45	1.35
23	V		1.31	1.35
24	Cr		1.25	1.40
25	Mn		1.37	1.40
26	Fe		1.24	1.40
27	Co		1.25	1.35
28	Ni		1.25	1.35
29	Cu		1.28	1.35
30	Zn		1.33	1.35
31	Ga		1.22	1.30
32	Ge		1.23	1.25
33	As	1.85	1.25	1.15
34	Se	1.90	1.16	1.15
35	Br	1.85	1.14	1.15
36	Kr	2.02	$(1.10)^d$	
37	Rb		2.48 (metal)	2.35
38	Sr		2.15	2.00
39	Y		1.78	1.80
40	Zr		1.59	1.55
41	Nb		1.43	1.45
42	Mo		1.36	1.45

Table 2-18 *Continued*)

Z	Element Symbol	Radius (Å) van der Waals $(r_W)^a$	Covalent-metallic $(r_{CM})^b$	Bragg-Slater $(r_{BS})^c$
43	Tc		1.35	1.35
44	Ru		1.33	1.30
45	Rh		1.35	1.35
46	Pd		1.38	1.40
47	Ag		1.45	1.60
48	Cd		1.49	1.55
49	In		1.63	1.55
50	Sn		1.41	1.45
51	Sb		1.45	1.45
52	Te	2.06	1.43	1.40
53	I	1.98	1.33	1.40
54	Xe	2.16	$(1.30)^d$	
55	Cs		2.65 (metal)	2.60
56	Ba		2.17	2.15
57	La		1.87	1.95
58	Ce		1.83	1.85
59	Pr		1.82	1.85
60	Nd		1.81	1.85
61	Pm		(ca. 1.81)	1.85
62	Sm		1.80	1.85
63	Eu		1.99	1.85
64	Gd		1.79	1.80
65	Tb		1.76	1.75
66	Dy		1.75	1.75
67	Ho		1.74	1.75
68	Er		1.73	1.75
69	Tm		(ca. 1.73)	1.75
70	Yb		1.94	1.75
71	Lu		1.72	1.75
72	Hf		1.56	1.55
73	Ta		1.43	1.45
74	W		1.37	1.35
75	Re		1.37	1.35
76	Os		1.34	1.30
77	Ir		1.36	1.35
78	Pt		1.39	1.35
79	Au		1.44	1.35
80	Hg		1.50	1.50
81	Tl		1.70	1.90
82	Pb		1.75	
83	Bi		1.55	1.60
84	Po		1.18	1.90

Table 2-18 (*Continued*)

Z	Element Symbol	van der Waals $(r_W)^a$	Covalent-metallic $(r_{CM})^b$	Bragg-Slater $(r_{BS})^c$
			Radius (Å)	
85	At			
86	Rn		$(1.45)^d$	
87	Fr			
88	Ra			2.15
89	Ac		1.88	1.95
90	Th		1.80	1.80
91	Pa		1.61	1.80
92	U		1.39	1.75
93	Np			1.75
94	Pu		1.51	1.75
95	Am		1.30–1.32	1.75
96	Cm			
97	Bk			
98	Cf			
99	Es			

^{a} A. Bondi, *J. Phys. Chem.*, **68**, 441 (1964).

^{b} L. E. Sutton (Science Ed.), *Tables of Interatomic Distances and Configuration in Molecules and Ions,* Spec. Publ. No. 18, pp. 53s–513s, The Chemical Society, London (1965). Largely; some scattered data from other sources.

^{c} J. C. Slater, *J. Chem. Phys.*, **41**, 3199 (1964).

^{d} See J. E. Huheey, *J. Chem. Educ.*, **45**, 791 (1968); *Inorganic Chemistry*, 2nd ed., pp. 232–233, Harper & Row, New York (1978).

Trends and Variations in Atomic Radii. Comparisons of the data in Table 2-18 with the ground-state electronic configurations in Table 2-14 indicate that, although some trends can be discerned, it is clearly impossible to establish quantitative relationships to either configuration or nuclear charge alone. Qualitatively, it is apparent that where ground-state configurations have the same maximum *n* values atoms of the metals are larger than those of the nonmetals; that as the number of outer electrons of a given *n* value increases, atomic size decreases; and that as the maximum *n* value increases with increasing *Z* for atoms of the same type of outermost electronic configuration, atomic size increases. None of these trends is completely regular, however, and each reflects the combined effects of nuclear charge, screening, and increases of spatial extension of orbitals with increasing *n*. There is a marked periodicity in size with increasing *Z* that brings together atoms of the same overall type of ground-state electronic configuration (Fig. 3-10).

Significant also are the general decreases in atomic size with increasing *Z* as $(n-1)d$ or $(n-2)f$ orbitals are successively occupied. Each of these reflects incomplete screening of nuclear charge effects by electrons in the *d* and *f* orbitals. These contractions in size are small, but they affect significantly both the properties of the d- and f-transition elements and also those of the elements that immediately follow them in atomic number.

After each d-transition series is completed, there is no complete regularity in atomic radii, but the sizes of these atoms (with underlying 18-electron cores) never quite equal those of atoms with comparable 8-electron cores (e.g., compare K with Cu, Y with In).

Parallel small decreases that occur as f orbitals are progressively occupied (i.e., the *lanthanide* and *actinide contractions*) are responsible for slight variations in certain physical and chemical properties with increasing nuclear charge in each series. The lanthanide contraction is responsible for two seeming anomalies among the elements: the marked parallel between the chemistry of yttrium ($Z = 39$) and those of the heavier lanthanides (especially dysprosium-erbium, $Z = 66$ to 68) and the close parallels between the chemistries of the elements that immediately follow the lanthanides in atomic number and their immediate congeners of lower atomic number [e.g., Hf and Zr, Ta and Nb, W and Mo (pp. 149–152)]. Reference to Table 2-18 will indicate clearly close similarities in atomic radii. Within the lanthanides, the seemingly anomalously large radii of the Eu ($Z = 63$) and Yb ($Z = 70$) atoms are associated with the presence of the dipositive state in these metals as opposed to the tripositive state in other lanthanide metals.

As noted in subsequent discussions, all properties that measure the ease of the loss or gain of electrons by atoms are related to atomic size. In later chapters, variations in the sizes of atoms and derived ions are related to bonding, molecular and crystal structures, and chemical reactions.

4-3. Electron Loss and Gain by Atoms

Inasmuch as the chemical reactions of the elements are believed to proceed through involvement of electrons in the highest quantum levels of their atoms, electron-gain, electron-loss, and electron-sharing processes are important in the necessary rupture of bonds and formation of new bonds (Ch. 4). Thus properties of atoms that measure the ease with which electrons are lost or gained must be considered. Those discussed in this section are *ionization energy, electron affinity,* and *electronegativity.*

Ionization Energy (IE or ΔH_i*).* As previously indicated, successive increments of energy must be absorbed to promote an electron to successively higher atomic energy levels. Ultimately, the electron is removed to an infinite quantum level ($n = \alpha$), effecting ionization and the formation of a cation. The process is indicated graphically for the hydrogen atom in Fig. 2-23. Similar energy-level diagrams can be constructed for other atomic species.

The *first ionization energy* (I) of an atom is defined as the quantity of energy required to remove completely the most loosely bound electron from that

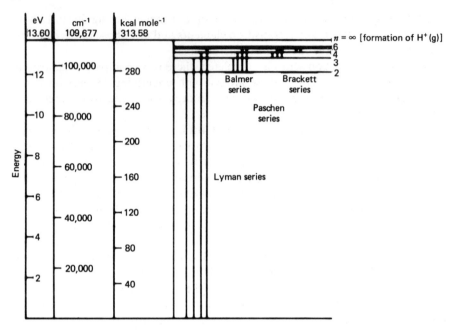

Figure 2-23 Energy-level diagram for hydrogen atom. [A. W. Adamson, *A Textbook of Physical Chemistry*, Fig. 16-1, p. 747, Academic Press, New York (1973), modified.]

neutral, gaseous atom (E) at $0°K$, that is, the energy associated with a process formulated as

$$E(g) \xrightarrow{0°K} E^+(g) + e^-$$

Second (II), third (III), and so on, ionization energies are similarly defined in terms of the removal of the second, third, and so on, electrons from gaseous unipositive, bipositive, and so on, ions. Inasmuch as it becomes progressively more difficult to remove an electron as the positive charge of the ion increases, ionization energies increase in magnitude as first < second < third < ···.[70] Since an ionization energy is usually measured and expressed in electron volts, the term *ionization potential* is also used, particularly in older treatments. It is understood that the unqualified term "ionization energy" (or potential) applies to the removal of one electron from a neutral atom.

Ionization energies are summarized in Table 2-19.[71] Ionization energies are always endothermic in character and are thus assigned positive values in accordance with thermodynamic convention (Sec. 9.2-1). The numerical magnitude of an ionization energy is influenced by a variety of factors, among which the more obvious ones are the actual nuclear charge, the penetration and shielding effect of inner electronic shells, and the atomic radius. Inasmuch as the measured ionization energy is determined by a combination of these factors, it is not always possible to ascribe a greater influence to a single factor. Thus although a given electron may be shielded from the nucleus by underlying shells, as soon as it penetrates close to the nucleus, shielding becomes less important and attraction by the nucleus more important.

Inasmuch as an ionization energy in excess of ca. 15 eV (ca. 350 kcal, or 1464 kJ mole^{-1}) is impossible to achieve by chemical means alone, many of the states of ionization indicated in Table 2-19 are observable only spectro-scopically. Correspondingly, the formation of simple cations with charges in excess of 2 is unfavorable. However, the formation of a gaseous cation, as measured by ionization energy, is not always the controlling factor in the energetics of compound formation (Sec. 4.3-4).

The comparative stabilities with respect to removal of electrons of the noble gas ($1s^2$ and $ns^2\,np^6$) configurations, and to a lesser extent of the pseudo noble gas $[(n-1)d^{10}]$ configurations, are emphasized by the "stair-step" designations in Table 2-19. The periodic variations in first ionization energies with atomic numbers apparent in this table are shown graphically in Fig. 3-11. First ionization energies have maximum values for the noble gas atoms. Addition of the next electron (ns^1, adjacent alkali metals) is accompanied by a sharp decrease in ionization energy, reflecting the increased "size" of the higher quantum level. Increase with increasing atomic number from alkali metal to halogen is associated with increasing nuclear or effective nuclear charge without compensation by increase in size. The significant decreases characterizing the elements B, Al, Ga, In, and Tl are associated with the appearances of the first np electrons, which are less penetrating than the ns electrons. Within the d- and f-transition series, small increases in first ionization energies with increasing Z reflect incomplete shielding of outermost electrons by those in these d or f shells. Within a family of elements (e.g., the alkali metals or the halogens), general decrease in first ionization energies with increasing nuclear charge is obviated among the elements immediately following the lanthanides (i.e., Hf through Pb) as consequences of the large increase of nuclear charge and corresponding contraction in the region $Z = 57$ to 71. Further useful information on regularities and relationships among ionization energies is available in a paper by Liebman.[72]

Ionization energies are particularly useful in systematizing the reducing properties of the elements and in determining the natures of the bonds that their atoms form (Ch. 4).

Electron Affinity (EA or ΔH_a). The first, or atomic, electron affinity is defined as the quantity of energy *released* when an electron is added to a gaseous atom at $0°K$ to form a gaseous uninegative ion, that is, the energy released in the process

$$E(g) + e^- \overset{0°K}{\to} E^-(g)$$

Its inverse relationship to ionization energy is apparent in terming the energy associated with the reverse of the foregoing process the "zeroth ionization

Table 2-19 Ionization Energies of the Elements

Z	Symbol	Outermost Configuration	Ionization Energy (eV)							
			I	II	III	IV	V	VI	VII	VIII
1	H	$1s^1$	13.598							
2	He	$1s^2$	24.587	54.416						
3	Li	$2s^1$	5.392	75.638	122.451					
4	Be	$2s^2$	9.322	18.211	153.893	217.713				
5	B	$2s^22p^1$	8.298	25.154	37.930	259.368	340.217			
6	C	$2s^22p^2$	11.260	24.383	47.887	64.492	392.077	489.981		
7	N	$2s^22p^3$	14.534	29.601	47.448	77.472	97.888	552.057	667.029	
8	O	$2s^22p^4$	13.618	35.116	54.934	77.412	113.896	138.116	739.315	871.387
9	F	$2s^22p^5$	17.422	34.970	62.707	87.138	114.240	157.161	185.182	953.886
10	Ne	$2s^22p^6$	21.564	40.962	63.45	97.11	126.21	157.93	207.27	239.09
11	Na	$3s^1$	5.139	47.286	71.64	98.91	138.39	172.15	208.47	264.18
12	Mg	$3s^2$	7.646	15.035	80.143	109.24	141.26	186.50	224.94	265.90
13	Al	$3s^23p^1$	5.986	18.828	28.447	119.99	153.71	190.47	241.43	284.59
14	Si	$3s^23p^2$	8.151	16.345	33.492	45.141	166.77	205.05	246.42	303.17
15	P	$3s^23p^3$	10.486	19.725	30.18	51.37	65.023	220.43	263.22	309.41
16	S	$3s^23p^4$	10.360	23.33	34.83	47.30	72.68	88.049	280.93	328.23
17	Cl	$3s^23p^5$	12.967	23.81	39.61	53.46	67.8	97.03	114.193	348.28
18	Ar	$3s^23p^6$	15.759	27.629	40.74	59.81	75.02	91.007	124.319	143.456
19	K	$4s^1$	4.341	31.625	45.72	60.91	82.66	100.0	117.56	154.86
20	Ca	$4s^2$	6.113	11.871	50.908	67.10	84.41	108.78	127.7	147.24
21	Sc	$3d^14s^2$	6.54	12.80	24.76	73.43	91.66	111.1	138.0	158.7
22	Ti	$3d^24s^2$	6.82	13.58	27.491	43.266	99.22	119.36	140.8	168.5
23	V	$3d^34s^2$	6.74	14.65	29.310	46.707	65.23	128.12	150.17	173.7
24	Cr	$3d^54s^1$	6.766	15.50	30.96	49.1	69.3	90.56	161.1	184.7
25	Mn	$3d^54s^2$	7.435	15.640	33.667	51.2	95	119.27	196.46	221.8
26	Fe	$3d^64s^2$	7.870	16.18	30.651	54.8	75.0	99	125	151.06

Z	Element	Configuration	I_1	I_2	I_3	I_4	I_5	I_6	I_7	I_8
27	Co	$3d^7 4s^2$	7.86	17.06	33.50	51.3	79.5	102	129	157
28	Ni	$3d^8 4s^2$	7.635	18.168	35.17	54.9	75.5	108	133	162
29	Cu	$3d^{10} 4s^1$	7.726	20.292	36.83	55.2	79.9	103	139	166
30	Zn	$3d^{10} 4s^2$	9.394	17.964	39.722	59.4	82.6	108	134	174
31	Ga	$3d^{10} 4s^2 4p^1$	5.999	20.51	30.71	64				
32	Ge	$3d^{10} 4s^2 4p^2$	7.899	15.934	34.22	45.71	93.5	127.6	155.4	192.8
33	As	$3d^{10} 4s^2 4p^3$	9.81	18.633	28.351	50.13	62.63	81.70	103.0	126
34	Se	$3d^{10} 4s^2 4p^4$	9.752	21.19	30.820	42.944	68.3	88.6	111.0	136
35	Br	$3d^{10} 4s^2 4p^5$	11.814	21.8	36	47.3	59.7	78.5	99.2	122.3
36	Kr	$3d^{10} 4s^2 4p^6$	13.999	24.359	36.95	52.5	64.7	84.4	106	129
37	Rb	$5s^1$	4.177	27.28	40	52.6	71.0	90.8	116	
38	Sr	$5s^2$	5.695	11.030	43.6	57	71.6	93.0	125	
39	Y	$4d^1 5s^2$	6.38	12.24	20.52	61.8	77.0	102.6		
40	Zr	$4d^2 5s^2$	6.84	13.13	22.99	34.34	81.5			
41	Nb	$4d^3 5s^2$	6.88	14.32	25.04	38.3	50.55	68		
42	Mo	$4d^5 5s^1$	7.099	16.15	27.16	46.4	61.2			
43	Tc	$4d^5 5s^2$	7.28	15.26	29.54					
44	Ru	$4d^7 5s^1$	7.37	16.76	28.47					
45	Rh	$4d^8 5s^1$	7.46	18.08	31.06					
46	Pd	$4d^{10}$	8.34	19.43	32.93					
47	Ag	$4d^{10} 5s^1$	7.576	21.49	34.83					
48	Cd	$4d^{10} 5s^2$	8.993	16.908	37.48					
49	In	$4d^{10} 5s^2 5p^1$	5.786	18.869	28.03	54				
50	Sn	$4d^{10} 5s^2 5p^2$	7.344	14.632	30.502	40.734	72.28	108	126.8	153
51	Sb	$4d^{10} 5s^2 5p^3$	8.641	16.53	25.3	44.2	56			
52	Te	$4d^{10} 5s^2 5p^4$	9.009	18.6	27.96	37.41	58.75	70.7	137	
53	I	$4d^{10} 5s^2 5p^5$	10.451	19.131	33					

Table 2.19 (*Continued*)

Z	Symbol	Outermost Configuration	Ionization Energy (eV)							
			I	II	III	IV	V	VI	VII	VIII
54	Xe	$4d^{10}5s^25p^6$	12.130	21.21	32.1					
55	Cs	$6s^1$	3.894	23.1						
56	Ba	$6s^2$	5.212	10.004						
57	La	$5d^16s^2$	5.577	11.06	19.175					
58	Ce	$4f^15d^16s^2$	5.47	10.85	20.20	36.72				
59	Pr	$4f^36s^2$	5.42	10.55	21.62	38.95	57.45			
60	Nd	$4f^46s^2$	5.49	10.72						
61	Pm	$4f^56s^2$	5.55	10.90						
62	Sm	$4f^66s^2$	5.63	11.07						
63	Eu	$4f^76s^2$	5.67	11.25						
64	Gd	$4f^75d^16s^2$	5.85	11.52						
65	Tb	$4f^96s^2$	5.85	11.52						
66	Dy	$4f^{10}6s^2$	5.93	11.67						
67	Ho	$4f^{11}6s^2$	6.02	11.80						
68	Er	$4f^{12}6s^2$	6.10	11.93						
69	Tm	$4f^{13}6s^2$	6.18	12.05	23.71					
70	Yb	$4f^{14}6s^2$	6.254	12.17	25.2					
71	Lu	$4f^{14}5d^16s^2$	5.426	13.9	23.3					
72	Hf	$4f^{14}5d^26s^2$	7.0	14.9	23.3	33.3				
73	Ta	$4f^{14}5d^36s^2$	7.89							
74	W	$4f^{14}5d^46s^2$	7.98							
75	Re	$4f^{14}5d^56s^2$	7.88							
76	Os	$4f^{14}5d^66s^2$	8.7							
77	Ir	$4f^{14}5d^76s^2$	9.1							

Z		Configuration	I	II	III	IV	V	VI
78	Pt	$4f^{14}5d^96s^1$	9.0	18.563				
79	Au	$4f^{14}5d^{10}6s^1$	9.225	20.5				
80	Hg	$4f^{14}5d^{10}6s^2$	10.437	18.756	34.2			
81	Tl	$4f^{14}5d^{10}6s^26p^1$	6.108	20.428	29.83			
82	Pb	$4f^{14}5d^{10}6s^26p^2$	7.416	15.032	31.937	42.32	68.8	
83	Bi	$4f^{14}5d^{10}6s^26p^3$	7.289	16.69	25.56	45.3	56.0	88.3
84	Po	$4f^{14}5d^{10}6s^26p^4$	8.48					
85	At	$4f^{14}5d^{10}6s^26p^5$						
86	Rn	$4f^{14}5d^{10}6s^26p^6$	10.748					
87	Fr	$7s^1$						
88	Ra	$7s^2$	5.279	10.147				
89	Ac	$6d^17s^2$	6.9	12.1				
90	Th	$6d^27s^2$		11.5	20.0	28.8		
91	Pa	$5f^26d^17s^2$						
92	U	$5f^36d^17s^2$						
93	Np	$5f^46d^17s^2$						
94	Pu	$5f^67s2$	5.8					
95	Am	$5f^77s^2$	6.0					
96	Bk	$5f^76d^17s^2$						

energy."[73] Second, third, and so on, electron affinities are associated with the addition of two, three, and so on, electrons. Although experimentally determined ionization energies have been available for many years,[72] prior to 1970 only the atomic electron affinities of hydrogen, fluorine, chlorine, bromine, iodine, carbon, oxygen, and sulfur had been measured. New techniques have yielded values for many other elements.[74, 75] Summarized in Table 2-20 are the "best available" values.

Unfortunately, the accepted definition noted above has led to assignment of positive signs to exothermic electron affinities, in direct opposition to thermodynamic convention (Sec. 9.2-1). It is necessary, therefore, to change the sign when using an electron affinity as an enthalpy change in a calculation (e.g., in Born–Fajans–Haber cycle applications; Sec. 4.3-4). The atomic electron affinities of the active nonmetals are indeed exothermic quantities. However, because of repulsions of like charges, the second, third, and so on, electron affinities of these elements are endothermic quantities, and indeed the *total* electron affinities for the formation of ions such as O^{2-}, S^{2-}, and Se^{2-} are endothermic. The scarcity of simple ions with charges larger than -1 is a reflection of this characteristic. The seemingly anomalously small electron affinity of the fluorine atom is probably due to the small dissociation energy of the gaseous F_2 molecule, which in turn results from repulsions among nonbonded electrons in this small species.[76]

Additional data are available in several review references.[74, 75, 77–79]

The data in Table 2-20 show periodic variations with nuclear charge (Fig. 3-12). As may be expected, minimum values characterize the noble gas atoms and maximum values the halogen atoms. The chalcogen family atoms (0, S, Se, Te, Po) also have large first electron affinities, corresponding to their observed tendency to form anions. Values for many atoms appear to cluster around 1.0 eV. The large negative values for atoms of the beryllium family are unique and indicate clearly a marked tendency not to form anions. Electron affinities are useful in systematizing the oxidizing properties of the elements and, in conjunction with ionization energies, in determining the natures of the bonds that their atoms form (Ch. 4).

Electronegativity (*x*). Both the ionization energy and the electron affinity are quantitative measures of the attraction, or the lack of attraction, a gaseous atom has for an electron. In more general terms, and in the words of L. Pauling,[80] these properties measure "the qualitative property that the chemist calls *electronegativity*, the power of an atom in a molecule to attract electrons to itself." It is apparent that, other factors being equal, a small atom attracts electrons more strongly than a larger atom and is thus more electronegative. It is apparent also that larger electronegativities may be expected for atoms in which the available orbitals are nearly completely occupied than for atoms in which the available orbitals are sparsely populated. Strictly speaking, electronegativity is not a property of an isolated atom, although it is often so considered and related to some of the properties of an isolated atom. Rather, it is a property of

Table 2-20 Atomic Electron Affinities

Z	Symbol	Electron Affinity (eV)[a]	Z	Symbol	Electron Affinity (eV)[a]
1	H	0.754	34	Se	2.02
2	He	(−0.22)	35	Br	3.363
3	Li	0.620	36	Kr	(−0.40)
4	Be	(−2.5)	37	Rb	0.4859
5	B	0.86	38	Sr	(−1.74)
6	C	1.270	42	Mo	1.0
7	N	0.0	47	Ag	1.303
8	O	1.465	49	In	0.35
9	F	3.339	50	Sn	1.25
10	Ne	(−0.30)	51	Sb	1.05
11	Na	0.548	52	Te	1.90
12	Mg	(−2.4)	53	I	3.061
13	Al	(0.52)	54	Xe	(−0.42)
14	Si	1.24	55	Cs	0.472
15	P	0.77	56	Ba	(−0.54)
16	S	2.077	73	Ta	0.8
17	Cl	3.614	74	W	0.5
18	Ar	(−0.36)	75	Re	0.15
19	K	0.501	78	Pt	2.128
20	Ca	(−1.62)	79	Au	2.3086
22	Ti	0.391	81	Tl	0.5
23	V	0.937	82	Pb	1.05
24	Cr	0.66	83	Bi	1.05
26	Fe	0.582	84	Po	(1.8)
27	Co	0.936	85	At	(2.8)
28	Ni	1.276	86	Rn	(−0.42)
29	Cu	1.276	87	Fr	(0.456)
31	Ga	(0.37)			
32	Ge	1.20			
33	As	0.80			

[a] All values experimental except those in parentheses.

an atom in a molecule or complex ion where that atom is bonded to, and affected by, one or more other atoms.

Several scales of numerical values of electronegativities have been devised.[80-85] Four of these are included in Table 2-21, and where the values in those scales are not equivalent to those of the Pauling scale,[80] equivalent values are given also. The periodicity apparent from the values in any one of these scales is shown graphically for the Allred–Rochow values in Fig. 3-13.

Pauling's original scale of electronegativities[86] was based upon variations in the difference Δ between the measured energy of a bond between two atoms A and B and the energy expected for a purely covalent single bond (p. 236) between these two atoms. Values of Δ increase as the difference in electronegativity

Table 2-21   Electronegativities of the Elements

Z	Symbol	Electronegativity, x			
		Pauling[a]	Mulliken–Jaffé (eV)[b]	Allred–Rochow[c]	Sanderson[d]
1	H	(2.20)	7.17 (2.21) s	2.20	3.55 (2.31)
2	He			[3.2]	
3	Li	(0.98)	3.10 (0.84) s	0.97	0.74 (0.86)
4	Be	(1.57)	4.78 (1.40) sp	1.47	1.99 (1.42)]
5	B	(2.04)	6.33 (1.93) sp^2	2.01	[2.93 (1.93)]
6	C	(2.55)	7.98 (2.48) sp^3	2.50	3.79 (2.47)
7	N	(3.04)	11.54 (3.68) sp^3	3.07	4.49 (2.93)
8	O	(3.44)	15.25 (4.93) sp^3	3.50	5.21 (3.46)
9	F	(3.98)	12.18 (3.90) p	4.10	5.75 (3.92)
10	Ne			[5.1]	
11	Na	(0.93)	2.80 (0.74) s	1.01	0.70 (0.85)
12	Mg	(1.31)	4.09 (1.17) sp	1.23	[1.56 (1.21)]
13	Al	(1.61)	5.47 (1.64) sp^2	1.47	[2.22 (1.54)]
14	Si	(1.90)	7.30 (2.25) sp^3	1.74	[2.84 (1.88)]
15	P	(2.19)	8.90 (2.79) sp^3	2.06	[3.43 (2.22)]
16	S	(2.58)	10.14 (3.21) sp^3	2.44	[4.12 (2.67)]
17	Cl	(3.16)	9.38 (2.95) p	2.83	4.93 (3.28)
18	Ar			[3.3]	
19	K	(0.82)	2.90 (0.77) s	0.91	[0.42 (0.74)]
20	Ca	(1.00)	3.30 (0.99) sp	1.04	1.22 (1.06)
21	Sc	(1.36)		1.20	1.30 (1.09)
22	Ti	(1.54)		1.32	1.40 (1.13)
23	V	(1.63)		1.45	1.60 (1.24)
24	Cr	(1.66)		1.56	1.88 (1.35)
25	Mn	(1.55)		1.60	2.07 (1.44)
26	Fe	(1.83)		1.64	2.10 (1.47)
27	Co	(1.88)		1.70	2.10 (1.47)
28	Ni	(1.91)		1.75	2.10 (1.47)
29	Cu	(1.90)	[4.31] (1.36) s	1.75	2.60 (1.74)
30	Zn	(1.65)	[4.71] (1.49) sp	1.66	[2.98 (1.96)]
31	Ga	(1.81)	6.02 (1.82) sp^2	1.82	[3.28 (2.13)]
32	Ge	(2.01)	8.07 (2.50) sp^3	2.02	3.59 (2.31)
33	As	(2.18)	8.30 (2.58) sp^3	2.20	[3.90 (2.53)]
34	Se	(2.55)	9.76 (3.07) sp^3	2.48	[4.21 (2.72)]
35	Br	(2.96)	8.40 (2.62) p	2.74	4.53 (2.96)
36	Kr			[3.1]	4.81 (3.17)
37	Rb	(0.82)	2.09 (0.50) s	0.89	[0.36 (0.72)]
38	Sr	(0.95)	3.14 (0.85) sp	0.99	[1.06 (0.98)]
39	Y	(1.22)		1.11	1.05 (0.98)
40	Zr	(1.33)		(1.22)	1.10 (1.00)
41	Nb	1.6		(1.23)	1.36 (1.12)
42	Mo	(2.16)		(1.30)	1.62 (1.24)

Table 2-21 (*Continued*)

Z	Symbol	Pauling[a]	Mulliken–Jaffé (eV)[b]	Allred–Rochow[c]	Sanderson[d]
			Electronegativity, x		
43	Tc	1.9		(1.36)	1.80 (1.33)
44	Ru	2.2		(1.42)	1.95 (1.40)
45	Rh	(2.28)		(1.45)	2.10 (1.47)
46	Pd	(2.20)		(1.35)	2.29 (1.57)
47	Ag	(1.93)		1.42	2.57 (1.72)
48	Cd	(1.69)		1.46	2.59 (1.73)
49	In	(1.78)	5.28 (1.57) sp^2	1.49	[2.84 (1.88)]
50	Sn	(1.96)	7.90 (2.44) sp^3	1.72	[3.09 (2.02)]
51	Sb	(2.05)	8.48 (2.64) sp^3	1.82	[3.34 (2.16)]
52	Te	2.1	9.66 (3.04) sp^3	2.01	[3.59 (2.31)]
53	I	(2.66)	8.10 (2.52) p	2.21	3.84 (2.50)
54	Xe			[2.4]	4.06 (2.62)
55	Cs	(0.79)		0.86	[0.28 (0.69)]
56	Ba	(0.89)		0.97	0.78 (0.93)
57	La	(1.10)		1.08	0.88 (0.92)
58	Ce	(1.12)		(1.08)	0.90 (0.92)
59	Pr	(1.13)		(1.07)	0.91 (0.92)
60	Nd	(1.14)		(1.07)	0.92 (0.93)
61	Pm			(1.07)	0.93 (0.94)
62	Sm	(1.17)		(1.07)	0.94 (0.94)
63	Eu			(1.01)	0.95 (0.94)
64	Gd	(1.20)		(1.11)	0.96 (0.94)
65	Tb			(1.10)	0.97 (0.94)
66	Dy	(1.22)		(1.10)	0.97 (0.94)
67	Ho	(1.23)		(1.10)	0.98 (0.96)
68	Er	(1.24)		(1.11)	0.98 (0.96)
69	Tm	(1.25)		(1.11)	0.99 (0.96)
70	Yb			(1.06)	0.99 (0.96)
71	Lu	(1.27)		(1.14)	1.00 (0.96)
72	Hf	1.3		(1.23)	1.05 (0.98)
73	Ta	1.5		(1.33)	1.21 (1.04)
74	W	(2.36)		(1.40)	1.39 (1.13)
75	Re	1.9		(1.46)	1.53 (1.19)
76	Os	2.2		(1.52)	1.67 (1.26)
77	Ir	(2.20)		(1.55)	1.78 (1.33)
78	Pt	(2.28)		(1.44)	1.91 (1.36)
79	Au	(2.54)		(1.42)	2.57 (1.72)
80	Hg	(2.00)		(1.44)	2.93 (1.92)
81	Tl	(2.04)		(1.44)	3.02 (1.96)
82	Pb	(2.33)		(1.55)	3.08 (2.01)
83	Bi	(2.02)		(1.67)	3.16 (2.06)
84	Po	2.0		(1.76)	

Table 2-21 (*Continued*)

			Electronegativity, x		
Z	Symbol	Pauling[a]	Mulliken–Jaffé (eV)[b]	Allred–Rochow[c]	Sanderson[d]
85	At	2.2		(1.90)	
86	Rn				
87	Fr	0.7		(0.86)	
88	Ra	0.9		(0.97)	
89	Ac	1.1		(1.00)	
90	Th	1.3		(1.11)	
91	Pa	1.5		(1.14)	
92	U	(1.38)		(1.22)	
93	Np	(1.36)		(1.22)	
94	Pu	(1.28)		(1.22)	
95	Am	1.3		(1.2)	
96	Cm	1.3		(1.2)	
97	Bk	1.3		(1.2)	
98	Cf	1.3		(1.2)	
99	Es	1.3		(1.2)	
100	Fm	1.3		(1.2)	
101	Md	1.3		(1.2)	
102	No	1.3		(1.2)	
103	Lr	1.3			

[a] Values in parentheses from A. L. Allred, Ref. 89; other values from L. Pauling, Ref. 80, p. 93.

[b] First values, for orbital or hybrid indicated, from H. H. Jaffé et al., Ref. 90 and 91; values in brackets from H. O. Pritchard and H. A. Skinner, Ref. 82; values in parentheses the equivalent ones on the Pauling scale.

[c] Values in parentheses from E. J. Little, Jr., and M. M. Jones, Ref. 92; values in brackets from J. E. Huheey, *Inorganic Chemistry*, 2nd ed., pp. 162–164, Harper & Row, New York (1978); other values from A. L. Allred and E. G. Rochow, Ref. 84, except Pauling's value used for hydrogen.

[d] First values on relative compactness scale; those in parentheses the Pauling equivalent ones as calculated by R. T. Sanderson, Ref. 85(b), pp. 77–78. Bracketed values from Sanderson, Ref. 85(c), p. 41, with conversions to Pauling scale equivalents calculated from Eq. 2-38.

between the two atoms increases, but these values do not satisfy any relationship amounting to a difference between numerical properties of the atoms involved. Pauling found, however, that the square roots of the Δ values do approach such a relationship and evaluated a term $0.208\sqrt{\Delta}$ for a number of bonds, this relationship being derived from the conversion of Δ values for kcal mole^{-1} to eV by the ratio $\Delta/23.06$. Numerical values of electronegativities were so chosen that their differences were approximately equal to $\sqrt{\Delta}$ in electron volts, and these values were arbitrarily adjusted by adding a constant factor to bring the electronegativity values of the atoms C to F to 2.5 to 4.0. The general relationship

$$\Delta'(A\text{—}B) = 30(x_A - x_B)^2 \qquad (2\text{-}33)$$

where $\Delta'(A{-}B)$ is the mean (finally geometric) of extra ionic resonance energies, and x_A and x_B are the electronegativities of atoms A and B, resulted. Such a procedure allowed the evaluation of electronegativities of H, C, N, O, F, Si, P, S, Cl, Ge, As, Se, Br, and I, values for other atoms then being calculated from thermochemical data for compounds containing one of those elements and one of the above-listed elements. Details of the procedure are best reviewed in Pauling's classic monograph.[80]

The Pauling approach has been extended by M. Haïssinsky[87] and M. L. Huggins.[88] More recently, an extensive set of values has been calculated from thermochemical data by A. L. Allred.[89] The "Pauling" values listed in Table 2-21 combine those given by Pauling and Allred.

Shortly after Pauling's initial proposal, R. S. Mulliken[81] suggested that the electronegativity of an atom is measured by an average of its ionization energy and its electron affinity and is thus expressible in energy units such as electron volts. The expression used was

$$x_M = 0.168(IE + EA - 1.23) \qquad (2\text{-}34)$$

Inasmuch as a plot of x_P (Pauling electronegativity) vs. x_M yields a straight line with a slope of 62.5, conversion to the Pauling scale is effected as

$$x_P = \frac{x_M}{62.5} \qquad (2\text{-}35)$$

In terms of the same general approach, H. H. Jaffé and coworkers[90,91] have emphasized that the electronegativity of an atom in a given bond depends upon the particular pure orbital or hybrid of orbitals involved in the bond and is thus an orbital electronegativity. The "Mulliken" values given in Table 2-21 are designated in this way, and their Pauling equivalents are given.

In a generally very reasonable approach, A. L. Allred and E. G. Rochow[84] defined electronegativity as the electrostatic force exerted on its outermost, or valence, electrons by the nucleus of the atom in question. Expressed in terms of Z_{eff}, as calculated by Slater's rules (Sec. 2.4-1), electronegativity (x_{AR}) was obtained on the Pauling scale as

$$x_{AR} = \frac{0.359 Z_{eff}^2}{r^2} + 0.744 \qquad (2\text{-}36)$$

where r is the single-bond covalent radius, for some 43 elements. Many additional values were calculated in the same way by E. J. Little, Jr., and M. M. Jones.[92] These data are summarized in Table 2-21, and the periodicity in values is apparent in Fig. 3-13.

The relative compactness approach of R. T. Sanderson[85] also emphasizes the effects of size and nuclear charge. However, it stresses the qualitative concept that if an atom attracts its own electronic atmosphere strongly, it will attract an electron from another atom strongly if a vacant orbital is available to accommodate that electron. If, on the other hand, interelectronic repulsions are sufficient within an atom to weaken its hold on its own electronic atmosphere, that atom will not strongly attract an external electron even though orbital vacancies exist. The average electron density (ED), defined by the relationship

$$ED = \frac{3Z}{4\pi r^3} \tag{2-37}$$

where r is again the covalent radius (in Å), that is, the number of electrons per cubic angstrom, is considered to represent the average degree of compactness of the electronic sphere about the nucleus. Electronegativity is then expressed as a *stability ratio*, SR, defined as the ratio of the average electron density of the atom to that of "an inert atom having the same number of electrons" (e.g., a noble gas atom). Stability ratios vary from less than unity for the alkali metal atoms to 5.75 for the fluorine atom. These electronegativities, together with their Pauling equivalents calculated in terms of their observed linearity with $\sqrt{x_P}$ from the relationship

$$x^{1/2} = 0.21SR + 0.77 \tag{2-38}$$

are also summarized in Table 2-21.

Comparison of the values in Table 2-21 shows generally excellent agreement on a common basis, irrespective of the method of approach. The Allred–Rochow values are generally accepted as the most useful.* The Sanderson values have not been widely accepted, although in Sanderson's hands they have provided many interesting correlations among the elements and their compounds.[85] A particularly significant conclusion by Sanderson is that in a bond between two atoms electron density is so shifted that the electronegativities of these atoms become equalized (i.e., the *electronegativity equalization principle*). Although not shown by the data in Table 2-21, there is ample evidence that the electronegativity of a given element in combination with another element is dependent upon its oxidation state [e.g., Pauling values of 1.62 for Tl(I) vs. 2.04 for Tl(III)].

Irrespective of multitudinous arguments that decry the quantitative aspects of the electronegativity concept, electronegativities are extremely useful in determining bond type and character (Ch. 4), in approximating bond energies, in establishing qualitatively the thermal stabilities of compounds, and in rationalizing observed similarities and differences in chemical behavior. Many of these applications appear in subsequent chapters.†

4-4. Magnetic Properties

Since it is well known that a magnetic effect is produced by a moving electrical charge, it is to be expected that nuclear "spin" and the spin and orbital motions

---

* The very severe criticism by R. S. Drago [*J. Inorg. Nucl. Chem.*, **15**, 237 (1960)] of alternation of electronegativity in the C–Si–Ge–Sn–Pb sequence has been rationally and reasonably refuted by A. L. Allred and E. G. Rochow [*J. Inorg. Nucl. Chem.*, **20**, 167 (1961)].

† The reader is referred to the excellent summary and extension of ideas in J. E. Huheey, *Inorganic Chemistry*, 2nd ed., Ch. 4, Harper & Row, New York (1978), which includes a good many of Huheey's own significant contributions to this area.

Table 2-22 Classification of Magnetic Behaviors

Type	Origin of Behavior	Magnitude at 20°C (χ_M, cgs units)	Temperature Dependence of Susceptibility	Examples
Diamagnetism	Interaction of magnetic field with closed electronic shells	Negative, very small (ca. 10^{-5}–10^{-6})	None	NaBr, He, $[Fe(CN)_6]^{4-}$
Normal paramagnetism	Resultant spin and/or orbital angular momentum of incomplete electronic shells	Positive, small (ca. 10^{-2}–10^{-3})	$1/T$ or $1/(T + \Delta)$	d- or f-transition metal ions, such as $[Fe(CN)_6]^{3-}$, Nd^{3+}; some other species, such as O_2, NO_2
Temperature-independent paramagnetism (Van Vleck)	Upper electronic level separated from ground state by energy interval large vs. kT	Positive, very small (ca. 10^{-6})	None	MnO_4^-, MoO_4^{2-}, some cobalt(III) ammines
Metallic paramagnetism (Pauli)	Uncompensated spins of metallic conduction electrons	Positive, very small (ca. 10^{-6})	Very slight	K, Na, Cu, Ag; some interstitial carbides, nitrides
Ferromagnetism	Parallel orientation of molecular magnets over microscopic regions of solid	Positive, very large (ca. 10^0–10^2)	Passes into normal paramagnetism at Curie point	Fe, Heusler alloys, Gd
Ferrimagnetism	Interpenetrating crystal lattices containing unequal numbers of electrons with antiparallel spins	Positive, very small (ca. 10^{-5})	Positive	Fe_3O_4

of electrons will result in magnetic moments.* The nuclear contribution, although important (Sec. 2.2-3), is small in comparison with the electronic contribution. As a consequence, the measured magnetic properties of pure substances are usually those associated with electrons. Each moving electron is an individual micromagnet, and the total moment exhibited by a substance is the resultant of the moments of the individual electrons.

When placed in a magnetic field, every substance exhibits some degree of orientation or polarization. The nature of the orientation, together with its relationships to temperature and the strength of the applied field, permit one to distinguish a number of types of magnetic behavior, as indicated in Table 2-22.[93] Of these behaviors, two are of particular importance in inorganic chemistry. Thus a substance may be less permeable to the magnetic lines of force than a vacuum and, as a consequence, tend to move from the stronger part of the field to the weaker part. The field induced in the substance is in effect opposite to the external field. Such a substance is said to be *diamagnetic*. On the other hand, a substance may exhibit essentially the opposite behavior and align itself parallel to the lines of force in the external magnetic field. Such a substance is said to be *paramagnetic*. The relationship between diamagnetic and paramagnetic behavior is indicated in Fig. 2-24.

When a substance is placed in a magnetic field of strength H (in gauss or oersted), the total magnetic flux B within the substance is given by the relationship

$$B = H + 4\pi I \tag{2-39}$$

where I is the intensity of magnetization. The *magnetic susceptibility* κ, per unit volume, is then

$$\kappa = \frac{I}{H} \tag{2-40}$$

or per unit mass, κ/d or I/Hd, where d is the density. Magnetic susceptibility is commonly expressed per gram as χ or per mole as χ_M. Thus

$$\chi_M = \frac{M\kappa}{d} \tag{2-41}$$

where M is the molecular weight. The *magnetic permeability* of the substance, B/H, is the ratio of the density of magnetic lines of force within a substance in a magnetic field to the density of lines of force in that field in the absence of the substance. It is apparent that Eq. 2-39 can be rewritten

$$\frac{B}{H} = 1 + 4\pi\kappa \tag{2-42}$$

The volume susceptibility (κ) is the quantity that is commonly measured experimentally.

---

*   In a simple bar magnet, the magnetic moment is given by the product of the pole strength and the distance between the poles. Similar considerations apply on the atomic scale.

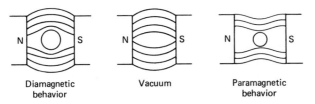

Diamagnetic behavior   Vacuum   Paramagnetic behavior

Figure 2-24 Changes in magnetic lines of force when substances are placed in a magnetic field. [T. Moeller, *Inorganic Chemistry*, Fig. 5.10, p. 165, John Wiley & Sons, New York (1952), modified.]

The molar magnetic susceptibility is related to the *permanent magnetic moment* μ_B of a pure substance in terms of the Langevin expression

$$\chi_M = \frac{N\mu_B^2}{3kT} \qquad (2\text{-}43)$$

where N is Avogadro's number, k is the Boltzmann constant, and T is the temperature in °K. The permanent moment is expressed in Bohr magnetons. The Bohr magneton, having the magnitude 5564 gauss-cm mole^{-1}, is calculated from the ratio $he/4\pi mc$, where h is Planck's constant, e the electronic charge, m the electronic mass, and c the velocity of light.

Conveniently, Eq. 2-43 can be rearranged to

$$\mu_B = \left(\frac{3kT\chi_M}{N}\right)^{1/2} \qquad (2\text{-}44)$$

which, on evaluation of the constants $3k/N$, becomes

$$\mu_B = 2.84(T\chi_M)^{1/2} \qquad (2\text{-}45)$$

Instead of the permanent moment, it is more nearly accurate to use the *effective magnetic moment* μ_{eff}, which is obtained by correction for diamagnetic susceptibility (see below).

Somewhat empirically, the temperature dependence indicated by Eq. 2-43 is given for many magnetically dilute substances* by the Curie law (after P. Curie),

$$\chi_M = \frac{C}{T} \qquad (2\text{-}46)$$

or the Curie–Weiss law,

$$\chi_M = \frac{C}{T - \Theta} \qquad (2\text{-}47)$$

where C is a constant and Θ is the Curie temperature (Table 2-22).

---

* A magnetically dilute substance is a substance wherein the paramagnetic species represent so small a proportion of the whole that their magnetic interactions can be neglected. Extensive solvation or complexation often renders a paramagnetic solid magnetically dilute.

Diamagnetic Behavior.   Diamagnetism is a function of the distribution of electron density within an atom, molecule, or ion and arises from the interaction of the applied magnetic field with filled orbitals. All pure substances are inherently diamagnetic. Even paramagnetic materials also exhibit diamagnetism, and correction for this property must be made in a precise measurement of a permanent magnetic moment. The precession of electron pathways resulting from the applied magnetic field induces a directionally opposed field in the substance. For this reason, the diamagnetic susceptibility is always negative. It is also independent of the temperature and the strength of the magnetic field (Table 2-22).

Diamagnetic ions are those having noble gas (8 electron), pseudo-noble gas (18 electron), or pseudo-noble gas $+2$ ($18 + 2$ electron) configurations, as summarized in Table 2-23. Molecular diamagnetism is characteristic of nonionic substances in which all electron spins are paired. Many organic compounds and a variety of covalent inorganic compounds are typical examples. Because the pairing of electron spins is essential, a first criterion for diamagnetic behavior is usually the presence of an even number of electrons. It is not an absolute criterion, however, since the principle of maximum multiplicity (p. 59) can preclude pairing, and a limited number of other exceptions (e.g., the paramagnetic O_2 molecule) also exist.

Paramagnetic Behavior.   Normal paramagnetism (Table 2-22) results when the magnetic effects of the individual electrons are not mutually neutralized, a condition that pertains when there is at least one unpaired electron present in the atom, molecule, or ion in question. When such a substance is placed in a

Table 2-23   Diamagnetic Ionic and Pseudoionic Species

NOBLE GAS TYPE

$H^- \rightarrow He \leftarrow Li^+, Be^{2+}, B(III), C(IV), N(V)$

$C(-IV), N^{3-}, O^{2-}, F^- \rightarrow Ne \leftarrow Na^+, Mg^{2+}, Al^{3+}, Si(IV), P(V), S(VI), Cl(VII)$

$Si(-IV), P(-III), S^{2-}, Cl^- \rightarrow Ar \leftarrow K^+, Ca^{2+}, Sc^{3+}, Ti(IV), V(V), Cr(VI), Mn(VII)$

$Ge(-IV), As(-III), Se^{2-}, Br^- \rightarrow Kr \leftarrow Rb^+, Sr^{2+}, Y^{3+}, Zr(IV), Nb(V), Mo(VI), Tc(VII)$

$Sb(-III), Te^{2-}, I^- \rightarrow Xe \leftarrow Cs^+, Ba^{2+}, La^{3+}, Ce^{4+}$

$Po^{2-}, At^- \rightarrow Rn \leftarrow Fr^+, Ra^{2+}, Ac^{3+}, Th^{4+}$

PSEUDO-NOBLE GAS TYPE

$Ni \leftarrow Cu^+, Zn^{2+}, Ga^{3+}, Ge(IV), As(V), Se(VI), Br(VII)$

$Pd \leftarrow Ag^+, Cd^{2+}, In^{3+}, Sn(IV), Sb(V), Te(VI), I(VII)$

$Pt \leftarrow Au^+, Hg^{2+}, Tl^{3+}, Pb(IV), Bi(V), Po(VI)$

PSEUDO-NOBLE GAS $+2$ TYPE

$Zn \leftarrow Ga^+, Ge^{2+}, As(III), Se(IV), Br(V)$

$Cd \leftarrow In^+, Sn^{2+}, Sb^{3+}, Te(IV), I(V)$

$Hg \leftarrow Tl^+, Pb^{2+}, Bi^{3+}, Po(IV)$

magnetic field, these micromagnets are subjected to two opposing forces: (1) the magnetic field strength (H), which tends to align them in the direction of the field; and (2) the thermal forces of vibration, rotation, and translation (kT effect), which tend to randomize their directions. As indicated by Eqs. 2-39 to 2-43, the observed molar susceptibility is determined by both field strength and temperature.

Using the quantum mechanics, J. H. Van Vleck[94] derived a modification of the Langevin equation,

$$\chi_M = \frac{N\bar{\mu}_B^2}{3kT} + N\bar{\alpha} \qquad (2\text{-}48)$$

in which $\bar{\mu}_B$ represents the low-frequency part of the magnetic moment vector (in Bohr magnetons) and $\bar{\alpha}$ represents the combined temperature-independent contributions of the high-frequency part and the diamagnetic susceptibility. The observed moment results from an appropriate coupling of the *spin* and *orbital* components of electron motion. Coupling is usefully considered in terms of the Russell–Saunders pattern (Sec. 2.3-6), leading, in terms of Eq. 2-30, to values of J, the *inner quantum number*.

The permanent magnetic moment is determined by the magnitudes of energy separation between levels corresponding to successive J values (so-called *multiplet intervals*) as compared with kT (Eq. 2-43). The following cases can be distinguished:

1. Multiplet intervals small compared to kT. Here the high-frequency component and the diamagnetic contribution (Eq. 2-48) can be neglected. In the general case where both spin and angular momenta are operative, the permanent moment is then given by

$$\bar{\mu}_B = [4S(S+1) + L(L+1)]^{1/2} \qquad (2\text{-}49)$$

and Eq. 2-48 reduces to

$$\chi_M = \frac{N[4S(S+1) + L(L+1)]}{3kT} \qquad (2\text{-}50)$$

The observed moments of such d-transition metal ions as Fe^{2+} and Co^{2+} are in agreement with these considerations.

In certain instances, notably where a subshell is half-filled (e.g., p^3, d^5, f^7 configurations) or where the unpaired electrons responsible for the magnetic moment are in the valence shell, the orbital angular momentum is effectively quenched, and the permanent moment is determined solely by spin, as

$$\bar{\mu}_B = [4S(S+1)]^{1/2} = 2[S(S+1)]^{1/2} \qquad (2\text{-}51)$$

or in terms of the number of unpaired electrons n, as

$$\bar{\mu}_B = [n(n+2)]^{1/2} \qquad (2\text{-}52)$$

Thus the expected moment, in Bohr magnetons, for *one* unpaired electron

calculates to $\sqrt{3}$ or 1.73, for two electrons to $\sqrt{8}$ or 2.83, and for 3, 4, 5, ...
unpaired electrons to 3.87, 4.90, 5.92, ... The data summarized in Fig. 2-25
indicate that this situation pertains for the selected $3d$-transition metal ions and
thus confirms for these species operation of the principle of maximum multiplicity. *Spin-only* considerations apply much less rigorously to ions derived from
the $4d$- and $5d$-transition elements.[95]

2.   Multiplet intervals large compared to kT.   Here both spin and orbital
angular momentum contributions are important, and the high-frequency
component cannot be neglected. The permanent magnetic moment is calculated
as

$$\bar{\mu}_B = g[J(J+1)]^{1/2} \qquad (2\text{-}53)$$

where g, the Landē splitting factor, is defined as

$$g = 1 + \frac{S(S+1) + J(J+1) - L(L+1)}{2J(J+1)} \qquad (2\text{-}54)$$

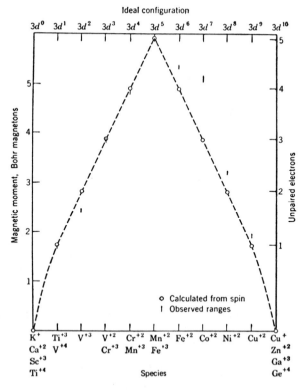

Figure 2-25   Magnetic moments of ions of $3d$-transition elements. [T. Moeller,
Inorganic Chemistry, Fig. 5.11, p. 169, John Wiley & Sons, New York (1952).]

Equation 2-48 then becomes

$$\chi_M = \frac{Ng^2 J(J+1)}{3kT} + N\bar{\alpha} \tag{2-55}$$

The shielding of unpaired electrons is essential. Experimentally, the permanent magnetic moments of all of the lanthanide(III) ions (Ln^{3+}) except Sm^{3+} and Eu^{3+} (see below) agree with those calculated from Eq. 2-55. These variations are shown in Fig. 2-26.

Substances of this type and those of the preceding type obey the Curie law.

3. Multiplet intervals comparable to kT. Here a summation of contributions based upon different J values is required, but the theoretical evaluation of the permanent moment is complicated by the necessity for considering a Boltzmann distribution of the various energy states.[93] The observed moments of the Sm^{3+} and Eu^{3+} ions and of the nitrogen(II) oxide molecule agree closely with those calculated on these bases. Large departures from the Curie law are common.

An observed paramagnetic moment is chiefly useful to the inorganic chemist as an indicator of the number of unpaired electrons in an atom, molecule, or ion. Subsequent discussions indicate how this kind of information can be interpreted in terms of oxidation states, bonding and bond types, and stereochemistry.

Other Magnetic Behaviors. The other behaviors indicated in Table 2-22 are much less significant to a broad understanding of inorganic chemistry as defined

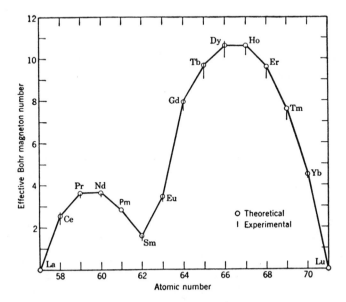

Figure 2-26 Magnetic moments of terpositive lanthanide ions. [T. Moeller, *Inorganic Chemistry*, Fig. 5.12, p. 170, John Wiley & Sons, New York (1952).]

in Chapter 1 and need not be considered further in this treatment. All of the discussions above can be supplemented to advantage by consultation of more detailed references.[94, 96-100] Nuclear magnetic characteristics are of particular importance to the technique of nuclear magnetic resonance spectroscopy, which is discussed later (Sec. 6.5-4).

EXERCISES

2-1. For a given nuclear charge, radioactive nuclides of larger mass numbers commonly decay by negatron (β^-) emission, whereas those of smaller mass number commonly decay by either positron (β^+) emission or electron capture (EC). Explain.

2-2. A diamagnetic ion always contains an even number of electrons, but a paramagnetic ion does not always contain an odd number of electrons. Explain.

2-3. Rationalize each of the following seemingly "irregular" ground-state electronic configurations for the gaseous atoms: $_{24}$Cr: [Ar] $4s^1 3d^5$; $_{46}$Pd: [Kr] $4d^{10}$; $_{70}$Yb: [Xe] $4f^{14}6s^2$; $_{97}$Bk: [Rn] $5f^9 7s^2$.

2-4. Lithophilic ions that are chemically either distantly related or unrelated (e.g., K^+, Sr^{2+}; Eu^{2+}, Ca^{2+}) commonly occur together in minerals. Why?

2-5. Clearly identify or define each of the following: a g atomic orbital; the Paschen series in the atomic emission spectrum of H(g); an s process in nuclear synthesis; dualistic properties of electromagnetic radiation; hydrogenic wave function; atomic term symbol; effective nuclear charge; electronegativity.

2-6. Using such supplementary references as are necessary, compare and contrast Pauling and Sanderson electronegativities as to meaning, evaluation, and applicability.

2-7. Assuming that the types of electronic configurations that characterize the lighter gaseous atoms are continued in the *trans*-actinide region, give the atomic number of eka-lead (EkPb) (i.e., the next heavier congener of lead), the probable ground-state electronic configuration of its gaseous atom, estimates of the radii of its atom and dipositive cation, and the formulas of two probable water-soluble and two water-insoluble salts.

2-8. Account for the observation (Table 2-19) that although the first ionization energies of the periodic group IIIB atoms are less than those of atoms of immediately smaller and larger nuclear charges, the corresponding second ionization energies are larger.

2-9. What is the significance of an atomic, or Russell–Saunders, term symbol? Determine the term symbol for each of the following electronic configurations:

[He] $2s^2 2p^1$; [Ne] $3s^1$; [Ar] $4s^2$;[Ar] $3d^1 4s^2$; [Ar] $3d^{10}4s^2$.

2-10. Illustrate the effects of the lanthanide contraction for the terpositive ions by as many sets of tabulated numerical data as you can find. Include data for the Y^{3+} ions. If yttrium does not occupy a constant position, suggest a reason, or reasons.

REFERENCES

1. A. J. Ihde, *The Development of Modern Chemistry*, Chs. 4–6, 9, 18, 19, Harper & Row, New York (1964).

2. J. J. Thomson, *Philos. Mag.*, [5], **44**, 293 (1897).

3. R. A. Millikan, *Philos. Mag.*, [6], **19**, 209 (1910); *Phys. Rev.*, [1], **32**, 349 (1911); **2**, 109 (1913).

4. L. de Broglie, *Ann. Phys.*, **3**, 22 (1925).

5. F. S. Dainton, *Chem. Soc. Rev.*, **4**, 323 (1975).

6. E. Goldstein, *Ber. Preuss. Akad. Wiss.*, **39**, 691 (1886).

7. J. Chadwick, *Proc. R. Soc. Lond. Ser. A*, **136**, 692 (1932).

8. E. Rutherford, *Philos. Mag.*, [6], **21**, 669 (1911).

9. H. G. J. Moseley, *Philos. Mag.*, [6], **26**, 1024 (1913); [6], **27**, 703 (1914).

10. G. Friedlander, J. W. Kennedy, and J. M. Miller, *Nuclear and Radiochemistry*, 2nd ed., John Wiley & Sons, New York (1964).

11. B. G. Harvey, *Nuclear Chemistry*, Prentice-Hall, Englewood Cliffs, N.J. (1965).

12. H. A. C. McKay, *Principles of Radiochemistry*, Butterworth, London (1971).

13. A. G. Maddock (Ed.), *MTP International Review of Science, Inorganic Chemistry, Series Two*, Vol. 8, Butterworth, London (1975).

14. O. Hahn and F. Strassmann, *Naturwissenschaften*, **27**, 11, 89 (1939).

15. H. D. Smyth, *Atomic Energy for Military Purposes*, Princeton University Press, Princeton, N.J. (1945).

16. S. Glasstone, *Sourcebook on Atomic Energy*, 2nd ed., Ch. 13, D. Van Nostrand, Princeton, N.J. (1958).

17. M. G. Mayer and J. H. D. Jensen, *Elementary Theory of Nuclear Shell Structure*. John Wiley & Sons, New York (1955).

18. F. W. Clarke, *The Data of Geochemistry*, U.S. Geol. Surv. Bull. 770, 5th ed. (1924).

19. V. M. Goldschmidt, *Geochemistry*, A. Muir (Ed.), Clarendon Press, Oxford (1954).

20. D. T. Gibson, *Q. Rev. (Lond.)*, **3**, 263 (1949).

21. M. Fleischer, *J. Chem. Educ.*, **31**, 446 (1954).

22. H. E. Suess and H. C. Urey, *Rev. Mod. Phys.*, **28**, 53 (156).

23. W. D. Ehmann, *J. Chem. Educ.*, **38**, 53 (1961).

24. A. G. W. Cameron, *Space Sci. Rev.*, **15**, 121 (1973).

25. B. Mason and C. B. Moore, *Principles of Geochemistry*, 4th ed., Ch. 2. John Wiley & Sons, New York (1982).

26. W. D. Harkins, *Chem. Rev.*, **5**, 371 (1928).

27. R. A. Alpher and R. C. Herman, *Annual Reviews of Nuclear Science*, Vol. 2, pp. 1–40, Annual Reviews, Palo Alto, Calif. (1953).

28. F. A. Hoyle, W. A. Fowler, G. R. Burbidge, and E. M. Burbidge, *Science*, **124**, 611 (1956).

29. E. M. Burbidge, G. R. Burbidge, W. A. Fowler, and F. Hoyle, *Rev. Mod. Phys.*, **29**, 547 (1957).

30. M. Burbidge and G. Burbidge, *Science*, **128**, 387 (1958).

31. T. P. Kohman, *J. Chem. Educ.*, **38**, 73 (1961).

32. C. D. Coryell, *J. Chem. Educ.*, **38**, 67 (1961).

33. W. A. Fowler, *Science*, **135**, 3508 (1962); *Proc. Natl. Acad. Sci.*, **52**, 524 (1964); *Chem. Eng. News*, Mar. 16, 1964, p. 90.

34. J. Selbin, *J. Chem. Educ.*, **50**, 306, 380 (1973). Detailed summary.

35. B. Mason and C. B. Moore, Ref. 25, Ch. 3.

36. V. M. Goldschmidt, *Chem. Ber.*, **60**, 1263 (1927).

37. K. Rankama and T. G. Sahama, *Geochemistry*, University of Chicago Press, Chicago (1950).

38. K. B. Krauskopf, *Introduction to Geochemistry*, McGraw-Hill Book Company, New York (1967).

39. B. Mason and C. B. Moore, *Principles of Geochemistry*, 4th ed., John Wiley & Sons, New York (1982).

40. M. Planck, *Ann. Phys. (Leipzig)*. **1**, 69 (1900).

41. N. Bohr, *Philos. Mag.*, [6], **26**, 1, 476, 857 (1913).

42. W. Heisenberg, *Z. Phys.*, **43**, 172 (1927).

43. A. Sommerfeld, *Phys. Z.*, **17**, 491 (1916); *Ann. Phys. (Leipzig)*, **51**, 1 (1916).

44. E. C. Stoner, *Philos. Mag.*, **48**, 719 (1924).

45. J. D. Main-Smith, *Chem. Ind. (Lond.)*, **43**, 323 (1924).

46. L. de Broglie, *Philos. Mag.*, **47**, 446 (1924); *Ann. Phys.*, **3**, 22 (1925).

47. C. J. Davisson and L. H. Germer, *Phys. Rev.*, **30**, 705 (1927).

48. E. Schrödinger, *Ann. Phys. (Leipzig)*, **79**, 361, 489, 734 (1926); **80**, 437 (1926); **81**, 109 (1926). *Naturwissenschaften*, **14**, 664 (1926). *Phys. Rev.*, **28**, 1049 (1926).

49. E. A. Ogryzlo and G. B. Porter, *J. Chem. Educ.*, **40**, 256 (1963).

50. D. T. Cromer, *J. Chem. Educ.*, **45**, 626 (1968).

51. Graphical representations of this type for many atomic and molecular systems are illustrated very effectively in J. R. Van Wazer and I. Absar, *Electron Densities in Molecules and Molecular Orbitals*, Academic Press, New York (1975).

52. H. G. Friedman, Jr., G. R. Choppin, and D. G. Feuerbacher, *J. Chem. Educ.*, **41**, 354 (1964).

53. C. Becker, *J. Chem. Educ.*, **41**, 358 (1964).

54. A. W. Adamson, *J. Chem. Educ.*, **42**, 140 (1965).

55. E. A. Ogryzlo, *J. Chem. Educ.*, **42**, 150 (1965).

56. D. R. Hartree, *Proc. Camb. Philos. Soc.*, **24**, 89 (1928); *Proc. R. Soc. Lond. Ser. A*, **141**, 282 (1933); *Rep. Prog. Phys.*, **11**, 113 (1946–1947).

57. V. Fock, *Z. Phys.*, **61**, 126 (1930).

58. J. C. Slater, *Phys. Rev.*, **35**, 210 (1930); **36**, 57 (1930).

59. M. Gorman, *J. Chem. Educ.*, **50**, 189 (1973).

60. O. K. Rice, *Electronic Structure and Chemical Bonding*, p. 96, McGraw-Hill Book Company, New York (1940).

61. W. E. Duncannon and C. A. Coulson, *Proc. R. Soc. Edinb.*, **62**, 37 (1944).

62. R. McWeeny, *Coulson's Valence*, 3rd ed., pp. 41–45, Oxford University Press, London (1979).

63. E. Clementi and D. L. Raimondi, *J. Chem. Phys.*, **38**, 2686 (1963).

64. J. E. Huheey, *J. Chem. Educ.*, **45**, 791 (1968).

65. L. Pauling, *The Nature of the Chemical Bond*, 3rd ed., pp. 257–264, Cornell University Press, Ithaca, N.Y. (1960).

66. A. Bondi, *J. Phys. Chem.*, **68**, 441 (1964).

67. L. Pauling, Ref. 65, pp. 256–257.

68. W. L. Bragg, *Philos. Mag.*, **40**, 169 (1920).

69. J. C. Slater, *J. Chem. Phys.*, **41**, 3199 (1964).

70. J. Sherman, *Chem. Rev.*, **11**, 93 (1932).

71. C. E. Moore, *Ionization Potentials and Ionization Limits Derived from the Analyses of Optical Spectra*, NSRDS-NBS 34, National Bureau of Standards, Washington, D.C. (1970).

72. J. F. Liebman, *J. Chem. Educ.*, **50**, 831 (1973).

73. D. W. Brooks, E. A. Meyers, F. Sicilio, and J. C. Nearing, *J. Chem. Educ.*, **50**, 487 (1973).

74. E. C. M. Chen and W. E. Wentworth, *J. Chem. Educ.*, **52**, 486 (1975).

75. R. S. Berry, *Chem. Rev.*, **69**, 533 (1969).

76. A. G. Sharpe, *Q. Rev. (Lond.)*, **11**, 49 (1957).

77. H. O. Pritchard, *Chem. Rev.*, **52**, 529 (1953).

78. A. F. Kapustinskii, *Q. Rev. (Lond.)*, **10**, 283 (1956).

79. B. Moiseiwitsch, *Adv. At. Mol. Phys.*, **1**, 61 (1965).

80. L. Pauling, *The Nature of the Chemical Bond*, 3rd ed., Ch. 3, Cornell University Press, Ithaca, N.Y. (1960). Quotation from p. 88.

81. R. S. Mulliken, *J. Chem. Phys.*, **2**, 782 (1934); **3**, 573 (1935).

82. H. O. Pritchard and H. A. Skinner, *Chem. Rev.*, **55**, 745 (1955).

83. W. Gordy and W. J. O. Thomas, *J. Chem. Phys.*, **24**, 439 (1956).

84. A. L. Allred and E. G. Rochow, *J. Inorg. Nucl. Chem.*, **5**, 264, 269 (1958).

85. R. T. Sanderson, (a) *J. Chem. Educ.*, **29**, 539 (1952); **31**, 2, 238 (1954); **32**, 140 (1955); (b) *Inorganic Chemistry*, Ch. 6, Reinhold Publishing Corp., New York (1967); (c) *Chemical Bonds and Bond Energies*, 2nd ed., Academic Press, New York (1976).

86. L. Pauling, *J. Am. Chem. Soc.*, **54**, 3570 (1932).

87. M. Haïssinsky, *J. Phys. Radium*, **7**, 7 (1946).

88. M. L. Huggins, *J. Am. Chem. Soc.*, **75**, 4123 (1953).

89. A. L. Allred, *J. Inorg. Nucl. Chem.*, **17**, 215 (1961).

90. J. Hinze and H. H. Jaffé, *J. Am. Chem. Soc.*, **84**, 540 (1962); *J. Phys. Chem.*, **67**, 1501 (1963); *Can. J. Chem.*, **41**, 1315 (1963).

91. J. Hinze, M. A. Whitehead, and H. H. Jaffé, *J. Am. Chem. Soc.*, **85**, 148 (1963).

92. E. J. Little, Jr., and M. M. Jones, *J. Chem. Educ.*, **37**, 231 (1960).

93. R. S. Nyholm, *J. Inorg. Nucl. Chem.*, **8**, 401 (1958).

94. J. H. Van Vleck, *The Theory of Electric and Magnetic Susceptibilities*, 2nd ed., Oxford University Press, London (1948).

95. B. Figgis, *J. Inorg. Nucl. Chem.*, **8**, 476 (1958).

96. P. W. Selwood, *Magnetochemistry*, 2nd ed., Interscience Publishers, New York (1956).

97. L. N. Mulay, *Magnetic Susceptibility*, Interscience Publishers, New York (1963).

98. J. B. Goodenough, *Magnetism and the Chemical Bond*, Interscience Publishers, New York (1963).

99. A. Earnshaw, *Introduction to Magnetochemistry*, Academic Press, New York (1968).

100. F. E. Mabbs and D. J. Machin, *Magnetism and Transition Metal Complexes*, Chapman & Hall, London (1973).

SUGGESTED SUPPLEMENTARY REFERENCES

A. S. Chakravarty, *Introduction to the Magnetic Properties of Solids*, John Wiley & Sons, New York (1980).

D. A. Dingee, *Chem. Eng. News*, Apr. 2, 1979, pp. 32–47. Nuclear fusion.

A. Earnshaw, *Introduction to Magnetochemistry*, Academic Press, New York (1968).

G. Friedlander, J. W. Kennedy, and J. M. Miller, *Nuclear and Radiochemistry*, 2nd ed., John Wiley & Sons, New York (1964).

S. Glasstone, *Sourcebook on Atomic Energy*, 2nd ed., D. Van Nostrand, Princeton, N.J. (1958). Development.

G. Herzberg, *Atomic Spectra and Atomic Structure*, Dover Publications, New York (1944). Largely electronic properties of elements.

A. G. Maddock (Ed.), *MTP Internaional Review of Science, Inorganic Chemistry, Series Two*, Vol. 8, Butterworth, London (1975). Nuclear stability.

B. Mason and C. B. Moore, *Principles of Geochemistry*, 4th ed., John Wiley & Sons, New York (1982). Sources, abundances, distributions of elements.

L. Pauling, *The Nature of the Chemical Bond*, 3rd ed., Ch. 3, Cornell University Press, Ithaca, N.Y. (1960). Electronegativities, etc.

J. Selbin, *J. Chem. Educ.*, **50**, 306, 380 (1973). Origins of the elements.

P. W. Selwood, *Magnetochemistry*, 2nd ed., Interscience Publishers, New York (1956).

CHAPTER THREE

THE PERIODIC TABLE AND PERIODIC RELATIONSHIPS

Even a cursory examination of the ground-state electronic configurations of the atoms of the various elements reveals the periodic recurrence of configurations of particular types as nuclear charge (i.e., atomic number) is regularly increased (Sec. 2.3-5). When this observation is coupled with our realization that the chemical properties and many of the physical properties of the elements are determined to a large degree by the electronic configurations of their atoms, it is immediately obvious that it is possible to arrange the elements in any one of a number of regular ways, each of which will reflect periodic similarities in properties. Unless the matter is pursued in more detail, however, it is tempting to conclude that the familiar periodic table is merely a consequence of the investigations that led to extranuclear atomic structures and is only a convenient way of expressing these results. Such is not the case, however, for the development of the periodic table preceded the elucidation of modern atomic theory chronologically. Indeed, early periodic tables, derived as they were solely as the results of meticulous observations of the properties of the elements and their compounds, provided the bases for those elucidations of electronic configurations that ultimately provided sound theoretical support for their construction.

The periodic table is, in the light of our present-day knowledge, so obvious a systematization that we are often inclined to accept it at face value without further inquiry and to overlook the scientific thought and experimental study from which it developed. It is, therefore, both important and instructive to examine, in roughly chronological order, a number of the significant developments that led to this classification before discussing its construction and uses in terms of modern concepts. The brief summary that follows can be supplemented by reference to more comprehensive discussions.[1-5]

1. SIGNIFICANT DEVELOPMENTS IN THE SYSTEMATIC CLASSIFICATION OF THE ELEMENTS

Although it is conceivable that early investigators may have attempted to classify the elements after their discovery and recognition as fundamental substances, modern effort can be traced directly or indirectly to John Dalton's atomic theory (Sec. 2.1). Dalton's atomic theory contained nothing relating to any systematic grouping, but it did stimulate thought and led to speculation that atoms of the various elements, although apparently different in properties, might be composed of the same fundamental substance and that the marked similarities observed among certain elements might be traceable to their atoms.

The first of these considerations was expressed in 1815 in what is now known as *William Prout's hypothesis*: that the weights of all atoms are simple multiples of the weight of the hydrogen atom and thus that hydrogen is the primary substance from which all other elements derive. Fundamental as was Prout's hypothesis, it was discredited by experimentally observed deviations of atomic weights from whole numbers and ultimately abandoned in 1860 when precise atomic weights were determined by J. S. Stas. Our presently held concepts on the nuclear syntheses of the elements (Sec. 2.2-4) indicate the true significance of this hypothesis. The second of these considerations was not adequately elucidated until early in the twentieth century.

In 1817, J. W. Döbereiner noted that within a group of three elements closely related to each other in chemical properties atomic weights are nearly the same (e.g., Fe 55.8, Co 58.9, Ni 58.7), or the atomic weight of the middle element is the approximate mean of those of the other two (e.g., Cl 35.5, Br 79.9, I 126.9; mean of 35.5 and 126.9 = 81.2). The *triads of Döbereiner* represent the first reported attempt at systematic classification of the elements. As a modification of Döbereiner's ideas, M. von Pettenkofer, in 1850, suggested that among chemically similar elements successive differences in atomic weights amount either to some constant or to a multiple of some constant, which suggestion is, in fact, a statement that among such elements atomic weights can be derived from a modified arithmetical progression involving the smallest atomic weight and multiples of an integer. Thus in the series O(16), S(32), Se(80), and Te(128), the difference between the first two atomic weights is 16, and between any other two it is 48, or 3 × 16. The same general concept is implicit in J. P. Cooke's statement (in 1854) that triads are merely portions of series for which the increase in atomic weights of the members of the series follows an algebraic law.

William Odling's arrangement (in 1857) of the known elements into 13 groups, based upon similarities in chemical and physical properties, listed the elements in each group in order of increasing atomic weight. However, it showed no periodic relationships between these properties and atomic weight and thus could really not be called a periodic table. Rather, his bringing together elements, the compounds of which have similar solubilities, resulted in arrangements more nearly comparable to those used in qualitative analysis.

The first periodic classification, in the sense of modern usage of the term, was the *telluric screw*, or *helix*, proposed by A.-E. B. de Chancourtois in 1862.* Using a cylinder as a geometric basis, de Chancourtois divided its surface into 16 equal segments (because the atomic weight of oxygen was taken as 16) and plotted atomic weights as ordinates on the generatrix of a helix that descended the surface of the cylinder at an angle of 45° to its top. Elements differing from each other in atomic weight by 16 units fell into the same segments and resembled each other strikingly in their properties. In referring to this periodicity, de Chancourtois stated that the properties of the elements are the properties of numbers and observed that atomic weights commonly followed the formula $n + 16n'$, where n was frequently 7 or 16 and n' an integer.

Of probable equal importance as a forerunner of the periodic table is the arrangement proposed in 1864–1866 by J. A. R. Newlands, who noted that when the known elements were placed in the order of increasing atomic weight, similarities in chemical and physical properties reappeared after each interval of

---

* For an interesting account on the search for the discoverer of the periodic law, see H. Cassebaum and G. B. Kauffman, *Isis*, **62**, Pt. 3, No. 213, pp. 314–327 (1971).

eight elements. Because of a fancied resemblance to the musical scale, Newlands termed this concept the *law of octaves*, an unfortunate choice because it resulted in ridicule by his peers. However, the fundamental importance of this arrangement was appreciated and finally recognized by the award of the Davy Medal to Newlands in 1887. The Newlands tabulation is given as Fig. 3-1. Although inaccuracies in atomic weight data and the subsequent discovery of elements unknown in Newlands' time produced inconsistencies, his arrangement bears a striking resemblance to some periodic tables now in use. Indeed, if the Newlands table is rearranged into vertical families and the column beginning with hydrogen (H) is displaced one position upward, a striking similarity to the Mendeleev table is apparent.[6]

The evolution of the periodic table reached a plateau in 1869 as a consequence of independent efforts by D. I. Mendeleev and J. Lothar Meyer. Mendeleev's approach was based largely upon considerations of chemical properties of the elements, whereas that of Lothar Meyer stressed their physical properties. Yet the two tables are remarkably similar, and both emphasize the periodicity of recurrence of similar properties as atomic weight increases. In clarity of presentation and in fundamental understanding of the importance of periodicity, of the significance of odd and even series of elements, and of the place of the transition elements, Mendeleev's considerations went beyond those of Lothar Meyer. As a consequence, greater credit is usually given to Mendeleev. The initial publications of Mendeleev and Lothar Meyer were communicated in March[7] and December[8] of 1869, respectively.*

Lothar Meyer based the tabulation given in Fig. 3-2 upon periodic variations noted when properties such as atomic volume, melting point, and boiling point were plotted against atomic weight. His arrangement of 55 elements is strikingly similar to that of Mendeleev given in Fig. 3-4. Because of the importance of atomic volumes in the development of concepts of structure and bonding, Lothar Meyer is better remembered for his atomic volume curve than for his periodic table. Subsequent corrections in atomic weights and realization that atomic number (i.e., nuclear charge) is a better basis for classification of the

H	F	Cl	Co, Ni	Br	I	Pt, Ir
Li	Na	K	Cu	Rb	Cs	Os
G(Be)	Mg	Ca	Zn	Sr	Ba, V	Hg
Bo(B)	Al	Cr	Y	Ce, La	Ta	Tl
C	Si	Ti	In	Zr	W	Pb
N	P	Mn	As	Di, Mo	Nb	Bi
O	S	Fe	Se	Ro(Rh), Ru	Au	Th

Figure 3-1 The classification of Newlands.

* Credit must be given also to L. R. Gibbes for his very similar and apparently independently conceived "synoptical table of the chemical elements," which was formulated in the period 1870–1874 but not published until 1886. [See W. H. Taylor, *J. Chem. Educ.*, **18**, 403 (1941).] The work of Gibbes and other American forerunners is interestingly reviewed in G. B. Kauffman, *J. Chem. Educ.*, **46**, 128 (1969).

I	II	III	IV	V	VI	VII	VIII	IX
	B = 11.0	Al = 27.3				?In = 113.4		Tl = 202.7
	C = 11.97	Si = 28				Sn = 117.8		Pb = 206.4
	N = 14.01	P = 30.9	Ti = 48	As = 74.9	Zr = 89.7	Sb = 122.1	Ta = 182.2	Bi = 207.5
	O = 15.96	S = 31.98	V = 51.2	Se = 78	Nb = 93.7	Te = 128?	W = 183.5	
	F = 19.1	Cl = 35.38	Cr = 52.4	Br = 79.75	Mo = 95.6	I = 126.5		
			Mn = 54.8		Ru = 103.5		Os = 198.6?	
			Fe = 55.9		Rh = 104.1		Ir = 196.7	
			Co = Ni = 58.6		Pd = 106.2		Pt = 196.7	
Li = 7.01	Na = 22.99	K = 39.04	Cu = 63.3	Rb = 85.2	Ag = 107.66	Cs = 132.7	Au = 196.2	
?Be = 9.3	Mg = 23.9	Ca = 39.9	Zn = 64.9	Sr = 87.0	Cd = 111.6	Ba = 136.8	Hg = 199.8	

Figure 3-2 Lothar Meyer's periodic table of the elements.

elements than atomic weight have had little effect on altering the general shape of the atomic volume curve. A curve based upon the latest data given as Fig. 3-9 has the same general appearance and emphasizes the same points as Lothar Meyer's original curve.

Because of their importance in influencing subsequent chemical progress, Mendeleev's proposals merit detailed consideration. The original Mendeleev table (Fig. 3-3) was altered from its vertical form to a horizontal form (Fig. 3-4) in his later publications. It is this latter form, the so-called "short form," modified only to the extent of adding an additional group to accommodate the noble gas elements, of including elements discovered since it was originally published, and of providing a more reasonable place for the lanthanide (or rare earth) elements,[9] that was the common and most widely used periodic table through at least the 1920s. The resemblances between the vertical form (Fig. 3-3) and various of the "long forms" now in use (pp. 109–110) are apparent.

In the light of the limited amount, and often limited accuracy, of information available to him, the breadth of Mendeleev's understanding of the system of the chemical elements was remarkable. This understanding is apparent from the following statements, as summarized from his early publications:[7, 10] periodicity of properties is inherent in the arrangement; the arrangement corresponds with the valencies of the elements; the characteristics of the elements are determined by the magnitudes of their atomic weights; errors in atomic weights may be

I	II	III	IV	V	VI
			Ti = 50	Zr = 90	? = 180
			V = 51	Nb = 94	Ta = 182
			Cr = 52	Mo = 96	W = 186
			Mn = 55	Rh = 104.4	Pt = 197.4
			Fe = 56	Ru = 104.4	Ir = 198
			Ni = Co = 59	Pd = 106.6	Os = 199
H = 1			Cu = 63.4	Ag = 108	Hg = 200
	Be = 9.4	Mg = 24	Zn = 65.2	Cd = 112	
	B = 11	Al = 27.4	? = 68	Ur = 116	Au = 197?
	C = 12	Si = 28	? = 70	Sn = 118	
	N = 14	P = 31	As = 75	Sb = 122	Bi = 210?
	O = 16	S = 32	Se = 79.4	Te = 128?	
	F = 19	Cl = 35.5	Br = 80	I = 127	
Li = 1	Na = 23	K = 39	Rb = 85.4	Cs = 133	Tl = 204
		Ca = 40	Sr = 87.6	Ba = 137	Pb = 207
		? = 45	Ce = 92		
		?Er = 56	La = 94		
		?Yt = 60	Di = 95		
		?In = 75.6	Th = 118?		

Figure 3-3 Mendeleev's original periodic table (1869).

Series	Group I — R_2O	Group II — RO	Group III — R_2O_3	Group IV RH_4 RO_2	Group V RH_3 R_2O_5	Group VI RH_2 RO_3	Group VII RH R_2O_7	Group VIII — RO_4
1	H = 1							
2	Li = 7	Be = 9.4	B = 11	C = 12	N = 14	O = 16	F = 19	
3	Na = 23	Mg = 24	Al = 27.3	Si = 28	P = 31	S = 32	Cl = 35.5	
4	K = 39	Ca = 40	= 44	Ti = 48	V = 51	Cr = 52	Mn = 55	Fe = 56, Co = 59, Ni = 59, Cu = 63
5	(Cu = 63)	Zn = 65	= 68	= 72	As = 75	Se = 78	Br = 80	
6	Rb = 85	Sr = 87	?Yt = 88	Zr = 90	Nb = 94	Mo = 96	= 100	Ru = 104, Rh = 104, Pd = 106, Ag = 108
7	(Ag = 108)	Cd = 112	In = 113	Sn = 118	Sb = 122	Te = 125	I = 127	
8	Cs = 133	Ba = 137	?Di = 138	?Ce = 140				
			?Er = 178	?La = 180	Ta = 182	W = 184		Os = 195, Ir = 197, Pt = 198, Au = 199
11	(Au = 199)	Hg = 200	Tl = 204	Pb = 207	Bi = 208			
12				Th = 231		U = 240		

Figure 3-4 Mendeleev's periodic table of 1872.

corrected from the positions the elements occupy in the table; elements with very similar properties have atomic weights that are either nearly the same (e.g., Fe, Co, Ni) or increase regularly (e.g., Li, Na, K); elements that are most widely dispersed in nature are those with small atomic weights. Even more striking was Mendeleev's boldness both in predicting the existence of then undiscovered elements and in summarizing their properties and those of their compounds, all on the basis of unoccupied positions in his table. In every case, these predictions were subsequently verified. Thus his eka-aluminum (atomic weight = 68) was gallium, discovered by P.-E. L. de Boisbaudran in 1875; his eka-boron (atomic weight = 44) was scandium, discovered by L. F. Nilson in 1879; and his eka-silicon (atomic weight = 72) was germanium, discovered by C. A. Winkler in 1886. The comparison of predicted and measured properties for the last of these substances, as summarized in Table 3-1, is particularly revealing of Mendeleev's insight.

Acceptance of the periodic table marked the beginning of a true renaissance of chemical thought and practice. For the first time, variations among the properties of the elements and their compounds could be fitted into a logical pattern, and it was no longer necessary to treat each chemical element as a species detached from and unrelated to the other elements. Modern developments in inorganic chemistry are largely traceable to the periodic table, in particular those relating to atomic structure (Ch. 2) and the relationships between the electronic configurations of atoms, ions, and molecules and their properties.

Developments in the 40-odd years subsequent to the initial publications of Mendeleev and Lothar Meyer centered very largely in extensions, expansions, and some geometrical modifications of the Mendeleev table, but with atomic weight as the sole basis for ordering the elements. It is of particular interest that in an attempt to resolve the problems associated with including widely different elements (e.g., Na and Cu, Cl and Mn) in subgroups within main groups in this table, P. J. F. Rang in 1893 produced in simple form an extended table that is a forerunner of the currently widely utilized "long form," a type of tabulation extended by A. Werner[11] (Fig. 3-5) and others.

It was not until the development of the concept of the nuclear atom by E. Rutherford[12] (Sec. 2.1-2), the proposal of atomic numbers by H. G. J. Moseley[13] (Sec. 2.1-2), the elucidation of ground-state electronic configurations of the atoms stemming from N. Bohr's[14] theory of the structure of the hydrogen atom (Sec. 2.3-1), and the periodicity exhibited by these electronic configurations that it was realized that the periodic table brings together elements the atoms of which have the same types of electronic configurations. An arrangement based upon atomic numbers accomplishes this result.

Inasmuch as periodicity is a function of electronic configuration rather than mass, one may well ask why a table based upon atomic weights could have been devised. The answer lies in the nature of the "Aufbau process' (Sec. 2.3-5). As protons are added (increasing the nuclear charge), neutrons are also added (increasing the mass) in order to yield nuclei of maximum stabilities. Although

Table 3-1 Comparison of Properties Predicted by Mendeleev and Those Measured Experimentally

Predicted for Eka-silicon (Es) (Mendeleev, 1871)	Property	Found for Germanium (Ge) (Winkler, 1886)
72	Atomic weight	72.32
5.5	Specific gravity	5.47 (at 20°C)
13	Atomic volume (cm^3 $mole^{-1}$)	13.22
Dirty gray	Color	Grayish white
0.073	Specific heat	0.076
EsO_2 (white)	Heating in air gives	GeO_2 (white)
Slight	Action of acids	None by HCl
EsO_2 heated with Na	Preparation	GeO_2 heated with C
K_2EsF_6 heated with Na		K_2GeF_6 heated with Na
Refractory, sp. gr. 4.7, molecular volume 22 cm^3 $mole^{-1}$	Dioxide	Refractory, sp. gr. 4.703, molecular volume 22.16 cm^3 $mole^{-1}$
B.p. 100°C, sp. gr. 1.9, molecular volume 113 cm^3 $mole^{-1}$	Tetrachloride	B.p. 86°C, sp. gr. 1.887, molecular volume 113.35 cm^3 $mole^{-1}$
B.p. 160°C, sp. gr. 0.96	Tetraethyl derivative	B.p. 160°C, sp. gr. < 1.00

...																														...	...
H																														...	He
Li																									Be	B	C	N	O	Fl	Ne
Na																									Mg	Al	Si	P	S	Cl	A
K	Ca															Sc	Ti	V	Cr	Mn	Fe	Co	Ni	Cu	Zn	Ga	Ge	As	Se	Br	Kr
Rb	Sr															Y	Zr	Nb	Mo	...	Ru	Rh	Pd	Ag	Cd	In	Sn	Sb	Te	I	X
Cs	Ba	La	Ce	Nd	Pr	...	...	Sa	Eu	Gd	Tb	Ho	Er	Tu	Yb	...	...	Ta	W	...	Os	Ir	Pt	Au	Hg	Tl	Pb	Bi	...	...	...
...	Ra	Laα	Th	...	...	...	...	...	U	...	...	...	...	Ac	...	...	...	...	...	...	...	...	...	...	...	...	Pbα	Biα	Teα	...	...

Figure 3-5 Alfred Werner's periodic table of 1905.

the rate of increase of mass exceeds the rate of increase of nuclear charge, the increases are parallel, and an arrangement based upon atomic weight is exactly parallel to one based upon atomic number, except in those few instances (Ar and K, Co and Ni, Te and I, and Th and Pa) where abundances of heavier isotopes effect small reversals in average atomic weights.

2. THE MODERN PERIODIC TABLE

Bewildering arrays of arrangements and geometries, none of which is completely satisfactory, have been used to bring together in two or three dimensions those elements the atoms of which have the same types of outermost and (usually) underlying ground-state electronic configurations. To some degree, each of these tables obviates problems inherent in the Mendeleev table. Among these problems are inconsistencies in predictions of oxidation states; marked differences in properties of elements placed in the same group; incompleteness in the separation of metals from nonmetals; inconsistencies in the grouping together of elements giving colorless, diamagnetic ions with elements giving colored, paramagnetic ions; and lack of reasonable positions for hydrogen, the lanthanide elements, and the actinide elements. Some of these tables are designed primarily to reflect the occupancy of s, p, d and f orbitals, whereas others are designed primarily to collect together elements by broad types or classes. No table can reflect completely those minor differences in ground-state electronic configurations that result from small differences in orbital energies (Sec. 2.3-5).

It is not pertinent to this discussion to include a variety of these periodic tables.* Rather a typical version of the "long form," the advantages of which are legion,[15] is included as Fig. 3-6† Fundamentally, this periodic table (1) reflects the "outermost," or valence-shell, ground-state electronic configuration of an atom of each of the elements in terms of the distinguishing maximum principal quantum number n and such $n-1$ and/or $n-2$ quantum numbers as are essential, and (2) classifies the elements, on the basis of similarities in electronic configurations of their atoms, into three broad types. Although this table will be broadly useful in relationship to subsequent discussions, it suffers from its inability to classify the lanthanide,[16] actinide,[16] and many possible "superheavy"[17] elements logically. Another version of the long form, which does effect these classifications, is presented as Fig. 3-8 later in this chapter.

---

* The interested reader can examine with profit Ref. 2 and 5, the many pertinent articles that have appeared since 1934 in the *Journal of Chemical Education*, and R. T. Sanderson, *Inorganic Chemistry*, Reinhold Publishing Corp., New York (1967).

† Contrary to the recommendations of the International Union of Pure and Applied Chemistry [*Nomenclature of Inorganic Chemistry, Definitive Rules 1970*, p. 11, Butterworth, London (1971)] but in keeping with long-standing American practice, the representative or main-group elements (p. 111) are designated A, and the d-transition or subgroup elements (p. 111) are designated B, both in this periodic table and throughout this textbook.

Representative, or main-group, elements ns^{1-2}

d-Transition, or Subgroup, elements $(n-1)d^{1-10}ns^2$

Representative, or main-group, elements ns^2np^{1-6}

	IA	IIA	IIIB	IVB	VB	VIB	VIIB	VIIIB			IB	IIB	IIIA	IVA	VA	VIA	VIIA	VIIIA
$n=1$	1 H $1s^1$																	2 He $1s^2$
$=2$	3 Li $2s^1$	4 Be $2s^2$											5 B $2s^2p^1$	6 C $2s^2p^2$	7 N $2s^2p^3$	8 O $2s^2p^4$	9 F $2s^2p^5$	10 Ne $2s^2p^6$
$=3$	11 Na $3s^1$	12 Mg $3s^2$											13 Al $3s^2p^1$	14 Si $3s^2p^2$	15 P $3s^2p^3$	16 S $3s^2p^4$	17 Cl $3s^2p^5$	18 Ar $3s^2p^6$
$=4$	19 K $4s^1$	20 Ca $4s^2$	21 Sc $3d^14s^2$	22 Ti $3d^24s^2$	23 V $3d^34s^2$	24 Cr $3d^54s^1$	25 Mn $3d^54s^2$	26 Fe $3d^64s^2$	27 Co $3d^74s^2$	28 Ni $3d^84s^2$	29 Cu $3d^{10}4s^1$	30 Zn $3d^{10}4s^2$	31 Ga $4s^24p^1$	32 Ge $4s^24p^2$	33 As $4s^24p^3$	34 Se $4s^24p^4$	35 Br $4s^24p^5$	36 Kr $4s^24p^6$
$=5$	37 Rb $5s^1$	38 Sr $5s^2$	39 Y $4d^15s^2$	40 Zr $4d^25s^2$	41 Nb $4d^45s^1$	42 Mo $4d^55s^1$	43 Tc $4d^55s^2$	44 Ru $4d^75s^1$	45 Rh $4d^85s^1$	46 Pd $4d^{10}$	47 Ag $4d^{10}5s^1$	48 Cd $4d^{10}5s^2$	49 In $5s^25p^1$	50 Sn $5s^25p^2$	51 Sb $5s^25p^3$	52 Te $5s^25p^4$	53 I $5s^25p^5$	54 Xe $5s^25p^6$
$=6$	55 Cs $6s^1$	56 Ba $6s^2$	57 La $5d^16s^2$	72 Hf $5d^26s^2$	73 Ta $5d^36s^2$	74 W $5d^46s^2$	75 Re $5d^56s^2$	76 Os $5d^66s^2$	77 Ir $5d^76s^2$	78 Pt $5d^96s^1$	79 Au $5d^{10}6s^1$	80 Hg $5d^{10}6s^2$	81 Tl $6s^26p^1$	82 Pb $6s^26p^2$	83 Bi $6s^26p^3$	84 Po $6s^26p^4$	85 At $6s^26p^5$	86 Rn $6s^26p^6$
$=7$	87 Fr $7s^1$	88 Ra $7s^2$	89 Ac $6d^17s^2$	104 Rf $6d^27s^2$	105 Ha $6d^37s^2$													

f-Transition, or inner transition, elements

Lanthanides $4f^{1-14}5d^{0-1}6s^2$	58 Ce $4f^15d^16s^2$	59 Pr $4f^36s^2$	60 Nd $4f^46s^2$	61 Pm $4f^56s^2$	62 Sm $4f^66s^2$	63 Eu $4f^76s^2$	64 Gd $4f^75d^16s^2$	65 Tb $4f^96s^2$	66 Dy $4f^{10}6s^2$	67 Ho $4f^{11}6s^2$	68 Er $4f^{12}6s^2$	69 Tm $4f^{13}6s^2$	70 Yb $4f^{14}6s^2$	71 Lu $4f^{14}5d^16s^2$
Actinides $5f^{1-14}6d^{0-1}7s^2$	90 Th $6d^27s^2$	91 Pa $5f^26d^17s^2$	92 U $5f^36d^17s^2$	93 Np $5f^46d^17s^2$	94 Pu $5f^67s^2$	95 Am $5f^77s^2$	96 Cm $5f^76d^17s^2$	97 Bk $5f^97s^2$	98 Cf $5f^{10}7s^2$	99 Es $5f^{11}7s^2$	100 Fm $5f^{12}7s^2$	101 Md $5f^{13}7s^2$	102 No $5f^{14}7s^2$	103 Lr $5f^{14}6d^17s^2$

Figure 3-6 Typical long-form periodic table.

2-1. Types of Elements

Three types of broadly different elements can be distinguished among known species, namely representative, or main-group, elements; d-transition, transition, or subgroup elements; and f-transition, or inner transition, elements. Some authors prefer to include the noble gas elements (He, Ne, Ar, Kr, Xe, Rn) as a fourth type, basing their arguments upon those special properties of reduced chemical reactivity that result from the closed "outermost" electron shells in the atoms in question. However, since each of these elements represents the completion of a series of representative elements and since those compounds of the noble gas elements that have been described are often comparable with those of the elements of immediately preceding atomic numbers, it appears more logical to classify the noble gas elements with the representative, or main-group elements. In Fig. 3-6, the main groups are designated as IA through VIIIA and the subgroups as IB through VIIIB. The f-transition elements are collected at the bottom of the periodic table.

Representative, or Main-Group, Elements. Atoms of these elements are characterized by "outermost" ground-state configurations varying from ns^1 through ns^2np^6, with all underlying orbitals that are present filled to capacity. Included among elements of this broad type are the very strongly reducing alkali (IA) and alkaline earth (IIA) metals, the somewhat less strongly reducing metals of groups IIIA and IVA, the strongly oxidizing nonmetals of the halogen (VIIA) and oxygen (VIA) families, the less strongly oxidizing nonmetals of groups IIIA to VIIA, the chemically inert or nearly inert nonmetals of group VIIIA, and the metalloids, which lie along the stair-step line that roughly separates the metals from the nonmetals. The elements in groups IA and IIA are sometimes referred to as the "s-block" elements (ns^{1-2}), whereas those in groups IIIA to VIIIA are termed the p-block elements (np^{1-6}). The placing of hydrogen and helium is somewhat arbitrary.

Within each horizontal series (given n value), there is a regular trend from strongly reducing character (easy loss of electrons) through decreasing reducing strength to increasing oxidizing strength and then ultimately to near or complete inertness that parallels addition of electrons to the nth quantum level. Many of the representative elements are abundant, familiar, and of substantial importance either as such or in compounds. Resemblances within a particular main group (or family) are close. Oxidation states are fixed in the IA (as $+1$) and IIA (as $+2$) families but variable in the other families. Where variability in oxidation states of the representative elements is found, individual states for a given element commonly differ from each other by two units, reflecting electron pairing.

d-Transition, or Subgroup, Elements. Atoms of these elements are characterized by "outermost" ground-state configurations varying from $(n-1)d^1ns^2$ through $(n-1)d^{10}ns^2$ ($n=4$ and larger), with all underlying orbitals present filled to capacity. The inclusion of the Group IB and Group IIB elements in this category is opposed by those persons who believe that the criterion for classification should be an incompletely filled $(n-1)d$ orbital arrangement. However, parallels in overall chemistry between these elements and those that immediately precede them are sufficient to make it advantageous to include them as d-transition elements. The interposing of the d-transition elements between the s-block and the p-block representative elements is a consequence of the occurrence of $(n-1)d$ orbitals at lower energy than np orbitals.

All of these elements are metals, which vary from the naturally very abundant (e.g., Ti, Fe) to the scarce (e.g., Pt metals, Re) to the terrestrially probably nonexistent (e.g., Tc). Many of the subgroup elements are important as structural metals, in alloys of various types, as protective or decorative plates, in jewelry, as catalysts, and (in compounds) as catalytically and electronically useful substances. The metals vary from the readily oxidized IIIB elements to the relatively inert platinum metals and gold. Except for the group IIIB metals, for which $+3$ is the only oxidation state encountered in stable compounds, all of the transition metals exist in at least two oxidation states when in chemical combination. The $+2$ state, corresponding to the involvement of the ns^2 electrons, is particularly common. For a given element, other oxidation states usually differ from each other by single units, corresponding to the involvement of the $(n-1)d$ electrons one by one. Compounds of the d-transition elements are, probably without exception, complexes (Chs. 6–8, 11). As a consequence of the presence of unpaired $(n-1)d$ electrons, many of these compounds are colored and paramagnetic.

f-Transition, or Inner Transition, Elements. Atoms of these elements are characterized by "outermost" ground-state configurations varying from $(n-2)f^1(n-1)d^{0-1}ns^2$ through $(n-2)f^{14}(n-1)d^1ns^2$, where $n = 6$ or 7, with all underlying orbitals present filled to capacity. The term "outermost" is applied advisedly since electrons in $4f$ orbitals are relatively well buried within other electron shells and thus not really in the valency shell. On the other hand, electrons in $5f$ orbitals are less buried and are in the valency shell. Two f-transition series can be distinguished among the known elements, namely the $4f$, or *lanthanide* series ($_{58}$Ce through $_{71}$Lu) and the $5f$, or *actinide* series ($_{90}$Th through $_{103}$Lr). There is no argument that the first $4f$ electron characterizes the cerium atom. However, the orbital energies of the $5f$ and $6d$ orbitals are so nearly the same that the first $5f$ electron appears in the $_{91}$Pa atom, not in the $_{90}$Th atom (Sec. 2.3-5). However, an overall comparison of properties justifies including thorium as a member of the actinide series.

All of these elements are metals, none of which is extremely abundant. Only traces of $_{61}$Pm, $_{93}$Np, and $_{94}$Pu have been detected in the crust of the earth, each species as a product of a nuclear reaction.* The elements americium $(Z = 95)$

---

* It appears to be probable [R. West, *J. Chem. Educ.*, **53**, 336 (1976)] that natural nuclear reactors, evidences for which were discovered at the Oklo uranium mine in Gabon, produced $^{239}_{94}$Pu, which then decreased in quantity by both alpha decay and nuclear fission.

through lawrencium ($Z = 103$) have been obtained only as products of nuclear bombardment reactions. Of the lanthanides, only cerium and neodymium are even moderately abundant; of the naturally occurring actinides, only thorium and uranium. Major applications of the elements or their compounds are in alloys, in electronic devices (e.g., as chromophoric compounds in color television screens), as catalysts, and in nuclear reactions (e.g., $^{235}_{92}U$ or $^{239}_{94}Pu$ as fuel, various isotopes of Eu and Gd as moderators of neutron flux). Resemblances between compounds of elements in the two series (e.g., Nd and U, Gd and Cm) are less striking than within a given series, each in a particular oxidation state (e.g., Pr^{3+} and Nd^{3+}, UO_2^{2+} and PuO_2^{2+}). Again, oxidation states for a given element differing by single units are common, as is the occurrence of colored and paramagnetic species. Complexation is less significant with ions derived from these elements than with d-transition metal ions. Operationally and because of similarities in overall chemistry, it is not uncommon to include lanthanum and actinium with the f-transition elements.

2-2. Extending the Periodic Table. Superheavy Elements*

Prior to about 1940, the heaviest known element, uranium, was quite generally believed to terminate the periodic table, although a number of somewhat speculative papers dealing with possible transuranium elements had appeared.[18] In the years following the recognition of fission of the $_{235}U$ atom with slow neutrons,[19] neutron capture, followed by β decay, and other nuclear bombardment reactions have produced transuranium elements with atomic numbers at least through 105. Elements 93(Np) through 103(Lr) are accommodated nicely

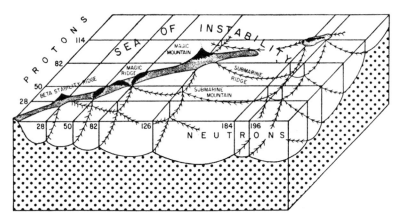

Figure 3-7 Known and predicted regions of nuclear stability within a "sea of instability." [Reproduced with permission from G. T. Seaborg, *J. Chem. Educ.*, **46**, 631 (1969).]

* So-called "SHES."

Table 3-2 Comparison of Properties of Known Elements with Those Predicted for Unknown "Superheavy" Congeners[17]

Property	Thallium (Z = 81)	Eka-thallium (Z = 113)	Lead (Z = 82)	Eka-lead (Z = 114)
Atomic weight (amu)	204.37	297	207.19	298
Density (g cm^{-3})	11.85 (20°C)	16	11.34 (20°C)	14
Atomic volume, (cm^3 mole^{-1})	17.3 (20°C)	19	18.3 (20°C)	21
Most stable oxidation state	+1	+1	+2	+2
Oxidation potential for:	$Tl(s) \rightarrow Tl^+(aq) + e^-$	$M(s) \rightarrow M^+(aq) + e^-$	$Pb(s) \rightarrow Pb^{2+}(aq) + 2e^-$	$M(s) \rightarrow M^{2+}(aq) + 2e^-$
Volts	+0.336	−0.6	+0.126	−0.8
First ionization energy (eV)	6.108	7.4	7.416	8.5
Second ionization energy (eV)	20.428		15.032	16.8
Ionic radius (M^{n+}) (Å)	1.50 (Tl^+)	1.48 (M^+)	1.18 (Pb^{2+})	1.31 (M^{2+})
Metallic radius (C.N. = 12) (Å)	2.0	1.75	2.0	1.85
Melting point (°C)	303	430	327	70
Boiling point (°C)	1457	1100	1737	150
Enthalpy of vaporization (kcal mole^{-1})	39	31	42.5	9
Entropy at 25°C (cal mole-deg^{-1})	15.35	70	15.49	46

as members of the actinide ($5f$-transition) series[17,20,21] and elements 104(Rf) and 105(Ha) as congeners of hafnium and tantalum, respectively, in the $6d$ transition series.

More recently, theoretical considerations of nuclear stability in terms of neutron (N) and proton (Z) numbers (Sec. 2.2-2) suggest the possible existence of an "island of stability" in a "sea of instability," centered at the "magic numbers" (Sec. 2.2-2) $Z = 114$ and $N = 184$ (i.e., the doubly magic nuclide $A = 298$, $Z = 114$) and the closed nuclear shell at $N = 196$,[17,21] as shown in Fig. 3-7.[17] A second region of potentially enhanced nuclear stability can be expected at $Z = 164$. Nuclear charges of 114 and 164, together with those of 50(Sn) and 82(Pb) which appear as a "mountain" and a "ridge" in the "peninsula of stability" in Fig. 3-7, are those of the group IVA elements. Following Mendeleev's procedure, elements 114 and 164 have been termed "eka-lead" and "eka-eka-lead," respectively. Species in these regions are conceived to have half-lives of sufficient potential lengths to allow either their persistence in nature from primordial time or their formation by bombardment of existing heavy nuclei with massive atomic particles (e.g., ^{48}Ca, ^{238}U), the capture of which might bridge the gap imposed by the "sea of instability."[22] Various schemes for the chemical separation of the products of bombardment have been suggested.[23] As of 1979–80, reports of success must be viewed with skepticism.

By extrapolation, certain properties of some of the elements in the "superheavy" region have been predicted.[17,24] Typical are the data given for eka-thallium ($Z = 113$) and eka-lead ($Z = 114$) in Table 3-2, where corresponding

Table 3-3 Calculated Ground-State Electronic Configurations of Gaseous Atoms

Atomic Number	Ground-State Configuration	Atomic Number	Ground-State Configuration
104	[Rn]^a$5f^{14}6d^27s^2$	119	[118]^a$8s^1$
105	[Rn]$5f^{14}6d^37s^2$	120	[118]$8s^2$
106	[Rn]$5f^{14}6d^47s^2$	121	[118]$8s^28p^1$
107	[Rn]$5f^{14}6d^57s^2$	122	[118]$7d^18p^18s^2$
108	[Rn]$5f^{14}6d^67s^2$	123	[118]$6f^17d^18p^18s^2$
109	[Rn]$5f^{14}6d^77s^2$	124	[118]$6f^37d^08p^18s^2$
110	[Rn]$5f^{14}6d^87s^2$	125	[118]$5g^16f^37d^08p^18s^2$
111	[Rn]$5f^{14}6d^97s^2$	126	[118]$5g^26f^27d^18p^18s^2$
112	[Rn]$5f^{14}6d^{10}7s^2$	127	[118]$5g^36f^37d^18s^2$
113	[Rn]$5f^{14}6d^{10}7s^27p^1$	128	[118]$5g^46f^37d^18s^2$
114	[Rn]$5f^{14}6d^{10}7s^27p^2$	129	[118]$5g^56f^27d^28s^2$
115	[Rn]$5f^{14}6d^{10}7s^27p^3$	130	[118]$5g^66f^27d^28s^2$
116	[Rn]$5f^{14}6d^{10}7s^27p^4$	131	[118]$5g^76f^17d^38s^2$
117	[Rn]$5f^{14}6d^{10}7s^27p^5$	132	[118]$5g^86f^17d^38s^2$
118	[Rn]$5f^{14}6d^{10}7s^27p^6$		

^{a} Underlying completed noble gas atom core.

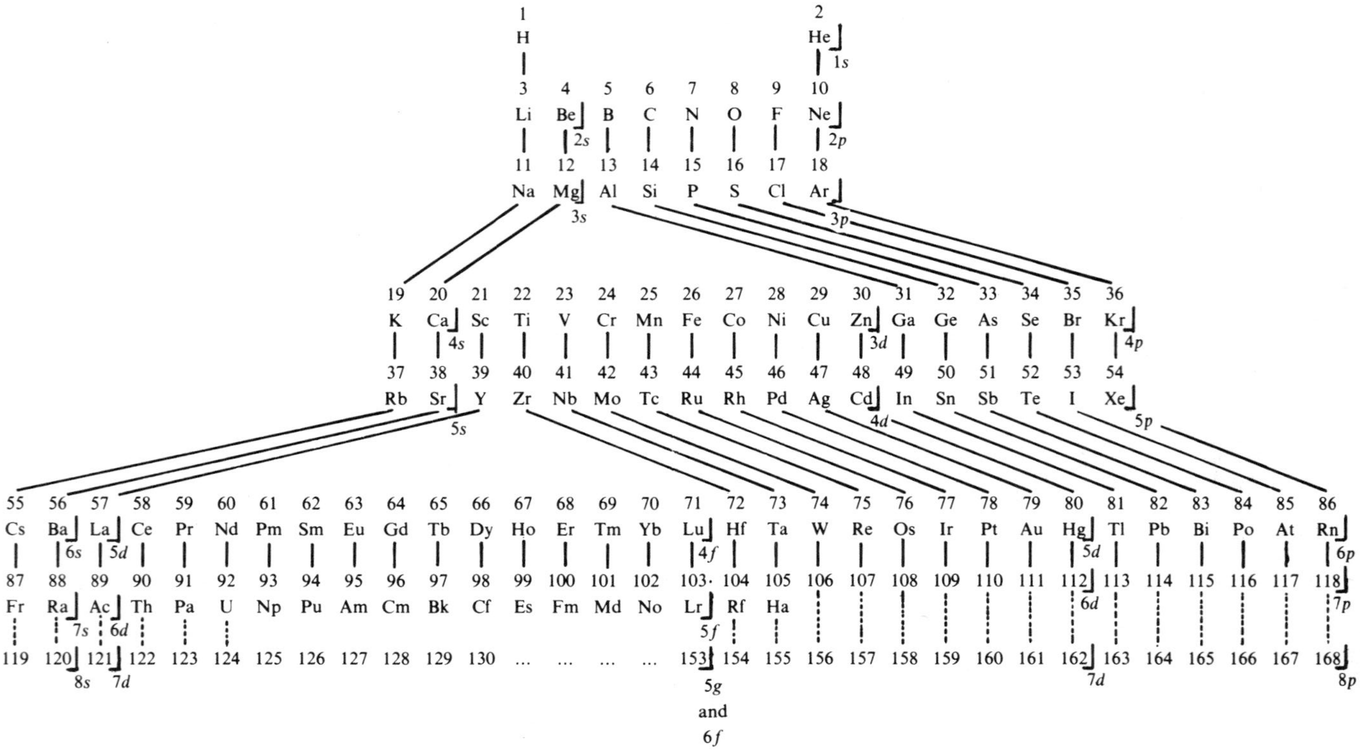

Figure 3-8 Modified long-form periodic table, with completions of orbital occupancies indicated. [Redrawn from G.T. Seaborg, *J. Chem. Educ.*, **46**, 626 (1969).]

data for thallium and lead are also included for comparison. Perhaps even more pertinent are computer-generated predictions of ground-state electronic configurations, some of which involve both $6f$ and $5g$ orbitals, as given in Table 3-3.[17] A third f-transition series ($6f$) apparently merges with a g-transition series somewhere in the vicinity of $Z = 125$, and these two series are then completed at $Z = 154$.

An extension of the long form of the periodic table, which both accommodates the $4f$ and the $5f$ series and also provides reasonable positions for the known (Rf and Ha) and possible transactinide elements, is given as Fig. 3-8. Presumably the physical properties of these elements, if and when the latter are obtained, will correlate in terms of this table with those of their known lighter congeners, but differences in chemical properties may be expected.

3. PERIODIC AND NONPERIODIC PROPERTIES

Only those properties of the elements and their compounds that are determined by the *arrangements* of electrons can vary periodically with atomic number. Those properties that are determined solely by the *numbers* of electrons cannot vary in this way. Only a few properties (e.g., x-ray spectra) are of the nonperiodic type. Among those properties that do vary periodically are the following: atomic volume, metallic or covalent radius, ionic radius (species of same underlying structure type), ionization energy, electron affinity, electronegativity, standard electrode potential (free element), oxidation number or state, ionic mobility, melting point, boiling point, enthalpy of formation (particular compound type), compressibility, optical spectrum, magnetic behavior, hardness, refractive index, and acidity or basicity and many other chemical characteristics.[25]

Some typical properties are discussed in subsequent paragraphs, but largely only from the periodic point of view. Detailed discussion of all of these properties appear in subsequent chapters or in Chapter 2.

Atomic Volumes. The *atomic volume* of an element, as calculated by dividing its gram atomic weight (g mole^{-1}) by its density (g cm^{-3}), is the volume occupied, at the temperature at which the density is measured, by 1 mole (i.e., 6.023×10^{23} atoms) of that element. A modern version of Lothar Meyer's atomic volume curve, as given in Fig. 3-9, is still very useful in emphasizing periodicity. Although atomic volumes are calculated for the same number of atoms of each element, they cannot be used to compare the relative volumes of individual atoms because of variability among the elements in packing and molecular complexity (e.g., Cl_2, S_8, P_4) and have no direct utility in determining crystal structures or dimensions.

Metallic and Covalent Radii. Internally consistent single-bond metallic and covalent radii, as summarized in Table 2-18, are plotted vs. atomic numbers in Fig. 3-10. Periodic variations are clearly apparent.

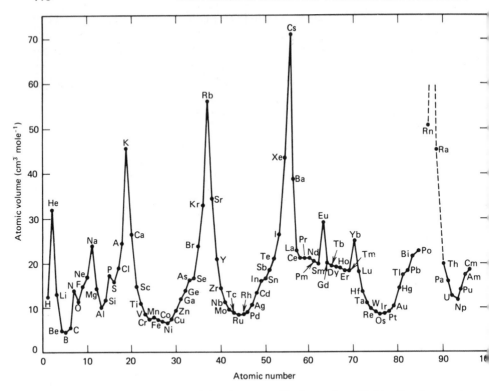

Figure 3-9   Atomic volumes of the elements. [Redrawn from T. Moeller, *Inorganic Chemistry*, pp. 130–131, John Wiley & Sons, New York (1952).]

The trends and variations already pointed out (Sec. 2.4-2) are evident and should be reviewed.

Ionization Energies.   Numerical values of first ionization energies, as summarized in Table 2-19, are plotted vs. atomic numbers in Fig. 3-11. Periodic trends are clearly evident, but operative in essentially the reverse direction of those observed for radii since, other factors being equal, it is more difficult to remove an electron from a small atom than from a larger one. Again the trends discussed earlier should be reviewed from a periodic point of view.

Electron Affinities.   Numerical values of first electron affinities, as summarized in Table 2-20, are plotted vs. atomic numbers in Fig. 3-12. Again periodic trends are noted, although they are much less regular than those in Fig. 3-11 and are less well defined because of the smaller number of data points available. As noted in Sec. 2.4-3, the regularity in trends associated with ionization energies is often absent. Cases in point are essential constancy among heavier members of most of the families of representative elements. A striking exception is found for the elements in group IIA. Atoms with completed outer-

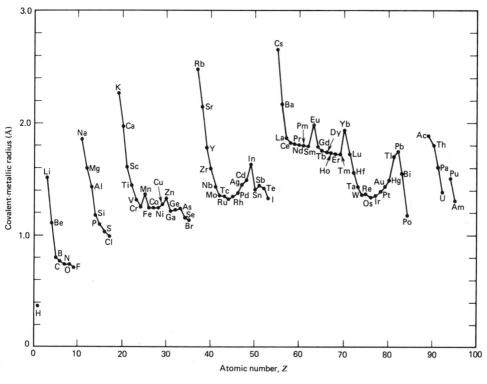

Figure 3-10 Metallic and covalent radii of the elements.

shell electronic arrangements (e.g., s^2 with He and the group IIA atoms, $s^2 p^6$ with the noble gas atoms) or with half-filled valence-shell arrangements (e.g., p^3 with the group IIIA atoms and d^5 with Re) have smaller electron affinities than do adjacent atoms.

Electronegatives. Numerical values of electronegatives of the Allred–Rochow type (Table 2-21) are plotted vs. atomic numbers in Fig. 3-13. Periodicity is clearly evident, as might be expected since the electronegativity of an atom is influenced by its ionization energy and electron affinity, and thus by its ground-state electronic configuration.

Other Physical Properties. Among other physical properties that vary periodically are crystal radii (Table 4-4) and melting points. Variations among the latter are shown in Fig. 3-14, where it is evident that periodicity is less striking and less regular than for the properties already discussed. The melting point of a crystalline element, one indicator of the quantity of energy required to destroy the regular ordering of the crystal and produce the comparative randomness of the liquid, is determined by many factors, one of which is apparently the atomic electronic configuration.

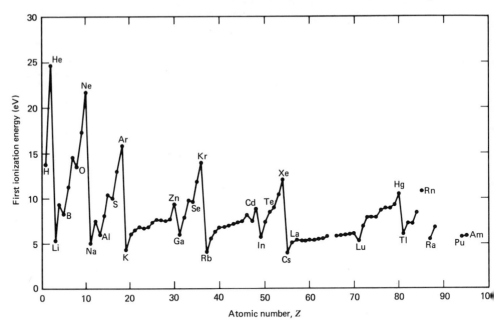

Figure 3-11 First ionization energies of the elements. [Redrawn from T. Moeller, *Inorganic Chemistry*, p. 160, John Wiley & Sons, New York (1952).]

Chemical Properties. Although the chemical properties of the elements do not readily lend themselves to graphic comparisons, they are determined by such factors as ionization energies, electron affinities, and electronegativities and can, therefore, be expected to vary periodically with atomic number. For each subgroup, the same oxidation states prevail, the same types of compounds are produced by chemical reactions, comparable reactions occur with a given reagent, within compounds of the same type bonds are fundamentally the same, comparable compounds have the same types of molecular and crystal structures, and a knowledge of the chemistry of one member allows one to predict the chemistries of the other members. From subgroup to subgroup, there is a regular progression in chemical characteristics.

4. IN CONCLUSION

No attempt has been made to make this chapter all inclusive, either in terms of types of periodic tables or in terms of comprehensiveness in applications. Rather, the objectives of this chapter have been to develop a sound basis for classification of the elements and to emphasize the potential of the periodic table in systematizing a very large segment of inorganic chemistry. Throughout the discussions in subsequent chapters, periodicity and the significance of the

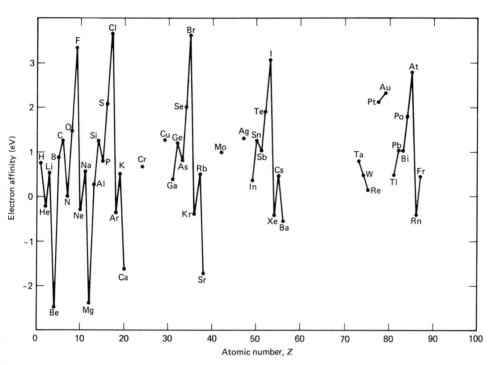

Figure 3-12 First electron affinities of the elements. [Redrawn from data of E. C. M. Chen and W. E. Wentworth, *J. Chem. Educ.*, **52**, 486 (1975).]

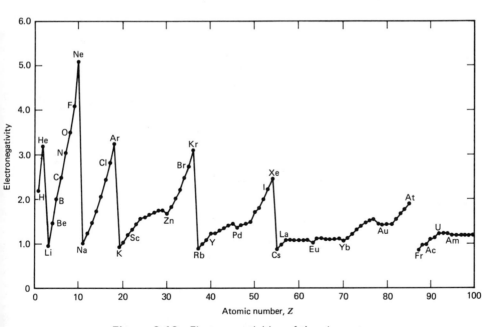

Figure 3-13 Electronegativities of the elements.

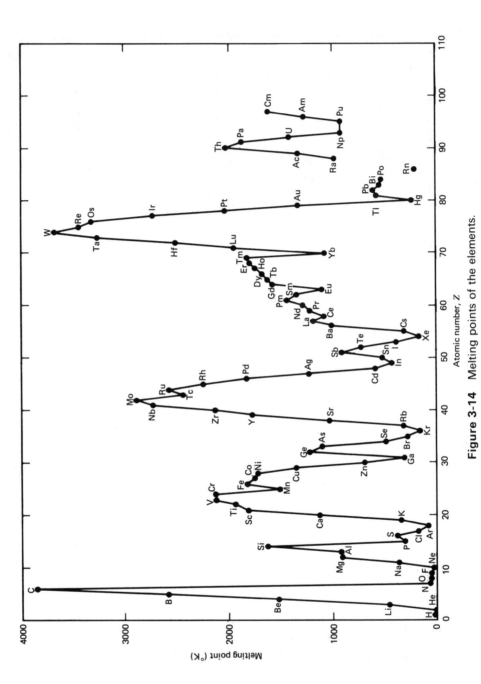

Figure 3-14 Melting points of the elements.

periodic classification are emphasized. The periodic table remains as the single most powerful tool available to us for understanding and systematizing inorganic chemistry.

EXERCISES

3-1. From the list of general periodic properties given on pp. 117–120, select several that were not considered in Figs. 3-9 through 3-14 and construct comparable plots of appropriate numerical values.

3-2. Review the periodic table proposed by L. M. Simmons [*J. Chem. Educ.*, **24**, 588 (1947); **25**, 658 (1948)] and construct an expanded version that will include also the 5g and 6f elements.

3-3. Select a nonperiodic numerical property of the elements and construct a plot comparable with those in Figs. 3-9 through 3-14.

3-4. On the basis of known properties and trends in properties, predict the atomic weight, density, most stable oxidation state, first ionization energy, atomic radius, melting point, and enthalpy of vaporization of eka-thalluim $(Z = 113)$. Compare your values with those given by G. T. Seaborg [*J. Chem. Educ.*, **46**, 627 (1969)].

3-5. Refer to Fig. 3-7 and interpret the designations "submarine mountain" and "submarine ridge" in terms of both nuclear stability and possible periodicity in the super-heavy-element region.

3-6. Indicate clearly why observed periodicity in atomic volumes (Lothar Meyer and Fig. 3-9) may be expected in terms of presently held concepts of atomic structure.

3-7. Implement the brief section on *chemical properties* (p. 120) by citing specific examples that support the general properties listed.

3-8. After appropriate consultation of the literature, provide some data in support of consideration of the species of $Z = 1$, $A = 0$ (positronium) and $Z = 0$, $A = 0$ (light isotope of the neutron, or product of oxidation of the hydrated electron) as "exotic elements." [See R. Rich, *Periodic Correlations*, pp. 109–110, W. A. Benjamin, New York (1965).]

REFERENCES

1. F. P. Venable, *The Development of the Periodic Law*, Chemical Publishing Co., Easton, Pa. (1896).

2. G. N. Quam and M. B. Quam, *J. Chem. Educ.*, **11**, 27, 217, 288 (1934).

3. A. H. Ihde, *The Development of Modern Chemistry*, Ch. 9, Harper & Row, New York (1965).

4. M. E. Weeks and H. M. Leicester, *Discovery of the Elements*, 7th ed., Ch. 14, Journal of Chemical Education Circulation Services, Easton, Pa. (1968).

5. E. G. Mazurs, *Graphic Representations of the Periodic System During One Hundred Years*, University of Alabama Press, University, Ala. (1974).

6. W. H. Taylor, *J. Chem. Educ.*, **26**, 491 (1949).

7. D. Mendeléev, *J. Russ. Phys.-Chem. Soc.*, **1**, 60 (1869); *Z. Chem.*, **5**, 405 (1869).

8. L. Meyer, *Ann. Chem.*, Supple. **7**, 354 (1870).

9. B. Brauner, *Z. Anorg. Chem.*, **33**, 1 (1902).

10. D. Mendeleev, *Ann. Chem.*, Suppl. **8**, 133 (1872).

11. A. Werner, *Chem. Ber.*, **38**, 914 (1905).

12. E. Rutherford, *Philos. Mag.*, [6], **21**, 669 (1911).

13. H. G. J. Moseley, *Philos. Mag.*, [6], **26**, 1024 (1913); [6], **27**, 703 (1914).

14. N. Bohr, *Philos. Mag.*, [6], **26**, 1, 476, 857 (1913).

15. L. S. Foster, *J. Chem. Educ.*, **16**, 409 (1939).

16. T. Moeller, *J. Chem. Educ.*, **47**, 417 (1970).

17. G. T. Seaborg, *J. Chem. Educ.*, **46**, 626 (1969); *Chem. Eng. News*, Apr. 16, 1979, pp. 46–52; *Science*, **203**, 711 (1979); *Am. Sci.*, **68**, 279 (1980). "Required reading" references.

18. L. L. Quill, *Chem. Rev.*, **23**, 87 (1938). General review.

19. O. Hahn and F. Strassmann, *Naturwissenschaften*, **27**, 11, 89 (1939).

20. G. T. Seaborg, *Man-Made Transuranium Elements*, Prentice-Hall, Englewood Cliffs, N.J. (1963).

21. B. Fricke, *Struct. Bonding*, **21**, 89 (1975).

21. G. T. Seaborg, *Annu. Rev. Nucl. Sci.*, **18**, 68 (1968).

22. R. J. Otto, D. J. Morissey, D. Lee, A. Ghiorso, J. M. Nitschke, G. T. Seaborg, M. M. Fowler, and R. J. Silva, *J. Inorg. Nucl. Chem.*, **40**, 589 (1978).

23. E. P. Horwitz and C. A. A. Bloomquist, *J. Inorg. Nucl. Chem.*, **37**, 425 (1975).

24. R. A. Penneman, J. B. Mann, and C. Klixbüll Jørgensen. *Chem. Phys. Lett.*, **8**, 321 (1971). (Typical of many other references that can be consulted.)

25. R. Rich, *Periodic Correlations*, 2nd ed., Addison-Wesley, Reading, Mass. (1972).

SUGGESTED SUPPLEMENTARY REFERENCES

D. G. Cooper, *The Periodic Table*, Butterworth, Washington, D.C. (1964). Elements and compounds by groups.

B. Fricke, *Structure and Bonding*, Vol. 21, pp. 89–144, Springer-Verlag, New York (1975). Superheavy elements.

A. H. Ihde, *The Development of Modern Chemistry*, Ch. 9, Harper & Row, New York (1965). Historical.

E. G. Mazurs, *Graphic Representations of the Periodic System during One Hundred Years*, University of Alabama Press, University, Ala. (1974). Detailed representations of proposed arrangements.

R. Rich, *Periodic Correlations*, 2nd ed., Addison-Wesley, Reading, Mass (1972). Correlation between positions in table and properties of elements and compounds.

J. W. van Spousen, *The Periodic System of the Chemical Elements*, American Elsevier, New York (1969). Detailed summary.

M. E. Weeks and H. M. Leicester, *Discovery of the Elements*, 7th ed., Ch. 14, Journal of Chemical Education Circulation Services, Easton, Pa. (1968). Historical.

CHAPTER FOUR

THE CHEMICAL BOND I: INTRODUCTION, MAJOR ELECTROSTATIC INTERACTIONS

Valency and the *chemical bond* are terms and concepts that permeate and apply to all of chemistry. They are used universally in all types of discussions of chemistry — but unfortunately with a variety of meanings and interpretations.

In its broadest sense, valency implies the power or ability of one atom or ion to interact chemically with another. In a more restricted sense, it may be either a numerical indication of combination or a description of the type of union between two species. On the other hand, a chemical bond can be clearly, and often quantitatively, defined. Paraphrasing L. Pauling,[1] we may say that a chemical bond exists between two atoms or groups of atoms when the forces acting between them are large enough to yield an aggregate of sufficient stability to be distinguishable as an independent molecular or ionic species. Thus a chemical bond is a manifestation of a force of attraction. In terms of different types of these forces of attraction, one can then distinguish different types of chemical bonds.

It is apparent from the discussions in Chapter 2 that bonding is a consequence of the interaction of electrons in the "outermost" shells of the atomic or ionic species involved. Two important characteristics of a bond in general emerge from a consideration of the energy changes that result when two gaseous atoms or ions A and B, which are capable of chemical interaction, approach each other, as indicated in Fig. 4-1. As the internuclear distance (d) decreases, a *force of repulsion* (resulting from electron–electron, nucleus–nucleus, and/or like ionic charge interactions), measured qualitatively in terms of the potential energy of the system A–B, increases exponentially. By contrast, as d decreases, a *force of*

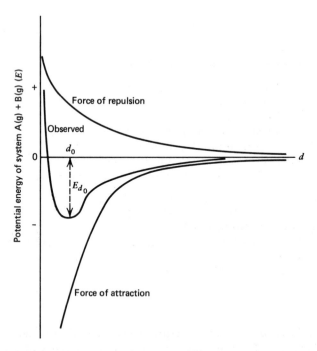

Figure 4-1   Changes in potential energy paralleling the approach of A(g) and B(g). *d*, internuclear, distance; d_0, *equilibrium internuclear distance*; E_{d_0}, minimum energy of system.

attraction (resulting from electron–nucleus and/or unlike ionic charge interactions) increases exponentially. The observed potential energy of the system, amounting to a combination of the forces of repulsion and attraction, passes through a minimum at an internuclear distance d_0 (i.e., the bond length in the species AB). The magnitude of the energy (E_{d_0}) at this bond distance is a measure of the strength of the bond. Quantitatively, bond strength is expressed as either the *bond dissociation energy* or the *bond energy*, both of which are enthalpy terms (Sec. 5.1-7) expressed in kcal mole^{-1} or kJ mole^{-1}.

The bond dissociation energy is the energy required to rupture a specific bond in a molecule, that is, to split the molecule in question into two fragments that were originally connected by that specific bond. The bond energy is a quantity so chosen that the sum of all bond energies for a given molecule is numerically equal to the observed enthalpy of formation of that species from its constituent atoms in their normal states. For a diatomic species, the bond dissociation energy and the bond energy are identical. For a polyatomic species derived from two elements, the bond energy is the mean energy required for complete disruption or atomization. Thus for the $H_2O(g)$ molecule, two dissociation processes occur,

$$H_2O(g) \rightarrow H(g) + OH(g) \qquad \Delta H_1 = 119.7 \text{ kcal (500.8 kJ)}$$

$$OH(g) \rightarrow H(g) + O(g) \qquad \Delta H_2 = 101.5 \text{ kcal (424.7 kJ)}$$

from which a mean of ca. 110.6 kcal (462.8 kJ) mole^{-1} represents the O—H bond energy. The enthalpy change associated with the formation of a water molecule as

$$2 H(g) + O(g) \rightarrow H_2O(g)$$

is -221.5 kcal mole^{-1} (-926.8 kJ), that is, essentially twice the O—H bond energy. Unfortunately, bond and bond dissociation energy are sometimes used synonymously for polyatomic species.

Tables of bond energies and applications are presented in Chapter 5 (pp. 237–238).

1. DEVELOPMENT OF ELECTRONIC CONCEPTS OF VALENCY AND BOND FORMATION

Many detailed accounts of the development of these concepts are available,[2-6] but only a limited discussion covering the more significant contributions is essential to placing subsequent discussions in this chapter in proper perspective. The early proposal by J. J. Berzelius (1812) that the forces of attraction between atoms in chemical compounds are purely electrostatic in character, although strongly supported by M. Faraday's laws of electrochemical equivalence (1834) and subsequent experiments in electrochemistry, was not compatible with such observations from organic chemistry as the lack of profound alteration in properties upon substitution of positive hydrogen in a hydrocarbon molecule by negative chlorine. The extremely useful structural approach to organic chemistry, which ignored the forces between atoms, produced a conflict that continued until (1) the theory of electrolytic dissociation of S. Arrhenius (1887) and (2) the correlation of theory with experiment in the elucidation of the properties of coordination compounds by A. Werner (1893) indicated clearly the existence of two "kinds" of valency. The kinds of valency correlated well with the later development of the concepts of ionic and covalent bonding. It is an interesting commentary that resolution of this conflict was aided materially by a person trained in organic chemistry who received a Nobel Prize for his work in inorganic chemistry: Alfred Werner.

In developing his periodic table, D. Mendeleev had indicated a relationship between valence number and group number (p. 104). This relationship was expanded by R. Abegg[7] into a so-called *rule of eight* by assignment to each element maximum positive and negative valence numbers such that their numerical sum (signs not included) is eight and the positive number is the periodic group number (e.g., $C \pm 4$, $N + 5$ and -3). The significance of Abegg's rule was aptly summarized and put into essentially modern terminology by P. Drude, who stated that "Abegg's positive valency number v signifies the number of loosely attached negative electrons in the atom; his negative valency number v' means that the atom has the power of removing v' negative electrons from other atoms, or at least of attaching them more firmly to itself."[8] The fundamental importance of this statement became apparent only after the elucidation of atomic numbers in 1913 (Sec. 2.1).

Modern concepts of valency, and bonding, have their origin in two significant and independently conceived papers that appeared nearly simultaneously in 1916.[9,10] Inherent in both of these is an appreciation of the chemical stability associated with the noble gas atom "outermost" electronic configuration ($1s^2$ and ns^2np^6, where $n > 1$) and of the tendency of an atom of an element close to a noble gas element to achieve or approach such a configuration through appropriate involvement of its "outermost" electronic shell.

W. Kossel[9] directed attention to the strongly electropositive characteristics of elements immediately following the noble gas elements in atomic number and to the strongly electronegative characteristics of elements immediately preceding the noble gas elements, as reflected, respectively, in positive and negative valence numbers in their compounds. On this basis, he proposed that atoms of these elements combine with each other by losing or gaining sufficient electrons to achieve the appropriate noble gas atom electronic arrangements. The valence number, or the ionic charge, is then the number of electrons lost or gained per atom. Thus the bonding in such compounds involves attractions between oppositely charged ions. In support of this concept, Kossel included a plot, a modified version of which is given as Fig. 4-2, in which the constancy of total number of electrons in ions derived from elements in the noble gas element regions is apparent. It is also apparent that 18-electron and (18 + 2)-electron numbers are also found with considerable frequency. Kossel recognized,

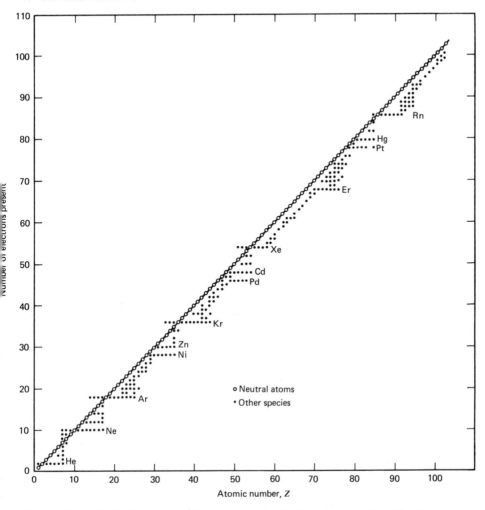

Figure 4-2 Oxidation states of the elements in Kossel-type plot. [T. Moeller, *Inorganic Chemistry*, Fig. 6-1, p. 174, John Wiley & Sons, New York (1952), updated.]

however, that many compounds, in particular those for which decision as to the positive and negative elements is difficult to make, could not be described in these terms.

The somewhat broader approach of G. N. Lewis[10] suggested that the comparatively stable noble gas atom "outermost" arrangement of electrons could be achieved either by the transfer of sufficient electrons from one atom to another to form ions in the Kossel sense or by the sharing of sufficient electrons, in pairs, between the atomic species with the formation of molecules (e.g., four chlorine atoms with one carbon atom). Each of the bonded atoms was assumed to provide one electron to the shared pair, and the shared electrons were then assumed to belong simultaneously to both bonded atoms. In presenting this concept, Lewis assumed the existence in an atom of (1) an inert *kernel* consisting of the nucleus and electrons in inner shells, and (2) an outer *valence shell* within which the electrons were located spatially at the corners of an imaginary circumscribed cube.

Much of the credit for popularizing the Lewis concept must be given to I. Langmuir,[11] who extended the approach to many additional cases, recognized the improbability of a stationary cubical arrangement of electrons, and suggested the term *covalence* for the sharing of electron pairs, as opposed to *electrovalence* for the transfer of electrons. That both electrons of a shared pair may come from the same atom was apparently suggested by G. A. Perkins,[12] not by Langmuir as is generally believed. The familiar symbolism, indicated by the formulations

$$Na^+ \qquad :\overset{..}{\underset{..}{Cl}}:^- \qquad H:\overset{..}{\underset{..}{Br}}:$$

where the symbol of an element represents the kernel and the dots (crosses, etc.) represent the "outer-shell electrons," is a direct consequence of the Lewis–Langmuir valency model.

Subsequent developments have centered mostly in attempts, largely by the application of the quantum-mechanical approaches suggested in Chapter 2, to provide quantitative understanding of the mechanisms by means of which electrons form bonds.

2. CLASSIFICATION OF CHEMICAL BONDS

Very broadly, the physical properties of pure substances allow us to divide the latter into three broad classes:

1. The relatively high-melting, high-boiling, hard, brittle, crystalline compounds that do not conduct the electric current as solids but do conduct it when melted or when dissolved in solvents of large dielectric constants (e.g., H_2O, HF, SO_2—Ch. 10). These properties are consistent with the presence of essentially immobile ions in the crystalline solids and of relatively mobile ions in melts and solutions.

2. The generally much lower-melting, lower-boiling, softer elements and compounds that are nonconductors as solids or liquids and in most of the solvents in which they can be dissolved. Some substances of this class have very high melting points and are hard and brittle (e.g., SiC, BN). Some become conducting as temperature is raised (e.g., Si, GaAs). A limited number react chemically with solvents of large dielectric constants (e.g., NH_3, HCl, SO_2) to yield conducting solutions. These properties are all consistent with the presence of nonionic, molecular groups in the pure substances.

3. The elements, and certain of their derivatives, which although widely variable in melting and boiling points and hardness, are excellent conductors either as liquids or solids, have high luster, are soluble without apparent chemical reaction only in each other, and have the mechanical properties of ductility and malleability. In total, these properties are consistent with the presence of relatively mobile electrons in the crystalline solids and their melts.

It is reasonable to conclude that within substances of the first and third classes the bonding is electrostatic, but of fundamentally different types, whereas with substances of the second class, interactions must be primarily nonelectrostatic.

However, as subsequent discussions indicate, each of these classes, as described, represents a limiting case, there being many substances the properties of which lie in all regions between any two of these extremes. More realistically, the interactions should be designated as primarily or predominantly electrostatic or nonelectrostatic. Furthermore, there are also electrostatic interactions that do not involve ions or mobile electrons, and there exist also solids in which three-dimensional, nonelectrostatic interactions result in molecules the sizes of which are limited only by the sizes of the samples themselves.

Table 4-1 presents a broad classification of chemical bonds reflecting these considerations in terms of formation, properties, and examples. This table should be consulted frequently in relationship to subsequent sections of this chapter and Chapters 5 and 7.

3. IONIC CRYSTALS AND THE IONIC BOND

Of the interactions given in Table 4-1, the ionic bond is the easiest to visualize and the best suited to simple treatment. This type of bond is substantially limited to the crystalline state, although fusion of such crystals renders the ions present mobile, and there are evidences for the existence of ion pairs in the vapors of such compounds. The ionic bond is a consequence of electrostatic interaction between the ions that result from the transfer of one or more electrons from a less electronegative atom to a more electronegative atom or group of atoms. In the simplest case, it may be expected to form most readily, therefore, when an element of small ionization energy (p. 73) combines with an element of large electron affinity (p. 75). However, even under the most optimum of such conditions (e.g., the combination of an atom of cesium of ionization energy 3.894 eV with an atom of fluorine of electron affinity 3.339 eV) there is insufficient energy available from this source to allow the formation of an isolated ionic bond [e.g., for the case in point $Cs^+F^-(g)$]. However, the energy gained through electrostatic attractions in the formation of a crystal [i.e., the crystal or lattice energy (Sec. 4.3-4)] is sufficient to permit the formation of ions.

Ionic crystals are orderly, three-dimensional arrays of charged particles.

Table 4-1 Classification of Chemical Bonds

	Electrostatic				Nonelectrostatic			
	Major		Minor		Normal Covalent			Coordinate Covalent
					Nonpolar		Polar	
Description	Ionic	Metallic	Hydrogen	Dipolar[a]	Small Molecules	Giant Molecules		
Formation	Electron transfer	Single electron for several cations	H-atom bridge between two electro-negative atoms		Electron-pair sharing—one electron from each atom			Electron-pair sharing—both electrons from one atom
					Atoms of same electronegativity or compensating different electronegativities		Atoms of different electro-negativities	
Bond energy (kcal mole^{-1})	ca. 100	Variable	ca. 5–7		ca. 100			ca. 100
Examples	NaCl, CaO, TlOH, La(ClO$_4$)$_3$, UCl$_4$	Na, Cu, Nd, Pu, alloys	(NH$_3$)$_n$, (H$_2$O)$_n$, (HF)$_n$	(NF$_3$)$_n$, Ar(l)	H$_2$, Cl$_2$, CO$_2$, SiCl$_4$, Ge$_2$H$_6$, PF$_5$, XeF$_6$	C (diamond or graphite), BN, SiC, SiO$_2$, AlPO$_4$	HF, OF$_2$, H$_2$O, SF$_4$, PCl$_3$, ClF$_3$	F$_3$B: NH$_3$, [Co(C$_2$O$_4$)$_3$]$^{3-}$, Fe$_2$(CO)$_9$

Properties	Hard, brittle, high m.p. and b.p., insulators as solids, conductors as melts or aqueous solutions	Malleable, ductile, opaque, lustrous, conductors as solids or melts	Moderate m.p. and b.p.	Generally low m.p., b.p.	Soft, low m.p. and m.p., insulators	Hard, brittle, high m.p. and b.p., insulators or semi-conductors	Like non-polar small molecules but with higher m.p. and b.p.	Variable

^{a} Including London or van der Waals interactions.

Inasmuch as there are no distinguishable molecular groups present, the chemical formula of an ionic compound is only an expression of the stoichiometry of combination and is amenable in a quantitative sense only to an expression of a formula weight. The numerical magnitudes of the ionic charges* of the species present are limited, as indicated for simple species by ionization energy and electron affinity data (Tables 2-19, 2-20). In ionic crystals, *simple* cations with charges larger than *two* are uncommon and are probably limited to species such as Y^{3+}, Ln^{3+} or Ln^{4+} (Ln = any lanthanide element), and An^{3+} or An^{4+} (An = any actinide element). Correspondingly, *simple* anions with charge larger than *two*, except possibly the N^{3-} ion in a few nitrides, are unknown. The simple cationic species are thus M^+ (M = Li, Na, K, Rb, Cs, Tl, and sometimes Ag); M^{2+} (M = Mg, Ca, Sr, Ba, Ra, and sometimes Zn, Cd, a few *d*-transition metals, and Ln^{2+}); Y^{3+}, Ln^{3+}, and An^{3+}; and Ln^{4+} and An^{4+} (e.g., Ce^{4+}, Th^{4+}, U^{4+}). The simple anionic species are, correspondingly, H^-, F^-, Cl^-, Br^-, I^-, O^{2-}, S^{2-}, Se^{2-}, Te^{2-}, and N^{3-}). On the other hand, both complex cationic and anionic species with larger charges are not uncommon, as a consequence of the operation of other energy factors. It is apparent that many, but not all simple ions exemplify the rule of eight.

3-1. Structures of Ionic Crystals†

The internal structure of an ionic crystal is given by the repeating three-dimensional arrangement of the component ions. The smallest distinguishable repeating combination of ions is called the *unit cell.*‡ Of necessity, the unit cell must reflect the total symmetry of the crystal in question, but it can be established in a variety of ways. Perhaps the simplest approach is to characterize the unit cell in terms of (1) the distances (*a*, *b*, *c*) from an arbitrary origin in a Cartesian coordinate system at which the experimentally measured planes of ions intercept the *X*, *Y*, and *Z* coordinates, and (2) the angles (α, β, γ) between these planes, as indicated in Fig. 4-3. On this basis, the unit cells shown in Fig. 4-4 and described in Table 4-2 can be distinguished. The rhombohedral unit cell is sometimes called trigonal. Both rhombohedral and hexagonal unit cells can be regarded as subclasses of the trigonal cell.

All points, or ionic sites, within a crystal with identical environments lie in a three-dimensional framework known as a *crystal lattice*. All crystals, ionic or otherwise, are based upon the 14 types of lattices (the Bravais lattices, after M. A. Bravais, who first noted this in 1850) shown in Fig. 4-4. The Bravais

---

* For a simple ion, the ionic charge is also the *oxidation number*, the number of electrons that must be added to a cation to give a neutral atom or the number electrons that must be removed from an anion to give a neutral atom.

† Although the discussion is presented in terms of ionic crystals, it is equally applicable to nonionic crystals, where the fundamental particles present are atoms or molecules.

‡ In more sophisticated terminology, the unit cell is the smallest possible subdivision that has the properties of the visible macrocrystal and that, by repetitive translation of itself in all directions, forms the macro crystal.

Table 4-2 Crystal Systems

System	Unit-Cell Description	Bravais Lattices	Examples
Cubic	$a = b = c$ $\alpha = \beta = \gamma = 90°$	P, I, F	NaCl, CsCl, CaO, Cu
Tetragonal	$a = b \neq c$ $\alpha = \beta = \gamma = 90°$	P, I	SnO_2, K_2PtCl_4, Sn
Orthorhombic	$a \neq b \neq c$ $\alpha = \beta = \gamma = 90°$	P, C, I, F	K_2SO_4, $HgCl_2$, I_2
Rhombohedral	$a = b = c$ $\alpha = \beta = \gamma \neq 90° < 120°$	R	Al_2O_3, Bi
Hexagonal	$a = b \neq c$ $\alpha = \beta = 90°; \gamma = 120°$	P	SiO_2 (quartz), AgI, Zn
Monoclinic	$a \neq b \neq c$ $\alpha = \gamma = 90°; \beta \neq 90°$	P, C	$KClO_3$, $K_3Fe(CN)_6$, S_8
Triclinic	$a \neq b \neq c$ $\alpha \neq \beta \neq \gamma \neq 90°$	P	$CuSO_4 \cdot 5 H_2O$, $K_2S_2O_8$

lattices are characterized as *primitive* (P or R) and *nonprimitive* (C, I, F), the distinction between the two classes being that in the former there is only one lattice point per unit cell, whereas in the latter there are more than one. The designations used are P (primitive), I (body-centered), C (centered on one face only), F (centered on all faces), and R (rhombohedral) (Table 4-2), in accordance with Fig. 4-4.

Because these classifications appear to reflect crystal *shapes*, they may be misleading, for a crystal may have a higher internal symmetry that its physical

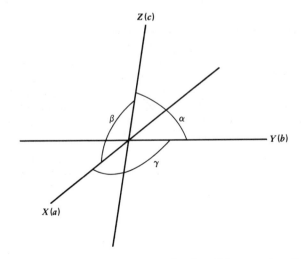

Figure 4-3 Coordinate system for describing a unit cell.

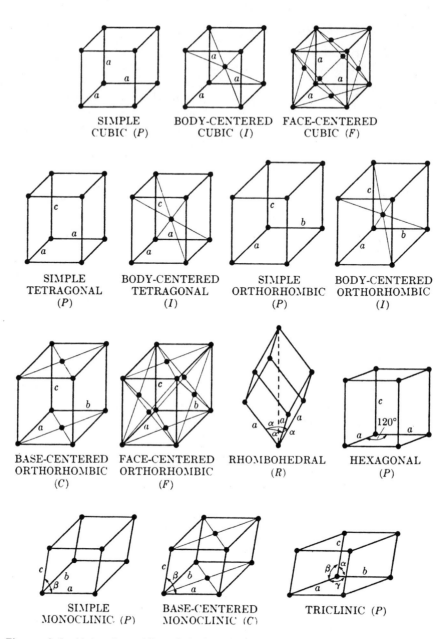

Figure 4-4 Unit cells and Bravais lattices for crystals. [B. D. Cullity, *Elements of X-ray Diffraction*, 2n ed., Fig. 2-3, p. 36, Addison-Wesley, Menlo Park, Calif. (1978).]

shape suggests.[13] Crystal symmetry is an expression of the fact that the crystal has equal properties in different directions. It is not pertinent to our discussion to describe lattice symmetry in detail. It is pertinent to note, however, that a particular combination of translational operations involving screw axes and glide planes with the symmetry operations used to establish point groups in molecular species (Sec. 6.4-1) yields a *space group*. The 32 point groups combine with the 14 Bravais lattices to yield 230 space groups. The result is that a given crystal shape may not agree with the space group that defines that crystal. Thus although a unit cell with $a \neq b \neq c$ and $\alpha = \beta = \gamma = 90°$ suggests orthorhombic symmetry, many apparently monoclinic crystals have unit cells of this type. A comparison of space-group and point-group notations is given in Chapter 6 (p. 326). For more comprehensive coverage of crystal chemistry, standard references can be consulted.[14–18]

3-2. Ionic Radii (Crystal Radii)

As suggested by Fig. 4-1, the equilibrium positions occupied by a cation and an anion in an ionic crystal are determined by a balance between forces of attraction and forces of repulsion. Largely by means of x-ray and neutron diffraction techniques (Sec. 6.5-1), it is possible to determine experimentally internuclear anion–cation distances. Somewhat broadly, these distances are termed *interionic distances*, suggesting that they are the distances between the ions themselves. It is unlikely, however, that the anions and cations in a given crystal are completely isolated from each other, and interaction between electronic atmospheres must occur—interactions that increase, of course, as the degree of covalent interaction increases. Notwithstanding, it is convenient to assume that each ion is a hard sphere, to assume contacts between these spheres, and then to determine radii for these spheres, by appropriate apportionment of measured internuclear distances. Depending upon the method of apportionment used, differing values of the radius of a given ion may result. The hoped for ultimate objective is a set of ionic radii that are additive to give measured internuclear distances in ionic crystals.

The apportionment procedure commonly followed necessitates the adoption of some independently determined ionic radius as a standard and the assumptions that (1) this radius remains unchanged in a series of crystals containing that ion, and (2) ionic radii are indeed additive. This is the substitutional procedure, first used by W. L. Bragg[19] and subsequently adopted by other prominent investigators such as J. A. Wasastjerna,[20] V. M. Goldschmidt,[21] and W. H. Zachariasen.[22]

Based upon the assumption that in a crystal of lithium iodide the relatively small Li^+ ion cannot prevent the much larger I^- ions from contacting each other, A. Landé[23] obtained the radius of the I^- ion as 2.13 Å from half the measured I^-—I^- internuclear distance. From this value and the constancies of differences among internuclear distances in crystals of the alkali metal halides, as shown in Table 4-3, it is then possible to assign radii to the other halide ions and to the

Table 4-3 Internuclear Distances in Crystals of Alkali Metal Halides[a]

Ion	Distance (Å)						
M^+ \ X^-	F^-	$\Delta = Cl^- - F^-$	Cl^-	$\Delta = Br^- - Cl^-$	Br^-	$\Delta = I^- - Br^-$	I^-
Li^+	2.01	0.56	2.57	0.18	2.75	0.28	3.03
$\Delta = Na^+ - Li^+$	0.30		0.24		0.23		0.20
Na^+	2.31	0.50	2.81	0.17	2.98	0.25	3.23
$\Delta = K^+ - Na^+$	0.35		0.33		0.31		0.30
K^+	2.66	0.48	3.14	0.15	3.29	0.24	3.53
$\Delta = Rb^+ - K^+$	0.16		0.15		0.14		0.13
Rb^+	2.82	0.47	3.29	0.14	3.43	0.23	3.66
$\Delta = Cs^+ - Rb^+$	0.19		0.27		0.28		0.29 $\left(\genfrac{}{}{0pt}{}{NaCl}{CsCl}\right)^b$
Cs^+	3.01	0.55	3.56	0.15	3.71	0.24	3.95

[a] A. F. Wells, *Structural Inorganic Chemistry*, 3rd ed., p. 357, Clarendon Press, Oxford (1962).

[b] NaCl-type crystal structure (CsCl-type crystal structures within ... boundary) mean differences (for NaCl types): $\Delta_{Na^+ - Li^+} = 0.25$ Å, $\Delta_{K^+ - Na^+} = 0.32$ Å, $\Delta_{Rb^+ - K^+} = 0.15$ Å, $\Delta_{Cl^- - F^-} = 0.50$ Å, $\Delta_{Br^- - Cl^-} = 0.16$ Å, $\Delta_{I^- - Br^-} = 0.25$ Å.

alkali metal ions. By the use of similar methods,[21,24] the radius of the O^{2-} ion was established as 1.33 Å, and this value was used to develop extensive tables of ionic radii. The values obtained differ only slightly from those currently used (Table 4-4). It is of interest that on the assumptions that the observed molecular refractivity of a solution is the sum of the component ionic refractivities and that a theoretical proportionality exists between the refractivity of an ion with a noble gas atom electronic configuration and the cube of its radius, values of ionic radii agreeing closely with those determined by substitutional procedures have been obtained.[20]

Based upon the knowledge that internuclear distances depend upon the types of electronic distributions in the ions involved, the structure of the crystal in question, and the ratio of cationic to anionic radius [i.e., the *radius ratio*, ρ (p. 152)], and assuming that the radius of an ion should be inversely proportional to effective nuclear charge, L. Pauling[25] devised a theoretical approach to the apportionment of internuclear distances. To minimize difficulties, Pauling used, as points of reference, the so-called standard crystals NaF, KCl, RbBr, and CsI, with internuclear distances of 2.31, 3.14, 3.43, and 3.95 Å, respectively. These crystals have about equal ionic character, the same radius ratio (ca. 0.75), and, except for CsI, the same sodium chloride–type structure;* and they contain, in each case, anions and cations of the same nonpolarizable noble gas type, but with different nuclear charges.

In the Pauling treatment, ionic radius and effective nuclear charge are related as

$$r = \frac{C_n}{Z_{\text{eff}}} = \frac{C_n}{Z - S} \tag{4-1}$$

where C_n is a constant, the value of which is determined by the maximum principal quantum number, and S is a screening constant (see Eq. 2-31). Pauling evaluated screening constants both theoretically and from molar refraction and x-ray term values,[25,26] but essentially the same values can be obtained by application of Slater's rules (Sec. 2.4-1). Thus, applying Pauling's values to NaF(c), where by experiment the internuclear distance is 2.31 Å, or

$$r_{\text{Na}^+} + r_{\text{F}^-} = 2.31 \text{ Å} \tag{4-2}$$

and the screening constant is 4.52, and using Eq. 4-1,

$$\underset{(Na^+)}{\frac{C_n}{11 - 4.52}} + \underset{(F^-)}{\frac{C_n}{9 - 4.52}} = 2.31 \text{ Å} \tag{4-3}$$

whence $C_n = 6.14$ Å. Using this value and Eq. 4-1, Pauling then calculated

---

* Both types are cubic (Table 4-2). In the NaCl type each M^+ ion is surrounded octahedrally by 6 X^- ions and each X^- ion by 6 M^+ ions, but in the CsCl type, 8 X^- ions surround each Cs^+ ion cubically. Pauling was able to correct the measured Cs^+—I^- distance to what it would be in an NaCl type crystal.

$$r_{Na^+} = 0.95 \text{ Å} \quad \text{and} \quad r_{F^-} = 1.36 \text{ Å}$$

Similarly, from the other crystals, he obtained

$$r_{K^+} = 1.33 \text{ Å} \quad r_{Cl^-} = 1.81 \text{ Å}$$

$$r_{Rb^+} = 1.48 \text{ Å} \quad r_{Br^-} = 1.95 \text{ Å}$$

$$r_{Cs^+} = 1.69 \text{ Å} \quad r_{I^-} = 2.16 \text{ Å}$$

By utilization of these values in conjunction with the measured internuclear distances for other crystals and the principle of additivity, he then obtained radii for other simple ions or pseudoions of the noble gas atom (8-electron) and pseudo-noble gas (18-electron) types. The term pseudoion is applied to species such as N^{5+}, S^{6+}, Cl^{7+}, and so on, which have no real existence as ions.

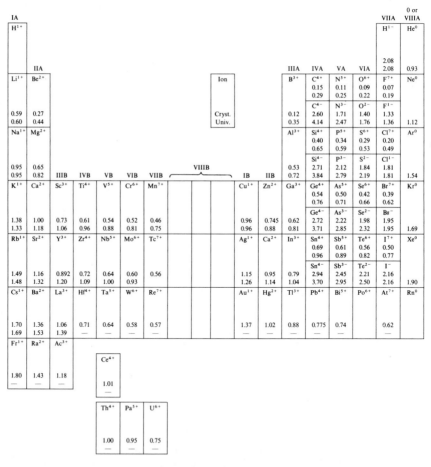

Figure 4-5 Univalent and crystal radii. [T. Moeller, *Inorganic Chemistry*, Fig. 5-1, p. 139, John Wiley & Sons, New York (1952)]

Table 4-4 Selected Crystal Radii

Z	Ion	Coordination Number	Crystal Radius (Å)		
			Pauling[25]	Shannon–Prewitt[30]	Other (Reference)
1	H^+	1		−0.38	
		2		−0.18	
	H^-		2.08		
3	Li^+	4		0.59	
		6	0.60	0.74	0.68 (27)
4	Be^{2+}	4	0.31	0.27	
		6			0.35 (27)
5	B^{3+}	3		0.02	
		4		0.12	
		6	0.20		0.23 (27)
6	C^{4+}	4			
		6	0.15		0.16 (27)
7	N^{5+}	3		−0.12	
		6	0.11		0.13 (27)
	N^{3-}	6	1.71		
8	0^{6+}	6	0.09		0.10 (27)
	O^{2-}	2		1.35	
		4		1.38	
		6	1.40	1.40	
9	F^{7+}	6	0.07		0.08 (27)
	F^-	6	1.36	1.33	
11	Na^+	6	0.95	1.02	0.97 (27)
12	Mg^{2+}	6	0.65	0.72	0.66 (27)
13	Al^{3+}	4		0.39	
		6	0.50	0.53	0.51 (27)
14	Si^{4+}	4		0.26	
		6	0.41	0.40	0.42 (27)
	Si^{4-}	6	2.71		
15	P^{5+}	4		0.17	
		6	0.34		0.35 (27)
	P^{3+}	6			0.44 (27)
	P^{3-}	6	2.12		
16	S^{6+}	4			
		6	0.29	0.12	0.30 (27)
	S^{4+}	6			0.37 (27)
	S^{2-}	6	1.84		
17	Cl^{7+}	4	0.26	0.20	
		6			0.27 (27)
	Cl^{5+}	3		0.12	
		6			0.34 (27)
	Cl^-	6	1.81		1.81 (22)
19	K^+	6	1.33	1.38	1.33 (27)
20	Ca^{2+}	6	0.99	1.00	0.99 (27)

Table 4-4 (*Continued*)

Z	Ion	Coordination Number	Crystal Radius (Å)		
			Pauling[25]	Shannon–Prewitt[30]	Other (Reference)
21	Sc^{3+}	6	0.81	0.73	0.81 (27)
22	Ti^{4+}	6	0.68	0.605	0.68 (27)
	Ti^{3+}	6		0.67	0.76 (27)
	Ti^{2+}	6		0.86	0.80 (21)
23	V^{5+}	6	0.59	0.54	0.59 (27)
	V^{4+}	6		0.59	0.63 (27)
	V^{3+}	6		0.64	0.74 (27)
	V^{2+}	6		0.79	0.88 (27)
24	Cr^{6+}	6	0.52		0.52 (27)
	Cr^{4+}	6		0.55	
	Cr^{3+}	6		0.615	0.63 (27)
	Cr^{2+}	6		0.73	0.83 (21)
25	Mn^{7+}	4		0.26	
		6	0.46		0.46 (27)
	Mn^{6+}	4		0.27	
	Mn^{4+}	6	0.50	0.54	0.60 (27)
	Mn^{3+}	6		0.58	0.66 (27)
	Mn^{2+}	6		0.67	0.80 (27)
26	Fe^{3+}	4		0.49	
		6		0.55	0.64 (27)
	Fe^{2+}	4		0.63	
		6		0.77	0.74 (27)
27	Co^{3+}	6		0.525	0.63 (27)
	Co^{2+}	6		0.735	0.72 (27)
28	Ni^{3+}	6		0.56	
	Ni^{2+}	6		0.70	0.69 (27)
29	Cu^{2+}	4 (SQ)[a]		0.62	
		6		0.73	0.72 (27)
	Cu^{+}	2		0.46	
	Cu^{+}	6	0.96		0.96 (27)
30	Zn^{2+}	4		0.60	
		6	0.74	0.745	0.74 (27)
31	Ga^{3+}	4		0.47	
		6	0.62	0.62	0.62 (27)
32	Ge^{4+}	4		0.40	
		6	0.53	0.54	0.53 (27)
	Ge^{2+}	6			0.73 (27)
	Ge^{4-}	6	2.72		
33	As^{5+}	4		0.335	
		6	0.47	0.50	0.46 (27)
	As^{3+}	6			0.58 (27)
	As^{3-}	6	2.22		

Table 4-4 (*Continued*)

Z	Ion	Coordination Number	Crystal Radius (Å)		
			Pauling[25]	Shannon–Prewitt[30]	Other (Reference)
34	Se^{6+}	4		0.29	
		6	0.42		0.42 (27)
	Se^{4+}	6			0.50 (27)
	Se^{2-}	6	1.98		1.91 (21)
35	Br^{7+}	6	0.39		0.39 (27)
	Br^{5+}	6			0.47 (27)
	Br^-	6	1.95		1.96 (21)
37	Rb^+	6	1.48	1.49	1.47 (27)
38	Sr^{2+}	6	1.13	1.16	1.12 (27)
39	Y^{3+}	6	0.93	0.892	0.92 (27)
		8		1.015	
		9		1.10	
40	Zr^{4+}	6	0.80	0.72	0.79 (27)
		8		0.84	
41	Nb^{5+}	4		0.32	
		6	0.70	0.64	0.69 (27)
		7		0.66	
	Nb^{4+}	6		0.69	0.74 (27)
	Nb^{3+}	6		0.70	
42	Mo^{6+}	4		0.42	
		6	0.62	0.60	0.62 (27)
	Mo^{5+}	6		0.63	
	Mo^{4+}	6		0.65	0.70 (27)
	Mo^{3+}	6		0.67	
43	Tc^{7+}	6			0.56 (27)
	Tc^{4+}	6		0.64	
44	Ru^{4+}	6		0.62	0.67 (27)
	Ru^{3+}	6		0.68	
45	Rh^{4+}	6		0.615	
	Rh^{3+}	6		0.665	0.68 (27)
46	Pd^{4+}	6		0.62	0.65 (27)
	Pd^{2+}	4 (SQ)		0.64	
		6		0.86	0.80 (27)
47	Ag^{3+}	4 (SQ)		0.65	
	Ag^{2+}	6			0.89 (27)
	Ag^+	2		0.67	
		4 (SQ)		1.02	
		6	1.26	1.15	1.26 (27)
48	Cd^{2+}	4		0.84	
		6	0.97	0.95	0.97 (27)
49	In^{3+}	6	0.81	0.79	0.81 (27)

Table 4-4 (*Continued*)

Z	Ion	Coordination Number	Crystal Radius (Å)		
			Pauling[25]	Shannon–Prewitt[30]	Other (Reference)
50	Sn^{4+}	6	0.71	0.69	0.71 (27)
	Sn^{2+}	6			0.93 (27)
	Sn^{4-}	6	2.94		
51	Sb^{5+}	6	0.62	0.61	0.62 (27)
	Sb^{3+}	6			0.76 (27)
	Sb^{3-}	6	2.45		
52	Te^{6+}	6	0.56		0.56 (27)
	Te^{4+}	6			0.70 (27)
	Te^{2-}	6	2.21		2.18 (27)
53	I^{7+}	6	0.50		0.50 (27)
	I^{5+}	6		0.95	0.62 (27)
	I^-	6	2.16		2.20 (21)
55	Cs^+	6	1.69	1.70	1.67 (27)
56	Ba^{2+}	6	1.35	1.36	1.34 (27)
		8		1.42	
57	La^{3+}	6	1.15	1.061	1.14 (27), 1.061 (32)
		8		1.18	
		9		1.20	1.101 (33)
		10		1.28	
58	Ce^{4+}	6	1.01	0.80	0.94 (27), 0.92 (35)
	Ce^{3+}	6		1.034	1.07 (27), 1.034 (32)
		8		1.14	
		12		1.29	
59	Pr^{4+}	6		0.78	0.92 (27), 0.90 (35)
		8		0.99	
	Pr^{3+}	6		1.013	1.06 (27), 1.013 (32)
		8		1.14	
60	Nd^{3+}	6		0.995	1.04 (27), 0.995 (32)
		8		1.12	
		9		1.09 (?)	1.035 (33)
61	Pm^{3+}	6		0.979	1.06 (27), 0.979 (32)
62	Sm^{3+}	6		0.964	1.00 (27), 0.964 (32)
		8		1.09	
		9			1.000 (33)
	Sm^{2+}	6			1.11 (35)
63	Eu^{3+}	6		0.95	0.98 (27), 0.950 (32)
		8		1.07	
	Eu^{2+}	6		1.17	1.09 (35)
		8		1.25	
64	Gd^{3+}	6		0.938	0.97 (27), 0.938 (32)
		8		1.06	
		9			0.974 (33)

Table 4-4 (*Continued*)

		Coordination	Crystal Radius (Å)		
Z	Ion	Number	Pauling[25]	Shannon–Prewitt[30]	Other (Reference)
65	Tb^{4+}	6		0.76	0.81 (27), 0.84 (35)
		8		0.88	
	Tb^{3+}	6		0.923	0.93 (27), 0.923 (32)
		8		1.04	
		9			0.957 (33)
66	Dy^{3+}	6		0.908	0.92 (27), 0.908 (32)
		8		1.03	
		9			0.942 (33)
67	Ho^{3+}	6		0.894	0.91 (27), 0.894 (32)
		8		1.02	
		9			0.929 (33)
68	Er^{3+}	6		0.881	0.89 (27), 0.881 (32)
		8		1.00	
		9			0.916 (33)
69	Tm^{3+}	6		0.869	0.87 (27), 0.869 (32)
		8		0.99	
	Tm^{2+}	6			0.94 (35)
70	Yb^{3+}	6		0.858	0.86 (27), 0.858 (32)
		8		0.98	
	Yb^{2+}	6			0.93 (35)
71	Lu^{3+}	6		0.848	0.85 (27), 0.848 (32)
		8		0.97	
72	Hf^{4+}	6		0.71	0.78 (27)
		8		0.83	
73	Ta^{5+}	6		0.64	0.68 (27)
		8		0.69	
	Ta^{4+}	6		0.66	
	Ta^{3+}	6		0.67	
74	W^{6+}	4		0.41	
		6		0.58	0.62 (27)
	W^{4+}	6	0.66	0.65	0.70 (27)
75	Re^{7+}	4		0.40	
		6		0.57	0.56 (27)
	Re^{6+}	6		0.52	
	Re^{5+}	6		0.52 (?)	
	Re^{4+}	6		0.63	0.72 (27)
76	Os^{4+}	6		0.63	0.69 (27)
77	Ir^{4+}	6		0.63	0.68 (27)
	Ir^{3+}	6		0.73	
78	Pt^{4+}	6		0.63	0.65 (27)
	Pt^{2+}	6			0.80 (27)
79	Au^{3+}	4 (SQ)		0.70	
		6			0.85 (27)

Table 4-4 (*Continued*)

Z	Ion	Coordination Number	Crystal Radius (Å)		
			Pauling[25]	Shannon–Prewitt[30]	Other (Reference)
	Au^+	6	1.37		1.37 (27)
80	Hg^{2+}	2		0.69	
		4		0.96	
		6	1.10	1.02	1.10 (27)
		8		1.14	
	Hg^+	3		0.97	
81	Tl^{3+}	6	0.95	0.88	0.95 (27)
		8		1.00	
	Tl^+	6		1.50	1.47 (27)
		8		1.60	
82	Pb^{4+}	6	0.84	0.775	0.84 (27)
		8		0.94	
	Pb^{2+}	4		0.94	
		6		1.18	1.20 (27)
		8		1.29	
	Pb^{4-}	6	2.15		
83	Bi^{5+}	6	0.74		0.74 (27)
	Bi^{3+}	5		0.99	
		6		1.02	0.96 (27)
		8		1.11	
84	Po^{6+}	6			0.67 (27)
	Po^{4+}	6			1.04
		8		1.10	
	Po^{2+}	6			2.30
85	At^{7+}	6			0.62 (27)
87	Fr^+	6			1.80 (27)
88	Ra^{2+}	6			1.43 (27)
89	Ac^{3+}	6			1.18 (27), 1.076 (34)
90	Th^{4+}	6		1.00	1.02 (27), 0.984 (34)
		8		1.06	
	Th^{3+}	6			1.08 (34)
91	Pa^{5+}	6			0.89 (27), 0.90 (34)
		9		0.95	
	Pa^{4+}	6			0.98 (27), 0.944 (34)
		8		1.01	
	Pa^{3+}	6			1.13 (27)
92	U^{6+}	2		0.45	
		4		0.48	
		6		0.75	0.80 (27), 0.83 (34)
	U^{5+}	6		0.76	0.88 (34)
	U^{4+}	6			0.97 (27), 0.929 (34)
		8		1.00	
	U^{3+}	6		1.06	1.005 (34)

Table 4-4 (*Continued*)

Z	Ion	Coordination Number	Pauling[25]	Crystal Radius (Å) Shannon–Prewitt[30]	Other (Reference)
93	Np^{5+}	6			0.87 (34)
	Np^{4+}	6			0.95 (27), 0.913 (34)
		8		0.98	
	Np^{3+}	6		1.04	1.10 (27), 0.986 (34)
94	Pu^{5+}	6			0.87 (34)
	Pu^{4+}	6		0.80	0.93 (27), 0.896 (34)
		8		0.96	
	Pu^{3+}	6		1.00	1.08 (27), 0.974 (34)
95	Am^{4+}	6			0.92 (27), 0.888 (34)
		8		0.95	
	Am^{3+}	6		1.01	1.07 (27), 0.962 (34)
96	Cm^{4+}	8		0.95	0.886 (34)
	Cm^{3+}	6		0.98	0.946 (34)
97	Bk^{4+}	6			0.870 (34)
		8		0.93	
	Bk^{3+}	6		0.96	0.935 (34)
98	Cf^{3+}	6		0.95	
99					
100					
101					
102					
103					
104					
105					
106					

[a] SQ indicates square-planar geometry.

Radii so calculated are relative to those of the alkali metal and halide ions, but they are not absolute radii in the sense that for non-unipositive and non-uninegative ions their sums do not agree with measured internuclear distances (e.g., $r_{Mg^{2+}} = 0.82$ Å and $r_{O^{2-}} = 1.76$ Å, with a sum substantially different from the measured Mg—O distance of 2.10 Å). For multivalent ions, radii so calculated are those that these ions should have if they were to retain their electronic configurations but behave in Coulombic attraction as if they were univalent. Thus they are really *univalent radii*. Pauling showed, however, that univalent radii can be converted to *crystal radii*, which are additive to give measured internuclear distances, by the relationship

$$r_x = r_u Z^{-2/(n'-1)} \tag{4-4}$$

where x and u refer to crystal and univalent, respectively, and n' is the Born repulsion exponent, resulting from forces of repulsion due to the interpenetration of ionic atmospheres and having a specific value for each noble gas or pseudo-noble gas configuration ($=5$ for He, 7 for Ne, 9 for Ar or Cu^+, 10 for Kr or Ag^+, 12 for Xe or Au^+). Comparative values for univalent and crystal radii, as calculated by Pauling, are summarized in Fig. 4-5, where periodic trends are clearly apparent.

Numerous extensions of Pauling's approach and revisions of various sets of crystal radii have been published.[27-30] A comparatively recent extension of the Goldschmidt technique has yielded a very useful set of cationic radii based upon 1000 evaluations of internuclear distances in oxides and fluorides, the reference values $r_{F^-} = 1.33$ Å and $r_{O^{2-}} = 1.40$ Å, and an essentially linear relationship between ionic volume and unit-cell volume.[30] These approaches are applicable only to simple and spherical ions. The only radii available for nonspherical ions (e.g., NO_3^-, CO_3^{2-}, CrO_4^{2-}, NCS^-) are calculated thermochemical values[31] which, except possibly for tetrahedral species (e.g., SO_4^{2-}, ClO_4^-, MnO_4^-, CrO_4^{2-}), have little utility in the structural sense.

The crystal radius of a given ion is not necessarily the same in all crystals. As the crystal coordination number (C.N.) of an ion (i.e., the number of nearest neighbors it has in the lattice) decreases, the crystal radius decreases. Both Goldschmidt[21] and Pauling[25] recognized this situation and gave similar factors for converting from the common C.N. $= 6$ to less common C.N. $= 4, 8$. Change in oxidation number of a neighboring ion also alters the crystal radius. These effects are apparent in the values given in Table 4-4.

Trends in crystal radii are comparable with those noted earlier for atomic radii (Sec. 2.4-2). Periodic similarities and differences among selected crystal radii are apparent in Fig. 4-6. These reflect in general the major influence of (1) expansion by addition of electron shells, (2) contraction with increasing nuclear charge, or (3) the effect of a combination of the two in terms of effective nuclear charge (Sec. 2.4-1). In general for ions of the same charge derived from elements in a specific periodic family, ionic radius increases with increasing nuclear charge, thus reflecting the greater importance of increasing principal quantum number. This trend is more apparent among the representative, or main-group elements than among the d-transition elements. Paralleling an increase in nuclear charge, without alteration in principal quantum number, there is a general decrease in the sizes of ions with comparable electronic configurations in any horizontal periodic series.

Within the d-transition series, size decreases of this type are smaller because addition of electrons to "inner" d orbitals results in smaller shielding of "outer-shell" electrons from the nuclear charges. But the radii of ions with 18-electron cores are never as large as those of comparable ions with 8-electron cores (e.g., Ag^+, Kr core $+ 4d^{10}$, $r = 1.26$ Å vs K^+, Kr core, $r = 1.47$ Å). Within a given d-transition series, the relatively small variations in the radii of ions of the same charge (e.g., Ti^{2+}, 0.80 Å through Ni^{2+}, 0.69 Å) account for similarities in crystal structure, solubility, and so on, of their compounds.

Parallel small decreases in size of ions of the same charge with increasing

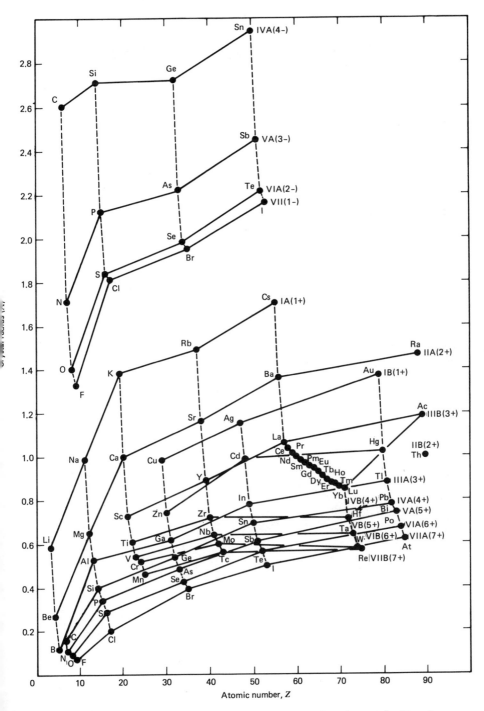

Figure 4-6 Periodic variations in crystal radii. [T. Moeller, *Inorganic Chemistry*, Fig. 5-6, pp. 148–149, John Wiley & Sons, New York (1952), redrawn and updated.]

Table 4-5 Size Relationships among Periodic Group IIIB and 4f- Transition Metal Species

Symbol	Z	Atomic Volume (cm³ mole⁻¹)	Molecular Volume (cm³ mole⁻¹)				Radius of M³⁺ (Å)
			$M_2(SO_4)_3 \cdot 8H_2O$	$A—Ln_2O_3$	$B—Ln_2O_3$	$C—Ln_2O_3$	
Sc	21	15.028	—	—	—	35.53	0.81
Y	39	19.786	240.8	—	—	45.17	0.92
La	57	22.501	—	50.28	—	ca. 47	1.061
Ce	58	20.695	—	47.89	—	—	1.034
Pr	59	20.778	253.9	46.65	—	—	1.013
Nd	60	20.60	252.4	46.55	ca. 51	—	0.995
Pm	61	—	—	—	—	—	0.979
Sm	62	19.950	247.9	—	4.69	48.38	0.964
Eu	63	28.91ᵃ	247.3	—	4.65	48.28	0.950
Gd	64	19.88	246.4	—	ca. 43	47.58	0.938
Tb	65	19.245	—	—	—	46.38	0.923
Dy	66	19.032	242.8	—	—	45.49	0.908
Ho	67	18.742	241.1	—	—	44.89	0.894
Er	68	18.473	239.3	—	—	44.38	0.881
Tm	69	18.151	—	—	—	44.11	0.869
Yb	70	24.80ᵃ	235.1	—	—	42.5	0.858
Lu	71	17.779	234.7	—	—	42.25	0.848
Ac	89	—	—	—	—	—	1.18

[a] Reflect $+2$ oxidation state in Eu and Yb. Others exist in $+3$ state.

Table 4-6 Variations in Size (Radius, Å) as Related to Lanthanide Contraction (by Periodic Family)

IIIB	IVB	VB	VIB	VIIB	VIIIB		
Sc 1.61	Ti 1.45	V 1.31	Cr 1.25	Mn 1.37	Fe 1.24	Co 1.25	Ni 1.25
Sc^{3+} 0.81	Ti^{4+} 0.68	V^{5+} 0.59	Cr^{6+} 0.52	Mn^{7+} 0.46			
Y 1.78	Zr 1.59	Nb 1.43	Mo 1.36	Tc 1.35	Ru 1.33	Rh 1.35	Pd 1.38
Y^{3+} 0.92	Zr^{4+} 0.79	Nb^{5+} 0.69	Mo^{6+} 0.62	Tc^{7+} 0.56	Ru^{4+} 0.67	Rh^{4+} 0.62	Pd^{4+} 0.65
La 1.87	Hf 1.56	Ta 1.43	W 1.37	Re 1.37	Os 1.34	Ir 1.36	Pt 1.39
La^{3+} 1.061	Hf^{4+} 0.78	Ta^{5+} 0.68	W^{6+} 0.62	Re^{7+} 0.56	Os^{4+} 0.69	Ir^{4+} 0.68	Pt^{4+} 0.65

nuclear charge among the f-transition elements are due to the effects of incomplete shielding by electrons in f orbitals. Among the $4f$ species, the so-called *lanthanide contraction* (Sec. 2.4-2) is apparent not only in the radii of ions of the same charge ,but also in atomic radii (Table 2-18), atomic volumes, and molecular volumes of isomorphous compounds (Table 4-5).[36] The effects of the lanthanide contraction are significant in affecting the chemistries of both the lanthanide elements and those that immediately follow them in atomic numbers. The classic difficulties in the separation of the terpositive ions (Ln^{3+}) from each other result from small differences in the degree to which properties such as solubility or complexation change with the small decrease in ionic radius from La^{3+} ion through Lu^{3+} ion. The contraction is sufficient that ionic radius drops to that of the Y^{3+} ion ($Z = 39$, $4d$ transition series) in the vicinity of Dy^{3+} and then decreases still more. The preferential occurrence of yttrium in nature with the heavier lanthanides and the difficulty of separating yttrium from these elements are consistent with similarities in ionic radii rather than complete similarities in electronic configurations. Furthermore, as a consequence of the lanthanide contraction, the radii of ions of elements that follow lutetium in atomic number (i.e., Hf and on) are decreased to those of comparable ions of congeners in the $4d$-transition series, again resulting in occurrence together in nature and difficulty of separation. Thus although zirconium ($Z = 40$) was discovered in 1789 by M. H. Klaproth, it was not until 1923 that its congener hafnium ($Z = 72$) was identified by D. Coster and G. von Hevesy, even though all zirconium minerals contain hafnium [average $Hf/(Hf + Zr) = 0.02$ (Sec. 1.1-7)]. The separation of hafnium from zirconium is one of classic difficulty and, of necessity, always requires a fractionation technique. Comparable resemblances also characterize such other pairs as Nb–Ta, Mo–W, and Tc–Re, as indicated by the data in Table 4-6.

The effects of the parallel *actinide contraction* are apparent in similarities in properties of compounds containing these ions in a constant oxidation state and in the necessity for using fractionation techniques to effect separations from each other.

3-3. Radius Ratio and the Packing of Ions in Crystals

On the assumption that each ion is effectively a hard sphere, the optimum arrangement of ions for the most thermodynamically stable crystal will be that involving the "touching" of the largest number of oppositely charged ions without crowding. The resulting type of packing (i.e., *closest packing*) is thus dependent upon the *radius ratio* (i.e., r_+/r_-) (p. 152). Limiting radius ratios can be calculated from simple geometrical relationships. For example, in a crystal of sodium chloride (Fig. 4-7), each Na^+ ion ($r_+ = 0.97$ Å) contacts six Cl^- ions ($r_- = 1.81$ Å), four in a plane and two others perpendicular to that plane in an overall octahedral array. The arrangement within the plane is indicated in Fig. 4-7. It is apparent that for the designated 45° right triangle

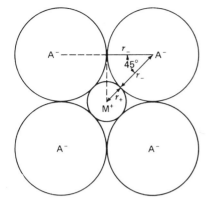

Figure 4-7 Planar projection of four A^- ions contacting one M^+ ion and each other in a plane.

$$\cos 45° = \frac{r_-}{(r_- + r_+)} = \frac{\sqrt{2}}{2} \tag{4-5}$$

which solves to

$$\frac{r_+}{r_-} = 0.414 \tag{4-6}$$

Thus 0.414 is the lower limit of radius ratio for the packing of four anions in a square-planar arrangement about a cation or for the packing of six anions in an octahedral arrangement about a cation. This calculation and other similar ones yield the limiting results given in Table 4-7. The lattices noted in Table 4-7 are illustrated in Fig. 4-8.

The observed packings of ions in real crystals also reflect the facts that (1) anions are, on the average, larger than cations, and (2) Coulombic repulsions among anions must be minimized.[14,15,37] As shown in Fig. 4-9(a), closest packing of *neutral* spheres of the same radius is achieved in *two dimensions* when each sphere contacts *six* other spheres (C.N. = 6). If, however, the spheres have

Table 4-7 Packing of Rigid Anion Spheres (A) about a Cation Sphere (C)

Radius Ratio: r_+/r_- or r_C/r_A	Crystal Coordination Number	Arrangement of A's around a C	Typical Ionic Lattices
Up to 0.155	2	Collinear[a]	
0.155–0.225	3	Trigonal planar[a]	
0.225–0.414	4	Tetrahedral	ZnS (wurtzite, zinc blende)
0.414–0.732	4	Square planar[a]	
	6	Octahedral	NaCl, TiO_2 (rutile)
0.732–	8	Cubic	CsCl, CaF_2 (fluorite)

[a] Uncommon among ionic crystals; often found in coordination entities, where the same geometrical considerations apply (Ch. 7).

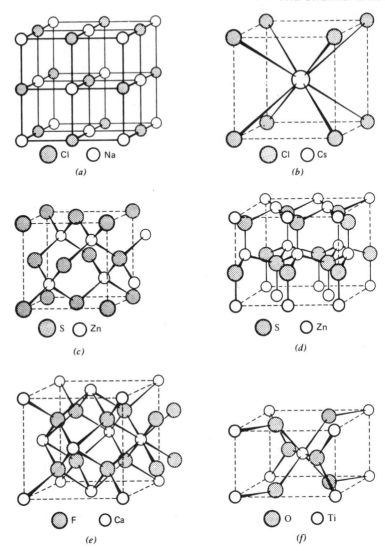

Figure 4-8 Common crystal structures. (a) NaCl; (b) CsCl; (c) ZnS–zinc blende; (d) ZnS–wurtzite; (e) CaF₂ (fluorite); (f) TiO₂ rutile). [Composite from A. F. Wells, *Structural Inorganic Chemistry*, 3rd ed., Clarendon Press, Oxford (1953), as done in J. E. Huheey, *Inorganic Chemistry*, 2nd ed., p. 52, Harper & Row, New York (1978).]

the same radius but half are anions and half cations, this two-dimensional arrangement [Fig. 4–9(b)] is energetically less stable than one in which each anion contacts *four* cations and each cation contacts *four* anions (i.e., C.N. = 4) [Fig. 4–9(c)].

If now we consider a three-dimensional array of spheres, closest packing of neutral spheres is achieved by fitting one layer upon another in such a way that a

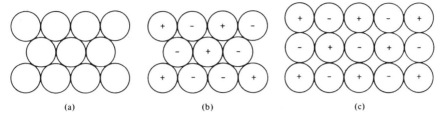

Figure 4-9 Packing of spheres of same radius in two dimensions. Two-dimensional lattices. (*a*) Stable, 6-coordinate, closest-packed lattice of uncharged atoms; (*b*) unstable, 6-coordinate lattice of charged ions; (*c*) stable, 4-coordinate lattice of charged ions. [J. E. Huheey, *Inorganic Chemistry*, 2nd ed., Fig. 3.8, p. 76, Harper & Row, New York (1978).]

given sphere contacts *six* others in the same layer, *three* more in the layer below its own, and another *three* in the layer above [i.e., *twelve* contacts (C.N. = 12)]. Two such arrangements are possible, as shown in Fig. 4-10. In the first of these, the arrangement in the initial layer (A) is repeated in the third, fifth, seventh, and so on, layer, whereas the arrangement in the second layer (B) is repeated in the fourth, sixth, eighth, and so on, layer. This arrangement, ABABAB ..., gives *cubic close packing* (ccp). In the second arrangement (Fig. 4-10), the first three layers are different (i.e., ABC), the arrangement in the first layer then being repeated in the fourth, that in the second layer in the fifth, and so on. This arrangement, ABCABC ..., gives *hexagonal close packing* (hcp).* Crystal structures of these two types are characteristic of metals, alloys, intermediate phases, and interstitial solid solutions (Sec. 4.4-3) but are found only infrequently among ionic substances.

In the layering process delineated in the preceding paragraph, holes surrounded by *four* contacting spheres (i.e., tetrahedral, or T_d, holes) develop between each two layers, and somewhat larger holes surrounded by *six* contacting spheres (i.e., octahedral, or O_h, holes) develop within a given layer and between this layer and the two layers that contact it (Fig. 4-11). In a perfect close-packed lattice, for N spheres there are N octahedral holes and $2N$ tetrahedral holes. In a metallic lattice, small nonmetal atoms (H, B, C, N) may be trapped within these holes to give interstitial solid solutions. In an ionic lattice, there is no absolute parallel, but if the anion is substantially larger than the cation, the close packing of anion spheres provides holes to accommodate the charge-balancing cation spheres.[38] The lithium iodide crystal, as treated by Landé (Sec. 4-3.2),[23] is a case in point, with each Li^+ ion surrounded octahedrally by I^- ions. This lattice is of the NaCl type, and the same considerations apply to sodium chloride crystals.

Although the bonding in crystalline zinc sulfide is largely covalent, it is

* It is recommended that the reader use either commercially available three-dimensional models or construct them from suitable cork or plastic balls to illustrate and verify this discussion.

(a)

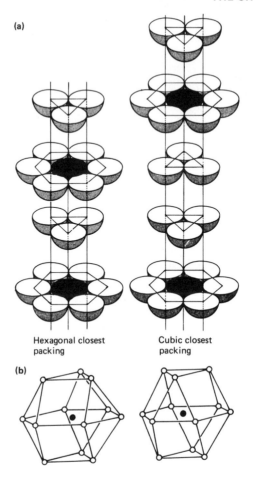

Hexagonal closest   Cubic closest
packing   packing

(b)

Figure 4-10 Cubic and hexagonal close packing of spheres of same radius. (a) Exploded views of stacking of planes of spheres; (b) arrangement of the 12 nearest neighbors about one sphere. In both types of packing, only the spheres in the A layers are directly above one another. [T. Moeller, J. C. Bailar, J. Kleinberg, C. O. Guss, M. E. Castellion, and C. Metz, *Chemistry with Inorganic Qualitative Analysis*, Fig. 12.13, p. 337, Academic Press, New York (1980).]

convenient to consider the crystals as close-packed arrays of relatively large S^{2-} ions ($r = 1.84$ Å) which accommodate the smaller Zn^{2+} ions ($r = 0.60$ Å) in tetrahedral holes. If the sulfide-ion lattice is cubic close packed, the crystal has the zinc blende structure; if it is hexagonal close packed, the crystal has the wurtzite structure (Fig. 4-8).

The radius ratio determines exactly the number of anions that fit around a cation if the two ions have the same numerical charge, and the stoichiometry is thus 1 : 1 (e.g., NaCl, CaO). If the stoichiometry is not 1 : 1, both crystal coordination numbers of the ions and stoichiometry must be considered in

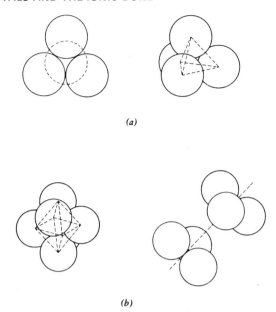

(a)

(b)

Figure 4-11 (a) Tetrahedral and (b) octahedral holes in close-packed crystal structures. [J. J. Lagowski, *Modern Inorganic Chemistry*, Fig. 2-15, p. 73, Marcel Dekker, New York (1973).]

determining the nature of the lattice. For example, for the oxide Na_2O, calculations of radius ratios for both the Na^+ and O^{2-} ions as

$$\frac{r_{Na^+}}{r_{O^{2-}}} = \frac{0.97 \text{ Å}}{1.40 \text{ Å}} = 0.69 \quad \text{and} \quad \frac{r_{O^{2-}}}{r_{Na^+}} = \frac{1.40 \text{ Å}}{0.97 \text{ Å}} = 1.44$$

indicate maximum crystal coordination numbers of four or six for Na^+ and eight for O^{2-}. In keeping with the stoichiometry of the compound, C.N. = 4 is selected for the Na^+ ion, and the crystal structure is guessed (and correctly) to be of the antifluorite types, that is, with anions and cations reversed in the fluorite type depicted in Fig. 4-8.

3-4. Energetics of Ionic Crystals

Inasmuch as many of the characteristics of the ionic bond are associated with the formation and destruction of ionic crystals, the energy change involved in the formation of such a crystal from appropriate numbers of ions is important. The Coulombic energy (E) of attraction between a *cation* (C) of charge z_C and an anion (A) of charge z_A separated by an internuclear distance d is given by

$$E = \frac{z_C z_A e^2}{d} \tag{4-7}$$

where e is the electronic charge. In terms of Fig. 4-1, as d decreases the energy of repulsion increases, and a balance between the energy of attraction and that of repulsion results at the equilibrium internuclear distance (d_0) in the crystal. Inasmuch as the energy of repulsion between ions varies as some power (n) of d, the potential energy V_{CA} for the system involving only two ions becomes[25]

$$V_{CA} = -\frac{z_C z_A e^2}{d_{CA}} + \frac{be}{d_{CA}^n} \tag{4-8}$$

where b is a constant and the minus sign indicates release of energy. For the three-dimensional array in a crystal, the total potential energy (V_0) then becomes

$$V_0 = -\frac{Me^2 z_{CA}^2}{d_0} + \frac{Be^2}{d_0^n} \tag{4-9}$$

where z_{CA} is the highest common factor of z_C and z_A, B is a repulsion coefficient, and M is the *Madelung constant*.

The Madelung constant is a geometrical correction factor that depends upon the type of crystal lattice. It can be evaluated from crystal structure data via the numerical solution of an infinite series* characteristic of that particular lattice.[39-42] Values of the Madelung constant for various structure types and the d_0 condition are summarized in Table 4-8.

From an experimentally determined value of d_0, the constant B can be evaluated as

$$B = \frac{d_0^{n-1} M z_{CA}^2}{n'} \tag{4-10}$$

where n' is the Born exponent (Eq. 4-4 and values, Sec. 3-2). Combining Eq. 4-9 with Eq. 4-10 and simplifying gives

---

* In a crystal of sodium chloride, as depicted in Fig. 4-12, a given Na^+ ion is surrounded by 6 Cl^- ions at a distance d, 12 Na^+ ions at a distance $d\sqrt{2}$, 8 Cl^- ions at a distance of $d\sqrt{3}$, 6 Na^+ ions at a distance $2d$, 24 Cl^- ions at a distance $d\sqrt{5}$, and so on. Thus the Na^+ ion is subjected alternately to forces of attraction and repulsion that are related to the crystal structure and lattice parameters. In terms of Eq. 4-7, the total energy associated with all of these interactions becomes

$$E = -\frac{6(z_{NA^+})(z_{Cl^-})e^2}{d} + \frac{12(z_{Na^+})^2 e^2}{d\sqrt{2}} - \frac{8(z_{Na^+})(z_{Cl^-})e^2}{d\sqrt{3}}$$

$$+ \frac{6(z_{Na^+})^2 e^2}{2d} - \frac{24(z_{Na^+})(z_{Cl^-})e^2}{d\sqrt{5}} + \cdots$$

$$= -\frac{(z)^2 e^2}{d}\left[6 - \frac{12}{\sqrt{2}} + \frac{8}{\sqrt{3}} - \frac{6}{2} + \frac{24}{\sqrt{5}} + \cdots\right]$$

where the infinite series in square brackets reduces to 1.74756, which then is the numerical value of the Madelung constant for sodium chloride or any other crystal of the same structure.

$$V_0 = -\frac{Me^2 z_{CA}^2}{d_0}\left(1 - \frac{1}{n'}\right)$$

(4-11)

or for 1 mole of ion pairs (an Avogadro number, N),

$$U_0 = -NV_0 = -\frac{NMe^2 z_{CA}^2}{d_0}\left(1 - \frac{1}{n'}\right)$$

(4-12)

where U_0 is the *lattice*, or *crystal*, *energy*. This relationship is termed the Born–Landé equation.

The lattice energy is defined as the energy that is released when the gaseous ions required to form 1 mole of product are brought from infinite separation to the positions they occupy in the crystal lattice of that product. Thus the lattice energy is the *enthalpy of formation* of an ionic solid from the gaseous ions. It is most commonly expressed in kcal mole^{-1} or kJ mole^{-1}. The numerical value of the lattice energy for a given substance is a measure of the thermochemical stability of that substance. It is apparent from Eq. 4-12 that the numerical magnitude of the lattice energy is determined by both the oxidation states and the radii of the ions present. Increase in oxidation state of either or both ions or decrease in the radius of either or both ions increases the lattice energy by enhancing attractions between cations and anions.

Modification of Eq. 4-12 to[43]

$$U_0 = -\frac{NMe^2 z_{CA}^2}{d_0}\left(1 - \frac{\rho}{d_0}\right)$$

(4-13)

where ρ describes quantum-mechanical forces of repulsion acting between

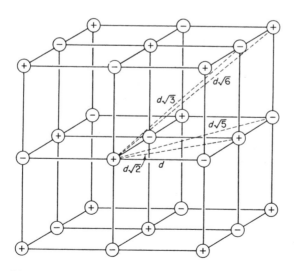

Figure 4-12 Distances between neighboring ions in NaCl-type crystal structure. [J. J. Lagowski, *Modern Inorganic Chemistry*, Fig. 2-19, p. 79, Marcel Dekker, New York (1973).]

Table 4-8   Values of Madelung Constant[a]

Structure Type	Coordination Numbers	M
Sodium chloride (NaCl), C^+A^-	6, 6	1.74756
Cesium chloride (CsCl), C^+A^-	8, 8	1.76267
Zinc blende (ZnS), C^+A^- or $C^{2+}A^{2-}$	4, 4	1.63085
Wurtzite (ZnS), C^+A^- or $C^{2+}A^{2-}$	4, 4	1.64132
Fluorite (CaF$_2$), $C^{2+}A_2^-$	8, 4	5.03878
Cuprite (Cu$_2$O), $C_2^+A^{2-}$	2, 4	4.44248
Rutile (TiO$_2$), $C^{4+}A_2^{2-}$	6, 3	4.7701
β-Quartz (SiO$_2$), $C^{4+}O_2^{2-}$	4, 2	4.402
Corundum (Al$_2$O$_3$), $C_2^{3+}O_3^{2-}$	6, 4	24.242

[a] Values M are consistent with Eq. 4-12. Other commonly cited values are based upon the product $z_C z_A$ rather than the highest common factor z_{CA}. For all 1 : 1 combinations, the two sets of values are identical. For others they are not. Comparisons of values and suggested precautions in their use are given by D. Quane, *J. Chem. Educ.*, **47**, 396 (1970).

electron shells of the ions and is essentially 0.345 Å for most crystals, allows improved calculation of lattice energies but requires that for each crystal the type of structure be either known or assumed. Modification of Eqs. 4-12 and 4-13 and introduction of other factors led A. F. Kapustinskii[42,44] to the relationship

$$U_0 = -287.2\left(\frac{v z_C z_A}{r_C + r_A}\right)\left(1 - \frac{0.345}{r_C + r_A}\right) \qquad (4\text{-}14)$$

where v is the number of ions in the "chemical" molecule and the constant 287.2 derives from NMe^2 for the sodium chloride type and the assumption that each crystal type can be transformed to the sodium chloride type without change in U_0 if $M/(v/2)$ and d_0 are simultaneously modified to values for this crystal type.

Equation 4-14 allows the calculation of many lattice energies.[42,45] However, it is often both more common and more convenient to calculate a lattice energy from other known thermochemical quantities in terms of a Born–Fajans–Haber cycle[46-49] by application of the law of Hess.* This cycle considers that a given ionic crystal may be formed either by direct combination of the elements or by a sequence of processes in which the elements are vaporized, the gaseous atoms ionized, and the gaseous ions combined. The two total changes in heat energy are then equal.

For a simple substance such as an alkali metal halide (MX), the cycle can be formulated as

* The law of Hess summarizes experimental observations that the total change in enthalpy (or free energy) for a given chemical process is the same regardless of the reaction pathway (Sec. 9.2-2).

$$\tfrac{1}{2}X_2(g) + M(s) \xrightarrow{-\Delta H_f} M^+X^-(c)$$

$$+\tfrac{1}{2}\Delta H_d \quad\Big\downarrow\quad +\Delta H_s \qquad \Big\uparrow\quad -U_0$$

$$M(g) \xrightarrow{+\Delta H_i} \overbrace{M^+(g)}$$
$$+$$
$$X(g) \xrightarrow[-\Delta H_a]{} X^-(g)$$

where ΔH_f is the ordinary enthalpy of formation of $M^+X^-(c)$, ΔH_d is the enthalpy of dissociation of $X_2(g)$, ΔH_s is the enthalpy of sublimation of $M(s)$, ΔH_i is the enthalpy of ionization (ionization energy) of $M(g)$, and ΔH_a is the enthalpy of ionization (electron affinity) of $X(g)$. Exothermic quantities are indicated by *minus* signs; endothermic quantities by *plus* signs (Sec. 9.2-1). By application of

Table 4-9 Calculation of Lattice Energies of Alkali Metal Halides by Born–Fajans–Haber Technique

| | Enthalpy Change (kcal mole^{-1}) | | | | | U_0 |
Formula	ΔH_f	ΔH_s	$\tfrac{1}{2}\Delta H_d$	ΔH_i	ΔH_a	(kcal mole^{-1})
LiF	− 145.7	37.1	18.9	124.4	− 77.0	− 249.1
NaF	− 136.3	26.0	18.9	118.5	− 77.0	− 222.7
KF	− 134.5	21.5	18.9	100.1	− 77.0	− 198.0
RbF	− 131.3	20.5	18.9	96.3	− 77.0	− 190.0
CsF	− 126.9	18.8	18.9	89.8	− 77.0	− 177.4
LiCl	− 97.9	37.1	28.5	124.4	− 83.4	− 204.6
NaCl	− 98.2	26.0	28.5	118.5	− 83.4	− 187.8
KCl	− 104.2	21.5	28.5	100.1	− 83.4	− 170.9
RbCl	− 102.9	20.5	28.5	96.3	− 83.4	− 164.8
CsCl	− 103.5	18.8	28.5	89.8	− 83.4	− 157.2
LiBr	− 83.7	37.1	22.7	124.4	− 77.6	− 190.3
NaBr	− 86.0	26.0	22.7	118.5	− 77.6	− 175.6
KBr	− 93.7	21.5	22.7	100.1	− 77.6	− 160.4
RbBr	− 93.0	20.5	22.7	96.3	− 77.6	− 154.9
CsBr	− 94.3	18.8	22.7	89.8	− 77.6	− 148.0
LiI	− 64.8	37.1	17.7	124.4	− 70.6	− 173.4
NaI	− 68.8	26.0	17.7	118.5	− 70.6	− 160.4
KI	− 78.3	21.5	17.7	100.1	− 70.6	− 147.0
RbI	− 78.5	20.5	17.7	96.3	− 70.6	− 142.4
CsI	− 80.5	18.8	17.7	89.8	− 70.6	− 136.2

the law of Hess (Sec. 9.2-2) and neglecting signs

$$\Delta H_f = \tfrac{1}{2}\,\Delta H_d + \Delta H_s + \Delta H_a + \Delta H_i + U_0 \tag{4-15}$$

or

$$U_0 = \Delta H_f - (\tfrac{1}{2}\,\Delta H_d + \Delta H_s + \Delta H_a + \Delta H_i) \tag{4-16}$$

When substitutions are made in Eq. 4-15 or 4-16, each numerical quantity must be given its proper sign.* Values so calculated are given in Table 4-9.

A lattice energy calculated from the Born–Fajans–Haber cycle is an *experimental* value and is thus independent of any assumption as to the effects of crystal structure or nature of the bond in the crystal. The influence of these factors is at once apparent when values calculated from Eq. 4-14 are compared with the Born–Fajans–Haber values, as in Table 4-10. For the essentially completely ionic alkali metal halides, hydrides, and many fluorides, agreement is quite good. For other halides and oxides deviations between the two reflect the effects

Table 4-10 Comparison of Some Calculated and Born–Fajans–Haber Lattice Energies

	U_0 (kcal mole^{-1})			U_0 (kcal mole^{-1})	
Crystal	Calculated	Table 4-9	Crystal	Calculated	Born–Fajans–Haber[42]
LiF	−227.7	−249.1	LiH	−234.0	−216.5
NaF	−211.5	−222.7	NaH	−202.0	−193.8
KF	−188.5	−198.0	KH	−177.2	−170.7
RbF	−181.7	−190.0	RbH	−168.6	−164.4
CsF	−170.4	−177.4	CsH	−162.0	−156.1
LiCl	−192.1	−204.6	AgF	−220	−228
NaCl	−179.9	−187.8	AgCl	−199	−216
KCl	−162.7	−170.9	AgBr	−195	−214
RbCl	−158.2	−163.8	AgI	−186	−211
CsCl	−149.4	−157.2			
LiBr	−189.5	−190.3	MgF$_2$	−696.8	−695
NaBr	−175.5	−175.6	MgBr$_2$	−513	−575
KBr	−161.3	−160.4	MgI$_2$	−486	−548
RbBr	−149.7	−154.9	MnF$_2$	−656.3	−662
CsBr	−143.9	−148.0	FeF$_2$	−657.7	−696
LiI	−170.4	−173.5	CoF$_2$	−688	−708
NaI	−161.0	−160.4	NiF$_2$	−697.1	−728
KI	−146.8	−147.0	Ag$_2$O	−585	−714.1
RbI	−141.0	−142.4	Cu$_2$O	−682	−777.7
CsI	−134.7	−136.2			

---

* Attention is again directed to the problem of sign convention for electron affinities. Here the thermodynamic convention noted above is used.

of covalency. Where unfilled d-shells are present among transition metal ions, deviations also reflect *crystal field stabilization* (Sec. 8.1-2).[50] Waddington's review[42] includes many additional examples.

Applications of Crystal Energetics. The Born–Fajans–Haber cycle has been chiefly useful for the evaluation of (1) the enthalpy of formation of a gaseous ion or an electron affinity, where the assumption that the crystal lattice is ionic is reasonable; (2) the nonionic contribution of a lattice by comparison of a value so obtained with that calculated theoretically (e.g., by use of Eq. 4-12); and (3) the standard enthalpy of formation of a hypothetical compound, and thus some probability of its existence, provided that reasonable assumptions as to structure and bonding can be made. Theoretically calculated lattice energies permit the estimation of enthalpies of formation of gaseous ions (e.g., Table 4-11), proton affinities (Table 4-12[51] and Sec. 9.6-1), and enthalpies of formation

Table 4-11 Calculated Standard Enthalpies of Formation of Gaseous Ions

Ion	ΔH_f^0 (kcal mole^{-1})	Ion	ΔH_f^0 (kcal mole^{-1})
HF_2^-	-150	NH_2^-	-15
O^{2-}	-217	NH^{2-}	-261
O_2^-	19	N_3^-	-35
O_2^{2-}	-171	NO_3^-	81
OH^-	-50	C_2^{2-}	245
S^{2-}	-152	CN^-	-7
SH^-	31	CNO^-	63
Se^{2-}	-165	BH_4^-	23
Te^{2-}	-145	BF_4^-	406

Table 4-12 Calculated Proton Affinities

Group	Proton Affinity (kcal mole^{-1})
O^{2-}	-554
S^{2-}	-550
NH^{2-}	-613
OH^-	-375
SH^-	-342
NH_2^-	-393
H_2O	-182
NH_3	-209

of complex ions. They also allow estimation of electrostatic contributions to bonds in crystals and evaluation of thermochemical stabilities of hypothetical compounds or compounds encompassing unusual oxidation states. Of the many applications cited,[45, 52-54] those that follow are typical.*

1. Exchange reactions involving halide ions. In the absence of an ionizing solvent (Sec. 10.1-1), a halogen exchange reaction involving a covalent halide (RX) and an ionic alkali metal halide (MX') and formulated as

$$RX + MX' \rightarrow RX' + MX$$

depends largely upon the difference in lattice energies between MX and MX' and is favored by a larger value for MX. Thus if $X = Cl$ and $X' = F$, CsF would be the best fluorinating agent and LiF the poorest, but if $X = F$ and $X' = Cl$, the reverse would be true. The large lattice energy of lithium fluoride is probably an important reason why covalent fluorides react more readily with lithium aryls than do the analogous chlorides. A case in point is noted in cyclic phosphazene chemistry (Sec. 6.2-1), where stepwise substitution of fluoride by phenyl can be effected as[55]

$$N_3P_3F_6 + x\,LiC_6H_5 \xrightarrow{(C_2H_5)_2O} N_3P_3F_{6-x}(C_6H_5)_x + x\,LiF$$

2. Formation of ionic fluorocomplexes. Certain heterocations,[56] for example, NO^+, NO_2^+, NF_4^+, ClO_2^+, IO_2^+, ClF_2^+, BrF_2^+, and BrF_6^+, and the O_2^+, ion, which do not form ionic fluorides do form salts with fluoroanions such as BF_4^-, PF_6^-, AsF_6^-, and PtF_6^-. These differences in behavior can be rationalized by comparison of energy changes. Using BF_3 as a typical anion precursor and M as a cation source, a Born–Fajans–Haber cycle comparable with that describing the formation of an ionic fluoride can be constructed, as

Comparison of these two cycles indicates major differences involving the total process[52]

$$\tfrac{1}{2}F_2(g) + BF_3(g) + e^- \rightarrow BF_4^-(g) \qquad \Delta H = -153\ \text{kcal}$$

Thus although the lattice energy of $M^+BF_4^-(s)$ is probably smaller than that of

---

* Inasmuch as these and other applications are based upon changes in enthalpy rather than in free energy (Sec. 9.2), the conclusions drawn are qualitatively correct, but not always quantitatively so.

M^+F^-(s), the formation of the ion BF_4^- (g) is sufficiently exothermic to favor the formation of the ionic heterocation tetrafluoroborate. A case in point is $O_2^+[PtF_6]^-$, for which a Born–Fajans–Haber calculation of the lattice energy supports the formula given.[57]

3. Prediction of thermochemical stabilities of unknown compounds or oxidation states. Calculation of the lattice energy for a hypothetical compound by means of Eq. 4-14, followed by a Born–Fajans–Haber summation to give an enthalpy of formation, is a useful approach to the thermochemical stability of both the compound and an oxidation state. Of course, all other pertinent enthalpy terms must be known. Typically, thermochemical stabilities of chlorides and oxides of the $3d$ transition metals in the $+1$, $+2$, and $+3$ states have been shown to agree well with experimental isolability.[58, 59]

The ubiquitous $+1$ and $+2$ oxidation states for the Group 1A and Group IIA elements, respectively, are reconcilable in the same fashion. To take a specific case, the observation that RbF(s) can be isolated but RbF_2(s) cannot is in accordance with the inability of the larger lattice energy of RbF_2 to compensate for the very large (529.2 kcal mole^{-1}) endothermic second ionization energy term in the cyclic treatment.

4. Rationalization of high oxidation states in metal fluorides. Equation 4-16 indicates that to a substantial degree the enthalpy of formation is dependent upon the lattice energy. Since for the halogen atoms the combination of $\frac{1}{2}\Delta H_d$ and ΔH_a is about -60 kcal (-2.51 kJ) mole^{-1}, ΔH_f values for a series of halides are largely determined by lattice energies. Inasmuch as the lattice energies of the fluorides of a given metal are larger than those of the other halides, the highest oxidation states are noted among the fluorides. To a degree, the same is true for oxides, but the larger lattice energy of a given oxide vs. a comparable fluoride is lost because of the endothermic nature [ca. $+220$ kcal ($+9.20$ kJ) per mole of O atoms] of the combination of $\frac{1}{2}\Delta H_d$ and ΔH_a.

3-5. Nonstoichiometric Ionic Crystals and Defect Lattices

One of the basic tenets of chemistry is the *law of definite proportions*, or *constant composition*, one statement of which appeared in Dalton's original atomic theory (Sec. 2.1). Its validity is indicated both by a wealth of stoichiometric data and by the restrictions placed upon bond formation involving electrons. There can be little, if any question as to its applicability to pure gaseous or liquid substances, but as more and more exacting measurements have been made, we have been forced to conclude that its rigorous applicability to crystalline solids may be the exception rather than the rule. Thus numerous solid compounds, the exact analytical compositions of which deviate significantly from those predicted by valency theory have been, and are being, described.*

---

* Termed Berthollide compounds [N. S. Kurnakov, *Z. Anorg. Chem.*, **88**, 109 (1914)] in recognition of C. L. Berthollet's early contention that the composition of a compound may vary. By contrast, the rigorously stoichiometric compounds are termed Daltonide compounds. [See A. H. Ihde, *The Development of Modern Chemistry*, pp. 98–101, Harper & Row, New York (1964)].

Lack of absolute stoichiometry is particularly common among hydrides (e.g., $VH_{0.56}$, $CeH_{2.69}$), oxides (e.g., $TiO_{0.69-1.33}$, $FeO_{1.055-1.19}$, $CeO_{1.65-1.812}$), sulfides (e.g., $Fe_{0.88}S$, $Cu_{1.7}S$), selenides (e.g., $Cu_{1.6}Se$, $CrSe_{1.33-1.50}$), tellurides (e.g., $Cu_{1.65}Te$, $CrTe_{1.3}$), arsenides (e.g., $Ni_{>1.00}As_2$), various ternary compounds (e.g., $K_{2-x}Ti_8O_{16}$, $CuFeS_{1.94}$), and certain minerals, most of which are synthesized at elevated temperatures. [60-67]

Departure from exact stoichiometry in an ionic crystal implies a change in charges of some of the ions, usually the cations, and is associated with defects in the crystal structure. Such a departure is commonly related to the existence of another oxidation state of the cation. Thus copper(I) oxide is metal deficient because copper(II) ions can be incorporated into the crystal lattice to compensate for vacant cation sites. Departures from stoichiometry involving metal deficiency are common because the incorporation of cations of larger charge and smaller size increases Coulombic attractions and thus lattice energies. Correspondingly, departures involving metal excess are usually small because inclusion of metal atoms or cations of smaller charge reduces lattice energy. An exception is γ-Fe_2O_3 which readily converts to Fe_3O_4 because its incomplete lattice can accommodate iron(II) ions.

Thus for a binary compound AB_n, departure from true stoichiometry with respect to B may result from (1) *substitutional* replacement of A ions by B ions in the crystal lattice, (2) *interstitial* inclusion of B ions in "holes" in the lattice, or (3) *subtractive* omission of some A ions while retaining all B sites occupied. In the sense that each of these processes results in a departure from an ideal crystal where every lattice site is completely and correctly occupied, each results in a *crystal defect*. Careful consideration suggests that defects are not unexpected from the points of view that various atoms or ions of appropriate size can be accommodated and that thermal vibrations at any temperature above absolute zero can render omissions possible.

Defects in crystals have been classified as follows:[68]

1. Imperfect crystal.[69, 70] Here the composition is stoichiometric, and electroneutrality in the crystal is preserved. *Point defects* result from either (a) the omission of ion pairs (or atoms in a covalent crystal), a so-called *Schottky defect*, as illustrated in Fig. 4-13(a), or (b) the migration of ions (or atoms) from normal lattice sites leaving "holes," a so-called *Frenkel defect*, as illustrated in Fig. 4-13(b). Such defects are energy-dependent equilibrium situations, and for a given crystal the population of point defects is temperature controlled. Point defects affect the chemistry and electrochemistry of crystals.

A *dislocation*, or *lattice*, *defect* amounts to an organization of point defects. It is not thermodynamically controlled. Rather, it is determined by the mode of formation of the crystal and the subsequent mechanical treatment to which it is subjected. A lattice defect is an *edge*, or *line*, *dislocation* if an extra lattice half-plane is envisioned, that is, a slip-plane on one side of which the crystal is in tension and on the other side of which it is in compression. A *screw dislocation* results if a dislocation line slips from one plane to another giving the effect of a spiral ramp. Dislocations do not affect crystal chemistry.

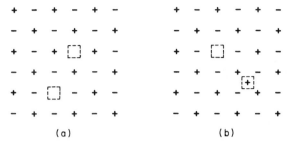

Figure 4-13 Diagrammatic representation of Schottky (*a*) and Frenkel (*b*) defects. [A. W. Adamson, *A Textbook of Physical Chemistry*, 2nd ed., Fig. 19.36, Academic Press, New York (1979).]

2. Nonstoichiometric crystals. As noted earlier, an excess of one component, or a deficiency of the other, is involved. For example, when crystalline sodium chloride is heated in sodium vapor, sodium atoms enter the lattice, where Na^+ ions that occupy lattice sites are formed. The released electrons are trapped in corresponding newly created anion sites (*F* centers), where they impart color and electrical conductance to the crystal. At elevated temperatures, stoichiometric zinc oxide loses oxygen or reacts with zinc vapor and becomes both colored and conducting.[71] The lattice contains relatively free electrons. Compounds such as iron(II) and copper(I) oxide add the anionic component, thereby producing lattices with vacant cation sites. These "positive holes" (*V* centers) function as centers of effective positive charge and migrate through the crystal when a potential is applied. In other instances, notably with δ-TiO (see above), an excess of either cation or anion can develop. This compound, even when stoichiometric, has about 15% unoccupied crystal sites and may conduct the electric current by either positive holes (if O/Ti > 1) or by electrons (if O/Ti < 1). More information is found in the subsequent treatment of semiconduction (Sec. 4.4-2).

3. Impurity systems. These systems result when impurity atoms enter crystal lattices to form solid solutions. Cases in point include (a) the substitution of a group IIIA element (e.g., Ga) or a group VA element (e.g., As) for a germanium atom in the production of doped semiconductors; (b) the formation of mixed crystals when the Cd^{2+} ion occupies Ag^+ ion sites in a silver chloride crystal; and (c) the incorporation of Li_2O into an NiO lattice to produce a controlled number of compensating Ni^{3+} cations and yield a dark-colored, conducting solid.

4. Dual-valency defects. Defects of this type result when ions of the same element, but in different oxidation states, are present in the crystal in stoichiometric proportions. Magnetic iron oxide, Fe_3O_4, is a classic example. In crystals of this compound, all of the Fe^{2+} ions and half of the Fe^{3+} ions are distributed statistically over octahedral lattice sites, and the remaining Fe^{3+} ions occupy tetrahedral sites. This crystal is electrically conducting because of easy transfer of an electron from one oxidation state to the other.

4. THE METALLIC STATE AND THE METALLIC BOND

In a strictly chemical sense, a metal is an element the atoms of which yield simple cations by reaction with highly electronegative atoms. On this basis it is possible to define a metal in terms of the ground-state electronic configuration of its atom and, as we have done earlier, to place the element in the periodic table. It becomes difficult, however, to distinguish on this basis between metals and nonmetals within the same periodic family or to compare metals classified into widely divergent periodic families.

A more nearly exact distinction between metals and nonmetals can be made on the basis of those physical properties that are unique to metals and define thereby a *metallic state*, which is characterized by bonding that is neither ionic nor covalent but has some of the characteristics of each. These unique properties, which are restricted to the condensed phases (i.e., liquid and solid), include:

1. Comparatively large densities, which are indicative of close-packed crystal structures.

2. Excellent electrical and thermal conduction. The former cannot involve material transfer, as opposed to electrical conduction in solution, since no observable changes in either physical or chemical properties either accompany or result from the passage of the electric current. Electrical conductivity, which is observed to be a periodic function of atomic number, is maximal on an atomic basis with metals that give univalent cations and minimal with metals that give multivalent cations. The inverse variation of electrical conductivity with temperature for a given metal suggests that the flow of electrons is impeded by increase in atomic vibrations with increase in temperature.

3. Emission of electrons under conditions of excitation. Both exposure to radiation of frequency larger than a characteristic threshold value and elevation of the temperature above another threshold value cause a given metal to emit electrons. The processes are termed *photoelectric* and *thermionic emission*, respectively.

4. Reversible dissolution without chemical reaction in strongly basic solvents. Strongly electropositive metals (e.g., Li, Na, K, Rb, Cs, Ca, Sr, Ba, Eu, Yb) dissolve readily in liquid ammonia and other comparable solvents, yielding dilute blue solutions that conduct electrolytically and concentrated, bronze-colored solutions that conduct like metals (Sec. 10.1-3). Removal of solvent allows ultimate recovery of the metal.

5. High luster when surfaces are free from oxidation products. Luster is associated with absorption and immediate radiation of visible radiation of a variety of frequencies. Only copper and gold are colored; other metals are silvery to gray in appearance.

6. Very   extensive   three-dimensional   deformability   without   fracture.

Malleability, ductility, and flow under pressure indicate an omnidirectional bonding that involves strong cohesive forces but presents little resistance to deformation when stress is applied.

The first of these properties is completely structural in character and conveniently discussed separately from the others. The others are all related to bonding and suggest strongly the presence of mobile and relatively free electrons in liquid and solid metals. Each of these types may be discussed to advantage.

4-1. Crystal Structures of Metals[72]

The concept of the packing of rigid atomic spheres is particularly useful to the description of the crystal structures of the metals in general because for a given metal all spheres have the same size, and interionic attraction–repulsion effects are absent. The data given in Table 4-13 indicate that the majority of the metals (ca. 60%) have close-packed (cubic or hexagonal) crystal structures, whereas roughly half of the remaining 40% have the more open body-centered cubic (bcc, Fig. 4-14) crystal structures. A limited number of other crystal structures have been described.

In both of the close-packed structures, ca. 74% of the available space is occupied by metal atoms. The body-centered cubic arrangement is characterized by coordination number 8, with ca. 67% of the available space occupied. It is apparent that in 12-coordination not all internuclear distances are the same, whereas in body-centered 8-coordination they are. Relative radii for the various coordination numbers stand in the ratios (C.N. = 12) : (C.N. = 8) : (C.N. = 6) as 1.00 : 0.97 : 0.96. The tetrahedral sites in the close-packed crystal structures (Sec. 4.3-2) can accommodate atoms of radii no larger than 0.23 times that of the metal atom. There are $2N$ tetrahedral sites for every N metal atoms. The octahedral sites can accommodate atoms of radii no larger than 0.41 times that of the metal atom. For every N metal atoms, there are N octahedral sites. The importance of these sites is emphasized in a subsequent section on alloys.

The exact structural orders characteristic of these crystal lattices commonly extend over regions of ca. 10^{-5} cm, that is, over the distances represented by some hundreds of unit cells. A normal single crystal of a metal is thus a mosaic in which unit cell combinations are slightly displaced. The size and distribution of mosaic units affect the physical properties, accounting for observations that the

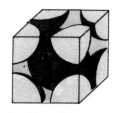

Body-centered cubic (e.g., Li, Na, Ba)

Figure 4-14 Body-centered cubic array of spheres of same radius [T. Moeller, J. C. Bailar, J. Kleinberg, C. O. Guss, M. E. Castellion, and C. Metz, *Chemistry with Inorganic Qualitative Analysis*, Fig. 12.18, p. 339, Academic Press, New York (1980).]

physical properties of a metal depend markedly upon the previous history of the sample and the impurities that it contains.

4-2. Bonding in Metals

The nature of the metallic bond is dictated both by those properties that indicate the presence of relatively mobile electrons and the fact that in a crystal the number of nearest neighbors for each atom is larger than the number of possible bonding electrons. Thus complete electron-pair bonding is impossible, and an extreme case of delocalization of electronic charge density must be explained. Any theoretical approach to the metallic bond must account for at least the following[73]:

1. Ability to act simultaneously between both identical and widely different atoms.
2. Lack of direction, as shown by retention of properties in the liquid state, and lack of saturation to permit large numbers of close neighbors.
3. Forces of attraction that vary inversely as some large power of internuclear distance.
4. Equilibrium forces of repulsion that are atomic in nature.
5. Ability to permit essentially unlimited electron transfer from atom to atom.

Strengths of bonds in the various metals are indicated by the large enthalpies of atomization, or cohesive energies (ΔH^0_{atm}, Table 4-13). Although widely variable, these cohesive energies are substantially larger than those that characterize molecular crystals. They reflect clearly the effects of large numbers of nearest neighbor (C.N. = 12 for ccp and hcp structures) interactions. Periodicity and the changes within particular families are evident in Fig. 4-15.

The modern electronic theory of the metallic bond may be traced to P. Drude,[74] who postulated that in metals free electrons move in spaces among atoms like the molecules of an ideal kinetic-molecular gas. Subsequent expansion of this concept led H. A. Lorentz to the conclusion that in the solid state a metal consists of a lattice of rigid cationic spheres, with free electrons moving in the interstices.[75] These electrons were considered to be the ordinary valence electrons. With the development of quantum-mechanical treatments and enunciation of the Pauli exclusion principle, this view was modified to the extent that although electrons in a metal may be considered to be "free," they are the property of all the atoms in the aggregated structure and occupy discrete energy states, each of which can hold no more than two electrons.[76, 77] Thus these electrons are restricted to limited energy *bands* (Brillouin zones) within the bulk of the material, with transfer between two allowed bands requiring the expenditure of only minimal energy. These concepts are inherent in the modern free-electron, energy-band model of a metal.[14, 74-82]

If a metal is envisioned as a lattice of cationic spheres permeated by freely

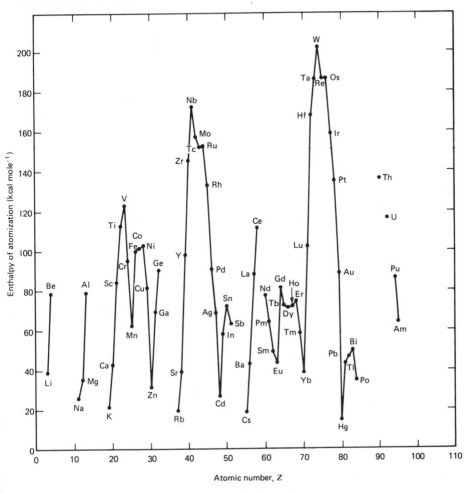

Figure 4-15 Enthalpies of atomization of the elements.

moving electrons, the result is an energy well that is essentially flat throughout the metal itself but rises sharply at its boundaries, that is, the metal is a continuum of energy levels (or bands), each of which can contain two oppositely moving electrons. Conduction then results when applied energy causes these electrons to move in the same direction. However, this concept predicts that even for atoms of the largest ionization energy, electrical conductivity should decrease on heating by about 10^2. Yet in insulators, this decrease is by 10^{24} or more. It is concluded, therefore, that the lattice contains energy gaps, or forbidden regions, that separate the energy bands rather than amounting to a continuum.

In a solid metal, the band width for inner-shell electrons is small, and the energies of these electrons are comparable with those for electrons in an isolated gaseous atom. However, the overlap of higher-energy atomic orbitals at the

Table 4-13 Some Physical Constants of the Metals

Z	Symbol	Density [g cm^{-3}(°C)]	Melting Point (°K)	Normal Boiling Point (°K)	ΔH^0_{atm} at 298.1°K^a (kcal mole^{-1})	Crystal Structure^b Type	a (Å)	c (Å)
3	Li	0.53	452	1590	38.4 ± 0.4	bcc (α)	3.509	
11	Na	0.9674(25°)	370.93	1156	25.8 ± 0.1	bcc	4.2906	
19	K	0.856(18°)	336.5	1030	21.3 ± 0.2	bcc	5.344	
37	Rb	1.532(20°)	312	961	19.5 ± 1	bcc	5.703	
55	Cs	1.90(20°)	301.6	978	18.7	bcc	6.141	
4	Be	1.85	1556	3243	78.25 ± 0.5	hcp	2.2857	3.5839
12	Mg	1.74(20°)	923	1383	35.1 ± 0.4	hcp	3.2093	5.2107
20	Ca	1.54(20°)	1124	1755	42.81 ± 0.02	ccp	5.588	
38	Sr	2.6(20°)	1043	1653	39.1	ccp	6.086	
56	Ba	3.6(20°)	1002	1910	42.5	bcc	5.023	
88	Ra	5.50	973	1413		bcc	5.148	
21	Sc	2.99	1812	3000	88 ± 4	hcp	3.3088	5.2675
39	Y	4.472(25°)	1782	3473	98 ± 2	hcp (α)	3.6471	5.7318
57	La	6.166(25°)	1193	3727	88	hcp' (α)	3.760	12.143
71	Lu	9.84	1948	3588	102.2	hcp	3.5044	5.5504
89	Ac		ca. 1323	ca. 3573		ccp	5.311	
103	Lr							
22	Ti	4.507(20°)	1941	3533	112.5	hcp (α)	2.9503	4.6836
40	Zr	6.506(20°)	2123	4650	145.4 ± 0.4	hcp (α)	3.2317	5.1476
72	Hf	13.29	2503	5473	168	hcp (α)	3.1940	5.0511

23	V	6.11	2163	3273	123	bcc	3.0238	
41	Nb	8.57	2741	5200	172.53(288°K)	bcc	3.3007	
73	Ta	16.654	3269	5700	186.8 ± 1	bcc	3.3031	
24	Cr	7.19(20°)	2148	2472	95 ± 1	bcc	2.8840	
42	Mo	10.22(20°)	2883	5833	157.5	bcc	3.1470	
74	W	19.3(20°)	3683	6203	201.8 ± 2	bcc	3.1651	
25	Mn	7.44(20°)	1517	2370	61.7	bcc (α)	8.9129	
43	Tc	11.49	2443	5303	152	hcp	2.738	4.394
75	Re	21.04	3443	5903	187 ± 2	hcp	2.7608	4.4580
26	Fe	7.8733(20°)	1809.6	3273	99.5	bcc (α)	2.86638	
44	Ru	12.45(20°)	2583	ca. 4392	153 ± 2	hcp	2.7053	4.280
76	Os	22.61(20°)	3323	ca. 4500	187 ± 2	hcp	2.7348	4.3193
27	Co	8.90	1766	3373	101.6	hcp (β)	2.5070	4.0698
45	Rh	12.41(20°)	2233	ca. 4000	138 ± 2	ccp	3.8032	
77	Ir	22.65(20°)	2716	4662	159 ± 2	ccp	3.8900	
28	Ni	8.908(20°)	1726	ca. 3003	102.8	ccp	3.5241	
46	Pd	12.02(20°)	1825	ca. 3020	91 ± 1	ccp	3.8895	
78	Pt	21.45(20°)	2042.5	ca. 4100	135.2	ccp	3.9233	
29	Cu	8.94(20°)	1356	2855	81.1	ccp	3.6148	
47	Ag	10.5(20°)	1233.9	2483	68.4	ccp	4.0857	
79	Au	19.32(20°)	1336	3081	88.3 ± 0.9	ccp	4.0782	
30	Zn	7.133(25°)	692.6	1180	31.2 ± 0.5	hcp	2.6644	4.9454
48	Cd	6.83(0°)	594	1040	26.75 ± 0.2	hcp	2.9788	5.6164
80	Hg	13.546(20°)	234.23	630	14.65 ± 0.02	Rhombohedral (α)	3.0057	

Table 4-13 (*Continued*)

Z	Symbol	Density [g cm^{-3}(°C)]	Melting Point (°K)	Normal Boiling Point (°K)	ΔH^0_{atm} at 298.1°K[a] (kcal mole^{-1})	Crystal Structure[b]		
						Type	a (Å)	c (Å)
13	Al	2.6989(20°)	933.3	2740	78.0 ± 0.4	ccp	4.04953	
31	Ga	5.904(29.6°)	302.88	2676	69.0	Orthorhombic	4.5192	7.6586[c]
49	In	7.31(20°)	429.7	2348	58 ± 2	Tetragonal	3.2520	4.9470
81	Tl	11.85(20°)	576	1730	43.0 ± 1	hcp	3.4563	5.5263
32	Ge	5.323(25°)	1210.5	3103	90 ± 2	Cubic (α)	5.65739	
50	Sn	7.29	505	2960	72.0 ± 2	Tetragonal (β)	5.8316	3.1815
82	Pb	11.34(20°)	600	2024	46.75 ± 0.13	ccp	4.9502	
51	Sb	6.697(26°)	903.6	1908	63 ± 2	Rhombohedral (α)	4.5065	
83	Bi	9.8(20°)	548.4	1833	49.5 ± 1	Rhombohedral (α)	4.7460	
84	Po	9.196	527	1235	34.5	Cubic (α)	3.352	
58	Ce	6.773(25°)	1071	3530	111.6	ccp (γ)	5.1608	
59	Pr	6.475(25°)	1208	3290		hcp (α)	3.6717	11.833
60	Nd	7.003(25°)	1289	3400	77.3	hcp (α)	3.6570	11.7995
61	Pm	7.22(25°)	1441	ca. 2733	ca. 64	hcp	3.65	11.65
62	Sm	7.536(25°)	1345	2173	49.3	Rhombohedral (α)	8.988	
63	Eu	5.245(25°)	1099	1712	43.1	bcc	4.5821	
64	Gd	7.886(25°)	1585	3273	81.22	hcp (α)	3.6333	5.7794
65	Tb	8.253(25°)	1629	2753	72	hcp (α)	3.6041	5.6961
66	Dy	8.559(25°)	1680	2608	71.2	hcp	3.5918	5.6518
67	Ho	8.78(25°)	1743	2993	71.7	hcp	3.5769	5.6169
68	Er	9.045(25°)	1795	2783	74.5	hcp	3.5592	5.5885

69	Tm	9.318(25°)	1818	1998	58.3	hcp	3.5376	5.5543
70	Yb	6.972(25°)	1089	1466	38.2	ccp (α)	5.4843	
90	Th	11.72	2023	ca. 4073	136.6	ccp	5.0851	
91	Pa	15.37	<1873			Tetragonal	3.945	3.240
92	U	18.97(25°)	1405	4091	115 ± 3	Orthorhombic (α)	2.8538	4.9557[d]
93	Np	20.25(20°)	913	ca. 4175		Orthorhombic (α)	4.723	6.663[e]
94	Pu	17.70	914	3600	86 (?)	Monoclinic (α)	6.183	10.963[f]
95	Am	13.67(20°)	ca. 1273		64 (?)	hcp (α)	3.4681	11.241
96	Cm	13.51	1613 (±40)					
97	Bk					hcp	3.496	11.331
98	Cf					hcp	3.416	11.069
99	Es							
100	Fm							
101	Md							
102	No							

[a] Data largely from H. A. Skinner and G. Pilcher, Q. Rev. (Lond.), **17**, 264 (1963).

[b] Modification stable at room temperature. Data largely from J. Donohue, *The Structures of the Elements*, John Wiley & Sons, New York (1974). Values commonly averages.

[c] $b = 4.5258$ Å.

[d] $b = 4.8680$ Å.

[e] $b = 4.887$ Å.

[f] $b = 4.822$ Å.

characteristic internuclear distance is so extensive that wider bands which result can encompass a range of orbital energies and within which electrons are free to move. Such bands are *conduction bands*. By contrast, in a solid nonmetal, overlap of higher-energy orbitals at the characteristic internuclear distance is minimal as a consequence of the localization of electronic charge density in atom–atom bonds, and no conduction band can result except possibly at extremely high energies.

The difference is indicated graphically by comparison of Figs. 4-16 and 4-17, which represent, respectively, the orbital energies in solid sodium and solid carbon (as diamond), both as functions of internuclear distance r. Of the eleven electrons in the sodium atom, 10 are in the $1s$, $2s$, and $2p$ atomic orbitals, which are separated from each other at the equilibrium internuclear distance r_0. However, overlap of higher-energy orbitals at this distance allows the $3s$ electron to move freely in the resulting conduction band. By contrast, the four outer-shell electrons of the carbon atom are, at the equilibrium internuclear distance, in overlapping $2s$ and $2p$ bands and cannot be promoted to the higher-

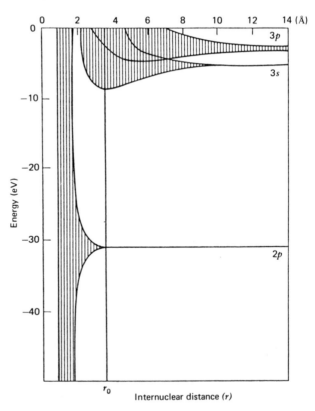

Figure 4-16 Orbital energies in crystalline sodium. [C. S. G. Phillips and R. J. P. Williams, *Inorganic Chemistry*, Vol. 1, Fig. 6.15, p. 203, Oxford University Press, London (1965).]

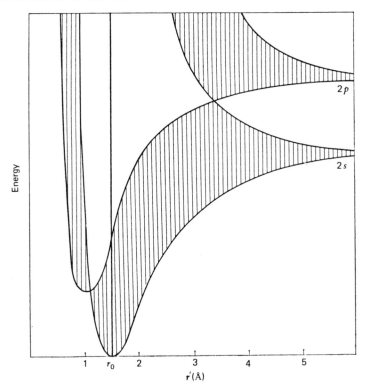

Figure 4-17   Orbital energies in diamond. [C. S. G. Phillips and R. J. P. Williams, *Inorganic Chemistry*, Vol. 1, Fig. 6.22, p. 214, Oxford University Press, London (1965).]

energy orbitals required for conduction unless considerable extra energy is supplied. The fundamental difference between a metal and a nonmetal is, therefore, in the former the conduction band merges into the valence band, whereas in the latter it is separated by an energy gap (Fig. 4-18 and Table 4-14).

A distinction can be made between an *insulator* and a *semiconductor*[83, 84] in terms of the size of the energy gap. If the energy gap for a pure substance is small enough ($=$ca. kT, where k is the Boltzmann constant), heating can promote an electron into the conduction band, giving *intrinsic* semiconduction (e.g., with Si, Ge). Doping an element such as Si or Ge appropriately decreases the energy gap and allows conduction at lower temperatures. If an atom of the dopant (e.g., As) can fit into the crystal lattice of silicon or germanium and contains one *more* electron in its valence shell than an atom of the host, the impurity level is closer to the conduction band, and promotion of an electron to this band is effected at a lower temperature. Here conduction is by negative electrons (*n*-type). If an atom of the dopant (e.g., Ga in Ge or Si) can fit into the crystal lattice of the host and has one or more *fewer* electrons in its valence shell than an atom of the host, the impurity level will lie just above the occupied level, where it can accept

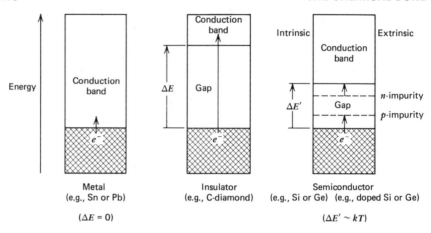

Figure 4-18 Figurative comparison of energy bands in solid insulators, semi-conductors, and metals. (Crosshatched areas represent occupied orbitals in the atoms in question.)

electrons from the latter and leave a corresponding number of positive "holes". Conduction in the host is then by these positive holes (p-type). Semiconduction resulting from impurities is of the *extrinsic* type.[70]

This concept of metallic bonding resembles the molecular-orbital treatment of the covalent bond (Sec. 5.1-1) in associating the bonding electrons with all atoms in a metallic crystal.

An alternative approach involving an extension of the valence-bond approximation to covalent bonding (Sec. 5.1-1) has been developed by L. Pauling.[85] By proposing the existence of normal electron-pair covalency with resonance involving all the possible equivalent pairs of atoms, this approach accounts for the fact that the number of nearest neighbors of an atom in a metallic crystal exceeds the number of valence electrons. Thus for elemental lithium (bcc crystal, Table 4-13), for which each atom has a single valence electron, resonance involves a given Li atom and each of its eight nearest-neighbor Li atoms. Inasmuch as only a single atomic orbital (2s) is available per atom, no Li atom

Table 4-14   Energy Gaps for Some Elements

Symbol	Energy Gap, ΔE (eV)
C(diamond)	5.6 (insulator)
Si	1.1 (semiconductor)
Ge	0.7 (semiconductor)
Sn(α, gray)	0.08 (metal)
Sn(β, white)	0 (metal)
Pb	0 (metal)
Metals in general	0

can form more than one bond at the same time, and synchronous resonance, as postulated for the benzene molecule (Sec. 5.1-1), is required. The inclusion of ionic structures such as

$$Li\!-\!Li^-$$

$$Li^+\!-\!Li$$

would increase resonance stabilization but would require the use of an additional, *metallic* orbital by the atom that accepts the electron. The availability of such an orbital is a necessary condition for metallic bonding. For lithium, a $2s\,2p$ hybrid would provide such an orbital.

In this approach, hybridization provides a link between metallic and covalent bonding, particularly for the d-transition elements.[86] For an atom of one of these elements, Pauling proposes that *nine* orbitals, namely $(n-1)d^5ns^1np^3$, participate in metallic bonding, but in *three* separate sets, namely (1) hybridized $d^x\mathrm{sp}^3$ bonding orbitals, (2) pure md orbitals with no bonding properties, and (3) $5-(m+x)$ metallic orbitals for resonance. Both m and x may be fractional, as indicated by Pauling's assignment of 5.78 bonding orbitals, 2.44 atomic orbitals, and 0.78 metallic orbital to elemental iron.

This general concept is in agreement with observed internuclear distances for many metals and with close-packing involving large coordination numbers. Ductility and malleability are expected because the number and direction of bonds are not restricted, and the bonds remaining after deformation of a sample are as strong as the initial bonds. It is advantageous to reconsider this approach after covering the valence-bond approximation to covalency (Sec. 5.1-1).

4-3. Other Crystalline Solids with Metallic Properties

Metallic properties are characteristic also of alloys, including intermediate phases; solid metal ammonates, such as $Ca(NH_3)_6$ (Sec. 10.1-3); amalgams involving free radicals, such as NH_4 or $(CH_3)_4N$; species containing metal–metal bonds, such as $(C_2H_5Hg)_n$, PbS, FeS_2, Na_xWO_3, $[Ta_6Cl_{12}]Cl_2$; and polyplumbides, polystannides (Sec. 10.1-4), and so on. Only alloys and intermediate phases are considered in the present discussion. References to other species appear in later chapters.

Alloys. The products of blending together two or more metals may be physical mixtures, solid solutions, or intermetallic compounds. Mixtures result when melts that contain two or more metals that have no solubility in the solid state are cooled. All eutectic compositions are mixtures. Solid solutions and intermetallic compounds, unlike mixtures, are homogeneous.

Solid solutions result when atoms of one kind fit into the crystal lattice formed by atoms of another kind. Solid solutions involving metals may be either *substitutional* or *interstitial* in nature.

1. **Substitutional solid solution.** Atoms of one metal randomly occupy positions in the crystal lattice of another metal (i.e., metal atoms of one element substitute for those of the other). Complete solid solubility is observed only when the radii of the two kinds of atoms differ from each other by less than 15%, the crystal structures of the two metals are identical, and the chemical properties of the two elements are closely similar. Thus nickel (ccp, $a = 3.5168$ Å, $r = 1.25$ Å) and copper (ccp, $a = 3.615$ Å, $r = 1.28$ Å) form a complete series of solid solutions, or are completely soluble in each other in the solid state. On the other hand, sodium (bcc, $a = 4.28$ Å, $r = 1.86$ Å) and potassium (bcc, $a = 5.24$ Å, $r = 2.27$ Å) from no solid solutions, and zinc (hcp, $a = 2.664$ Å, $r = 1.33$ Å) and copper form limited series of solid solutions (α, up to 35% Zn in Cu; and η, up to 3% Cu in Zn—both atomic percentages; see Fig. 4-19). Occasionally, ordered distribution resulting in one type of atom having only the other type of atom as nearest neighbors (i.e., a *superlattice*) results.

2. **Interstitial solid solution.*** Small nonmetal atoms (namely H, B, C, N) occupy interstitial positions in the crystal lattices of d-transition metals.[87, 88] The interstitial positions may be either the octahedral sites or the tetrahedral sites already described. On a simple geometrical basis, nonmetal atoms could fit into octahedral holes without lattice distortion if the ratio of the radius of the nonmetal atom (r_A) to the radius of the metal atom (r_M) does not exceed 0.41. On this basis the limiting radii of metal atoms for octahedral interstitial solid solutions are: H($r_H = 0.37$ Å), 0.90 Å; B($r_B = 0.80$ Å), 1.95 Å; C($r_C = 0.77$ Å), 1.88 Å; and N($r_N = 0.74$ Å), 1.80 Å. The most common interstitial solid solutions are formed by the group IVB, VB, and VIB (except Cr) elements, the metallic radii of which (Table 2-15) are in the range 1.31 to 1.59 Å—values too small to accommodate any atoms except those of hydrogen without major distortions of crystal lattices. Similarly, nonmetal atoms might be expected to

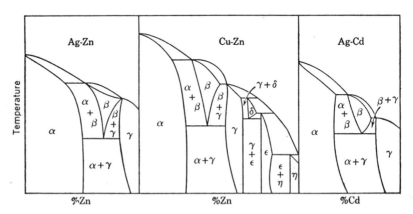

Figure 4-19 Phase diagrams of some typical binary alloy systems. [T. Moeller, *Inorganic Chemistry*, Fig. 6-8, p. 222, John Wiley & Sons, New York (1952).]

---

*   Sometimes referred to as an interstitial compound.

occupy tetrahedral holes without distortion only if r_A/r_M does not exceed 0.23, and corresponding limiting radii for metal atoms would range from 1.60 Å (for H) to 3.48 Å (for B). Thus except possibly for hydrogen atoms, tetrahedral holes cannot be significant in accounting for trapping nonmetal atoms.

As is indicated in Table 4-15, the values of r_A/r_M for known compounds are well below the limiting values based upon geometry alone, and the alterations in r_M that result are not major. Indeed, based upon structural studies the critical value of r_M for interstitial carbide formation appears to be closer to 1.3 Å than to 1.88 Å. Since the nonmetals are accommodated as atoms, substantial quantities of energy must have been expended to break bonds in the original polyatomic nonmetal species. On this basis, it is apparent that the simple picture of physical trapping is complicated by metal-to-nonmetal bonding.

Complete saturation of octahedral holes by boron, carbon, or nitrogen atoms gives species of 1 : 1 molar stoichiometry with the ideal NaCl-type lattice (fcc, or ccp), irrespective of the metal present (Table 4-15). These substances are characterized by primarily metallic properties, but they are commonly much harder and higher melting than the parent metals. Many other essentially metallic carbide and nitride phases, within which incomplete occupancy of octahedral sites occurs, can be distinguished also (e.g., $Cr_{23}C_6$-ccp; Mo_2C-orthorhombic; Fe_3C-hcp; Mn_4N, Fe_4N). Most of the other borides,[89] and many carbides[90] and nitrides,[91] have smaller metal to nonmetal molar ratios and many of the properties of valency compounds rather than of metals.

The metal-like hydrides include derivatives of nearly all of the metals in the periodic families IVB through IIB, but they are less well defined than the other interstitial substances.[92] Compositions are variable; it is not always possible to distinguish between absorption and adsorption of hydrogen in their formation; the uptake and release of hydrogen by a metal is reversible, depending on the temperature and pressure; the enthalpies of formation are not uncommonly comparable with those of ionic hydrides (e.g., Na^+H^-, $Ba^{2+}H_2^-$); and the crystal structure of the metal may change significantly on reaction with hydrogen. Metallic appearance and conductivity, small changes in atomic volume, and commonness of the ccp crystal structure are in accord with the interstitial trapping of hydrogen atoms, presumably in the tetrahedral holes, but there are evidences for the presence of ionized hydrogen, both H^+ and H^-, in some compounds.

Intermediate phases, intermetallic compounds, or *electron compounds*—all terms are used—are metal–metal combinations of variable, but reproducible, stoichiometry that are stable over limited compositional ranges and may be structurally unrelated to the parent metals. That they are not valency compounds in the usual sense is indicated by compositions such as Cu_3Al, $CoZn_3$, $Na_{31}Pb_8$, Ag_5Al_3, and Na_5Zn_{21}. Lack of complete utilization of valence electrons in bonding is indicated by their metallic properties. Such species seldom form between members of the same periodic group and never between members of the same periodic subgroup. A given metal either forms intermediate phases with *all* members of a given periodic subgroup or with *none*.

Table 4-15 Comparison of Selected Properties of Metals with Those of 1:1 Interstitial Solid Solutions

| Metal | | | | | | Interstitial Solid Solution | | | | | | | |
Symbol	Density (g cm⁻³)	Melting Point (°K)	r_M (Å)	Lattice Type	a (Å)	Formula	Density (g cm⁻³)	Melting Point (°K)	r_M (Å)	r_A/r_M	Lattice Type	a (Å)	Hardness (Mohs[a])
Ti	4.507	1941	1.45	hcp	2.9504	TiB		2900	1.48	0.55	ccp	4.20	
						TiC		3410	1.53	0.53	ccp	4.32	8⁺
						TiN	5.18	3220	1.50	0.51	ccp	4.24	8–9
Zr	6.506	2123	1.59	hcp	3.2321	ZrB			1.64	0.50	ccp		
						ZrC		3810	1.66	0.48	ccp	4.69	8–9
						ZrN	6.93	3520	1.61	0.47	ccp	4.56	8⁺
Hf	13.29	2503	1.59	hcp	3.2321	HfB			1.63	0.50	ccp	4.62	
						HfC		4160	1.64	0.48	ccp	4.65	
V	6.11	2163	1.31	bcc	3.026	VC		3080	1.47	0.59	ccp	4.18	9–10
						VN	5.63	2570	1.48	0.57	ccp	4.17	9–10
Nb	8.57	2741	1.43	bcc	3.294	NbC		4100	1.57	0.54	ccp	4.44	9⁺
						NbN	8.40	2570		0.52			8⁺
Ta	16.654	3269	1.43	bcc	3.2959	TaB				0.56			
						TaC		4150	1.57	0.54	ccp	4.46	9⁺
						TaN		3360		0.52			8⁺
W	19.3	3683	1.37	bcc	3.1585	WC		3140		0.56			9⁺

[a] Hardness scale of Mohs: 1 (talc), 2 (rock salt or gypsum), 3 (calcite), 4 (fluorite), 5 (apatite), 6 (feldspar), 7 (quartz), 8 (topaz), 9 (ruby or sapphire), 10 (diamond). Metals: 0.2(Cs)–9(Cr); mostly under 5.

In many binary systems, the addition of a metal (M′) of larger inherent oxidation number to another metal (M) of smaller inherent oxidation number results in the same general succession of alloy phases, which can be indicated roughly as

α	α + β	β bcc	β + γ	γ c*	γ + ε	ε hcp	ε + η	η

M   M′

composition as mole percent

where α and η are simple substitutional solid solutions and β, γ, and ε are intermediate phases. The phase diagram for the system Cu–Zn (Fig. 4-19) is typical.

The data in Table 4-16 indicate that intermediate phases of the same crystal structure may vary widely in composition. In an attempt to systematize formulations of β-phase alloys, W. Hume-Rothery[93] pointed out that for this phase the *ratio of total valence electrons to total atoms* is $\frac{3}{2}$ (or 21/14), provided that the valence electrons are considered to be only those in the highest principal quantum level and atoms of the group VIII elements are assumed to have no valence electrons. This principle was extended to the γ phase,[94] where the corresponding ratio is 21/13, and to the ε phase,[95] where the ratio is 7/4 (or 21/12). These ratios are the *Hume–Rothery* ratios. The typical examples given in Table 4-16 indicate that many, but not all, β, γ, and ε phases adhere to these rules.

The Hume–Rothery rules are sufficiently broadly applicable that they must have more than empirical significance. It has been suggested[60,96] that when electrons are fed into a metallic lattice, the last electrons to enter a given Brillouin zone many overflow into the next allowed zone, or the lattice may change to accommodate more electronic states within the given zone (i.e., a new phase may result). The theoretical β-phase and γ-phase boundaries are at 1.480 and 1.538 electrons per atom, respectively. These values compare well with the Hume–Rothery ratios of 1.500 and 1.615, in particular since each phase can cover a range of compositions (e.g., 1.63 to 1.77 for the γ phase in the Cu–Zn system, or 1.59 to 1.63 for the γ phase in the Ag–Cd system). Furthermore, phase boundaries do not change regularly with temperature (Fig. 4-19), and occupation of a few more energy states than those representing a filled zone may be essential to effecting a change in crystal structure.

Reference must be made also to *Laves*[97] and *Zintl*[98] *phases*. Laves phases are characterized by constant radius ratio, irrespective of composition, and thus are dependent in composition solely upon metallic radii. A 20% smaller radius is a requirement for one metal. Examples for radius ratio 1.25 include Cu_2Mg, KBi_2, $MgNi_2$, and Au_2Bi. Zintl phases are combinations of the highly electropositive metals of periodic groups IA or IIA with the less electropositive elements of periodic groups IB, IIB, or IIIA to VA that are characterized by mixed metallic

---

* This structure is a complex cubic one with 52 atoms per unit cell.

Table 4-16 Some Typical Binary Intermediate Phases

β Structures		γ Structures		ε Structures	
Formula	Valence Electrons / Atoms	Formula	Valence Electrons / Atoms	Formula	Valence Electrons / Atoms
CuBe	$\frac{3}{2}$	Cu_5Zn_8	$\frac{21}{13}$	$CuBe_3$	$\frac{7}{4}$
CuZn	$\frac{3}{2}$	Cu_9Al_4	$\frac{21}{13}$	$CuZn_3$	$\frac{7}{4}$
Cu_3Al	$\frac{3}{2}$	Cu_9Ga_4	$\frac{21}{13}$	Cu_3Sn	$\frac{7}{4}$
Cu_5Sn	$\frac{3}{2}$	$Cu_{31}Sn_8$	$\frac{21}{13}$	Cu_3Ge	$\frac{7}{4}$
Cu_5Si	$\frac{3}{2}$	Ag_5Zn_8	$\frac{21}{13}$	$AgZn_3$	$\frac{7}{4}$
AgZn	$\frac{3}{2}$	Fe_5Zn_{21}	$\frac{21}{13}$	$AgCd_3$	$\frac{7}{4}$
Ag_3Al	$\frac{3}{2}$	Co_5Zn_{21}	$\frac{21}{13}$	Ag_5Al_3	$\frac{7}{4}$
Au_3Al	$\frac{3}{2}$				
FeAl	$\frac{3}{2}$	Ni_5Cd_{21}	$\frac{21}{13}$	Au_3Sn	$\frac{7}{4}$
$CoZn_3$	$\frac{3}{2}$	$Na_{31}Pb_8$	$\frac{21}{13}$	$FeZn_7$	$\frac{7}{4}$
CoAl	$\frac{3}{2}$				
LiAg	$\frac{2}{2}$	$Li_{10}Pb_3$	$\frac{22}{13}$		
LiTl	$\frac{4}{2}$	$Li_{10}Ag_3$	$\frac{13}{13}$		
MgTl	$\frac{5}{2}$				
AlNd	$\frac{6}{2}$				
TlSb	$\frac{8}{2}$				

and ionic lattices and compositions that are often in accord with usual valencies. Typical examples include Mg_2Sn, Sr_2P, NaTl, and LiCd. Lattice dimensions are dependent upon the elements present. Zintl phases of type MA have simple cubic lattices with alternating M and A atoms but with A atoms bonded into networks that incorporate the M species in interstices. Thus in crystals of the substance NaTl, the Tl atoms form a diamond-type lattice (Sec. 5.1-4) involving four bonds per Tl atom. The extra electron needed for each bond must come from a Na atom, giving at least some Na^+ and Tl^- ions. The lattice of the phase $CaZn_2$, on the other hand, consists of sheets of Zn-atom hexagons, with Ca species inserted between sheets. Bonding involving sp^2 hybrids formed by three-electron Zn^- ions may result when Ca^{2+} ions are formed.

4-4. Noncrystalline Metals

As noted earlier, metallic properties such as luster and conductivity are characteristic of liquid metals and liquid alloys. Much shorter-range structural ordering is of course more evident in the liquid state than in the solid, but it is likely that the same general principles of bonding apply. Reference to a detailed monograph[99] is recommended.

The glassy state can be achieved in metals by extremely rapid cooling of the liquids, for example, by impinging tiny molten globules at high velocities on cold surfaces or by injecting molten metal between counter-rotating cooled drums.[100] The latter process produces threadlike metallic ribbons characterized by unusual combinations of high strength and plasticity, corrosion resistance, and novel magnetic and acoustic properties.

EXERCISES

4-1. Distinguish clearly between *bond energy* and *bond dissociation energy*, giving typical numerical values and units for the energy terms in each instance. Give five examples of bond energy and five examples of bond dissociation energy where the two energies are not identical.

4-2. Given the complex salt $[Co(NH_3)_6]^{3+}[Cr(CN)_6]^{3-}$, indicate clearly where and how Werner's "two types of valency" are exemplified.

4-3. Using three-dimensional models or suitable drawings, determine the number of nearest neighbors for each cation in (a) a face-centered cubic crystal, and (b) a body-centered cubic crystal.

4-4. Suggest, as a result of calculations using data for crystal radii from Table 4-4, the probable type of structure for each of the following ionic crystals: Li^+F^-, Na^+Br^-, K^+Cl^-, Cs^+I^-, $Mg^{2+}O^{2-}$, Tl^+Cl^-.

4-5. Using appropriate data from Table 4-5, determine whether isomorphism is probable or improbable in each of the following pairs of ionic crystals:

ScF$_3$ and LaF$_3$, ScF$_3$ and LuF$_3$, YCl$_3$ and YbCl$_3$, LaF$_3$ and LaI$_3$, PmCl$_3$ and PmBr$_3$. Give reasons for your answers.

4-6. Based upon appropriate data from Table 4-6, which pair(s) of solid salts would you expect to be isomorphous: KMnO$_4$–KReO$_4$, KMnO$_4$–KTcO$_4$, KReO$_4$–KTcO$_4$? Explain.

4-7. Using data from Tables 4-4 and 4-5, predict which of the following ions would be most likely to carry, and thus concentrate, the Cm^{3+} ion in a crystallization process: Ba^{2+}, Ce^{3+}, Sm^{3+}, Er^{3+}, or Y^{3+}. Explain.

4-8. In your own words, *define* or *clearly identify* each of the following: Hume–Rothery ratio; Born–Fajans–Haber thermal cycle; semiconduction; Berthollide compound.

4-9. Calculate the limiting radius of a spherical atom that can be accommodated in (a) a tetrahedral hole and (b) an octahedral hole in a close-packed, three-dimensional array of spheres of radius r.

4-10. By application of the Born–Fajans–Haber technique to whatever data are necessary, check the U_0 values given in Table 4-10 for the following ionic solids: KH, AgF, MgBr$_2$, Ag$_2$O. Account for any significant differences.

4-11. Refer to Figures 4-15 and 3-14 and compare periodicity in enthalpies of atomization and melting points of the metals. Account for similarities and differences. Account also for the minima at europium and ytterbium among the lanthanides.

4-12. Consult suitable references and summarize in detail the chemistry involved in preparing ultrapure silicon from quartz.

4-13. Using data from whatever sources you choose, summarize differences in physical properties between α-Sn and β-Sn (Table 4-14). Relate these, where possible, to differences in crystal structures.

4-14. Consult more recent literature sources than Ref. 100 and summarize briefly material on the preparation and properties of noncrystalline metals.

REFERENCES

1. L. Pauling, *The Chemical Bond*, p. 5, Cornell University Press, Ithaca, N.Y. (1967).

2. O. K. Rice, *Electronic Structure and Chemical Binding*, Ch. 11, McGraw-Hill Book Company, New York (1940).

3. W. G. Palmer, *Valency: Classical and Modern*, 2nd ed., Cambridge University Press, Cambridge (1959).

4. E. Cartmell and G. W. A. Fowles, *Valency and Molecular Structure*, 4th ed., Butterworth, Boston (1977).

5. C. A. Coulson, *Valence*, 2nd ed., Oxford University Press, Oxford (1961); R. McWeeny, *Coulson's Valence*, 3rd ed., Oxford University Press, London (1979).

6. A. J. Ihde, *The Development of Modern Chemistry*, Chs. 4, 5, 20, Harper & Row, New York (1964).

7. R. Abegg, *Z. Anorg. Chem.*, **39**, 330 (1904).

8. P. Drude, *Ann. Phys. (Leipzig)*, [4], **14**, 677 (1904).

9. W. Kossel, *Ann. Phys. (Leipzig)*, [4], **49**, 229 (1916).

10. G. N. Lewis, *J. Am. Chem. Soc.*, **38**, 762 (1916).

11. I. Langmuir, *J. Am. Chem. Soc.*, **41**, 868 (1919).

12. G. A. Perkins, *Philippine J. Sci.*, **19**, 121 (1921).

13. K. J. Mysels, *J. Chem. Educ.*, **34**, 10 (1957).

14. A. F. Wells, *The Third Dimension in Chemistry*, Clarendon Press, Oxford (1956); *Structural Inorganic Chemistry*, 4th ed., Clarendon Press, Oxford (1975).

15. R. W. G. Wyckoff, *Crystal Structures*, 2nd ed., Vols. 1–5, Interscience Publishers, New York (1963–1966).

16. L. Bragg, *The Crystalline State*, Cornell University Press, Ithaca, N.Y. (1965).

17. G. L. Clark, *Applied X-Rays*, 4th ed., Chs. 13–17, McGraw-Hill Book Company, New York (1955).

18. T. R. P. Gibb, Jr., and A. Winnerman, *J. Chem. Educ.*, **35**, 579 (1958); **36**, 46 (1959).

19. W. L. Bragg, *Philos. Mag.*, [6], **40**, 169 (1920).

20. J. A. Wasastjerna, *Z. Phys. Chem.*, **101**, 193 (1922); *Soc. Sci. Fenn. Commentat. Phys.-Math.*, **1**(38) (1923).

21. V. M. Goldschmidt, *Skr. Norske Vidensk.-Akad. Oslo*, No. 2 (1926); No. 8 (1927); *Trans. Faraday Soc.*, **25**, 253 (1929).

22. W. H. Zachariasen, *Z. Kristallogr., Kristallgeom., Kristallphys., Kristallchem.*, **80**, 137 (1931).

23. A. Landé, *Z. Phys.*, **1**, 191 (1920).

24. W. L. Bragg and J. West, *Proc. R. Soc. Lond. Ser. A*, **114**, 450 (1927).

25. L. Pauling, *J. Am. Chem. Soc.*, **49**, 765 (1927); *Proc. R. Soc. Lond. Ser. A*, **114**, 181 (1927); *The Nature of the Chemical Bond*, 3rd ed., Ch. 13, Cornell University Press, Ithaca, N.Y. (1960).

26. L. Pauling and J. Sherman, *Z. Kristallogr., Kristallgeom., Kristallphys., Kristallchem.*, **81**, 1 (1932).

27. L. H. Ahrens, *Geochim. Cosmochim. Acta*, **2**, 155 (1952).

28. K. H. Stern and E. S. Amis, *Chem. Rev.*, **59**, 1 (1959).

29. L. E. Sutton (Ed.), *Tables of Interatomic Distances and Configuration in Molecules and Ions*, Spec. Publ. No. 18, The Chemical Society, London (1965).

30. R. D. Shannon and C. T. Prewitt, *Acta Crystallogr.*, **B25**, 925 (1969).

31. K. B. Yatsimirskii, *Izv. Acad. Nauk SSSR, Otd., Khim. Nauk*, 453 (1947); 398 (1948). A. F. Kapustinskii and K. B. Yatsimirskii, *Zh. Obshch. Khim.*, **19**, 2191 (1949).

32. D. H. Templeton and C. H. Dauben, *J. Am. Chem. Soc.*, **76**, 5237 (1954).

33. G. W. Beall, W. O. Milligan, and H. A. Wolcott, *J. Inorg. Nucl. Chem.*, **39**, 65 (1977).

34. S. Ahrland, J. O. Liljenzin, and J. Rydberg, in *Comprehensive Inorganic Chemistry*, A. F. Trotman-Dickenson (Executive Ed.), Vol. 5, p. 466, Pergamon Press, Oxford (1973).

35. W. H. Zachariasen, in *The Actinide Elements*, G. T. Seaborg and J. J. Katz (Eds.), p. 775, McGraw-Hill Book Company, New York (1954).

36. T. Moeller, *The Chemistry of the Lanthanides*, pp. 6–8, Pergamon Press, Oxford (1975).

37. A. F. Wells, *Models in Structural Inorganic Chemistry*, Chs. 3–6, Oxford University Press, London (1970). A lucid and informative treatment.

38. For greater detail, see S.-M. Ho and B. E. Douglas, *J. Chem. Educ.*, **45**, 474 (1968).

39. E. Madelung, *Phys. Z.*, **19**, 524 (1918).

40. M. Born, *Z. Phys.*, **7**, 124 (1921).

41. D. H. Templeton, *J. Chem. Phys.*, **21**, 2097 (1953); **23**, 1826 (1955).

42. T. C. Waddington, in *Advances in Inorganic Chemistry and Radiochemistry*, H. J. Emeléus and A. G. Sharpe (Eds.), Vol. 1, pp. 157–221, Academic Press, New York (1959).

43. M. Born and J. E. Mayer, *Z. Phys.*, **75**, 1 (1932).

44. A. F. Kapustinskii, *Q. Rev. (Lond.)*, **10**, 283 (1956).

45. G. J. Moody and J. D. R. Thomas, *J. Chem. Educ.*, **42**, 204 (1965).

46. M. Born, *Verh. Dtsch. Phys. Ges.*, **21**, 679 (1919).

47. K. Fajans, *Verh. Dtsch. Phys. Ges.*, **21**, 714 (1919).

48. F. Haber, *Verh. Dtsch. Phys. Ges.*, **21**, 750 (1919).

49. D. F. C. Morris and E. L. Short, *Nature*, **224**, 950 (1969). An interesting historical commentary noting that the articles by Born, Fajans, and Haber, all of which appeared in the December 5, 1919, issue of the journal, were submitted, respectively, on October 9, October 27, and November 14 of that year. For some obscure reason, Fajans has seldom been associated with the development of the thermochemical cyclic approach.

50. N. S. Hush and M. N. L. Pryce, *J. Chem. Phys.*, **28**, 244 (1958).

51. D. Holtz, J. L. Beauchamp, and J. R. Eyler, *J. Am. Chem. Soc.*, **92**, 7045 (1970).

52. A. G. Sharpe, *Endeavor*, **27**, 120 (1968).

53. J. L. Holm, *J. Chem. Educ.*, **51**, 460 (1974).

54. G. P. Haight, *J. Chem. Educ.*, **45**, 420 (1968).

55. C. W. Allen and T. Moeller, *Inorg. Chem.*, **7**, 2177 (1968).

56. A. W. Woolf, in *Advances in Inorganic Chemistry and Radiochemistry*, H. J. Emeléus and A. G. Sharpe (Eds.), Vol. 9, pp. 217–314, Academic Press, New York (1966).

57. N. Bartlett and D. H. Lohmann, *J. Chem. Soc.*, **1962**, 5253.

58. M. Barber, J. W. Linnett, and N. H. Taylor, *J. Chem. Soc.*, **1961**, 3323.

59. G. J. Moody and J. D. R. Thomas, *J. Chem. Soc.*, **1964**, 1417.

60. J. M. Honig, *J. Chem. Educ.*, **34**, 224, 343 (1957).

61. W. D. Johnson, *J. Chem. Educ.*, **36**, 605 (1959).

62. L. Brewer, *J. Chem. Educ.*, **38**, 90 (1961).

63. H. W. Etzel, *J. Chem. Educ.*, **38**, 225 (1961).

64. W. J. Moore, *J. Chem. Educ.*, **38**, 233 (1961).

65. N. B. Hannay, *Solid-State Chemistry*, Prentice-Hall, Englewood Cliffs, N.J. (1967).

66. N. N. Greenwood, *Ionic Crystals, Lattice Defects, and Non-Stoichiometry*, Butterworth, London (1968).

67. A. F. Wells, *Structural Inorganic Chemistry*, 4th ed., Clarendon Press, Oxford (1975).

68. A. L. G. Rees, *Chemistry of the Defect Solid State*, John Wiley & Sons, New York (1954).

69. H. G. Van Bueren, *Imperfections in Crystals*, 2nd ed., North-Holland, Amsterdam (1961).

70. P. F. Weller, *J. Chem. Educ.*, **48**, 831 (1971).

71. W. J. Moore and E. L. Williams, *Disc. Faraday Soc.*, **28**, 86 (1959).

72. A. F. Wells, *Structural Inorganic Chemistry*, 4th ed., Ch. 29, Clarendon Press, Oxford (1975).

73. J. D. Bernal, *Trans. Faraday Soc.*, **25**, 367 (1929).

74. P. Drude, *Ann. Phys. (Leipzig)*, [4], **1**, 566 (1900); [4], **3**, 369 (1900).

75. H. A. Lorentz, *The Theory of Electrons*, G. E. Stechert and Co., New York (1923).

76. A. Sommerfeld, W. V. Houston, and C. Eckart, *Z. Phys.*, **47**, 1 (1928).

77. F. Bloch, *Z. Phys.*, **52**, 555 (1928).

78. J. C. Slater, *Rev. Mod. Phys.*, **6**, 209 (1934).

79. N. F. Mott and H. Jones, *Theory of the Properties of Metals and Alloys*, Clarendon Press, Oxford (1936).

80. F. Seitz, *Modern Theory of Solids*, McGraw-Hill Book Company, New York (1940).

81. W. Hume-Rothery and G. V. Raynor, *The Structure of Metals and Alloys*, Institute of Metals, London (1954).

82. W. Hume-Rothery, *Electrons, Atoms, Metals, and Alloys*, Metal Industry, London (1955).

83. E. Mooser, *Science*, **132**, 1285 (1960).

84. E. F. Gurnee, *J. Chem. Educ.*, **46**, 80 (1969).

85. L. Pauling, Ref. 1, Ch. 11.

86. W. Hume-Rothery, H. M. Irving, and R. J. P. Williams, *Proc. R. Soc. Lond. Ser. A* **208**, 431 (1951).

87. K. Becker, *Phys. Z.*, **34**, 185 (1933).

88. R. Kiessling, *Fortschr. Chem. Forsch.*, **3**, 41 (1954).

89. R. Thompson, in *Progress in Boron Chemistry*, R. J. Brotherton and H. Steinberg (Eds.), Vol. 2, pp. 173–230, Pergamon Press, Elmsford, N.Y. (1970).

90. W. A. Frad, in *Advances in Inorganic Chemistry and Radiochemistry*, H. J. Eméleus and A. G. Sharpe (Eds.), Vol. 11, pp. 153–247, Academic Press, New York (1968).

91. R. Juza, in *Advances in Inorganic Chemistry and Radiochemistry*, H. J. Eméleus and A. G. Sharpe (Eds.), Vol. 9, pp. 81–131, Academic Press, New York (1966). Restricted to $3d$ metals.

92. W. M. Mueller, J. P. Blackledge, and G. P. Libowitz (Eds.), *Metal Hydrides*, Academic Press, New York (1968).

93. W. Hume-Rothery, *J. Inst. Met.*, **35**, 295 (1926).

94. A. J. Bradley and J. Thewlis, *Proc. R. Soc. Lond. Ser. A*, **112**, 678 (1926).

95. A. F. Westgren and G. Phragmén, *Metallwirtschaft*, **7**, 700 (1928); *Trans. Faraday Soc.*, **25**, 279 (1929).

96. H. Jones, *Proc. R. Soc. Lond. Ser. A*, **144**, 225 (1934); **147**, 396 (1934).

97. O. W. Liebiger, *Am. Fiat. Rev. Ger. Sci.*, **31**, 33 (1950).

98. E. Zintl. *Angew. Chem.*, **52**, 1 (1939).

99. M. Shimoji, *Liquid Metals*, Academic Press, New York (1978).

100. R. J. M. Cotterill, *Am. Sci.*, **64**, 430 (1976).

SUGGESTED SUPPLEMENTARY REFERENCES

E. C. Baughan, *Structure and Bonding*, Vol. 15, pp. 53–71, Springer-Verlag, New York (1973). Radii of various types.

L. Bragg, *The Crystalline State*, Cornell University Press, Ithaca, N.Y. (1965). Interesting treatment by pioneer in the field.

E. Cartmell and G. W. A. Fowles, *Valency and Molecular Structure*, 4th ed., Butterworth, Boston (1977). General approach. Less detailed than McWeeny.

J. Donohue, *The Structures of the Elements*, John Wiley & Sons, New York (1974). Crystal structural data for elements.

N. N. Greenwood, *Ionic Crystals, Lattice Defects, Non-stoichiometry*, Butterworth, London (1968). Self-explanatory title.

N. B. Hannay, *Solid-State Chemistry*, Prentice-Hall, Englewood Cliffs, N.J. (1967). Classic introduction.

W. Hume-Rothery, *Electrons, Atoms, Metals, Alloys*, Metal Industry, London (1955). Classic discussion of structure and bonding.

R. McWeeny, *Coulson's Valence*, 3rd ed., Oxford University Press, London (1979). Modern treatment of valency and related topics.

D. F. C. Morris, *Structure and Bonding*, Vol. 4, pp. 63–82, Springer-Verlag, New York (1968). Ionic radii and enthalpies of hydration of ions. Also, Appendix, Vol. 6, pp. 157–159 (1969).

L. Pauling, *The Nature of the Chemical Bond*, 3rd ed., Cornell University Press, Ithaca, N.Y. (1959); or *The Chemical Bond: A Brief Introduction to Modern Structural Chemistry*, Cornell University Press, Ithaca, N.Y. (1967). Classical reference.

A. F. Wells, *Structural Inorganic Chemistry*, 4th ed., Clarendon Press, Oxford (1975). Classic volume on structures.

W. A. Weyl and E. Chostner Marloe, *The Constitution of Glasses: A Dynamic Interpretation*. Interscience Publishers, New York (1962). Glasses and the glassy state.

CHAPTER FIVE

THE CHEMICAL BOND, II NONELECTROSTATIC INTERACTIONS, MINOR ELECTROSTATIC INTERACTIONS

1. THE COVALENT BOND

The terms nonelectrostatic and covalent[1] are used somewhat interchangeably, but with preference given to the latter. As suggested in Chapter 4 (pp. 129–130), the covalent bond can be, somewhat noncommittally but nevertheless usefully, formulated as a shared pair of electrons in the Lewis kernel notation.[2,*] A formulation of this type indicates only the importance of a shared pair of electrons but gives no information as to how sharing is effected or as to the nature of the bond.

---

* Variations in this notation can be illustrated, for example, for the NH_4^+ ion, as

$$\left[\begin{array}{c} H \\ H\!\cdot\!\!\overset{\times}{N}\!\!\cdot\!H \\ H \end{array}\right]^{+} \quad \left[\begin{array}{c} H \\ H\!:\!\overset{\cdot\cdot}{N}\!:\!H \\ H \end{array}\right]^{+} \quad \left[\begin{array}{c} H \\ | \\ H\!-\!N\!-\!H \\ | \\ H \end{array}\right]^{+} \quad \left[\begin{array}{c} H \\ | \\ N \\ H\diagup \;|\;\diagdown H \\ H \end{array}\right]^{+}$$

The first formulation, by suggesting the origins of the bonding electrons, is useful in "electron bookkeeping," although once a given bond has formed, the electrons involved are indistinguishable from each other. The last two formulations substitute a bond line for each shared pair of electrons (single covalent bond), and the last formulation is useful in depicting the observed tetrahedral geometry of the ion. Double and triple covalent bonds can be indicated similarly, for example,

$$:\!\overset{\cdot\cdot}{O}\!:\!:\!C\!:\!:\!\overset{\cdot\cdot}{O}\!: \qquad\qquad :\!N\!:\!:\!:\!N\!:$$

$$\text{or} \qquad\qquad\qquad \text{or}$$

$$:\!O\!=\!C\!=\!\overset{\cdot\cdot}{O}\!: \qquad\qquad :\!N\!\equiv\!\!\equiv\!N\!:$$

However, several simplistic, but useful, interpretations which do not depend upon the nature of the bond result and can be summarized as follows:

1. **Bond order.** The number of pairs of electrons shared between two atoms: for example, 1 for each bond in the species NH_3, NH_4^+, $GeCl_4$, SF_6; 2 for each bond in the molecule CO_2; 3 for the bond in either the CO or the N_2 molecule.

2. **Covalency.** The number of atoms bonded to the specific atom being described: for example, 2 for the O atom in the H_2O molecule; 3 for the N atom in the NH_3 molecule or the NO_3^- ion; 4 for the S atom in the SO_4^{2-} ion or the Pt atom in the $[Pt(NH_3)_2Cl_2]$ molecule; 5 for the Fe atom in the $[Fe(CO)_5]$ molecule; 6 for the Co atom in the $[Co(CN)_6]^{3-}$ ion.

3. **Oxidation state.** An assigned indication of the relative charge borne by a specific atom as determined by both the number of electrons it shares with other bonded atoms and the relative electronegativities of the bonded atoms: for example, $+1$ for the H atom, -2 for each O atom, and $+6$ for the S atom in the H_2SO_4 molecule. It will be recalled that of necessity the algebraic sum of the *total* oxidation numbers must be zero for a neutral molecule or equal to the ionic charge for an ion. Oxidation state is sometimes indicated by a Roman numeral and ionic charge by an Arabic numeral as

$$+VI-II \qquad +IV \quad -I \qquad\qquad +V-II-I \qquad +I+VII-II$$
$$Se\ O_4^{2-} \qquad [Mo(CN)_8]^{4-} \qquad P\ \ S\ \ Cl_3 \qquad H\ \ Cl\ \ O_4$$

4. **Formal charge.** The relative charge on a given atom as calculated from the relationship

$$\text{formal charge} = (\text{number of valence-shell electrons}) - \tfrac{1}{2}(\text{number of electrons shared with other atoms}) - (\text{number of unshared valence-shell electrons}) \qquad (5\text{-}1)$$

for example, for the oxo acids of chlorine,

$$H\!-\!O\!-\!\ddot{\underset{..}{C}l}: \text{ formal charge on } Cl = 7 - \tfrac{1}{2}(2) - 6 = 0$$

$$H\!-\!O\!-\!\underset{..}{\ddot{C}}l\!-\!O = 7 - \tfrac{1}{2}(4) - 4 = +1$$

$$\begin{array}{c} O \\ \| \\ H\!-\!O\!-\!\underset{..}{C}l\!-\!O \end{array} = 7 - \tfrac{1}{2}(6) - 2 = +2$$

$$\begin{array}{c} O \\ \| \\ H\!-\!O\!-\!Cl\!-\!O \\ | \\ O \end{array} = 7 - \tfrac{1}{2}(8) - 0 = +3$$

Formal charges are indicated in formulas by Arabic numbers, as $HOC\overset{+3}{l}O_3$.

1-1. Theoretical Approaches and Interpretations

Any useful theoretical interpretation of the covalent bond must take into consideration and be consistent with at least the following characteristics of nonionic substances and provide answers to the questions raised:

1. Inasmuch as these substances are electrical insulators, electronic charge density must be shared between atoms rather than transferred. How can this sharing be rationalized in terms of orbital descriptions of electronic configurations?

2. Experimental diffraction (electron, x-ray, neutron) and spectroscopic (infrared, Raman, microwave) techniques (Ch. 6) indicate definite molecular geometries which require that the covalent bond be unidirectional (i.e., directed in space relative to the atom in question) rather than omnidirectional (as in ionic crystals and metals). How can this directional property be related to the sharing of electronic charge density?

3. Covalent substances not uncommonly persist through the solid, liquid, and gaseous states, with some persisting without decomposition at exceedingly high temperatures (e.g., CO, N_2, SiC, BN, diamond, graphite), indicating that the covalent bond can be a very strong bond. In the absence of electrostatic attraction between bonded atoms, why can this bond be so strong?

These characteristics, and other less striking ones, suggest that an increase in electronic charge density between bonded atoms occurs. This increase has been rationalized in terms of two somewhat different theoretical approaches, each of which is the result of approximate solutions of appropriate wave equations. These approaches can be characterized as

1. The *valence-bond, or directed-bond, approximation*, originally proposed by W. Heitler and F. London[3] and developed extensively by J. C. Slater[4] and L. Pauling.[5] In brief, this approximation considers covalent interaction to occur between two atomic species as a consequence of the overlap of two atomic orbitals of the same, or nearly the same, energy and proper spatial orientation and each containing a single electron. The covalent bond then amounts to the pairing of electron "spins" (p. 42), with essential retention of the properties of each atomic orbital in its ground state. The significance of the shared pair in the Lewis sense is thus apparent.

2. The *molecular-orbital, or undirected-bond, approximation*, as developed by F. Hund,[6] J. E. Lennart-Jones,[7] and, especially, R. S. Mulliken.[8] In brief, this approximation considers covalent interaction to occur between two atomic species as a consequence of the occupancy by electrons (originally in atomic orbitals of appropriate energies and spatial orientations) of new *polycentric molecular orbitals* peculiar to the species formed. In the process, the original atomic orbitals lose their identities. Inasmuch as excited states

are essential to the formation of molecular orbitals, excited states, rather than ground states, are significant.

By contrast, the valence-bond approximation is modular in concept and lends itself to representation by models or drawings depicting models, whereas the molecular-orbital approximation is energy-oriented and describes bonding in terms of energy states.

The Interaction, or Overlap, of Atomic Orbitals. Implicit in both of these approximations is the necessity for the overlap of atomic orbitals of comparable energies and proper symmetries. In quantum-mechanical terms, for the covalent interaction of atoms A and B the extent of this overlap can be evaluated in terms of an overlap integral

$$S = \int \psi_A \psi_B \, dv \tag{5-2}$$

where ψ_A and ψ_B are appropriate wave functions for the orbitals in question (p. 42), dv is an element of volume, and S is a measure of the bond strength in the resulting molecular species. The magnitude of the overlap integral depends upon the magnitudes and signs of the wave functions and may have all values, including zero, between -1 and $+1$. Qualitatively, if $S = 0$, there is no overlap and no bonding (i.e., the orbitals are *nonbonding*). If $S > 0$, there is attraction and bonding interaction (*bonding orbitals*), and if $S < 0$, there is repulsion and opposition to bonding (*antibonding orbitals*). In a pictorial fashion, orbital interaction that concentrates electronic charge density between two atomic nuclei results in covalent bonding. Conversely, orbital interaction that concentrates electronic charge density away from between the two nuclei opposes covalent bonding.

Allowed (bonding) and forbidden (antibonding) overlaps in terms of orbital symmetry, as indicated by the signs of appropriate wave functions (Secs. 2.3-3, 2.3-4), involving s, p, and d atomic orbitals, are depicted in two-dimensional projection in Fig. 5-1. Also indicated are situations of zero (nonbonding) overlap. It is apparent that covalent bonding based upon *pure* atomic orbitals can be of two types:

1. Along the *bond*, or *internuclear*, axis, taken arbitrarily as the z axis, where the orbitals are then rotationally symmetrical about the axis. Such orbitals are termed *sigma* (σ) orbitals, and a resulting covalent bond is a *sigma bond*. In Fig. 5-1, sigma bonds are exemplified by s-s, s-p_z, and p_z-p_z overlaps.

2. At right angles to the *bond*, or internuclear, axis, where the atomic orbitals that are involved are characterized by nodal planes that contain that axis. Such orbitals are termed *pi* (π) orbitals, and a resulting covalent bond is a *pi bond*. In Fig. 5-1, pi bonds are exemplified by p_x-p_x, p_y-p_y, d_{xz}-p_x, d_{yz}-p_y, d_{yz}-d_{yz}, and d_{xz}-d_{xz} overlaps.

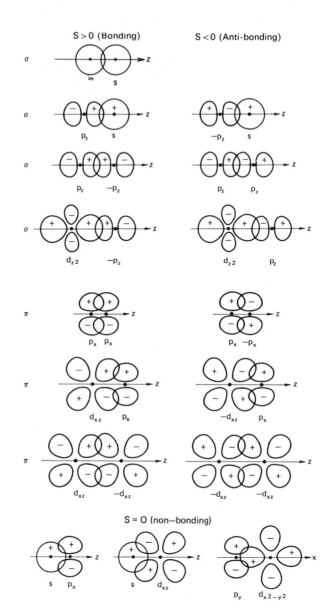

Figure 5-1 Overlaps of atomic orbitals that lead to positive (bonding), negative (antibonding), and zero (nonbonding) overlap integrals (S). [s-s and Ralf Steudel, *Chemistry of the Non-Metals, With an Introduction to Atomic Structure and Chemical Bonding*; English Edition by F. C. Nachod, J. J. Zuckerman. Walter de Gruyter, Berlin, New York (1977).]

It is also apparent from Fig. 5-1 that certain overlaps (e.g., s-p_y, s-p_x, d_{xy}-d_{yz}, d_{xz}-d_{yz}, d_{xy}-d_{xz}, s-d_{yz}) are impossible and represent nonbonding combinations. Atomic orbitals involved are, of necessity, valence-shell (i.e., maximum principal quantum number for the type in question) orbitals. Inner orbitals are not involved in bond formation except under unusual circumstances.

For a given molecular species, more than one covalent bond may exist between two atoms. Thus in the dinitrogen molecule, three types of orbital overlap exist, namely $2p_z$ with $2p_z$ (σ bond), $2p_y$ with $2p_y$ (π bond), and $2p_x$ with $2p_x$ (π bond), giving a triple covalent bond (Fig. 5-2). Similarly, an atom of one kind can combine with more than one atom of another kind to give a nonlinear molecular species. Molecular geometries and observed deviations from predictions based upon *pure* atomic orbital overlaps are discussed in Chapter 6.

The Valence-Bond, or Directed-Bond, Approximation.   In its primitive form,[3] this approximation required the following:

1.   Interaction of electrons of opposed spin from two atoms, that is, of two appropriate singly occupied atomic orbitals.
2.   In the absence of initially nonpaired electrons, unpairing via transfer of a previously paired electron to a vacant orbital belonging to the same atom.
3.   The same principal quantum number for the orbitals involved in bond formation.
4.   The possible formation of but one covalent bond for each stable orbital of the same principal quantum number.

The Slater–Pauling contribution[4, 5] was essentially an expansion of these ideas, which accounts also for the directional characteristics of covalent bonds.

It appears that maximum chemical stability characterizes a chemical species

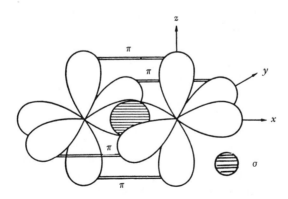

Figure 5-2 Overlap representation of $N_2(g)$ molecule. [E. Cartmell and G. W. A. Fowles, *Valency and Molecular Structure*, 2nd ed., Fig. 27, p. 82, Butterworths, London (1961).]

when every stable atomic orbital has been used either in forming covalent bonds or in accommodating nonbonding electrons. The valency normally exhibited by a given atom is determined by the number of nonpaired electrons present. Although participation in bond formation by electrons that are already paired is formally forbidden, it is allowable if the unpairing process does not require the expenditure of an excessively large quantity of energy and can occur without change in the principal quantum number. Thus the nitrogen atom, with the "outermost" ground-state electronic configuration $2s^2 2p_x^1 2p_y^1 2p_z^1$, can share its $2p$ electrons with three hydrogen atoms or three fluorine atoms to give the covalent species NH_3 or NF_3, but it cannot form either NH_5 or NF_5 because there is no additional atomic orbital in the second principal quantum level to accommodate one of the unpaired $2s$ electrons. However, the phosphorus atom, with the "outermost" ground-state electronic configuration $3s^2 3p_x^1 3p_y^1 3p_z^1$, does form the covalent molecule PF_5, presumably because one of the $3s$ electrons can be promoted to a potentially available $3d$ atomic orbital at only slightly higher energy.

The Concept of Resonance in Covalent Bonds.[9,10]   Not uncommonly, it is possible to construct more than one Lewis formulation for a specific covalent substance while maintaining the same arrangement of constituent atoms. Whenever this situation pertains, it is essential to determine which of the possible molecular arrangements can account best for the observed properties of that species. Often those properties are most consistent not with one of these specific molecular formulations but rather by an arrangement of electrons that partakes of the natures of all of the possible arrangements and lies between the extreme ones.

A case in point is the molecule $HNO_3(g)$, for which one can construct the conventional bond and electronic formulations

$$\text{(I)}$$

and

$$\text{(II)}$$

In this planar molecule, the measured bond angles O(1)—N—O(2) and O(1)—N—O(3) are equal (114 to 116°, within experimental error), and the measured bond lengths N—O(3) and N—O(2) are equal (1.22 Å) but between the single N—O bond length (ca. 1.36 Å) and the double N=O bond length (1.19 Å). Thus the true molecular structure is represented by neither formulation I nor formulation II but rather by one that lies between these limiting extremes and partakes of the nature of both. It is postulated, therefore,[9] that *resonance** occurs between these extreme structures and that the true molecular structure is then a *resonance hybrid* of the extreme structures. To depict this situation, the formulation

$$\text{H:O:N} \overset{\oplus}{\underset{}{}} \begin{matrix} \text{O} \\ \\ \text{O} \end{matrix} \quad \longleftrightarrow \quad \text{H:O:N} \overset{\oplus}{\underset{}{}} \begin{matrix} \text{O} \\ \\ \text{O} \end{matrix}$$

may be used, *provided that it is not interpreted as representing an equilibrium between limiting molecular structures.*

From the quantum mechanical point of view, a given chemical species may exist in any one of several stationary states to each of which there can be assigned a definite energy describable by an appropriate wave function, ψ_n. Of these states, the one of lowest energy is the *normal* state. For a given substance in its normal state, to which two conceivable molecular structures (I and II) can be assigned, the overall energy can be expressed in terms of a wave function ψ_0, as

$$\psi_0 = a\psi_I + b\psi_{II} \tag{5-3}$$

where a and b are numerical mixing coefficients, with $a + b = 1$. Only the ratio b/a is significant in altering ψ_0. Evaluation of the minimum energy of the system as a function of b/a gives a good indication of the relative importance of molecular structures I and II. If minimum energy is not achieved for either a large or a small value of b/a, ψ_0 represents significant contributions from ψ_I and ψ_{II}, the normal state involves both molecular structures, and the true molecular structure is represented as a resonance hybrid.

Although analogies have been drawn between mechanical and chemical systems, it must be emphasized that in the latter individual canonical structures do not exist as such. Two energetically comparable individual molecular structures can interact to give two new energy states, one of lower energy (the coupled state) and one of higher energy (the uncoupled state) than the energies of the separate structures. The coupled state is stabilized by the difference in energy between it and the uncoupled state. This quantity of energy is the *resonance energy*. Of course, more than two initial molecular structures can be involved.

A resonance energy can be evaluated as the difference between the observed enthalpy of formation or reaction of a given species and that calculated for an idealized depicted molecular structure.[9] For example, the enthalpy of formation of benzene (ΔH_f^0) calculated (Sec. 9.2-2) from its enthalpy of combustion and the enthalpies of formation of its combustion products is -1323 kcal (-5535 kJ) mole^{-1}, whereas the value calculated from the summation of bond energies, e.e., for 6 C—H + 3 C—C + 3 C=C bonds, is -1286 kcal (-5381 kJ) mole^{-1}. The benzene molecule is thus said to be stabilized by a resonance energy of 37 kcal

---

*   The alternative term *mesomerism* has been suggested (C. K. Ingold, *J. Chem. Soc.*, **1933**, 1120) but less widely used. Limiting resonance structures are called *canonical forms*.

(154 kJ) mole^{-1}. Similarly, the carbon monoxide molecule, the molecular structure of which may be regarded as a hybrid of the Lewis structures

$$:\overset{\oplus}{C}:\overset{\ominus}{\ddot{O}}: \qquad :C::\ddot{O}: \text{ (or } :C::\underset{\cdot\cdot}{O}: \text{)} \qquad :\overset{\ominus}{C}:::\overset{\oplus}{O}:$$

$$\text{I} \qquad\qquad\qquad \text{II} \qquad\qquad\qquad \text{III}$$

has with respect to II a resonance energy of 83 kcal (347 kJ) mole^{-1}. Resonance energies and resonance stabilization have been applied much more extensively to organic compounds than to inorganic compounds.

The ease with which the concept of resonance permits graphic representation of valence-bond molecular structures has resulted in abuses that have raised questions as to the validity of the concept. Although for a given covalent species it may be possible to depict a number of electronic arrangements, only those formulations that satisfy the following rules are admissible:

1. All must have the same relative arrangement of atoms.
2. All must be of comparable energy.
3. Each must indicate the same number of unpaired electrons.
4. Like formal charges should not reside on atoms close to each other; unlike formal charges should not be widely separated.
5. Negative formal charges should reside on more electronegative atoms; positive formal charges on the less electronegative atoms.
6. The larger the total number of covalent bonds, the more important the contributing structure.

These rules can be illustrated for the dinitrogen(I) oxide molecule, a colinear species containing a nitrogen–nitrogen bond, for which possible canonical forms can be depicted as

$$:\overset{\ominus}{\ddot{N}}=\overset{\oplus}{N}=\ddot{O}: \qquad :\overset{\ominus}{N}=\overset{\oplus}{N}=\overset{\cdot\cdot}{\ddot{O}}: \qquad :\overset{\oplus}{N}\equiv\overset{\ominus}{N}-\ddot{O}: \qquad :\overset{2\ominus}{\ddot{N}}-\overset{\oplus}{N}\equiv\overset{\oplus}{O}:$$

$$\text{I} \qquad\qquad \text{II} \qquad\qquad \text{III} \qquad\qquad \text{IV}$$

$$:\overset{\ominus}{\ddot{N}}=\overset{\oplus}{\ddot{N}}-\ddot{O}: \qquad :\ddot{N}-\ddot{N}=\ddot{O}:$$

$$\text{V} \qquad\qquad\qquad \text{VI}$$

Forms I and II differ only in that the N=N bond in I involves p_x orbitals and the N=O bond p_y orbitals, whereas in II these orbitals are reversed. Forms I and II are, therefore, completely equivalent. Forms I, II, and III all conform to the above-noted rules and are likely arrangements. Form IV is unlikely because of the arrangement of formal charges, and forms V and VI are ruled out because of the presence of fewer bonds. Forms I, II, and III should contribute about equally.

Experimental evidences for the existence of resonance include:

1. Measured internuclear distances do not always conform to summations of appropriate single-, double-, and triple-bond radii. Rather, they commonly have intermediate values. Thus the measured C to O distance in the molecule CO(g) is 1.128 Å, compared with expected values of 1.43, 1.21, and 1.07 Å for formulations I, II, and III, respectively, on p. 200. Pauling[9] interprets these data in terms of contributions of 10% (I), 20% and 20% (II), and 50% (III) to the total molecular structure.

2. The true energy content of a molecule in question is less (by the resonance energy) than that predicted by any contributing molecular structure.

3. The dipole moment (Sec. 5.1-8) of a molecular species is less than that predicted in terms of contributing structures. For example, the measured dipole moment of N_2O(g) is only 0.17×10^{-18} esu, although that for each canonical form (p. 200) should be sizable.

4. The chemical properties of a given species are not those of a specific canonical form but rather those of a hybrid of all forms. Thus none of the cyclic species C_6H_6, $B_3N_3H_6$ (borazine, Sec. 5.1-3), $N_3P_3Cl_6$ (hexachloro-cyclotriphosphazene, Secs. 6.2-1 and 10.4-2), or $N_3S_3O_3Cl_3$ (trimeric sulfanuric chloride, Secs. 6.2-1 and 10.4-2) has properties consistent with a ring system based upon alternating single and double bonds.

Hybridization of Atomic Orbitals in Covalent Bonding. All overlap considerations offered earlier were based upon *pure* atomic orbitals. Interpretations of the observed properties of certain species are better made, however, in terms of a mixing, or hybridization, of atomic orbitals. This approach may be illustrated as follows:

1. The covalent molecule $BeCl_2$(g) at ca. 750°C has a linear geometry with both Be—Cl bond distances being the same (1.7 Å). The ground-state electronic configuration of the gaseous Be atom, $1s^22s^2$, would preclude bond formation. The excited-state configuration $1s^22s^12p^1$ would suggest the formation of two types of bonds, at variance with experimental observation. The apparent problem can be solved by assuming that in the excited state the 2s and 2p orbitals combine linearly to give two sp hybrid orbitals (Fig. 5-3), each with a capacity of two electrons and, because of interelectronic repulsions, directed at 180° to each other. Overlap with the $3p_z^1$ orbitals of two Cl atoms then gives the linear $BeCl_2$(g) molecule.

2. The covalent molecules BX_3(X = F, Cl, Br, I) have regular trigonal planar geometries, with all B—X bond distances equal for a given halogen and all X—B—X bond angles 120°. This geometry is accounted for by assuming (1) excitation from the ground-state electronic configuration $1s^22s^22p^1$ to $1s^22s^12p^12p^1$; (2) interaction of the 2s and 2p orbitals to give three sp^2 hybrid orbitals, each containing one electron, and directed as far apart in

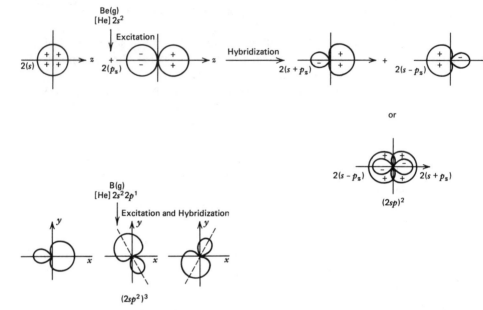

Figure 5-3 Formation of directional sp and sp^2 hybrid orbitals. [Ralf Steudel, *Chemistry of the Non-Metals, With an Introduction to Atomic Structure and Chemical Bonding*, English Edition by F. C. Nachod, J. J. Zuckerman. Walter de Gruyter, Berlin, New York (1977).]

    space as possible; and (3) ultimate overlap with np_z^1 orbitals from the halogen atoms (Fig. 5-3).

3.   The covalent molecules $CX_4(g)$ (X = H, F, Cl, Br, I) have regular tetrahedral geometry with all C—X bond distances being equal for a hydrogen or for a given halogen atom. These properties are accounted for by assuming (1) excitation of the ground-state electronic configuration of the carbon atom $1s^2 2s^2 2p^1 2p^1$ to $1s^2 2s^1 2p^1 2p^1 2p^1$; (2) combination of the $2s$ and $2p$ orbitals to give four sp^3 hybrid orbitals, each containing a single electron and so repelled by the others to as large a spatial separation as possible (i.e., to the apices of a circumscribed tetrahedron); and (3) ultimate overlap with $1s^1$ or np_z^1 orbitals of hydrogen or the halogen atoms (Fig. 5-3).

These hybrids, and those involving d orbitals, are treated in more detail in Chapter 6 in relationship to molecular geometries (Sec. 6.1-1).

    Valence-Bond Interpretation of Bonding in the Hydrogen Molecule. Following the approach suggested in Fig. 4-1, change in potential energy as two hydrogen atoms H_a and H_b approach each other is plotted as curve 1 in Fig. 5-4. If the two electrons present are labeled (1) and (2), the H_2 molecule can be assumed to form by the sharing of an electron pair in terms of either of the formulations

$$\begin{matrix} (1) & & (2) \\ H_a^{.} . H_b & \text{or} & H_a^{.} . H_b \\ (2) & & (1) \end{matrix}$$

$$\text{I} \qquad\qquad\qquad \text{II}$$

These two species can be described by the wave functions

$$\psi_I = \psi_a(1)\psi_b(2) \qquad \text{and} \qquad \psi_{II} = \psi_a(2)\psi_b(1) \tag{5-4}$$

which, since structures I and II are completely equivalent, can be combined to

$$\psi_{cov} = c_1\psi_I + c_2\psi_{II} \tag{5-5}$$

where c_1 and c_2 are mixing coefficients. Application of Eq. 5-5, using true nuclear charges (Z) led to curve 2 in Fig. 5-4.[3] Recognition of screening effects and the consequent use of Z_{eff} gave curve 3,[11] which approaches more closely the observed curve 0. A further assumption that the ionic resonance forms

$$\begin{matrix} H_a^+ & : H_b^- & \text{and} & ^-H_a : & H_b^+ \end{matrix}$$

$$\text{III} \qquad\qquad\qquad \text{IV}$$

for which

$$\psi_{ion} = c_3\psi_{III} + c_4\psi_{IV} \tag{5-6}$$

may exist permits combination of covalent and ionic interactions,[12] as

$$\psi = (1 - 2\lambda + 2\lambda^2)^{-\frac{1}{2}}[(1 - \lambda)\psi_{cov} + \psi_{ion}] \tag{5-7}$$

where λ is another mixing coefficient. Additional refinements, including polarization correction[12] and ultimately interelectron distance,[13, 13a] gave essentially complete agreement in both binding energy and equilibrium bond distance, as shown in Table 5-1.

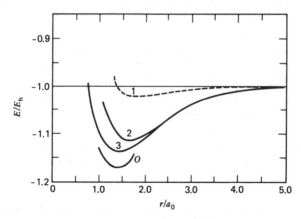

Figure 5-4 Potential-energy curves for interaction of atoms H_a and H_b (true compared with various approximations). [R. McWeeny, *Coulson's Valence*, 3rd ed., Fig. 5-1, p. 116, Oxford University Press, London (1979), modified.]

**Table 5-1   Comparison of Observed Ground-State Energy and
Bond Length for H_2 Molecule with Valence-Bond
Approximations**[a]

| | Binding Energy | | Bond Distance |
Type of Data	eV	kcal mole^{-1}	(Å)
Observed	4.7467	109.5	0.7412
Heitler–London[3]	3.14	72.4	0.740
Heitler–London with screening[11]	3.78	87.2	0.743
Heitler–London with screening plus ionic[12]	4.00	92.2	0.749
Heitler–London with screening plus ionic plus polarization[12]	4.12	95.0	0.749
Inclusion of interelectronic distance (13-term equation)[13]	4.72	108.5	0.740
100-term wave function[13a]	4.7467	109.5	0.7412

[a] Adapted from R. A. McWeeny, *Coulson's Valence*, 3rd ed., p. 120, Oxford University Press, London (1979).

Valence-Bond Interpretations of More Complex Molecules. The quantum-mechanical approach outlined for the interpretation of bonding in the H_2 molecule can be extended to more complicated species, but as the total number of electrons increases, it becomes progressively more difficult to obtain even approximate solutions to the appropriate wave equations.

However, covalent bonding results only when the antiparallel spins of electrons can be coupled. If in addition the previously noted principles are accepted, (1) that appropriate wave functions for two atoms can be combined only if they represent similar energy states, (2) that extensive overlap of appropriate atomic orbitals can occur, and (3) that these orbitals have the same symmetry with respect to the molecular axis, one can derive a valence-bond model for the molecule in question. Unfortunately, although useful, the picture is only qualitative. A case in point is the O_2 molecule, the paramagnetic properties ($\chi_M = 0.993/T$) of which are consistent with the presence of two unpaired electrons but the bond energy (118 kcal mole^{-1}) and bond distance (1.207 Å) of which are consistent with the presence of a double covalent bond. Thus a valence-bond approach leading to the formulation

$$:\ddot{O}{=}\ddot{O}:$$

is not completely satisfactory. Nor is the assumption of the presence of two three-electron covalent bonds,[14] as

$$:O\vdots\vdots O:$$

The Molecular-Orbital, or Undirected-Bond, Approximation[15-19] The molecular-orbital treatment has some similarity to the atomic-orbital treatment, as is apparent from the following principles:

1. Each electron occupies a molecular orbital, which is described by an appropriate combination of quantum numbers. The molecular orbital is associated with all of the atomic species involved and is, therefore, polycentric. It is described by a wave function ψ, such that $\psi^2 dv$ is a measure of the probability of finding the electron in a tiny element of volume dv.

2. The total energy associated with the molecular or ionic species in question is the sum of the energies associated with the individual wave functions, corrected for interelectronic repulsions.

3. Each molecular orbital can accommodate no more than two electrons, the "spins" of which must be paired.

4. The Aufbau principle (p. 57) is obeyed.

As molecular orbitals are occupied, series of closed electronic shells are conceived to be built up in much the same fashion as those that characterize isolated atoms. The quantum numbers n and l have the same connotations for both atomic and molecular orbitals, but it is useful to substitute for m_l the quantum number λ, which measures, in units of $h/2\pi$, the component of angular momentum along the bond, or molecular, axis (z). Permitted values of the quantum numbers, their designations, and comparisons between atomic and molecular orbitals are summarized in Table 5-2. A *sigma* (σ) molecular orbital is one that is cylindrically symmetrical about a bond axis. A *pi* (π) molecular orbital is one that is characterized by a nodal plane in which the bonded nuclei lie and through which the bond axis passes. A *delta* (δ) molecular orbital is one that is so oriented that the bond axis lies in two mutually perpendicular planes Sec. 8.2-8.3.

The σ, π, and δ designations also characterize bonds formed by combinations of atomic orbitals and described by molecular orbitals. Sigma bonds, formed in effect by "end-on" orbital overlap along the bond (z) axis, may involve s or p_z orbitals. Pi bonds, formed in effect by "sidewise" overlap at 90° to the bond axis, may involve p_x, p_y, d_{xz}, or d_{yz} orbitals. Delta bonds may, of necessity, involve

Table 5-2   Comparison of Atomic and Molecular Orbitals

Completed Electronic Group in Isolated Atom	l	λ	Completed Electronic Groups in Molecular Species
s^2	0	0	σ^2
p^6	1	$-1, 0, +1$	$\sigma^2 \quad \pi^4$
d^{10}	2	$-2, -1, 0, +1, +2$	$\sigma^2 \quad \pi^4 \quad \delta^4$

only d_{xy} or $d_{x^2-y^2}$ orbitals. For a given value of $n = 3$ or larger, no more than *six* sigma bonds, *eight* pi bonds, and *four* delta bonds can be formed. Most covalent species are describable in terms of sigma bonds or combinations of sigma and pi bonds. Metal–metal interactions involving d-transition metal ions [e.g., cluster compounds and ions (Sec. 8.2-3)] are describable in terms of delta bonds.

Although the general quantum-mechanical approach developed for atomic systems can be extended to molecular systems, multinuclear-polyelectron systems do not lend themselves to exact solutions of the Schrödinger equation (Sec. 2.3-3). Of the several approximations to solution of this equation, a particularly useful one is the *linear combination of atomic orbitals* (LCAO) technique, first proposed by R. S. Mulliken. Only atomic orbitals of appropriate symmetries can be so combined. The general approach and the results achieved are best illustrated by treating first the H_2^+ and H_2 species and then proceeding through other homonuclear diatomic species to multinuclear polyatomic species.

Molecular-Orbital Treatments of the Gaseous H_2^+ Ion and the Gaseous H_2 Molecule. The H_2^+ ion involves a single electron moving in the charge field generated by two fixed positively charged nuclei (H_a and H_b) separated at a distance R, as depicted in Fig. 5-5. At a given instant, this electron will be found at distances r_a and r_b from the two nuclei. For this system, the Hamiltonian operator (Sec. 2.3-3) is

$$\mathcal{H}_e = -\frac{h^2}{2m}\nabla^2 - \frac{e^2}{r_a} - \frac{e^2}{r_b} + \frac{e^2}{R} \tag{5-8}$$

A rigorous solution of the Schrödinger equation corresponding to this operator is possible, although it is much more difficult to effect than that for the hydrogen atom. However, a useful approximate solution can be obtained by a linear combination of wave functions for the two atomic orbitals, $\psi(1s)_{H_a}$ and $\psi(1s)_{H_b}$, each of which describes the case where the $1s$ electron is associated with a specific nucleus. The result is

$$\psi_b = \frac{1}{\sqrt{2}}[\psi(1s)_{H_a} + \psi(1s)_{H_b}] \tag{5-9}$$

where $1/\sqrt{2}$ is a normalization factor that obviates overlap of the atomic functions. Another suitable solution is

$$\psi_a = \frac{1}{\sqrt{2}}[\psi(1s)_{H_a} - \psi(1s)_{H_b}] \tag{5-10}$$

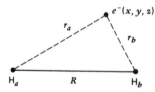

Figure 5-5 One-electron $H_aH_b^+$ ion, showing H_a, H_b, e^- distances and coordinates.

It can be shown that the energy associated with ψ_b (i.e., E_b) can be represented, for an element of volume dv, by the expression

$$E_b = \int \psi(1s)_{H_a} \mathscr{H} \psi(1s)_{H_a} \, dv + \int \psi(1s)_{H_a} \mathscr{H} \psi(1s)_{H_b} \, dv \qquad (5\text{-}11)$$

where the first term is called an atomic integral (α) and the second term a resonance integral (β), or

$$E_b = \alpha + \beta \qquad (5\text{-}12)$$

Similarly, a difference, as suggested by Eq. 5-10, results in

$$E_a = \alpha - \beta$$

Graphically, these energies can be shown as functions of R in ψ and ψ^2 plots, as in Fig. 5-6.

It is evident that ψ_b implies a concentration of electronic charge density between the two atoms (i.e., a bond). Thus ψ_b describes a *bonding molecular*

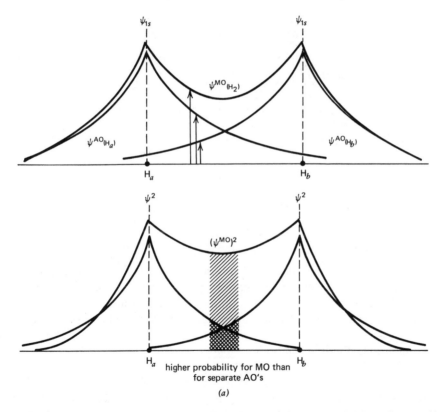

Figure 5-6 Molecular-orbital development for attraction and repulsion of atoms H_a and H_b. (a) Bonding molecular orbital (σ_{1s}); (b) antibonding molecular orbital (σ_{1s}^*). [W. W. Porterfield, *Concepts of Chemistry*, Figs. 6-8, 6-9, 6-10, pp. 264, 266, W. W. Norton, New York (1972)]

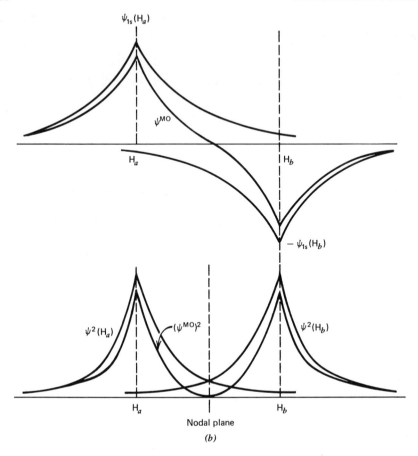

Figure 5-6   (*Continued*)

orbital. Similarly, ψ_a implies a concentration of electronic charge density in opposition to bonding and is, therefore, a description of an *antibonding molecular orbital.* Both of these apply to the ground state of the species H_2^+. Similar bonding and antibonding molecular orbitals could be constructed for the various excited states that would appear in the observed emission spectrum of this species.

Comparable bonding and antibonding molecular orbitals can be constructed similarly for the two-electron molecule $H_2(g)$, again in terms of the two atoms H_a and H_b. The formation of these orbitals in terms of electronic charge clouds is shown diagrammatically in Fig. 5-7. Computer-generated electron-density diagrams for pseudo-bonding and antibonding molecular orbitals resulting from the linear combination of two $1s$ atomic orbitals are shown in Fig. 5-8(*a*). Inasmuch as the molecular orbitals in question are cylindrically symmetrical about the bond axis, they are σ orbitals. The designations (σ_{1s} and σ_{1s}^*) indicate, respectively, the bonding and antibonding molecular orbitals for the H_2 molecule.*

The molecular orbital description of the H_2 molecule is then written $(\sigma_{1s})^2$, since there are only two electrons involved and they must occupy preferentially the lower-energy bonding molecular orbital. The formation of this species is indicated in terms of an energy-level diagram in Fig. 5-9. The *bond order*, that is, half the difference between the number of bonding electrons and the number of antibonding electrons, is then $(2-0)/2 = 1$, in agreement with the observation that a single covalent bond exists in this molecule.

Molecular orbitals can be described also in terms of a symmetry operation called *inversion*, which consists in reflecting a point in an orbital through the center of the molecule, that is, changing its spatial coordinates from $+x$, $+y$, $+z$ to $-x$, $-y$, $-z$. If, as a consequence of this reflection, a new point for which the wave function has the same sign and amplitude results, the orbital is said to be *symmetrical* with respect to inversion and is designated g (for *gerade*, German meaning even). If an orbital is antisymmetric with respect to inversion, it is designated u (for *ungerade*, uneven). The inversion operations involved for s, p, and d orbitals are shown in Fig. 5-10. The σ_s and σ_s^* orbitals are *gerade* and *ungerade*, respectively.

Other Diatomic s-Electron Species. Using comparable energy-level diagrams, one can construct molecular-orbital descriptions of the gaseous 1s species He_2^+ (from He and He^+) and He_2 (from He and He). Data for the 1s species are summarized and compared in Table 5-3. It is significant that this approach predicts zero bond order (i.e., no bond) for the species He_2, which prediction is in accord with the experimental observation that, in the gaseous state, helium is always monoatomic. Similarly, but invoking the overlap of 2s atomic orbitals, an energy-level diagram can be constructed for the molecule $Li_2(g)$ (Fig. 5-11). That the bond distance and bond energy of the $Li_2(g)$ molecule (Table 5-3) are, respectively, larger and smaller than those of the $H_2(g)$ molecule reflects both the large spatial extension of the 2s orbital, resulting in interaction at a larger internuclear distance, and the enhanced interelectronic repulsions involving the (σ_{1s}) and (σ_{1s}^*) orbitals. Comparison of the (σ_{2s}) orbitals with the (σ_{1s}) orbitals in Fig. 5-8(b) emphasizes this point. In agreement with observation, equality in numbers between bonding and antibonding electrons precludes the formation of the molecule $Be_2(g)$. The thermal stabilities of these gaseous species are indicated

---

* This type of notation, indicating both the type of bond and the specific atomic orbitals involved, is used in this book. Other notations for bonding molecular orbitals derived from 1s atomic orbitals include (σ_1), (σ_s), (σ_s^b), (σ_{ss}^b), and so on. Antibonding orbitals may be indicated also by a superscript a [e.g., (σ_1^a), (σ_s^a), ...]. For the second quantum level, the following equivalences appear in many publications and in certain of the figures in this and later chapters:

σ_{2s}, σ_1, σ_s^b	σ_{2s}^*, σ_2, σ_s^*
σ_{2p_z}, σ_3, σ_z^b	$\sigma_{2p_z}^*$, σ_4, σ_z^*
π_{2p_x} or π_{2p_y}, π_1, π_x^b or π_y^b	$\pi_{2p_x}^*$ or $\pi_{2p_y}^*$, π_2, π_x^* or π_y^*

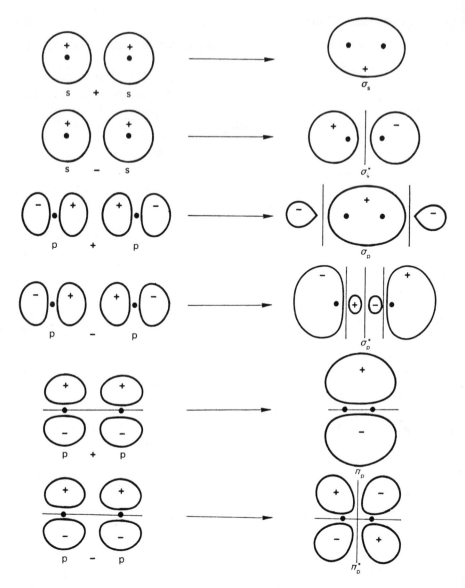

Figure 5-7 Formation of σ and π molecular orbitals from combinations of atomic orbitals of different symmetries. [R. Steudel, *Chemistry of the Non-Metals, With an Introduction to Atomic Structure and Chemical Bonding*; English Edition by F. C. Nachod, J. J. Zuckerman. Walter de Gruyter, Berlin, New York (1977), modified.]

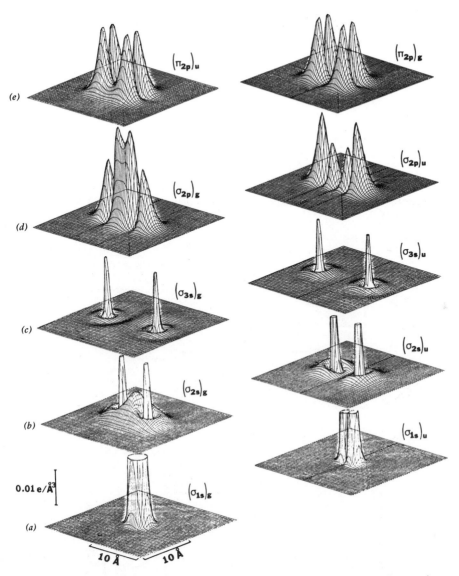

Figure 5-8 Three-dimensional electron-density maps for *s-s* and *p-p* interactions. The base plane of these diagrams passes through the two hydrogen nuclei, which have been placed so as to have maximum overlap of the atomic wave functions. [J. R. Van Wazer and I. Absar, *Electron Densities in Molecules and Molecular Orbitals*, Fig. 1.2, p. 5, Academic Press, New York (1975).]

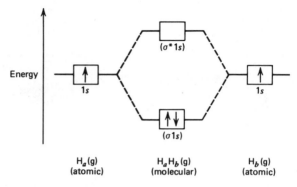

Figure 5-9 Molecular-orbital, energy-level diagram for $H_aH_b(g)$ molecule.

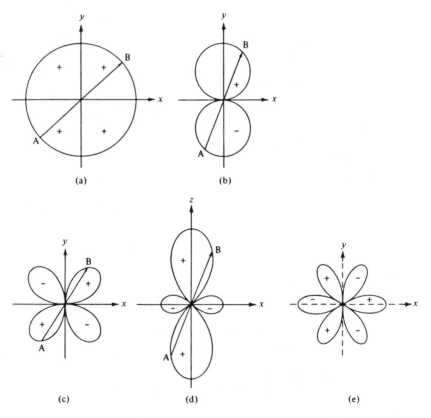

Figure 5-10 Explanation for *gerade* and *ungerade* symmetries. (*a*) *s* orbital, *gerade*; (*b*) *p* orbital, *ungerade*; (*c*) d_{xy} orbital, *gerade*; (*d*) d_z^2 orbital, *gerade*; (*e*) f_z^3 orbital, *ungerade*. [J. E. Huheey, *Inorganic Chemistry*, 2nd ed., Fig. 2.12, p. 23, Harper & Row, New York (1979).]

Table 5-3 Molecular-Orbital Descriptions and Properties of Gaseous Diatomic s Electron Species

Formula	Molecular-Orbital Formulation[a]	Bond Properties		
		Order	Length (Å)	Energy (kcal mole^{-1})
H_2^+	$(\sigma_{1s})^1$	$\frac{1}{2}$	1.06	61
H_2	$(\sigma_{1s})^2$	1	0.74	104.2
He_2^+	$(\sigma_{1s})^2(\sigma_{1s}^*)^1$	$\frac{1}{2}$	1.08	60
He_2	$(\sigma_{1s})^2(\sigma_{1s}^*)^2$	0	—	—
Li_2	$(\sigma_{1s})^2(\sigma_{1s}^*)^2(\sigma_{2s})^2$	1	2.67	25
Be_2	$(\sigma_{1s})^2(\sigma_{1s}^*)^2(\sigma_{2s})^2(\sigma_{2s}^*)^2$	0	—	—

[a] Since the configuration $(\sigma_{1s})^2(\sigma_{1s}^*)^2$ represents completion of the innermost, or K, shells of the two atoms, it is often designated as KK. Thus the molecular-orbital formulation of the Li_2 molecule could be written $KK(\sigma_{2s})^2$.

by their bond energies. Comparable bonds involving (σ_{3s}) orbitals are less stable thermally [Fig. 5-8(c)].

Gaseous Homonuclear Diatomic p-Electron Species. For these species, consideration must be given to the facts that for a given principal quantum number the three p orbitals (1) are spatially oriented at $90°$ to each other and (2) are described by wave functions that change sign as nodal planes are crossed. If the z axis is taken as the molecular (or bond) axis in a diatomic molecular species, interaction of atomic p_z orbitals can give the molecular orbitals (σ_{npz}) and (σ_{npz}^*).[†] As shown in Figs. 5-1, 5-7, and 5-8, however, the atomic p_x and p_y orbitals must overlap in the π fashion to give the molecular orbitals (π_{npx}), (π_{npx}^*), (π_{npy}), and (π_{npy}^*).[‡] As a consequence of the equivalence between np_x and np_y atomic orbitals, we have the energy equivalences

$$(\pi_{npx}) = (\pi_{npy}) \qquad \text{and} \qquad (\pi_{npx}^*) = (\pi_{npy}^*)$$

that is, both the (π_{np}) and (π_{np}^*) molecular orbitals are *doubly degenerate*. In terms of inversion operations, the (σ_{npz}) and (π_{np}^*) molecular orbitals are *gerade*, and the (σ_{npz}^*) and (π_{np}) orbitals are *ungerade*.

The relative energies of the molecular orbitals for gaseous diatomic species decrease with increasing atomic electronegativity in the series boron through fluorine, as indicated in Fig. 5-12. It is significant that for the gaseous B_2, C_2, and N_2 molecules the (π_{2px}) and (π_{2py}) orbitals lie at lower energies than the (σ_{2pz}) orbital, whereas for the gaseous O_2, F_2, and Ne_2 molecules the reverse is true. The first circumstance is sometimes said to reflect the presence of hybridization and the second its absence. Bond orders, distances, and energies

[†] Other designations include (σ_z^b) and (σ_z^*) or (σ_z^a).
[‡] Other designations include (π_x^b), (π_x^*), (π_x^a), (π_y^*), and (π_y^a).

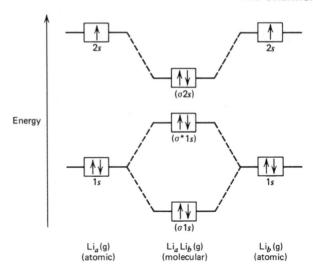

Figure 5-11   Molecular-orbital, energy-level diagram for $Li_2(g)$ molecule.

and magnetic properties, together with molecular-orbital descriptions, are summarized in Table 5-4.

Both the gaseous B_2 and O_2 molecules are paramagnetic. It is of particular interest that the molecular-orbital formulation of the O_2 molecule agrees both with its double-bond character and the presence of two unpaired electrons. The nonexistence of the Ne_2 molecule is also predicted correctly. Bond distances

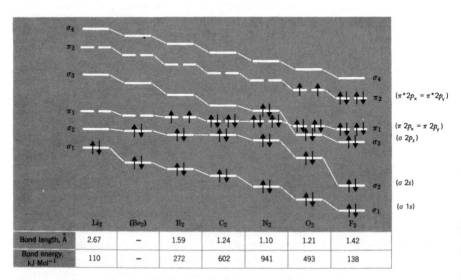

Figure 5-12   Energy sequence for second quantum level. B_2 and O_2 have one electron in each of two degenerate molecular orbitals. [F. A. Cotton and G. Wilkinson, *Basic Inorganic Chemistry*, Fig. 3-12, p. 66, John Wiley & Sons, Inc., New York (1976).]

Table 5-4 Molecular-Orbital Descriptions and Properties of Gaseous Homonuclear Diatomic p-Electron Species

Formula	Molecular-Orbital Formulation	Magnetic Properties (Observed)[a]	Bond Properties		
			Order	Length Å	Energy (kcal mole^{-1})
B_2	$KK(\sigma_{2s})^2(\sigma^*_{2s})^2(\pi_{2p_x})^1=(\pi_{2p_y})^1$	para (2)	1	1.59	69
C_2	$KK(\sigma_{2s})^2(\sigma^*_{2s})^2(\pi_{2p_x})^2=(\pi_{2p_y})^2$	dia	2	1.24	150
N_2	$KK(\sigma_{2s})^2(\sigma^*_{2s})^2(\sigma_{2p_z})^2(\pi_{2p_x})^2=(\pi_{2p_y})^2$	dia	3	1.10	226
O_2	$KK(\sigma_{2s})^2(\sigma^*_{2s})^2(\sigma_{2p_z})^2(\pi_{2p_x})^2=(\pi_{2p_y})^2(\pi^*_{2p_x})^1=(\pi^*_{2p_y})^1$	para (2)	2	1.21	118
F_2	$KK(\sigma_{2s})^2(\sigma^*_{2s})^2(\sigma_{2p_z})^2(\pi_{2p_x})^2=(\pi_{2p_y})^2(\pi^*_{2p_x})^2=(\pi^*_{2p_y})^2$	dia	1	1.42	36
Ne_2	$KK(\sigma_{2s})^2(\sigma^*_{2s})^2(\sigma_{2p_z})^2(\pi_{2p_x})^2=(\pi_{2p_y})^2(\pi^*_{2p_x})^2=(\pi^*_{2p_y})^2(\sigma^*_{2p_z})^2$	dia	0	—	—
S_2	$KKLL(\sigma_{3s})^2(\sigma^*_{3s})^2(\sigma_{3p_x})^2(\pi_{3p_x})^2=(\pi_{3p_y})^2(\pi^*_{3p_x})^1=(\pi^*_{3p_y})^1$	para (2)	2	1.887	101

[a] dia = diamagnetic; para (·) = paramagnetic (number of unpaired spins).

vary inversely with, and bond energies directly with, bond order. Extension of the approach to the $3s$ $3p$ combination is shown for the molecule $S_2(g)$.

Gaseous Heteronuclear Diatomic Species. The same general approach can be extended to gaseous heteronuclear diatomic species, except that knowledge of both the symmetries and the energies of all valence orbitals are required to determine those atomic orbitals that can interact, whether or not there is hybridization, and the presence or absence of strong orbital overlap. The highest occupied atomic orbitals have energies determined by the negative ionization energies (p. 73). Because the ionization energies of the two atoms are different, the molecular-orbital energy diagrams are appropriately distorted.

This situation can be illustrated using the gaseous HF molecule, as shown in Fig. 5-13. The ionization energies of the hydrogen and fluorine atoms are 13.598 eV (324 kcal mole^{-1}) and 17.422 eV (402 kcal mole^{-1}), respectively (Table 2-19), giving the distortion shown. Only the $2p_z$ orbital of the fluorine atom has the proper symmetry and energy to interact with the $1s$ orbital of the hydrogen atom, giving a single sigma bond. The bonding pair of electrons is largely localized near the fluorine atom. The σ^* molecular orbital differs but little in energy from the atomic $1s$ orbital of the hydrogen atom. The completely occupied $2s$, $2p_x$, and $2p_y$ atomic orbitals yield *nonbonding* molecular orbitals at the same energies.

The 10-valence electron, diamagnetic, gaseous species NO$^+$, CO, and CN$^-$ are all isoelectronic with the N_2 molecule and may be expected to be describable in rather similar molecular-orbital terms. The molecules N_2 and CO, which are characterized by the same number of atoms, the same total number of electrons, the same arrangement of electrons, and the same sum of nuclear charges, are said to be *isosteric*[20] It is not surprising that the physical constants of these two substances are remarkably similar (Table 5-5). The same is true of the molecular-

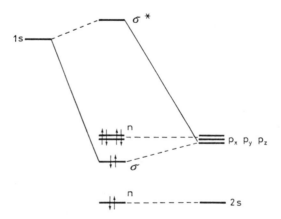

Figure 5-13 Molecular-orbital, energy-level diagram for HF(g) molecule. [Ralf Steudel, *Chemistry of the Non-Metals, With an Introduction to Atomic Structure and Chemical Bonding*, English Edition by F. C. Nachod, J. J. Zuckerman. Walter de Gruyter, Berlin, New York (1977).]

Table 5-5 Comparison of Physical Constants of Isosteric Molecular Species

Property	N_2	CO	N_2O	CO_2
Total electrons	14	14	22	22
Sum of nuclear charges	14	14	22	22
Molecular weight (amu)	28.014	28.0104	44.0134	44.0098
Melting point (°C)	−209.97	−205.06	−90.84	−56 (press.)
Normal boiling point (°C)	−195.81	−191.50	−88.48	−78 (subl.)
Density of liquid [g ml^{-1} (°C)]	0.808	0.793	0.856 (10)	0.858 (10)
Density of gas at 1 atm, 273.14°K (g liter^{-1})	1.25	1.25	1.79	1.79
Critical temperature (°C)	−146.89	−139	35.4	31.35
Critical pressure (atm)	33.54	35	75	73
Enthalpy of fusion (kcal mole^{-1})	0.1732	0.1997	1.563	1.90
Enthalpy of vaporization (kcal mole^{-1})	1.3329		3.958	3.08 (−28.9°C)

orbital energy-level diagrams (Fig. 5-14) except, of course, for the actual orbital energies. Similarity in properties can be related to a continuous change of the molecular orbitals for the N_2 molecule into those for the CO molecule as positive charge is shifted from one nitrogen nucleus to the other to give, respectively, carbon and oxygen nuclei. The isoelectronic species NO^+ and CN^- are similarly described, again with appropriate differences in orbital energies. Observed similarities in the coordination chemistries of these four species (Sec. 8.2-1) are relatable to similarities in bonding.

Of interest also is the paramagnetic, 11 valence-electron species NO(g), the general properties of which are consistent with the molecular-orbital formulation

$$KK(\sigma_{2s})^2(\sigma_{2s}^*)^2(\pi_{2p_x})^2 = (\pi_{2p_y})^2(\sigma_{2p_z})^2(\pi_{2p_x}^* = \pi_{2p_y}^*)^1$$

The unpaired electron is believed to be in an antibonding π orbital. Loss of this electron gives the NO^+ ion of larger bond energy. The molecular-orbital configuration of the NO molecule can be rationalized in terms of either the removal of a $\pi_{2p_x}^*$ or $\pi_{2p_y}^*$ electron from an O_2 molecule or the addition of an electron to the N_2 molecule, where of necessity it must occupy a π_{2p}^* orbital.

Bond-related properties of various species discussed in this section are summarized in Table 5-6.

Gaseous Heteronuclear Multiatomic Species. Gaseous triatomic molecules may have either linear (e.g., BeH_2, $BeCl_2$, CO_2, N_2O) or angular (e.g., H_2O, H_2S, OF_2, Cl_2O, NO_2) geometry (Sec. 6.1-2). In developing a molecular-orbital description of any one of these species, it is important to realize that for one atom two identical sets of atomic orbitals are involved. It is apparent in the examples that follow that molecular geometry does indeed affect the energies of resulting molecular orbitals.

In Fig. 5-15, molecular-orbital energy-level diagrams for the linear s-electron molecule BeH_2(g) and the s- and p-electron molecule $BeCl_2$(g) are compared. In Fig. 5-16 similar comparisons are given for the angular s- and p-electron molecules H_2O(g) and OF_2(g). In Fig. 5-17 the linear molecule CO_2(g) and the angular molecule NO_2(g) are compared. Inasmuch as the N_2O molecule is isosteric with the CO_2 molecule (Table 5-5), its molecular-orbital discription might be expected to resemble that for CO_2, modified of course because the structural arrangement of atoms is NNO, as opposed to OCO.

Extensions to more complex systems result in the development of even more involved molecular-orbital treatments. Applications to d-transition metal complex ions are described in Chapter 8.

Interpretations — A Summary. Both the valence-bond and the molecular-orbital approximations consider the electrons in a covalent bond to be associated simultaneously with both nuclei, predict an increase in the density of the electronic charge cloud between the two nuclei as a consequence of suitable overlap of atomic orbitals, and distinguish between sigma and pi overlap in

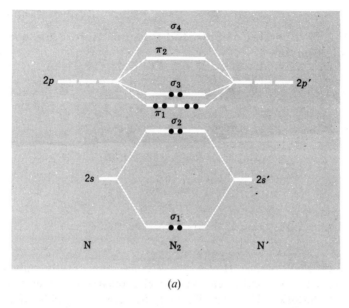

(a)

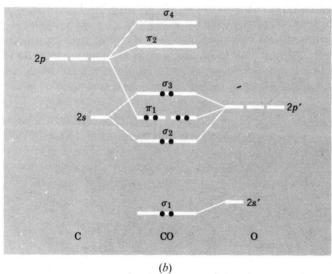

(b)

Figure 5-14 Comparison of molecular-orbital, energy-level diagrams for (a) $N_2(g)$ and (b) $CO(g)$ molecules. [F. A. Cotton and G. Wilkinson, *Basic Inorganic Chemistry*, Fig. 3-13, p. 68, John Wiley & Sons, New York (1976).]

Table 5-6   Properties of Selected Gaseous Heteronuclear Diatomic Species

| Formula | Order | Bond Properties | |
		Length (Å)	Energy (kcal mole^{-1})
N$_2$	3	1.10	225
CN$^-$	3	1.14	
CO	3	1.128	256
NO$^+$	3	1.062	
NO	2$\frac{1}{2}$	1.15	162

multiple covalent bonds. However, each describes exactly only a limiting condition, and thus neither can always give reliable interpretations without suitable modification. For a specific case, the necessary modification may even make the two approaches identical. The valence-bond approximation, through its use of resonance, is particularly useful for predicting molecular structures and reaction mechanisms. The molecular-orbital approximation, however, provides better descriptions of simple species such as H$_2^+$ and O$_2$ than do the one- and three-electron bonds that must be involved in the valence-bond approach. The molecular orbital method has the additional advantages of better describing

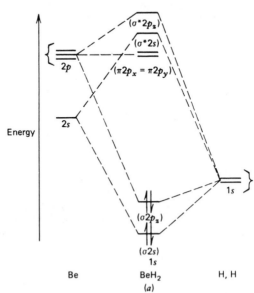

Figure 5-15 Comparison of molecular-orbital, energy-level diagrams for (a) BeH$_2$(g) and (b) BeCl$_2$(g) molecules.

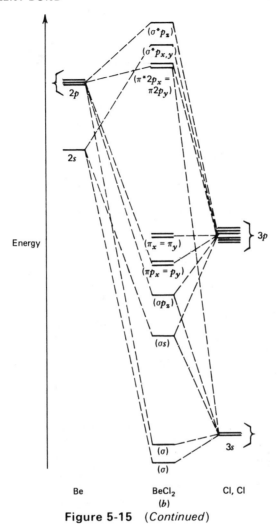

Figure 5-15 (*Continued*)

excited states, multiple bonding, and bond weakening via occupancy of antibonding orbitals. Furthermore, the energies of molecular orbitals are derivable from molecular spectra, as indicated, for example, in Fig. 5-18 for the $N_2(g)$ molecule.[21] There is no comparable experimental approach to verification of valence-bond formulations.

In summary, each approximation has its own specific uses and advantages, and it is often desirable to use the approach that is better suited to the specific situation in question. It is patently restrictive, and probably unfair, to conclude that the molecular-orbital approximation should replace completely its valence-bond counterpart.

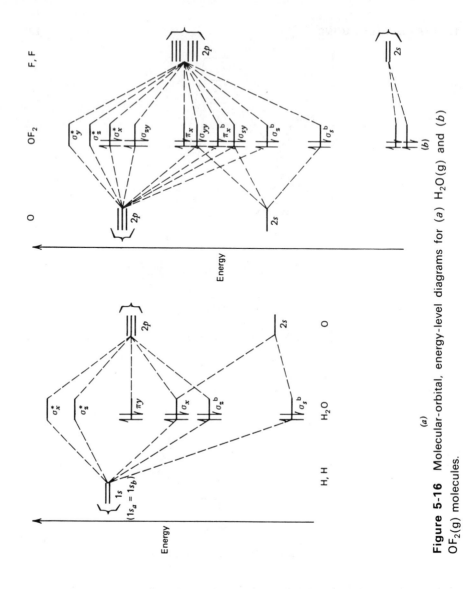

Figure 5-16 Molecular-orbital, energy-level diagrams for (a) $H_2O(g)$ and (b) $OF_2(g)$ molecules.

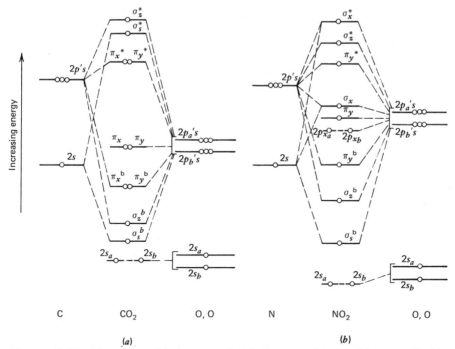

Figure 5-17 Molecular-orbital, energy-level diagrams for $CO_2(g)$ and $NO_2(g)$ molecules. [H. B. Gray, *Electrons and Chemical Bonding*, 1st ed., Fig. 3-11, p. 99 and Fig. 7-11, p. 151, W. A. Benjamin, New York (1965).]

1-2. Directional Characteristics of Covalent Bonds

In terms of previous emphasis on both the spatial distributions of electronic charge densities in atoms and molecules and the formation of covalent bonds as a consequence of the overlaps of appropriate atomic orbitals, the directional characteristics that relate covalent bonds to observed molecular structures are both reasonable and expected. In Chapter 6, these relationships are discussed in greater detail.

1-3. Delocalization Phenomena

As thus far discussed, both the valence-bond and the molecular-orbital approaches have dealt with the concentration of electronic charge density as being *localized* between two bonded atoms. On this basis, single, double, and triple covalent bonds are characterizable in terms of the electrons involved, bond lengths, and bond energies. Cases in point are most usefully organic compounds. In particular, we may consider the two-carbon atom molecules $H_3C—CH_3$ (four sigma bonds, in sp^3 hybridization, to each C atom, C—C bond length = 1.534 Å, C – C bond energy = 82.6 kcal mole^{-1}); $H_2C=CH_2$

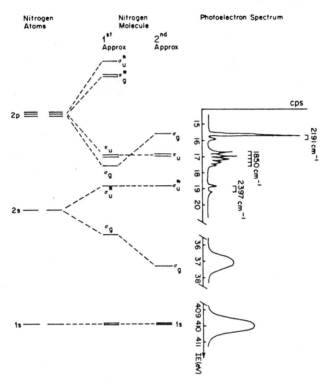

Figure 5-18 Comparison of qualitative molecular-orbital, energy-level scheme for $N_2(g)$ molecule with its photoelectron spectrum. [H. Bock and P. D. Mollère, *J. Chem. Educ.*, **51**, 506 (1974), Fig. 5, p. 508.]

(three sigma bonds, in sp^2 hybridization, and one pi bond to each C atom, C—C bond length = 1.337 Å, C=C bond energy = 145.8 kcal mole^{-1}); and HC≡CH (two sigma bonds, in sp hybridization, and two pi bonds to each C atom, C≡C bond length = 1.205 Å, C=C bond energy = 199.6 kcal mole^{-1}), where *localization* of electronic charge density between carbon atoms is essential.

By contrast, consider the cyclic planar molecule C_6H_6, the molecular structure of which is sometimes described in terms of a resonance hybrid involving the limiting canonical forms of the Kekulé type:

but more frequently in terms of delocalization of charge density as

That all C-to-C bonds in this molecule have the same length (1.395 Å) indicates that they are identical. That the C-to-C bond lengths are between those noted above for double and single bonds indicates an order between 1 and 2 for each such bond. These facts can all be rationalized by assigning to each C atom three sigma bonds in sp^2 hybridization and one pi bond, but suggesting that the π orbital for a given C atom overlaps those from both adjacent C atoms, as indicated in Fig. 5-19(b). The result is continuing overlap around the C_6 ring, complete *delocalization* of the π electronic charge density over all six C atoms, and the formation of a diffuse cloud of charge density above and below the plane of the cyclic structure [Fig. 5-19(c)]. Thermodynamic stabilization, as measured by the resonance energy, is then a consequence of delocalization, and *delocalization energy* and resonance energy are synonymous for cyclic systems of this type.

The $p_\pi\text{-}p_\pi$ overlap delocalization ascribed to the C_6H_6 molecule is a criterion for its so-called "aromaticity." Aromaticity and delocalization of this type are in accord with the Hückel rule,

$$\text{number of } \pi \text{ electrons} = 4n + 2 \qquad (5\text{-}13)$$

where n is an integer ($=1$ for C_6H_6). Delocalized cyclic ligands of particular interest in coordination chemistry (Ch. 8) include the benzene molecule and the cyclopentadienide ion ($C_5H_5^-$) for $n = 1$ and the cyclooctatetraenide ion ($C_8H_8^{2-}$) for $n = 2$. The dicarbollide ion, $B_9C_2H_{11}^{2-}$, as a ligand, and the borazine molecule, $B_3N_3H_6$, and its derivatives also exemplify $p_\pi\text{-}p_\pi$ delocalization. Borazine is of particular interest since it is isosteric with benzene and very closely comparable in many properties, the BN group being isoelectronic with the CC group. On the other hand, cyclic molecules in the phosphazene, or phosphonitrilic, series [i.e., $(NPX_2)_n$, where $X = F$, Cl, Br, etc., $n = 3$, 4, etc.] and the sulfanuric series [i.e., $(NSOX)_3$, where $X = F$, Cl, etc.] exemplify $p_\pi\text{-}d_\pi$ delocalization involving p orbitals from the N atoms and d orbitals from the P and S atoms.[22] (Sec. 6.2-1)

1-4. Atomic Crystals, or Giant Molecules

In crystals such as diamond, silicon, germanium, gray tin, silicon carbide (SiC), cubic boron nitride (BN), gallium arsenide (GaAs), and zinc blende or wurtzite (ZnS), completely repetitive, tetrahedrally oriented covalent bonds link all atoms into regular three-dimensional arrays, within which no individual molecular groups can be distinguished (Fig. 5-20). As a consequence, any

(a) σ-bonded C—H skeleton

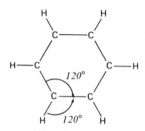

(b) p atomic orbitals available for molecular orbitals

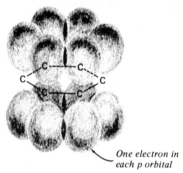

*One electron in
each p orbital*

(c) Sum of delocalized π orbitals

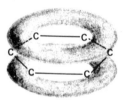

(d) Symbol for benzene that emphasizes
delocalized bonding

Figure 5-19 Delocalization of electronic charge density in C_6H_6 molecule. [T. Moeller, J. C. Bailar, J. Kleinberg, C. O. Guss, M. E. Castellion, and C. Metz, *Chemistry with Inorganic Qualitative Analysis*, Fig. 21.17, p. 654, Academic Press, New York (1980.]

fragment of a substance of this type, irrespective of its size, can be termed a molecule—literally a *giant molecule*. The somewhat less descriptive term *atomic crystal* has been used also.

By contrast, crystals of graphite and hexagonal boron nitride are built up, respectively, of stacked layers of covalently bonded carbon atoms and alternating covalently bonded boron and nitrogen atoms, as indicated in Fig. 5-21. Within each layer, infinite fused six-membered ring systems are distinguishable, and the arrangements therein are those of two-dimensional giant molecules. Bond distances between the layers are larger than those within the layers, and

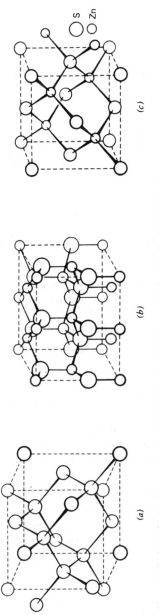

Figure 5-20 Crystal structures of (*a*) diamond, (*b*) wurtzite (ZnS), and (*c*) zinc blende (ZnS). [A. F. Wells, *Structural Inorganic Chemistry*, 4th ed., Fig. 3.35, p. 102; Fig. 21.1, p. 727, Clarendon Press, Oxford (1975).]

S

Zn

(*a*)

(*b*)

(*c*)

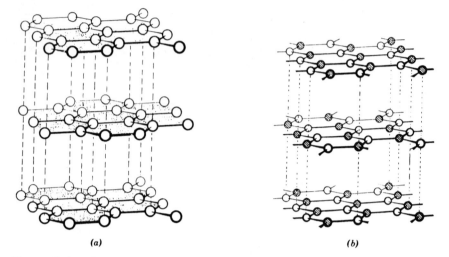

Figure 5-21 Crystal structures of (a) graphite and (b) hexagonal boron nitride. [A. F. Wells, *Structural Inorganic Chemistry*, 4th ed., Fig. 21.3, p. 735; Fig. 24.9, p. 848, Clarendon Press, Oxford (1975).]

bond energies between layers are reduced. Still another type of lattice is found in crystals of black phosphorus (Fig. 5-22), wherein each phosphorus atom is covalently bonded to three other phosphorus atoms in what appears to be a puckered form of a hexagonal network. Again, the arrangement results in a giant molecule.

As a consequence of the large number of bonds that must be broken to rupture a continuing three-dimensional crystal lattice, many of these substances are hard, high-melting solids which are difficult to vaporize, have limited solubilities in all solvents, and are chemically unreactive except at high temperatures (e.g., diamond, SiC, cubic BN). Gray tin is a striking exception, being the stable allotrope only at comparatively low temperatures (transition to white tin at 13.2°C), powdery rather than crystalline, and readily oxidized by hydronium ion. These substances are either insulators or semiconductors (Sec. 4.4-2). The two-dimensional species have somewhat similar properties,

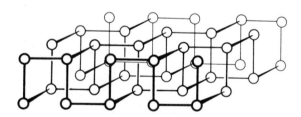

Figure 5-22 Crystal structure of black phosphorus. [A. F. Wells, *Structural Inorganic Chemistry*, 4th ed., Fig. 19.1, p. 674, Clarendon Press, Oxford (1975).]

except that rupture between layers is relatively easy to effect. Upon appropriate application of force, the layers slide over each other, thereby imparting lubricating characteristics. Graphite is an excellent electrical conductor, as a consequence of continuing π delocalization of electronic charge density within layers.

1-5. Electron-Deficient Covalent Bonding

Certain atoms having fewer valence electrons than valence-shell orbitals (e.g., Li, Be, B, Al, Pt) form covalent species in which there are insufficient electrons to allow for an electron-pair bond between each pair of bonded atoms. Typical of these species are elemental boron; the boron hydrides and their derivatives, including the carboranes; the lower boron halides; the polymeric alkyls of metals such as lithium, beryllium, aluminum, and platinum; certain interstitial compounds (Sec. 4.4-3); and a limited number of other substances [e.g., a few polynuclear metal carbonyls (Sec. 8.2-1)]. Bonding in these species is said to be *electron deficient* and is rationalized in terms of multinuclear interactions that involve at least three atomic centers.

Three-centered, two-electron covalent bonds are the most common. Two types, *open* and *closed*, can be distinguished geometrically, as indicated in Fig. 5-23. In an open-type bond [Fig. 5-23(a)], atoms Y_1 and Y_2 are sufficiently close to atom X to allow interaction of orbitals from all three atoms but sufficiently far removed from each other to preclude interaction with each other. Atoms X, Y_1, and Y_2 may be identical (e.g., all B atoms), or Y_1 and Y_2 may be identical, with X different (e.g., two B atoms and one H atom, or two Al atoms and one C atom). In a closed-type bond [Fig. 5-23(b)], all atoms are commonly identical and, ideally, in a planar, equilateral, triangular arrangement. This type of bonding is most characteristic of boron atoms. The designations of bonds included in Fig. 5-23 are those used by W. N. Lipscomb[23-25] in his semitopological representations of localized bonding interactions in the boron hydride molecules and their derivatives.

The open-type bond can be approached in terms of the structure of the diborane (B_2H_6) molecule, as depicted in Fig. 5-24.[26,27] The two boron atoms and four of the hydrogen atoms lie in a plane, with the other two hydrogen atoms symmetrically located in a perpendicularly bisecting plane but at slightly longer B—H bond distances. The arrangement of hydrogen atoms about each boron atom is roughly tetrahedral, suggesting that each boron atom is involved in sp^3 hybridization. Two of the sp^3 hybrid orbitals for each boron atom are then so oriented spatially that they can overlap simultaneously with the $1s$ orbitals of the two bridging hydrogen atoms, as depicted in Fig. 5-25. Some additional interaction of these orbitals to give a partial B—B bond is suggested by the fact that the measured B-to-B internuclear distance is only slightly larger than twice the covalent radius of the boron atom (0.80 Å). This type of overlap is better shown for one of the bridging hydrogen atoms in Fig. 5-26.

One sp^3 hybrid orbital from each of the boron atoms and the $1s$ orbital from a

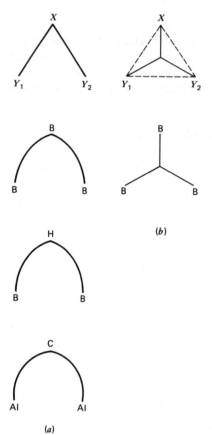

(a)

Figure 5-23 Geometries of three-centered, two-electron bonds. (*a*) Open; (*b*) closed.

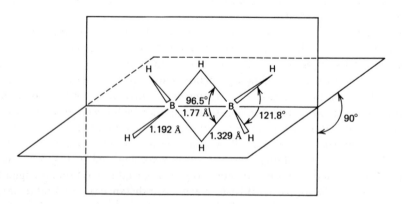

Figure 5-24 Diagrammatic representation of structure of $B_2H_6(g)$ molecule.

230

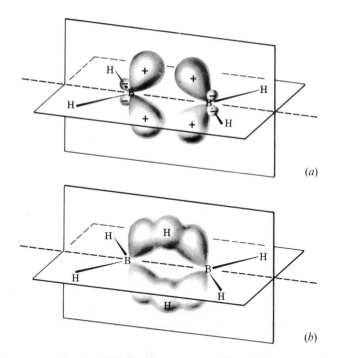

Figure 5-25 Overlap of 1s orbital from H atom with two sp^3 hybrid orbitals from two B atoms (three-dimensional). (a) Orientation of two coplanar H_2B groups with each B atom in sp^3 hybridization; (b) continuous overlap with 1s orbitals of two H atoms [F. A. Cotton and G. Wilkinson, *Basic Inorganic Chemistry*, Fig. 3-24, p. 80, John Wiley & Sons, New York (1976).]

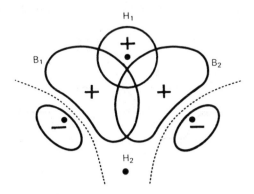

Figure 5-26 Overlap of 1s orbital from H atom with two sp^3 hybrid orbitals from two B atoms (two-dimensional). [Ralf Steudel, *Chemistry of the Non-Metals, With an Introduction to Atomic Structure and Chemical Bonding*; English Edition by F. C. Nachod, J. J. Zuckerman. Walter de Gruyter, Berlin, New York (1977).]

231

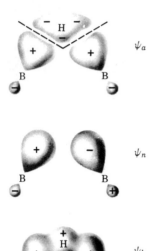

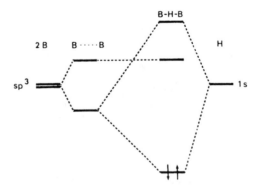

ψ_b **Figure 5-27** Formation of molecular orbitals for $B_2H_6(g)$ molecule. [F. A. Cotton and G. Wilkinson, *Basic Inorganic Chemistry*, Fig. 3-25, p. 81, John Wiley & Sons, New York (1976).]

hydrogen atom can be combined to yield, in order of increasing energy, one bonding molecular orbital (Ψ_b), one nonbonding molecular orbital (Ψ_n), and one antibonding molecular orbital (Ψ_a), as depicted in Fig. 5-27. The two available electrons then occupy the bonding molecular orbital (Fig. 5-28), giving a bond order of $\frac{1}{2}$.

Similar approaches, as indicated in Fig. 5-29, lead to descriptions of the open and closed three-center, two-electron bonds involving only boron atoms that characterize the molecular structures of elemental boron and the higher boron hydrides and their derivatives. The molecular structure of dimeric aluminum

Figure 5-28 Molecular-orbital, energy-level diagram for B—H—B system. [Ralf Steudel, *Chemistry of the Non-Metals, With an Introduction to Atomic Structure and Chemical Bonding*, English Edition by F. C. Nachod, J. J. Zuckerman. Walter de Gruyter, Berlin, New York (1977).]

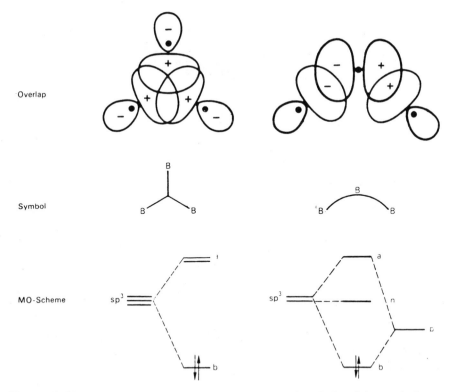

Overlap

Symbol

MO-Scheme sp^3

Figure 5-29 Types of three-centered, two-electron bonds involving only boron atoms. [Ralf Steudel, *Chemistry of the Non-Metals, With an Introduction to Atomic Structure and Chemical Bonding*; English Edition by F. C. Nachod, J. J. Zuckerman. Walter de Gruyter, Berlin, New York (1977).]

trimethyl $\{[Al(CH_3)_3]_2\}$, as depicted in Fig. 5-30,[28] can be similarly rationalized in terms of two bridging CH_3 groups, for each of which an sp^3 hybrid orbital of a carbon atom interacts with an sp^3 hybrid orbital of each of the aluminum atoms. Similar bonding explanations can be offered for methyl bridges in the infinite chains that characterize the molecular structures of $[Be(CH_3)_2]_\infty(s)$[29] and $[Mg(CH_3)_2]_\infty(s)$[30, 31] and the species $MgAl_2(CH_3)_8$[32] and $Al_2(CH_3)_2X_4$ (X = Cl, Br, I).[33]

Other more involved multicenter, electron-deficient bonds are known, for example, four-center, two-electron bonds in the molecular species $Rh_6(CO)_{16}(s)$ and $[Li(CH_3)]_4(s)$. In molecules of the former, three triangularly arranged Rh atoms apparently share electronic charge density with the C atom in a single CO molecule.[34] In crystals of the latter, four Li atoms are tetrahedrally arranged with a CH_3 group located symmetrically in each triangular face, as depicted in Fig. 5-31.[35] If each Li atom is regarded as exemplifying an sp^3 hybrid, extending from each Li atom and directed toward a center above each face, there is a hybrid orbital that can overlap one of the sp^3 hybrid orbitals of the carbon atom

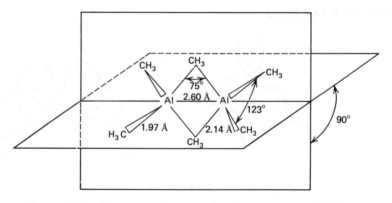

Figure 5-30 Representation of molecular structure of $Al_2(CH_3)_6$.

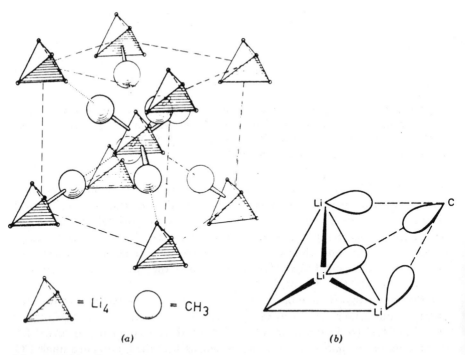

(a)   *(b)*

Figure 5-31 Representations of *(a)* molecular structure and *(b)* orbital overlap description of $Li_4(CH_3)_4(s)$ — formation of bonding molecular orbital on each face of Li_4 tetrahedrun by overlap of sp^3 hybrid C orbital with symmetric terminal Li orbitals. [E. Weiss and E. A. C. Lucken, *J. Organomet. Chem.*, **2**, 197 (1964).]

in a methyl group. The result is a four-center, two-electron bond in each triangular face.

It has been noted by R. E. Rundle[36] that in all species that have electron-deficient molecular structures, an atom with more low-energy orbitals than valence electrons is combined with atoms or groups of atoms that contain no unshared electron pairs. Thus electron-deficient covalent bonding tends to involve all low-energy orbitals by the use of one electron pair and one unused excess low-energy orbital per atom.

A measure of electron deficiency is the electron-pair accepting, or Lewis acid (Sec. 9.6-1), behavior of each such species when in contact with an electron-pair donor species such as water.[37] Information on other properties associated with electron deficiency is available in several summaries.[38-41]

1-6. Odd Molecules and Ions

Only a few molecules and ions, other than species derived from certain d- and f-transition metal species, that contain odd numbers of electrons have been characterized.[40-42] These are the so-called *odd molecules* and *ions*. Included in this small group are such species as He_2^+, $NO(g)$, $NO_2(g)$, O_2^- (superoxide ion), $ON(SO_3)_2^{2-}$(aq) (nitrosyldisulfonate ion), $CN(g)$, and $ClO_2(g)$. All of these species are paramagnetic.* Some [e.g., NO_2, ClO_2, and $ON(SO_3)_2^-$(aq)] are colored; others (e.g., NO) are not. Some (e.g., NO_2, O_2^-, ClO_2) are chemically very reactive; others (e.g., NO) somewhat less so. Some [e.g., CN, $ON(SO_3)_2^{2-}$, NO_2] dimerize readily; others (e.g., NO, ClO_2) either do so reluctantly or not at all.

In the valence-bond terminology, these species are described in terms of resonance-stabilized, three-electron bonds.[40] Molecular-orbital treatments are not inconsistent with this broad concept [cf. treatments of NO(g) with bond order 2.5 and $NO_2(g)$ with bond order 0.5]. That the NO molecule forms a dimer only at relatively low temperatures is attributable to distribution of the π^* orbital containing the unpaired electron over the entire molecule. The much greater ease with which the NO_2 molecule forms a dimer, at just below room temperature, is associated with the presence of the unpaired electron in a σ_x orbital and thus its ability to pair with a similar electron in a second molecule. However, the observation that both the NO(g) and $NO_2(g)$ molecules form dimers much less readily than does, for example, the CN(g) molecule can be readily and more simply accounted for [43] in terms of G. Herzberg's principle[44] that a molecule is chemically metastable or unstable if it can change into another species with a net increase in total number of bonds.

---

* Crystalline potassium nitrosyldisulfonate (Fremy's salt) is yellow and dimagnetic, but it dissolves in water to give a violet-blue, paramagnetic solution. The solid is believed to be dimeric, the aqueous ion monomeric.[42]

1-7. Covalent Bond Energies

The distinction between bond energy and bond dissociation energy has been noted earlier (p. 127).[45] Bond dissociation energies for various diatmic species are summarized in Table 5-7. From the enthalpy of formation of a particular covalent species and the appropriate bond dissociation energies, a bond energy can be calculated by means of a thermochemical cycle (Sec. 4.3-4). Thus having given the ΔH^0 values at 25°C for the following

$$H_2(g) + \tfrac{1}{2}O_2(g) \rightarrow H_2O(g) \qquad \Delta H_f^0 \quad = -57.80 \text{ kcal}$$

$$H_2(g) \rightarrow 2\,H(g) \qquad \Delta H_{H_2}^0 = 104.2 \text{ kcal}$$

$$O_2(g) \rightarrow 2\,O(g) \qquad \Delta H_{O_2}^0 = 118.9 \text{ kcal}$$

one calculates for

$$H_2O(g) \rightarrow 2\,H(g) + O(g) \qquad \Delta H_{H_2O}^0 = 221.5 \text{ kcal}$$

that is, an energy of 221.5 kcal (926.8 kJ) mole^{-1} to rupture both O—H bonds in the $H_2O(g)$ molecule. Half of this value, or 110.8 kcal (463.4 kJ) mole^{-1}, is then the *mean thermochemical energy* of the O—H bond at 25°C. Since the individual bond dissociation energies are, respectively, 119.7 and 101.5 kcal mole^{-1}, this value is indeed a mean bond energy (p. 127).

Although the energies are known for the individual bonds in this molecule, this information is not generally available for polyatomic species. What is done for such species is to use mean values so determined that they will agree, within minimal mean deviation, with appropriate atomization energies of a large number of related molecules. A number of mean thermochemical covalent bond energies at 25°C are summarized in Tables 5-8 and 5-9.[45-48] Only a few values for multiple bonds are available.[5(b), 49]

It is apparent that in general, for a given combination of atoms, mean covalent bond energy increases, as expected, with bond multiplicity. For single bonds involving a specific atom, bond energy increases as the electronegativity of the second atom increases. The mean energies of the single covalent bonds F—F, O—O, and N—N are somewhat smaller than might be expected. These smaller energies are the consequences of repulsions among unshared electron pairs in these relatively small atoms.[50] That similar small values are not evident for atoms of the heavier elements in these periodic families can be related to the larger fractions of d character and the lesser orbital overlaps, as compared with those characteristic of the first-row atoms. Mean covalent bond energies are usually additive for individual species. Certain of the many applications of covalent bond energies[45, 51] are discussed in subsequent chapters.

It is of interest that, by application of approaches already noted (Sec. 2.4-3), R. T. Sanderson has calculated many bond energies that agree well with thermochemical values.[52]

Table 5-7 Bond Dissociation Energies at 25°C

Molecule	Energy (kcal mole^{-1})	Molecule	Energy (kcal mole^{-1})	Molecule	Energy (kcal mole^{-1})
H$_2$	104.2	BS	119	NBr	67
HB	79	BF	186	OO	119.2
HC	81	BCl	119	OSi	192
HSi	74.6	BBr	104	OGe	159
HGe	76.5	CC	143	OP	141
HPb	43	CSi	104	OAs	114
HN	86	CN	174	OS	125
HO	102.4	CP	140	OSe	101
HS	85	CO	256.9	FF	38
HSe	73	CS	175	FCl	61
HTe	64	CSe	116	FBr	60
HF	135.8	CF	116	ClCl	58
HCl	103.0	CCl		ClBr	52
HBr	87.5	CBr	95.6	ClI	50
HI	71.3	NN	226	BrBr	46
HLi	56.91	NP	139	BrI	42.8
HCs	42	NAs	116	II	36.1
BB	70	NO	151	AtAt	(19)
BN	152	NS	116	LiLi	26.3
BO	173	NF	62.6	CsCs	11.3

Table 5-8 Mean Thermochemical Single Covalent Bond Energies at 25°C (kcal mole^{-1})

	H	B	C	Si	Ge	N	P	As	Sb	O	S	Se	F	Cl	Br	I
H	104		99	76	68	93	78	71	61	111	88	73	136	103	88	71
B		72								125			154	106	88	65
C			83	73	61	73	63	48		86	65	58	117	78	65	51
Si				54		80				111			143	96	79	56
Ge					45	61				86			113	81	67	51
N						38				39			67	45		
P							50			88			119	79	64	44
As								43		79			116	74	61	43
Sb									34				75	63		
O										34			51	49		48
S											63		78	65	51	
Se												38	68	58		
F													38	60	57	
Cl														58	52	50
Br															46	43
I																36

Table 5-9 Mean Thermochemical Multiple Covalent Bond Energies at 25°C

Bond	Bond Energy (kcal mole^{-1})	Bond	Bond Energy (kcal mole^{-1})
C=C	146	N=N	100
C≡C	200	N≡N	226
C=O	178	N=O	142
C≡O	257	P≡P	125
C=S	114		
C=N	147	O=O	119
C≡N	213	S=S	103
Si=O	153	S=O	125
Si≡O	192		

1-8. Polarity in Covalent Bonds

Although it is easy to imply from the preceding discussions that, in a given covalent bond, electrons are shared equally between the bonded atoms, differences in electronegativity of those bonded atoms (Sec. 2.4-3) suggest that equal sharing, or pure covalency, is uncommon. The difference in electronegativity is a measure of departure from pure covalency, but it is also improbable that 100% covalency can always be assigned to a bond between two atoms of the same electronegativity (e.g., the H_2 molecule; Sec. 5.1-1). However, it is convenient, and commonly very useful, either to describe covalent bonds between atoms of different electronegativity in terms of *partial ionic character* or to indicate the *degree of covalency* in ionic bonds where intermediate differences in electronegativity exist. Covalent bonds having partial ionic character are referred to as *polar covalent bonds*; those having little or no partial ionic character as *nonpolar* covalent bonds.*

The degree of polarity in a covalent bond can be expressed in terms of the *bond moment*, that is, the product of the partial charge on either atom by the internuclear distance, since each bond amounts to a dipole.[53-55] A number of typical bond moments are summarized in Table 5-10 as functions of differences in electronegativities of the bonded atoms.

Bond polarity determines the overall polarity of a molecular species. The average polarity of a given species over a period of time is measured by its *permanent dipole moment* (μ), which is often designated as a vector pointing from the center of positive charge to the center of negative charge ($\leftrightarrow$). The permanent dipole moments of a number of molecules in the gaseous state are summarized in Table 5-11. The permanent dipole moments of polar diatomic molecules are the same as their bond moments. The permanent dipole moments of multiatomic molecules, however, are determined by both the bond moments and the stereochemical arrangements of the bonded atoms. Thus although the moment of the C=O bond is 2.3 D, the permanent dipole moment of the CO_2 molecule is zero Debye. The obvious conclusion is that the CO_2 molecule has a linear geometry, with mutual internal neutralization of the two oppositely directed C=O bonds. Similarly, compensation of the sizable moments of the four C—Cl bonds in the tetrahedral CCl_4 molecule results in zero dipole moment. However, the tetrahedral $CHCl_3$ molecule has a dipole moment because the C—H bond moment differs markedly from the C—Cl bond moment (Table 5-10), thereby obviating compensation. The large dipole moment of the $H_2O(g)$ molecule (Table 5-11) suggests that its molecular geometry is angular (experimental bond angle $= 104°41'$). Since the negative charge center is the oxygen atom, the positive charge center must be at a resultant position between the hydrogen atoms, as

where $\delta-$ and $\delta+$ indicate partial charges.

Because of dipolar properties, neutral molecules may interact with other molecules and ions to give stoichiometric combinations that are not explicable

* In the older literature, *ionic* and *polar* were often used synonymously. Modern practice (Table 4-1) restricts the terms *polar* and *nonpolar* to covalent bonds. See S. J. French, *J. Chem. Educ.*, **13**, 122 (1936).

Table 5-10 Difference in Electronegativity and Bond Moment

Bond	$x_A - x_B{}^a$	Moment $(D)^b$	Bond	$x_A - x_B{}^a$	Moment $(D)^b$
C—H	0.30	0.4	F—C	1.60	1.41
N—H	0.87	1.31	Cl—C	0.33	1.46
N—D		1.30	Br—C	0.24	1.38
H—P	0.14	0.36	C—I	0.29	1.19
H—As	0	0.10	O—N	0.43	0.3
H—Sb	0.38	0.08	O=N		2.0
O—H	1.30	1.51	F—N	1.03	0.17
O—D		1.52	O=P		2.7
S—H	0.24	0.68	S=P		3.1
F—H	1.90	1.94	Cl—P	0.77	0.81
Cl—H	0.63	1.08	Br—P	0.68	0.36
Cl—D		1.09	I—P	0.15	0
Br—H	0.54	0.78	O=As		4.2
I—H	0.01	0.38	F—As	1.90	2.03
C—C	0	0	Cl—As	0.63	1.64
C=C		0	Br—As	0.54	1.27
C≡C		0	I—As	0.01	0.78
N—C	0.57	0.22	Cl—Sb	1.01	2.6
N=C		0.9	Br—Sb	0.92	1.9
N≡C		3.5	I—Sb	0.39	0.8
O—C	1.00	0.74	O=S		2.8
O=C		2.3	Cl—S	0.39	0.7
C—S	0.06	0.9	O—Cl	0.67	0.7
C=S		2.6	F—Cl	1.27	0.88
C—Se	0.02	0.8	F—Br	1.36	1.3
C—Te	0.49	0.6	Cl—Br	0.09	0.57
			Br—I	0.53	1.2

^{a} Allred–Rochow electronegativities, Table 2-18. More electronegative atom given first in each bond.
^{b} D = Debye, 1 D = 10^{-18} esu.

in terms of bonding by either electron transfer or electron pairing. Interactions of this type may be classified as:

1. Dipole–dipole interactions. This type of interaction results in the alignments of molecules that are characteristic of the associated liquids that commonly function as *electrolytic* or *ionizing solvents* (Secs. 10.1-1, 2). The potential energy of interaction, which may be expressed as

$$E \sim - \frac{\mu_A \mu_B}{d^3} \tag{5-14}$$

Table 5-11 Dipole Moments of Selected Gaseous Molecules

Molecule	Dipole Moment (D)	Molecular Geometry	Molecule	Dipole Moment (D)	Molecular Geometry
H_2, N_2, O_2, F_2, Cl_2, Br_2, I_2	0	Linear	BCl_3	0	Trigonal planar
			CCl_4	0	Tetrahedral
HF	1.94	Linear	$CHCl_3$	1.02	Tetrahedral
HCl	1.08	Linear	$SiCl_4$	0	Tetrahedral
HBr	0.78	Linear	$SnCl_4$	0	Tetrahedral
HI	0.38	Linear	$TiCl_4$	0	Tetrahedral
H_2O	1.84	Angular	CO	0.12	Linear
H_2S	0.89	Angular	CO_2	0	Linear
H_2Se	ca. 0.4	Angular	CS_2	0	Linear
H_2Te	ca. 0.2	Angular	N_2O	0.166	Linear
NH_3	1.45	Pyramidal	PF_3	1.025	Pyramidal
PH_3	0.55	Pyramidal	PCl_5	0	Trigonal bipyramidal
AsH_3	0.15	Pyramidal	AsF_3	2.815	Pyramidal
SbH_3	ca. 0.1	Pyramidal	SO_2	1.62	Angular
CH_4	0	Tetrahedral	SF_6	0	Octahedral
SiH_4	0	Tetrahedral	SO_2Cl_2	1.81	Tetrahedral
GeH_4	0	Tetrahedral	ClF_3	0.554	T-shaped
SnH_4	0	Tetrahedral	ClF	0.88	Linear

where d is the interparticle distance, is never large, but it may be sufficient to decrease vapor pressure and increase normal boiling point. If interactions of this type involve polar hydrogen atoms, they are conveniently treated separately in terms of the *hydrogen bond* (Sec. 5.2-5).

2.   Ion–dipole interactions.   Strongly dipolar species (e.g., NH_3, H_2O molecules) interact extensively with both cations and anions. The potential energy of interaction, expressed as

$$E \sim -\frac{(Z^{\pm})\mu}{4\pi d^2} \tag{5-15}$$

where $Z^{\pm}$ is the numerical charge borne by the ion, is much smaller than that for ion–ion interactions, since dipolar charges are smaller than electronic charges, and is more sensitive to interparticle distance. The solvation of ions and the related phenomena of the dissolution of ionic solids in polar solvents (Secs. 10.1-1, 10.1-2) are dependent upon interactions of this type. Certain coordination compounds (Chs. 7, 8) may owe their existence to ion–dipole interactions.

3.   Induced dipole interactions.   Highly charged ions when brought into the neighborhoods of comparatively easily polarized molecules so disturb the electronic atmospheres of the latter as to induce dipole moments into them. The potential energy of the resulting interaction between ion and induced dipole may be expressed as

$$E \sim -\frac{(Z^{\pm})^2 \alpha}{2d^4} \tag{5-16}$$

where α represents the inherent polarizability ("softness"; Sec. 9.6-4) of the molecule. Similarly, a strongly dipolar molecule may induce a dipole in a polarizable species, with a potential energy of interaction

$$E \sim -\frac{\mu^2 \alpha}{d^6} \tag{5-17}$$

where μ is the moment of the *inducing* dipole. Both ion–induced dipole and dipole–induced dipole interactions are comparatively weak, and, because d is involved at a relatively large power, they are significant only at small interparticle distances. The dissolution of ionic or polar compounds in nonpolar solvents involves interactions of these types.

4.   Instantaneous dipole–induced dipole interactions.   As a consequence of continuous changes in the intensities of charge concentration in their electronic atmospheres, instantaneous and fluctuating dipoles exist in molecules that have no permanent dipole moments. These dipoles induce dipoles in adjacent molecules, giving interactions for which the potential energies may be expressed as

$$E \sim -\frac{\bar{\mu}\alpha}{d^6} \tag{5-18}$$

where $\bar{\mu}$ is the mean instantaneous dipole moment. These interactions, commonly called London or dispersion forces or interactions,* are extremely weak and are operative only at very short interparticle distances. They increase rapidly with molecular weight (or, better, molecular volume) among comparable species. The liquefaction of a noble gas element occurs as a consequence of London forces. Observed increase in normal boiling point with molecular weight among these elements (Fig. 5-36) reflects increase in polarizability with increasing molecular size.

Partial Ionic Character in Covalent Bonds. The exact evaluation of the percentage of ionic character in a given covalent bond is difficult. In principle, this evaluation might involve the ratio of observed bond moment to theoretical bond moment. By evaluating the ratios of the then accepted dipole moments of the molecules HCl(g), HBr(g), and HI(g) and an assumed value for HF(g) to the values calculated for purely ionic bonding through multiplication of unit charge by observed internuclear distances, L. Pauling[56] determined the ionic characters of these molecules to be HF, 60%; HCl, 19%; HBr, 11%; and HI, 4%. In terms of differences in Pauling electronegativities, $x_A - x_B$, these values then led to the empirical relationship

$$\text{fraction of ionic character} = 1 - e^{-1/4(x_A - x_B)^2} \tag{5-19}$$

as shown graphically in Fig. 5-32. On this basis, any bond for which the difference in electronegativity exceeds 1.7 is more than 50% ionic and can be termed an ionic bond. Other evaluations of dipole moments[57,58] led to the second curve in Fig. 5-32 and to the relationship

$$\text{fraction of ionic character} = 0.16(x_A - x_B) + 0.035(x_A - x_B)^2 \tag{5-20}$$

in terms of which bonds of 50% or more ionic character can be expected if $x_A - x_B$ is 2.1 or larger.

The amount of ionic character in a covalent bond can be estimated also by assuming contributions from ionic resonance forms and the concomitant conclusion that the measured bond dissociation energy D_{A-B} of a molecule AB(g) is the sum of covalent and ionic contributions,[56] as

$$D_{A-B} = E_{cov} + E_{ion} \tag{5-21}$$

On the further assumption that the covalent bond dissociation energy of AB(g) is the geometric mean of the bond dissociation energies of the parent molecules $A_2(g)$ and $B_2(g)$, it follows that

$$E_{cov} = [(D_{A-A})(D_{B-B})]^{\frac{1}{2}} \tag{5-22}$$

and by substitution in Eq. 5-21

$$E_{ion} = D_{A-B} - [(D_{A-A})(D_{B-B})\}^{\frac{1}{2}} \tag{5-23}$$

---

* Dipolar and induced dipolar interactions all exemplify van der Waals forces, that is, interactive forces that are not due to bond formation or simple ionic attractions. A London force is thus a specific type of van der Waals force.

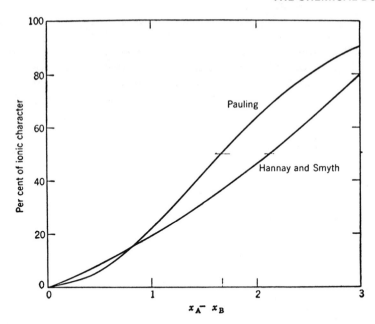

Figure 5-32 Partial ionic character in bond A—B vs. electronegativity difference. [T. Moeller, *Inorganic Chemistry*, Fig. 6-6, p. 207, John Wiley & Sons, New York (1952).]

The fraction of ionic character is then

$$\text{fraction of ionic character} = \frac{E_{\text{ion}}}{D_{\text{A--B}}} \tag{5-24}$$

Comparative values based upon Pauling's data are summarized in Table 5-12. Substantial variability is evident.

A comparable approach developed by R. T. Sanderson[52] assumes that in a bond ionic energy substitutes partially for covalent energy. Thus for a molecule AB(g)

$$D_{\text{A--B}} = t_{\text{cov}}E_{\text{cov}} + t_{\text{ion}}E_{\text{ion}} \tag{5-25}$$

where t_{cov} and t_{ion} are weighting coefficients such that their sum is unity, with t_{ion} being calculated from Sanderson's values of electronegativity (Table 2-21).

Irrespective of the approach used, bonds between fluorine or oxygen atoms and metal atoms have more than 50% ionic character. Only a few other bonds involving the fluorine atom (e.g., C—F, S—F, Cl—F, I—F) have less than 50% ionic character. Sizable differences in partial ionic character exist between bonds of fluorine and chlorine atoms to common atoms and of oxygen and sulfur atoms to common atoms. Differences within the series Cl—Br—I and S—Se—Te are less striking. Most bonds involving the alkali metal, barium, strontium, calcium, magnesium, the lanthanide, and the actinide atoms are predominantly

Table 5-12 Partial Ionic Character in Bonds in Gaseous Hydrogen Halide Molecules

Molecule	From Eq. 5-19 (%)	From Bond Dissociation Energies				From Dipole Moments (%)
		D_{A-B} (kcal mole^{-1})	E_{cov} [a] (kcal mole^{-1})	E_{ion} [b] (kcal mole^{-1})	Percent	
HF	60	134.6	61.8	72.8	54	45
HCl	19	103.2	77.7	25.5	25	17
HBr	11	87.5	69.3	18.3	21	12
HI	4	71.4	61.3	10.1	14	5

[a] From Eq. 5-22, using $D_{H-H} = 104.2$, $D_{F-F} = 36.6$, $D_{Cl-Cl} = 58.0$, $D_{Br-Br} = 46.1$, $D_{I-I} = 36.1$, each in kcal mole^{-1}.

[b] From Eq. 5-23.

ionic. All bonds, except those to oxygen or fluorine atoms, involving the beryllium and aluminum atoms are predominantly covalent. The only bonds involving the boron atom that are more than 50% ionic are those to fluorine atoms. All bonds involving carbon atoms are predominantly covalent. With silicon, the Si—F bond is predominantly ionic, but the Si—O bond is borderline, being 50% ionic by Eq. 5-20 but less so by Eq. 5-21.* Predictions for bonds involving hydrogen atoms are complicated by the fact that the small hydrogen atom so closely approaches each other atom that, except in the ionic hydrides and hydrogen fluoride, covalency predominates.

Polarization, or Deformation, of Ions. A less quantitative, but particularly useful, approach to bond polarity is based upon the transition from ionic character to covalent character as a consequence of *polarization*, or *deformation*, of ions. As a cation and a quite generally larger anion approach each other, the attraction of the cation for the electronic charge cloud of the anion and the simultaneous repulsion between the two nuclei result in deformation, distortion, or polarization of that charge cloud, as shown in Fig. 5-33. The electronic charge cloud of the cation undergoes comparable polarization, but because of the smaller radius of the cation, the effect is less pronounced. Thus the anion is said to be the polarizable species, and the cation is regarded as the polarizing agent.

In terms of the *rules of Fajans*,[59, 60] polarization of ions, and thus the transition from ionic to covalent bonding, is favored by:

1. Large charge borne by either cation or anion.
2. Small cation.
3. Large anion.
4. Cation with non-noble gas electronic configuration.

The first three require no comment, although the first assumes perfect shielding. The last reflects the general observation that shielding of the nucleus by electrons in the $(n - 1)d$ orbitals is less effective than shielding by $(n - 1)s^2(n - 1)p^6$ arrangements. Thus cations with $(n - 1)d^{10}$, or 18-electron underlying

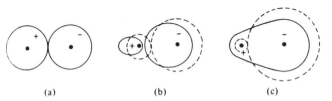

(a)　　　　　(b)　　　　　(c)

Figure 5-33 Polarization of an anion on approach of a cation. (*a*) Ion pair without polarization; (*b*) mutually polarized ion pair; (*c*) sufficient polarization to give covalent bond. [J. E. Huheey, *Inorganic Chemistry*, 2nd ed., Fig. 3.21, p. 91, Harper & Row, New York (1979.]

* In describing the molecular structures of silicates, it is convenient to consider this bond as more than 50% ionic.

arrangements are more polarizing than cations of the same size and charge with $(n-1)s^2(n-1)p^6$, or 8-electron, underlying arrangements.

Within the limitations of the predictive power of melting points in indicating the degree of ionic bonding in solids, the data summarized in Tables 5-13 and 5-14 illustrate and support the rules of Fajans. Since it is manifestly impossible to divorce from each other the individual effects on polarization of increasing cationic charge and decreasing cationic radius, it is both reasonable and useful to combine the two. The *ionic potential* (ϕ) of G. H. Cartledge,[61] defined as

$$\phi = \frac{\text{cationic charge}}{\text{cationic radius}} \qquad (5\text{-}26)$$

combines these effects nicely. As supported by the data in Tables 5-13 and 5-14, it is found that those anhydrous halides for which $\sqrt{\phi} < 2.2$ are electrolytic conductors in the molten state, whereas those for which $\sqrt{\phi} > 2.2$ are nonconductors when molten and are thus covalently bonded. Ionic-potential data correlate well with the hydration of ions and the formation of other complex species (Chs. 7, 8) and also with acid–base behavior (Sec. 9.6). With reference to the periodic table, the classic diagonal similarities between the Li^+ ion ($\phi = 1.35$, $\sqrt{\phi} = 1.16$) and the Mg^{2+} ion ($\phi = 2.78$, $\sqrt{\phi} = 1.67$) and between the Be^{2+} ion ($\phi = 5.70$, $\sqrt{\phi} = 2.48$) and the Al^{3+} ion ($\phi = 5.66$, $\sqrt{\phi} = 2.34$) are correlated. Other combined relationships, such as (cationic charge)/(cationic radius)^2, can be used with equal effectiveness.*

Some Consequences of Decreasing Bond Polarity. Decrease in bond polarity is accompanied by decreases in electrical conductivity in the molten state, in boiling point, in melting point, and in solubility in polar solvents (Sec. 10.1-2). It is of interest also that deviation of the color of a solid compound from the colors of its component ions or intensification or deepening of color may also indicate enhanced covalent character in bonding.[56,62] This situation is particularly striking among the anhydrous halides, for which change from colorless (or white) fluorides and chlorides (e.g., SnF_4, $PbCl_2$, SbF_3, $BiCl_3$, $AgCl$, $HgCl_2$) through palely colored bromides to intensely colored iodides (e.g., red SnI_4, SbI_3, BiI_3, HgI_2; yellow PbI_2, AgI) is not uncommon. Color associated with strongly polarized iodide ions is to be contrasted with that associated with partial occupancy of d or f orbitals [e.g., violet Ti^{3+}(c or aq), green Pr^{3+}(c or aq) or U^{4+}(c or aq), yellow $Co(NH_3)_6^{3+}$ (c or aq)]; with unpaired electrons in certain odd molecules and ions [e.g., $ClO_2(g)$, $NO_2(g)$, $ON(SO_3)_2^{2-}$(aq) (Sec. 5.1-6)]; or with the presence of two oxidation states of a metal in a crystalline solid [e.g., black $\overset{II\ III}{FeFe_2O_4}$, red $\overset{II\ IV}{Pb_2PbO_4}$, black $\overset{I\ \ I\ III}{M_2AuAuCl_6}$].

---

* For an interesting and detailed discussion of ionic models for the bonding in metal chlorides, see R. G. Pearson and H. B. Gray, *Inorg. Chem.*, **2**, 358 (1963).

Table 5-13 Melting Points and Ionic Potentials of Anhydrous Halides: Three Rules of Fajans

Halide	Anion (and Radius) (Å)	Cation					Melting Point (°C)	Rule Illustrated
		Formula	Charge	Radius (Å)	ϕ	$\sqrt{\phi}$		
NaCl	Cl⁻ (1.81)	Na^+	+1	1.02	0.98	0.99	801	→ 1
CaCl₂		Ca^{2+}	+2	1.00	2.00	1.41	772	
MgCl₂		Mg^{2+}	+2	0.72	2.78	1.67	708	→ 1
AlCl₃		Al^{3+}	+3	0.53	5.66	2.34	190 (2.5 atm)	
BeCl₂	Cl⁻ (1.81)	Be^{2+}	+2	0.35	5.70	2.48	405	←
MgCl₂		Mg^{2+}		0.72	2.78	1.67	708	
CaCl₂		Ca^{2+}		1.00	2.00	1.41	772	2
SrCl₂		Sr^{2+}		1.16	1.72	1.31	873	
BaCl₂		Ba^{2+}		1.36	1.47	1.21	963	→
CaF₂	F⁻ (1.33)	Ca^{2+}	+2	1.00	2.00	1.41	1360	
CaCl₂	Cl⁻ (1.81)						772	3
CaBr₂	Br⁻ (1.95)						730	
CaI₂	I⁻ (2.16)						740	→

Table 5-14 Melting Points and Ionic Potentials of Anhydrous Chlorides: Fourth Rule of Fajans

	Underlying 8-Electron Arrangement					Underlying 18-Electron Arrangement			
Cation	Radius (Å)	ϕ	$\sqrt{\phi}$	Melting Point (°C)	Cation	Radius (Å)	ϕ	$\sqrt{\phi}$	Melting Point (°C)
Na^+	1.02	0.98	0.99	801	Cu^+	0.96	1.04	1.02	430
K^+	1.38	0.72	0.85	776	Ag^+	1.15	0.87	0.93	455
Rb^+	1.49	0.67	0.82	715	Au^+	1.37	0.73	0.85	170 (dec.)
Mg^{2+}	0.72	2.78	1.67	708	Zn^{2+}	0.745	2.68	1.63	283
Ca^{2+}	1.00	2.00	1.41	772	Cd^{2+}	0.95	2.10	1.45	568
Sr^{2+}	1.16	1.72	1.31	873	Hg^{2+}	1.02	1.96	1.40	276

2. THE HYDROGEN BOND

The conclusion that the hydrogen atom can be attracted simultaneously to two or more atoms of other elements and thereby serve as a bridging atom is usually ascribed to W. M. Latimer and W. H. Rodebush,[63] who used this concept to account for the association of both water and liquid hydrogen fluoride, the large dielectric constants of these two liquids (Table 10-1), the small electrolytic dissociation of aqueous ammonia, and the existence of dimeric molecules of acetic acid. However, the existence of a hydrogen-atom bond between the nitrogen and oxygen atoms in trimethylammonium hydroxide had been postulated earlier,[64] and this type of linkage had been suggested to account for the reduced acidity of phenolic hydrogen atoms *ortho* to carbonyl groups in benzenoid compounds.[65] Furthermore, M. L. Huggins[66] had already used the concept to explain tautomerism in acetoacetic ester, the interaction of ammonia with water, and the molecular structure of sulfuric acid.[67] Indeed, Latimer and Rodebush referred to unpublished work by Huggins, and G. N. Lewis stated: "The idea was first suggested by Dr. M. L. Huggins, and was also advanced by Latimer and Rodebush, who showed the great value of the idea in their paper...."[68]

It is now recognized that the hydrogen bond is the most common and strongest of all intermolecular interactions and, except for ionic and covalent interactions, is probably the most important single structural principle in chemistry, physics, biology, crystallography, mineralogy, geology, meteorology, and so on.[67] The brief treatment that follows should be supplemented by reference to the excellent summaries that are available.[67, 69-75] The hydrogen bond is not an electron-deficient bond and must be clearly distinguished from the three-centered, two-electron hydrogen bridge discussed earlier (Sec. 5.1-5). Hydrogen bonding, although normally discussed in terms of the ^1H isotope, is of course characteristic of deuterium (^2H) and tritium (^3H) as well.

2-1. Some Characteristics of the Hydrogen Bond

Hydrogen bonds are ordinarily restricted to the highly electronegative atoms fluorine, oxygen, nitrogen, and chlorine, although there are well-documented examples involving carbon atoms bonded to strongly electronegative substituents (e.g., in $CHCl_3$, $CHCl_2F$, HCN)[76-78] or aromatic nuclei.[79]

It is evident from bond-energy data,[72] some typical values of which are summarized in Table 5-15, that, although hydrogen bonds vary widely in strength, they are relatively weak compared with most ionic and covalent bonds but relatively strong compared with other intermolecular interactions. Except for the $F—H\cdots F^-$ entity in crystals, bond energies usually range between 3 and 10 kcal mole^{-1}. Bonds below this range are weak; those above this range strong.[72,*] Hydrogen bonds in the weak and normal range form and break so

---

* For additional examples of strong hydrogen bonds, see J. Emsley, *Chem. Soc. Rev.*, **9**, 91 (1980).

Table 5-15   Energies for Some Common Hydrogen Bonds

Type of Bond	System	$-\Delta H$ (kcal mole^{-1})
F—H⋯F	HF(g)	6.8
	$HF + (CH_3)_4N^+F^-$	37
F—H⋯O	$HF + CH_3COCH_3$	11
O—H⋯O	CH_3CO_2H(g)	7.0
	$CCl_3CO_2H + (C_6H_5)_3PO$	16
	H_2O(g)	5.0
N—H⋯O	$CH_3CONHCH_3$(in CCl_4)	3.9
C—H⋯O	$CHCl_3(l) + CH_3COCH_3(l)$	0.95
C—H⋯N	HCN(g)	3.3

readily that they are of particular importance in biological systems where energy changes are minimal. Enhancement of the electronegativity of a particular atom to which a hydrogen atom is bonded increases the strength of hydrogen bonds that it can form. Thus bond strength increases from aliphatic alcohols to phenols, from the ammonium ion to substituted—but not quaternary— ammonium ions, and from methyl chloride to methylene chloride to chloroform. Size of the bonded atom is apparently of importance also, since the nitrogen atom forms stronger hydrogen bonds than the larger chlorine atom, although the two electronegativities are comparable.

The length of a hydrogen bond is expressed as the internuclear distance between the two bonded electronegative atoms. A number of bond lengths are listed in Table 5-16.[74] Substantial variability is apparent for both F—H⋯F and O—H⋯O bonds; less for others. Longer O—H⋯O bonds that characterize certain crystalline metal hydroxides and alkali metal hydroxide hydrates[80] were at one time called *hydroxyl bonds* to indicate an assumed retention of the hydrogen atom by the oxygen atom to which it was bonded originally. This distinction is no longer recognized, since variations of this type probably reflect differences in packing in the crystal lattices.[70] The data in Table 5-17 indicate that hydrogen-bond distances are commonly less than van der Waals contacts.[81] Indeed, such a shortening may be taken as an indication of the presence of a hydrogen bond, as opposed to a definitive van der Waals interaction.

The data in Table 5-17 on H⋯B distances indicate that hydrogen bonds in general are unsymmetrical; that is, the hydrogen atom is not located midway between the other two atoms. Although it was proposed originally that hydrogen bonds are symmetrical, early experiments by J. D. Bernal and R. H. Fowler[82] indicated clearly that in water such is not the case. The resulting implication that, to a reasonable degree, water molecules in ice retain their identities is consistent with L. Pauling's conclusion that the residual entropy of ice at sufficiently low temperatures prevents a "switchover" from one bond length to another.[83] Every long O—H⋯O bond in which the position of the

Table 5-16 Lengths of Some Common Hydrogen Bonds

Bond	Compound	Length (Å)	Bond	Compound	Length (Å)
F—H···F	NaHF$_2$(c)	2.27	N—H···O	NH$_4$OOCH$_3$	2.81-2.89
	KH$_4$F$_5$(c)	2.45		CO(NH$_2$)$_2$	2.99-3.01
	HF(l)	2.49	N—H···F	NH$_4$F, N$_2$H$_6$F$_2$	2.61-2.82
O—H···O	Acid salts	2.40-2.55	N—H···Cl	(CH$_3$)$_3$NHCl, (CH$_3$)$_2$NH$_2$Cl	3.00, 3.11
	Ice, hydroxy compounds, hydrates	2.70-2.90		NH$_3$OHCl	3.20
			N—H···N	NH$_4$N$_3$,	2.94
O—H···F	CuF$_2$·2H$_2$O, [Fe(H$_2$O)$_6$][SiF$_6$]	2.65-2.72		NCNC(NH$_2$)$_2$	3.15
O—H···Cl	HCl·nH$_2$O	2.92-2.95		NH$_3$	3.35
	[NH$_3$OH]Cl	3.04	S—H···S	H$_2$S(c)	3.94
O—H···Br	CsBr·$\frac{1}{3}$(H$_3$OHBr$_2$)	3.04	C—H···O	organic species	2.92
O—H···N	N$_2$H$_4$·4CH$_3$OH	2.68	Cl—H···Cl	CsCl·$\frac{1}{3}$(H$_3$OHCl$_2$)	3.14
	N$_2$H$_4$·H$_2$O	2.79	Br—H···Br	CsBr·$\frac{1}{3}$(H$_3$OHBr$_2$)	3.35

Table 5-17 Comparison of Observed Bond Distances in Species A—H···B with Distances Calculated for van der Waals Contacts[a]

Bond Type	Bond Distance (Å)			
	A···B		H···B	
	Observed	Calculated	Observed	Calculated
[F—H···F]⁻	2.4	2.7	1.2	2.6
O—H···O	2.7	2.8	1.7	2.6
O—H···F	2.7	2.8	1.7	2.6
O—H···N	2.8	2.9	1.9	2.7
O—H···Cl	3.1	3.2	2.2	3.0
N—H···O	2.9	2.9	2.0	2.6
N—H···F	2.8	2.9	1.9	2.6
N—H···Cl	3.3	3.3	2.4	3.0
N—H···N	3.1	3.0	2.2	2.7
N—H···S	3.4	3.4	2.4	3.1
C—H···O	3.2	3.0	2.3	2.6

[a] Based upon radii as listed by L. Pauling, Ref. 5(b).

hydrogen atom has been established is not symmetrical. On occasion, non-symmetrical hydrogen bonds of more than one length have been found in the same system e.g., 2.53 Å and 2.74 Å in crystalline $Na_3H(CO_3)_2 \cdot 2\ H_2O$.[84] It appears that short O—H···O bonds (ca. 2.4 Å) are symmetrical. In the ion F—H···F⁻, the bond has been reported to be symmetrical,[85] but a single-crystal, neutron-diffraction investigation of *p*-toluidinium bifluoride indicated F—H distances of 1.025 and 1.235 Å in the same ion.[86]

Although it may be assumed, as a first approximation, that the atoms in a hydrogen bond are collinear, deviations from absolute linearity are observed. Thus in the O—H···O bonds linking the HCO_3^- and CO_3^{2-} ions in crystalline $Na_3H(CO_3)_2 \cdot 2\ H_2O$, there is a departure of 3.5° from true linearity,[84] and in the F—H···F bond in crystalline *p*-toluidinium bifluoride the departure is 1.9°.[86]

2-2. Types of Hydrogen Bonds

Hydrogen bonds can be classified, in terms of structural arrangements, as:[69]

1. Intermolecular, extending over many molecules. Ice-I$_h^*$ is a typical example involving a tetrahedral array of oxygen atoms around each oxygen atom, as predicted by M. L. Huggins and verified by W. H. Bragg[87] [see Fig. 5-34(a)].

* Ice-I$_h$ is the hexagonal polymorph that normally crystallizes at 1 atm. Some nine other polymorphs have been described (see A. F. Wells, Ref. 74, Ch. 15).

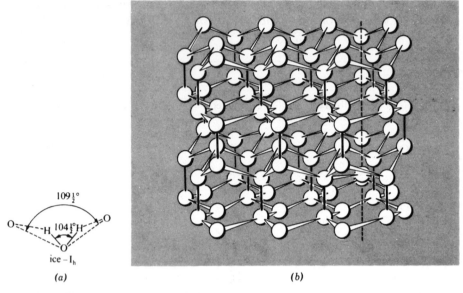

Figure 5-34 Crystal structure and hydrogen bond array of ice-I_h. Only the oxygen atoms are shown. The O---O distances are 2.75 Å. [(a) A. F. Wells, *Structural Inorganic Chemistry*, 4th ed., p. 305, Clarendon Press, Oxford (1975); (b) F. A. Cotton and G. Wilkinson, *Basic Inorganic Chemistry*, Fig. 9-2, p. 213, John Wiley & Sons, New York (1976).]

2. Intermolecular, extending over only two molecules in a dimeric species. The crystalline aliphatic carboxylic acids are typical examples. Thus dimeric formic acid can be formulated, without reference to structural details, as

$$H-C\underset{\displaystyle O-H\cdots O}{\overset{\displaystyle O--H-O}{\big<}}C-H$$

3. Intramolecular, involving a hydrogen atom bonded to two atoms in the same molecule. The *cis*-isomer of *o*-chlorophenol, formulated as

may be an example. Conclusions that *bifurcated* bonds exist in structures where a covalently bonded hydrogen atom is close to two other electropositive atoms are open to question.

4. The specific case of the FHF⁻ ion, which exists as a linear or nearly linear entity of relatively short bond distances (Table 5-16). By contrast, the longer bond in HF(l) is found in a zigzag structure, as

The same type of arrangement probably characterizes the ions $H_2F_3^-$, $H_3F_4^-$, $H_4F_5^-$, $H_6F_7^-$, and $H_7F_8^-$.[88]

A comparable, but slightly different, classification based upon crystal structures[74] lists hydrogen-bonded systems among solids as:

1. Finite groups: for example, $[H(CO_3)_2]^{3-}$, $(RCOOH)_2$, FHF^-
2. Infinite chains: for example, HCO_3^-, HSO_4^-
3. Infinite layers: for example, $B(OH)_3$, H_2SO_4, $N_2H_6F_2$
4. Three-dimensional networks: for example, ice, H_2O_2, $Te(OH)_6$, $H_2PO_4^-$ ion in KH_2PO_4

2-3. Theoretical Considerations [69–72,89]

To be completely satisfactory, a theoretical treatment of the hydrogen bond should yield bond energies of the correct order of magnitude, should be consistent with the observed directional properties of that bond, and should interpret correctly other observed characteristics of hydrogen-bonded systems.[89] Although electrostatic, valence-bond, and molecular-orbital treatments have been developed, none of these is completely useful for all systems.

The simple electrostatic model, based upon ion–dipole or dipole–dipole interactions, as supported by the observed limitation of hydrogen bonds to cases involving strongly electronegative atoms (p. 251), is particularly attractive because it avoids the problem of a covalency for the hydrogen atom larger than one and accounts well for many measured bond energies and experimentally determined geometries. However, the shortness of hydrogen bonds (Tables 5-16, 5-17) indicates considerable van der Waals overlap, which should result in forces of repulsion that would decrease bond strength markedly. Furthermore, symmetrical bonds would not be expected if the hydrogen atom were covalently bonded to one electronegative atom and electrostatically attracted to the other atom.

The simple valence-bond approach, based upon the fact that in the bond A—H···:B an unshared electron pair on atom B is directed toward the H atom would impose a covalency of two upon that atom. Resonance structures may contribute, as

$$A—H\cdots B \leftrightarrow \bar{A}—\overset{+}{H}\cdots B \leftrightarrow \bar{A}\cdots H—B^+ \leftrightarrow A\cdots H—B$$

which would require delocalization of the covalent bond on both sides of the H atom and ionic contributions. However, a molecular-orbital treatment involving a three-centered, four-electron bond effects the same result in terms of a "smear" of the covalent interaction over all three atoms. This approach is

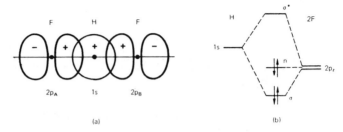

Figure 5-35 Molecular-orbital description of symmetrical hydrogen bond. (*a*) Atomic orbital overlaps, (*b*) energy-level diagram. [Ralf Steudel, *Non-Metal Chemistry, With an Introduction to Atomic Structure and Chemical Bonding*; English Edition by F. C. Nachod, J. J. Zuckerman. Walter de Gruyter, Berlin, New York (1977).]

depicted for the symmetrical bond in the FHF^- ion in Fig. 5-35, where both orbital overlap and an energy-level diagram are shown. Two electrons are then in a bonding σ orbital and two in a nonbonding orbital. For a symmetrical bond, the electronic charge density is the same for both linkages; for an asymmetrical bond, it is larger for the shorter linkage.

In spite of its deficiencies, the electrostatic model is still the most generally useful.

2-4. Establishment of the Presence of Hydrogen Bonding

The existence of hydrogen bonding is inferred by changes in certain physical properties that reflect association or lack of association. Changes ascribable to hydrogen bonding include increases in normal boiling point and enthalpy of fusion or vaporization and decreases in vapor pressure and normal freezing point. Proof of its existence, however, requires the use of a structure-dependent instrumental technique.[72] The structure-indicating techniques of refined x-ray and/or neutron diffraction (Sec. 6.5-1) permit the exact location of H atoms relative to other atoms.[71] Vibrational absorption spectroscopy (Sec. 6.5-3) establishes hydrogen bonding in terms of the displacements of absorption bands characteristic of the A—H bond to lower frequencies because of the weakening of that bond when a hydrogen bond is formed also.[72] A down-field shift in proton nuclear magnetic resonance absorption (Sec. 6.5-4) associated with an A—H bond indicates participation of the proton in hydrogen bonding.[90]

2-5. Hydrogen Bonding in Inorganic Systems

Among the more important examples of hydrogen bonding are the following:

1. Association among simple covalent hydrides. The seemingly abnormally high normal melting and boiling points and large enthalpies of vaporization of the simple covalent hydrides of fluorine, oxygen, and nitrogen, as indicated in Figs. 5-36 and 5-37, together with the sharp decreases to those of the chlorine,

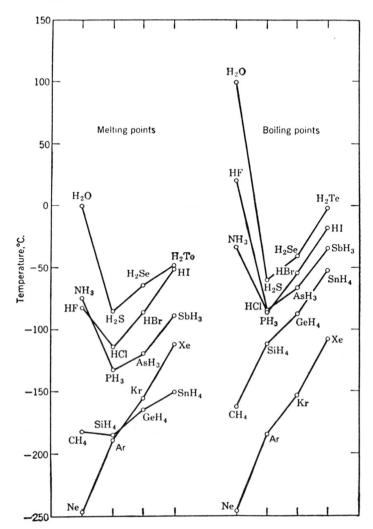

Figure 5-36 Melting points and boiling points of simple covalent hydrides and noble gas elements. [T. Moeller, *Inorganic Chemistry*, Fig. 6-3, p. 190, John Wiley & Sons, New York (1952).]

sulfur, and phosphorus analogs are interpreted as consequences of molecular association through hydrogen bonding involving these derivatives of the strongly electronegative nonmetals. By contrast, the subsequent regular increases in these constants with increasing molecular weight, paralleling those for the noble gas elements, suggest the operation of only weak London forces (Sec. 5.1-8) and the absence of any strong intermolecular interactions. The symmetrical, nonpolar CH_4 molecule gives no evidences of association.

2. Association among other polar hydrogen compounds. Comparable association through hydrogen bonding also characterizes the molecular species HCN, H_2O_2, and NH_2OH.

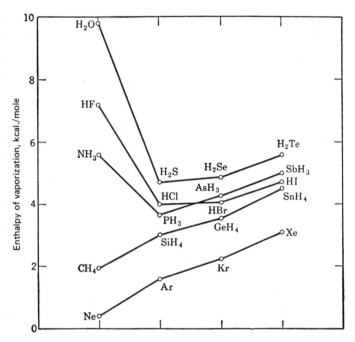

Figure 5-37 Enthalpies of vaporization of simple covalent hydrides and noble gas elements. [T. Moeller, *Inorganic Chemistry*, Fig. 6-4, p. 191, John Wiley & Sons, New York (1952).]

3. **Anionic solvation.** The HF_2^- ion has been described. Other related species in the ionic solids $K^+H_2F_3^-$, $K^+H_4F_5^-$, and $K_2^+H_5F_7^{2-}$ exemplify more extensive solvation of the F^- ion by HF molecules to give molecular arrangement such as [91,92]

In crystals of the salt $FeSiF_6 \cdot 6\,H_2O$, the six H_2O molecules bond to the Fe^{2+} ion through the O atoms to form the octahedral $[Fe(OH_2)_6]^{2+}$ ion, each of the H atoms is linked by a hydrogen bond, and each F atom forms two hydrogen bonds to H_2O molecules.[93] Salts such as $[N(CH_3)_4]^+HCl_2^-$, $[N(C_2H_5)_4]^+HBr_2^-$, and $[N(C_4H_9)_4]^+HI_2^-$ are thermally stable at room temperature and apparently contain halide ions solvated via hydrogen bonds by hydrogen halide molecules.[94] Crystals of the compound $HBr \cdot 2\,H_2O$ at $-40°C$

contain the ions $[H_5O_2]^+$ and Br^-, with each cation associated with four Br^- ions through hydrogen bonds. The central O—H—O group is slightly asymmetric with a total length of 2.44 Å and an OHO angle of 174.7°.[95] In crystals of $CuSO_4 \cdot 5H_2O$, four H_2O molecules bond through O atoms to a Cu^{2+} ion, each forming two hydrogen bonds. The fifth H_2O molecule lies between the $[Cu(H_2O)_4]^{2+}$ and SO_4^{2-} ions and forms four hydrogen bonds, two of which are to O atoms in the SO_4^{2-} ion and the other two to H atoms in coordinated water molecules.[96] In crystals of $Na_4XeO_6 \cdot 6H_2O$, alternating layers of compositions $(XeO_6^{4-} \cdot 2Na^+)$ and $(2Na^+ \cdot 6H_2O)$ are held together by hydrogen bonds between H_2O molecules and the O atoms of the XeO_6^{4-} ions.[97] For crystals of $CsH(NO_3)_2$, neutron-diffraction data indicate the presence of the $O_2NO\text{-}H \cdots ONO_2^-$ ion, characterized by a short [$= 2.468(8)$ Å] $O \cdots O$ distance and a symmetrical hydrogen bond.[98]

4. The hydrated proton. Inasmuch as the tiny proton represents an extremely high concentration of positive charge, it is reasonable to expect that it will interact with polar molecules and solvate in polar solvents such as water. The monohydrate H_3O^+, universally termed the *hydronium* ion, but probably more properly called the *oxonium* ion, is a non-hydrogen-bonded species that exists in a number of crystals, for example, $H_3O^+ClO_4^-$,[99] $(H_3O^+)_2SO_4^{2-}$,[100] $(H_3O^+)(HSeO_4)^-$,[100a] $H_3O^+AsF_6^-$ and $H_3O^+SbF_6^-$,[101] and $H_3O^+NO_3^-$.[101a] It is commonly assumed to be present in acidic aqueous solutions (p. 587). Consideration of theoretical potential energy curves for OHO bond systems prompted M. L. Huggins to suggest that in aqueous solutions the proton is at the center of a short, strong, symmetrical bond involving two H_2O molecules (i.e., $H_2O \cdots \overset{+}{H} \cdots OH_2$), or stoichiometrically $H_5O_2^+$.[102] The large mobility of the proton in aqueous solutions can then be accounted for in terms of slight shifts of the positions of two H nuclei within this ion, resulting in the addition of an H_2O molecule at one end and the simultaneous removal of an H_2O molecule at the other end with displacement of the positive charge through the solution.

The $H_5O_2^+$ ion, or its deuterium analog, has been characterized in a number of crystalline salts, among them $(H_5O_2)^+ClO_4^-$,[103] $(H_5O_2)^+Cl^-$,[104] $(H_5O_2)_2^+SO_4^{2-}$ and $(D_5O_2)_2^+SO_4^{2-}$,[105] $(H_5O_2)^+[Ln(C_2O_4)_2]^- \cdot H_2O$ (Ln = Y, Er),[106] *trans*- $[Co(en)_2Cl_2]^+Cl^- \cdot (H_5O_2)^+Cl^-$ (en = ethylenediamine),[107] $(H_5O_2)^+AuCl_4^- \cdot 2H_2O$,[108] and $(H_5O_2)^+CF_3SO_3^-$.[108a] An asymmetric $H_5O_2^+$ ion is present in crystals of $(H_5O_2)^+C_6H_2(NO_2)_3SO_3^- \cdot 2H_2O$, picrylsulfonic acid 4-hydrate.[109] The ions $H_7O_3^+$ and $H_9O_4^+$ are also present in a number of crystalline salts, for example, $(H_7O_3)^+(H_9O_4)^+Br_2^- \cdot H_2O$,[110] 2,5-dichloro-benzenesulfonic acid 3-hydrate,[111] *o*-sulfobenzoic acid 3-hydrate,[112] and $H_7O_3^+ClO_4^-$ and $H_7O_3^+NO_3^-$ (i.e., $HClO_4 \cdot 3H_2O$ and $HNO_3 \cdot 3H_2O$).[113, 113a]

The related peroxonium ion, H_2OOH^+, has been characterized by vibrational and nuclear magnetic resonance spectroscopic techniques (Ch. 6) in the salts $(H_3O_2)^+Sb_2F_{11}^-$, $(H_3O_2)^+SbF_6^-$, and $(H_3O_2)^+(AsF_6)^-$.[113b]

5. Hydrogen-bonded ferroelectric compounds.[71, 114] In a number of substances, ferroelectric transitions, that is, spontaneous polarizations amounting to displacements between centroids of positive and negative electric charges, are associated with reorientations of hydrogen bonds. Examples include $NaKC_4H_4O_6 \cdot 4\,H_2O$ (Rochelle salt), $CaB_3O_4(OH)_3 \cdot H_2O$ (colemanite), $(NH_4)_2SO_4$, $K_4[Fe(CN)_6] \cdot 3\,H_2O$, and KH_2PO_4. The principle can be illustrated in terms of a diagrammatic representation of the crystal structure of KH_2PO_4,[114] where the uniform polarization of all the hydrogen bonds is

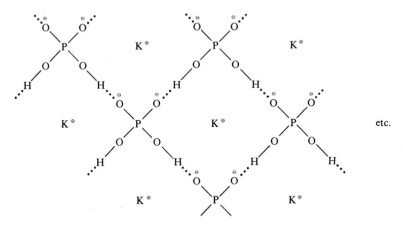

readily reversed; that is, H atoms in O—H···O bonds move to give O···H—O bonds.

3. INCLUSION COMPOUNDS

An *inclusion compound* is a *solid* substance in which molecules of one type (the *guest*) are trapped in appropriate voids in the crystal lattice formed by molecules of a second type (the *host*) in such a way that their escape is precluded.[115-123] They are most commonly formed when the host is crystallized in the presence of the guest. Their properties are very roughly those of the host. Each type has an idealized stoichiometry based upon the limitations of crystal structure, but this stoichiometry is more often approached than exactly achieved. Inclusion compounds are usually stable with respect to decomposition into their components as long as the trapping structure is maintained, but destruction of the structure by melting or dissolution permits the guest to escape. Although the formation of inclusion species is governed primarily by the combination of a crystal structure with voids of suitable size with molecules of the correct size to be trapped within these voids, it is not unlikely that van der Waals interactions contribute to some degree to both their formation and their stability.

The nature of the cavity allows classification in terms of *layer*, *channel*, and *cage* structures.[116] For each of these structures, conditions for formation are limited and highly specific. The following conditions are uniformly applicable:

1. An open crystal structure of the host, which is a consequence of directed bonds, sufficient extension of groups to form cavities of suitable size, and rigid structure.

2. Small access holes to enclosing cavities, resulting from either proper distribution of groups in the formation of the crystal or sufficient area in the enclosing groups.

3. Ready availability of the guest species when the cavity is closed.

3-1. Layer Structures

In inclusion compounds of this type, the guest species are held between layers of atomic or molecular thickness within crystals of the host species. The lamellar compounds of graphite (Fig. 5-21),[119, 124-126] hexagonal boron nitride (Fig. 5-21),[126, 127] and certain silicates such as micas, vermiculites, and montmorillonite[119] are typical examples. Inclusion compounds of this type are termed *intercalation* compounds.[128]

3-2. Channel Structures

Channels in crystalline solids can be long and essentially parallel tubes or short, wide "windows" between relatively large, three-dimensional cavities. The first arrangement is characteristic of hexagonal urea and thiourea.[129, 130] The second arrangement is characteristic of the synthetic crystalline zeolites that function as *molecular sieves*.[119, 131]

Each channel in crystalline urea is formed by three interpenetrating spirals of molecules held together by hydrogen bonding. Each O atom is hydrogen bonded to four N atoms and each N atom to two O atoms. The hydrogen bonds are essentially coplanar with the urea molecules. Two N—H$\cdots$O bond lengths are found, 2.93 and 3.04 Å. The spirals are randomly right- or left-handed. Each channel has the correct size to accommodate a zigzag, unbranched hydrocarbon chain.[130, 131] Channels in crystals of thiourea are comparable but, as a consequence of the larger size of the sulfur atom, have larger cross-sectional areas and can trap branched-chain, alicyclic, and other molecules of similar dimensions.[130] Again, the spiral channels are right- or left-handed.

The zeolites, although differing substantially from each other in chemical compositions, all have three-dimensional crystal lattices based upon networks of bonded SiO_4^{4-} and AlO_4^{5-} tetrahedra and containing channels that readily trap small molecules such as H_2, O_2, N_2, CO, CO_2, H_2O, NH_3, and CH_4.[119]

3-3. Cage Structures

Compounds in which guest molecules are trapped in three-dimensional cavities with openings too small to allow their escape have cage structures and are called *clathrate* compounds.[132] (Latin *clathratus*, enclosed by cross bars of a grating). Typical compounds include the β-quinol compounds of the type $3 C_6H_4(OH)_2 \cdot n G$ (G = HCl, HBr, H_2S, CH_3OH, SO_2, CO_2, HCN, CH_3CN,

HCO_2H, Ar, Kr, Xe; $n = 1$ ideally, but actually ca. 0.3 to 0.97;[115,133,134] the related phenol compounds of the type $12\,C_6H_5OH \cdot nG$ ($n = 2$ for G = CS_2; = 4 for SO_2; = 5 for HCl, HBr);[135-137] the products derived from Dianin's compound, $nC_{18}H_{20}O_2 \cdot G$ ($n = 2$ for G = CH_3OH; = 3 for SO_2, CCl_4; = 6 for C_6H_5Br, Ar);[138-141] the Hofmann-type dicyanomonoamminenickel(II) compounds, $Ni(NH_3)(CN)_2 \cdot G$ (G = C_6H_6, C_6H_5OH);[142-144] the related compounds $MNi(CN)_4 \cdot nG$ and $MNi(NH_3)_2(CN)_4 \cdot nG$ (M = Cu, Zn, Cd; G = C_6H_6);[145-147] the tetra-4-methylpyridinemetal(II) thiocyanates, $M(4\text{-}Mepy)_4(SCN)_2 \cdot G$ (M = Fe, Co, Ni; G = xylenes, p-disubstituted benzenes), and related compounds;[148,149] and the hydrates, ideally $46\,H_2O \cdot 6G$ (G = AsH_3, C_2H_4, C_2H_6, N_2O, Cl_2, COS, Br_2, CO_2, SO_2, ClO_2, $(CH_3)_2O$, etc.), $46\,H_2O \cdot 8G$ (G = Ar, Kr, Xe, N_2, O_2, H_2S, H_2Se, PH_3, CH_4, C_2H_4, CH_3F), $136\,H_2O \cdot 8G$ (G = CH_3I, CH_2Cl_2, C_2H_5Cl, CF_2Cl_2, $CFCl_3$, CH_3CHCl_2, C_3H_8, $(CH_3)_2CO$, $(CH_3)_2O$, $(CH_2)_4O$, $(CH_2)_4O_2$), and $136\,H_2O \cdot 16\,H_2S \cdot 8G$ (G = CH_3I, COS, CS_2, $CFCl_3$, CCl_4, SF_6, C_6H_6, $CHCl_3$, CCl_3NO_2, etc.).[71,120,150-154] Other reported clathrate compounds are polyhedral water–fluoride cluster anions in the tetramethylammonium fluoride hydrates;[155] a series of clathrochelates derived from iron(II), dimethylglyoxime, and either boric acid or boron trifluoride;[156] benzene clathrates of compositions $Cd(en)Cd(CN)_4 \cdot 2\,C_6H_6$ and $Cd(en)Hg(CN)_4 \cdot 2\,C_6H_6$;[157] and other Hofmann-type compounds of compositions $M(NH_3)_2M'(CN)_4 \cdot 2G$ (M = Mn, Fe, Co, Ni, Cu, Zn, Cd; M' = Ni, Pd, Pt; G = C_6H_6, $C_6H_5NH_2$, pyrrole, thiophene).[158] In each of these types of clathrate compounds, the cage apparently results from appropriate hydrogen bonding.

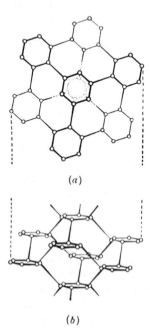

(a)

(b)

Figure 5-38 Representation of crystal structure of β-quinol. (a) In plane. Each regular hexagon denotes six hydrogen bonds between oxygen atoms. Hexagons at different levels are represented by different line thickness. The tapered lines, representing the O---O axis of a hydroquinone molecule, show the method of linking to form an infinite three-dimensional cagework. Each taper points downward from the observer. (b) Perspective drawing corresponding to the above. The hexagons denote the hydrogen bonds; the longer lines connecting different hexagons denote the O—O axis of the hydroquinone molecule. The complete structure as found in the molecular compounds consists of the cagework shown, together with a second identical interpenetrating cagework, which is displaced vertically halfway between the top and bottom hexagons. [H. M. Powell, in *Non-Stoichiometric Compounds*, L. Mandelcorn (Ed.)., Fig. 2, p. 456, Academic Press, New York (1966).]

Only the β-quinol and water clathrate compounds need be discussed in greater detail. β-Quinol is a normally metastable crystalline modification that is apparently stabilized by the inclusion of other species. In crystals of β-quinol, the OH groups from three quinol molecules are hydrogen-bonded to each other, giving a six-membered O—H—O type ring, with the C_6H_4 groups projecting alternately above and below its plane. Inasmuch as the two OH groups in a given quinol molecule are *para* to each other, both ends of the molecule are similarly involved, and a three-dimensional network results. Interpenetration of two networks of this type gives vertical alternation of the O—H—O rings and yields cavities that are further enclosed by the linking C_6H_4 groups. The result is a structure comparable to that depicted in Fig. 5-38. Each roughly spherical cavity has a diameter of ca. 7.9 Å and can accommodate without excessive strain a spherical species of radius up to ca. 2.25 Å. Thus neither very small molecules (e.g., He) nor large molecules (e.g., CCl_4) can be trapped. The idealized composition $3\,C_6H_4(OH)_2 \cdot G$ corresponds to single occupancy of every cavity present (Fig. 5-39). The β-quinol clathrates of the heavier noble gas elements were the first inclusion compounds of these elements to be investigated in detail.[134]

The structural feature common to all but a few clathrate hydrates is a pentagonal dodecahedron of oxygen atoms formed as a consequence of the linking of H_2O molecules via hydrogen bonding.[153, 154, 159] This polyhedron, corresponding to a unit $H_{40}O_{20}$ and having 12 regular pentagonal faces, 20 vertices, and 30 edges, is shown diagrammatically in Fig. 5-40. It is apparent that this polyhedron encompasses a void. The interaction of pentagonal dodeca-

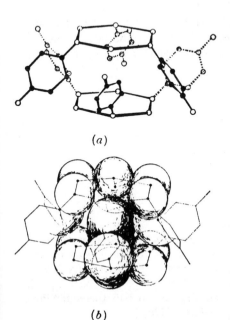

(a)

(b)

Figure 5-39 Representation of β-quinol-argon clathrate. (a) The positions of the six hydroquinone molecules that form the immediate surroundings of an argon atom are shown in perspective drawing. This is deceptive because the centers of the atoms are shown, and the square occupied by the atoms cannot be shown at the same time. (b) The spaces occupied by the atoms are indicated in a drawing on the same scale as that above. A few only of the atoms surrounding the central argon atom are shown. They are distinguished from those that have been omitted by small circles drawn around their centers. Broken lines are used to indicate that a part of the structure so represented lies behind some portion drawn with full lines. [H. M. Powell, in *Non Stoichiometric Compounds*, L. Mandelcorn (Ed.), Fig. 4, p. 459, Academic Press, New York (1966).]

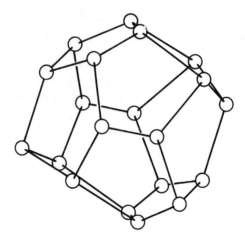

Figure 5-40   Pentagonal dodecahedron of oxygen atoms in $H_{40}O_{20}$ cluster.

hedra with tetrakaidecahedra (14-hedra, with 14 pentagonal faces, 24 vertices, and 36 edges) gives unit cells of the type depicted in Fig. 5-41. Within this unit cell ($a = $ ca. 12 Å), 46 water molecules enclose two decahedral voids and six somewhat larger 14-hedral voids. When all of these voids are occupied, the hydrate has the composition $46\,H_2O \cdot 8\,G$, or $G \cdot 5.75\,H_2O$ (e.g., if G = Ar, Kr, CH_4, H_2S). When only the larger voids are occupied, the composition is $46\,H_2O \cdot 6\,G$, or $G \cdot 7.67\,H_2O$ (e.g., if G = Cl_2).

The interaction of pentagonal dodecahedra with hexakaidecahedra (16-hedra, with 16 pentagonal faces, 28 vertices, and 42 edges) gives the unit cell depicted in Fig. 5-42. Within this unit cell ($a = $ ca. 17.2 Å), 136 water molecules enclose 24 voids, 8 of which are larger than the other 16. If only the larger voids are occupied, the composition is $136\,H_2O \cdot 8\,G$ or $G \cdot 17\,H_2O$ (e.g., with G = $CHCl_3$,

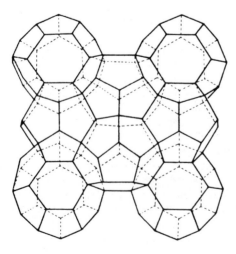

Figure 5-41   Oxygen-atom framework of unit cell in crystal of composition $46\,H_2O \cdot 6$ or 8 G. [A. F. Wells, *Structural Inorganic Chemistry*, 4th ed., Fig. 15.6, p. 545, Clarendon Press, Oxford (1975).]

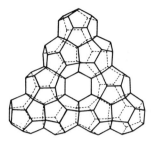

Figure 5-42 Oxygen-atom framework of unit cell in crystal of composition 136 $H_2O \cdot 8$ G. [A. F. Wells, *Structural Inorganic Chemistry*, 4th ed., Fig. 15.7, p. 546, Clarendon Press, Oxford (1975).]

CHI_3). However, if molecules of two types and appropriate sizes are trapped, both types of holes are occupied (e.g., $CHCl_3 \cdot 2 H_2S \cdot 17 H_2O$). For additional structural details, available excellent summaries[153, 154, 160] can be consulted.

Another class of cage compounds that has received attention more recently is that of the *crown* and *cryptand* ligands, that is, multichelating ligands in which donor atoms such as N, O, or S are so located as to trap metal ion acceptors within cages which may be cryptlike in nature and function.[161-163] Certain of these are discussed later in conjunction with donor species (Sec. 7.3-2) trapping alkali metal cations and anions (Sec. 10.1-3).

Although it is highly unlikely that the clathrate compounds owe their existence solely to the physical trapping of species, the nature of the interactions between host and guest is not clearly understood.[164] Alteration in the magnetic susceptibility of trapped O_2 molecules suggests some interaction,[165, 166] but the magnetic properties of trapped NO molecules are those expected for the free gas, but below its normal temperature of liquefaction.[167] The enthalpies of formation of a series of β-quinol clathrates correlate well with the polarizabilities of the trapped molecules[168] (Table 5-18), indicating that dipole–dipole interactions are unimportant but suggesting that London dispersion forces and forces of repulsion may be operative. However, enthalpy of formation data for noble gas–phenol clathrates (Ar, -9.85; Kr, -8.97; Xe, -8.8 kcal mole^{-1})[169] do not follow this order. A statistical mechanical treatment in terms of the potential field in

Table 5-18 Some Properties of β-Quinol Clathrates

Guest	Enthalpy of Formation of Clathrate (kcal mole^{-1} at 25°C)	Polarizability of Guest (cm$^3 \times 10^{24}$)
Ar	-6.0	1.63
O_2	-5.5	1.57
N_2	-5.8	1.74
HCl	-9.2	2.63
HBr	-10.2	3.58
CH_3OH	ca. -11	3.26(?)
HCOOH	-12.2	3.40(?)

which the gas molecule moves in a β-quinol clathrate does give excellent agreement between calculated and measured[168] enthalpies of formation where $G = Ar$, O_2, N_2, and HCl.[170-172] Measured heat capacities of β-quinol clathrates of argon and methane,[173] as well as other measured and calculated values, are in accord with the van der Waals model. All of these data, the small entropy values characterizing the clathrates,[173] and dielectric and infrared data[164] suggest minimal interactions among trapped molecules.

3-4. Applications

Past and continuing interest in the crystal and molecular structures of inclusion compounds is augmented by possible practical applications. Among these are the purification of gaseous species such as argon and the development of an ^{85}Kr β-source using β-quinol clathrates; the separation of hydrocarbons using urea and the thiocyanato complexes; and the desalination of seawater through the formation of the clathrate hydrate of propane and its subsequent decomposition.

EXERCISES

5-1. For each of the following species, give for the atom marked with an asterisk (a) bond order for each bond involving that atom, (b) covalency toward each bonded atom, (c) oxidation state, and (d) formal charge: $\overset{*}{N}O_3^-$, $\overset{*}{H}PO_3^{2-}$, $\overset{*}{N}_2O_4$, $\overset{*}{I}Cl_4^-$, $H_5\overset{*}{I}O_6$, $\overset{*}{B}_2Cl_4$, $(\overset{*}{S}O_3)_3$, $\overset{*}{S}_2O_3^{2-}$, $\overset{*}{P}Br_4^+$, $\overset{*}{W}_2Cl_9^{3-}$.

5-2. In terms of formal charges, suggest a reason why NNO is a more reasonable array of atoms than NON for the $N_2O(g)$ molecule.

5-3. Compare and contrast the valence-bond and the molecular-orbital descriptions for the paramagnetic molecule $S_2(g)$, S—S bond distance $= 1.89$ Å. Criticize the statement "the valence-bond description for this molecule is incorrect."

5-4. Indicate clearly the orbital hybridization and molecular geometry characteristic of each of the following species: $NF_3(g)$, $PF_3(g)$, $PF_5(g)$, SbF_6^-. Why is NF_5 a nonexistent species?

5-5. Indicate clearly the differences in bonding between the two dimeric species $[Al_2Cl_6]$ and $[Al_2(CH_3)_6]$. Why is there a difference?

5-6. Account for the difference in molecular complexity in the series $[Al(CH_3)_3]_2 - [Li(CH_3)]_4 - [Be(CH_3)_2]_n$ ($n =$ very large number).

5-7. For each of the following heteronuclear diatomic gaseous molecules, construct a molecular-orbital energy-level diagram and predict both the bond order and the number of unpaired electrons: HCl, NO, BN, CO_2, ClF.

5-8. In the linear carbon suboxide molecule, OCCCO, the C to O and C to C bond distances are 1.19 Å and 1.28 Å, respectively. Describe electronic configurations and bonding for this species.

5-9. Indicate clearly how hydrogen-bond strength varies, at constant temperature, among $H_2O(l)$, $D_2O(l)$, and $T_2O(l)$. Explain your answer.

5-10. Suggest a reason, or reasons, why ease of dimerization of the two odd molecules $NO(g)$ and $NO_2(g)$ is markedly different.

5-11. With reference to Figs. 5-36 and 5-37, melting points decrease in the order $H_2O > NH_3 > HF$, normal boiling points decrease in the order $H_2O > HF > NH_3$, and enthalpies of vaporization decrease in the order $H_2O > HF > NH_3$. Suggest an explanation.

5-12. Using the data in Table 5-4 and adding the values for the $Li_2(g)$ and $Be_2(g)$ molecules, plot bond energy against periodic group number. Discuss any apparent trends in terms of the molecular-orbital electronic configurations.

5-13. Using data from other chapters in this book or from other sources, support the statement that the molecules $B_3N_3H_6$ and $P_3N_3F_6$ exemplify delocalized covalent bonding.

5-14. Supplement the data in Table 5-5 by comparing comparable properties of the isosteric pair $C_6H_6 — B_3N_3H_6$. Discuss fundamental differences in covalent bonding within each of the three pairs.

5-15. Select from the literature a crystalline salt, other than those discussed in this chapter, for which anionic hydration is well established and, if possible, list significant O---O distance(s).

5-16. After consultation of suitable references, list and summarize the details of some practical applications of clathrate formation.

REFERENCES

1. I. Langmuir, *J. Am. Chem. Soc.*, **41**, 868 (1919).

2. G. N. Lewis, *Valence and the Structure of Atoms and Molecules*, Chemical Catalog Co., New York (1923).

3. W. Heitler and F. London, *Z. Phys.*, **44**, 455 (1927).

4. J. C. Slater, *Phys. Rev.*, **37**, 481 (1931); **38**, 1109 (1931).

5. L. Pauling, (a) *J. Am. Chem. Soc.*, **53**, 1367 (1931); *Phys. Rev.*, **40**, 891 (1932); (b) *The Chemical Bond*, Chs. 1–10, Cornell University Press, Ithaca, N.Y. (1967).

6. F. Hund, *Z. Phys.*, **51**, 788, 793 (1928); *Z. Elektrochem.* **34**, 437 (1928).

7. J. E. Lennart-Jones, *Trans. Faraday Soc.*, **25**, 668 (1929).

8. R. S. Mulliken, *Chem. Rev.*, **9**, 347 (1931); *Rev. Mod. Phys.*, **4**, 1 (1932); *Science*, **157**, 13 (1967).

9. L. Pauling, Ref. 5(b), Chs. 1, 6, 8.

10. C. K. Ingold, *Nature*, **141**, 314 (1938).

11. S. C. Wang, *Phys. Rev.*, **31**, 579 (1928).

12. S. Weinbaum, *J. Chem. Phys.*, **1**, 593 (1933).

13. H. M. James and A. S. Coolidge, *J. Chem. Phys.*, **1**, 825 (1933).

13a. W. Kolos and L. Wolniewicz, *J. Chem. Phys.*, **49**, 404 (1968); **51**, 1417 (1969).

14. L. Pauling, *J. Am. Chem. Soc.*, **53**, 3225 (1931).

15. R. McWeeny, *Coulson's Valence*, 3rd ed., Oxford University Press, London (1979).

16. H. B. Gray, *Electrons and Chemical Bonding*, W. A. Benjamin, New York (1965).

17. C. J. Ballhausen and H. B. Gray, *Molecular-Orbital Theory*, W. A. Benjamin, New York (1964).

18. J. A. Pople and D. L. Beveridge, *Approximate Molecular Orbital Theory*, McGraw-Hill Book Company, New York (1970).

19. H. F. Schaeffer III, *The Electronic Structure of Atoms and Molecules*, Addison-Wesley, Reading, Mass. (1972).

20. I. Langmuir, *J. Am. Chem. Soc.*, **41**, 1543 (1919).

21. H. Bock and P. D. Mollère, *J. Chem. Educ.*, **51**, 506 (1974).

22. D. P. Craig, *J. Chem. Soc.*, **1959**, 997; *Suom. Kemistil.*, **A33**, 142 (1960).

23. R. E. Dickerson and W. N. Lipscomb, *J. Chem. Phys.*, **27**, 212 (1957).

24. W. N. Lipscomb, in *Advances in Inorganic Chemistry and Radiochemistry*, H. J. Emeléus and A. G. Sharpe (Eds.), Vol. 1, pp. 117–156, Academic Press, New York (1959).

25. W. N. Lipscomb, *Boron Hydrides*, W. A. Benjamin, New York (1963).

26. W. C. Price, *J. Chem. Phys.*, **15**, 614 (1947); **16**, 894 (1948).

27. K. Hedberg and V. Schomaker, *J. Am. Chem. Soc.*, **73**, 1482 (1951).

28. R. G. Vranka and E. L. Amma, *J. Am. Chem. Soc.*, **89**, 3121 (1967).

29. A. I. Snow and R. E. Rundle, *J. Chem. Phys.*, **18**, 1125 (1950).

30. G. D. Stucky and R. E. Rundle, *J. Am. Chem. Soc.*, **86**, 4825 (1964).

31. E. Weiss, *J. Organomet. Chem.*, **2**, 314 (1964).

32. J. L. Atwood and G. D. Stucky, *J. Am. Chem. Soc.*, **91**, 2538 (1969).

33. G. P. van der Kelen and M. A. Herman, *Bull. Soc. Chim. Belges*, **65**, 362 (1956).

34. E. R. Corey, L. F. Dahl, and W. Beck, *J. Am. Chem. Soc.*, **85**, 1202 (1963).

35. E. Weiss and E. A. C. Lucken, *J. Organomet. Chem.*, **2**, 197 (1964).

36. R. E. Rundle, *J. Am. Chem. Soc.*, **69**, 1327 (1947); *J. Chem. Phys.*, **17**, 671 (1949).

37. H. C. Longuet-Higgins, *Q. Rev. (Lond.)*, **11**, 121 (1957).

38. F. G. A. Stone, *Endeavor*, **20**, 61 (1961).

39. K. Wade, *Electron-Deficient Compounds*, Nelson, London (1971).

40. L. Pauling, Ref. 5(b), Ch. 10.

41. M. Green and J. W. Linnett, *J. Chem. Soc.*, **1960**, 4959.

42. R. W. Asmussen, *Z. Anorg. Allgem. Chem.*, **212**, 317 (1933).

43. E. A. Guggenheim, *J. Chem. Educ.*, **43**, 474 (1966).

44. G. Herzberg, *The Structure of Molecules*, p. 174, Blackie & Sons, London (1932).

45. S. W. Benson, *J. Chem. Educ.*, **42**, 502 (1965). Excellent summary of bond energies.

46. T. L. Cottrell, *The Strengths of Chemical Bonds*, 2nd ed., Butterworth, London (1958).

47. D. A. Johnson, *Some Thermodynamic Aspects of Inorganic Chemistry*, Cambridge University Press, Cambridge (1968).

48. B. Darwent, *Natl. Bur. Stand. Ref. Data Ser.*, *NRDS-NBS31*, National Bureau of Standards, Washington, D.C. (1970).

49. J. Goubeau, *Angew. Chem.*, **69**, 77 (1957).

50. E. Politzer, *Inorg. Chem.*, **16**, 3350 (1977).

51. R. A. Howald, *J. Chem. Educ.*, **45**, 163 (1968).

52. R. T. Sanderson, *Chemical Bonds and Bond Energy*, 2nd ed., Academic Press, New York (1976).

53. P. Debye, *Polar Molecules*, Chemical Catalog Co., New York (1929).

54. R. J. W. LeFevre, *Dipole Moments: Their Measurement and Application in Chemistry*, 3rd ed., Chs. 1, 4, John Wiley & Sons, New York (1953).

55. C. P. Smyth, *Dielectric Behavior and Structure*, Chs. 8, 12, McGraw-Hill Book Company, New York (1955).

56. L. Pauling, Ref. 5(b), Ch. 3.

57. N. B. Hannay and C. P. Smyth, *J. Am. Chem. Soc.*, **68**, 171 (1946).

58. R. A. Oriani and C. P. Smyth, *J. Chem. Phys.*, **16**, 1167 (1948).

59. K. Fajans, *Naturwissenschaften*, **11**, 165 (1923); *Chimia*, **13**, 349 (1954); *Struct. Bonding*, **2**, 88 (1967).

60. K. Fajans and G. Joos, *Z. Phys.*, **23**, 1 (1924).

61. G. H. Cartledge, *J. Am. Chem. Soc.*, **50**, 2855, 2863 (1928); **52**, 3076 (1930).

62. K. S. Pitzer and J. H. Hildebrand, *J. Am. Chem. Soc.*, **63**, 2472 (1941).

63. W. M. Latimer and W. H. Rodebush, *J. Am. Chem. Soc.*, **42**, 1419 (1920).

64. T. S. Moore and T. F. Winmill, *J. Chem. Soc.*, **101**, 1635 (1912).

65. P. Pfeiffer, *Ann. Chem.*, **398**, 137 (1913).

66. M. L. Huggins, Thesis, University of California at Berkeley (1919).

67. M. L. Huggins, *Angew. Chem., Int. Ed. Engl.*, **10**, 147 (1971). An interesting and very readable account of 50 years of involvement in hydrogen bonding by Huggins.

68. G. N. Lewis, Ref. 2, p. 109.

69. C. A. Coulson, *Research*, **10**, 159 (1957); R. McWeeny, *Coulson's Valence*, 3rd ed., pp. 358–363, Oxford University Press, London (1979).

70. G. C. Pimentel and A. L. McClelland, *The Hydrogen Bond*, W. H. Freeman, San Francisco (1960).

71. W. C. Hamilton and J. A. Ibers, *Hydrogen Bonding in Solids*, W. A. Benjamin, New York (1968).

72. M. D. Joesten and L. J. Schaad, *Hydrogen Bonding*, Marcel Dekker, New York (1974). Ca. 2700 literature citations.

73. Anonymous, *Hydrogen Bond Studies 100*, Institute of Chemistry, University of Uppsala, Uppsala, Sweden (1975). Summary of 174 papers in a still continuing series.

74. A. F. Wells, *Structural Inorganic Chemistry*, 4th ed., pp. 301–312, Clarendon Press, Oxford (1975).

75. P. Schuster, G. Zundel, and C. Sandorfy (Eds.), *The Hydrogen Bond: Recent Developments in Theory and Experiments*, Vol. 1 (Theory), Vol. 2 (Structure and Spectroscopy), Vol. 3 (Dynamics, Thermodynamics, Special Systems), North-Holland, New York (1976).

76. L. Pauling, Ref. 5(b), Ch. 12.

77. G. F. Zellhoefer, M. J. Copley, and C. S. Marvel, *J. Am. Chem. Soc.*, **60**, 1337 (1938).

78. B. Zaslow, *J. Chem. Educ.*, **37**, 578 (1960).

79. M. Tamres, *J. Am. Chem. Soc.*, **74**, 3375 (1952).

80. I. Gennick and K. M. Harmon, *Inorg. Chem.*, **14**, 2214 (1975).

81. Ref. 71, p. 16.

82. J. D. Bernal and R. H. Fowler, *J. Chem. Phys.*, **1**, 515 (1933).

83. L. Pauling, *J. Am. Chem. Soc.*, **57**, 2680 (1935).

84. G. E. Bacon and N. A. Curry, *Acta Crystallogr.*, **9**, 82 (1956).

85. R. M. Bozorth, *J. Am. Chem. Soc.*, **45**, 2128 (1923).

86. J. M. Williams and L. F. Scheemeyer, *J. Am. Chem. Soc.*, **95**, 5780 (1973).

87. W. H. Bragg, *Proc. R. Soc. Lond.*, **34**, 98 (1922).

88. I. Gennick, K. M. Harmon, and M. M. Potvin, *Inorg. Chem.*, **16**, 2033 (1977).

89. S. H. Lin, in *Physical Chemistry: An Advanced Treatise*, H. Eyring, D. Henderson, and W. Jost (Eds.), Vol. 5, pp. 439–482, Academic Press, New York (1970).

90. J. C. Davis, Jr., and K. K. Deb, in *Advances in Magnetic Resonance*, Vol. 4, (J. S. Waugh (Ed.), pp. 201–270, Academic Press, New York (1970).

91. J. D. Forrester, M. E. Senko, A. Zalkin, and D. H. Templeton, *Acta Crystallogr.* **16**, 58 (1963).

92. B. A. Coyle, L. W. Schroeder, and J. A. Ibers, *J. Solid State Chem.*, **1**, 386 (1970).

93. W. C. Hamilton, *Acta Crystallogr.*, **15**, 353 (1962).

94. G. E. Bacon and N. A. Curry, *Proc. R. Soc. Lond. Ser. A*, **266**, 95 (1962).

95. R. Attig and J. M. Williams, *Angew. Chem.*, **88**, 507 (1976).

96. C. A. Beevers and H. Lipson, *Proc. R. Soc. Lond. Ser. A*, **146**, 570 (1934).

97. A. Zalkin, J. D. Forrester, and D. H. Templeton, *Inorg. Chem.*, **3**, 1417 (1964).

98. J. Roziere, M.-T. Roziere-Bories, and J. M. Williams, *Inorg. Chem.*, **15**, 2490 (1976).

99. M. Volmer, *Ann. Chem.*, **440**, 200 (1924).

100. I. Taesler and I. Olovsson, *J. Chem. Phys.*, **51**, 4213 (1969).

100a. J.-O. Lundgren and I. Taesler, *Acta Crystallogr.*, **B35**, 2384 (1979).

101. K. O. Christe, C. J. Schack, and R. D. Wilson, *Inorg. Chem.*, **14**, 2224 (1975).

101a. R. G. Delaplane, I. Taesler, and I. Olovsson, *Acta Crystallogr.*, **B31**, 1486 (1975).

102. M. L. Huggins, *J. Phys. Chem.*, **40**, 723 (1936).

103. I. Olovsson, *J. Chem. Phys.*, **49**, 1063 (1968).

104. J.-O. Lundgren and I. Olovsson, *Acta Crystallogr.*, **23**, 966 (1967).

105. T. Kjällman and I. Olovsson, *Acta Crystallogr.*, **B28**, 1692 (1972).

106. R. R. Ryan and R. A. Penneman, *Inorg. Chem.*, **10**, 2637 (1971).

107. J. M. Williams, *Inorg. Nucl. Chem. Lett.*, **3**, 297 (1967).

108. J. M. Williams and S. W. Peterson, *J. Am. Chem. Soc.*, **91**, 776 (1969). Reviews evidences for $H_5O_2^+$ in several crystals.

108a. R. G. Delaplane, J.-O. Lundgren, and I. Olovsson, *Acta. Crystallogr.*, **B31**, 2202 (1975).

109. J.-O. Lundgren and R. Tellgren, *Acta. Crystallogr.*, **B30**, 1937 (1974).

110. J.-O. Lundgren and I. Olovsson, *J. Chem. Phys.*, **49**, 1068 (1968).

111. J.-O. Lundgren and P. Lundin, *Acta. Crystallogr.*, **B28**, 486 (1972).

112. R. Attig and J. M. Williams, *Inorg. Chem.*, **15**, 3057 (1976).

113. J. Roziere and J. Potier, *J. Inorg. Nucl. Chem.*, **35**, 1179 (1973).

113a. I. Taesler, R. G. Delaplane, and I. Olovsson, *Acta Crystallogr.*, **B31**, 1489 (1975).

113b. K. O. Christe, W. W. Wilson, and E. C. Curtis, *Inorg. Chem.*, **18**, 2578 (1979).

114. M. L. Huggins, *J. Chem. Educ.*, **34**, 480 (1957).

115. H. M. Powell, *Endeavor*, **9**, 154 (1950).

116. F. D. Cramer, *Angew. Chem.*, **64**, 437 (1952); *Rev. Pure Appl. Chem. (Aust.)*, **5**, 143 (1955).

117. L. Mandelcorn, *Chem. Rev.*, **59**, 827 (1959).

118. Sr. M. M. Hagan, *J. Chem. Educ.*, **40**, 643 (1963); *Clathrate Inclusion Compounds*, Reinhold Publishing Corp., New York (1962).

119. R. M. Barrer, in *Non-Stoichiometric Compounds*, L. Mandelcorn (Ed.), Ch. 6, Academic Press, New York (1964).

120. H. M. Powell, in Ref. 119, Ch. 7.

121. L. A. K. Staveley, in Ref. 119, Ch. 10.

122. J. E. D. Davies, *J. Chem. Educ.*, **54**, 536 (1977).

123. D. D. MacNicol, J. J. McKendrick, and D. R. Wilson, *Chem. Soc. Rev.*, **7**, 65 (1978).

124. W. Rüdorff, in *Advances in Inorganic Chemistry and Radiochemistry*, H. J. Emeléus and A. G. Sharpe (Eds.), Vol. 1, pp. 223–266, Academic Press, New York (1959).

125. G. R. Hennig, in *Progress in Inorganic Chemistry*, F. A. Cotton (Ed.), Vol. 1, pp. 215–205, Interscience Publishers, New York (1959).

126. R. C. Croft, *Q. Rev. (Lond.)*, **14**, 1 (1960).

127. R. C. Croft, *Aust. J. Chem.*, **9**, 206 (1956).

128. F. R. Gamble and T. H. Geballe, in *Treatise on Solid State Chemistry*, N. B. Hannay (Ed.), Vol. 3, pp. 89–166, Plenum Press, New York (1976).

129. W. Schlenk, *Ann. Chem.*, **565**, 204 (1949).

130. L. C. Fetterly, Ref. 119, Ch. 8.

131. R. M. Barrer, *Q. Rev. (Lond.)*, **3**, 293 (1949).

132. H. M. Powell, *J. Chem. Soc.*, **1948**, 61.

133. D. E. Palin and H. M. Powell, *Nature*, **156**, 334 (1945); *J. Chem. Soc.*, **1947**, 208; **1948**, 571, 815.

134. H. M. Powell, *J. Chem. Soc.*, **1950**, 298, 300, 468.

135. O. J. Wehrhahn and M. von Stackelberg, *Naturwissenschaften*, **41**, 358 (1954).

136. M. von Stackelberg, *Rec. Trav. Chim.*, **75**, 902 (1956).

137. M. von Stackelberg, A. Hoverath, and C. Scheringer, *Z. Elektrochem.*, **62**, 129 (1958).

138. A. P. Dainen, *J. Russ. Phys.-Chem. Soc.*, **46**, 1310 (1914).

139. W. Baker, A. J. Floyd, J. F. W. Omie, G. Pope, A. S. Weaving, and J. H. Wild, *J. Chem. Soc.*, **1956**, 2010, 2018.

140. H. M. Powell and B. D. P. Wetters, *Chem. Ind. (Lond.)*, **1955**, 256.

141. L. Mandelcorn, N. N. Goldberg, and R. E. Hoff, *J. Am. Chem. Soc.*, **82**, 3297 (1960).

142. H. M. Powell and J. H. Rayner, *Nature*, **163**, 566 (1949).

143. J. H. Rayner and H. M. Powell, *J. Chem. Soc.*, **1952**, 319.

144. E. E. Aynsley, W. A. Campbell, and R. E. Dodd, *Proc. Chem. Soc.*, **1957**, 210.

145. R. Baur and G. Schwarzenbach, *Helv. Chim. Acta*, **43**, 842 (1960).

146. T. Iwamoto, *Inorg. Chim. Acta*, **2**, 269 (1968).

147. T. Iwamoto, T. Nakano, M. Morita, T. Miyoshi, T. Miyamoto, and Y. Sasaki, *Inorg. Chim. Acta*, **2**, 313 (1968).

148. W. D. Schaeffer, W. S. Dorsey, D. A. Skinner, and C. G. Christian, *J. Am. Chem. Soc.*, **79**, 5870 (1957).

149. F. V. Williams, *J. Am. Chem. Soc.*, **79**, 5877 (1957).

150. M. von Stackelberg, *Naturwissenschaften*, **36**, 327 (1949).

151. M. von Stackelberg and H. R. Müller, *J. Chem. Phys.*, **19**, 1319 (1951); *Z. Elektrochem.*, **58**, 25 (1954).

152. M. von Stackelberg and B. Meuthen, *Z. Elektrochem.*, **62**, 130 (1958).

153. G. A. Jeffrey and R. K. McMullan, in *Progress in Inorganic Chemistry*, F. A. Cotton (Ed.), Vol. 8, pp. 43–108, Interscience Publishers, New York (1967). Comprehensive review.

154. A. F. Wells, Ref. 74, pp. 543–548.

155. I. Gennick, K. M. Harmon, and J. Hartwig. *Inorg. Chem.*, **16**, 2241 (1977).

156. S. C. Jackels and N. J. Rose, *Inorg. Chem.*, **12**, 1232 (1973).

157. T. Iwamoto and D. F. Shriver, *Inorg. Chem.*, **11**, 2570 (1952).

158. R. Kuroda, *Inorg. Nucl. Chem. Lett.*, **9**, 13 (1973).

159. W. C. Hamilton and J. A. Ibers, Ref. 71, Ch. 6.

160. G. A. Jeffrey, *Acc. Chem. Res.*, **2**, 344 (1969).

161. C. J. Petersen and H. K. Frensdorff, *Angew. Chem., Int. Ed. Engl.*, **11**, 16 (1972).

162. B. Dietrich, J. M. Lehn, and J. P. Sauvage, *Acc. Chem. Res.*, **11**, 49 (1978).

163. J. J. Christensen, D. J. Eatough, and R. M. Izatt, *Chem. Rev.*, **74**, 351 (1974).

164. W. C. Child, Jr., *Q. Rev.* (*Lond.*), **18**, 321 (1964). Detailed summation of thermodynamic, dielectric, and infrared studies.

165. D. F. Evans and R. E. Richards, *Nature*, **170**, 246 (1952).

166. A. H. Cooke, H. Meyer, W. P. Woolf, D. F. Evans, and R. E. Richards, *Proc. R. Soc. Lond. Ser. A*, **225**, 112 (1954).

167. A. H. Cooke and H. J. Duffus, *Proc. Phys. Soc. Lond. Ser. A*, **67**, 525 (1954).

168. D. F. Evans and R. E. Richards, *Proc. R. Soc. Lond. Ser. A*, **223**, 238 (1954).

169. R. H. Lahr and H. L. Williams, *J. Phys. Chem.*, **63**, 1432 (1959).

170. J. H. van der Waals, *Trans. Faraday Soc.*, **52**, 184 (1956); *J. Phys. Chem. Solids*, **18**, 82 (1961).

171. J. H. van der Waals and J. C. Platteeuw, *Rec. Trav. Chim.*, **75**, 912 (1956).

172. J. H. van der Waals and J. C. Platteeuw, in *Advances in Chemical Physics*, L. Prigogine (Ed.), Vol. 2, pp. 1–57, Interscience Publishers, New York (1959).

173. N. G. Parsonage and L. A. K. Staveley, *Mol. Phys.*, **2**, 212 (1959); **3**, 59 (1960).

SUGGESTED SUPPLEMENTARY REFERENCES

C. J. Ballhausen and H. B. Gray, *Molecular-Orbital Theory*, W. A. Benjamin, New York (1964). In-depth treatment.

H. B. Gray, *Electrons and Chemical Bonding*, W. A. Benjamin, New York (1965). Useful and lucid introduction.

W. C. Hamilton and J. A. Ibers, *Hydrogen Bonding in Solids*, W. A. Benjamin, New York (1968). Restrictive summation.

M. D. Joesten and L. J. Schaad, *Hydrogen Bonding*, Marcel Dekker, New York (1974). Many literature citations.

J. C. Lockhart, *Redistribution Reactions*, Academic Press, New York (1970). Concise summation.

R. McWeeny, *Coulson's Valence*, 3rd ed., Oxford University Press, London (1979). Modern treatment.

L. Mandelcorn (Ed.), *Non-stoichiometric Compounds*, Academic Press, New York (1964). Extensive treatments by experts in various areas.

L. Pauling, *The Nature of the Chemical Bond*, 3rd ed., Cornell University Press, Ithaca, N.Y. (1959); or *The Chemical Bond: A Brief Introduction to Modern Structural Chemistry*, Cornell University Press, Ithaca, N.Y. (1967). Classical references.

J. A. Pople and D. L. Beveridge, *Approximate Molecular Orbital Theory*, McGraw-Hill Book Company, New York (1970). In-depth treatment.

G. C. Pimentel and A. L. McClelland, *The Hydrogen Bond*, W. H. Freeman, San Francisco (1960). General summation.

R. T. Sanderson, *Chemical Bonds and Bond Energy*, 2nd ed., Academic Press, New York (1976). Unique, nonclassical treatment.

K. Wade, *Electron-Deficient Compounds*, Nelson, London (1971). Summary of species and properties.

A. F. Wells, *Structural Inorganic Chemistry*, 4th ed., Clarendon Press, Oxford (1975). Classic volume on structures of all types.

CHAPTER SIX

MOLECULAR STRUCTURE AND STEREOCHEMISTRY; SIMPLE COVALENT ENTITIES

273

The directed nature of the covalent bond indicates clearly that, for a polyatomic molecule or ion, atoms which are covalently bonded to a central atom must occupy definite positions in space about that central atom. In other terms, a covalent species must have a spatial geometry relative to a centrally located atom. Experimentally determined geometries are well defined, limited in number, and directly dependent upon the covalencies (Sec. 5.1-2) of the central atoms. In describing a two-dimensional arrangement, all atoms are placed in a plane, and the angles between covalent bonds are given. In describing a three-dimensional arrangement, an assumed polyhedron commonly enclosing the central atom and placing the atoms covalently bonded to it at appropriate apices or within or above polyhedral faces is used.* A significant departure from this approach is in the trigonal pyramidal arrangement, wherein the "central" atom is at one apex, and the bonded atoms are at the other three apices.

Typical planar and three-dimensional geometries are depicted in Fig. 6-1 and summarized in Table 6-1, where they are exemplified by specific, experimentally determined molecular structures. Both relatively simple covalent species and complex, or coordination, entities are included. Molecular structures of species of the first type are considered in this chapter; with but few exceptions, those of the second type are better discussed in conjunction with other related characteristics in Chapters 7 and 8. Covalencies larger than six are almost invariably exemplified by coordination species.

1. THEORETICAL AND CONCEPTUAL APPROACHES

The observed molecular structures of simple covalent molecules and ions have been rationalized in terms of the valence-shell electronic configurations of the central atoms most commonly in two ways: through (1) directed bonding involving pure or hybridized atomic orbitals, and (2) repulsions among bonded and nonbonded electron pairs.

---

* Polyhedra are most specifically used to describe the molecular geometries of coordination entities (Ch. 7), but they are equally useful for simpler moieties. In general, polyhedra with triangular faces minimize atom–atom repulsions. Preferred polyhedra for various covalencies or coordination numbers have been derived from topological and group theoretical approaches.[1] Point-group designations are described later in this chapter (Sec. 6.4).

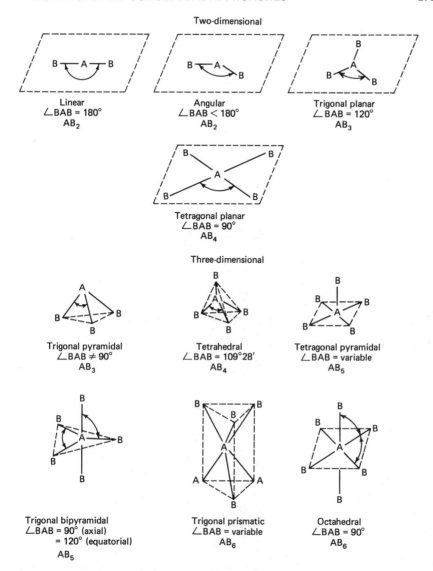

Figure 6-1 Geometries of covalently bonded species — spatial arrangements of atoms B around central atom A.

1-1. Directed Atomic Orbitals

Earlier discussions have emphasized the hybridization of ns and np orbitals as leading to bonding overlaps in accounting for observed linear [sp hybrid, e.g., $BeCl_2(g)$], trigonal planar [sp^2 hybrid, e.g., $BX_3(g)$], and tetrahedral [sp^3 hybrid, e.g., $CX_4(g)$] molecular geometries for covalent compounds of second-period elements (Sec. 5.1-1). These considerations can be extended by assuming that

Table 6-1 Some Experimentally Observed Molecular Geometries

Covalency of Central Atom	Geometrical Arrangement of Atoms Around Central Atom	Examples
2	Linear	$BeCl_2(g)$, $HgX_2(X = Cl, Br, I)(g)$
	Angular	$H_2O(g)$, $SnX_2(X = Cl, Br, I)(g)$
3	Trigonal planar	$BX_3(X = F - I)(g)$
	Trigonal pyramidal	$NX_3(X = H, F)$, NH_2Cl, $NHCl_2$, NR_3, $PX_3(X = H, F - I)$
4	Square planar	ICl_4^-, XeF_4, $Pd(NH_3)_2Cl_2$
	Tetrahedral	CX_4 or $SiX_4(X = H, F - I)$, NH_4^+, XeO_4
5	Tetragonal, or square, pyramidal	ClF_5, BrF_5, $(R_2NCS_2)_2Co(NO)^a$
	Trigonal bipyramidal	PF_5, $PCl_5(g)$, PF_3Cl_2, $SbCl_5$
6	Octahedral	SF_6, PX_6^- $(X = F, Cl)$, AlX_6^{3-} $(X = F, Cl)$, $Co(NO_2)_6^{3-}$, $PtCl_6^{2-}$, XeO_6^{4-}
	Trigonal prismatic	$(R_2C_2S_2)_3Re^b$
7	Pentagonal bipyramidal	IF_7
8	Tetragonal (Archimedes) antiprismatic	TaF_8^{3-}, $W(CN)_8^{4-}$, $U(NCS)_8^{4-}$
	D_{2d} docedahedral	$Mo(CN)_8^{4-}$, $Zr(C_2O_4)_4^{4-}$
9	Tricapped trigonal prismatic	ReH_9^{2-}, TcH_9^{2-}
10	Bicapped tetragonal antiprismatic	$[U(CH_3CO_2)_4]_n$
12	Icosahedral	$Pr(C_8H_6N_2)_6^{3+\,c}$

[a] $R_2NCS_2^- = N,N$-dialkyl (or aryl)dithiocarbamate ion.

[b] $R_2C_2S_2^{2-} = $ 1,2-dialkyl (or aryl)dithiolene ion.

[c] $C_8H_6N_2 = $ 1,8-naphthyridine.

nonbonded, valence-shell electron pairs may also participate in hybridization.[2,3] Thus the observed molecular geometries

$$NH_3(g), \quad <HNH = 106.8° \qquad H_2O(g), \quad <HOH = 104.5°$$

can be correlated with that of $CH_4(g)$ ($<HCH = 109.5°$ ideal tetrahedral angle, sp^3 bonding) by assuming that the comparatively small departures of bond angles from the true tetrahedral value are consequences of the inequalities produced in sp^3 hybridization by the presence of one or two nonbonded electron pairs.

The orbital energies of d electrons may be sufficiently close to those of s and p electrons to allow their participation in bonding hybrids.[4] Most commonly, these hybrids are useful in describing the complexes of d-transition metal atoms and ions (Ch. 7, 8) and thus involve $(n-1)d$-ns-np combinations. However, ns-np-nd hybrids have been invoked also to rationalize the observed molecular structures of covalent main-group species such as octahedral SiF_6^{2-}, $Ge(C_2O_4)_3^{2-}$, PCl_6^-, and SF_6 or cyclic $(NPX_2)_n$ ($n = 3, 4$ most commonly; Sec. 6.2-1) and $(NSOX)_3$ (Sec. 6.2-1).*

The relationships of directed atomic orbitals, both pure and hybridized, to spatial orientations are summarized in Table 6-2,[1,5] where specific d orbitals that may be involved in hybrids and postulated relative bond strengths[6] for commonly observed molecular geometries are included.

Although the use of directed atomic orbitals is a reasonable extension of earlier discussions based upon valence-shell electronic configurations, it is commonly difficult to determine the correct combinations of orbitals required for a specific complicated species and thus to predict with certainty even the approximate molecular geometry. It is for this reason that the electron-pair repulsion approach has proved to be both useful and popular.

1-2. Valence-Shell Electron-Pair Repulsions (VSEPR)

In 1940, N. V. Sidgwick and H. Powell[7] suggested that the arrangements in space of bonds from a multicovalent central atom are simply related to the size of the valence shell of electrons of that atom and that arrangements of pairs of electrons in this shell, either shared (bonding) or unshared (nonbonding), depend only upon their numbers (e.g., trigonal planar for three pairs, tetrahedral for four pairs, etc.). It was not until 1957, however, that these ideas were developed further and extended to a large number of covalently bonded species by R. J. Gillespie and R. S. Nyholm in what must be considered as a classic publication.[8] More recently, the concept has been utilized and popularized in a series of publications by Gillespie.[9-15]

---

* Arguments for and against d-orbital participation in σ and π bonds involving atoms of main-group elements are summarized, together with numerous literature citations, by T. B. Brill, *J. Chem. Educ.*, **50**, 392 (1973).

Table 6-2 Directional Characteristics of Covalent Bonds: Directed Pure or Hybrid Orbital Approach

Covalence	Bond Type	Spatial Arrangement of Atoms B Around Atom A	Relative Bond Strength
2	sp^a or dp	Linear	1.93
	p^2, ds, or d^2	Angular	
3	$sp^{2\,a}$, dp^2, d^2s, or d^3	Trigonal planar	1.99
	dsp	Unsymmetrical planar	
	p^3 or d^2p	Trigonal pyramidal	
4	$sp^{3\,a}$ or d^3s	Tetrahedral	2.00
	$dsp^{2\,a}$ or d^2p^2	Square planar	2.694
	d^2sp, dp^3, or d^3p	Irregular tetrahedral	
	d^4	Tetragonal pyramidal	
5	$d_{z^2}sp^{3\,a}$ or d^3sp	Trigonal bipyramidal	
	$d_{x^2-y^2}sp^{3\,a}$, d^2sp^2, d^4s, d^2p^3, or d^4p	Tetragonal pyramidal	
	d^3p^2	Pentagonal planar	
	d^5	Pentagonal pyramidal	
6	$d_{x^2-y^2}d_{z^2}sp^{3\,a}$	Octahedral	2.923
	$d_{xy}d_{yz}sp^{3\,a}$, d^4sp, or d^5p	Trigonal prismatic	
	d^3p^3	Trigonal antiprismatic	
		Mixed	
7	d^3sp^2, d^5s, d^4p^2	Pentagonal bipyramidal	
	$d_{xy}d_{x^2-y^2}d_{z^2}sp^{3\,a}$	Face-capped octahedral	
	$d_{x^2-y^2}d_{z^2}sp^3$, d_{xy}, $d_{xz}d_{z^2}sp^3$, or d^5sp	Face-capped trigonal prismatic	
	$d_{xy}d_{xz}d_{z^2}sp^3$, $d_{xy}d_{xz}d_{x^2-y^2}sp^3$, d^4sp^2, d^4p^3, or d^5p^2		
8	$d_{xy}d_{xz}d_{z^2}sp^{3\,a}$	D_{2d} dodecahedral	
	$d_{xy}d_{yz}d_{xz}d_{x^2-y^2}sp^{3\,a}$	D_{4d} tetragonal antiprismatic	
	d^5sp^2	Biface-capped trigonal prismatic	
9	d^5sp^3	Triface-capped trigonal prismatic	
10		Bicapped tetragonal antiprismatic	
12		Regular icosohedral	

General Principles. The Pauli exclusion principle requires that if two electrons have the same spin and the same space coordinates the describing wave function must vanish. Thus two electrons of the same spin cannot occupy the same position in space. It follows that two electrons of parallel spin have a very low probability of being close to each other in space and a very high probability of being far apart. Maximum separation of two electrons of parallel spin would be at 180°. Correspondingly, three electrons of parallel spin would achieve maximum separation when lying in a plane at angles of 120°, and four electrons of parallel spin would be most widely separated if at the apices of a regular tetrahedron at angles of 109° 28′. If now a central atom with four valence electrons of the same spin interacted with another species providing four electrons of opposite spins, electrons of opposed spins would be drawn into the same regions in space, bringing the two tetrahedra into coincidence and producing four doubly occupied orbitals tetrahedrally oriented about the atom. Similarly, three pairs of electrons would yield three localized orbitals at 120° to each other; two pairs of electrons would give two localized orbitals at 180° to each other; and so on.

The broad conclusion is thus that pairs of electrons in the valence shell of a central atom tend to occupy positions in space that minimize repulsions among the pairs and thus maximize distances among them. If, for convenience, the total valence shell is regarded as a sphere, the electron pairs, bonded or nonbonded, can be regarded as localizing on this spherical surface at maximum distances from each other. The result is shown diagrammatically in Fig. 6-2. On this basis, the predicted geometrical arrangements of electron pairs in the valence shell of a central atom are summarized in Table 6-3. An alternative approach, the tangent-sphere model described in detail by H. A. Bent[16] and based upon the packing of spherical orbitals in the valence shell about a central nucleus, leads to the same result.

These predicted geometries are the idealized ones that result if all of the repulsions among electron pairs are equal. Such a condition would exist only if the electrons in the valence shell were used in bonding to identical atoms (e.g., in CH_4, SF_6). Nonbonded electron pairs actually occupy more space in the valence shell and thus repel other electron pairs more strongly than bonded electron pairs. The result is deviation from idealized geometry and alteration of bond angles. The molecules CH_4, NH_3, and OH_2 are cases in point, as indicated in Fig. 6-3. Thus although overall geometry based upon both bonded and nonbonded electron pairs is essentially tetrahedral, the observed geometries of the molecules NH_3 and OH_2 in terms of bond angles depart from ideal tetrahedra. With respect to resulting spatial effects, repulsions decrease as

(nonbonded pair—nonbonded pair) > (nonbonded pair—bonded pair)
> (bonded pair—bonded pair)

In addition, repulsions among electron pairs in the valence shell of a central atom are altered by both differences in electronegativity and the presence of multiple bonding. Increasing the electronegativity of atom B, bonded to central

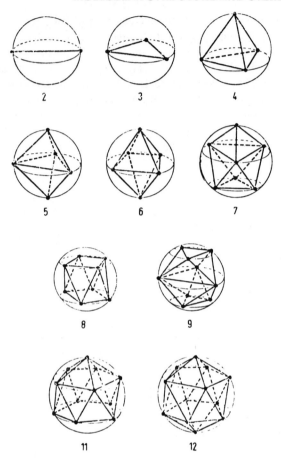

Figure 6-2 Locations of electron pairs at maximized distances on spherical surface around central atom A. [R. J. Gillespie, *Angew. Chem., Int. Ed. Engl.*, **6**, 819 (1967), Fig. 2.]

atom A, decreases the space occupied by a bonding electron pair in the valence shell of atom A as electronic charge density is displaced toward atom B, as indicated in Fig. 6-4, and decreases the bond angle. Thus the bond angle F—N—F of 102.1° in the NF_3 molecule is smaller than the angle H—N—H in the NH_3 molecule (Fig. 6-3). Other examples are given in Table 6-4.[13] Where a multiple bond is present, the increased space occupied by the additional pair(s) of bonding electrons enhances repulsions and alters bond angles. This effect is exemplified by the data in Table 6-4.

Predictions and Examples. Idealized molecular geometries for molecular or ionic species of compositions AB_mE_n, where E represents a nonbonded electron pair, are indicated in Fig. 6-5 in terms of the spherical valence-shell model. These geometries, together with some examples, are also summarized in Table 6-5.[12]

Table 6-3 Predicted Geoemetrical Arrangements of Electron Pairs in the Valence Shell of a Central Atom

Number of Electron Pairs	Arrangement of Electron Pairs Around Central Atom
2	Linear
3	Trigonal planar (equilateral triangle)
4	Tetrahedral
5	Trigonal bipyramidal
6	Octahedral
7	Monocapped octahedral
8	Tetragonal(square) antiprismatic
9	Tricapped trigonal prismatic
10	Bicapped tetragonal antiprismatic
11	Icosahedral, minus one apex
12	Icosahedral

Included in this table, in addition to simple covalent molecular and ionic species, are some transition-metal species of d^0- and d^{10}-configurations. These latter species both indicate the scope of the VSEPR approach and serve as a prelude to certain of the discussions in Chapter 7.

The Special Case of AB_mE_n *where* **m+n=5.** Although, as indicated in Table 6-5, trigonal bipyramidal geometry is most probable for these species, it must be recognized that even when $n = 0$ and all ligands B are identical the axial and equatorial positions are not identical. This situation is exemplified by the $PF_5(g)$ molecule, for which the structural diagram in Fig. 6-6(a) indicates the

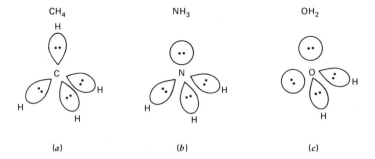

Figure 6-3 Effects of nonbonded electron pairs on tetrahedral geometry. (a) Regular tetrahedral geometry: four bonded pairs; zero nonbonded pairs; $\angle HCH = 109.5°$. (b) Trigonal pyramidal geometry (distorted tetrahedral): three bonded pairs; one nonbonded pair; $\angle HNH = 106.8°$. (c) Angular geometry (distorted tetrahedral): two bonded pairs; two nonbonded pairs; $\angle HOH = 104.5°$.

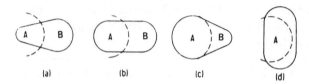

Figure 6-4 Effect of increasing electronegativity of atom B on space occupied by bonding electron pair. (a) $x_B > x_A$; (b) $x_B = x_A$; (c) $x_B < x_A$; (d) $x_B = 0$ (nonbonded pair). [R. J. Gillespie, *Angew. Chem., Int. Ed. Engl.*, **6**, 819 (1967), Fig. 5.]

axial P—F bond to be longer than the equatorial P—F bond.[13] Lack of equivalence in bond lengths arises from the facts that in this geometry each axial F atom has three nearest-neighbor F atoms at 90° bond angles, whereas each equatorial F atom has two nearest-neighbor F atoms at 120° bond angles, with the observed bond distances thus reflecting the equilibrium positions of the bonded F atoms. Comparable situations for other AB_5 species are exemplified in Table 6-6.

Substitution of methyl groups for fluorine atoms in the PF_5 molecule emphasizes the differences between the axial and equatorial positions in terms of

Table 6-4   Bond Angles in Simple Molecules Containing Double Bonds

Molecular Formula	Central Atom A	Bonded Atoms B	Bonded Atoms X	Bond Angles (deg) BAB	Bond Angles (deg) BAX
$F_2C=O$	C	F	O	108.0	126.0
$(CH_3)FC=O$	C	F, C	O	110	128, 122
$Cl_2C=O$	C	Cl	O	111.3	124.3
$H_2C=O$	C	H	O	115.8	122.1
$(H_2N)_2C=O$	C	N	O	118	121
$(H_2N)_2C=S$	C	N	S	116	122
$H_2C=CH_2$	C	H	C	116.8	122
$H_2C=CHF$	C	H	C	115.4	123.3, 120.9
$H_2C=CF_2$	C	H	C	109.3	125.3
$H_2C=CCl_2$	C	H	C	114.0	123
$F_2C=CH_2$	C	F	C	110.0	125
$F_2C=CFCl$	C	F	C	114.0	123
$F_2S=O$	S	F	O	92.8	106.8
$Br_2S=O$	S	Br	O	96	108
$(CH_3)_2S=O$	S	C	O	100	107
$(C_6H_5)_2S=O$	S	C	O	97.3	106.2
$F_3P=O$	P	F	O	102.5	
$Cl_3O=O$	P	Cl	O	103.6	
$Br_3P=O$	P	Br	O	106	

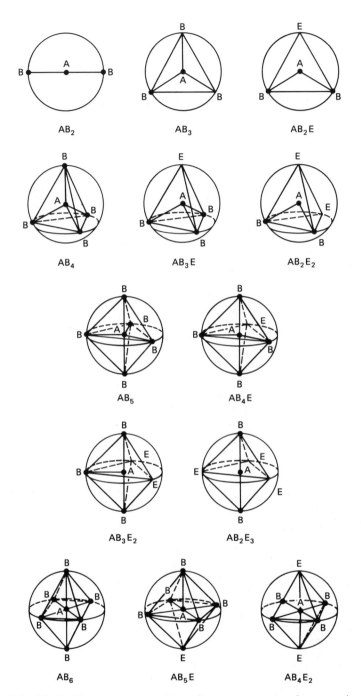

Figure 6-5 Idealized geometries predicted for arrangements of two to six electron pairs around central atom A (E = nonbonded electron pair). [R. J. Gillespie, *J. Chem. Educ.*, **47**, 18 (1970), Fig. 1.]

Table 6-5 Molecular Geometries and the VSEPR Concept

Type Formula	Electron Pairs in Valence Shell of Atom A			Predicted Arrangement Around A	Observed Molecular Geometry	Examples[a]
	Total	Bonded	Non-Bonded			
AB_2	2	2	0	Linear	Linear	$BeCl_2(g)$, ZnX_2, CdX_2, HgX_2, Hg_2Cl_2, $Ag(CN)_2^-$, $Au(CN)_2^-$
AB_3	3	3	0	Trigonal planar	Trigonal planar	BX_3, GaI_3, $In(CH_3)_3$
AB_2E	3	2	1	Trigonal planar	Angular (V-shaped)	SnX_2, PbX_2
AB_4	4	4	0	Tetrahedral	Tetrahedral	$(BeCl_2)_n(s)$, BeX_4^{2-}, BX_4^-, CR_4, CX_4, NH_4^+, NR_4^+, $OBe_4(RCO_2)_6$, $ZnO(s)$, Al_2Cl_6, SiX_4, GeX_4, SnX_4, AsR_4^+, ZnX_4^{2-}, HgX_4^{2-}, TiX_4, ZrX_4, HfX_4, ThX_4, SO_4^{2-}
AB_3E	4	3	1	Tetrahedral	Trigonal pyramidal	NH_3, NX_3, PX_3, AsX_3, SbX_3, P_4O_6, As_4S_6, H_3O^+, SO_3^{2-}
AB_2E_2	4	2	2	Tetrahedral	Angular (V-shaped)	OH_2, OX_2, SeX_2, TeX_2, ClO_2^-
AB_5	5	5	0	Trigonal bipyramidal	Trigonal bipyramidal	PF_5, $PCl_5(g)$, PF_3Cl_2, PF_2R_3, $SbCl_5(g)$, $NbCl_5(g)$, $NbBr_5(g)$, $TaCl_5(g)$, $TaBr_5(g)$, $V_2O_5(s)$, $(VO_3^-)_n$

Type		Geometry		Shape		Examples
AB$_4$E	5	Trigonal bipyramidal	4	1	Distorted tetrahedral	SF$_4$, SeF$_4$, R$_2$SeCl$_2$, R$_2$TeCl$_2$, R$_2$TeBr$_2$
AB$_3$E$_2$	5	Trigonal bipyramidal	3	2	T-shaped	ClF$_3$, BrF$_3$, C$_6$H$_5$ICl$_2$
AB$_2$E$_3$	5	Trigonal bipyramidal	2	3	Linear	ICl$_2^-$, I$_3^-$, XeF$_2$
AB$_6$	6	Octahedral	6	0	Octahedral	SF$_6$, SeF$_6$, TeF$_6$, MoF$_6$, WF$_6$, WCl$_6$, S$_2$F$_{10}$, Te(OH)$_6$, Sn(OH)$_6^{2-}$, Sb(OH)$_6^-$, H$_5$IO$_6$, (NbCl$_5$)$_2$, PF$_6^-$, PCl$_6^-$, SbF$_6^-$, SiF$_6^{2-}$, SnCl$_6^{2-}$, TaF$_6^-$, VF$_6^-$, FeF$_6^{3-}$, AlF$_6^{3-}$, GeF$_6^{2-}$, Cd(NH$_3$)$_6^{2+}$, Zn(NH$_3$)$_6^{2+}$
AB$_5$E	6	Octahedral	5	1	Tetragonal pyramidal	ClF$_5$, BrF$_5$, IF$_5$
AB$_4$E$_2$	6	Octahedral	4	2	Square planar	ICl$_4^-$, BrF$_4^-$, XeF$_4$

^{a} X = F, Cl, Br, or I.

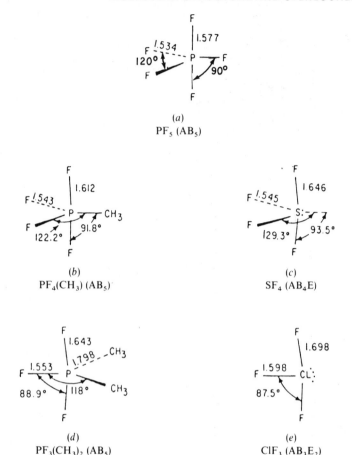

Figure 6-6. Molecular structures of species of type $MA_m E_n$ (where $m+n=5$).
[R. J. Gillespie, *J. Chem. Educ.*, **47**, 18 (1970), Fig. 5]

substantial changes in both distances and angles [Fig. 6-6(*b*) and (*d*)]. Even larger deviations from ideal trigonal bipyramidal geometry characterize the species AB_4E and AB_3E_2, as indicated for the SF_4 and ClF_3 molecules in Fig. 6-6(*c*) and (*e*).

The total electronic interaction on electron pairs in equatorial positions is less than that on those in axial positions, thus providing more space in the equatorial regions. As a consequence, nonbonding electron pairs are found in equatorial positions (e.g., in the SF_4, $SeCl_4$, ClF_3, BrF_3 molecules), and the most electronegative ligands appear in axial positions [e.g., in the PF_4CH_3 and $PF_3(CH_3)_2$ molecules]. Substitution of a methyl group for a fluorine atom in an equatorial position in the PF_5 molecule decreases the overall electronegativity of the phosphorus atom, causing all bonding electron pairs to move away from that atom and thus increasing all P—F bond lengths. Simultaneously, the

Table 6-6 Bond Distances and Angles in AB$_5$ Molecules

Formula	A—B Distance (Å)		BAB Angle (deg)	
	Axial	Equatorial	Axial	Equatorial
PF$_5$	1.577	1.534	90	120
PCl$_5$(g)	2.10	2.04		
P(C$_6$H$_5$)$_5$	1.987	1.850		
SbCl$_5$	2.34	2.29		

equatorial fluorine atoms are repelled sufficiently by the large electron-pair bonding the methyl group to increase the F—P—C bond angle and thus alter the ideal 180° F—P—F bond angle. The effects are enhanced in the SF$_4$ and ClF$_3$ molecules by the large equatorial nonbonding electron pairs.

d-*Transition Metal Complexes.* Although the molecular geometries of these species are discussed in detail in Chapter 7, it is useful to point out that the valence-shell electron-pair repulsion concept is applicable to those species that contain spherical d^0, d^5 (spin-free; Sec. 8.1-2), or d^{10} electron shells.[8, 13] For other species, the effects of unsymmetrical d shells upon the arrangements of electron pairs influence molecular geometry. Thus for an AB$_6$-type complex, the regular octahedral arrangement can be either compressed or elongated.

Three- and Four-Centered Covalent Bonds. The molecular geometries of these species can also be rationalized in terms of the VSEPR approach. Thus for the B$_2$H$_6$ molecule (Sec. 5.1-5), although the roughly tetrahedral geometry around each boron atom is reasonable, it may be assumed that the concentration of electronic charge density around a nucleus is less for a three-centered bond than for a two-centered bond and thus that the three-electron arrangement occupies less space than the two-electron arrangement. The result is that the terminal H—B—H angle (121.8°) is larger than the tetrahedral angle, whereas the bridging H—B—H angle (96.5°) is smaller. Similar approaches can be applied to account for the observed molecular structures of cluster species (Sec. 8.2-3) such as Mo$_6$Cl$_8^{4+}$, Ta$_6$Cl$_{12}^{2+}$, Re$_3$Cl$_{12}^{3-}$, and Re$_2$Cl$_8^{2-}$.[13]

Pros and Cons. The remarkable success enjoyed by the valence-shell electron-pair repulsion concept both in rationalizing observed molecular geometries and in predicting them without resort to detailed approximations involving hybrids, molecular orbitals, and so on, is a major factor in accounting for its widespread popularity. That one can, by application of a limited number of reasonable principles and with no recourse to involved calculations or sometimes difficult-to-remember rules, arrive rapidly at molecular geometries that are remarkably consistent with experimental observations is a compelling argument for use of this approach.

Yet there are exceptions, and thus this approach is not always accurate.[13, 17-20] The most serious problems appear where one or more nonbonded pairs of electrons are present in the valence shell of a central atom. In those cases already discussed, these pairs are essential in determining molecular geometry and can thus be termed "stereochemically active." There are, however, a number of species of the type AB_6E which, by experiment, have octahedral geometry but which, in terms of the electron-pair repulsion concept should have other geometries (Table 6-3).[13, 18] For each of these species, the nonbonded electron pair is said to be "stereochemically inactive." A limited number of AB_6E species do have nonoctahedral geometry, wherein the nonbonded pairs appear to be stereochemically active. Thus in the molecular structure of the $Sb(C_2O_4)_3^{3-}$ ion, a nonbonded electron pair occupies an apical position in a distorted pentagonal bipyramidal arrangement, with oxygen atoms from the oxalate ions occupying the other six positions.[21] In the gaseous state, the XeF_6 molecule is close to octahedral but not exactly so.[22] Examples of both types are summarized in Table 6-7.[23-32] The D_{2d}-dodecahedral* array of eight sulfur atoms around the tellurium atom in the compound tetrakis(diethyldithiocarbamato)tellurium(IV), $Te[S_2CN(C_2H_5)_2]_4$, exemplifies the presence of a stereochemically inactive nonbonding electron pair in a compound of the type AB_8E.[33]

Although a difference between observed molecular geometry and that predicted by the electron-pair repulsion approach may be a consequence of ligand–ligand repulsions,[13] it has been suggested[17] that it is a consequence of changing from the packing of valence-shell electron pairs to the packing of closely spaced peripheral atoms. Two empirical rules, using the terminology of R. G. Pearson's concept of hard and soft acids and bases (Sec. 9.6-4),[34] have been suggested by K. J. Wynne,[18] as:

1. Only the hardest donor atoms (O, F) generally lead to the presence of stereochemically active nonbonding electron pairs in species of the type AB_6E.
2. Soft donor atoms generally lead to the presence of stereochemically inactive nonbonded electron pairs in these species.

The second rule is explained in terms of enhanced ligand–ligand repulsions between soft donor species, availability of orbitals in soft donor atoms into which nonbonded electron pairs can be delocalized, and decreased bonded electron-pair—nonbonded electron-pair repulsions resulting from longer bond lengths and more diffuse electronic charge density distribution in soft donor species. The first rule is explained by the converse of these ideas.

It is pertinent, and of interest, that a number of solid compounds containing AB_6E arrangements of regular octahedral symmetry, and thus stereochemically inactive ns^2 nonbonded electron pairs,[35-39] where A is a lower-valent p-block species [e.g., Sn(II), Pb(II), As(III), Sb(III), Bi(III), Se(IV), Te(IV), Po(IV)], are

_____

* Point-group symmetry symbols are discussed in Sec. 6.4.

Table 6-7 Classification of AB_6E Species

Nonbonded Electron Pair Stereochemically Active[a]	Nonbonded Electron Pair Stereochemically Inactive[a]
$Sb(C_2O_4)_3^{3-}$ [21]	$SbBr_6^{3-}$ [24,25]
$XeF_6(g)$ [22]	$[BiBr_4]_n^{n-}$ [26]
II	
$Sn(EDTA)$ unit in $Sn_2(EDTA) \cdot 2 H_2O$ [23]	$[BiBr_5]_n^{2n-}$ [27]
	$SeCl_6^{2-}$, $SeBr_6^{2-}$ [28]
	$SeCl_4(C_5H_5N)_2$ [29]
	$TeCl_6^{2-}$ [30]
	$TeBr_6^{2-}$ [31]
	$TeX_4(tmtu)_2 (X = Cl, Br)$ [32]

[a] EDTA = ethylenediaminetetraacetate; tmtu = tetramethylthiourea.

colored and often exhibit metallic or semiconducting properties.[40] It has been suggested that for these species the distorting effects of the nonbonding ns^2 electron pairs are reduced by their tendencies to occupy lower-energy de-localized (or conduction) bands (Sec. 4.4-2).[40]

For species AB_nC_{6-n}, where B and C are different ligands, there may be departures from the regular octahedral geometry predicted by the electron-pair repulsion concept. Thus an electron-diffraction, microwave investigation of $IOF_5(g)$ indicated a distorted octahedral geometry, with bond distances I—O = 1.715(4) Å, I—F (equatorial) = 1.817(2) Å, and I—F (axial) = 1.863 Å and the bond angle O—I—F (equatorial) = 98.0(3)°.[20] The authors suggest that if the change from undistorted octahedral TeF_6 to IOF_5 is regarded as removal of a proton from an F nucleus (giving an O nucleus) and its subsequent addition to a Te nucleus (giving an I nucleus), with resulting deformation from octahedral geometry by altered nuclear charges or electron-pair repulsions (a "primary" effect), the bond along which the alteration occurs (the I—O bond) assumes some double bond character. The resulting larger volume of space occupied around the I atom should then lengthen the adjacent equatorial I—F bonds relative to the more remote axial I—F bond and simultaneously open the O—I—F (equatorial) bond angle beyond 90°. Then as the equatorial F atoms become closer to the axial F atom, primary deformations tend to lengthen the axial I—F bond as the molecule relaxes to the observed equilibrium structure ("secondary effect"). Experimental data suggest that for this compound the secondary effect overrides the primary effect.

For the molecule SF_5Cl, the same technique yielded the bond distances S—Cl = 2.047(3) Å, S—F (mean) = 1.570(1) Å, S—F (axial) = 1.588(9) Å, and S—F (equatorial) = 1.566 Å, and the bond angle Cl—S—F (equatorial) = 90.7(0.2)°.[41] The same situation of an overiding "secondary" effect is believed to exist, but to a lesser degree. These observations do not negate the valence-shell electron-pair repulsion approach; rather they indicate the significance of secondary effects in accounting for observed bond distances and angles.[20]

Memorization of a set of those rules, presumed to be based upon facts and suggested to be "intellectually more satisfying," has been suggested by R. S. Drago[42] as a means of predicting molecular geometry. In an extensive and particularly thorough comparison, R. J. Gillespie has convincingly and conclusively refuted this suggestion and clearly established both the simplicity and utility of the VSEPR approach, while recognizing that it cannot be infallible.[43]

The valence-shell electron-pair repulsion approach is intended only to predict basic geometry or direction (but not magnitude) of departure from ideal symmetry and is thus always only qualitative in nature.[17] It is simulated by Hückel molecular-orbital models which neglect explicit electron–electron electrostatic repulsions and ignore completely nonbonded interactions. The VSEPR approach applies best to molecular or ionic species containing large central atoms; it is less successful for internally crowded molecular or ionic species. For the latter, an approach based upon the packing of repelling atoms about a central atom, as opposed to the packing of electron pairs, is more useful, but in turn this approach fails for large central atoms.[17] These two models are compared and contrasted by L. S. Bartell,[17] who also introduces a conceptual model that gives some insight into the quantum mechanical basis for the VSEPR concept.

1-3. The Rules of Walsh

In a series of detailed, but seldom quoted, papers on the electronic orbitals, molecular shapes, and spectra of simple polyatomic molecules,[44] A. D. Walsh related probable molecular shapes for various types of simple covalent compounds in their ground states to numbers of valence electrons, as summarized in Table 6-8. These predictions are in rather good accord with observed geometries and are consistent with the view that in the qualitative prediction of molecular geometry via molecular orbitals only those forces that are related to the highest occupied molecular orbital (termed HOMO) are necessary.[45-48] Table 6-9 summarizes predictions of the molecular shapes of AB_4 and AB_5 species based upon fragments and an extension of these HOMO ideas.[48] Walsh's rules amount to emphasis of a general rule as to the significance of a closed shell.[49]

2. STEREOCHEMISTRIES OF SELECTED p-BLOCK ELEMENTS IN COVALENT SPECIES

The general principles emphasized in earlier sections of this chapter are implemented usefully by considering in more detail the molecular geometries of a variety of covalently bonded compounds of selected p-block elements. Data of this type are summarized in Tables 6-10 through 6-15, where many additional examples beyond those appearing in earlier tables in this chapter are included.

Table 6-8   Molecular Geometry and Walsh's Rules

Type of Molecule	Number of Valence Electrons	Expected Shape in Ground State
AH_2	4	Linear
	5–8	Bent
AB_2	Not over 16	Linear
	17–20	Bent (apex angle decreases markedly from 16 to 17 or 17 to 18, less markedly from 18–19 or 19–20
	22	Linear or nearly linear
HAB	10 or less	Linear
	10–14	Bent
	16	Linear
HAAB	10	Linear
	12	Bent (planer, with *cis* and *trans*)
	14	Bent (nonplanar)
AH_3	6	Planar
	7 or 8	Pyramidal
AB_3	Not over 24	Planar
	25 or 26	Pyramidal (angle decreasing with number of electrons)

It is apparent that the observed geometries are well in accord with the predictions of the valence-shell electron-pair repulsion concept, although exceptions and examples that require some liberality of interpretation are evident. A parallelism among the molecular structures of isoelectronic species derived from elements in adjacent periodic families is also evident (e.g., for those derived from Xe, I, and Te). Typical examples include: AB_4E type ($XeOF_3^+$, $IO_2F_2^-$, R_2TeBr_2), AB_5E type ($XeOF_4$, IF_5, $CH_3TeI_4^-$), and AB_6 type [XeO_6^{4-}, IO_6^{5-}, $Te(OH)_6$].

That geometrical data for noble gas compounds are largely restricted to those of xenon is a consequence of the fact that only compounds of this element have been studied in detail.[50, 51] The fluorine atom, unlike the other less electronegative halogen atoms, appears as a central atom in only a limited number of species, e.g., in the $Xe_2F_3^+$ ion, which has the molecular geometry[52]

In periodic group VIA, the oxygen atom (Table 6-12) has a somewhat different stereochemistry from the sulfur, selenium, and tellurium atoms

Table 6-9 Shapes of AB$_4$ and AB$_5$ Species as Related to AB$_2$ Fragments[48]

Fragment				Corresponding Composition	Larger Species		
Composition	Valence Electrons	Geometry[a]	Examples		Valence Electrons	Geometry[a,b]	Examples
AB$_2$	18	Bent	CCl$_2$	AB$_4$	32	Tetrahedral or D_{2d}	CCl$_4$, IO$_4^-$, SeO$_4^{2-}$, TiCl$_4$, FeCl$_4$, ZnCl$_4^{2-}$, AlCl$_4^-$, HgCl$_4^{2-}$, Cu(CN)$_4^{3-}$, BF$_4^-$, BeF$_4^{2-}$, XeO$_4$, POCl$_3$, NF$_4^+$, F$_3$NO, PO$_4^{3-}$
	20	Bent	SF$_2$		34	C_{2v}	SF$_4$, XeO$_2$F$_2$, IO$_2$F$_2^-$
	22	Linear	XeF$_2$		36	Square planar	XeF$_4$, BrF$_4^-$
AB$_2$	20	Bent	SCl$_2$	AB$_5$	40	Trigonal bipyramidal	PCl$_5$, Sb(CH$_3$)$_3$Br$_2$, NbCl$_5$, P(CH$_3$)F$_4$, PCl$_2$F$_3$, SOF$_4$, SnCl$_5^-$
	22	Linear	BrF$_2^-$		42	Tetragonal pyramidal	BrF$_5$, SbF$_5^{2-}$, TeF$_5^-$, XeOF$_4$

[a] For ground state.

[b] For an interpretation of point-group symmetry designations, see p. 313.

Table 6-10 Stereochemistry of Covalently Bonded Xenon

Type Formula	Valence-Shell Electron Pairs		σ Bonds	Geometry Around Central Atom[a]	Apparent Hybrid	Examples
	Total	Bonded				
AB_3E	4	3	3	Trigonal pyramidal (tetrahedral)	sp^3	XeO_3
AB_4	4	4	4	Tetrahedral	sp^3	XeO_4
AB_2E_3	5	2	2	Linear (trigonal bipyramidal)	dsp^3	XeF_2
AB_3E_2	5	3	3	T-shaped (trigonal bipyramidal)	dsp^3	XeF_3^+
AB_4E	5	4	4	Distorted tetrahedral (trigonal bipyramidal)	dsp^3	$XeOF_3^+$
AB_4E_2	6	4	4	Square planar (octahedral)	d^2sp^3	XeF_4
AB_5E	6	5	5	Tetragonal pyramidal (octahedral)	d^2sp^3	$XeOF_4$, XeF_5^+
AB_6	6	6	6	Octahedral	d^2sp^3	XeO_6^{4-}
AB_6E	7	6	6	Distorted octahedral		$XeF_6(g)$[b]
AB_8E	9	8	8	Distorted Archimedean antiprismatic	d^5p^3	XeF_8^{2-} in $(NO^+)_2XeF_8^{2-}$

[a] Geometries in parentheses are idealized ones, including nonbonded electron pairs on central Xe atoms.

[b] Crystalline XeF_6 is ionic, $XeF_5^+F^-$. In the gaseous state at 75°C, three electronic isomers are proposed — one (30% of molecules) as a ground state (O_h point-group symmetry), a second (55%) at 450 cm^{-1} above ground state (D_{3d}), a third (15%) at 1230 cm^{-1} above ground state (also D_{3d}). See G. L. Goodman, *J. Chem. Phys.*, **56**, 5038 (1972).

Table 6-11 Stereochemistry of Covalently Bonded Chlorine, Bromine, and Iodine

Type Formula	Valence-Shell Electron Pairs		σ Bonds	Geometry Around Central Atom[a]	Apparent Hybrid	Examples[b]
	Total	Bonded				
ABE_3	4	1	1	[c]		HX, X_2, XX', XO^-
AB_2E_2	4	2	2	Angular (tetrahedral)	sp^3	$HOXO, XO_2^-, BrF_2^+,$ ClF_2^+, ICl_2^+, I_3^+
AB_3E	4	3	3	Trigonal pyramidal (tetrahedral)	sp^3	$HOXO_2, XO_3^-$
AB_2E_3	5	2	2	Linear (trigonal bipyramidal)	sp^3d	$I_3^-, ICl_2^-, BrICl^-$
AB_3E_2	5	3	3	T-shaped (trigonal bipyramidal)	sp^3d	$ClF_3, BrF_3, ICl_3(g), RICl_2$
AB_4E	5	4	4	Distorted tetrahedral (trigonal bipyramidal)	sp^3d	$IF_4^+, I_5^+, IO_2F_2^-$
AB_4E_2	6	4	4	Square planar (octahedral)	sp^3d^2	BrF_4^-, ICl_4^-
AB_5E	6	5	5	Tetragonal pyramidal (octahedral)	sp^3d^2	ClF_5, BrF_5, IF_5
AB_6	6	6	6	Octahedral	sp^3d^2	$IOF_5, (HO)_5IO, IO_6^{5-}$
AB_6E	7	6	6	Distorted octahedral	sp^3d^3	IF_6^-
AB_7	7	7	7	Pentagonal bipyramidal	sp^3d^3	IF_7

[a] Geometries in parentheses are idealized ones, including nonbonded electron pairs associated with central atoms.
[b] X = Cl, Br, or I; R = alkyl group.
[c] No central atom.

Table 6-12 Stereochemistry of Covalently Bonded Oxygen

Type Formula	Valence-Shell Electron Pairs		σ Bonds	Geometry Around Central Atom[a]	Apparent Hybrid	Examples
	Total	Bonded				
AB_2E_2	4	2	2	Linear	sp	$[Cl_5RuORuCl_5]^{4-}$
AB_2E_2	4	2	2	Angular (tetrahedral)	sp^3	OH_2, OF_2, OCl_2, O_2H_2
AB_3E	4	3	3	Trigonal planar	sp^2	$[O(HgCl)_3]^+$
AB_3E	4	3	3	Trigonal pyramidal (tetrahedral)	sp^3	H_3O^+
AB_4	4	4	4	tetrahedral	sp^3	$[Be_4O(OCR)_6]$
						$[Zn_4O(OCR)_6]$

[a] Geometries in parentheses are idealized ones including nonbonded electron pairs on central O atoms.

295

Table 6-13 Stereochemistries of Covalently Bonded Sulfur, Selenium, and Tellurium

Type Formula	Valence-Shell Electron Pairs Total	Valence-Shell Electron Pairs Bonded	σ Bonds	Geometry Around Central Atom[a]	Apparent Hybrid	Examples[b]
AB_2E	3	2	2	Angular (trigonal planar)	sp^2	SO_2
	3	3	3	Trigonal planar	sp^2	$SO_3(g)$
AB_2E_2	4	2	2	Angular (tetrahedral)	sp^3	H_2Z, H_2S_2, S_2^{2-}, SCl_2, S_nCl_2, $O_3S(S)_nSO_3^{2-}$, $O_3SSZSSO_3^{2-}$
AB_3E	4	3	3	Trigonal pyramidal (tetrahedral)	sp^3	OSF_2, $OSCl_2$, $OSeCl_2$, $SeO_2(s)$, H_2SeO_3, R_3Te^+
AB_4	4	4	4	Tetrahedral	sp^3	H_2ZO_4, HZO_4^-, ZO_4^{2-}, O_2SCl_2, $O_3SNH_2^-$, O_3SS^{2-}
AB_4E	5	4	4	Distorted tetrahedral (trigonal bipyramidal)	sp^3d	ZF_4, $SeCl_4$, $TeCl_4$, R_2SeCl_2, R_2TeBr_2
AB_5E	5	5	5	Distorted tetragonal pyramidal (octahedral)	sp^3d	CH_3TeI^-, $OSeCl_3^-$, $OSeCl_2(NC_5H_5)_2$
AB_6	6	6	6	Octahedral	sp^3d^2	ZF_6, Te_2F_{10}, TeF_5Cl, $Te(OH)_6$, $(F_5S)_2O$, $O_6Te_2(OH)_4^{4-}$, SeX_6^{2-}, TeX_6^{2-}
AB_6E	7	6	6	Distorted octahedral		$SeCl_4(NC_5H_5)$

[a] Geometries in parentheses are idealized ones, including nonbonded electron pairs on central Z atoms.

[b] $Z = S$, Se, or Te for general cases. R = alkyl, aryl.

Table 6-14 Stereochemistry of Covalently Bonded Nitrogen

Type Formula	Valence-Shell Electron Pairs		σ Bonds	Geometry Around Central Atom[a]	Apparent Hybrid	Examples[b]
	Total	Bonded				
ABE_3	4	1	1	[c]		NH^{2-}
AB_2E_2	4	2	1	Angular (tetrahedral)	sp^3	NH_2^-
ABE	4	3	1	[c]		N_2
AB_2E_2	4	3	3	Angular (trigonal planar)	sp^2	$ONCl$
AB_3E	4	3	3	Trigonal pyramidal (tetrahedral)	sp^3	NH_3, NH_2X, NX_3, N_2H_4
AB_2	4	4	2	Linear	sp	NNO
AB_3	4	4	3	Trigonal planar	sp^2	NO_3^-, O_2NCl, O_2NNO_2
AB_4	4	4	4	Tetrahedral	sp^3	NH_4^+, NRH_3^+, $NR_2H_2^+$, NR_3H^+, NR_4^+

[a] Geometries in parentheses are idealized one, including nonbonded electron pairs on central N atoms.

[b] X = F, Cl, Br; R = allyl or aryl group.

[c] No central atom.

Table 6-15 Stereochemistry of Covalently Bonded Phosphorus

Type Formula	Valence-Shell Electron Pairs		σ Bonds	Geometry Around Central Atom[a]	Apparent Hybrid	Examples[b]
	Total	Bonded				
AB_2E_2	4	2	2	Angular (tetrahedral) [c]	sp^3	PH_2^-
ABE	4	3	1	[c]		$P_2(g)$
AB_3E	4	3	3	Trigonal pyramidal (tetrahedral)	sp^3	$P_4(g)$, PH_3, P_2H_4, PX_3, P_2I_4, $P(OR)_3$, P_4O_6, PR_3, etc.
AB_4	4	4	4	Tetrahedral	sp^3	PH_4^+, PR_4^+, PCl_4^+, PBr_4^+, OPX_3, OPR_3, H_3PO_4, $OP(OR)_3$, $H_2PO_4^-$, HPO_4^{2-}, PO_4^{3-}, $P_2O_7^{4-}$, $(PO_3)_n^{n-}$, HPO_3^{2-}, $H_2PO_2^-$, $(NPX_2)_m$, PO_3S^{3-}, etc.
AB_5	5	5	5	Trigonal bipyramidal	sp^3d	PF_5, $PCl_5(g)$, PF_3Cl_2, $P(CH_3)F_4$, PR_5
AB_6	6	6	6	Octahedral	sp^3d^2	PF_6^-, PCl_6^-

[a] Geometries in parentheses are idealized ones including nonbonded electron pairs on central P atoms.
[b] X = F, Cl, Br; R = alkyl, aryl.
[c] No central atom.

(Table 6-13) because of both its small size and the absence of d orbitals. The structural chemistry of polonium is not well delineated, largely because the relatively short half-life (138.4 days) and the intense α-activity (5.4 MeV) of the longest lived and only obtainable nuclide ($A = 210$) preclude the isolation of sizable quantities, complicate the handling of compounds, and result in both rapid atom displacements and rapid thermal decompositions.[53] Furthermore, polonium is essentially a metal.

In periodic group VA, the nitrogen atom (Table 6-14), like the oxygen atom, is limited to four covalent bonds and thus differs from the phosphorus atom (Table 6-15), where the availability of $3d$ orbitals at easily obtainable energies allows covalencies of 5 and 6. The molecular geometries in arsenic and antimony compounds are comparable with those in phosphorus compounds, with somewhat greater tendencies toward covalencies of 5 and 6. Bismuth is again primarily a metal.

Atoms of the periodic group IVA elements are less versatile geometrically, with a marked preference for the four-covalent condition. In carbon compounds, the situation is complicated by double and triple bonding and both reduction in covalency and the introduction of trigonal planar and linear geometries (Sec. 5.1-1). A limited number of six-bonded species have been described [e.g., SiF_6^{2-}, GeF_6^{2-}, $SnCl_6^{2-}$, $Sn(OH)_6^{2-}$].

Molecular geometry in covalent species derived from the periodic group IIIA elements is complicated with boron in particular and aluminum and gallium to lesser extents by the formation of three-centered, electron-deficient bonds (Sec. 5.1-5). However, tetrahedral [e.g., in BF_4^-, $Al(OH)_4^-$, $GaCl_4^-$, $InBr_4^-$] and octahedral [e.g., in $Al(C_2O_4)_3^{3-}$, $InCl_6^{3-}$] geometries are well known.

The stereochemistries of many of the p-block elements in their lower positive oxidation states are influenced by the presence of ns^2 nonbonded electron pairs.[54] These are the "inert pairs" of N. V. Sidgwick[55] that are responsible for less than maximum oxidation states (e.g., $+1$ for IIIA atoms, $+2$ for IVA atoms, $+3$ for VA atoms, $+4$ for VIA atoms, $+5$ for VIIA atoms). The effects of these pairs on geometry are apparent in earlier discussions and tabulations in this chapter. However, as noted in Table 6-7, there are exceptions in octahedral geometry. That the solid compounds containing ns^2 p-block elements in high-symmetry sites are colored and often metallic conductors or semiconductors suggests that the distorting effects of these nonbonded pairs are reduced by delocalization into low-energy bands (Sec. 4-4.2).[40] In a discussion of the stereochemical activity of ns^2 nonbonded electron pairs, it is pointed out that in terms of extended Hückel-theory calculations for molecules of the type AH_n ($n = 2$ to 6) the only repulsions of the Pauli type that are important are those between bonded electron pairs.[56] Thus the H_2O molecule bends not as a consequence of nonbonded electron-pair repulsions but rather as a consequence of $2s$-$2p$ energy separation that lowers the energy of the $2s$ orbital to such a degree that its occupation is maximized only by bonding.

2-1. Heterocyclic *p*-Block Species

Discrete heterocyclic molecules are common in organic chemistry. To a much more limited extent, they have parallels in silicon chemistry. Among other *p*-block elements, they are particularly characteristic of phosphorus and sulfur in combination with oxygen and nitrogen.[57-59]

Oxygen-Derived Heterocyclic Species. Sulfur(VI) oxide is obtainable in three solid modifications, (1) α-SO_3, an asbestoslike solid consisting of infinite chains

$$(SO_3)_n, \rightarrow \overset{\overset{O}{|}}{\underset{\underset{O}{|}}{S}}O \rightarrow \overset{\overset{O}{|}}{\underset{\underset{O}{|}}{S}}O \rightarrow , \text{ joined in a layerlike arrangement; (2) } \beta\text{-}SO_3, \text{ an asbestos-}$$

like solid consisting of infinite chains $(SO_3)_n$; and (3) γ-SO_3, an icelike solid (orthorhombic) in which trimeric cyclic molecular species $(SO_3)_3$ of arrangement

can be distinguished. The arrangement of O atoms around each S atom is tetrahedral, giving the ring a staggered geometry.[60]

Phosphorus(III) oxide is dimeric both in solution in naphthalene and in the vapor state. Electron-diffraction data (p. 331) show the four phosphorus atoms in the molecule P_4O_6 to lie at the apices of a tetrahedron with the six oxygen atoms just outside the center points of the edges,[61,62] giving the arrangement

with angles P—O—P and O—P—O 127.5° and 99°, respectively. The arrangement of O atoms relative to each P atom is thus distorted trigonal pyramidal. Phosphorus(V) oxide is also dimeric in the vapor state, with a molecular structure exactly comparable with that of the P_4O_6 molecule but with an additional O atom bonded to each P atom in such a way that the arrangement of the four oxygen atoms about each P atom is tetrahedral.[61] In the solid state, hexagonal, orthorhombic, and tetragonal modifications have been distinguished. In the volatile hexagonal form, the molecular structure of four interlocking PO_4 tetrahedron found for the vapor is present.[63] Crystals of the orthorhombic modification are built of interlocking rings based upon PO_4 tetrahedra,[64] and crystals of the tetragonal modification contain corrugated sheets of P_4O_{10} molecules.[65] Crystals of the cyclic metaphosphates, $(PO_3)_n^{n-}$ ($n = 3, 4$), contain the heterocyclic structures[66,67]

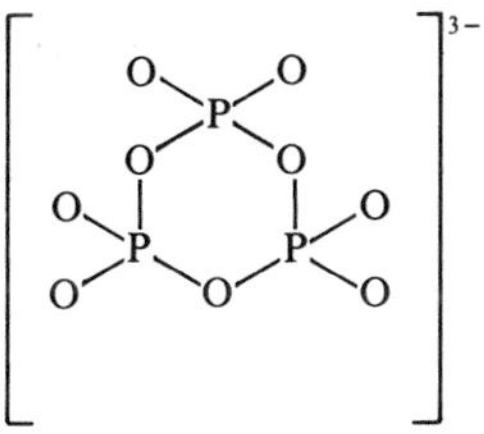

and

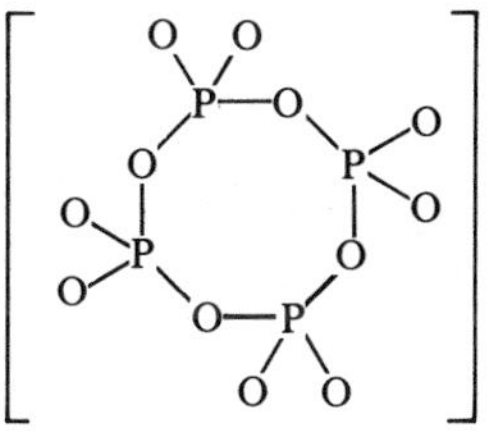

wherein each P atom is bonded tetrahedrally to four O atoms. Both the phosphorus(III) structures (three bonded pairs, one nonbonded pair on each P atom) and the phosphorus(V) structures (four bonded pairs on each P atom) are in agreement with the valence-shell electron-pair repulsion approach.

Related compounds are spirophosphoranes of graphic[67a] formulations

$R = H, CH_3$

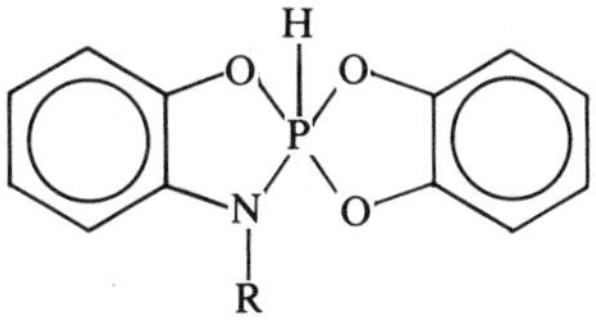

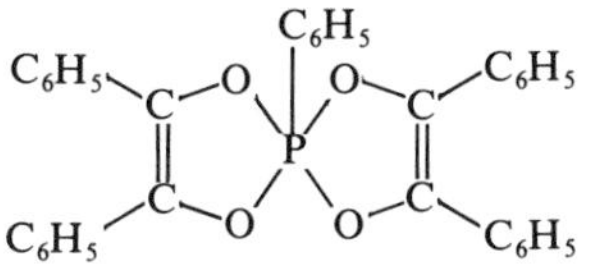

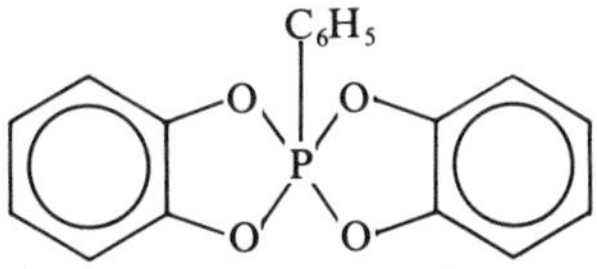

The first of these species exemplifies nearly perfect trigonal bipyramidal geometry, the second 86% displacement toward tetragonal pyramidal geometry, and the third nearly perfect tetragonal pyramidal geometry, each relative to the central phosphorus atom.

Nitrogen-Derived Heterocyclic Species. Although one can formulate the cyclic sulfur–oxygen and phosphorus–oxygen species just discussed in terms of localized electron-pair bonds, substitution of a nitrogen atom $(2s^2 2p^3)$ for an oxygen atom $(2s^2 2p^4)$ to give cyclic sulfur–nitrogen or phosphorus–nitrogen species requires either another atom covalently bonded to the nitrogen atom or localized or delocalized multiple bonding. As a consequence, a great variety of such compounds has been prepared,[58-60] some of which are indicated in Tables 6-16 and 6-17.

The compounds listed in Table 6-16 are representative of an active area of investigation in inorganic chemistry.[58, 59, 68-73] Cyclotetrathiazene, S_4N_4, is an

Table 6-16 Some Heterocyclic Sulfur–Nitrogen Species

Stoichiometric Formula	Name(s)	Graphic Formula(s)[a]
S_4N_4	Cyclotetrathiazene, tetrasulfur tetranitride	(etc.)
$(NSX)_n$	Cyclothiazenes ($n = 3, 4$; $X = F, Cl$)	
$(NSOX)_3$	Trioxocyclotrithiazenes, or sulfanuric compounds ($X = F, Cl$)	

$S_{8-n}(NH)_n$ Cyclothiazanes, or sulfur imides
($n = 1, 2, 3, 4$)

$(HNSO_2)_n$ Dioxocyclothiazanes,
or sulfimides ($n = 3, 4$)

[a] Formulas represent only arrangements of atoms and are not to be interpreted in terms of natures of bonds or complete molecular structures. Possible resonance forms are indicated, in part only, for S_4N_4.

303

endothermic solid [$\Delta H_f^0 = 110.0$ kcal (460.2 kJ) mole^{-1}] that explodes when subjected to shock; hydrolyzes in water or alkaline media; undergoes reduction to $S_4(NH)_4$; undergoes oxidation to $(OSNH)_4$, $S_3N_3Cl_3$, and so on; forms a variety of N-substituted organic derivatives; and behaves as both a Lewis acid and a Lewis base (Sec. 9.6-1).[71,72] In both the gaseous and crystalline states, the molecule has the tublike structure

in which the four nitrogen atoms are in a square plane and the sulfur atoms are in a distorted tetrahedral arrangement.[73] Important data are: all S—N bonds of equal length (average 1.62 Å), N—S—N bond angles ca. 103° or ca. 113°, S⋯S (a) distances 2.58 Å. The S—N bond length is intermediate between those expected for single (1.74 Å) and double (1.54 Å) bonds, suggesting cyclic p_π-d_π delocalization in bonding. The S⋯S distance is larger than the normal single S—S bond distance (ca. 2.05 Å), but much less than the van der Waals radius sum (3.70 Å), and not materially larger than the S—S distance in the ion O_2S—SO_2^{2-}. Molecular-orbital calculations[74] lead to the conclusions: (1) coplanar N atoms are favored over coplanar S atoms; (2) appreciable bonding between S atoms on the same side of the plane of four N atoms is expected; (3) negative ions, $N_4S_4^{(1-4)-}$, should exist and have π-delocalized electronic structures; and (4) the role played by d-orbitals of the S atoms is a contributing factor, but not a main factor, in describing the electronic structure of the molecule.

The trioxocyclotrithiazenes, or sulfanuric compounds, also exemplify cyclic delocalization of electronic charge density and thus inorganic aromaticity.[75] The crystalline chloride, which is isoelectronic with trichlorocyclotriphosphazene, exists in two isomeric forms: δ, m.p. 144 to 145°C, orthorhombic; and β, m.p. 45 to 47°C, needles. The molecular structure of the δ isomer,[76] has a

chairlike conformation with bond distances S—N: 1.57 Å, S—Cl = 2.00 Å, S—O = 1.41 Å; and bond angles N—S—N = 113°, S—N—S = 122°, N—S—O = 120°, N—S—Cl = 106°, O—S—Cl = 108°. The arrangement of bonds around each S atom is roughly tetrahedral, and the intermediate S—N bond distance again suggests d_π-p_π delocalization. The smaller dipole moment of the β-isomer (1.91 D vs. 3.88 D for the α-isomer) is consistent with a more symmetrical structure, probably also a chairlike configuration, but with at least one Cl atom in an equatorial position.[77]

Unlike these compounds, molecules of the cyclothiazanes and the dioxocyclothiazanes do not exemplify extensive delocalization of electronic charge density. As discussed in detail in Sec. 10.1-3, the NH group is electronically comparable with both the O and S atoms. The cyclothiazanes can thus be regarded as being derived from cyclooctasulfur (Table 6-18) by replacement of one or more S atoms by NH groups. Indeed, there are many resemblances between these compounds and cyclooctasulfur. Orthorhombic crystals of S_7NH (m.p. 113.5°C) contain puckered eight-membered rings with average bond distances S—S = 2.02Å and S—N = 1.68 Å and average S—S—S bond angle = 106°,[78,79]

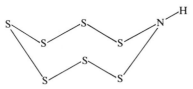

Orthorhombic crystals of $S_4(NH)_4$ (d.p. 100 to 148°C) contain comparable eight-membered rings with alternating S and N atoms

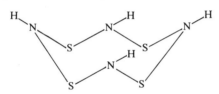

amounting to a square plane of N atoms lying above (or below) a square plane of S atoms, but displaced from coincidence by 90°.[80,81] The average bond angles are S—N—S = 122° and N—S—N = 108.4°. The S—N bond distance 1.674 Å is only slightly less than the single-bond distance, suggesting only small delocalization in bonding. Other pseudocyclooctasulfur species have comparable molecular structures.

The dioxocyclothiazanes, or sulfimides, are aquo ammono sulfur compounds[82] relatable to the aquo sulfuric acids in terms of the ammonation–deammonation schemes presented in Chapter 10 (Table 10-10). Trisulfimide, $(HNSO_2)_3$, is obviously related to $(SO_3)_3$ by replacement of cyclic O atoms by NH groups.

Of the compound types listed in Table 6-17, the cyclophosphazenes are undeniably the most widely investigated and potentially the most useful.[58,59,83-87] Bonding in these compounds and their reactions are discussed in other sections of this book (Sec. 10.4-2). Crystal and molecular structure data for numerous trimeric compounds and a limited number of higher homologs support planar or puckered cyclic arrangements of alternating phosphorus and nitrogen atoms, with endocyclic P—N distances essentially equal for a given ring and intermediate between single- and double-bond distances, indicating

Table 6-17 Some Heterocyclic Phosphorus–Nitrogen Species

Stoichiometric Formula	Name(s)	Graphic Formula(s)[a]
$(NPX_2)_n$	Cyclophosphazenes, or phosphonitriles ($n = 3, 4, 5, \ldots$; $X = F$, Cl, Br, NCS, OR, OAr, NH_2, NHR, NR_2, R, Ar, ...)[b]	
$(HO)_n(PONH')_n$	Cyclophosphazanes(V) ($n = 3, 4$; HO replaceable by RO, ArO, Cl; H' replaceable by R; oxidation state of $P = +5$; coordination number of $P = 4$)	
$(RNPX_3)_2$	Tetracyclophosphazanes(III) (R = alkyl, aryl; X = F, Cl; oxidation state of $P = +3$; coordination number of $P = 5$)	

$(RNPX)_n$

Cyclophosphazanes(III) (R = R, Ar; X = Cl, OR, OAr, NHR; oxidation state of $P = +3$; coordination number of $P = 3$)

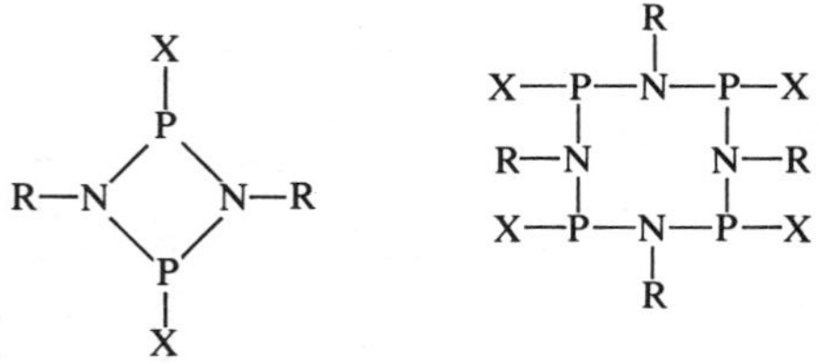

$X_3P_4(NR)_6$

Closocyclotetraphosphazanes(V) ($X = O$, S; oxidation state of $P = +5$; coordination number of $P = 4$)

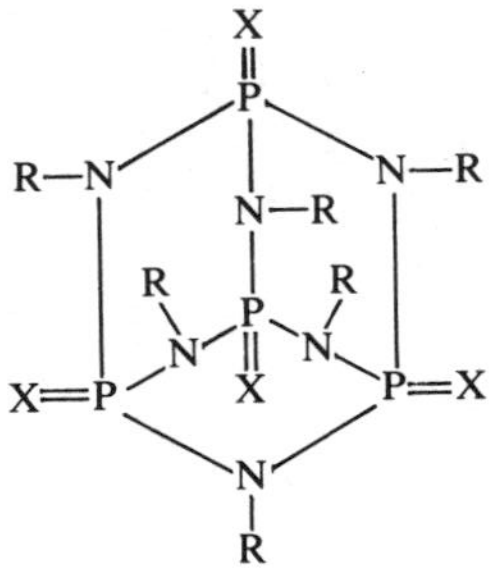

$P_4(NR)_6$

Closocyclotetraphosphazanes(III) (oxidation state of $P = +3$; coordination number of $P = 3$)

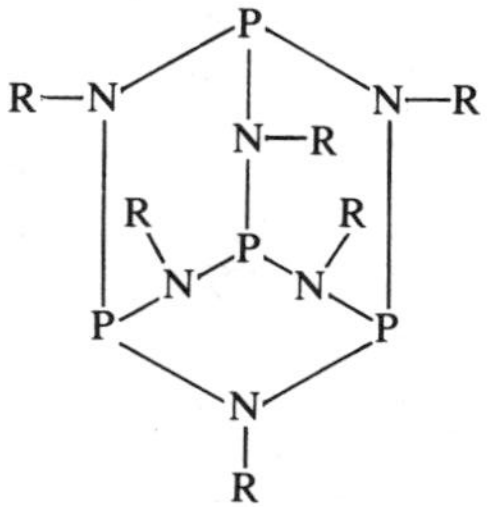

[a] Intended to show only arrangements of atoms and bond multiplicity.

[b] R = alkyl, except were otherwise designated; Ar = aryl.

307

delocalization of electronic charge density (or resonance).[86,88] Thus in the molecule $(NPF_2)_3$, the ring conformation is planar, with N—P—N angles 119.5° and 119.3°, P—N—P angles 119.6 and 121°, F—P—F angles < 99.9°, P—N distances 1.56 Å, and P—F distances 1.52 Å.[89] In the molecule $(NPCl_2)_3$, the ring conformation is also planar, with N—P—N angles 118.33 to 120.93°, P—N—P angles 118.48 to 120.38°, Cl—P—Cl angles 101.77 to 102.05°, P—N distances 1.59 Å, and P—Cl distances 1.98 Å.[90] Two crystalline modifications of the compound $(NPCl_2)_4$ exist — the stable T form and the metastable K form,[91,92] but in each the molecule $(NPCl_2)_4$ has a puckered ring conformation with N—P—N angles 121.2 to 122°, P—N—P angles 131.3 to 135°, Cl—P—Cl angles 102.8 to 103°, P—N distances 1.56 to 1.57 Å, and P—Cl distances 1.99 Å. The molecule $(NPCl_2)_5$ has a planar ring conformation with N—P—N angles 118.4°, P—N—P angles 148°, Cl—P—Cl angles 102.0°, P—N distances 1.52 Å, and P—Cl distances 1.985 Å.[93] In these molecular structures, and others for variously substituted cyclophosphazenes, the geometry around each P atom is distorted tetrahedral and that around each N atom roughly planar. These molecular structures, and those of related species, are depicted in Fig. 6-7.

Of interest also is a hybrid compound of molecular arrangement

which has a chairlike conformation with N—P—N angle 115.3°, N—S—N angles 115.0°, S—N—S angle 120.3°, P—N—S angles 123.5° and 120.6°, average N—P distance 1.585 Å, average P—Cl distance 1.957 Å, average S—O distance 1.421 Å, average S—Cl distance 2.018 Å, and average N—S distances 1.578 Å (for S—N—S) and 1.540 Å (for S—N—P).[94]

The parent cyclophosphazane(V) (Table 6-17) compounds are better known as phosphimic acids, which yield phosphimate salts. In crystals of the salt $[NaOP(O)NH]_3 \cdot 4H_2O$, the ion $[OP(O)NH]_3^{3-}$ has a chairlike molecular conformation, with average N—P—N angles 104.5°, P—N—P angles 123°, O—P—O angles 118°, O—P—N angles 106 to 109°, P—N distances 1.68 Å, and P—O distances 1.50 Å.[95,96] Molecules of the acid $[HOP(O)NH]_4 \cdot 2H_2O$ have a conformation with average N—P—N angles 107.3°, P—N—P angles 125.6°, O—P—O angles 116.1°, O—P—N angles 105 to 111°, P—N distances 1.66 Å, and P—O distances 1.48 to 1.52 Å.[97,98] The longer endocyclic P—N bond distances suggest the absence of delocalized π bonding in these heterocyclic species.

Molecules of *closo*cyclotetraphosphazanes(V)[99,100] and *closo*cyclotetraphosphazanes(III)[101,102] are structurally comparable with those of P_4O_{10} and P_4O_6, respectively, with the —N— group substituting for the bridging —O— group.

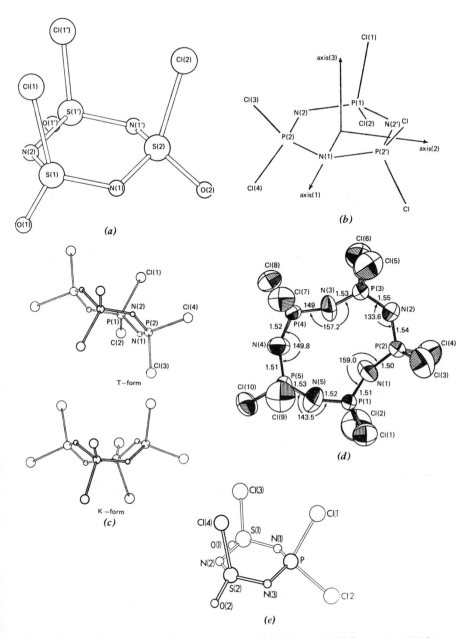

Figure 6-7 Molecular structures of (*a*) α-(NSOCl)$_3$, (*b*) (NPCl$_2$)$_3$; (*c*) (NPCl$_2$)$_4$, (*d*) (NPCl$_2$)$_5$, and (*e*) (NPCl$_2$)(NSOCl)$_2$. [(*a*), A. C. Hazell, G. A. Wiegers, and A. Vos, *Acta Crystallogr.*, **20**, 186 (1966); (*b*), G. J. Bullen, *J. Chem. Soc. A*, **1971**, 1450, Reproduced by permission of the Royal Society of Chemistry; (*c*), A. J. Wagner and A. Vos, *Acta Crystallogr.*, **B24**, 707 (1968); (*d*), A. W. Schlueter and R. A. Jacobson, *J. Chem. Soc. A*, **1968**, 2317, Reproduced by permission of the Royal Society of Chemistry; (*e*) J. C. van de Grampel and A. Vos, *Acta Crystallogr.*, **B25**, 651 (1969).]

Molecular structure data, based upon single-crystal x-ray diffraction measurements, for the species $O_4P_4(NCH_3)_6$ and $S_4P_4(NCH_3)_6$[100] are as follows

	$O_4P_4(NCH_3)_6$	$S_4P_4(NCH_3)_6$
N—P—N angle (deg)	104.1 (15)	103 (2)
P—N—P angle (deg)	119.2 (13)	120.6 (5)
P—N—C angle (deg)	116 (3)	119 (1)
N—P—X angle (deg)	114.4 (11)	116 (4)
P—X distance (Å)	1.45 (1)	1.912 (24)
P—N distance (Å)	1.67 (2)	1.656 (14)
N—C distance (Å)	1.52 (5)	1.47 (3)

The geometry around each P atom is essentially tetrahedral and that around each N atom planar. Again delocalized π bonding is absent.

3. DIHEDRAL, OR TORSIONAL, ANGLES AND THEIR GEOMETRICAL SIGNIFICANCE

In a number of their compounds, atoms of the periodic group VIA elements bond covalently either to atoms of the same element or to atoms of other elements of the family to produce chainlike molecular structures. Some typical species of this type are described in Table 6-18.

In a chain of this type, each atom, except the terminal atoms in certain cases, is two-bonded, with two nonbonded electron pairs in its valency shell. Electron-pair repulsions result in an overall tetrahedral (sp^3) geometry about each such atom, with bond angles approaching the tetrahedral angle, in accordance with earlier discussions. In addition, however, repulsions between nonbonded electron pairs on the two adjacent atoms cause the orbitals in question to assume positions in space ideally at 90° to each other, as illustrated for a molecule of the type X—Z—Z—X in Fig. 6-8(a). The net result is that the bonded groups X are forced out of plane — ideally at 90° to each other. The actual angle between the intersecting planes then occupied by the X atoms [Fig. 6-8(b)] is the dihedral, or torsional, angle.

The existence of a dihedral angle eliminates colinearity in molecules and ions of all catenated group VIA species, except the simple diatomic molecules (Z_2), and requires that all molecules of this type, irrespective of whether Z and X are atoms of the same or different elements, have staggered or puckered molecular structures. In ions of the polythionate type or of the seleno- or telluro-polythionate types, dihedral angles also account for observed geometrical isomerism.[103]

Figure 6-9 depicts the molecular structures of several catenated species of various types. Table 6-18 includes measured dihedral angles. Differences among substituted groups (X) cause variations among dihedral angles and deviations from 90°. In unsubstituted catenated species, such as cyclic S_8 and Se_8 or linear

Table 6-18 Dihedral (Torsional) Angles in Catenated Periodic Group VIA Species

Species Formula	Name	Structural Grouping[b]	Bond ZZZ	Torsional ZZZZ
H_2O_2	Hydrogen peroxide	H—O—O—H	96°52'	93°51'
H_2S_2	Disulfane	H—S—S—H	ca. 95	ca. 90
S_6^{2-}	Hexasulfide ion	S—S—S—S—S—S	109	98, 101, 119
S_2Cl_2	Dichlorodisulfane	Cl—S—S—Cl	104.5	92
S_n	Polycatenasulfur	S—S—S—S⋯	106	85.3
S_8	Cyclooctasulfur (orthorhombic)	S—S—S—S⋯	108.0 ± 0.7 (av.)	98.3 ± 2.1 (av.)
S_{12}	Cyclododecasulfur	S—S—S⋯	106.5 ± 1.4 (av.)	86.1 ± 5.5 (av.)
Se_8	Cyclooctaselenium (α-monoclinic)	Se—Se—Se—Se⋯	105.7 ± 1.6 (av.)	101.3 ± 3.2 (av.)
$(C_6H_5)_2Se_2$	Diphenyldiselenide	C—Se—Se—C	106	82 ± 3
$S_4O_6^{2-}$	Tetrathionate ion	S—S—S—S	103.8 ± 0.5	90.4 ± 1
$S_5O_6^{2-}$	cis-Pentathionate ion (Ba²⁺ salt)	S—S—S—S—S	105, 106	110
$(O_3SS)Se(SSO_3)^{2-}$	cis-Selenopentathionate ion (Ba²⁺ salt)	S—S—Se—S—S	103(SSSe) 101(SSeS)	109
$(O_3SS)Te(SSO_3)^{2-}$	cis-Telluropentathionate ion (Ba²⁺ salt)	S—S—Te—S—S	106(SSTe) 101(STeS)	103
$(O_3SS)Te(SSO_3)^{2-}$	trans-Telluropentathionate ion (NH₄⁺ salt)	S—S—Te—S—S	104, 105(SSTe) 103(STeS)	90° (av.)

[a] $Z = O$, S, Se, Te.
[b] ⋯ = additional S or Se atoms as required.

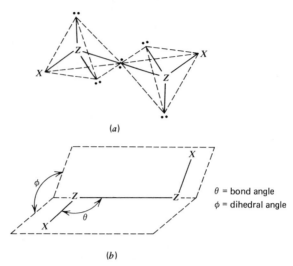

Figure 6-8 Dihedral, or torsional, angle developed when two periodic group VIA atoms (Z) are bonded (*a*) to each other and (*b*) to other atoms (X) also.

S_n or S_x^{2-}, successive dihedral angles result in staggered rings or helical chains.[104-107] The effects of dihedral angles in determining the molecular geometries of polyhomoatomic cations such as S_{16}^{2+} and S_8^{2+} are apparent in later discussions (Table 10-2).

4. MOLECULAR SYMMETRY AND POINT-GROUP SYMBOLISM

The importance of symmetry in the description of crystals has been emphasized in Chapter 4. Symmetry is also an important property of a polyatomic molecule or ion, not only in terms of the geometrical arrangement of its atoms but also as a description of its structure as determined by diffraction and spectroscopic methods (Sec. 6.5). A significant difference between crystals and polyatomic molecules or ions is that the former are rigidly restricted in their symmetries, whereas the latter are not. It is our purpose here to present only information that will be generally useful. To this end, supplementary reference to useful general discussions will be helpful.[108-111] For more detailed information, particularly in relationship to the applications of *group theory*, other excellent detailed treatments are available.[112-119]

4-1. Symmetry Operations and Symmetry Elements

A *symmetry operation* is either an actual or an envisioned process which when carried out on an object (here a molecule or an ion) leaves that object in an indistinguishable, but not necessarily identical, situation with reference to its initial situation. Five such symmetry operations, as delineated in Table 6-19, can be distinguished. Associated with each symmetry operation is a *symmetry element*, which is designated by a particular symbol (Table 6-19).

Before attempting to determine the symmetries of specific molecules and ions by application of the symmetry operations, it is desirable to characterize the symmetry elements more extensively. More than one *proper axis of rotation* may describe a given species, depending upon its complexity and molecular geometry. The *angle of rotation* is of necessity the angle that gives an integral value of n. Depending on that value, this element is designated C_2, C_3, C_4, and so on. The *proper axis* of highest symmetry is referred to as the principal C_n axis. A *mirror plane* may be *horizontal* and perpendicular to the principal C_n axis (σ_h), *vertical* containing the principal C_n axis (σ_v), or *diagonal* containing the principal C_n axis and bisecting the angle formed by two horizontal C_2 axes perpendicular to the principal C_n axis (σ_d). An *improper axis of rotation* is an axis involved when a C_n rotation is followed by reflection across a σ_h mirror plane. The *identity element*, although defined by a rotation operation of 360°, is a symmetry element that can be associated with no symmetry operation.

Let us now illustrate some of these points by considering a few simple species of known geometries. In the H_2O molecule (H—O—H bond angle = 104.5°), all three atoms lie in a plane, which may be considered as the plane of the paper in Fig. 6-10(a). Rotation (counterclockwise by convention) may occur around a vertical axis C bisecting the bond angle and lying in the plane of the three atoms. Rotation through 180° returns the molecule to a position that is indistinguishable from its original position since the two H atoms are identical. Thus n (Table 6-19) = 2, and a symmetry element C_2 exists. In addition, a mirror plane (σ_v) can be visualized as bisecting the bond angle and being perpendicular to the molecular plane. Although reflection exchanges the hydrogen atoms, the initial and final states are again indistinguishable from each other. This operation gives another symmetry element σ_v.

The HOCl molecule is similar to the HOH molecule [Fig. 6-10(b)], but rotation through 180° does not return the molecule to an indistinguishable position. Rotation through 360° returns the molecule to an identical position; so $n = 1$, giving the symmetry element C_1. Since a C_1 axis is characteristic of all molecules, it is unimportant in describing molecular symmetry. Like the H_2O molecule and all other planar molecules, the HOCl molecule has a mirror plane that coincides with the molecular plane (σ_v). However, a mirror plane bisecting the valence angle is absent.

The pyramidal NH_3 molecule (bond angle = 106.8° is depicted in Fig. 6-10(c). The H atoms lie in three planes, each containing the N atom and lying at 120° angles to the other two. Rotation of 120° about the proper axis C_n returns the molecule to an indistinguishable position. Three such rotations return the molecule to its original position. Thus the C_3 symmetry element exists. A vertical plane containing the N atom and one H atom bisects the remaining H—N—H bond angle and is a vertical mirror plane (σ_v). Two other vertical mirror planes (σ_v', σ_v''), each containing one H atom and bisecting an H—N—H bond angle, can be distinguished [Fig. 6-10(c)].

By contrast, the trigonal planar BCl_3 molecule (bond angles = 120°) is depicted in Fig. 6-11. A principal proper axis C_n perpendicular to the molecular plane and extending through the boron atom exists. As with the NH_3 molecule, this axis is designated C_3 since rotation around it by 120° returns the molecule to an indistinguishable position. Within the molecular plane, however, three additional proper C_2 axes exist, each along a B—Cl bond. Each of these axes is perpendicular to the C_3 axis. The molecular plane is a horizontal mirror plane (σ_h). Correspondingly, a vertical mirror plane (σ_v), perpendicular to the molecular plane, exists along each B—Cl bond (Fig. 6-11).

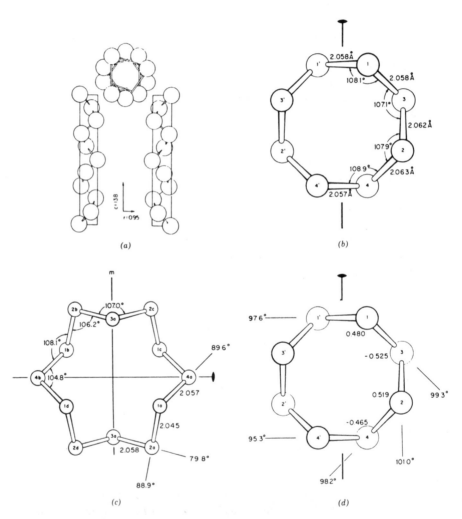

(a)

(b)

(c)

(d)

Figure 6-9 Typical molecular structures illustrating dihedral, or torsional, angles. (a) S_n helix; (b) bond distances and angles in the S_8 molecule in orthorhombic sulfur; (c) bond distances, bond angles, and torsion angles in the S_{12} molecule; (d) torsion angles and distances from the mean molecular plane in the S_8 molecule in orthorhombic sulfur (e) bond distances and bond angles in the Se_8 molecules in α-monoclinic selenium; (f) torsion angles and distances from the mean plane (in Å) in the Se_8 molecules in α-monoclinic selenium; (g) the tetrathionate ion as seen along its

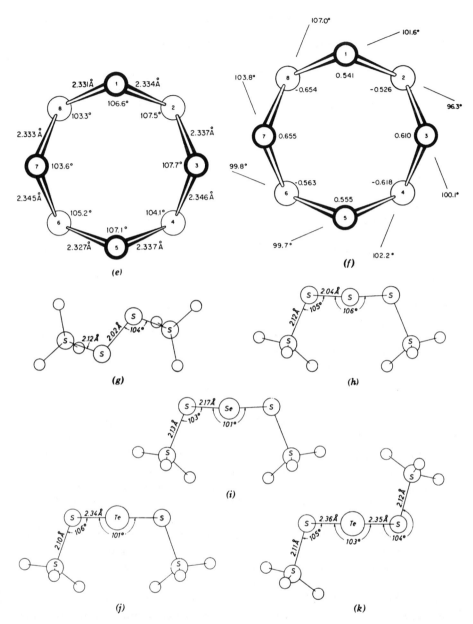

twofold axis in the sodium salt; (h) the pentathionate ion as it occurs in the barium salt; (i) the selenopentathionate ion in orthorhombic barium selenopentathionate dihydrate; (j) the *cis* form of the telluropentathionate ion as it occurs in monoclinic barium telluropentathionate dihydrate; (k) the *trans* form of the telluropentathionate ion as found in the ammonium salt.[(a), B. Meyer, *Chem. Rev.*, **76**, 367 (1976), Fig. 3, Reprinted with permission from the American Chemical Society; (b)–(f), J. Donohue, *The Structures of the Elements*, Figs. 16-7, 16-8, 16-21, 34-5, 34-6, John Wiley & Sons, New York (1974); (g)–(k), O. Foss, in *Advances in Inorganic and Radiochemistry*, Vol. 2, pp. 257, 260, 161, Academic Press, New York (1960).]

Table 6-19   Symmetry Operations and Symmetry Elements

Symmetry Operation	Symmetry Element	
	Description	Symbol[a]
Rotation, counterclockwise, about an axis of symmetry by an order $n(= 360°/$angle of rotation)	*Proper axis of rotation*[b]	C_n
Reflection through a *plane* of symmetry	*Mirror plane*	σ
Reflection through a *center* of symmetry	*Center of symmetry, or inversion center*	i
Rotation, counterclockwise, followed by *reflection* in a plane perpendicular to the axis	*Improper axis of rotation*	S_n
Rotation, by 360°, about an axis of symmetry	*Identity*	E

[a]The symbolism is that of A. Schoenflies [*Krystallsyteme und Krystallstruktur*, Leipzig (1891)].
[b]Species having proper axes of rotation only are not superposable on their mirror images and exhibit optical isomerism (Sec. **7-4**).

Other symmetrical planar molecules and ions have comparable symmetries. For example, in the planar $[PdCl_4]^{2-}$ ion, the principal proper axis is a C_4 axis perpendicular to the molecular plane and extending through the Pd atom. Four C_2 axes, one along each Pd—Cl bond, are perpendicular to the C_4 axis. A horizontal mirror plane (σ_h) and four vertical mirror planes (σ_v) also exist. Similarly, the planar C_6H_6 molecule has a C_6 proper axis perpendicular to the molecular plane and passing through the center of the planar hexagon of carbon atoms (Fig. 6-12). Six C_2 axes can then be distingished in the molecular plane, three through opposite C atoms and three through the centers of opposite edges of the hexagon. Since each C_2 axis is perpendicular to the C_6 axis, an overall symmetry element of D_6 describes the molecule.

Four types of specialized molecular symmetry can be distinguished: linear, tetrahedral, octahedral, and icosahedral. Linear polyatomic molecules and ions are unique in that no rotation about a proper axis containing the atoms in such a species or perpendicular to the molecular axis can produce a distinguishable difference in atomic arrangements. Thus an infinite number of n values exists, corresponding to a symmetry element of $C_{\infty n}$ or $D_{\infty n}$.

The other special types exemplify very high symmetries, with the icosahedral arrangement being the highest of all. Each of these molecular symmetries is characterized by a specific symmetry symbol: T_d (tetrahedral), O_h (octahedral), and I_h (icosahedral). Representative proper rotation axes for these symmetries are indicated in Fig. 6-13. It can be shown that in exact T_d symmetry there are 24

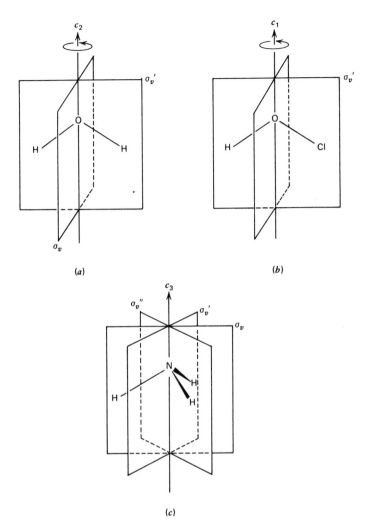

Figure 6-10 Symmetry elements in (a) H_2O, (b) HOCl, and (c) NH_3 molecules.

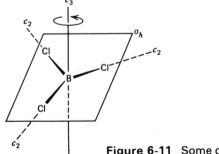

Figure 6-11 Some of the symmetry elements in BCl_3 molecule.

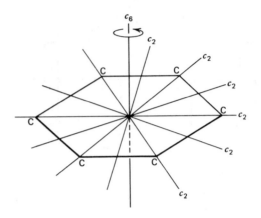

Figure 6-12 Symmetry elements in C_6H_6 molecule.

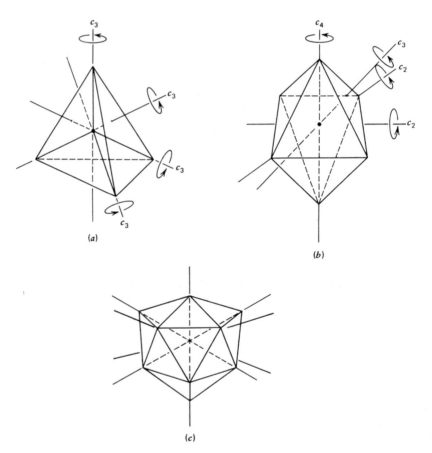

(a)

(b)

(c)

Figure 6-13 Some proper rotation axes for *(a)* T_d, *(b)* O_h, and *(c)* I_h point groups.

possible symmetry operations involving the elements C_2, C_3, S_4, and σ_d. Removal of all improper rotations leaves 12 proper rotations, symbolized T. Similarly, in exact O_h symmetry, there are 48 symmetry operations, and removal of improper rotations yields the symmetry symbol O. Icosahedral symmetry I_h includes both the icosahedron and the pentagonal dodecahedron as polyhedra for which there are 120 possible symmetry operations, of which 60 are proper (symbol I). Full I_h symmetry is exhibited by the $B_{12}H_{12}^{2-}$ anion, the molecular structure of which is indicated in Fig. 6-14.

The improper axis of rotation element (S_n) is less easy to visualize than the other symmetry elements. If the tetrahedral CH_4 molecule is considered as lying within a circumscribed cube with the C atom at the center and the H atoms at four of the corners (Fig. 6-15), an axis of rotation bisecting two H—C—H bond angles is an S_4 axis [Fig. 6-15(a)]. Rotation by 90° about this axis alters the positions of the H atoms [Fig. 6-15(b)], but reflection through a mirror plane perpendicular to the axis returns the molecule to a configuration indistinguishable from the original one [Fig. 6-15(c)]. The symmetry element is then symbolized S_4.

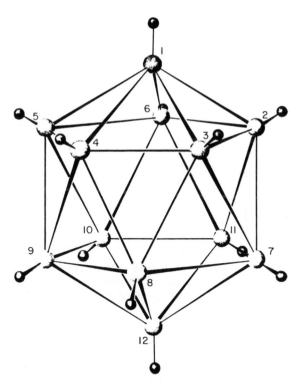

Figure 6-14   Icosahedral molecular structure of $B_{12}H_{12}^{2-}$ ion. [E. L. Muetterties, *The Chemistry of Boron and Its Compounds*, Fig. 5.2b, p. 233, John Wiley & Sons, New York (1967).]

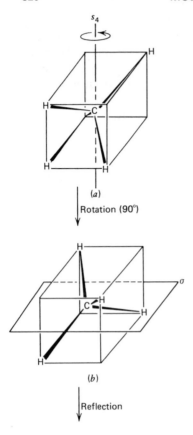

(a)

Rotation (90°)

(b)

Reflection

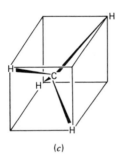

(c)

Figure 6-15 Representation of an improper axis of rotation. Change from (a) to (b) by 90° rotation, to (c) by reflection.

Although the center of symmetry, or inversion center, element should be apparent from Fig. 6-15, this type of center can be defined as a point in a molecule or ion having the Cartesian coordinates (O, O, O) such that changing the coordinates (x, y, z) of all bonded atoms to $(-x, -y, -z)$ converts that molecule or ion to an indistinguishable configuration. Geometrically, this

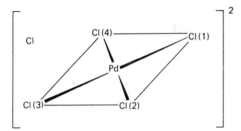

Figure 6-16 Inversion center in the planar ion $[PdCl_4]^{2-}$

situation exists when a line projected from a particular atom in the species through this point (i.e., center) will, if continued in the same direction, encounter an identical atom at the same distance.

Thus with the square-planar ion $[PdCl_4]^{2-}$ (p. 316), depicted in Fig. 6-16, the center of the Pd^{2+} ion is the center of symmetry since the correct symmetry operation would transfer Cl(1) through that center to Cl(3) and Cl(2) to Cl(4) without altering the geometry of the ion as a whole. Similarly, an inversion center characterizes the *trans* isomer of the octahedral dichlorotetraammine-cobalt(III) ion, $[Co(NH_3)_4Cl_2]^+$, but not the *cis* isomer, as indicated in Fig. 6-17.

4-2. Point Groups and Point-Group Symbols

Although innumberable polyatomic molecules and ions of a number of geometries have been identified, all can be characterized in terms of molecular symmetry by sets or combinations of symmetry elements. Each of these combinations is called a *point*

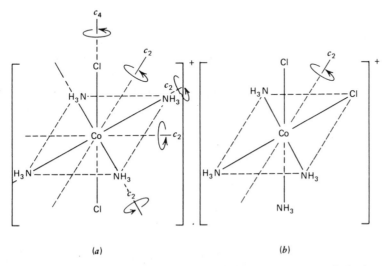

Figure 6-17 Comparison of (a) *trans*- and (b) *cis*-$[Co(NH_3)_4Cl_2]^+$ ions.

group, since the necessary combination of symmetry elements must be such as to leave a specific point in the species unchanged. No translation motions are involved in defining a point group. In considerations of a crystal structure, translational motions generate a *space lattice.*

Each point group is described by a point-group symbol, which in turn consists of a capital letter and a subscript that is a number, a lowercase letter, or a combination of a number and a letter, or a combination of two letters. In arriving at a point-group symbol for a particular species, one can to advantage follow the sequence:

1. Determine whether or not one of the special groups (i.e., linear, $C_{\infty n}$ or $D_{\infty n}$; tetrahedral, T_d; octahedral, O_h; or icosahedral, I_h) exists.

2. Determine whether or not a principal proper rotation axis exists. If only one i present, then the possible point-group symbols may be
   (a) C_n, if no other symmetry features exist.
   (b) C_{nh}, if a σ_h plane also exists.
   (c) C_{nv}, if no σ_h plane exists but a σ_v plane does. Then n planes of this type must be present.

3. Determine if more than one proper rotation axis exists. The principal axis is the one of highest order $(=n)$. The symbolism then becomes
   (a) D_n, if no other symmetry features exist.
   (b) D_{nh}, if a σ_h plane also exists.
   (c) D_{nd}, if no σ_h plane exists but a σ_d plane does.

4. Determine whether an S_{2n} axis is collinear with the principal C_n axis and there are no other symmetry features. The symbol is then S_{2n}. If only a plane of symmetry is found, the symbol is C_s; if only an inversion center, C_i; if no symmetry elements, C_1.

The general procedure is facilitated by the use of a graphic scheme,[108,111,113,119,120] one version of which is given in Fig. 6-18.[108] In this scheme, questions are raised in sequence, with the answers "yes" and "no" giving each time further reductions in total possible point-group symbols and leading ultimately to a final symbol that describes the point-group symmetry of the species in question.

This approach is illustrated by the following examples, assuming that in general the special symmetries are apparent:

1. **H_2O molecule [Fig. 6-10(a)].** Symmetry elements: C_2 axis, $2\sigma_v$, no other nontrivial ones. Point-group symbol: C_{2v}.

2. **NH_3 molecule [Fig. 6-10(c)].** Symmetry elements: C_3 axis, $3\sigma_v$, no other nontrivial ones. Point-group symbol: C_{3v}.

3. **BCl_3 molecule (Fig. 6-11).** Symmetry elements: C_3 axis, $3C_2$ axes $\perp C_3$ axis, $\sigma_h \perp C_3$ axis, $3\sigma_v \perp \sigma_h$, no other nontrivial ones. Point-group symbol: D_{3h}.

Spatial structural formula

(Possible point groups: $C_{\infty v}$, $D_{\infty h}$, T_d, O_h, I_h, C_i, C_1, C_s, C_n, C_{nv}, C_{nh}, D_n, D_{nh}, D_{nd}, S_n)

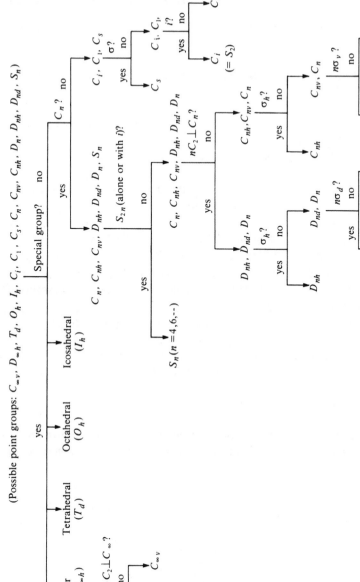

Figure 6-18. Scheme to select point-group symbol.

4. **[PdCl$_4$]$^{2-}$ ion (Fig. 6-16).** Symmetry elements: C_4 axis, $4C_2$ axes $\perp C_4$ axis, $\sigma_h \perp C_4$ axis, $4\sigma_v \perp \sigma_h$, no other nontrivial ones. Point-group symbol: D_{4h}.

5. ***trans*-[Co(NH$_3$)$_4$Cl$_2$]$^+$ ion [Fig. 6-17(*a*)].** Although this ion has overall octahedral geometry, its point-group symmetry is less than O_h because not all of the ligands bonded to the Co^{3+} ion are identical. Symmetry elements: C_4 axis, $4C_2$ axes $\perp C_4$ axis, $\sigma_h \perp C_4$ axis, no other nontrivial ones. Point-group symbol: D_{4h}.

6. ***cis*-[Co(NH$_3$)$_4$Cl$_2$]$^+$ ion [Fig. 6-17(*b*)].** The departure from O_h symmetry is even more pronounced. Symmetry elements: C_2 axis bisecting Cl—Co—Cl bond angle, $2\sigma_v$ collinear with C_2 axis and perpendicular to each other, no other nontrivial ones. Point-group symbol: C_{2v}.

7. ***trans*-[Pt(NH$_3$)$_2$Cl$_2$] molecule (Fig. 6-19).** Symmetry elements: principal C_2 axis $\perp$ molecular plane (σ_h), $2C_2$ axes $\perp$ principal C_2 axis, $2\sigma_v$ including minor C_2 axes, no other nontrivial ones. Point-group symbol: D_{2h}.

8. **PF$_5$ molecule (Fig. 6-20).** Symmetry elements: C_3 axis, $3C_2$ axes $\perp C_3$ axis, $\sigma_h \perp C_3$ axis, $3\sigma_v \perp \sigma_h$, no other nontrivial ones. Point-group symbol: D_{3h}.

9. **BrF$_3$ molecule (Fig. 6-21).** Symmetry elements: C_2 axis, C_1 axis $\perp C_2$ axis, $\sigma_v \perp C_1$ axis, no other nontrivial ones. Point-group symbol: C_{2v}.

Reference to physical models[121] or to symmetry operations in terms of a familiar object[122] may be helpful in visualizing the various symmetry elements.

4-3. Importance of Molecular Symmetry

The fundamental importance of molecular symmetry cannot be over-emphasized. Inasmuch as the symmetries of directed orbitals determine mole-

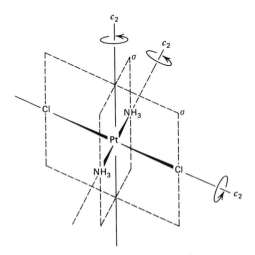

Figure 6-19 Symmetry of the square planar *trans*-[Pt(NH$_3$)$_2$Cl$_2$] molecule.

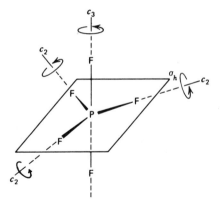

Figure 6-20 Symmetry of the trigonal pyramidal PF_5 molecule.

cular geometries, covalent bonding and symmetry are interdependent. Indeed, a strong case can be made for developing an interpretation of covalent bonding on the basis of molecular geometry.[115,116] Molecular symmetry is intimately related to and essential to the interpretation of vibrational spectra (Sec. 6.5-3). The help that it has given in indicating structures and geometries among coordination entities has aided materially in advancing coordination chemistry from substantial empiricism to a sophisticated combination of fact and theory (Ch. 7,8).

4-4. Comparison of Point-Group and Space-Group Notations

Reference is made to crystal symmetry, space groups, and Bravais lattices (Sec. 4.3-1). The same symmetry operations that relate to molecular structures apply also to crystal structures,[111] but in addition two translational operations may be involved. An operation in which rotation is followed by translation is referred to as a *screw axis*. An operation in which reflection in a mirror plane is followed by

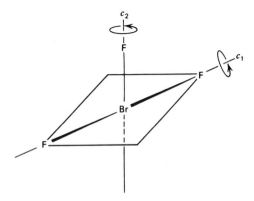

Figure 6-21   Symmetry of the T-shaped BrF_3 molecule.

translation parallel to that plane is termed a *glide plane*. A rotation axis is designated by an Arabic number giving its order. Thus space-group notations 1, 2, 3, and 4 correspond to point-group notations C_1, C_2, C_3, and C_4, respectively. Rotation about a specific axis, followed by inversion through a center of symmetry, is indicated by $\bar{1}$, $\bar{2}$, $\bar{3}$, and so on, depending on the axis involved. A mirror plane is designated *m*, and the symbolism $2/m$ then embraces the point-group symmetry elements C_2 and σ_h. A succession of *m*'s indicates a series of mutually perpendicular mirror planes (e.g., *m m m*). A screw axis is designated in terms of p_q, where *p* is the order of the axis (e.g., 1, 2, 3, 4, 6), and q/p is the fraction of unit cell length that the point is translated following rotation. A glide-plane operation is designated *a*, *b*, or *c* if translation is $a/2$, $b/2$, or $c/2$, *a*, *b*, and *c* being unit-cell dimensions.* The addition of translational symmetry operations results in combination of the 14 Bravais lattices with the 32 point groups to give 230 possible space groups.

Some comparisons involving Bravais lattices and space groups are apparent in Table 6-20,[111] which should be supplemented by Table 4-2. Molecular structures commonly result from crystal structure determinations wherein space groups are established and point groups deduced. For example, a single-crystal, x-ray diffraction investigation of crystals of α-P_4S_4 indicated a monoclinic unit cell, space group $C2/c$, with $a = 9.779(1)$ Å, $b = 9.055(1)$ Å, $c = 8.759(2)$ Å, and $\beta = 102.65°$, and within which each of four P_4S_4 molecules has approximately D_{2d} symmetry with mean P—S and P—P bond distances of 2.111 Å and 2.353 Å, respectively, and cage (p. 261) bond angles P—S—P, S—P—S, and S—P—P of 98.92°, 95.18°, and 100.42°, respectively.[123] Crystal and molecular structures of this compound are indicated in Fig. 6-22.

Table 6-20   Bravais Lattices and Space Groups

Crystal System	Lattices		Space-Group Symbol
	Number	Bravais Notation	
Cubic	3	*P, I, F*	*m 3 m*
Tetragonal	2	*P, I*	*4/m m m*
Orthorhombic	4	*P, C, I, F*	*m m m*
Rhombohedral	1	*P* or *R*	*3/m*
Hexagonal	1	*P*	*6/m m m*
Monoclinic	2	*P, C*	*2/m*
Triclinic	1	*P*	*m 3 m*

---

* For a particularly detailed discussion of space-group symmetry and symbolism, see G. L. Clark, *Applied X-Rays*, 4th ed., Ch. 13, McGraw-Hill Book Company, New York (1955).

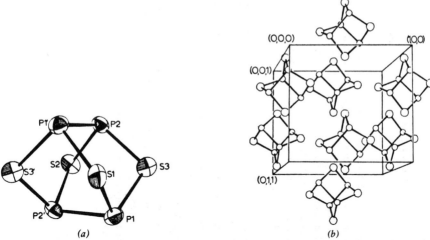

Figure 6-22 Molecular (*a*) and crystal (*b*) structures of α-P$_4$S$_4$. (*a*) Molecular structure; (*b*) unit cell; small spheres represent P atoms and the larger spheres represent S atoms. [C. C. Chang, R. C. Haltiwanger, and A. D. Norman, *Inorg. Chem.*, **17**, 2056 (1978), Figs. 3, 4. Reprinted with permission from the American Chemical Society.]

5. PHYSICAL METHODS FOR THE DETERMINATION OF MOLECULAR STRUCTURE AND GEOMETRY

Most of the experimental approaches to the determination of molecular structures are based upon either diffraction or spectroscopic measurements. Diffraction data can be interpreted in terms of the spatial arrangements of atoms in both crystals and polyatomic molecular or ionic species, together with internuclear distances and bond angles. Spectroscopic data can lead to energy relationships which, in turn, describe particular bonds and groups of atoms and thus molecular geometries. Other experimental approaches include magnetic measurements and dipole moment determinations. These approaches help to establish bond types, electronic structures, and molecular geometries.

The more useful experimental methods are summarized in Table 6-21.[124] It is inappropriate to discuss here the complete details of each technique. Some of the more useful techniques are described in general terms. Some additional information on the relationships between vibrational spectra and molecular symmetry is included also. For detailed information, available extensive treatments can be consulted with profit.[119, 125-128]

5-1. X-Ray and Neutron Diffraction[128,*]

X-ray diffraction, using single crystals, is a classical experimental approach to the determination of the structures of both crystalline solids and polyatomic

---

* A useful and informative summary is J. Wornald, *Diffraction Methods*, Clarendon Press, Oxford (1973).

Table 6-21 Physical Methods for Determining Molecular Structures

Method	Applicability	Information Obtained
Diffraction		
X-ray	Crystalline substances	Complete structural data, including accurate internuclear distances and bond angles; some limitations for light atoms and atoms of similar nuclear charge
Neutron	Crystalline substances	As with x-ray diffraction, plus data for light atoms, atoms of similar nuclear charge, and unpaired electrons
Electron	Commonly gaseous substances; sometimes liquids or solids	Complete structural data, but limited to species with limited numbers of structural parameters
Spectroscopy		
Microwave	Gaseous substances having dipole moments	Interatomic distances, nuclear interactions, dipole moments; limited to relatively simple species
Infrared	Gaseous, liquid, or crystalline substances	Interatomic distances, bond force constants, molecular geometries based upon symmetries of central atoms
Raman	Gaseous, liquid, or crystalline substances	As with infrared, but also applicable where energy changes not observable with infrared
Visible-ultraviolet	Gaseous, liquid, or crystalline substances	All information obtainable by spectroscopic techniques already noted, plus bond dissociation energies; particularly applicable to transition metal complexes
Optical rotatory dispersion and circular dichroism	Solutions, most commonly of transition metal complexes	Optical activity, indicating disymmetry as a consequence of molecular geometry

Nuclear magnetic resonance	Commonly solutions of diamagnetic compounds	Chemical environments of atoms, structural groups of atoms and relative numbers thereof; requires nuclei possessing magnetic moments (e.g., ^1H, ^2H, ^{11}B, ^{13}C, ^{14}N, ^{19}F, ^{31}P, ^{117}Sn, ^{129}Xe, ^{205}Tl)
Nuclear quadrupole resonance	Solids, usually as large samples	Different environments of the resonating nucleus (e.g., Cl, Br, I) in crystals
Electron spin resonance	Paramagnetic species	Ligand symmetries around metal ions
Mössbauer	Solids, γ source required	Symmetries or environments about specific nuclei (e.g., ^{57}Fe, ^{99}Ru, ^{119}Sn, ^{125}Te); solids only
X-ray absorption	Solids	Molecular environments of atoms or ions
Photoelectron	Solid or gas	Molecular environment of atoms or ions
Mass	Gases or reasonably volatile species	Fragmentation patterns as guides to molecular shapes
Miscellaneous		
Magnetic data	Solid or solution	Possible relationships to symmetries around transition metal ions
Dipole moment	Gas, liquid, or solution	Indications of molecular geometries

species present therein. Whereas up until only a few years ago the determination of a single structure could require a year or more of tedious experimentation and calculation, the advent of computer-controlled, single-crystal, x-ray diffracto-meters with computer processing of data now allows completion of a structure determination, with printout of data and computer mapping of molecular structure with thermal variations of given probability, in a few days. Indeed, given the availability of the necessary sophisticated equipment and a suitable crystal, a structure so determined may better characterize a new compound than a complete conventional analysis.[129] A case in point is the compound $La_2O_2[OP(C_6H_5)_3]_2[N\{Si(CH_3)_3\}_2]_4$, for which the observed oxidizing prop-erties were readily rationalized in terms of the peroxo group (O_2^{2-}) found in the crystal structure (Fig. 6-23).[130] Another is the reasonable accounting for the

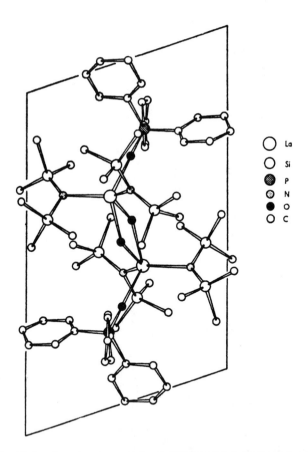

Figure 6-23 Molecular structure of $La_2O_2[OP(C_6H_5)_3]_2[N\{Si(CH_3)_3\}_2]_4$. [D. C. Bradley, J. S. Ghotra, F. A. Hart, W. B. Hursthouse, and P. R. Raithby, *J. Chem. Soc., Chem. Commun.*, **1974**, 40.]

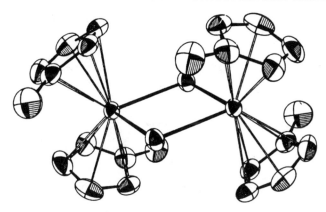

Figure 6-24   Molecular structure of $[Yb(C_5H_4CH_3)_2Cl]_2$ (H atoms not shown). [E. C. Baker, L. D. Brown, and K. N. Raymond, *Inorg. Chem.*, **14**, 1376 (1975), Fig. 2. Reprinted with permission from the American Chemical Society.]

dimeric formulation $[Yb(C_5H_4CH_3)_2Cl]_2$ involving bridging chlorine atoms, as indicated in Fig. 6-24.[131]

Neutron diffraction dates to initial work done on sodium hydride and deuteride in 1948.[132] The sizable neutron fluxes available from fission reactors have given particular impetus to the use of this technique. Neutrons are scattered by both atomic nuclei and unpaired electrons. In combination with x-ray diffraction, neutron diffraction can provide both a molecular structure and data on the distribution of bonding electrons.[133] Advantages over x-ray diffraction include less absorption and easier data collection, location of position and thermal motion of a nucleus, greater sensitivity to light atoms, distinction between elements adjacent to each other in the periodic table, and evaluation of magnetic properties.[133] Neutron diffraction is particularly useful in refining data from x-ray diffraction. A classical application is the determination of the position of the hydrogen atom in a hydrogen bond (Sec. 5.2-1). A case in point is the molecular structure of the ion $H_5O_2^+$ in the compound $HBr \cdot 2\,H_2O$, in which, as indicated in Fig. 6-25, the bonding hydrogen atom is essentially symmetrically located between the two oxygen atoms.[134]

5-2. Electron Diffraction

The number of molecules for which electron diffraction can yield good structural data is limited because of the requirement of relatively high symmetry, the impossibility of accurately locating light atoms in species containing substantially heavier ones, and the uncertainty of establishing positions of atoms because of thermal vibrations. An early success of the method was the establishment of the "H-bridge" structure (Fig. 5-24) of the B_2H_6 molecule.[135] Determinations of the molecular structures of the compounds IOF_5 and SF_5Cl[20, 41] have been discussed earlier in this chapter.

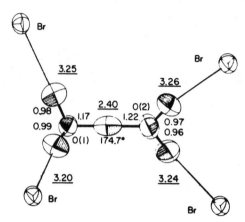

Figure 6-25 Structure of the $H(H_2O)_2^+$ ion in $HBr \cdot 2\ H_2O$ (underlined distances those between O and Br atoms). [R. Attig and J. M. Williams, *Angew. Chem., Int. Ed. Engl.*, **15**, 492 (1976), Fig. 1.]

5-3. Absorption Spectroscopy[128]

Some characteristics of the more common absorption techniques are summarized in Table 6-22.[136] Depending upon the energies involved, these methods give data on (1) rotational, vibrational, or rotational–vibrational deformations within a polyatomic molecule or ion; or (2) electronic transitions within molecular or ionic species. Electronic transitions are particularly important in elucidating molecular structures and bonding in d-transition metal species and, as such, are discussed in more detail in Chapter 8. The internal rotational energy states of polyatomic molecules and ions have no important symmetry characteristics, but the internal vibrational energy states do. For this reason, interpretations of microwave, infrared, and Raman data are useful in evaluating molecular symmetries, and known molecular symmetries are useful in the prediction of vibrational spectra.

Polyatomic molecules and ions undergo internal vibrations. The apparently complex vibrations that characterize a simple species are consequences of the superimposition of several simple vibratory motions known as *normal modes of vibration* or, more commonly, *normal vibrations*. The normal modes can be determined quite simply.[137] A given atom in a species more complicated than diatomic can move in the x direction from its original position independent of any motion in the y or z direction, in the y direction independent of any motion in the x or z direction, and in the z direction independent of any motion in the x or y direction. Thus the atom has 3 degrees of motional freedom. For n atoms in the species, there are then $3n$ degrees of motional freedom. However, all the n atoms can move simultaneously in the x direction, leading to a displacement of the center of mass without altering the internal dimensions of the species, and the same situation applies also to either the y or the z direction. Thus of the $3n$ degrees of freedom, only *three* are real vibrations. Similarly, concerted rotations

Table 6-22 Some Types of Absorption Spectroscopy

Type	Ranges of Energy Involved			Type of Molecular Energy Involved
	cycles sec^{-1}	cm^{-1}	kcal mole^{-1}	
Microwave	10^9–10^{11}	0.03–3	10^{-4} to 10^{-2}	Rotation—heavy molecules
Far infrared	10^{11}–10^{13}	3–300	0.01–1	Rotation—light molecules
				Vibration—heavy molecules
Infrared	10^{13}–10^{14}	300–3000	1–10	Vibrations—light molecules
				Vibrations–rotations
Raman	10^{11}–10^{14}	3–3000	0.01–10	Rotations–vibrations
Visible-ultraviolet	10^{14}–10^{16}	3×10^3 to 3×10^5	10–1000	Electronic transitions

in circular patterns around the x, y, and z axes are not real vibrations. Thus the allowed normal vibrations reduce to $3n - 6$. For a diatomic species, rotation can occur only about the two axes perpendicular to the molecular axis, giving $3n - 5$ normal modes. Each mode is characterized by an absorption at a wave number characteristic of the atoms in the bond in question. However, a real spectrum may contain more absorption bands than the normal ones.

A case in point is the molecule $SO_2(g)$ (bond angle = 119.5°) for which infrared absorption bands are noted at 519, 606, 1151, 1361, 1871, 2305, and 2499 cm^{-1}. The three fundamental vibrations are indicated in Fig. 6-26. Two of these vibrations, labeled *symmetric (a) stretching* and *asymmetric (b) stretching*, may be regarded as elongations of bonds; the other, a *bending mode*, as an alteration of the bond angle. The largest wave number associated with a symmetric vibration is designated v_1, the second largest v_2, and so on. After all symmetric vibrations have been accounted for, the asymmetric ones are designated (e.g., v_3 in Fig. 6-26). The band at 2305 cm^{-1} (i.e., $= 2v_1$) is an *overtone*; that at 606 cm^{-1} is a *difference* band ($= v_1 - v_2$); and those at 1871 and 2499 cm^{-1} are *combination* bands ($= v_2 + v_3$ and $v_1 + v_3$, respectively).* Assignments for this species are quite simple, but in many cases they are much more difficult to make. Fortunately, the absorptions characteristic of a given bond type repeat themselves fairly closely from species to species, thus allowing vibrational spectra to be used in a "fingerprinting" sense to identify particular arrangements of atoms in a variety of species. Typical fingerprinting regions are indicated in Fig. 6-27(a) to (e).[138]

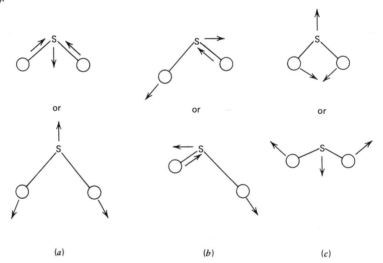

(a)   (b)   (c)

Figure 6-26   Normal vibrations of the $SO_2(g)$ molecule. (a) Symmetric stretching, v_1, 1151 cm^{-1}; (b) asymmetric stretching; v_3, 1361 cm^{-1}; (c) bending, v_2, 519 cm^{-1}.

* Other symbols used include v for stretching vibrations, δ for in-plane bending vibrations, and π for out-of-plane bending vibrations, with inclusion of the subscripts s (symmetric), as (asymmetric), and d (degenerate).

Examples are legion, but the approach is often particularly useful in identifying the bonding sites in such ambidentate ligands (Sec. 7.4-5) as the NCS^- and NCO^- ions, which can in principle bond through either the N or S atoms or the N and O atoms, respectively. Thus the presence of a strong absorption band at 528 cm^{-1} in the infrared spectrum of the cyclotriphosphazene $N_3P_3Cl_4(NCS)_2$[139] clearly indicates the presence of the P—NCS bond,[140] as do also the strong asymmetric N=C=S stretching vibration at 2010 cm^{-1} and the strong symmetric N=C=S stretching vibration at 1080 cm^{-1}.[141,142] Furthermore, in the crystalline lanthanide(III) complexes $[(C_2H_5)_4N]_3[Ln(NCO)_6]$ (Ln = Sc, Y, Eu, Gd, Dy, Ho, Er, Yb),[143] $[(C_2H_5)_4N]_3[Ln(NCO)_3(NO_3)_3]$ (Ln = La, Pr, Nd, Sm, Eu, Gd, Er, Yb),[144] and $[(C_2H_5)_4N]_3[Ln(NCO)_3X_3]$ (Ln = Y, Dy, Er, Yb; X = Cl, NCS),[145] the N-bonded isocyanato arrangement is indicated by observed increases in wave number for v_{as} and decreases in wave number for the δ_{NCO} modes with respect to the NCO^- ion, as predicted.[146] An even more reliable indication of the iso arrangement is in the displacement of the v_s mode from 1254 cm^{-1} for the NCO^- ion[147-149] to 1327 to 1337 cm^{-1}. Furthermore, the presence of the N-bonded isothiocyanato group is indicated for the last series of compounds (i.e., X = NCS).

The origins of vibrational absorptions bear evident relationships to molecular symmetry. It is not within the scope of this treatment to indicate how these relationships are derived.[119] However, some specific examples are instructive. Thus for both gallium and indium compounds of compositions GaX_2 and InX_2 (X = Cl, Br, I) the presence of the dipositive oxidation state is a possibility, although the diamagnetic natures of these compounds would preclude the presence of Ga^{2+} and In^{2+} ions. However, comparisons of Raman data for known tetrahedral halogen-containing species with those for these compounds (Table 6-23) prove conclusively the presence of species of tetrahedral symmetry. The known MX_2 stoichiometries then require the presence of the +3 oxidation state in the species GaX_4^- and InX_4^-.[150-156] Thus the +1 state must also be present, and the correct formulations are $\overset{I}{Ga}[\overset{III}{Ga}X_4]$ and $\overset{I}{In}[\overset{III}{In}X_4]$, both $\overset{I}{M}$ and $\overset{III}{M}$ being diamagnetic.

Change in point-group symmetry, as indicated by vibrational spectroscopy, has been particularly useful in deciding whether or not certain oxoanions (e.g., NO_3^-, ClO_4^-) are bonded to metal ions in various complex species. As indicated in Fig. 6-28, the planar NO_3^- ion has D_{3h} point-group symmetry, whereas if that ion is bonded to a metal ion (M^{n+}), symmetry is reduced to C_{2v}. Characteristic absorptions for these two symmetries are summarized in Tables 6-24[157] and 6-25.[158,159] For the NO_3^- ion, six normal modes of vibration, of which two are nondegenerate and two pairs are doubly degenerate, are predicted. The formation of an M—O—N covalent bond results in the splitting of the two doubly degenerate vibrations and the prediction of six nondegenerate vibrations. Ionic nitrate is usually distinguished from bonded nitrate by the

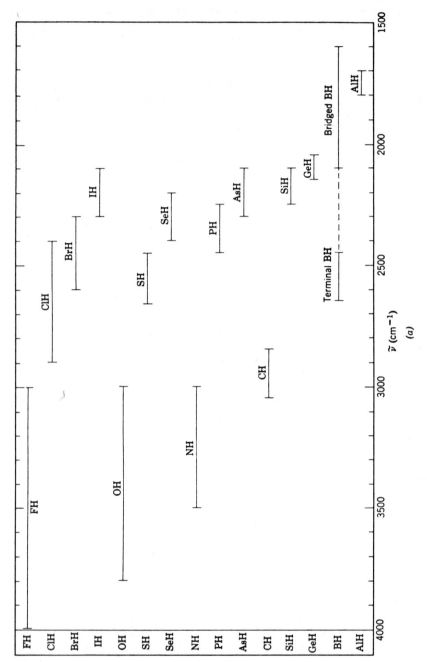

$\tilde{\nu}\ (cm^{-1})$

(a)

Figure 6-27

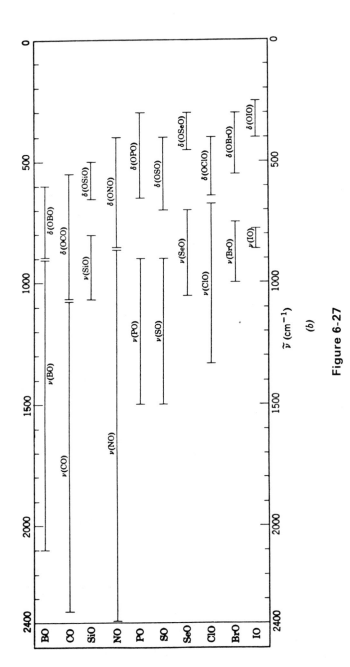

Figure 6-27

337

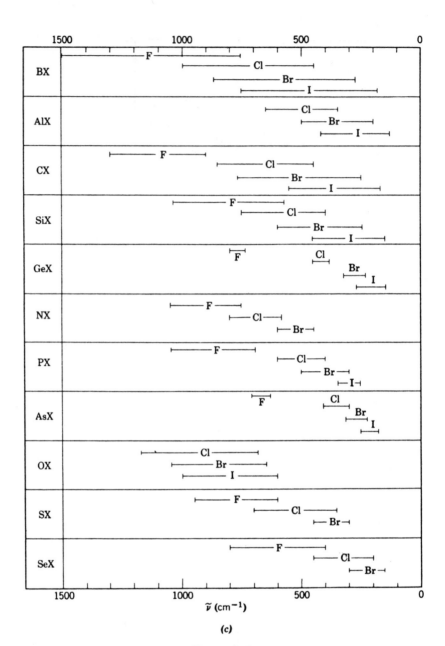

(c)

Figure 6-27

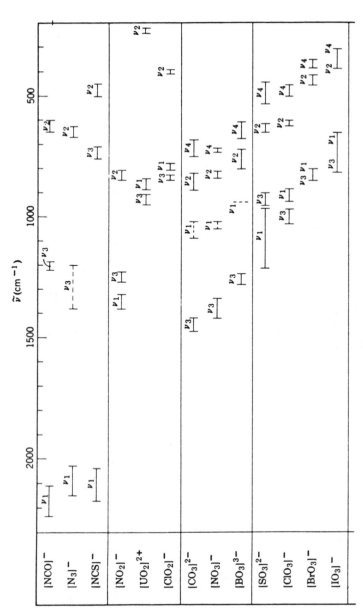

$\tilde{\nu}\,(\mathrm{cm}^{-1})$

Figure 6-27

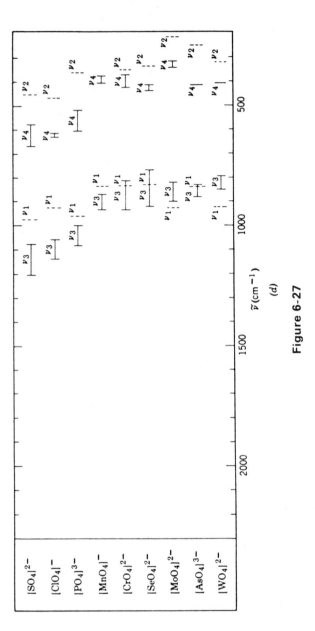

Figure 6-27

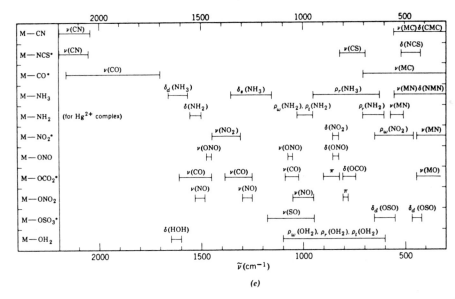

Figure 6-27   Fingerprinting regions in vibrational spectra. (a) Hydrogen stretching vibrations; (b) oxygen stretching and bending vibrations; (c) halogen (X) stretching vibrations; (d) characteristic frequencies of some inorganic ions (dashed lines indicate Raman active vibrations); (e) characteristic vibrations of metal complexes having simple ligands (wave-number ranges include bidentate and bridged complexes for the ligands marked by an asterisk). [K. Nakamoto, *Infrared and Raman Spectra of Inorganic and Coordination Compounds*, 3rd ed., pp. 436–441, John Wiley & Sons, New York (1978).]

Table 6-23   Raman Absorptions Characteristic of Tetrahedral MX_4 Species

Species	Absorption Band (cm^{-1})				Reference
	v_2	v_4	v_1	v_3	
GeI_4	60	80	159	264	150
ZnI_4^{2-}	44	62	122	170	150
$GaI_3 + HI(aq)$	52	73	145	222	150
$GaI_2(l)$	53	80	144	214	151
$GaBr_3 + HBr(aq)$	71	102	210	278	152
$GaBr_2(l)$	79	107	209	288	153
$Ga_2Cl_4(l)$	115	153	346	380	154
$(C_2H_5)_4NInBr_4(s)$			199	235	155
$InBr_4^-(aq)$	55	79	197	239	155
$InBr_2(s)$			197	228	155
$KInCl_4(l)$	87	114	320	346	156
$InCl_2(l)$	91	114	317	349	156

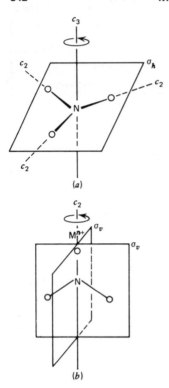

(a)

(b)

Figure 6-28 Symmetries of isolated and bonded NO_3^- groups. (a) NO_3^-, D_{3h}; (b) $M-ONO_2^{n-1}$, C_{2v}.

splitting of the v_3 1390 cm^{-1} band (D_{3h}) into two bands at ca. 1300 cm^{-1} (v_2, C_{2v}) and ca. 1500 cm^{-1} (v_4, C_{2v}). The v_4 720 cm^{-1} band (D_{3h}) is also split, but the resulting two absorptions are weak and may not be observable. Unfortunately, C_{2v} point-group symmetry characterizes both unidentate and bidentate bonded nitrate groups. Distinction between the two is impossible using infrared spectroscopy but possible using Raman spectroscopy. Table 6-26 summarizes a few of the many reported results.

Table 6-24 Vibrational Absorptions and Assignments for NO_3^- Ion (D_{3h})

				Activity[b]	
		Absorption			
Type[a]	cm^{-1}	Vibration	Assignment	Infrared	Raman
A$_1'$	1050	v_1	N—O stretching (symmetric)	ia	a
A$_2''$	831	v_2	NO_2 deformation	a	ia
E'	1390	v_3	NO_2 stretching (asymmetric)	a	a
E'	720	v_4	Planar rocking	a	a

[a] A = symmetric vibration with respect to highest-fold rotation axis; E = doubly degenerate vibration.

[b] a = active; ia = inactive.

Table 6-25 Vibrational Absorptions and Assignments for M—ONO_2^{n-1} Bonding (C_{2v})

Type[a]	cm^{-1}	Vibration	Assignment	Infrared	Raman
				Activity[b]	
A_1	1034–970	v_2	N—O stretching	a	a
B_2	800–781	v_6	Nonplanar rocking	a	a
B_1	1531–1481	v_4	Stretching (asymmetric)	a	a
A_1	1290–1253	v_1	NO_2 stretching (symmetric)	a	a
A_1	ca. 740	v_3	NO_2 bending	a	a
B_1	ca. 713	v_5	Planar rocking	a	a

[a] B = anti symmetric vibration with respect to highest-fold rotation axis.

[b] a = active.

Similar, but less extensive investigations, have shown that the T_d point-group symmetry of the ClO_4^- ion is altered to C_{3v} or C_{2v} when that ion is bonded to various metal ions, the bonding being either unidentate or bidentate.[161] Bidentate bonding is supported by infrared data for the compounds $M(ClO_4)_2 \cdot 2H_2O$ (M = Mn, Co, Ni),[162] unidentate bonding for the compounds $[Ni(R\text{-}py)_4(ClO_4)_2]$ (R-py = substituted pyridines),[163] and ionic ClO_4^- for the compounds $[Ln(phen)_4](ClO_4)_3$ (Ln = Nd, Dy, Er; phen = 1,10-phenanthroline).[164]

Microwave, infrared, and Raman spectroscopy are often mutually complementary (Table-6-21). This is particularly true of infrared and Raman measurements, where differences in *selection rules* may lead to characteristic absorptions in different spectral regions.[119, 157, 165] For a molecular vibration to be Raman active, a change in polarizability (p. 246) is necessary during that vibration. No fundamental absorptions are common to the infrared and Raman spectra of a polyatomic species if a center of symmetry is present in that species. Thus the infrared and Raman spectra of a given substance may be the same or different. Water and deuterium oxide are often useful solvents for Raman studies since they show transparency in regions where they exhibit infrared absorptions. Early experimental difficulties in obtaining sources of sufficient intensity for Raman investigations have been obviated by the development of laser sources. Nakamoto's treatise[119] is an excellent source for information on all aspects of vibrational spectroscopy.

A number of examples of structural studies using Raman absorption have been noted in preceding paragraphs. A typical example of less widely used microwave studies lies in determining for the molecule $(CH_3)_3NB(CH_3)_3$ the parameters: B—N distance, 1.698(1) Å; N—C distance, 1.470(10) Å; B—C distance, 1.69(4) Å; N—B—C angle, 180.0(15)°; and B—N—C angle, 111.6(5)°.[166]

Table 6-26 Infrared Data for Ionic and Bonded NO_3^- Groups

Compound	NO_3^- (D_{3h}) ν_1	ν_2	ν_3	ν_4	$-ONO_2$ ($C_{2v'}$) ν_1	ν_2	ν_3 or ν_5	ν_4	ν_6	Indicated Symmetry	Reference
$NaNO_3$	—	836 m	1358 vs							$D_{3h}(NO_3^-)$	159
KNO_3	—	824 m	1348 vs	733 m						$D_{3h}(NO_3^-)$	159
$Hg(NO_3)_2$					1376 s	1027 vs	750 m	1495 m	788 vs	$C_{2v'}(-ONO_2)$	158
$Mn(NO_3)_2$					1294 vs	1019 vs	759 m	1553 vs	805 s 799 vs	$C_{2v'}(-ONO_2)$	158
CH_3NO_3					1287 s	854 s		1672 vs	759 m	$C_{2v'}(-ONO_2)$	157
$Gd(NO_3)_3 \cdot 4en$[b]		832	1350							$D_{3h}(NO_3^-)$, or $[Gd(en)_4]^{3+}(NO_3^-)_3$	160
$Nd(NO_3)_3 \cdot 4en$		829	1345				725	1428	815,820	D_{3h} and $C_{2v'}$, or $[Nd(en)_4$ $(NO_3)]^{2+}(NO_3^-)_2$	160
$Gd(NO_3)_3 \cdot 3en$		832	1345				740	1505	815	D_{3h} and $C_{2v'}$, or $[Gd(en)_3$ $(NO_3)_2]^+NO_3^-$	160
$[(C_2H_5)_4N]_3$-$[Pr(NCO)_3(NO_3)_3]$					1312 vs	1038 vs	738 vs	1510 vs	823 s	$C_{2v'}(-ONO_2)$	144

[a] s = strong, m = medium, v = very.
[b] en = ethylenediamine.

5-4. Nuclear Magnetic Resonance Spectroscopy

Nuclear magnetic resonance spectroscopy is a type of absorption spectroscopy that involves the radio-frequency region of the electromagnetic spectrum.[127, 167-169] Samples under study are subjected simultaneously to an intense magnetic field and a weak radio-frequency field applied at right angles. When the intensities of the two fields satisfy certain requirements, an appropriate atomic nucleus characterized by "spin," or angular momentum (p. 23), changes its spin orientation with respect to the magnetic field with the simultaneous absorption of radio-frequency energy. Inasmuch as the magnetic environment of the nucleus is closely related to its molecular environment, that nucleus in different molecular environments in a chemical species will then resonate at different magnetic field strengths when the applied radio-frequency field strength is constant.

Differences dependent upon molecular environments are rendered quantitative by reference to a standard that is without interference: for example, $Si(CH_3)_4$ (symbolized commonly as TMS) for proton spectra, $CFCl_3$ for ^{19}F spectra, 85% phosphoric acid or triethylphosphate for ^{31}P spectra. If in measuring a spectrum at constant radio-frequency field strength v_0 (e.g., 60 MHz or 100 MHz for protons),* absorption is plotted against magnetic field strength (in Hz) with the standard taken as zero, the difference between zero and an absorption peak (i.e., Δ) can be measured. Each peak can then be characterized in terms of the relationship

$$\delta = \frac{\Delta \times 10^6}{v_0} \tag{6-1}$$

where δ is the *chemical shift*, expressed in parts per million (ppm). The chemical shift, which is independent of the probe frequency used, is characteristic of the chemical environment of the nucleus being investigated.

An early classical experiment using ethanol yielded three proton chemical shifts, resulting in areas under the absorption curve that stood in the ratios 3 : 2 : 1, assigned, respectively, to the methyl, methylene, and hydroxyl protons, in keeping with the molecular formula CH_3CH_2OH.[170, 171] A more refined set of experiments, using TMS as a reference, yielded the results shown for two different probe frequencies in Fig. 6-29.

Chemical shifts are sufficiently characteristic to be used in the "fingerprinting" sense of identifying particular groupings in compounds and thus providing useful structural indications, in particular for hydrogen-containing species.[172] A down-field shift (i.e., toward higher frequencies) is written with a positive sign; an up-field shift with a negative sign.

A second observable characteristic of a nuclear magnetic resonance spectrum is "spin-spin" coupling, which arises from the fact that the magnetic moments of nuclei within a given molecular species or ion allow interactions with electrons

* 1 Hz = 1 cycle sec^{-1}; 1 MHz = 10^6 Hz.

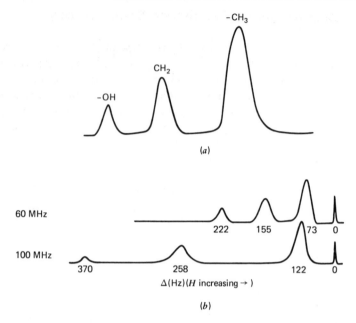

Figure 6-29 Proton nuclear magnetic resonance spectrum of C_2H_5OH. (*a*) Low resolution; (*b*) high resolution. [R. S. Drago, *Physical Methods in Chemistry*, Fig. 7-15, p. 205; Fig. 7-16, p. 206, W. B. Saunders, Philadelphia (1977).]

in bonds that link those nuclei. The result is that an otherwise sharply defined single absorption line in the spectrum is "split" into multiplets, the spacing and intensity distributions of which can be interpreted in terms of both the number of interacting nuclei and the bonds through which they are coupled. Splitting of this type involves only nonequivalent nuclei. In general, if an absorption band characteristic of a nucleus A is split as a consequence of the presence of a nonequivalent nucleus B, the number (*n*) of absorption peaks due to nucleus A is given by the relationship

$$n = 2 \, \Sigma \, S_B + 1 \tag{6-2}$$

where ΣS_B is the sum of the spins of the nonequivalent B nuclei. Thus for the molecule PF_3 ($S_P = \frac{1}{2}$, $S_F = \frac{1}{2}$), for the absorption due to the P atom, $n = 2(3 \times \frac{1}{2})$ $+ 1 = 4$ (a so-called "quartet"), and for that due to an F atom, $n = 2(1 \times \frac{1}{2}) + 1$ $= 2$ (a "doublet").

Spin-spin splitting is measured in terms of the *spin-spin coupling constant* J_{AB}, which is the separation between the absorption peaks comprising the fine structure and which is commonly expressed in hertz. Thus J_{HH} refers to $^1H - {}^1H$ splitting, J_{PH} to $^{31}P - {}^1H$ splitting, and so on. Although the magnitude of Δ is field dependent, the magnitude of J is field independent. Both Δ and J values are commonly given when a particular species is characterized by nuclear magnetic resonance studies. For example, in the identification of the compound *cis-(t-*

$C_4H_9PF_2)_2Mo(CO)_4$, the following data are given as useful: ^1H—$\delta_H = -1.01$ ppm (internal TMS), $^3J_{PH} = 16.6$ Hz and $^4J_{FH} = 1.2$ Hz; ^{19}F—$\delta_F = +65.4$ ppm (internal CCl_3F), $^1J_{PF} + ^3J_{PH} = 1104$ Hz; and ^{31}P—$\delta_P = -263.4$ ppm (external H_3PO_4).[173]

From a combination of chemical shift values, spin-spin coupling constants, and the observation that the integrated areas under absorption peaks are proportional to the number of atoms of a given type in the molecule or ion under investigation, it is often possible either to deduce or to approximate with a fair degree of certainty the molecular structure of that species. Nuclei that are particularly useful for the evaluation of molecular structures by way of nuclear magnetic resonance spectroscopy include ^1H, ^2H, ^{10}B, ^{11}B, ^{13}C, ^{14}N, ^{15}N, ^{19}F, ^{23}Na, ^{29}Si, ^{31}P, ^{35}Cl, ^{59}Co, ^{117}Sn, ^{119}Sn, ^{129}Xe, ^{203}Tl, and ^{205}Tl. Proton-resonance studies have been and are of particular value in evaluating the molecular structures of organic compounds, hydrogen bonds, and many inorganic compounds. Boron-10 or 11, carbon-13, nitrogen-14 or 15, fluorine-19, and phosphorus-31 nuclei are very useful in inorganic chemistry. Studies using phosphorus-31 have done much to elucidate the very involved structural chemistry of phosphorus.[174] Some other typical applications include establishment of a fluorine-bridged structure for the $Kr_2F_3^+$ ion, using ^{19}F resonance;[175] important contributions to the structural chemistry of xenon, using ^{129}Xe resonance;[176] confirmation of a planar molecular structure, for the $N_2F_3^+$ ion, using ^{19}F resonance;[177] and establishment of a rearrangement process in the syntheses of $B_9H_{12}S^-$ derivatives from $B_{10}H_{14}$, using ^{11}B resonance.[178]

Although much of the useful structural information obtained by means of nuclear magnetic resonance measurements is restricted to diamagnetic species, numerous paramagnetic complexes of the transition metals (Ch. 7) have been examined with profit, largely in terms of shifts in proton magnetic resonance spectra. The comparatively large isotropic shift in the diamagnetic resonance of a proton in a ligand that results from imposing electronic spin density on the ligand through bond formation with a paramagnetic cation is termed the *contact*, or *scalar, shift*.[179] An observed contact shift indicates covalence in the bond between a ligand atom and the metal ion, and the magnitude of this shift is taken as a measure of the degree of covalency in the bond. Another paramagnetic shift in proton resonance, similar to but distinct from the contact shift, results from the combined effects of coupling of the spin-orbit, orbit-nucleus, and spin-nucleus types. It is related to the molecular geometry of the species and is termed the *pseudocontact*, or *dipolar, shift*.[180] The observed proton shift in a given spectrum is the summation of the contact and pseudocontact terms, but the absolute contribution of each of the terms is not always known with accuracy. Contact shifts are of particular importance with complexes of the *d*-transition metals, but both appear to be significant for at least the $4f$-transition metal compounds.[181]

The utility of proton nuclear magnetic resonance spectroscopy in elucidating the molecular structures of certain organic compounds that contain donor oxygen atoms has been extended markedly by the use of *lanthanide shift*

Table 6-27 · β-Diketones Useful as Shift Reagents

Name	Molecular Formulation	Abbreviation
2,2,6,6-Tetramethyl-3,5-heptanedione (dipivaloylmethane)	$(CH_3)_3CC(O)CH_2C(O)C(CH_3)_3$	dpm, thd, tmhd
6,6,7,7,8,8,8-Heptafluoro-2,2-dimethyl-3,5-octanedione (1,1,1,2,2,3,3-heptafluoro-7,7-dimethyl-4,6-octanedione)	$CF_3CF_2CF_2C(O)CH_2C(O)C(CH_3)_3$	fod
1,1,1,5,5,6,6,7,7,7-Decafluoro-2,4-heptanedione (1,1,1,2,2,3,3,7,7,7-decafluoro-4,6-heptanedione)	$CF_3CF_2CF_2C(O)CH_2C(O)CF_3$	dfhd

reagents.[182] These reagents are most commonly tris(β-diketonates) of certain paramagnetic lanthanide ions (e.g., Pr^{3+}, Eu^{3+}, Yb^{3+}) which behave as Lewis acids (Sec. 9.6-1) in the sense that the bonded Ln^{3+} ion can, by facile increase in its coordination number, accept an electron pair from an oxygen atom in the organic moiety. The result is a general displacement and spreading out of absorption peaks in the proton nuclear magnetic resonance spectrum of that moiety and thus improvement in ease of interpretation.[182-184]

The first lanthanide shift reagent to be used was the dipyridine adduct of tris(dipivaloylmethanato)europium(III), $Eu(dpm)_3 \cdot 2 C_5H_5N$.[182] This reagent and other related substances are listed in Table 6-27.[184] A typical effect in clarifying a spectrum is apparent in Fig. 6-30.[185]

5-5. Other Methods[128]

Other methods listed in Table 6-21 are specifically applicable to determinations of electronic charge distributions in covalent entities. Thus only indirectly do they yield data indicative of molecular structure and geometry. Fundamentally, the same is true of nuclear magnetic resonance spectroscopy, but that approach is so versatile that it does give definitive data of a structural type in many cases.

Earlier discussions have included information on dipole-moment (Sec. 5.1-8) and magnetic susceptibility (Sec. 2.4-4) measurements. X-ray absorption spectroscopy is a particularly promising technique applicable to gases, liquids, crystalline solids, and amorphous solids, the scope of which can be remarkably extended through the use of synchrotron radiation.[186] X-ray photoelectron spectroscopy, or electron spectroscopy for chemical analysis (ESCA) is an application of this broad approach.[187, 188] Mössbauer spectroscopy[189-192] probes the electronic environments about nuclei in terms of fluorescence

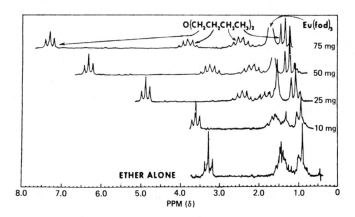

Figure 6-30 Effect of added $Eu(fod)_3$ on the proton nuclear magnetic resonance spectrum of di-*n*-butyl ether. [R. E. Rondeau and R. E. Sievers, *J. Am. Chem. Soc.*, **93**, 1522 (1971), Fig.2. Reprinted with permission from the American Chemical Society.]

produced from the absorption and emission of gamma radiation, yielding data in terms of an *isomer shift* which measure differences in the chemical environments of nuclei. Information derivable from Mössbauer spectra is somewhat analogous to, but much more restrictive than, that derivable from nuclear magnetic resonance spectra.

Mass spectroscopy, an ionization technique originally limited to isotope detection and mass evaluation (Sec. 2.1-1), yields, through fragmentation patterns which show relative intensities of positive ions of particular e/m (i.e., charge/mass) ratios, information as to groups of atoms present in a given molecule or ion, and in terms of appearance potentials, data on bond strengths.[193-195] Some of these data are often useful in confirming particular, previously determined or postulated molecular structures, although a more common use is in identification of species. For a series of closely related compounds, correlations between fragmentation patterns and molecular structures, which may permit assignments of molecular structures to other related compounds, may result. A case in point involves a series of phenylsubstituted chlorocyclophosphazenes, for which distinctions between geometrical isomers and the assignment to a trace quantity of a cyclotetraphosphazene of composition $N_4P_4Cl_6(C_6H_5)_2$ of a 4,4,6,6,8,8-hexachloro-2,2-diphenyl molecular structure have been reported.[196]

EXERCISES

6-1. For each of the following covalent species, determine the probable molecular geometry in terms of both the directed-orbital and the valence-shell electron-pair repulsion approaches: $BeCl_2(g)$, $SnCl_2(g)$, $PBr_3(g)$, $[PtCl_4]^{2-}$, $BrF_5(g)$. Compare your results with experimentally observed geometries.

6-2. Indicate clearly why in compounds of the type $X_3PO(X = F, Cl, Br)$ (*a*) the X—P—X bond angles approach but do not equal the ideal tetrahedral angle, and (*b*) this approach becomes closer as the atomic mass of the halogen atom increases (see Table 6-4).

6-3. Consult suitable references among those cited and summarize in more detail current interpretations of "stereochemically active" and stereochemically inactive" nonbonded electron pairs in determining molecular geometries.

6-4. Rationalize, as an underlying common principle, the observed molecular geometries of the molecular species $P_4(g)$, $P_4O_6(s)$, $P_4O_{10}(s)$, $P_4S_{10}(s)$, and $P_4S_4O_6(s)$.

6-5. Indicate clearly, with your reasons, which of the following triatomic species should have linear geometry and which should have angular geometry: F_2O, $CdBr_2$, COS, $ONCl$, S_3^{2-}.

6-6. The groups NNN in the HN_3 molecule and CNN in the CH_2N_2 molecule have linear geometry. Indicate electronic configurations and account for this geometry.

6-7. In alkaline media, the ions IO_6^{5-}, TeO_6^{6-}, $Sb(OH)_6^{-}$, and $Sn(OH)_6^{2-}$ can be distinguished. What structural feature is common to these species? On this basis, give reasonable formulas for the parent acids.

6-8. Consult the appropriate literature and summarize current interpretations of Sidgwick's "inert electron pair."

6-9. Why are S_n chains in covalent species nonlinear when $n > 2$?

6-10. For each of the following covalent species, indicate the characteristic symmetry elements: planar N_2O_4, pyramidal PF_3, tetrahedral GeH_4, angular H_2S, octahedral SeF_6.

6-11. Showing your method, determine the point group for each of the following covalent species: planar $[Pd(NH_3)_4]^{2+}$, SF_4, cis-$[Pt(NH_3)_2Cl_2]$, Fe-$(C_5H_5)_2]$, $[Cr(en)_3]^{3+}$, ClF_3.

6-12. Using the cited references and any others that you may find, indicate clearly how one can distinguish between —SCN and —NCS covalently bonded groups.

6-13. Consult suitable references and then indicate clearly how vibrational spectroscopy can be used to distinguish among ionic, unidentate covalent, and bidentate covalent bonding of the perchlorate group in metal complexes.

6-14. By means of suitable drawings, indicate possible isomers for a cyclophosphazene of molecular formula $N_3P_3Cl_4(OCH_3)_2$. How can one distinguish among these isomers?

6-15. Summarize the advantages of neutron diffraction over x-ray diffraction as a method of structure determination.

REFERENCES

1. R. B. King, *J. Am. Chem. Soc.*, **91**, 7211, 7217 (1969); **92**, 6455, 6460 (1970).

2. G. W. A. Fowles, *J. Chem. Educ.*, **34**, 187 (1957).

3. L. Pauling, *The Nature of the Chemical Bond*, 3rd ed., Ch. 4, Cornell University Press, Ithaca, N.Y. (1960).

4. L. Pauling, *The Nature of the Chemical Bond*, 3rd ed., Ch. 5, Cornell University Press, Ithaca, N.Y. (1960).

5. G. E. Kimball, *J. Chem. Phys.*, **8**, 188 (1940).

6. L. Pauling and J. Sherman, *J. Am. Chem. Soc.*, **59**, 1450 (1937).

7. N. V. Sidgwick and H. Powell, *Proc. R. Soc. Lond. Ser. A*, **176**, 153 (1940).

8. R. J. Gillespie and R. S. Nyholm, *Q. Rev. (Lond.)*, **11**, 339 (1957).

9. R. J. Gillespie, *Can. J. Chem.*, **38**, 818 (1960).

10. R. J. Gillespie, *J. Am. Chem. Soc.*, **82**, 5978 (1960).

11. R. J. Gillespie, *J. Chem. Educ.*, **40**, 295 (1963).

12. R. J. Gillespie, *Angew. Chem., Int. Ed. Engl.*, **6**, 819 (1967).

13. R. J. Gillespie, *J. Chem. Educ.*, **47**, 18 (1970).

14. R. J. Gillespie, *Molecular Geometry*, Van Nostrand Reinhold, London (1972).

15. R. J. Gillespie, *J. Chem. Educ.*, **51**, 367 (1974).

16. H. A. Bent, *J. Chem. Educ.*, **40**, 446 (1963); **40**, 523 (1963); **42**, 302 (1965); **42**, 348 (1965); **43**, 170 (1966); **45**, 768 (1968).

17. L. S. Bartell, *J. Chem. Educ.*, **45**, 754 (1968).

18. K. J. Wynne, *J. Chem. Educ.*, **50**, 328 (1973).

19. C. J. Marsden and L. S. Bartell, *Inorg. Chem.*, **15**, 3004 (1976).

20. L. S. Bartell, F. B. Clippard, and E. J. Jacob, *Inorg. Chem.*, **15**, 3009 (1976).

21. M. C. Poore and D. R. Russell, *J. Chem. Soc. Chem. Commun.*, **1971**, 8.

22. L. S. Bartell and R. M. Gavin, Jr., *J. Chem. Phys.*, **48**, 2460, 2466 (1968).

23. F. P. Remoortere, J. J. Flynn, F. P. Boer, and P. P. North, *Inorg. Chem.*, **10**, 1511 (1971).

24. S. L. Lawton, R. A. Jacobson, and R. S. Frye, *Inorg. Chem.*, **10**, 701 (1971).

25. S. L. Lawton and R. A. Jacobson, *Inorg. Chem.*, **5**, 743 (1966); **10**, 709 (1971).

26. B. K. Robertson, W. G. McPherson, and E. A. Meyers, *J. Phys. Chem.*, **71**, 3531 (1967).

27. W. G. McPherson and E. A. Meyers, *J. Phys. Chem.*, **72**, 532 (1968).

28. G. Engel, *Z. Kristallogr., Kristallgeom., Kristallphys., Kristallchem.*, **90**, 341 (1935).

29. I. R. Beattie, M. Milne, M. Webster, H. E. Blayden, P. J. Jones, R. C. G. Killean, and J. L. Lawrence, *J. Chem. Soc. A*, **1969**, 482.

30. A. C. Hazell, *Acta Chem. Scand.*, **20**, 165 (1966).

31. A. K. Das and I. D. Brown, *Can. J. Chem.*, **44**, 939 (1966).

32. S. Husebye and J. W. George, *Inorg. Chem.*, **8**, 313 (1969).

33. S. Esperås, S. Husebye, and S. E. Svaeren, *Acta Chem. Scand.*, **25**, 3539 (1971).

34. R. G. Pearson, *J. Chem. Educ.*, **45**, 581, 643 (1968).

35. J. Barrett, S. R. A. Bird, J. D. Donaldson, and J. Silver, *J. Chem. Soc. A*, **1971**, 3105.

36. J. D. Donaldson, D. R. Laughlin, S. D. Ross, and J. Silver, *J. Chem. Soc. Dalton*, **1973**, 1985.

37. D. C. Urch, *J. Chem. Soc.*, **1964**, 5775.

38. T. C. Gibb, R. Greatrex, N. N. Greenwood, and A. C. Sarma, *J. Chem. Soc. A*, **1970**, 212.

39. L. Atkinson and P. Day, *J. Chem. Soc. A*, **1969**, 2423.

40. J. D. Donaldson and J. Silver, *Inorg. Nucl. Chem. Lett.*, **10**, 537 (1974).

41. C. J. Marsden and L. S. Bartell, *Inorg. Chem.*, **15**, 3004 (1976).

42. R. S. Drago, *J. Chem. Educ.*, **50**, 244 (1973).

43. R. J. Gillespie, *J. Chem. Educ.*, **51**, 367 (1974).

44. A. D. Walsh, *J. Chem. Soc.*, **1953**, 2260, 2266, 2288, 2296, 2301.

45. B. M. Deb, *Rev. Mod. Phys.*, **45**, 22 (1973).

46. B. M. Deb, *J. Am. Chem. Soc.*, **96**, 2030 (1974).

47. B. M. Deb, P. N. Sen, and S. K. Bose, *J. Am. Chem. Soc.*, **96**, 2044 (1974).

48. B. M. Deb, *J. Chem. Educ.*, **52**, 314 (1975).

49. Y. Takahata, G. W. Schnuelle, and R. G. Parr, *J. Am. Chem. Soc.*, **93**, 784 (1971).

50. F. O. Sladky, in *MTP International Review of Science, Inorganic Chemistry, Series One*, H. J. Emeléus (Consultant Ed.), Vol. 3, V. Gutmann (Ed.), pp. 1–51, Butterworth, London (1972).

51. N. Bartlett and F. O. Sladky, in *Comprehensive Inorganic Chemistry*, A. F. Trotman-Dickinson (Executive Ed.), Vol. 1, Ch. 5, Pergamon Press, Elmsford, N.Y. (1973).

52. N. Bartlett, B. G. DeBoer, F. J. Hollander, F. O. Sladky, D. H. Templeton, and A. Zalkin, *Inorg. Chem.*, **13**, 780 (1974).

53. K. W. Bagnall, *The Chemistry of Selenium, Tellurium, and Polonium*, Elsevier, New York (1966); *The Chemistry of the Rare Radioelements*, pp. 3–94, Butterworth, London (1957).

54. L. E. Orgel, *J. Chem. Soc.*, **1959**, 3815.

55. N. V. Sidgwick, *The Electronic Theory of Valency*, p. 179, Clarendon Press, Oxford (1927); *Ann. Rep.*, **30**, 120 (1933).

56. M. B. Hall, *Inorg. Chem.*, **17**, 2261 (1978).

57. H. R. Allcock, *Heteroatom Ring Systems and Polymers*, Academic Press, New York (1967).

58. I. Haiduc, *The Chemistry of Inorganic Ring Systems*, Parts 1 and 2, Wiley-Interscience, New York (1970).

59. F. Armitage, *Inorganic Rings and Cages*, E. Arnold, London (1972).

60. R. Westric and C. H. MacGillavry, *Rec. Trav. Chim.*, **60**, 794 (1941).

61. G. C. Hampson and A. J. Stosick, *J. Am. Chem. Soc.*, **60**, 1814 (1938).

62. L. R. Maxwell, S. B. Hendricks, and L. S. Deming, *J. Chem. Phys.*, **5**, 626 (1937).

63. H. C. J. de Decker and C. H. MacGillavry, *Rec. Trav. Chim.*, **60**, 153 (1941).

64. H. C. J. de Decker, *Rec. Trav. Chim.*, **60**, 413 (1941).

65. C. H. MacGillavry, H. C. J. de Decker, and L. M. Nijland, *Nature*, **164**, 448 (1949).

66. V. Caglioti, G. Giacomello, and E. Bianchi, *Atti Accad. Ital. Rend.*, **3**, 761 (1942).

67. C. Romers, J. A. A. Ketelaar, and C. H. MacGillavry, *Acta Crystallogr.*, **4**, 114 (1951).

67a. T. E. Clark, R. O. Day, and R. R. Holmes, *Inorg. Chem.*, **18**, 1653, 1660, 1668 (1979).

68. M. Becke-Goehring, *Angew. Chem.*, **73**, 589 (1961).

69. O. Glemser, *Angew. Chem., Int. Ed. Engl.*, **2**, 530 (1963).

70. O. Glemser, *Endeavor*, **28**, 86 (1969).

71. H. J. Emeléus, *Endeavor*, **32**, 76 (1973).

72. C. W. Allen, *J. Chem. Educ.*, **44**, 38 (1967).

73. B. D. Sharma and J. Donohue, *Acta Crystallogr.*, **16**, 891 (1963).

74. A. G. Turner and F. S. Mortimer, *Inorg. Chem.*, **5**, 906 (1966).

75. T. Moeller and R. L. Dieck, in *Preparative Inorganic Reactions*, W. L. Jolly (Ed.), Vol. 6, pp. 63–79, Wiley-Interscience, New York (1971).

76. A. C. Hazell, G. A. Wiegers, and A. Vos, *Acta Crystallogr.*, **20**, 186 (1966).

77. A. Vandi, T. Moeller, and T. L. Brown, *Inorg. Chem.*, **2**, 899 (1963).

78. M. Goehring, H. Herb, and W. Koch, *Z. Anorg. Allg. Chem.*, **264**, 137 (1951).

79. J. Weiss, *Z. Anorg. Allg. Chem.*, **305**, 190 (1960).

80. E. W. Lund and S. R. Svendsen, *Acta Chem. Scand.*, **11**, 940 (1957).

81. R. L. Sass and J. Donohue, *Acta Crystallogr.*, **11**, 497 (1958).

82. L. F. Audrieth, M. Sveda, H. H. Sisler, and M. J. Butler, *Chem. Rev.*, **26**, 49 (1940).

83. R. A. Shaw, B. W. Fitzsimmons, and B. C. Smith, *Chem. Rev.*, **62**, 247 (1962).

84. C. D. Schmulbach, in *Progress in Inorganic Chemistry*, F. A. Cotton (Ed.), Vol. 4, pp. 275–379, Interscience Publishers, New York (1962).

85. H. R. Allcock, *Chem. Rev.*, **72**, 315 (1972).

86. H. R. Allcock, *Phosphorus–Nitrogen Compounds*, Academic Press, New York (1972).

87. H. R. Allcock, *Chem. Br.*, **10**, 118 (1974).

88. C. W. Allen, J. B. Faught, T. Moeller, and I. C. Paul, *Inorg. Chem.*, **8**, 1719 (1969).

89. M. W. Dougill, *J. Chem. Soc.*, **1963**, 3211.

90. A. Wilson and D. F. Carroll, *J. Chem. Soc.*, **1960**, 2548.

91. R. Hazekamp, T. Migchelsen, and A. Vos, *Acta Crystallogr.*, **15**, 539 (1962).

92. A. J. Wagner, A. Vos, J. L. de Boer, and T. Wichertjes, *Acta Crystallogr.*, **16**, A39 (1963).

93. A. W. Schleuter and R. A. Jacobson, *J. Am. Chem. Soc.*, **88**, 2051 (1966).

94. J. C. van de Grampel and A. Vos, *Acta Crystallogr.*, **B25**, 651 (1969).

95. R. Hazekamp and A. Vos, *Acta Crystallogr.*, **16**, A38 (1963).

96. R. Olthof, T. Migchelsen, and A. Vos, *Acta Crystallogr.*, **19**, 590 (1965).

97. T. Migchelsen and A. Vos, *Acta Crystallogr.*, **16**, A38 (1963).

98. T. Migchelsen, R. Olthof, and A. Vos, *Acta Crystallogr.*, **19**, 603 (1965).

99. R. R. Holmes and J. A. Forstner, *Inorg. Chem.*, **2**, 377 (1963).

100. F. Casabianca, F. A. Cotton, J. G. Riess, C. E. Rice, and B. R. Stults, *Inorg. Chem.*, **17**, 3232 (1978).

101. R. R. Holmes and J. A. Forstner, *J. Am. Chem. Soc.*, **82**, 5509 (1960).

102. R. R. Holmes, *J. Am. Chem. Soc.*, **83**, 1334 (1961).

103. O. Foss, in *Advances in Inorganic Chemistry and Radiochemistry*, H. J. Eméleus and A. G. Sharpe (Eds.), Vol. 2, pp. 237–278, Academic Press, New York (1960).

104. J. Donohue, *The Structures of the Elements*, pp. 324–392, John Wiley & Sons, New York (1974).

105. B. Meyer, *Chem. Rev.*, **76**, 367 (1976).

106. R. Rahman, S. Safe, and A. Taylor, *Q. Rev. (Lond.)*, **24**, 208 (1970).

107. T. Chivers and I. Drummond, *Chem. Soc. Rev.*, **2**, 233 (1973).

108. M. Zeldin, *J. Chem. Educ.*, **43**, 17 (1966).

109. N. C. Craig, *J. Chem. Educ.*, **46**, 23 (1969).

110. J. Donohue, *J. Chem. Educ.*, **46**, 27 (1969).

111. J. P. Fackler, Jr., *J. Chem. Educ.*, **55**, 79 (1978).

112. F. A. Cotton, *Chemical Applications of Group Theory*, Interscience Publishers, New York (1963).

113. M. Orchin and H. H. Jaffé, *J. Chem. Educ.*, **47**, 246, 372, 510 (1970).

114. J. P. Fackler, Jr., *Symmetry in Coordination Chemistry*, Academic Press, New York (1971); *Symmetry in Chemical Theory: Application of Group Theoretical Techniques to the Solution of Chemical Problems*, Dowden, Hutchinson & Ross Academic Press, New York (1973).

115 A. W. Adamson, *A Textbook of Physical Chemistry*, 2nd ed., Ch. 17, Academic Press, New York (1979).

116. F. Daniels and R. A. Alberty, *Physical Chemistry*, 4th ed., Ch. 13, John Wiley & Sons, New York (1975).

117. G. Davidson, *Introductory Group Theory for Chemists*, Elsevier, New York (1971).

118. G. W. King, *Spectroscopy and Molecular Structure*, Ch. 7, Holt, Rinehart and Winston, New York (1964).

119. K. Nakamoto, *Infrared and Raman Spectra of Inorganic and Coordination Compounds*, 3rd ed., John Wiley & Sons, New York (1978).

120. A. J. Krubsack, *J. Chem. Educ.*, **52**, 368 (1975).

121. N. C. Craig, *J. Chem. Educ.*, **46**, 23 (1969).

122. M. Herman and J. Lievin, *J. Chem. Educ.*, **54**, 597 (1977).

123. C.-C. Chang, R. C. Haltiwanger, and A. D. Norman, *Inorg. Chem.*, **17**, 2056 (1978).

124. J. E. Ferguson, *Stereochemistry and Bonding in Inorganic Chemistry*, Prentice-Hall, Englewood Cliffs, N.J. (1974).

125. H. A. O. Hill and P. Day (Eds.), *Physical Methods in Advanced Inorganic Chemistry*, Wiley-Interscience, New York (1968).

126. E. A. V. Ebsworth, *Chem. Br.*, **12**, 84 (1976).

127. R. S. Drago, *Physical Methods in Chemistry*, W. B. Saunders, Philadelphia (1977).

128. W. L. Jolly, *The Synthesis and Characterization of Inorganic Compounds*, pp. 263–427, Prentice-Hall, Englewood Cliffs, N.J. (1970).

129. F. A. Cotton and J. M. Troupe, *J. Am. Chem. Soc.*, **95**, 3798 (1973).

130. D. C. Bradley, J. S. Ghotra, F. A. Hart, M. B. Hursthouse, and P. R. Raithby, *J. Chem. Soc. Chem. Commun.*, **1974**, 40.

131. E. C. Baker, L. D. Brown, and K. N. Raymond, *Inorg. Chem.*, **14**, 1376 (1975).

132. C. G. Shull, E. O. Wollan, G. A. Morton, and W. L. Davidson, *Phys. Rev.*, **73**, 842 (1948).

133. H. Fuess, *Chem. Br.*, **14**, 37 (1978).

134. R. Attig and J. M. Williams, *Angew. Chem.*, **88**, 507 (1976).

135. K. Hedberg and V. Schomaker, *J. Am. Chem. Soc.*, **73**, 1482 (1951).

136. J. L. Hollenberg, *J. Chem. Educ.*, **47**, 2 (1970).

137. F. A. Cotton, Ref. 112, Ch. 9.

138. K. Nakamoto, Ref. 119, pp. 436–441.

139. R. L. Dieck and T. Moeller, *J. Inorg. Nucl. Chem.*, **35**, 75 (1973).

140. R. Stahlberg and E. Steger, *Spectrochim. Acta*, **23A**, 2185 (1967).

141. J. Goubeau and J. Reyling, *Z. Anorg. Allg. Chem.*, **294**, 96 (1958).

142. D. B. Sowerby, *J. Inorg. Nucl. Chem.*, **22**, 205 (1961).

143. R. L. Dieck and T. Moeller, *J. Inorg. Nucl. Chem.*, **35**, 3781 (1973).

144. R. L. Dieck and T. Moeller, *J. Inorg. Nucl. Chem.*, **36**, 2283 (1974).

145. R. L. Dieck and T. Moeller, *J. Less Common Met.*, **33**, 355 (1973).

146. A. H. Norbury and A. I. P. Simha, *J. Chem. Soc. A*, **1968**, 1598.

147. J. L. Burmeister, E. A. Deardorf, and C. E. Van Dyke, *Inorg. Chem.*, **8**, 170 (1969).

148. R. A. Bailey, S. L. Kozak, T. W. Michelsen, and W. N. Mills, *Coord. Chem. Rev.*, **6**, 407 (1971).

149. A. Sabatini and I. Bertini, *Inorg. Chem.*, **4**, 959 (1965).

150. L. A. Woodward and G. H. Singer, *J. Chem. Soc.*, **1958**, 716.

151. L. A. Woodward and I. J. Worrall, J. Inorg. Nucl. Chem., **35**, 1535 (1973).

152. L. A. Woodward and A. A. Nord, *J. Chem. Soc.*, **1955**, 2655.

153. L. A. Woodward, N. N. Greenwood, I. R. Hall, and I. J. Worrall, *J. Chem. Soc.*, **1958**, 1505.

154. L. A. Woodward and A. A. Nord, *J. Chem. Soc.*, **1956**, 3721.

155. L. A. Woodward and I. J. Worrall, *Inorg. Nucl. Chem. Lett.*, **8**, 123 (1972).

156. J. H. R. Clarke and R. E. Hester, *J. Chem. Soc. Chem. Commun.*, **1968**, 1042; *Inorg. Chem.* **8**, 1113 (1969).

157. G. Herzberg, *Infrared and Raman Spectra of Polyatomic Molecules*, p. 178, D. Van Nostrand, New York (1945).

158. B. M. Gatehouse, S. E. Livingstone, and R. S. Nyholm, *J. Chem. Soc.*, **1957**, 4222.

159. C. C. Addison and B. M. Gatehouse, *J. Chem. Soc.*, **1960**, 613.

160. J. H. Forsberg and T. Moeller, *Inorg. Chem.*, **8**, 883 (1969).

161. B. J. Hathaway and A. E. Underhill, *J. Chem. Soc.*, **1961**, 3091.

162. B. J. Hathaway, D. G. Holah, and M. Hudson, *J. Chem. Soc.*, **1963**, 4586.

163. L. E. Moore, R. B. Gayhart, and W. E. Bull, *J. Inorg. Nucl. Chem.*, **26**, 896 (1964).

164. R. C. Grandey and T. Moeller, *J. Inorg. Nucl. Chem.*, **32**, 333 (1970).

165. R. S. Tobias, *J. Chem. Educ.*, **44**, 2, 70 (1967).

166. P. M. Kusesof and R. L. Kuczkowski, *Inorg. Chem.*, **17**, 2308 (1978).

167. F. A. Bovey, *Chem. Eng. News*, Aug. 30, 1965, p. 98.

168. J. L. Holcomb and R. H. Sands, *Ind. Res.*, July 1965, p. 64.

169. E. L. Muetterties and W. D. Phillips, in *Advances in Inorganic Chemistry and Radiochemistry*, (H. J. Emeléus and A. G. Sharpe (Eds.), Vol. 4, pp. 231–292, Academic Press, New York (1962).

170. J. T. Arnold, S. S. Dharmatti, and M. E. Packard, *J. Chem. Phys.*, **19**, 507 (1951).

171. J. T. Arnold, *Phys. Rev.*, **102**, 136 (1956).

172. M. W. Dietrich and R. E. Keller, *Anal. Chem.*, **36**, 258 (1964).

173. O. Stelzer and R. Schmutzler, *Inorg. Synth.*, **18**, 176 (1978).

174. M. M. Crutchfield, C. H. Dungan, J. H. Letcher, V. Mark, and J. R. Van Wazer, in *Topics in Phosphorus Chemistry*, M. Grayson and E. J. Griffiths (Eds.), Vol. 5, Interscience Publishers, New York (1967).

175. R. J. Gillespie and G. J. Schrobilgen, *Inorg. Chem.*, **15**, 22 (1976).

176. G. J. Schrobilgen, J. H. Holloway, P. Granger, and C. Brevard, *Inorg. Chem.*, **17**, 980 (1978).

177. K. O. Christie and C. J. Schack, *Inorg. Chem.*, **17**, 2749 (1978).

178. A. R. Siedle, G. M. Bodner, A. R. Garber, and L. J. Todd, *Inorg. Chem.*, **13**, 1756 (1974).

179. D. R. Eaton and W. D. Phillips, *Adv. Magn. Reson.*, **1**, 103 (1965).

180. H. M. McConnell and R. E. Robertson, *J. Chem. Phys.*, **29**, 1361 (1958).

181. E. R. Birnbaum and T. Moeller, *J. Am. Chem. Soc.*, **91**, 7274 (1969).

182. C. C. Hinckley, *J. Am. Chem. Soc.*, **91**, 5160 (1969).

183. R. E. Sievers (Ed.), *Nuclear Magnetic Resonance Shift Reagents*, Academic Press, New York (1973).

184. K. A. Stone and R. E. Sievers, *Aldrichim. Acta*, **10**(4), 54 (1977).

185. R. E. Rondeau and R. E. Sievers, *J. Am. Chem. Soc.*, **93**, 1522 (1971).

186. K. O. Hodgson and S. Doniach, *Chem. Eng. News*, Aug. 21, 1978, p. 21.

187. J. M. Hollander and W. L. Jolly, *Acc. Chem. Res.*, **3**, 193 (1970).

188. D. N. Hendrickson, in Ref. 127, pp. 566–584.

189. N. N. Greenwood, *Chem. Br.*, **3**, 56 (1967).

190. G. M. Bancroft and R. H. Platt, in *Advances in Inorganic Chemistry and Radiochemistry*, H. J. Emeléus and A. G. Sharpe (Eds.), Vol. 15, pp. 59–258, Academic Press, New York (1972).

191. R. H. Herber, in *Progress in Inorganic Chemistry*, F. A. Cotton (Ed.), Vol. 8, pp. 1–41, Interscience Publishers, New York (1967).

192. V. I. Goldanskii and R. H. Herber, *Chemical Applications of Mössbauer Spectroscopy*, Academic Press, New York (1968).

193. M. R. Litzow and T. R. Spalding, *Mass Spectrometry of Inorganic and Organometallic Compounds*, Elsevier, New York (1973).

194. J. Lewis and B. F. G. Johnson, *Acc. Chem. Res.*, **1**, 245 (1968).

195. J. M. Miller and G. L. Wilson, in *Advances in Inorganic Chemistry and Radiochemistry*, H. J. Emeléus and A. G. Sharpe (Eds.), Vol. 18, pp. 229–286, Academic Press, New York (1976).

196. C. W. Allen, R. L. Dieck, P. Brown, T. Moeller, C. D. Schmulbach, and A. G. Cook, *J. Chem. Soc. Dalton*, **1978**, 173.

SUGGESTED SUPPLEMENTARY REFERENCES

F. A. Cotton, *Chemical Applications of Group Theory*, Interscience Publishers, New York (1963). Symmetry operations in depth.

J. Donohue, *The Structures of the Elements*, John Wiley & Sons, New York (1974). Crystal-structure data for elements.

R. S. Drago, *Physical Methods in Chemistry*, W. B. Saunders, Philadelphia (1977). Methodology and interpretation of data.

J. P. Fackler, Jr., *Symmetry in Chemical Theory: Application of Group Theoretical Techniques to the Solution of Chemical Problems*, Dowden, Hutchinson & Ross/Academic Press, New York (1973). Symmetry operations in depth.

R. J. Gillespie, *Molecular Geometry*, Van Nostrand Reinhold, London (1972). Valence-shell electron-pair repulsion.

H. A. O. Hill and P. Day (Eds.), *Physical Methods in Advanced Inorganic Chemistry*, Wiley-Interscience, New York (1968). Detailed descriptions of methods and interpretation of data.

H. H. Jaffé and M. Orchin, *Symmetry in Chemistry*, John Wiley & Sons, New York (1965).

W. L. Jolly, *The Synthesis and Characterization of Inorganic Compounds*, Prentice-Hall, Englewood Cliffs, N.J. (1970). Extensive treatment of physical methods, together with typical syntheses.

K. Nakamoto, *Infrared and Raman Spectra of Inorganic and Coordination Compounds*, 3rd ed., John Wiley & Sons, New York (1978). Detailed summation and interpretation of vibrational spectral data.

L. Pauling, *The Nature of the Chemical Bond*, 3rd ed., Cornell University Press, Ithaca, N.Y. (1959); or *The Chemical Bond: A Brief Introduction to Modern Structural Chemistry*, Cornell University Press, Ithaca, N.Y. (1967). Classical references.

A. F. Wells, *Models in Structural Inorganic Chemistry*, Oxford University Press, London (1970). Principles, model building, structural details.

A. F. Wells, *Structural Inorganic Chemistry*, 4th ed., Clarendon Press, Oxford (1975). Classical treatment of structures of all types.

A. F. Wells, *The Third Dimension in Chemistry*, Clarendon Press, Oxford (1956). Concise discussion of three-dimensional arrangements with stereoscopic photographs.

P. J. Wheatley, *The Determination of Molecular Structure*, 2nd ed., Oxford University Press, London (1967). Methodology.

CHAPTER SEVEN

COORDINATION CHEMISTRY, I PRINCIPLES; ISOMERISM; MOLECULAR STRUCTURES

The ability of certain ions and molecules, that are themselves capable of independent existence and apparently saturated with respect to valence, to combine in stoichiometric proportions to yield new and more complex species was first recognized prior to 1800. The resulting adducts, or *complex* species as they have been commonly termed,* vary widely in their physical and chemical properties. Some, for example, are stable only in the solid state and decompose into their components upon dissolution; others can be crystallized repeatedly from various solvents without change in composition.† Some dissociate completely into their components at room or slightly higher temperature; others are thermally stable at comparatively high temperatures and may even volatilize without change in composition. In some, bonding appears to be essentially ionic; in others, it is clearly covalent. Yet in spite of substantial differences in properties between extreme examples of adducts of this general type, there exist sufficient similarities in methods of synthesis, molecular structures, and overall chemical behaviors to justify our discussing them as a broad, single class of chemical substances.[4] That the vast majority of these species is derived from *d*-transition metal ions does not in any way require, or even recommend, that discussion be completely restricted to that group of complex substances.

It is difficult indeed to establish with absolute certainty the origin of this area of chemistry. It may well date to a description, by A. Libavius in 1597, of the

---

* The terms *coordination compounds or ions, complex ions or compounds,* and *Werner complexes* are also used widely.

† These two types are exemplified, respectively, by the compounds $CoCl_2 \cdot 6\,NH_3$ and $CoCl_3 \cdot 6\,NH_3$. On the assumption that equality in the molecular volumes of these two compounds is due to the compression of the ammonia molecules,[1] the latter substance and other similar species have been termed *penetration complexes* to distinguish them from the more labile *normal complexes.*[2] Inasmuch as correlations from molecular-volume data are subject to numerous errors, these terms have little significance.[3]

formation in aqueous solution of a blue species we now symbolize as $[Cu(NH_3)_4]^{2+}$. Prussian blue, now formulated as $KFeFe(CN)_6$, was first obtained accidentally in 1704 by Diesbach. However, the true origin is often ascribed to Citoyen Tassert,[5] who in 1798 observed the brownish mahogany color of what we now know as the $[Co(NH_3)_6]^{3+}$ ion but did not follow up his observation. Subsequent developments, encompassing a period ending in 1893, were associated with many prominent chemists, including Gustav Magnus, William Christoffer Zeise, Thomas Graham, Carl Ernst Claus, Christian Wilhelm Blomstrand, and Sophius Mads Jørgensen.[6, 7] Modern coordination chemistry dates to Alfred Werner (Sec. 7.1-3).

1. SYSTEMATIC APPROACH TO COMPLEX SUBSTANCES

1-1. Some Early Experimental Observations

The systematic study and interpretation of the properties of these species can be approached with profit by citing some significant experimental observations, dating indefinitely into history, that require explanation and justification. As typical of those involving chemically stable species, data for the ammonia containing derivatives of cobalt(III) chloride can be considered. Treatment of aqueous solutions of cobalt(II) chloride containing ammonium chloride with aqueous ammonia, followed by oxidation, yielded a number of isolable cobalt(III) compounds, among them the following:

1. **Luteocobaltic chloride, $CoCl_3 \cdot 6 NH_3$, an orange-yellow crystalline compound.*** Addition of 18 M sulfuric acid to this compound liberates all of the chlorine as hydrogen chloride, yielding the sulfate, $Co_2(SO_4)_3 \cdot 12 NH_3$. No ammonia is removed by hydrochloric acid, even at 100°C. Conductance data for aqueous solutions of the luteo chloride ($\Lambda_{1024} = 432$ ohm^{-1} cm^2 mole^{-1}) indicate the presence of four moles of ions per mole (p. 365), and all of the chloride in such solutions is immediately precipitated by silver(I) ion. Similarly, all of the chlorine is removed from the original solid by moist silver(I) oxide, leaving the strongly basic, water-soluble compound $CoOOH \cdot 6 NH_3$. This substance is, in turn, converted by acids to salts of the type $CoX_3 \cdot 6 NH_3$. Both stabilization of terpositive cobalt and the existence of very strong bonds to the ammonia molecules are indicated.

2. **Roseocobaltic chloride, $CoCl_3 \cdot 5 NH_3 \cdot H_2O$, a red crystalline compound.** The water molecules in this compound are firmly held at room temperature and lost only at 100°C or above, whereupon the new compound $CoCl_3 \cdot 5 NH_3$ results. Conductance measurements again show the presence of *four* moles of ions per mole ($\Lambda_{1024} = 394$ ohm^{-1} cm^2 mole^{-1}), and again silver(I) ion precipitates all the chlorine present. Even though this compound

---

* The "color-code" nomenclature was introduced by E. Frémy, *Ann. Chim. Phys.*, [3], 35, 257 (1852); *J. Prakt. Chem.*, **57**, 95 (1852).

contains less ammonia than the luteo chloride, its overall properties are closely similar, suggesting an essential equivalence between ammonia and water molecules in its molecular structure.

3. **Purpureocobaltic chloride, $CoCl_3 \cdot 5NH_3$, a purple compound.**   When treated with $18\,M$ sulfuric acid, this compound loses only two-thirds of its chlorine as hydrogen chloride, leaving as the solid product $CoClSO_4 \cdot 5NH_3$. An aqueous solution of the latter yields no immediate precipitate with silver(I) ion. Conductance measurements on solutions of the purpureo chloride show the presence of only *three* moles of ions per mole ($\Lambda_{1024} = 261$ ohm^{-1} cm^2 mole^{-1}), and from such solutions only *two-thirds* of the chlorine is precipitated immediately by silver(I) ion. The remaining chlorine is precipitated only slowly and after long standing or boiling. Two types of bonds involving chlorine are indicated.

In addition, but by another procedure, *praseocobaltic chloride*, $CoCl_3 \cdot 4NH_3$, a green compound, was described.* Conductance studies on its aqueous solution show the presence of *two* moles of ions per mole, and from such solutions only *one-third* of the chlorine is precipitated immediately by silver(I) ion.

1-2. Early Interpretations of Experimental Data

It is apparent, and worthy of particular emphasis, that the information noted above, together with a very substantial body of similar information relating to complex species containing, for example, platinum(II and IV), palladium(II), and chromium(III), was accumulated before definitive data on electronic configurations and molecular structures, and theoretical interpretations thereof, were available. Indeed, the successful interpretation of such data within the limited framework then available contributed significantly to modern views on valency and the nature of the chemical bond (Chs. 4–6). It is apparent that any interpretation required development of a theory that could both account for the existence of complex species and explain the striking changes in properties that accompany small changes in composition.

References already cited[6, 7] can be consulted for summaries of early approaches. The proposals of C. E. Claus,[8] C. W. Blomstrand,[9] and S. M. Jørgensen[10] require discussion as a prelude to that of A. Werner.[11]† The views of Claus, which closely anticipated those of Werner and appeared in modified form in Werner's theory, are summarized as:

1. When several equivalents of ammonia combine with one equivalent of a certain metal chloride to form a neutral substance, the basic property of

---

*   The isomeric *violeo* compound, violet $CoCl_3 \cdot 4NH_3$, was not reported until much later [A. Werner, *Chem. Ber.*, **40**, 4817 (1907)]. It was this synthesis, predictable in terms of Werner's theory (Sec. 7.1-3) that prompted S. M. Jørgensen, Werner's chief critic, to acknowledge the validity of Werner's ideas.

†   See Ref. 7, Part II, for English translations.

ammonia is destroyed, and ammonia can neither be detected by usual methods nor eliminated by double decomposition.

2. If the chlorine in these compounds is replaced by oxygen, strong bases result, the saturation capacities of which are always determined by the oxygen equivalents contained therein but not by the number of equivalents of ammonia present.

3. The number of equivalents of ammonia entering into these substances is not random. It is determined by the number of equivalents of water contained in hydrates of the metal oxides that enter into such compounds along with ammonia.

It is apparent that these views are conclusions and do not represent a theory. It is significant, however, that Claus assumed that in the compounds in question "ammonia occurs in a copulated compound and that the radicals of these compounds with ammonia are copulated metals." Formulas such as $PtNH_3 + Cl$ for Magnus's salt* were proposed.

Blomstrand's concept of the constitution of ammonia complexes reflected A. Kekulé's proposal[12] that successive numbers of an homologous series of organic compounds were related via the self-linking of carbon atoms through — CH_2— groups. Thus he proposed the self-linking of NH_3 molecules, as indicated by the formulations

$$Co\begin{cases} NH_3 \cdot NH_3 \cdot NH_3 \cdot Cl \\ NH_3 \cdot NH_3 \cdot NH_3 \cdot Cl \end{cases}$$

$$Pt\begin{cases} NH_3 \cdot NH_3 \cdot Cl \\ NH_3 \cdot NH_3 \cdot Cl \end{cases}$$

with the stability of the ammonia chain not dependent on its length. Jørgensen extended these ideas by proposing that in the cobalt(III) chloride complexes discussed above the nonprecipitable chlorine is bonded to the metal and the precipitable chlorine to ammonia, as

$$Co\begin{cases} NH_3 \cdot Cl \\ NH_3 \cdot NH_3 \cdot NH_3 \cdot NH_3 \cdot Cl \\ NH_3 \cdot Cl \end{cases} \qquad Co\begin{cases} Cl \\ NH_3 \cdot NH_3 \cdot NH_3 \cdot NH_3 \cdot Cl \\ NH_3 \cdot Cl \end{cases}$$

$$\text{luteo } (CoCl_3 \cdot 6 NH_3) \qquad\qquad \text{purpureo } (CoCl_3 \cdot 5 NH_3)$$

* The first platinum-ammonia complex, now formulated $[Pt(NH_3)_4][PtCl_4]$, to be discovered [G. Magnus, Poggendorf's *Ann. Phys. Chem.*, **14**, 239 (1828); English translation, Ref. 7, Part II].

A somewhat bitter conflict between Jørgensen's views and those of Werner continued for several years.*

1-3. The Theory of Alfred Werner

Werner's original interest in valency problems among organic compounds led him to consider in detail the adducts mentioned above. In 1893, these considerations were summarized in the revolutionary paper[11] that both outlined a theoretical approach to structure and properties and opened a new and fruitful field of investigation.† In subsequent years, Werner's investigative work, as described in a truly classic series of 129 publications, verified experimentally every postulate of his original theory. The award of the Nobel Prize in Chemistry in 1913‡ was a fitting recognition of his accomplishments. Sadly, he passed away just prior to his fifty-third birthday in 1919, to a substantial degree of victim of overwork in his science.[13]

The fundamental postulate in Werner's coordination theory is best stated in his own words[14]:

> Even when, to judge by the valence number, the combining power of certain atoms is exhausted, they still possess in most cases the power of participating further in the construction of complex molecules with the formation of very definite atomic linkages. The possibility of this action is to be traced back to the fact that, besides the affinity bonds designated as principal valencies, still other bonds on the atoms, called auxiliary valencies, may be called into action.

In other words, this postulate, and other dependent postulates, can be summarized as follows:

1. Metal atoms can possess two types of valency: *principal* (primary, ionizable) valency and *auxiliary* (secondary, nonionizable) valency.§

---

* For an informative account of this controversy, see P. S. Cohen, in *Werner Centennial*, Vol. 62, pp. 8–40, Advances in Chemistry Series, American Chemical Society, Washington, D.C. (1967). Papers from an American Chemical Society symposium honoring Alfred Werner and organized by G. B. Kauffman.

† It is reported reliably by Paul Pfeiffer, Werner's student, colleague, and "chief of staff" at the Chemical Institute of the Universität Zürich, that Werner awoke with a start at two o'clock one morning with the fundamental concepts of his theory in mind and that, writing furiously and without interruption, by five o'clock the following afternoon he had completed its summary [P. Pfeiffer, *J. Chem. Educ.*, 5, 1090 (1928)].

‡ It is of interest that not until 1973 was another Nobel Prize given to an inorganic chemist (G. Wilkinson and E. O. Fischer — shared, also for work with metal complexes).

§ In the original German, Hauptvalenz and Nebenvalenz, respectively. These terms were not mentioned — at least in print — until 1902 [A. Werner, *Ann. Chem.*, 322, 261 (1902)], although they were implicit in his first publication. Similarly, the possibility of optical isomerism was first noted in 1899 [A. Werner and A. Vilmos, *Z. Anorg. Chem.*, 21, 145 (1899)].

2. Every metal atom has a fixed number of auxiliary valencies (i.e., *coordination number*). Thus cobalt(III) and platinum(IV) were recognized as exhibiting coordination number *six*, whereas copper(II) and platinum(II) exhibit coordination number *four*.

3. Principal valencies are satisfied by negative ions, whereas auxiliary valencies may be satisfied by either negative ions or neutral molecules. In certain instances, a given negative ion may satisfy both, but in every case satisfaction of the coordination number appears to be essential.

4. The auxiliary valencies are directed in space about the central metal atom. Thus, for coordination number 6, these valencies were assumed to be directed to the apices of a circumscribed octahedron, whereas for coordination number 4, either a square planar or a tetrahedral distribution was assumed, depending upon the metal atom. A consequence of this postulation is the prediction of several types of isomerism, among them optical isomerism.

   The consistency of these postulations with the experimental observations noted above for the various cobalt(III) ammonia compounds illustrates their significance. By slight alteration of Werner's original notation, we have

$$CoCl_3 \cdot 6\,NH_3 \quad\quad CoCl_3 \cdot 5\,NH_3 \cdot H_2O \quad\quad CoCl_3 \cdot 5\,NH_3$$
$$\text{(luteo)} \quad\quad\quad \text{(roseo)} \quad\quad\quad\quad \text{(purpureo)}$$

$$CoCl_3 \cdot 4\,NH_3 \quad\quad\quad CoCl_3 \cdot 4\,NH_3$$
$$\text{(praseo)} \quad\quad\quad\quad \text{(violeo)}$$

Satisfaction of coordination number 6, the correct number of ions, and the correct number of chloride ions are indicated for each. That part of the chlorine in each of the last three species held by auxiliary valencies is assumed to be nonionic. The existence of two isomeric forms of composition $CoCl_3 \cdot 4\,NH_3$ is consistent with the limitations of octahedral geometry. Final *chemical* confirmation of the octahedral geometry of cobalt(III) was obtained in 1914 by the resolution into its predicted optical isomers of the ion

$$\left[Co\left(\frac{HO}{HO} > Co(NH_3)_4 \right)_3 \right]^{6+} \quad \text{(Sec. 7.4-10).}^{[15,*]}$$

Data summarized in Table 7-1 for a series of related platinum(IV) compounds clearly indicate satisfaction of coordination number six, effect of entry of anion into the coordination sphere on the oxidation number of the complex ion itself, and alteration of number of ions due to such entry. Conductance data reported are from classic papers by A. Werner and A. Miolati[16] and are representative of those reported for series of platinum(IV), platinum(II), cobalt(III), and chromium(III) complexes. At a dilution of 1000 to 1024 liters mole^{-1}, conductance values ($\Lambda_{1000-1024}$) of 520 to 560 ohm^{-1} cm^2 mole^{-1} are characteristic of *five* ions; 400 to 430 ohm^{-1} cm^2 mole^{-1} of *four* ions; 230 to 260 ohm^{-1} cm^2 mole^{-1} of *three* ions; and ca. 100 ohm^{-1} cm^2 mole^{-1} of *two* ions.† It is significant that a nonelectrolyte results when the charges of the *coordinated* anions exactly balance the oxidation number of the central metal ion. A graphic representation (Fig. 7-1) of the data from Table 7-1 is striking and characteristic.

Table 7-1 Some Data on Coordination Compounds of Platinum(IV)

Compositional Formula	Molar Conductance^a (ohm^{-1} cm^2 mole^{-1})	Number of Ions Indicated	Resulting Formulation
$PtCl_4 \cdot 6\,NH_3$	523	5	$[Pt(NH_3)_6]Cl_4$
$PtCl_4 \cdot 5\,NH_3{}^b$	404	4	$[Pt(NH_3)_5Cl]Cl_3$
$PtCl_4 \cdot 4\,NH_3$	228.9	3	$[Pt(NH_3)_4Cl_2]Cl_2$
$PtCl_4 \cdot 3\,NH_3$	96.75	2	$[Pt(NH_3)_3Cl_3]Cl$
$PtCl_4 \cdot 2\,NH_3$	0	0	$[Pt(NH_3)_2Cl_4]$
$PtCl_4 \cdot NH_3 \cdot KCl$	108.5	2	$K[Pt(NH_3)Cl_5]$
$PtCl_4 \cdot 2\,KCl$	256.8	3	$K_2[PtCl_6]$

^{a} At a dilution of 1000 liters mole^{-1} (i.e., Λ_{1000}) and 25°C.

^{b} Not known by Werner and Miolati. First synthesized by L. A. Chugaev and N. Vladmirov [*C. R. Acad. Sci.*, **160**, 840 (1915)] and found to have $\Lambda_{1024} = 404$ ohm^{-1} cm^2 mole^{-1}. (Known as Chugaev's salt since 1925.)

---

* By convention, the metal atom and all groups coordinated to it (i.e., the *coordination sphere*) are enclosed in formulas by square brackets, with balancing ions written outside the brackets, as shown in the diagrams above, or more simply as $[Co(NH_3)_6]Cl_3$, $[Co(NH_3)_5(H_2O)]Cl_3$, $[Co(NH_3)_5Cl]Cl_2$, *trans*-$[Co(NH_3)_4Cl_2]Cl$, and *cis*-$[Co(NH_3)_4Cl_2]Cl$.

† Conductance data of this type must be accepted with caution, for aquation and/or hydrolysis may occur upon the dissolution of complex compounds in water. Inasmuch as the extents to which these do take place are functions of both time and temperature, a single conductance value may be meaningless.[17,18]

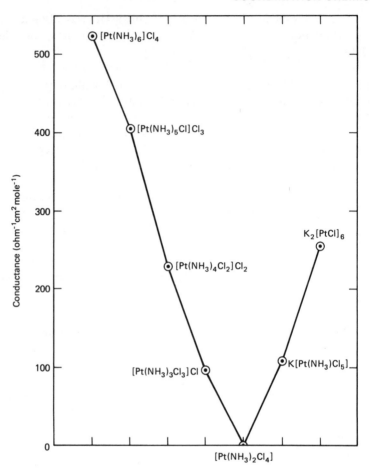

Figure 7-1 Molar conductance data for series of platinum(IV) complexes.

Early Electronic Extensions of the Werner Theory. Shortly after the importance of the electron in the formation of the chemical bond became apparent, N. V. Sidgwick[19] and T. M. Lowry[20] suggested, independently, that the principal valencies of Werner are involved when electron transfer occurs and the auxiliary valencies when electron-pair sharing takes place. Inasmuch as it was noted that the majority of the groups that coordinate to metal ions have in their molecular structures atoms with one or more nonbonded electron pairs, it was suggested that utilization of the auxiliary valencies requires donation of electron pairs from the coordinating species to the metal ion in question, with the formation of a *coordinate-covalent, semipolar,* or *dative* bond. These donor–acceptor bonds have often been designated by arrows pointing to the metal ion (e.g., $[Pt(\leftarrow NH_3)_6]Cl_4$).

The Sidgwick–Lowry approach was invaluable in systematizing rapidly accumulating information on coordination compounds. Sidgwick noted[19, 21] that in the formation of the most stable complex species the metal ion often added sufficient electrons in this fashion to give it the same total number of electrons (*effective atomic number*, or E.A.N.) as an atom of the next higher noble gas. Some examples, and exceptions, are indicated in Table 7-2. Both Sidgwick and A. A. Blanchard[22] emphasized this equality as a measure of the coordinating tendency of the metal ion and as an indication of the probable composition of a complex which that metal ion could form.

That there are many exceptions to this generality (Table 7-2) suggests that it is no more than an interesting correlation and thus lacks a fundamental basis. Those complexes that fit this generalization are diamagnetic (Sec. 2.4-4), whereas those that do not are often paramagnetic with observed magnetic moments that suggest equality between the number of unpaired electrons and the departure of the effective atomic number from the noble gas value. On the other hand, the effective atomic numbers of nickel(II), palladium(II), and platinum(II) in square-planar, diamagnetic, 4-coordinate complexes are not equal to the noble gas values (Table 7-2). The rule is obeyed with considerable frequency by metal carbonyls and nitrosyls and by many organometallic compounds (Sec. 8.2).

Although the concept of direct donation of electron pairs to a central cation has probably been more helpful than any other simple concept in systematizing the wealth of information that has been accumulated relative to complex species, it may be questioned on several bases. An important difficulty lies in the improbable accumulation of negative charge on the metal ion. Thus in the $[Co(NH_3)_6]^{3+}$ ion, simple donation of electrons should render the cobalt(III) ion negative and the nitrogen atom in each ammonia molecule positive. That the nitrogen atom is much more electronegative than the cobalt atom (Table 2-21) renders this situation even more unlikely. Furthermore, the nonbonded pairs of electrons that presumably form the coordinate covalent bond are commonly s^2 arrangements that have no bond-forming characteristics. Although excitation to give unpaired electrons may be considered, this process would normally require the expenditure of more energy than can be gained through bond formation and is thus unlikely. An added difficulty is presented by the remarkable chemical stabilities of complex species involving certain metal ions and olefinic or aromatic hydrocarbons (Sec. 8.2-3). Although the donor–acceptor approach is still useful for many Lewis acid–base interactions (Sec. 9.6-1) and systems involving electron-poor and/or electron-rich nonmetal atoms, it is apparent that the bonding in complex species involving metal ions requires more detailed consideration. Detailed and more modern approaches are more conveniently considered (Ch. 8), however, after the introduction of additional information about complex species.

Table 7-2 Effective Atomic Number Concept

Metal Ion	Atomic Number of Metal	Coordination Number of Metal Ion	Electrons Lost in Ion Formation	Electrons Added via Coordination	Effective Atomic Number of Metal Ion in Complexes
Fe^{2+}	26	6	2	12	36 (Kr)
Co^{3+}	27	6	3	12	36 (Kr)
Cu^{+}	29	4	1	8	36 (Kr)
Pd^{4+}	46	6	4	12	54 (Xe)
Ir^{3+}	77	6	3	12	86 (Rn)
Pt^{4+}	78	6	4	12	86 (Rn)
Cr^{3+}	24	6	3	12	33
Fe^{3+}	26	6	3	12	35
Ni^{2+}	28	6	2	12	38
Ni^{2+}	28	4	2	8	34
Pd^{2+}	46	4	2	8	52
Ir^{4+}	77	6	4	12	85
Pt^{2+}	78	4	2	8	84

2. THE NOMENCLATURE OF COORDINATION ENTITIES

An appreciation of systematic nomenclature requires the prior definition of a number of commonly used terms. Atoms or groups that are coordinated to the *central*, or *nuclear*, atom are called *ligands*. Any group containing more than one potential coordinating atom is referred to as a *multidentate ligand*, and the number of such coordinating sites is indicated by the terms *uni-*, *bi-*, *ter-*, *quadri-*, *quinque-*, *sexi-*, *septi-*, *octa-*, and so on, dentate. A *chelating* ligand is one that bonds to a *single* central atom through *two* coordinating atoms.* A *bridging* ligand (or group) is one that is bonded simultaneously to *more than one* central atom. *Multinuclear complexes* are those containing more than one central atom.

2-1. The Werner System of Nomenclature

Although the original system devised by Werner[23] has been modified substantially, it did serve as a basis for modern approaches, and it is, therefore, worthy of consideration. According to Werner's system, the cation and the anion are named in that order, as for simple compounds. Where coordinating groups are present, they are named before the metal is named, and in the order:

1. Negative groups, the suffix *-o* being added to the stem name of the group: for example, Cl^- (chloro), NO_2^- (nitro), $C_2O_4^{2-}$ (oxalato), CO_3^{2-} (carbonato), CN^- (cyano), O^{2-} (oxo), O_2^{2-} (peroxo), OH^- (hydroxo), and so on.
2. Water, the name *aquo* being used.
3. Derivatives of ammonia, the amine names being used directly.
4. Ammonia, the name *ammine* being used.

In every instance, the number of coordinating atoms or groups of each type is indicated by a prefix (e.g., *di-*, *tri-*, *tetra-*, etc.). The metal is then indicated by its *stem* name plus an appropriate *suffix*. If the metal appears in the cation, oxidation states of $1+$, $2+$, $3+$, and $4+$ are designated, respectively, by the suffixes *-a*, *-o*, *-i*, and *-e*; if the metal appears in the anion, these single-letter suffixes are followed by the ending *-ate*. If the complex species is a nonelectrolyte, the name of the metal is used as such without a suffix. The names of the bridging groups in polynuclear complexes are preceded by μ.

2-2. Modified Systems of Nomenclature

In order to broaden the Werner system and to render it more definitive, the *Nomenclature Committee* of the *International Union of Pure and Applied*

---

* The term *chelate* was first applied in 1920 by G. T. Morgan and H. D. K. Drew [*J. Chem. Soc.,* **117**, 1456 (1920)], who stated: "The adjective *chelate*, derived from the great claw or *chela* (*chely*) of the lobster or other crustaceans, is suggested for the caliperlike groups which function as two associating units and fasten to the central atom so as to produce heterocyclic rings."

Chemistry (IUPAC) initially suggested modifications[24] that differed only in recommending that the oxidation state of the metal be indicated by a Roman numeral placed in parentheses immediately after the name of the metal (for cationic or neutral complexes).* Subsequently, the problem was examined in detail by W. C. Fernelius and coworkers[25-27] and by R. V. G. Ewens and H. Bassett,[28] and a new series of recommendations was offered by the International Union.[29, 30] This series was subsequently modified in terms of definitive rules adopted in 1970.[31] Further recommendations covering specific cases are contained in a series of publications.[32]

The latest rules can be summarized as follows:

1.　The cation is named first and then the anion, but in each instance the central atom is named after the ligands.

2.　Within the complete name of the coordination entity, multiplying Greek prefixes *di-, tri-, tetra-*, and so on, are used with simple ligand names and Greek prefixes *bis-, tris-, tetrakis-*, and so on, with more complicated ligand names. Enclosing marks, in the sequence {[( )]}, are often used for clarity.

3.　The names for all anionic ligands end in *-o*. Although this rule usually amounts to replacing the final *-e* in the endings *-ide, -ite,* and *-ate* with *-o*, important variations include F^- (fluoro), Cl^- (chloro), Br^- (bromo), I^- (iodo), O^{2-} (oxo), OH^- (hydroxo), O_2^{2-} (peroxo), S^{2-} (thio), HS^- (mercapto), and CN^- (cyano). Cationic and neutral ligands are named as such except that the terms *aqua* (formerly *aquo*) and *ammine* are retained for H_2O and NH_3, respectively, and CO is termed *carbonyl* and NO *nitrosyl*.

4.　Anionic groups and neutral and cationic groups are often named in this order. Ligands within each charge group are listed in alphabetical order, regardless of the number or type of each.

5.　The oxidation state of the central atom in a coordination entity is indicated by a Roman numeral or an Arabic 0 (zero) placed in parentheses immediately following the name of the metal for a cation or following the *-ate* ending for an anion (Stock system, Appendix II, p. 765). Alternatively, the charge on the ion, expressed as a positive or negative number (zero omitted) in parentheses after the name of the ion (Ewens–Bassett number[28]), may be used.

6.　Structural information may be given in names and formulas by prefixes: for example, *cis-, trans-, fac-, mer-*.

7.　If a ligand can bond through more than one kind of atom, the atoms actually involved are designated by their italicized symbols placed after the name of the ligand. Thus the dithiooxalate ion $(C_2S_2O_2^{2-})$ is named dithiooxalato-*S,S* if it bonds through sulfur atoms and dithiooxalato-*O,O* if it bonds through oxygen atoms. In deference to established practice, —NCS is termed isothiocyanato, —SCN thiocyanato, —NO_2 nitro, and —ONO nitrito.

---

*　The Stock system [A. Stock, *Z. Angew. Chem.*, **32**, 1, 373(1919); **33**, 1, 79 (1920)].

8. A bridging ligand is indicated by inserting the Greek letter μ immediately before its name and separating the name from the remaining ligands by hyphens. More than one bridging group of the same kind is indicated as di-μ- or bis-μ-, tri-μ- or tris-μ, and so on.

9. Metals are abbreviated in general as M; ligands as L. Abbreviations for ligands should be brief and usually in lowercase letters: for example, en (ethylenediamine), bpy (bipyridine).*

Use of the IUPAC Rules is illustrated in general in Table 7-3 and specifically for many ions and compounds in subsequent discussions and in Appendix II. Where necessary [e.g., in discussions of isomerism (Sec. 7.4)], additional rules of nomenclature are applied and illustrated. Prefixes used are of both Greek and Latin origin. Ideally, Greek prefixes (e.g., mono, di, tri, tetra, penta, hexa, octa, poly, etc.) should be used with words of Greek origin (e.g., chelate), and Latin prefixes (e.g., uni, bi, ter, quadri, quinque, sexi, octa, multi, etc.) with words of Latin origin (e.g., dentate). In practice, however, this degree of purity is not always maintained.

3. ACCEPTORS AND DONORS IN COORDINATION ENTITIES

Certain combinations of metal ions and ligands are said to yield *stable* complexes, whereas others are said to yield *unstable* complexes. The distinction is more qualitative than quantitative, however. The term *stability* is perhaps used qualitatively and without definition with greater frequency than any other term in chemistry. As applied to a chemical substance in general, it may reflect resistance to complete or partial decomposition into its components or it may imply ease of formation from its components — always with reference to a specified set of conditions. Quantitatively, stability may be described in terms of (1) thermodynamic and kinetic data (Ch. 9) or (2) formation and other types of chemical reactions (Ch. 11). Here we shall restrict our considerations, when we speak of stability, to the natures of the metal ions and the ligands that are involved in the formation of chemically relatively stable complexes.

3-1. Acceptors

Although most of the known coordination entities are derived from metal ions, as acceptors in the broad Sidgwick–Lowry sense, nonmetal atoms and even molecules can function in the same way. Thus species such as PF_6^- [hexafluoro-phosphate(V) ion], BrF_4^- [tetrafluorobromate(III) ion], ICl_4^- [tetrachloro-iodate(III) ion], TeO_6^{6-} [hexaoxotellurate(VI) ion], XeO_3F^- [fluoro-

---

* Abbreviations should not be used in names, but they are common in formulas. Although a few (e.g., en, pn, bpy, acac- for acetylacetonate) are unambiguous, it is desirable always to state what each abbreviation in use actually means.

Table 7-3 Nomenclature of Coordination Entities

Formula	Werner Name	IUPAC Name
$[Co(NH_3)_6]Cl_3$	Hexamminecobalti chloride	Hexaamminecobalt(III) chloride
$[Co(NH_3)_4Cl_2]^+$	Dichlorotetramminecobalti ion	Tetraamminedichlorocobalt(III) or tetraamminedichlorocobalt(1 +) ion
$[Cr(H_2NC_2H_4NH_2)_3]^{3+}$	Triethylenediaminechromi ion	Tris(ethylenediamine)chromium(III) ion
$K_3[Fe(C_2O_4)_3]$	Potassium trioxalatoferriate	Potassium trioxalatoferrate(III)
$[PtCl_6]^{2-}$	Hexachloroplatineate ion	Hexachloroplatinate(IV) or hexachloroplatinate(2 −) ion
$[Pt(NH_3)_2Cl_2]$	Dichlorodiammineplatinum	Diamminedichloroplatinum(II)
$[Co(NH_3)_5(H_2O)]Cl_3$	Aquopentammminecobalti chloride	Pentaamminneaquacobalt(III) chloride
$[Co(en)_2Cl(NO_2)]^+$	Chloronitrodiethylenediaminecobalti ion	Chlorobis(ethylenediamine)nitrocobalt(III) ion
$[Cr(NH_3)_6][Co(CN)_6]$	Hexamminechromihexacyanocobaltiate	Hexaamminechromium(III) hexacyanocobaltate(III)
$[(NH_3)_5Co—O_2—Co(NH_3)_5]^{4+}$	Decammine-μ-peroxodicobalti ion	Decaammine-μ-peroxodicobalt(III) ion
$\left[Co\left\{ \genfrac{}{}{0pt}{}{HO}{HO} \!\!> Co(NH_3)_4 \right\}_3 \right] X_4$	Dodecammine-hexol-tetracobalti salts	Hexa-μ-hydroxo-{tetraamminedihydroxocobalt(III)}cobalt(III) salts

trioxoxenate(VI) ion], and $H_3B \cdot CO$ (borine carbonyl) are clearly definable as complexes. Indeed, atoms or ions of nearly all of the elements can, under appropriate circumstances, serve as central acceptors in complex species.

Among the metal ions, those that form the largest number of well-characterized complexes constitute a compact group at the center of the periodic table, as

										Al		
Sc	Ti	V	Cr	Mn	Fe	Co	Ni	Cu	Zn	Ga		
	Zr	Nb	Mo	Tc	Ru	Rh	Pd	Ag	Cd	In	Sn	
	Hf	Ta	W	Re	Os	Ir	Pt	Au	Hg	Tl	Pb	Bi

Most of these elements are d-transition elements; the others immediately follow the d-transition elements in atomic numbers. The f-transition elements — the lanthanide[33-34a] and actinide[35] elements — although originally believed to form only a limited number of coordination entities, are now known to form many with a variety of ligands. Noteworthy by their absence are the alkali (group IA) and alkaline earth (group IIA) metal ions, but it must not be assumed that no complexes of these cations exist.*

It is at once apparent that the cations that serve best as coordination centers are those that have small sizes, that have large ionic or effective nuclear charges, and/or that have incompletely occupied $(n-1)d$ orbitals. Each of these factors favors attraction by the central ion for electronic charge density. It is apparent also from the diversity of species that, except for electrostatic attraction effects, no single type of bond can characterize all types of complex species.

A considerable body of experimental data, obtained originally in comparing the solution stabilities of halide-containing complex ions of various cations[36-42] but later found to be characteristic of many other donor species,[43-45] suggests that acceptor cations can be divided into two broad classes: Type A — those that form the most stable complexes with the lightest ligand atom from a given periodic family (e.g., F in the halogen family) and Type B — those that form the most stable complexes with the second or subsequent ligand atom from a given periodic family (e.g., Cl or Br or I in the halogen family). The distribution of acceptor cations of these types, in their common states of oxidation, in the periodic table is shown in Fig. 7-2.[44] More detailed treatment as it pertains to the Hard–Soft Acid–Base Classification is found in Sec. 9.6-4.

3-2. Donors

The commonest donor atoms in ligands are undoubtedly oxygen and nitrogen. Less commonly, halogen, sulfur, phosphorus, arsenic, and carbon atoms act as

---

* Many have been identified in solution [e.g., D. Midgley, *Chem. Soc. Rev.*, **14**, 549 (1975)] and, largely by encapsulation, in solids (Secs. 7.3-2, 10.1-3).

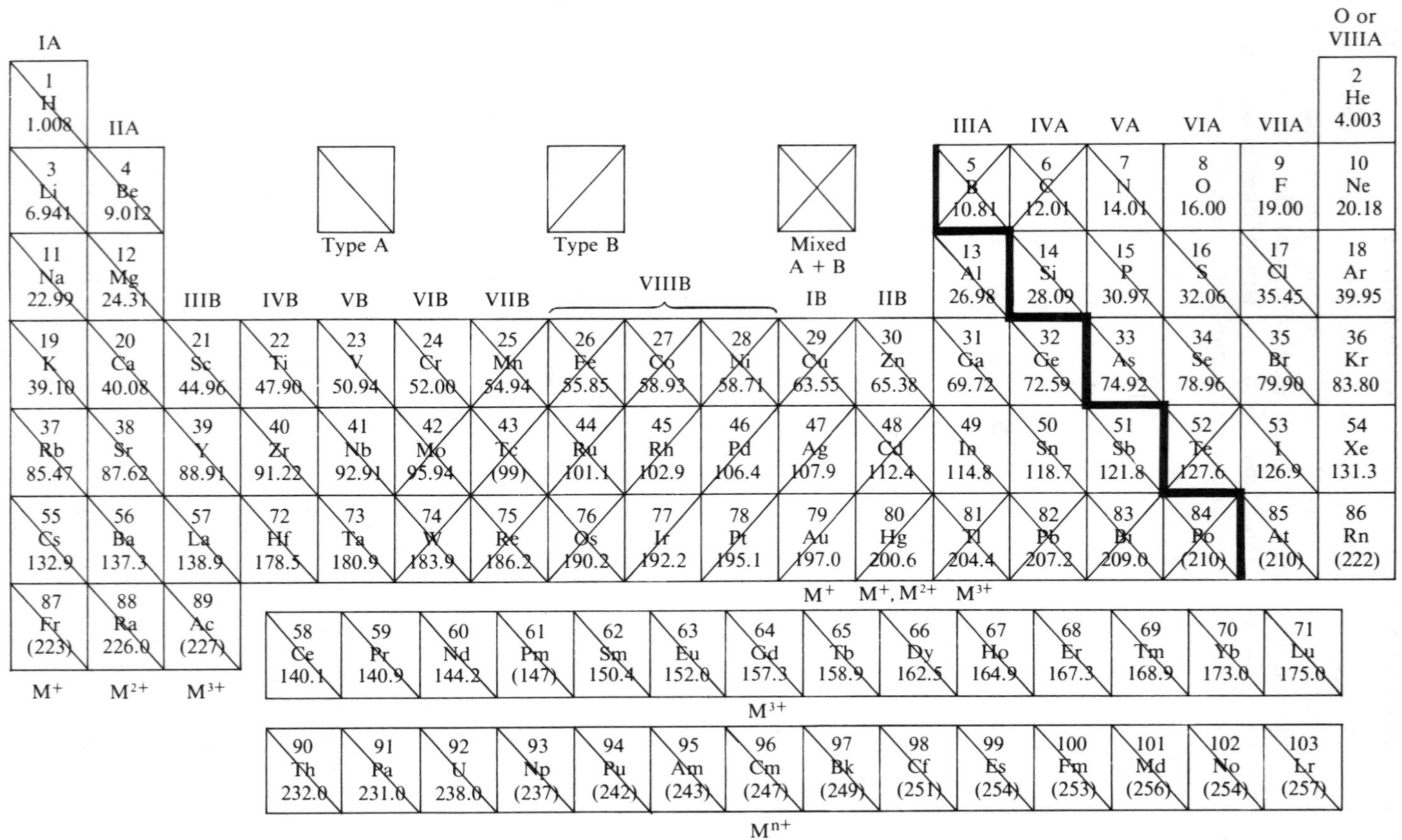

IA	IIA		IIIB	IVB	VB	VIB	VIIB		VIIIB		IB	IIB	IIIA	IVA	VA	VIA	VIIA	O or VIIIA
1 H 1.008																		2 He 4.003
3 Li 6.941	4 Be 9.012												5 B 10.81	6 C 12.01	7 N 14.01	8 O 16.00	9 F 19.00	10 Ne 20.18
11 Na 22.99	12 Mg 24.31												13 Al 26.98	14 Si 28.09	15 P 30.97	16 S 32.06	17 Cl 35.45	18 Ar 39.95
19 K 39.10	20 Ca 40.08		21 Sc 44.96	22 Ti 47.90	23 V 50.94	24 Cr 52.00	25 Mn 54.94		26 Fe 55.85	27 Co 58.93	28 Ni 58.71	29 Cu 63.55	30 Zn 65.38	31 Ga 69.72	32 Ge 72.59	33 As 74.92	34 Se 78.96	35 Br 79.90	36 Kr 83.80
37 Rb 85.47	38 Sr 87.62		39 Y 88.91	40 Zr 91.22	41 Nb 92.91	42 Mo 95.94	43 Tc (99)		44 Ru 101.1	45 Rh 102.9	46 Pd 106.4	47 Ag 107.9	48 Cd 112.4	49 In 114.8	50 Sn 118.7	51 Sb 121.8	52 Te 127.6	53 I 126.9	54 Xe 131.3
55 Cs 132.9	56 Ba 137.3		57 La 138.9	72 Hf 178.5	73 Ta 180.9	74 W 183.9	75 Re 186.2		76 Os 190.2	77 Ir 192.2	78 Pt 195.1	79 Au 197.0	80 Hg 200.6	81 Tl 204.4	82 Pb 207.2	83 Bi 209.0	84 Po (210)	85 At (210)	86 Rn (222)
87 Fr (223)	88 Ra 226.0		89 Ac (227)															

58 Ce 140.1	59 Pr 140.9	60 Nd 144.2	61 Pm (147)	62 Sm 150.4	63 Eu 152.0	64 Gd 157.3	65 Tb 158.9	66 Dy 162.5	67 Ho 164.9	68 Er 167.3	69 Tm 168.9	70 Yb 173.0	71 Lu 175.0
90 Th 232.0	91 Pa 231.0	92 U 238.0	93 Np (237)	94 Pu (242)	95 Am (243)	96 Cm (247)	97 Bk (249)	98 Cf (251)	99 Es (254)	100 Fm (253)	101 Md (256)	102 No (254)	103 Lr (257)

Type A Type B Mixed A + B

M^+ M^{2+} M^{3+} M^+, M^{2+} M^{3+} M^{n+}

Figure 7-2 Distributions of acceptor cations by types in the periodic table. [Variant from

donors. In other instances, π bonds appear to act as donors. Ligands based upon donor species have been usefully classified as follows:[46]

1. *Type A* — those ligands containing donor atoms having one or more nonbonded pairs of electrons. These are further classified as:
    (a) Having no vacant orbitals to receive electrons from the metal atom or ion (e.g., H_2O, NH_3, F^-, H^-, CH_3^-).
    (b) Having vacant orbitals or orbitals that can be vacated to receive π electronic charge density from the metal atom or ion (e.g., PR_3, I^-, CN^-, NO_2^-).
    (c) Having additional π electronic charge density that can be supplied to vacant orbitals of the metal atom or ion (e.g., OH^-, NH_2^-, F^-, I^-).
2. *Type B* — Those ligands containing donor atoms without nonbonded pairs of electrons but with π-bonding electronic charge density [e.g., C_2H_4, C_6H_6, $C_5H_5^-$ (cyclopentadienide ion)].

Although it is true that almost any ionic or molecular structure containing one or more electron-pair donor atoms or suitably arranged π bonds can act as a ligand, both the nature of the donor site and the other atoms present affect the chemical stability of the product.

Among Type A ligands, the general trends in coordinating affinity noted in Table 7-4 can be distinguished. With a given ligand atom, the orders $NH_3 > NH_2R > NHR_2 > NR_3$, $HOH > ROH > R_2O$, and $R_2S > RSH > HSH$ are also noted; and with anions, the order $CN^- > NCS^- > OH^-$ is observed. With Type A acceptors, chemical stability generally decreases as basicity of the donor

Table 7-4 Coordinating Affinities of Donor Atoms

Periodic Position of Donor Atom	Type of Metal Ion	Relative Coordinating Affinity of Donor Atom[a]
Group VA	A	$N. >> P > As > Sb > Bi$
	B	$N << P > As > Sb > Bi$
Group VIA	A	$O >> S > Se > Te$
	B	No regularity
Group VIIA	A	$F^- >> Cl^- > Br^- > I^-$
	B	$I^- > Br^- > Cl^- >> F^-$
Period II	A	$N > O > F$
	B	$N >> O >> F$
Period III	A or B	$P > S > Cl$
Period IV	A	$Se > As > Br$
	B	$As > Se > Br$
Period V	A or B	$Te > Sb > I$

[a] Donor atom bonded to H, alkyl, or aryl, except as otherwise indicated.

decreases. Type B ligands bond only to Type B acceptors, since π bonding is a necessity and only cations with appropriate d-electron properties can so bond. In general, a more strongly bonding ligand will displace a more weakly bonding one from the coordination sphere.

Uncommon Donor Species. Although the CO molecule is a well-known donor species, for example, in the metal carbonyls and their derivatives (Sec. 8.2-1), the isoelectronic N_2 molecule was not shown to act as a ligand until 1965, when the pentaammine(dinitrogen)ruthenium(II) cation, $[Ru(NH_3)_5(N_2)]^{2+}$, was isolated from the products of reduction of commercial ruthenium(III) chloride [containing some ruthenium(IV)] with hydrazine.[47] Numerous other dinitrogen complexes have since been described.[48-50] The tetraphosphorus molecule (P_4) much less commonly acts as a donor, giving species such as $[Rh(Cl)(P_4)\{P(C_6H_5)_3\}]_2$.[51] Many dioxygen complexes have been described in terms of both sigma and pi bonding.[52] However, at least some of these species contain the superoxide ion (O_2^-) as a sigma-bonded donor group.[53, 54] Polysulfide ions may act as donors, for example, in the compound $(NH_4)_3[Rh(S_5)_3]$.[55] Metal–metal acceptor–donor interactions apparently characterize certain metal-cluster (Sec. 8.2-3) species[56] [e.g., $[Os_3(CO)_{12}]$ and $[M_6X_{12}]^{2+}$ (M = Nb, Ta; X = Cl, Br)]. Of interest also are species containing tin(II) ligands [e.g., $[Pt(SnCl_3)_5]^{3-}$, $[(C_{10}H_8N_2)(CO)_3(Cl)Mo\{Sn(CH_3)Cl_2\}]$ ($C_{10}H_8N_2$ = bipyridine), and $[(C_4H_{10}S_2)(CO)_3(Cl)W\{Sn(CH_3)Cl_2\}]$ ($C_4H_{10}S_2$ = 2,5-dithiahexane)]. X-ray crystallographic studies on these compounds have indicated the presence of metal ion–tin(II) bonds based upon the donor groups $SnCl_3^-$ and $Sn(CH_3)Cl_2^-$, with 5-coordinate platinum(II) in a trigonal bipyramidal geometry[57] and 7-coordinate molybdenum and tungsten in capped octahedral geometry.[58] Many other tin(II) and comparable germanium(II) derivatives are known.[59]

Cations are, in general, either noncoordinating or very weakly coordinating ligands, reflecting electrostatic repulsion between metal ion and ligand. Other factors being equal, the distance between one donor atom in the ligand species and the donor atom carrying the positive charge is closely related to the overall donor strength of the ligand. Typical cationic donors are $\{(CH_3)_3\overset{+}{N}(CH_2)_2NH_2\}$ ($=L^+$) in $[M(L^+)_6](ClO_4)_8$ (M = $\overset{II}{Co}$, $\overset{II}{Ni}$)[60] and $\{H_2N\overset{+}{N}(CH_3)_3\}$ ($=L^+$) in $[Ni(L^+)_2Cl_4]$.[61]

"Noncoordinating" Anions. Although from time to time various anions have been believed to be noncoordinating and thus used, particularly in investigating solution chemistry, where no or minimal interaction with a cation was desired, it is now well established that all known anions do, to some degree, form complexes with many cations.[62] Of particular significance among these species are the nitrate (NO_3^-), perchlorate (ClO_4^-), perrhenate (ReO_4^-), hexafluorophosphate(V) (PF_6^-), tetrafluoroborate (BF_4^-), tetrahydroborate (BH_4^-), and tetraphenylborate $[B(C_6H_5)_4^-]$ ions.

Coordination by the NO_3^- group has been well established experimentally both in terms of vibrational spectroscopic indications of changes from D_{3h} to C_{2v} point-group symmetry (Sec. 6.5-3) and of molecular structures determined by diffraction methods. The factors that influence coordination by the NO_3^- ion and numerous individual molecular structures have been summarized.[63] Among the latter is the structure of the hexanitratocerate(IV) ion, in the compound $(NH_4)_2[Ce(NO_3)_6]$, in which structure each of the nitrate groups is bidentate, thus imparting a coordination number of 12 to the Ce^{4+} ion.[64] It is of interest also that x-ray diffraction studies on concentrated aqueous solutions of cerium(IV) ammonium nitrate also indicate coordination number = 12.[65] Early, but often disputed, qualitative proposals of the existence of the hexa-nitratocerate(IV) ion [66,67] are thus justified.

The perchlorate ion is a much weaker donor than the nitrate ion and is not uncommonly the anion of choice when complexation is to be minimized. Coordination by this species was first detected through changes in the infrared spectrum associated with reduction of the point-group symmetry of the ClO_4^- ion from T_d to C_{3v} (Table 7-5) as a consequence of interactions with cations.[68,69] Both unidentate[70] and bidentate[71] perchlorato groups have been identified by x-ray crystallographic studies. Use of perchlorate in systems containing organic substances should be accompanied by adequate precautions in the light of the numerous explosions that have been recorded.[72,73] The perrhenate ion, ReO_4^-, which imparts no explosion hazard, is closely comparable to the perchlorate ion in its coordinating ability, with infrared spectra interpreted in terms of lowering the point-group symmetry from T_d to C_{3v} (Sec. 6.5-3) as a consequence of unidentate coordination.[74,75]

Anionic coordination is still further minimized with perfluoro species such as PF_6^-, AsF_6^-, SbF_6^-, SiF_6^{2-}, and BF_4^-. However, there are

Table 7-5 Vibrational Absorptions of Perchlorate Group[a]

Ionic T_d (cm^{-1})	Covalently Bonded	
	Unidentate C_{3v} (cm^{-1})	Bidentate C_{2v} (cm^{-1})
$v_1(A)$ 932[b]	$v_2(A_1)$ 933	$v_2(A_1)$ 928
$v_2(E)$ 460[b]		
$v_3(T_2)$ 1110	$v_1(A_1)$ 1040	$v_1(A_1)$ 1038
	$v_4(E)$ 1122	$v_6(B_1)$ 1125
		$v_8(B_2)$ 1170
$v_4(T_2)$ 626	$v_3(A_1)$ 616	$v_3(A_1)$ 635
	$v_5(E)$ 632	$v_7(B_1)$ ⎫ 617, 623 $v_9(B_2)$ ⎭

[a] All Raman active

[b] Infrared inactive.

spectral evidences of at least weak coordination through fluorine atoms in the species $CuC_5H_5N)_4(PF_6)_2$,[76] $(CH_3)_3Sn(AsF_6)$, and $CH_3)_3Sn(SbF_6)$,[77] and $Ni(C_5H_5N)_4SiF_6$.[62] Spectral evidence indicates coordination by the BH_4^- ion in the species $NiA(BH_4)$ (ClO_4) where A = cyclic tetramine,[78] and a single-crystal, x-ray diffraction study of the compound $Y(BH_4)_3(C_4H_8O)_3$ (C_4H_8O = tetrahydrofuran) shows the presence of Y—H bonds with one BH_4^- group being bidentate and the other two terdentate.[79] The $B(C_6H_5)_4^-$ ion has no atom donor sites, but it does coordinate to d-transition metal ions through its π-bonding system, as indicated by x-ray diffraction[80] and infrared and nuclear magnetic resonance measurements.[81] On the other hand, in compounds such as $[Ln(pd)_4]$ $[B(C_6H_5)_4]_3$ (pd = pyramidone; Ln = Sc, Y, La, Nd, Gd, Er, Yb, Lu)[82] and $[Ln(ap)_6]$ $[B(C_6H_5)_4]_3$ (ap = antipyrine; Ln = La, Pr, Gd, Er, Yb),[83] the Ln^{3+} ions have so small a tendency to interact with π-donor systems under the aqueous conditions used for syntheses[84] that the $B(C_6H_5)_4^-$ ion is probably completely noncoordinating.

Olefinic and Related Ligands. The first olefin-containing species, originally formulated as $KCl \cdot PtCl_2 \cdot C_2H_4$ but better described as $K[PtCl_3(C_2H_4)]$, was obtained by W. C. Zeise as a precipitate formed by heating platinum(IV) chloride with ethanol under reflux and then adding potassium chloride.[85] A later study[86] indicated the reaction to proceed as

$$PtCl_4 \xrightarrow{C_2H_5OH} [PtCl_2(C_2H_4)]_2 \xrightarrow{KCl} K[PtCl_3(C_2H_4)]$$

Subsequently, comparable species were prepared from a variety of metal ions (e.g., those of Pt, Pd, Ir, Cu, Ag) and olefins, diolefins, and other compounds containing olefinic linkages (e.g., alcohols, acids, esters, etc.).[87-90] For a given ligand, the platinum(II) derivative is commonly the most stable thermodynamically. Acetylenes and π-allylic compounds form related complexes.

The molecular structures of typical species and interpretations of the donor properties of the multiple carbon-to-carbon double bonds are discussed in Secs. 8.1 and 8.2-3.

Carbocyclic Ligands. Although salts containing carbocyclic anions (Table 7-6) had been known for many years, it was not until 1951 that the first complex compound, di-π-cyclopentadienyliron(II), $Fe(C_5H_5)_2$, was first described— interestingly, quite accidentally by two independent groups.[91,92] The initial synthesis involved passing cyclopentadiene diluted with nitrogen over iron powder containing oxides of aluminum, molybdenum, and potassium at 300°C.[91] The other procedure involved the reaction of anhydrous iron(III) chloride with an excess of the Grignard reagent cyclopentadienylmagnesium bromide.[92] The orange-yellow crystalline compound was named *ferrocene*[93] and proposed to have a sandwich-type molecular structure,[93,94] a proposal subsequently verified by single-crystal, x-ray diffraction measurements.[95]

Subsequently, π-carbocyclic derivatives of a number of cations, most es-

Table 7-6 Carbocyclic Ligands

Ligand	Graphic Formula	Ligand Electrons	Hapto Notation[a]
Cyclopentadienide, $C_5H_5^-$		5	η^5
Methylcyclopentadienide, $C_6H_7^-$		5	η^5
Indenide, $C_9H_7^-$		5	η^5
Fluorenide, $C_{13}H_9^-$		5	η^5
Cycloheptatrienene, $C_7H_8^+$		7	η^7
Cyclooctatetraenide, $C_8H_8^{2-}$		10	η^8

[a] For a discussion, see p. 514.

pecially of the d-transition type, have been identified.[89,96,97] Again, molecular structure and bonding are discussed subsequently (Sec. 8.2-3).

Arene Ligands. Molecules of arenes (e.g., benzene, toluene, mesitylene, diphenyl, tetralin) interact to a more limited extent with certain d-transition metal ions to give somewhat less thermodynamically stable, sandwich-type complexes.[89,96,97] Of particular interest is dibenzenechromium(0), first obtained by F. Hein[98] in 1919 in admixture with other related species but not character-ized until 1956.[99-102] Dibenzenechromium is conveniently synthesized accord-ing to the reaction scheme[100]

$$CrCl_3 + AlCl_3 + Al + C_6H_6 \xrightarrow[\text{autoclave}]{180°C} [Cr(C_6H_6)_2][AlCl_4] \xrightarrow[\text{OH}^-]{S_2O_4^{2-}} Cr(C_6H_6)_2(s)$$

Bonding and molecular structures for these species are closely comparable with those for the carbocyclic species (Sec. 8.2-3).

Multidentate Ligands and Chelate Rings. By definition a bidentate ligand is a chelating ligand and forms a chelate ring that includes a metal ion. Early chelating ligands were only bidentate (e.g., ethylenediamine, oxalate ion, acetylacetonate ion), but modern studies have included ter-, quadri-, and in general multidentate ligands, in the molecular structures of which donor atoms are so located that more than a single chelate ring involving the same metal ion can form. Some typical chelating ligands and corresponding structural units are summarized in Table 7-7, following an early, but still useful, arbitrary classification.[103] It is apparent, of course, that the distinction between *acidic* (*anionic*) and *coordinating* (*neutral*) groups may be of no consequence once the chelated structure has formed. In the next section, chelating ligands that encapsulate the metal ions are discussed.

An outstanding characteristic of a chelated structure is its substantially larger thermal and solution stabilities over those of a nonchelated structure containing the same donor and acceptor atoms.[104-109] Stability is still more enhanced when a given metal atom or ion is located in a structure involving two or more fused rings. This *chelate effect** distinguishes these complexes from those derived from only unidentate ligand species and accounts for many of the unique and significant properties of the chelated species.

A few examples can be cited to advantage. Thus ethylamine forms with cobalt(III) and chromium(III) the ions $[M(H_2NCH_2CH_3)_6]^{3+}$, which are easily decomposed thermally or by chemical reactions. The corresponding complexes formed with ethylenediamine, and formulated as indicated in item 4 of Table 7-7, remain undecomposed at higher temperatures and are difficult to decompose chemically. Neither phenol nor benzaldehyde forms particularly thermally or thermodynamically (Ch. 9) stable complexes. However, combination of these functional groups in salicylaldehyde gives a bidentate donor that yields relatively stable complexes, as formulated in item 2, Table 7-7. Complexes derived from acetone (CH_3-C-CH_3) with carbonyl O are exceedingly difficult to prepare and preserve, but those derived from the enolic form of acetylacetone $(CH_3-C-CH=C-CH_3)$ with carbonyl O and OH and containing the ring system

---

*   See D. Munro, *Chem. Br.*, **13**, 100 (1977); E. L. Simmons, *J. Chem. Educ.*, **56**, 578 (1979).

can often be volatilized at elevated temperatures without thermal decomposition. Other β-hydroxocarbonyl ligands (Table 7-7) form complexes with comparable properties. Similarly, copper(II) acetate is an ionic compound, but copper(II) α-aminoacetate is a chelated complex compound (Table 7-7).

Even more enhancement of thermal and chemical stability results with fused-ring ligands. Thus the phthalocyanine derivatives of the dipositive metal ions

$$M = Cu, \; Zn, \; Co, \; Ni, \; Fe$$

are remarkably unreactive [e.g., the copper(II) compound is unaffected by hot, 18 M sulfuric acid], and the compound

is more resistant to reduction than the closely related compound

In general, these species are said to be "entropy stabilized" since the replacement of two unidentate ligands by a single chelating ligand increases the total number of ions or molecules in the system, and thus gives a positive increase in entropy (Sec. 9.2).

Table 7-7 Typical Chelating Groups and Graphic Structural Representations

Chelating Group: Classification	Chelating Group: Examples	Typical Chelated Structures
1. Two acidic groups — anions of:		
Dibasic inorganic acids	CO_3^{2-}, SO_3^{2-}, SO_4^{2-}	$[Co(NH_3)_4(CO_3)]X$
Dicarboxylic acids	$C_2O_4^{2-}$, o-$C_6H_4(CO_2)_2^{2-}$	$M^I_3[M^{III}(C_2O_4)_3](M^{III} = Co, Fe, Cr)$
Disulfonic acids	$CH_2(SO_3)_2^{2-}$	$[Co(en)_2\{CH_2(SO_3)_2\}_2]Br$
α-Hydroxycarboxylic acids	Salicylate, glycolate	
Diamides (acid amides)	$OC(NH_2)_2$, $HN(\overset{O}{\overset{\Vert}{C}}NH_2)_2$	
	$O_2S(NH_2)_2$	$Na[Rh(H_2O)_2\{(HN)_2SO_2\}_2]$
Acidic dihydroxy compounds	Glycols, pyrocatechol	
α-Hydroxy oximes	α-Acyloin oximes	

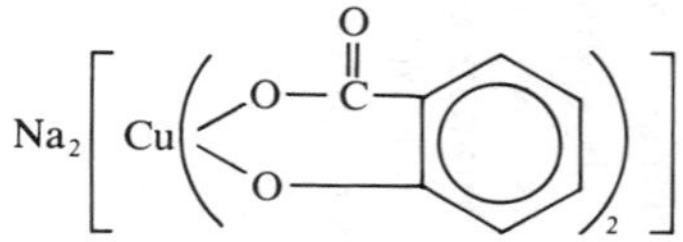

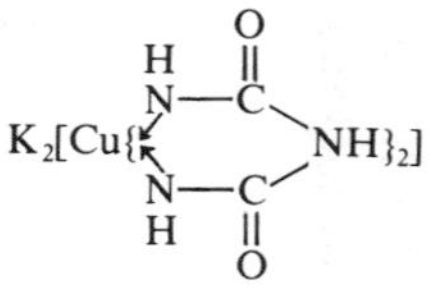

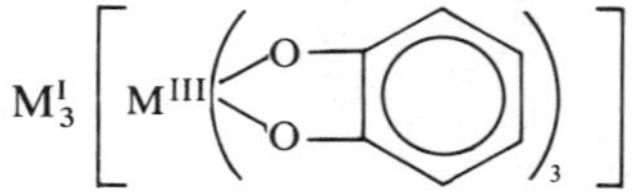

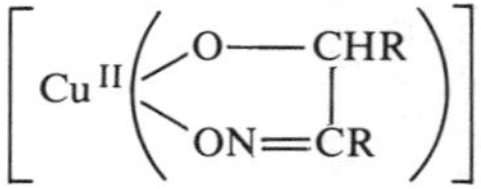

2. One acidic group, one coordinating group — anions of:

α-Aminocarboxylic acids

$RCH(NH_2)CO_2^-$

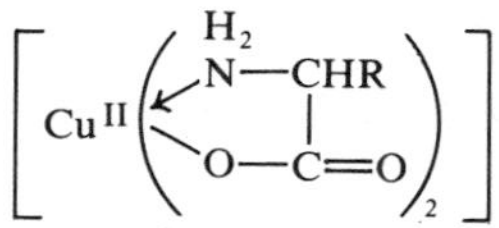

α-Hydroxycarboxylic acids

Lactate, glycolate

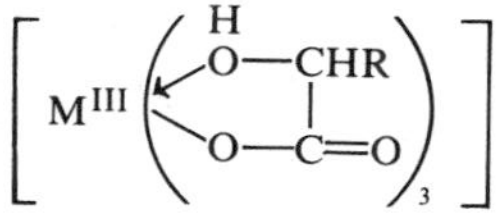

Certain dihydroxy compounds

Biphenol

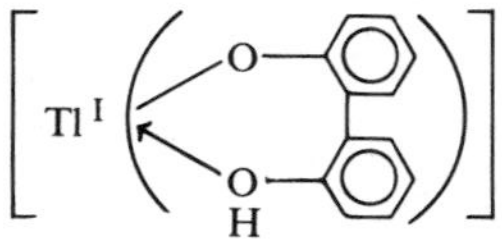

β-Hydroxy carbonyl compounds

β- or 1,3-Diketones,
o-hydroxyaldehydes,
o-hydroxyphenones

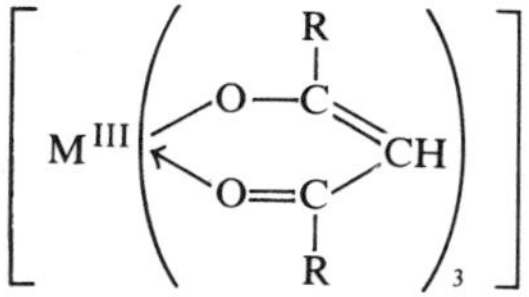

Hydroxy amines

Ethanolamines, o-aminophenol

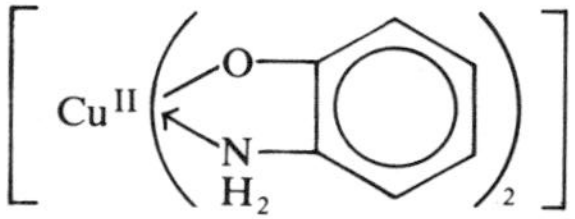

Table 7-7 (*Continued*)

Classification	Chelating Group Examples	Typical Chelated Structures
Hydroxy azo compounds	o-Hydroxyazobenzene, p-nitrobenzeneazo-resorcinol	
8-Quinolinols	8-Quinolinol, 5,7-dihalo-8-quinolinols	
α-Hydroxyoximes	α-Benzoin oximes, salicylaldoxime	

Ketoximes

Benzil oxime, *o*-nitrosophenol

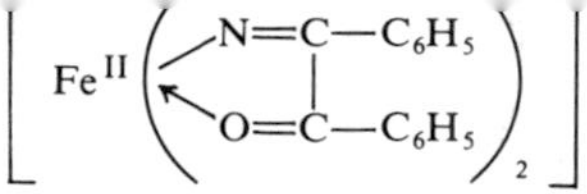

$$\left[Fe^{II} \left(\begin{array}{c} N=C-C_6H_5 \\ O=C-C_6H_5 \end{array} \right) \right]_2$$

Glyoximes

Dimethylglyoxime (anti)

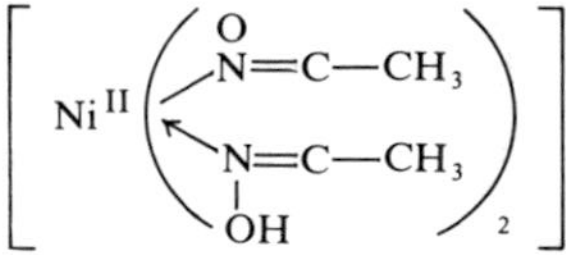

$$\left[Ni^{II} \left(\begin{array}{c} O \quad N=C-CH_3 \\ \quad \quad \| \\ N=C-CH_3 \\ \quad \quad OH \end{array} \right) \right]_2$$

3. Several acidic groups, one or several coordinating groups — anions of:

Aminepolyacetic acids

Nitrilotriacetate, $N(CH_2COO)^{-3}$

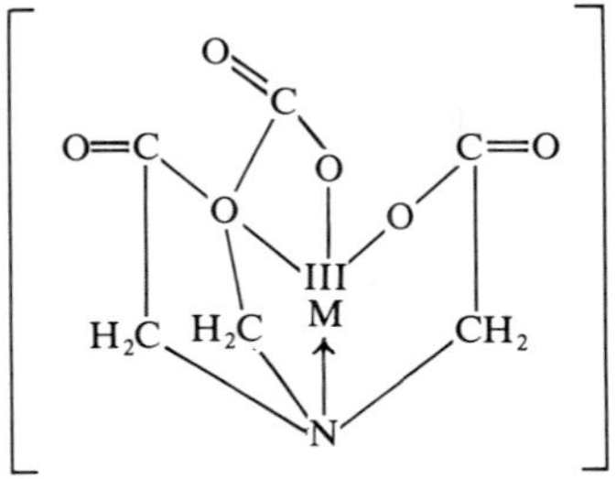

Polyaminepolycarboxylic acids

Ethylenetetraacetate (EDTA), diethylenetriaminepentaacetate (DTPA), triethylenetetraaminehexaacetate (TTHA)

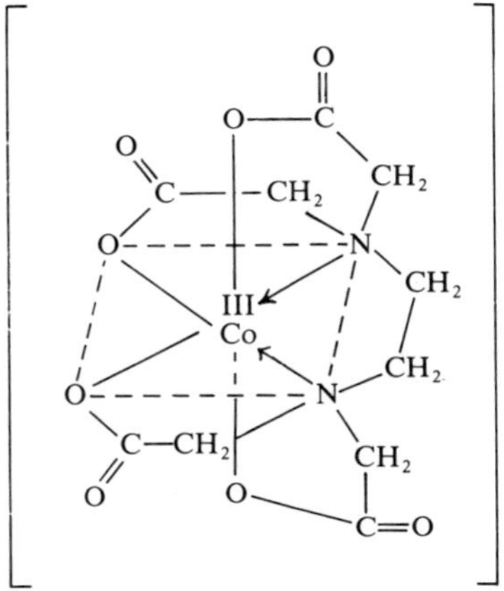

Table 7-7 (*Continued*)

Classification	Chelating Group	Typical
	Examples	Chelated Structures

4. Two or more coordinating groups

Diamines — Ethylenediamine (en), 1,2-diaminopropane (pn)

$$\left[\text{Co}^{III} \left(\begin{array}{c} H_2N-CH_2 \\ | \\ N-CHCH_3 \\ H_2 \end{array} \right)_3 \right]^{3+}$$

2,2'-Bipyridine (bpy), 1,10-phenanthroline (phen), 1,8-naphthyridine

$$\left[\text{Fe}^{II} \left(N \quad N \right)_3 \right]^{2+}$$

$$\left[\text{Pr}^{III} \left(N \quad N \right)_4 \right]^{3+}$$

Amino pyridyls — α-Pyridylhydrazine

$$\left[\text{Fe}^{II} \left(\begin{array}{c} N \quad NH \\ | \\ N \\ H_2 \end{array} \right)_2 \right]^{2+}$$

Triamines

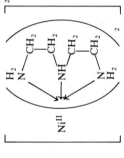

Diethylenetriamine

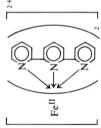

2,2',2''-Terpyridine (terpy)

Tetramines

Triethylenetetramine

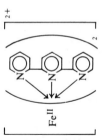

2,2'2''-Triaminotriethylamine (tren)

Dihydroxy compounds

1,2-Glycols, pinacol

Table 7-7 (*Continued*)

Classification	Chelating Group	Typical Chelated Structures
	Examples	
Disulfides	Dithioethers	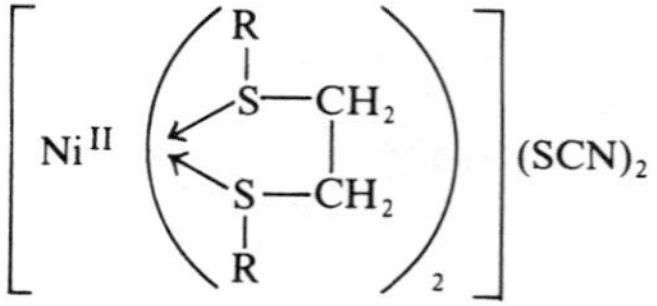
Diarsines	*o*-Phenylenebis(dimethylarsine)	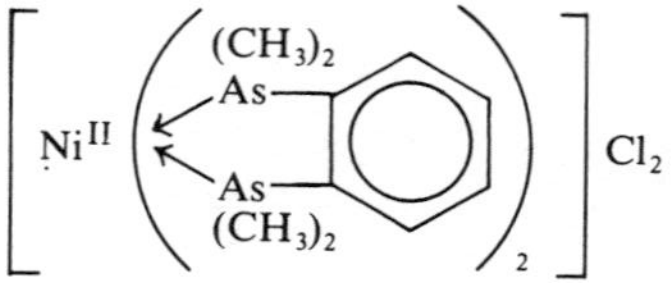
Triarsines	Methylbis(3-dimethylarsinopropyl)arsine	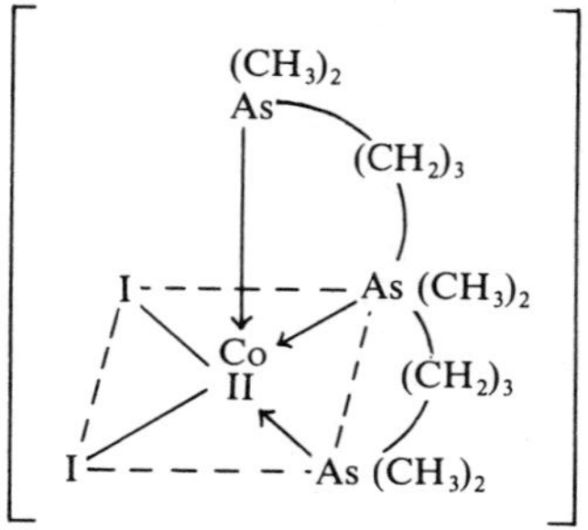
Macrocyclic ligands	Cyclic polyethers, cryptands, sepulchrands	See pp. 390–395

388

Ring size is of considerable importance. The broadly stated principle that maximum stability results with five- or six-membered rings, including the metal ion, is consistent with formation-constant data (Sec. 11.1 and Table 11.1).[104–109] However, careful examination of these data shows some important variations. Thus for saturated ligands of a given type, the five-membered ring is invariably more stable than the six-membered one. For example, 1,2-diaminopropane, although less basic than 1,3-diaminopropane, gives the more stable complex ions with dipositive zinc, cadmium, and copper[110] and with terpositive lanthanide ions.[111,112] Furthermore, 1,2,3-triaminopropanetetrachloroplatinum(IV) is optically active (Sec. 7.4-10),[113] indicating an asymmetry consistent with coordination as

$$
\left[
\begin{array}{c}
\quad\quad\quad\;\; H_2 \\
\quad\quad\quad N\!-\!CH_2 \\
Cl_4Pt{\Large\langle} \quad\quad\quad | \\
\quad\quad\quad N\!-\!CHCH_2NH_2 \\
\quad\quad\quad H_2
\end{array}
\right]
$$

rather than as

$$
\left[
\begin{array}{c}
\quad\quad\quad H_2 \\
\quad\quad\quad N\!-\!CH_2 \\
Cl_4Pt{\Large\langle} \quad\quad\quad\quad\; CHNH_2 \\
\quad\quad\quad N\!-\!CH_2 \\
\quad\quad\quad H_2
\end{array}
\right]
$$

A further decrease in stability with increasing ring size is common for a number of comparable saturated ligands, but it is observed that the stabilities of aliphatic diamine complexes of silver(I)[114] and of the diamine-N,N'-tetraacetato* complexes of mercury(II)[115] increase, whereas the stabilities of other diaminetetraacetato and dicarboxylato complexes become effectively constant after initial decreases.[114,115]

Steric factors are undoubtedly significant in many instances. Thus, as indicated in Fig. 7-3 and as can be shown by models, a five-membered saturated chelate ring can form with less strain than a six-membered one, whereas conjugated unsaturation in a six-membered diketonate chelate ring results in reduced strain. With a multidentate ligand, alteration in the metal ion and/or in the number of donor atoms used may affect the most stable ring size. It must be emphasized that meaningful direct correlations between formation constant data and ring size are possible only when the basicity (donor strength) of the ligand, the nature and multidentate character of the ligand, and resonance effects are constant.

Resonance within an unsaturated heterocyclic ring including a metal ion may increase or decrease stability, depending upon whether it is enhanced or interfered with by resonance in the overall molecular structure of the species.

* Diamine-N,N'-tetraacetates $(-OOCCH_2)_2N(CH_2)_nN(CH_2COO-)_2$, have six potential donor sites (Table 7-7).

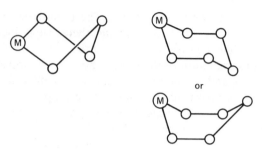

Figure 7-3 Conformations of five- and six-membered saturated chelate rings.

Thus although the stabilities of the metal phthalocyanions are associated with complete conjugation throughout their structures, the stability of a copper(II)-enolate ring

decreases from the acetylacetone to the salicylaldehyde compound because resonance in the latter is disturbed by participation in resonance of the benzene ring. Steric interference by groups in the proximity of a donor atom can also reduce stability. Thus 8-quinolinol (Table 7-7) gives a stable chelate with aluminum(III), but the closely related compound 2-methyl-8-quinolinol does not.[116]

Multidentate Macrocyclic Ligands. These are exemplified in nature as the porphyrin and carrin ring systems and in laboratory practice by the phthalocyanine complexes. The more recent, and very extensive interest in ligands of this type dates to the unexpected formation, as a by-product due to the presence of contaminating catechol in an attempt to prepare an alkyl phenolic ether, of a crystalline complex containing an Na^+ ion and the ligand 2,3,11,12-dibenzo-1,4,7,10,13,16-hexaoxacyclooctadeca-2,11-diene,[117]

(molecularly)   (graphically)

Figure 7-4 Complexation of Na^+ by dibenzene-18-crown-6. [from *duPont Innovation*, **1**(1), 18 (1969).]

Because of crownlike coordination by ether-type oxygen atoms (Fig. 7-4), the ligand was termed a "crown" ether, and designated as dibenzo-18-crown-6 (i.e., 18 atoms in the macrocycle, 6 donor oxygen atoms). Many other comparable polyethers were subsequently prepared in Pedersen's laboratory.[118] Interestingly, these ligands trap a variety of nontransition metal ions but form only a limited number of stable complexes with d-transition metal ions.[119-120a]

A related class of macrocyclic ligands involves donor atoms other than oxygen. Most commonly the other donor atom is nitrogen, either solely or in combination with oxygen atoms. Somewhat less commonly, sulfur atoms, and even phosphorus atoms, can act as donors. Ligands of these types typically form stable complexes with d-transition metal ions. Most of the nitrogen donors that have been studied are quadridentate.

A number of oxygen-, nitrogen-, and sulfur-donor ligands of these types are summarized in terms of graphic structures in Table 7-8.[119] It is apparent that both the size of the cation and the appropriate positioning of the donor atoms dictate whether or not a single ligand molecule can satisfy completely the coordination number of that cation in an unstrained geometry.

Multidentate macrocyclic ligand molecules that are appropriately internally cross-linked with donor atoms correctly positioned in the bridging groups can

Table 7-8 Crown Ether and Related Macrocyclic Ligands

Graphic Structure[a]	Typical Substituent Groups	Trivial Designation of Parent[117]
	1,2-Benzo; 2,4,6,8-methyl; 1,2-cyclohexyl	12-crown-4
	1,2-Cyclohexyl; 6,7-cyclohexyl; 3,4,5-naphthyl	16-crown-5
	1,2-Benzo; 3,4-benzo	30-crown-10
	1,2-Benzo; 3,4-benzo; 5,6-benzo	

Table 7-8 (*Continued*)

Graphic Structure[a]	Typical Substituent Groups	Trivial Designation of Parent[117]
	1-Methyl; 3-dimethyl; 6-methyl; 8-dimethyl	
	1,2-Benzo; 3,4-benzo	

[a] represents CH_2-CH_2 ; represents $CH_2 \overset{CH_2}{\diagdown} CH_2$

completely encapsulate metal ions in cagelike structures. That the cations are buried has prompted the terminologies *cryptands* (forming *cryptates*)[121] and *sepulchrands* (forming *sepulchrates*).[122] Some examples of ligands of this type are indicated graphically in Table 7-9.[122]

For detailed data on syntheses, general properties, and reactions of these ligands and the complexes they form, available excellent summations[119-122] can be consulted with profit. It is of particular interest that the syntheses of certain *d*-transition metal cryptates and sepulchrates are better effected by chemically altering organo ligands already bonded to the metal ion than by attempting to synthesize the macrocyclic ligand first and then incorporate the metal ion. This *template* type of synthesis[123] can be illustrated as follows:[122]

Macrocylic ligands and the resulting complexes are particularly important in many biological systems: for example, oxygen carriers, pigments, vitamin B_{12} and its analogs, and sodium and potassium ion transport and balance.[120, 124] The ready complexation of univalent cations[125] allows extraction of these species into nonaqueous media. It also allows for the stabilization of certain unusual anionic species by complexation of alkali metal counter cations: for example, Na^- *in ethylamine* through formation of $Na(C_{222})^+$ (Sec. 10.1-3);[126] and Te_3^{2-}, Sb_7^{3-}, Bi_4^{2-}, Sn_5^{2-}, and Pb_5^{2-}, again through formation of $Na(C_{222})^+$ or $K(C_{222})^+$ (Sec. 10.1-3).[127-129] In addition, encapsulation of *d*-transition metal ions has given much useful structural and electronic information.[119, 122, 130, 131] Sargeson[122] has noted that the large stabilities of sepulchrates over substantial pH ranges minimize hydrolysis and subsequent polymerization (Sec. 10.4-1), which has complicated investigations of solution chemistry for elements such as molybdenum and tungsten. He has noted also that encapsulating metal ions should allow chemical comparisons among a range of metal ions and studies of various oxidation states with constant stereochemistry and overall constitution.

4. ISOMERISM AND MOLECULAR STRUCTURE

Although isomerism is most commonly thought of as being associated with organic compounds and thus sometimes believed to be restrictive to those species, isomerism is a consequence of atomic arrangement or position and thus cannot be limited to the compounds of any one element or to any single class of compounds. Isomerism can be expected among appropriate inorganic sub-stances, but in practice most of the examples are among the complex species. With these substances, isomerism is so widespread and variable that a strong case can be made for introducing the general subject through inorganic, rather than organic, examples. Both variations in bond types and an increased number of possible geometrical arrangements combine to yield more types of inorganic isomerism than are recognized among the compounds of carbon.

Table 7-9   Encapsulating Ligands

Graphic Structure[a]	Typical Substituent Groups	Designation
		$C_{m,n,o}$ (e.g., $C_{2,2,2}$)
	1-Methyl, 2-methyl, 1-tosylate; 2-tosylate	

[a] or represents $\overset{CH_2-CH_2}{\diagup \qquad \diagdown}$

The broad classification of types of isomerism given by A. Werner[132] is generally useful and is followed here, with some appropriate modifications.[133, 134]

4-1. Polymerization Isomerism

Compounds are said to be polymerization isomers when, although they have the same stoichiometric composition, their molecular complexities are multiples of the simplest stoichiometry. The name is a misnomer in that a more complicated species does not result from polymerization of the simplest one. Differences among polymerization isomers are differences in arrangements of groups, not in numbers of repeating groups.

Werner distinguished between coordination polymerization and nuclear polymerization. The former can be illustrated by the two series of compounds based, respectively, on triamminetrinitrocobalt(III) and diamminedichloroplatinum(II), as summarized in Table 7-10. Other comparable series can be based upon the simple compounds $[Cr(NH_3)_3(NCS)_3]$, $[Pt(NH_3)_2(CN)_2]$, $[Pt(NH_3)_2(NCS)_2]$, $[Pt(NH_3)_2Cl_4]$, and $[Pt(C_5H_5N)_2Cl_2]$. Nuclear polymerization isomerism is much less common and is perhaps best illustrated by the examples

and

It is evident from differences in compositions of ionic and molecular species that substantial differences in physical and chemical properties exist among isomers of this type.

4-2. Coordination Isomerism

This type of isomerism results in compounds containing both coordinated cations and anions when differences in the distribution of the coordinating

Table 7-10 Coordination Polymerization Isomerism

Number of Stoichiometric Groups	Cobalt(III) Series	Platinum(II) Series
1	$[Co(NH_3)_3(NO_2)_3]$	$[Pt(NH_3)_2Cl_2]$
2	$[Co(NH_3)_6][Co(NO_2)_6]$	$[Pt(NH_3)_4][PtCl_4]$
	$[Co(NH_3)_4(NO_2)_2][Co(NH_3)_2(NO_2)_4]$	
3	$[Co(NH_3)_5(NO_2)][Co(NH_3)_2(NO_2)_4]_2$	$[Pt(NH_3)_4][Pt(NH_3)Cl_3]_2$
		$[Pt(NH_3)_3Cl]_2[PtCl_4]$
4	$[Co(NH_3)_6][Co(NH_3)_2(NO_2)_4]_3$	
	$[Co(NH_3)_4(NO_2)_2]_3[Co(NO_2)_6]$	
5	$[Co(NH_3)_5(NO_2)]_3[Co(NO_2)_6]_2$	

groups occur. If two different metal ions (M and M') are involved, the donor groups (A and B) may bond to give isomers of the extreme types $[MA_x][M'B_x]$ and $[M'A_x][MB_x]$ or any intermediate species between these extremes. Typical examples include

$[Co(en)_3][Cr(CN)_6]$ and $[Cr(en)_3][Co(CN)_6]$

$[Co(en)_2(pn)][Cr(CN)_6]$ and $[Cr(en)_2(pn)][Co(CN)_6]$

$[Co(NH_3)_6][Cr(C_2O_4)_3]$ and $[Cr(NH_3)_6][Co(C_2O_4)_3]$

$[Cu(NH_3)_4][PtCl_4]$ and $[Pt(NH_3)_4][CuCl_4]$

and the series of compounds

$[Co(en)_3][Cr(C_2O_4)_3]$, $[Co(en)_2(C_2O_4)][Cr(en)(C_2O_4)_2]$,

$[Cr(en)_2(C_2O_4)][Co(en)(C_2O_4)_2]$, and $[Cr(en)_3][Co(C_2O_4)_3]$

On the other hand, if the same metal ion appears in both the cation and the anion, either in the same oxidation state or in different oxidation states, coordination isomers exist if the ligands are differently distributed between cation and anion. Examples for which there is no difference in oxidation state include

$[Cr(NH_3)_6][Cr(NCS)_6]$ and $[Cr(NH_3)_4(NCS)_2][Cr(NH_3)_2(NCS)_4]$

$[Pt(NH_3)_4][PtCl_4]$ and $[Pt(NH_3)_3Cl][Pt(NH_3)Cl_3]$

$[Cr(en)_3][Cr(C_2O_4)_3]$ and $[Cr(en)_2(C_2O_4)][Cr(en)(C_2O_4)_2]$

$[Cr(en)_2(H_2O)_2]$- and $[Cr(en)_2(C_2O)_4]$-

$[Cr(C_2O_4)_3]\cdot 2\ H_2O$ $[Cr(H_2O)_2(C_2O_4)_2]\cdot 2\ H_2O$

Examples for which there is a difference in oxidation state include

$\overset{II}{[Pt(NH_3)_4]}\overset{IV}{[PtCl_6]}$ and $\overset{IV}{[Pt(NH_3)_4Cl_2]}\overset{II}{[PtCl_4]}$

$\overset{II}{[Pt(C_5H_5N)_4]}\overset{IV}{[PtCl_6]}$ and $\overset{IV}{[Pt(C_5H_5N)_4Cl_2]}\overset{II}{[PtCl_4]}$

Coordination isomers exhibit the same essential differences in properties that characterize polymerization isomers, and for the same reason.

4-3. Hydrate Isomerism

Combined water molecules may coordinate to metal ions or occupy lattice sites in a crystal without being closely associated with a metal ion. This difference can result in hydrate isomerism, a phenomenon classically illustrated by the existence of several different well-defined crystalline compounds of stoichiometric composition $CrCl_3 \cdot 6 H_2O$. On the basis of the ionic chloride content of aqueous solutions,[135, 136] three compounds were described as $[Cr(H_2O)_4Cl_2]Cl \cdot 2 H_2O$ (dark green), $[Cr(H_2O)_5Cl]Cl_2 \cdot H_2O$ (blue-green), and $[Cr(H_2O)_6]Cl_3$ (violet). A fourth compound, $[Cr(H_2O)_3Cl_3] \cdot 3 H_2O$ (brown), has been reported,[137] but its existence remains in doubt. Polarographic[138] and thermogravimetric[139, 140] data support the first three formulations. Commercially available $CrCl_3 \cdot 6 H_2O$ contains largely the trans-$[Cr(H_2O)_4Cl_2]^+$ ion, which slowly aquates in aqueous solution to $[Cr(H_2O)_5Cl]^+$ ion and ultimately to $[Cr(H_2O)_6]^{3+}$ ion, which species can be separated from each other by cationic exchange.[141] Several kinetic studies of these reactions have been reported.[142-144] Single-crystal x-ray diffraction data for the compound $[Cr(H_2O)_4Cl_2]Cl \cdot 2 H_2O$ indicate a trans arrangement of the two coordinated chloride ions, oxygen atoms from four water molecules completing octahedral geometry around the Cr^{3+} ion, and hydrogen-bonded cages of water molecules and chloride ions linking the octahedra in chains.[145]

Other examples of this type of isomerism are

$[Co(NH_3)_4(H_2O)Cl]Cl_2$ and $[Co(NH_3)_4Cl_2]Cl \cdot H_2O$

$[Co(NH_3)_3(H_2O)_2Cl]Br_2$ and $[Co(NH_3)_3(H_2O)(Cl)Br]Br \cdot H_2O$

$[Co(NH_3)_4(H_2O)Cl]Br_2$ and $[Co(NH_3)_4Br_2]Cl \cdot H_2O$

$[Cr(C_5H_5N)_2(H_2O)_2Cl_2]Cl$ and $[Cr(C_5H_5N)_2(H_2O)Cl_3] \cdot H_2O$

$[Co(NH_3)_5(H_2O)](NO_3)_3$ and $[Co(NH_3)_5(NO_3)](NO_3)_2 \cdot H_2O$

The chemical and physical properties of isomers of this type are understandably quite different.

4-4. Ionization Isomerism

Compounds that have the same stoichiometric composition but yield different ions in solution are called ionization isomers. Classical examples are red-violet $[Co(NH_3)_5Br]SO_4$ and red $[Co(NH_3)_5(SO_4)]Br$, which yield, respectively SO_4^{2-} and Br^- ions. Certain of the species used to exemplify hydrate isomerism also illustrate ionization isomerism. Other examples are

$[Co(NH_3)_5(NO_3)]SO_4$ and $[Co(NH_3)_5(SO_4)]NO_3$

$[Co(NH_3)_5Br]C_2O_4$ and $[Co(NH_3)_5(C_2O_4)]Br$

$[Co(NH_3)_4(NO_2)Cl]Cl$ and $[Co(NH_3)_4Cl_2]NO_2$

$[Co(en)_2(NO_2)Cl]NO_2$ and $[Co(en)_2(NO_2)_2]Cl$

$[Co(en)_2(NCS)Cl]NCS$ and $[Co(en)_2(NCS)_2]Cl$

$$[Pt(NH_3)_3Cl]I \qquad \text{and} \qquad [Pt(NH_3)_3I]Cl$$

$$[Pt(NH_3)_4Cl_2]Br_2 \qquad \text{and} \qquad [Pt(NH_3)_4Br_2]Cl_2$$

and the series $[Co(en)_2(NO_2)Cl]NCS$, $[Co(en)_2(NO_2)(NCS)]Cl$, and $[Co(en)_2(NCS)Cl]NO_2$. The existence of species in which a dinegative ion occupies a single coordination position is noteworthy.

4-5. Structural, or Linkage, Isomerism

Isomerism of this type may occur when more than a single atom in a coordinated group can function as a donor.* The classic example is S. M. Jørgensen's syntheses of two compounds of stoichiometric composition $Co(NH_3)_5(NO_2)Cl_2$, one red and the other yellow-brown.[146] The red compound was easily decomposed by acids; the yellow-brown compound was stable to acids. Either in solution or in the solid state, the red compound slowly transformed into the other one. Yet in aqueous solutions of either compound, silver ion precipitated all the chloride present. Since the source of isomerism had to be in the NO_2 group, the compounds were formulated as

<table>
<tr><td align="center">red</td><td></td><td align="center">yellow-brown</td></tr>
<tr><td align="center">$[Co(NH_3)_5(—ONO)]Cl_2$</td><td></td><td align="center">$[Co(NH_3)_5(—NO_2)]Cl_2$</td></tr>
<tr><td align="center">pentaamminenitritocobalt(III)</td><td align="center">and</td><td align="center">pentaamminenitrocobalt(III)</td></tr>
<tr><td align="center">chloride</td><td></td><td align="center">chloride</td></tr>
<tr><td align="center">(Co ← ONO bond)</td><td></td><td align="center">(Co ← NO_2 bond)</td></tr>
</table>

The similar isomers

$$[Co(en)_2(ONO)_2]X \qquad \text{and} \qquad [Co(en)_2(NO_2)_2]X$$

$$[Co(NH_3)_2(C_5H_5N)_2(ONO)_2]X \qquad \text{and} \qquad [Co(NH_3)_2(C_5H_5N)_2(NO_2)_2]X$$

again as red and yellow-brown solids, were obtained by A. Werner[147] and, much later, the mixed nitrito–nitro ion $[Co(en)_2(ONO)(NO_2)]^+$.[148] Closely related are the isomeric ions $[M(NH_3)_5(ONO)]^{n+}$ and $[M(NH_3)_5(NO_2)]^{n+}$ (M = Rh^{III}, Ir^{III}, Pt^{IV}).[149]

The authenticity of nitrito–nitro linkage isomerism is based upon considerable experimental evidence.[150] Spectrophotometric data indicate that the spontaneous conversion of Jørgensen's red compound to the yellow-brown isomer is a first-order process (Sec. 9.3-3), whereas direct formation of only the nitro isomer should be a second-order process.[151] Comparisons of the infrared spectra of these two pentaammines with those of known nitrites and nitro compounds show clearly that the red compound contains a nitrito group (M—ONO stretching vibrations at 1060 cm^{-1}) and that the other compound

---

* A ligand of this type is said to be ambidentate.

contains a nitro group ($M—NO_2$ deformation vibration at 820 cm^{-1}).[152,153] Convincing evidence for the presence of the nitrito arrangement is the retention of oxygen-18 (O*) in the complex when the species $[Co(NH_3)_5(O^*H_2)]^{3+}$ is allowed to react with the nitrite ion,[154]

$$[Co(NH_3)_5(O^*H_2)]^{3+} + NO_2^- \xrightarrow{(H_2O)} [Co(NH_3)_5(O^*—NO)]^{2+} + H_2O$$

This reaction can occur only if the original $Co—O$ bond is retained (i.e., if the product has the nitrito arrangement). Similar studies indicate that isomerization to the nitro ion is an intramolecular process that involves either a penta- or a hepta-coordinated intermediate.[154] Detailed single-crystal x-ray diffraction studies established without question the nitrito arrangement in *trans*-[Co(en)$_2$(NCS)(ONO)]X and the nitro arrangement in *trans*-[Co(en)$_2$(NCS)(NO$_2$)]X (X = ClO$_4^-$, I$^-$).[155] Crystals of both compound types are monoclinic, in space group $P2_1$, with $Z = 2$. Both structures are nearly identical, except for the bonding of the NO_2 group, with symmetry around the cobalt(III) ion nearly O_h. Interestingly, both photochemical and thermal isomerization of the nitro to the nitrito species occurs in the solid state, presumably via rotation within the NO_2 group. The low-temperature, solid-state, photochemical isomerization

$$[Co(NH_3)_5(NO_2)]Cl_2 \rightarrow [Co(NH_3)_5(ONO)]Cl_2$$

may involve the 7-coordinate intermediate[156]

Other examples of linkage isomerism include $M—NCS$ vs. $M—SCN$ bonds,[157-162] $M—CN$ vs. $M—NC$ bonds,[163-165] and $M—OSO_2S$ vs. $M—SSO_3$ bonds.[166]

Distinction between $M—NCS$ (isothiocyanato) and $M—SCN$ (thiocyanato) bonding is conveniently effected by infrared measurements (Sec. 6.5-3), the $C—N$ stretching absorption being sharp and above 2000 cm^{-1} for S-bonded complexes and broad and below 2000 cm^{-1} for N-bonded complexes.* Isolated linkage isomers of this type include [Pd(bpy)$_2$(—NCS)$_2$] and [Pd(bpy)$_2$(—SCN)$_2$] (bpy = 2,2'-bipyridine); [Pd{As(C$_6$H$_5$)$_3$}$_2$(—NCS)$_2$] and [Pd{As(C$_6$H$_5$)$_3$}$_2$(—SCN)$_2$]; and Pd(Et$_4$dien)(—NCS)]$^+$ and [Pd(Et$_4$dien)(—SCN)]$^+$ (Et$_4$dien = (C$_2$H$_5$)$_2$NC$_2$H$_4$NHC$_2$H$_4$N(C$_2$H$_5$)$_2$].[167,168] Other species have been detected in solution.[158,159]

Class A metal ions (Sec. 7.3-1) tend to form $M—NCS$ bonds, whereas Class B metal ions bond as $M—SCN$. A systematic study of the behavior in

---

* Less definitively, the $C—S$ stretching vibration is found at 780 to 860 cm^{-1} for N-bonding and at ca. 700 cm^{-1} for S-bonding.

solvents of varying polarity of complexes of the types $[ML_2X_2]$ (M = Pd^{II}, Pt^{II}; L = neutral ligand; $X^- =$ —SCN^-, —NCS^-, —$SeCN^-$, and —NCO^-), $[Rh\{P(C_6H_5)_3\}_2(CO)(NCX)_2]$ (X = O, S, Se), and $[Ir\{P(C_6H_5)_3\}_2(CO)(NCS)]$ indicated that in most instances the ambidentate character of thio- and selenocyanate ions was determined by the nature of the solvent, whereas when X = O it was not.[160] Solvents of relatively large dielectric constants favored M—S (or M—Se) bonds, whereas those of smaller dielectric constants favored M—N bonds. It has been noted[162] that under constant steric restraints S vs. N coordination is controlled by electronic effects. If the electronic factor is held constant in a series of species, the steric restraints control the donor site. It is of interest that both N- and S-bonded ligands are present in the compound[169]

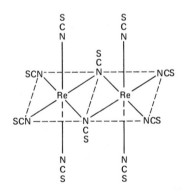

Indeed, in the molecular structures of compounds of the type [Pd(diphos) $(NCS)_2$], diphos = $(C_6H_5)_2P(CH_2)_nP(C_6H_5)_2$, both NCS groups are S-bonded if $n = 1$, one is S-bonded and the other N-bonded if $n = 2$, and both are N-bonded if $n = 3$.[169a] However, the molecular structure of the ion $[Re_2(NCS)_{10}]^{3-}$ contains only N-bonded groups arranged as[169b]

with two NCS groups serving as bridges and bringing the two Re atoms sufficiently close to each other [2.613(1) Å] to give a metal–metal bond (Sec. 8.2-3). Ambidentate character in the seleno- and oxo-analog complexes is much less well established.

Simple metal cyanides and most cyano complexes exemplify M—CN bonding. However, the CN group does function as a bridge in a variety of polymeric mixed-metal complexes such as K[Fe(—CN—)$_6$Mn]. Based upon ligand-field energies (Sec. 8.1-3),[170] site-preference energies calculated for compounds of this general type suggested that the arrangement K[Fe^{II}(—NC—)$_6Cr^{III}$] is less stable thermodynamically than $K[Fe^{II}(-CN-)_6Cr^{III}]$ and allowed syntheses of the two isomers according to the reaction sequence

$$[Cr(CN)_6]^{3-} \xrightarrow{K^+, Fe^{2+}} K[Fe^{II}(-NC-)_6Cr^{III}](s) \xrightarrow{100°C} K[Fe^{II}(-CN-)_6Cr^{III}](s)$$
$$\qquad\qquad\qquad\qquad\text{brick red}\qquad\qquad\qquad\qquad\quad\text{green}$$

Slow reaction of $[Cr(H_2O)_6]^{3+}$ with $[Co(CN)_6]^{3-}$ yielded the species $[H_2O)_5Cr(-NC-)Co(CN)_5]$, but rapid reaction of $[Co(CN)_5(N_3)]^{3-}$ with HNO_2 in the presence of $[Cr(H_2O)_5(CN)]^{2+}$ gave the linkage isomer $[(H_2O)_5Cr(-CN-)Co(CN)_5]$.[165] Separation and purification of each species were effected.

4-6. Coordination Position Isomerism

In multinuclear complex species, coordinating groups may be present in the same numbers but in different arrangements relative to the metal ions present. The result is coordination position isomerism. The following examples are typical:

$$\left[(H_3N)_4Co\underset{O_2}{\overset{\overset{H_2}{N}}{<}}Co(NH_3)_2Cl_2\right]Cl_2 \quad \text{and} \quad \left[Cl(H_3N)_3Co\underset{O_2}{\overset{\overset{H_2}{N}}{<}}Co(NH_3)_3Cl\right]Cl_2$$

$$\left[(H_3N)_4Co\underset{\underset{H}{O}}{\overset{\overset{H}{O}}{<}}Co(NH_3)_2Cl_2\right]SO_4 \quad \text{and} \quad \left[Cl(H_3N)_3Co\underset{\underset{H}{O}}{\overset{\overset{H}{O}}{<}}Co(NH_3)_3Cl\right]SO_4$$

$$\left[(R_3P)_2Pt\underset{Cl}{\overset{Cl}{<}}PtCl_2\right] \quad \text{and} \quad \left[Cl(R_3P)Pt\underset{Cl}{\overset{Cl}{<}}Pt(PR_3)Cl\right]$$

An analogy to the isomerism involving ethylene dichloride, $ClCH_2CH_2Cl$, and ethylidene chloride, CH_3CHCl_2, is apparent.

4-7. Valence Isomerism

A. Werner applied this terminology to substances of the same composition in which the same group is held in one instance by principal valency and in the other instance by secondary valency. Typical examples include

$$\left[(en)_2Co\underset{\underset{H_2}{N}}{\overset{\overset{O_2}{}}{<}}Co(en)_2\right]X_4 \quad \text{and} \quad \left[(en)_2Co\underset{\underset{H.HX}{N}}{\overset{\overset{O_2}{}}{<}}Co(en)_2\right]X_3$$

Werner also believed the pink and black compounds of empirical formulation $[Co(NH_3)_5(NO)]X_2$ to be examples of valence isomerism. However, although the black chloride is correctly formulated as $[Co(NH_3)_5(NO)]Cl_2$,[171a] the red nitrate appears to be a binuclear substance with an N_2O_4 bridge, $[(H_3N)_5Co(N_2O_4)Co(NH_3)_5](NO_3)_4$, and contains no NO donor.[171b]

4-8. Conformation Isomerism

In a limited number of cases, a different geometrical arrangement of the same donor atoms around a metal ion may occur. Examples are

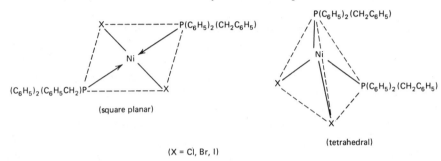

(square planar)

(tetrahedral)

(X = Cl, Br, I)

and a variety of related diphenylalkylphosphine compounds.[172, 173] Crystals of paramagnetic $[NiBr_2\{P(C_6H_5)_2(CH_2C_6H_5)\}_2]$ contain both the square-planar and the tetrahedral isomers in the same unit cell.[174]

4-9. Miscellaneous Types of Isomerism

Isomerism within a coordinating group leads to isomeric complexes: for example, the ions $[Co(en)_2(CH_3\underset{\underset{NH_2}{|}}{C}HCH_3)Cl]^+$ and $[Co(en)_2(CH_3CH_2CH_2NH_2)Cl]^+$ and the compounds $[Pt(H_2NC_6H_4COO)_2Cl_2]$, where the aminobenzoate ion may be *ortho*, *meta*, or *para*. Somewhat different are the compounds $\{(H_2N)_2CS\}_2$-$[Hg(SCN)_2]$ and $(NH_4)_2[Hg(SCN)_4]$, which reflect the isomerism between thiourea and ammonium thiocyanate.

Apparently feasible formulations such as

$$[Co(NH_3)_5(ClO_3)]Br_2 \quad \text{and} \quad [Co(NH_3)_5(BrO_3)]ClBr$$

$$\overset{IV}{[Pt(NH_3)_2(SO_3)_2]} \quad \text{and} \quad \overset{II}{[Pt(NH_3)_2(S_2O_6)]}$$

$$\overset{IV \; II}{Ce[Fe(CN)_6]} \quad \text{and} \quad \overset{III \; III}{Ce[Fe(CN)_6]}$$

are ruled out because the relative oxidizing and reducing strengths of appropriate components allow the existence of either no compounds or only one compound.

4-10. Stereoisomerism

Stereoisomerism is a consequence of differences in the spatial arrangements of groups about a central atom. It will be recalled that the existence of stereo-isomers of complex species of appropriate compositions is a fundamental corollary of Werner's original theory. Indeed, the syntheses of predicted stereoisomers, proof of their molecular structures, and the evaluations of their properties provide most convincing arguments for the acceptance of Werner's fundamental views. Of all the types of isomerism found among inorganic species, stereoisomerism is by a wide margin the most interesting and important.*

Among organic compounds, the tetrahedral arrangement of bonds around the carbon atom and the modifications resulting from multiple bonding impose an arbitrary limitation to stereoisomerism. Among inorganic species, however, the greater variety of possible spatial distributions of bonds (Ch. 6, pp. 274–290) complicates the subject. However, although more varied examples of ste-reoisomers can be found among the inorganic species, they are still classifiable as either *geometrical isomers* or *optical isomers*.[133, 175-180]

Geometrical and optical isomerism are most readily identified by con-sideration of the molecular structures of species in terms of *coordination polyhedra*, as described in Ch. 6. Geometrical isomerism is noted when two unidentate groups occupy either adjacent (*cis*) or opposite (*trans*) positions in the coordination polyhedron. In an octahedral geometry, this situation is exemplified by the violeo(*cis*) and praseo(*trans*) isomers of $CoCl_3 \cdot 4\,NH_3$,† as noted schematically on p. 364 and shown three-dimensionally for the ions $[Co(NH_3)_4Cl_2]^+$ as

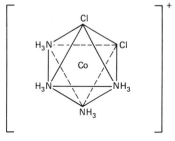   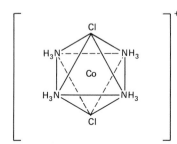

cis (violeo)   *trans* (praseo)

---

* The complexities in nomenclature induced by the great variety of stereochemical geometries, isomerism within these geometries, ligand donor sites, and so on, have prompted several detailed proposals for systematic extension of the IUPAC rules (Sec. 7.2-2). The scope of this discussion does not allow detailed presentation. Useful references that can be consulted with profit are: M. F. Brown, B. R. Cook, and T. E. Sloan, *Inorg. Chem.*, **14**, 1273 (1975) [mononuclear complexes, C.N. = 6]; M. F. Brown, B. R. Cook, and T. E. Sloan, *Inorg. Chem.*, **17**, 1563 (1978) [mononuclear complexes, C.N. = 7, 8, 9]; T. E. Sloan and D. H. Busch, *Inorg. Chem.*, **17**, 2043 (1978); and G. J. Leigh, *Inorg. Chem.*, **17**, 2047 (1978).

† As indicated in Fig. 6-17, the point-group symmetries of *cis*-$[Co(NH_3)_4Cl_2]^+$ and *trans*-$[Co(NH_3)_4Cl_2]^+$ ions are, respectively, C_{2v} and D_{4h}.

In square-planar geometry, *cis-trans* isomerism is illustrated schematically for the α and β forms of $PtCl_2 \cdot 2\,NH_3$ (Fig. 6-19) as

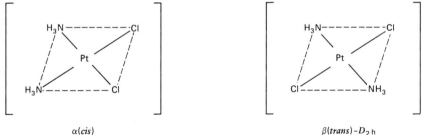

$\alpha(cis)$ $\beta(trans) - D_{2h}$

Optical isomerism characterizes a coordination species in which the arrangement of donor groups about the central metal ion is such that no plane or axis of symmetry (Sec. 6.4-1) can exist. This condition may result from either dissymmetry of the coordination entity as a whole or dissymmetry within the ligand itself. As is true in organic chemistry, optical isomers of a given species are distinguished from each other by opposite rotations of plane-polarized light.

Geometrical Arrangement and Coordination Number.* The variety of possible geometrical arrangements of donor atoms about a central atom or ion increases as the coordination number of the latter increases. It is convenient to explore these arrangements before discussing geometrical and optical isomerism in more detail. The geometrical arrangements are, of course, those discussed in Chapter 6 and summarized in Table 6-1.[179]

1. Coordination number 2. With species of this type (e.g., $[Ag(CN)_2]^-$, $[Ag(NH_3)_2]^+$, $[AuCl_2]^-$, $[Hg(CN)_2]$, $[UO_2]^{2+}$), the existence of stereoisomers is precluded by the ubiquitous linear geometry characteristic of this coordination number.

2. Coordination number 3. Well-established examples of this coordination number in isolable species are limited in number. Trigonal planar geometry, which precludes the existence of stereoisomers, characterizes species such as $[(CH_3)_2BeN(CH_3)_2]$,[181] $[Ag(PRR'_2)_2I]$,[182] $[HgI_3]^-$,[183] $[Cu\{(CH_3)_3PS\}Cl]_3$,[184] $[Pt\{P(C_6H_5)_3\}]$,[185] $[Cu\{SP(CH_3)_3\}_3]^+$,[186] and $[HgQ(CN)_2]$ (Q = quinoline, substituted quinolines, isoquinolines).[187] Stoichiometry is not in itself necessarily an indication of coordination number 3. Thus in crystals of K_2CuCl_3 and Cs_2AgCl_3, infinite, chlorine-bridged chains result in tetrahedral 4-coordination,[188] whereas in crystals of $KCu(CN)_2$ the chain $\cdots CN{-}Cu(CN){-}CN\cdots$ does exemplify 3-coordination.[189] A general summary[190] can be consulted with profit.

3. Coordination number 4. This second most common coordination number is exemplified in a large number of complex species.[176, 191-194] Two

---

* A useful general reference is M. Hargittai and J. Hargittai, *The Molecular Geometries of Coordination Compounds in the Vapour Phase*, Elsevier, Amsterdam (1977).

Table 7-11 Molecular Geometries of 4-Coordinate d-Transition Metal Complexes

Periodic Family	d^0	d^1	d^2	d^3	d^4	d^5	d^6	d^7	d^8	d^9	d^{10}
Square planar											
VIB				Cr(III)	Cr(II)						
VIIB						Mn(II)			Mn(-I)	Mn(-II)	
VIII							Fe(II)	Co(II)	Ni(II)[a]	Ni(I)	
									Pd(II)[a]		
									Pt(II)[a]		
								Rh(II)	Rh(I)[a]		
								Ir(II)	Ir(I)[a]		
IB									Cu(III)	Cu(II)[a]	
									Ag(III)	Ag(II)	
									Au(III)[a]		
Tetrahedral											
IVB	Ti(IV)[a]										
	Zr(IV)[a]										
VB	V(V)	V(IV)	V(III)								
VIB	Cr(VI)	Cr(V)	Cr(IV)	Cr(III)(?)							
	Mo(VI)										
	W(VI)										
VIIB	Mn(VII)	Mn(VI)	Mn(V)								
	Tc(VII)[a]										
	Re(VII)[a]										

406

Group										
VIII	Ru(VIII) Os(VIII)	Ru(VII) Os(VII)	Fe(VI) Ru(VI)	Fe(V)	Fe(III)	Fe(II)	Co(II)[a]	Co(I)	Co(0)	Fe(-II) Ru(-II) Os(-II)
						Co(III)		N(II)[a]	Ni(I)	Co(-I) Rh(-I) Ir(-I)
										Ni(0) Pd(0) Pt(0)
IB									Cu(II)	Cu(I)[a] Ag(I)[a] Au(I)

[a] Principal examples.

geometrical arrangements, the square-planar (dsp^2 hybrid), and the tetrahedral (sp^3 hybrid), characterize these complexes, but established examples of other geometries are not known. The distribution of the square-planar and tetrahedral geometries among d-transition elements in various oxidation states is indicated in Table 7-11.[194] Square-planar geometry is most common for d^n configurations where $6 < n \leq 9$, and it is related to the common octahedral geometry of coordination number 6 by removal of two ligands in *trans* positions. Tetragonal distortion of an octahedral arrangement can give a geometry between square planar and octahedral. 4-Coordinate complexes derived from p-block cations invariably exhibit tetrahedral geometry. The theoretical principles relating to these geometries and bonding within them are outlined in Chapter 8.

Although decision as to which type of geometrical arrangement of ligands better characterizes a particular species has been largely based upon x-ray diffraction, spectroscopic, or other numerical data (Sec. 6.5), a comparison of the number of isolable isomers with the number possible is often a useful approach to this type of problem. Regardless of whether the geometry is square-planar or tetrahedral, complexes of the compositions $[ML_4]$, $[ML_3L']$, or $[MLL'_3]$ (L and L' = different unidentate ligands) can have no stereoisomers, for every possible spatial distribution of the ligands for any of these compositions is the same for either geometry. However, as noted earlier in this section, for a substance of composition $[ML_2L'_2]$, *cis*- and *trans*-isomers are predicted for a square-planar geometry but no isomers for a tetrahedral geometry. Correspondingly, for a substance of composition $[MLL'L''L''']$, the square-planar geometry predicts three geometric isomers,

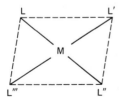 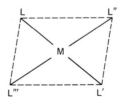 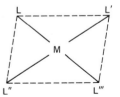

whereas the tetrahedral geometry predicts the two mirror-image, nonsuperimposable, optical isomers,

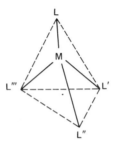 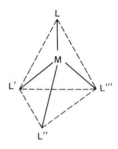

Determination of spatial arrangement via syntheses of predicted isomers is complicated by bond lability and the tendency toward rearrangement. Since this

behavior is not characteristic of platinum(II) compounds, they are useful in illustrating this approach.[191] For example, it was observed many years ago that whereas heating tetraammineplatinum(II) chloride, $[Pt(NH_3)_4]Cl_2$, yielded $[Pt(NH_3)_2Cl_2]$ (Reiset's second chloride,[195] or the *beta** compound), treating potassium tetrachloroplatinate(II), $K_2[PtCl_4]$, with aqueous ammonia gave a compound of the same composition but different chemical and physical properties (Peyrone's salt,[196] or the *alpha** compound). It appears, therefore, that the configurations around platinum(II) are square planar. Additional evidences were given by the syntheses of two isomers each for compositions $[Pt(NH_3)_2(C_5H_5N)_2]Cl_2$ and $K_2[PtCl_2(NH_2SO_3)_2]$ and three isomers each for compositions $[Pt(C_2H_4)(NH_3)(Cl)(Br)]$,[197] $[Pt(NH_3)(C_5H_5N)(Cl)(Br)]$,[198] and $[Pt(NH_3)(C_5H_5N)(NH_2OH)(NO_2)]^+$.[199]

Even more convincing proof that 4-coordinate platinum(II) has a square-planar geometry was given by the resolution into optical isomers of the *iso*-butylenediamine-*meso*-stilbenediamineplatinum(II) ion.[200] As suggested in Fig. 7-5, the square-planar arrangement, with no plane of symmetry, allows for the existence of mirror-image isomers, whereas the tetrahedral arrangement, with a plane of symmetry, does not. First x-ray diffraction proof of square-planar geometry involved the $[PtCl_4]^{2-}$ ion.[201]

4. Coordination number 5. Until comparatively recently, truly 5-coordinate complexes were believed to be rare.[202] Many of the apparent examples of this coordination number described in the early literature have been shown,

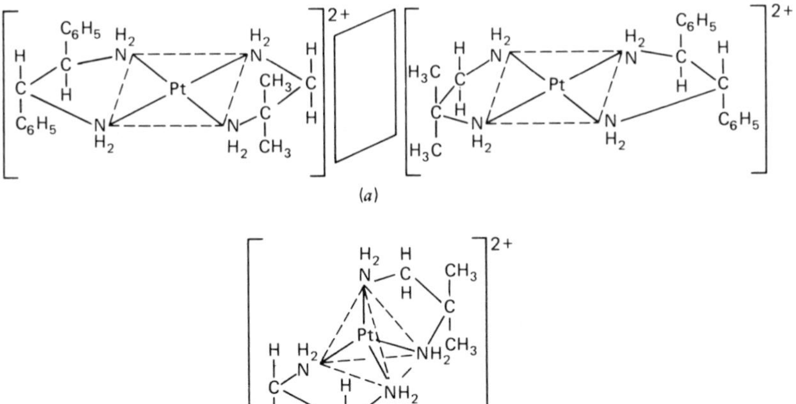

(a)

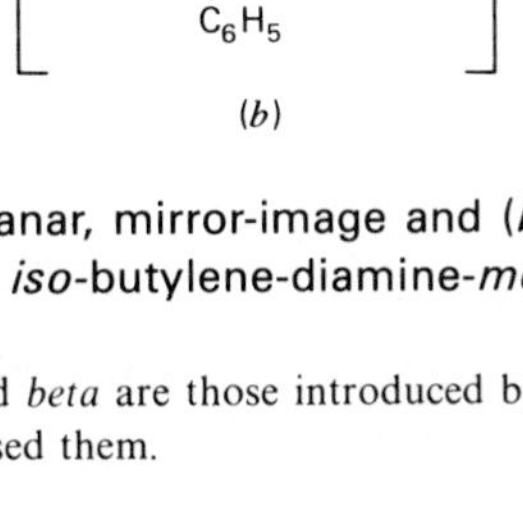

(b)

Figure 7-5 (a) Square-planar, mirror-image and (b) tetrahedral, nondissymmetric molecular structures for *iso*-butylene-diamine-*meso*-stilbenediamineplatinum(II) ion.

* The designations *alpha* and *beta* are those introduced by A. Werner [*Z. Anorg. Chem.*, **3**, 267 (1893)]. Later, Drew reversed them.

largely by x-ray diffraction methods, not to be 5-coordinate.[177, 202] For example, crystals of composition $(NH_4)_3ZnCl_5$ contain $[ZnCl_4]^{2-}$ and Cl^- ions;[203] those of compositions Rb_3CoCl_5 and Cs_3CoCl_5 contain $[CoCl_4]^{2-}$ and Cl^- ions;[204] and those of compositions $(NH_4)_2InCl_5 \cdot H_2O$[205] and $(NH_4)_2FeCl_5 \cdot H_2O$[206] contain, respectively, the octahedral ions $[InCl_5(H_2O)]^{2-}$ and $[FeCl_5(H_2O)]^{2-}$. A more recently described example has the composition $CoCl_2(dien)$ (dien = diethylenetriamine) but is really a salt containing the octahedral, 6-coordinate cation $[Co(dien)_2]^{2+}$ and the tetrahedral, 4-coordinate anion $[CoCl_4]^{2-}$.[207] Similarly, compounds of compositions $Co(NH_3)_6ZnCl_5$ and $Co(NH_3)_6$-$ZnCl_4NO_3$ contain $[ZnCl_4]^{2-}$ and either Cl^- or NO_3^- anions, as indicated by Raman and infrared data.[208] Single-crystal, x-ray diffraction data for the former compound confirm the presence of the ions $[Co(NH_3)_6]^{3+}$, $[ZnCl_4]^{2-}$, and Cl^-.[209]

Five-coordination is well established in terms of many completely authenticated examples.[202, 210-214] The trigonal-bipyramidal and distorted square-pyramidal geometries (Sec. 6.2) appear with about equal frequency. Typical examples are illustrated in Fig. 7-6. Completely regular trigonal-bipyramidal geometry is restricted to species in which all ligands are identical, for example, the $[CuCl_5]^{3-}$ ion in the compound $[Cr(NH_3)_6][CuCl_5]$, wherein the two axial Cu–Cl bonds have the length 2.2964(12)Å and the three equatorial Cu–Cl bonds the length 2.3912(13)Å.[215] However, quadridentate ligands occupying one axial position and three equatorial positions (e.g., in cyanotris(3-dimethyl-arsinopropyl)phosphinenickel(II) ion,[216] $[NiP\{(CH_2)_3As(CH_3)_2\}_3CN)]^+$) effect some distortion, with the metal ion not necessarily in the trigonal plane. The square-pyramidal geometry is usually distorted to the extent that the metal

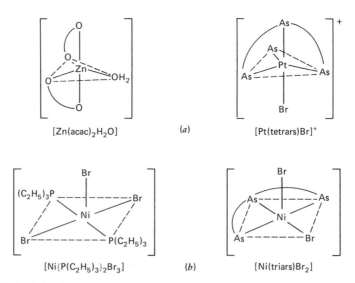

$[Zn(acac)_2H_2O]$ (a) $[Pt(tetrars)Br]^+$

$[Ni\{P(C_2H_5)_3\}_2Br_3]$ (b) $[Ni(triars)Br_2]$

Figure 7-6 Typical examples of (a) trigonal-bipyramidal and (b) tetragonal-pyramidal geometries for C.N. = 5.

ion lies above the plane of the four basal ligands, but this kind of distortion is favored by both hybridization and minimal steric repulsion approaches to molecular geometry. Regularity may exist, however, if the ligands are identical, as in the $[Ni(CN)_5]^{3-}$ ion in crystals of the compound $[Cr(tn)_3][Ni(CN)_5]$ $\cdot 2\,H_2O$ (tn = 1,3-propanediamine).[217]

Although for a particular species, a given geometry is commonly invariant, rearrangements do occur,[218-221] and in some instances (e.g., for the $[Ni(CN)_5]^{3-}$ ion in crystals of $[Cr(en)_3]\,[Ni(CN)_5]\cdot 1.5\,H_2O$),[222] both trigonal-bipyramidal and square-pyramidal geometries are present. That the two geometries differ but little in energy is exemplified by crystal-structure proof of tetragonal-pyramidal geometry for the red isomer and trigonal-bipyramidal geometry for the green isomer of $[Co(dpe)_2Cl]^+$ ion [dpe = 1,2-bis(diphenyl-phosphino)ethane].[223] A further complication is sometimes noted because of conversion of tetragonal-pyramidal geometry to octahedral; for example, the yellow $[Co(CN)_5]^{3-}$ ion in crystals of $[N(C_2H_5)_2(i-C_3H_7)_2]_3[Co(CN)_5]$ weakly aquates to a green species upon dissolution of the solid in water.[224]

Both geometrical and optical isomers are possible for coordination number 5, but only a limited number of examples have been described. Of course, the trigonal-bipyramidal and tetragonal-pyramidal forms of the $[Co(dpe)_2Cl]^+$ ion and of the $[Ni(CN)_5]^{3-}$ ion are geometrical isomers. Geometrical isomerism within tetragonal-pyramidal geometry is illustrated in Fig. 7-7, where the donor P and S atoms of the orange and violet forms of $[\{(C_6H_5)_3P\}_2\{(CF_3)_2C_2S_2\}-Ru(CO)]$ are differently arranged in the basal plane.[225]

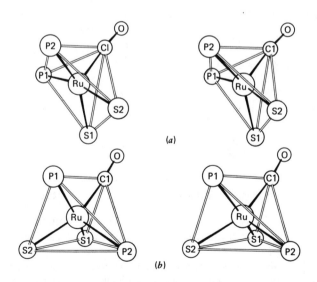

Figure 7-7 Stereoscopic projections of geometrical isomers of $[\{(C_6H_5)_3P\}_2-\{(CF_3)_2C_2S_2\}Ru(CO)]$. (a) Orange isomer; (b) violet isomer. [I. Bernal, A. Clearfield, and J. S. Ricci, Jr., J. Cryst. Mol. Struct., 4, 43 (174).]

5. Coordination number 6. As suggested earlier, this coordination number is the commonest and, therefore, the most widely investigated of all. In principal, six donor atoms can be arranged in equivalent positions about a central atom or ion at the corners of a hexagonal plane (D_{6h} symmetry), a trigonal prism (D_{3h}), or an octahedron (O).* These geometrical arrays, together with numbers designating positions for substitution, may be indicated as

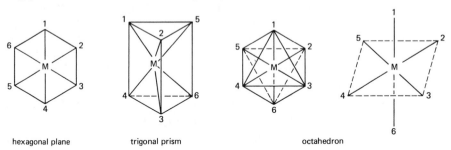

hexagonal plane   trigonal prism   octahedron

Again, decision as to the correct geometry for a particular composition can be based best upon structure-indicating data (Ch. 6). Classically, and in particular as related to Werner's conclusions, it was reached by comparison between the number of isolable isomers and the numbers predicted for each possible geometry. Predictions for species containing unidentate ligands and having compositions $[ML_5L']$ or $[ML_5'L]$, $[ML_4L_2']$ or $[ML_4'L_2]$, and $[ML_3L_3']$ are summarized in Table 7-12. It was because for species of either of the last two types no more than the two geometrical isomers predicted for octahedral geometry could be obtained that this geometry was early assigned to coordination number 6. Furthermore, for chelated species of the type $[M(LL)_3]$ (e.g.,

$$[Co(C_2O_4)_3]^{3-}, \quad [Cr(acac)_3], \quad [Rh(en)_3]^{3+}, \quad \left[Co\left\{ {HO \atop HO} \!\!> Co(NH_3)_4 \right\}_3 \right]^{6+} \text{) the}$$

predictions indicated are

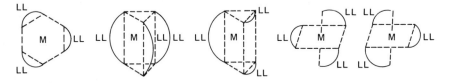

hexagonal plane   trigonal prism   octahedron
(1 arrangement)   (2 geometrical isomers)   (2 optical isomers)

with, until comparatively recently, only the octahedral geometry being in accord with experimental observation. Similar considerations based upon other com-

---

* These and other stereochemical arrangements are best visualized by the use of three-dimensional models. Thus in octahedral geometry, all adjacent positions are equivalent, with (1,3), (1,4), (1,5), (2,3), (3,4), (2,6), and so on, arrangements being the same. Similarly, (2,4) and (3,5) are identical with (1,6). In practice, the first substituent is usually considered to occupy the 1 position, and others are referred to it as simply as possible.

Table 7-12 Theoretically Possible Geometrical Isomers for Coordination Number 6

	Number of Isomers		
	Hexagonal Plane	Trigonal Prism	Octahedron
$[ML_5L']$ or $[ML_5'L]$	One	One	One
$[ML_4L_2']$ or $[ML_4'L_2]$	Three (1,2; 1,3; 1,4)	Three (1,2; 1,3; 1,4)	Two (1,2 or *cis*; 1,6 or *trans*)
$[ML_3L_3']$	Three (1,2,3; 1,2,4; 1,3,5)	Three (1,2,3; 1,2,5; 1,2,6)	Two (1,2,3 or *fac*; 1,2,6 or *mer*)

positions suggest that octahedral geometry is correct for coordination number = 6.* It is significant that the chemical data which led Werner to this conclusion have been supported by both crystal structure, spectroscopic, and magnetic data and theoretical interpretations.

Although exact octahedral symmetry (O_h) is found for many 6-coordinate complexes, particularly where all ligands are identical and unidentate, minor distortions result when small differences in kind and electronegativity exist among the ligands. Two major distortions, tetragonal and trigonal, are more significant. The natures of these distortions and their origins are indicated in Fig. 7-8.

Tetragonal distortion amounts to either compression or elongation along one of the fourfold rotational axes of the octahedron (see also the Jahn–Teller effect, Sec. 8.1-2). Elongation, by increasing the bond distances of two ligands in *trans* positions results in an approach to square-planar geometry involving the metal ion and the other four ligands. This effect is not uncommon in copper(II) complexes (e.g., α-$[Cu(NH_3)_2Br_2]$(s) with $2\,NH_3$ at 1.93 Å, 2 Br at 2.54 Å, and 2 Br at 3.08 Å or $[Cu(NH_3)_6]^{2+}$ with $4\,NH_3$ at 2.07 Å and $2\,NH_3$ at 2.62 Å).

Trigonal distortion amounts to elongation or compression along one of the four threefold rotational axes of the octahedron that pass through centers of faces. The end result is trigonal prismatic coordination.† Although this type of coordination geometry was long recognized in certain solid sulfides (e.g., MoS_2, WS_2), it was first established for a coordination entity, tris(*cis*-1,2-diphenylethane-1,2-dithiolato)rhenium, $[Re\{S_2C_2(C_6H_5)_2\}_3]$, only in 1965.[234] The mo-

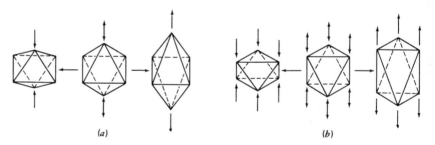

(a)　　　　　　　　　　　　　　(b)

Figure 7-8 Tetragonal (a) and trigonal (b) distortions of octahedral geometry. [J. E. Huheey, *Inorganic Chemistry*, 2nd ed., Fig. 10.19, p. 444, Harper & Row, New York (1978), modified.]

* Tables of possible geometrical isomers for a variety of compositions, uni- and multidentate ligands, and several coordination numbers date to J. D. Main Smith.[226] Approaches using models,[227] drawings,[228] and Polya's theorem[229,230] are useful, but the procedure of J. C. Bailar[231] and S. A. Mayper[232] for coordination number 6 has the advantages of simplicity and adaptability to computer use[233] in establishing both numbers of isomers and their molecular structures. Another variant applies to unidentate ligands.[233a]

† Summation of ligand–ligand repulsion energies for complexes of the type $[M(LL)_3]^{n+}$ suggests that a continuous change from regular octahedral geometry to trigonal prismatic geometry parallels progressively decreasing "bite" by the bidentate ligand.[234a]

lecular structure of this complex, as depicted in perspective for the donor and acceptor atoms in Fig. 7-9, amounts to essentially a perfect ReS_6 trigonal prism. Since then, an entire series of related trigonal prismatic compounds of composition $[M(S_2C_2R_2)_3]$ (M = V, Zr, Mo, W) has been described, and numerous other complexes of related geometry and derived from 1,2-dithiols, 1,2-dithiolenes, and other dithio ligands have been prepared.[235-239] The propensity for dithio ligands to give products of this geometry is striking, although restriction in rotation about the double bond in species such as

$$\left[\begin{array}{c} R \diagdown_{C}\diagup^{S} \\ \| \\ C \\ S\diagup \diagdown S \end{array} \right]^{2-}$$

aids materially in positioning the donor S atoms so that chelation can result along rectangular edges of the trigonal prism. Distorted trigonal prismatic geometry can result, also [e.g., in crystals of the tris(thiobenzoyldiazene)molybdenum(I) complex].[240] Mixed sulfur- and nitrogen-donor complexes have also been described, and the crystal and molecular structures of the very interesting pure oxygen-donor complex $[Co^{II}\{Co^{III}(OCH_2CH_2NH_2)_3\}_2]^{2+}$ indicate the cobalt(II) ion to be at the center of an almost perfect trigonal prism with oxygen atoms at the six apices and acting as bridges to the two cobalt(III) ions, as shown diagrammatically in Fig. 7-10.[241]

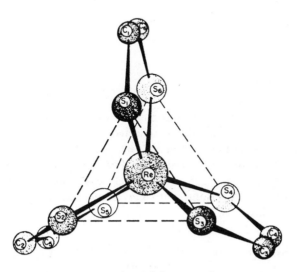

Figure 7-9 Perspective representation of donor and acceptor atoms in molecular structure of $[Re\{S_2C_2(C_6H_5)_2\}_3]$. [R. Eisenberg and J. A. Ibers, *Inorg. Chem.*, **5**, 411 (1966). Reprinted with permission from the American Chemical Society.]

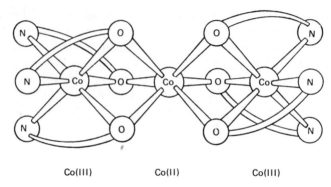

Co(III)   Co(II)   Co(III)

Figure 7-10 Representation of trigonal prismatic geometry of shared oxygen atoms around Co(II) ion in the species $[Co^{II} Co^{III}\{(OCH_2CH_2NH_2)_3\}_2]$. [J. A. Bertrand, J. A. Kelley, and E. G. Vassian, *J. Am. Chem. Soc.*, **91**, 2394 (1969). Reprinted with permission from the American Chemical Society.]

Although examples of geometrical isomerism in true trigonal prismatic geometry are limited, those in octahedral geometry are legion. The *cis* (violeo) and *trans* (praseo) isomers of the $[Co(NH_3)_4Cl_2]^+$ ion have been discussed. In the absence of any experimental evidence to the contrary, a bidentate group is assumed always to span *cis* positions. Thus the observed green and purple compounds of composition $[Co(en)_2Cl_2]Cl$ are assigned the arrangements

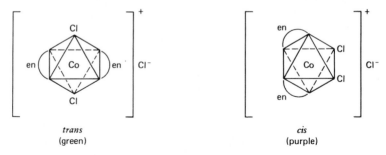

trans   *cis*
(green)   (purple)

For a given composition, the *cis* isomer is invariably less symmetrical than the *trans* isomer. Experimentally, this is reflected in the common observation that the *cis* isomer has a dipole moment, whereas the *trans* isomer does not. Furthermore, the *cis* isomer is in almost all cases in a higher-energy state than the *trans* isomer and may convert relatively easily in solution to the latter, whereas to convert a *trans* isomer into a *cis* isomer requires the input of energy. The conversion illustrated as

$$trans\text{-}[Co(en)_2Cl_2]Cl(s) \underset{\substack{aq.\ HCl,\\ room\\ temperature}}{\overset{\substack{heat\\ (NH_3)}}{\rightleftharpoons}} cis\text{-}[Co(en)_2Cl_2]Cl(s)$$

is a classic case in point.[242] If a ligand is sufficiently large, it may, because of excessive crowding, be impossible to prepare the *cis* isomer. Thus although both *cis*- and *trans*-$[Co(NH_3)_4Cl_2]Cl$ can be synthesized, only *trans*-$[Co(NH_3)_4Br_2]Br$ has been obtained. Replacement of a bidentate ligand by two unidentate ligands is a common procedure for the synthesis of a *cis* isomer, and substitution reactions involving either *cis* or *trans* isomers are sometimes used to assure the same configuration in a product, although isomerization can be a complication (Sec. 11.2-3).

Within these limitations, syntheses of *cis* and *trans* isomers are based upon conversion reactions involving species of known configurations. As illustrative of the general method, reactions resulting in the syntheses of various bis(ethylenediamine)cobalt(III) compounds are summarized in Table 7-13.[243] Similar reaction schemes characterize other series, for example, *cis*- and *trans*-tetraamminecobalt(III)[244] and diamminedichloroethylenediaminecobalt(III)[245] species. The last of these is of particular interest because rearrangement to the

Table 7-13   Geometrical Isomerism among Bis(Ethylenediamine)Cobalt(III) Species

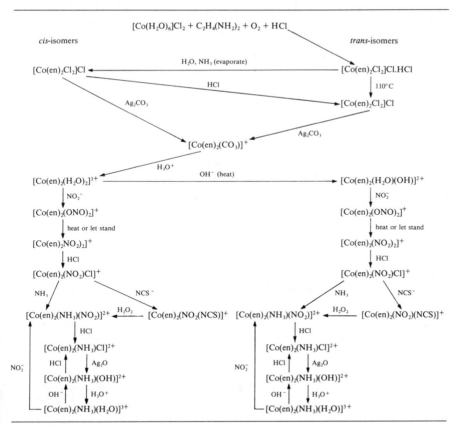

trans isomer is minimized by use of ethanolic hydrogen chloride solution. These reactions are summarized in Table 7-14.

Another type of geometrical isomerism is exemplified by the meridial (*mer*) and facial (*fac*) isomers of triamminetrinitrocobalt(III),[246] as indicated by the formulations

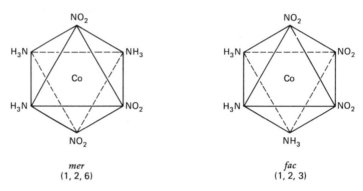

$$\begin{array}{cc} mer & fac \\ (1, 2, 6) & (1, 2, 3) \end{array}$$

The *fac* isomer shows a single peak (equivalent NH_3 groups) in a proton nuclear magnetic resonance spectrum; the *mer* isomer two peaks in the area ratio 2 : 1 (two NH_3 groups *trans* to one another, one NH_3 group *trans* to an NO_2 group). The infrared spectrum of the more symmetrical *fac* isomer is slightly less complex than that of the *mer* isomer, but distinction between the two on this basis is less reliable than for *cis-trans* isomers. Other examples of *mer-fac* isomerism include $[Ru(H_2O)_3Cl_3]$,[247,248] $[Pt(NH_3)_3Br_3]^+$ ion,[249] $[Pt(NH_3)_3I_3]^+$ ion,[250] $[Ir(H_2O)_3Cl_3]$,[251] and $[Rh(CH_3CN)_3Cl_3]$.[252]

6. Coordination number 7. Comparatively few clearly defined examples of 7-coordination have been reported.[177,253-257] The original belief, based solely upon known examples, that this coordination number is restricted to heavy-metal ions has been refuted by the isolation of a number of compounds derived from $3d$ metal ions.[253,254] Three idealized geometrical arrangements are usually considered: the pentagonal bipyramid (D_{5h} point-group symmetry), the rectangular faced monocapped trigonal prism (C_{2v}), and the triangular face monocapped octahedron (C_{3v}). A fourth geometry, the tetragonal base-trigonal base prism (C_s), is uncommon.[257a] The geometries are depicted in Fig. 7-11, and specific examples are summarized in Table 7-15.

It is apparent from Fig. 7-11 that these idealized geometries differ but little from each other and can be interconverted by only slight bending. Except in the crystalline state, the half-life for a given geometry can be relatively short,[258] since the energy difference between any two geometries is small with respect to the intramolecular forces.[253] Exact decision as to which geometry characterizes a particular complex requires a crystal-structure determination. Preferences among the pentagonal-bipyramidal, capped trigonal-prismatic, and capped octahedral geometries have been worked out in terms of a molecular-orbital analysis.[272] Although certain 7-coordinate complexes can be isolated from solution, definite characterization of species in solution is lacking. Because the

Table 7-14   Synthesis of Geometrical Isomers of [Co(en)(NH₃)₂Cl₂]*

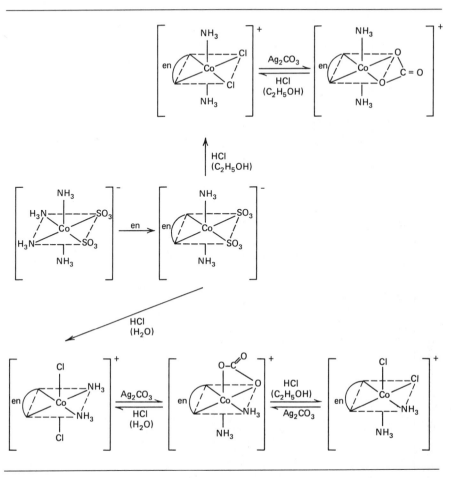

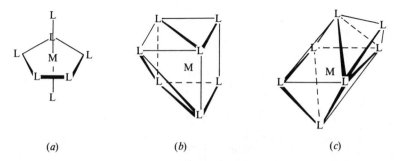

Figure 7-11   Polyhedra for C.N. = 7. (*a*) Pentagonal bipyramid D_{5h}; (*b*) capped trigonal prism C_{2v}; (*c*) capped octahedron C_{3v}. [R. K. Boggess, *J. Chem. Educ.*, **55**, 156 (1978).]

Table 7-15 Geometries in 7-Coordinate Complexes

Pentagonal Bipyramidal	Mono-Capped Trigonal Prismatic	Mono-Capped Octahedral
$[UO_2F_5]^{3-}$ [259]	$[NbF_7]^{2-}$ [264]	$[NbOF_6]^{3-}$ [266]
$[ZrF_7]^{3-}$ [260]	$[Mo(t\text{-}C_4H_9CN)_7]^{2+}$ [265]	$[W(CO)_4Br_3]^{-}$ [267]
$[V(CN)_7]^{4-}$ [261]		$[Mo\{P(CH_3)_2(C_6H_5)\}_3Cl_4]$ [268]
$[SnT_3X]^{262\,a}$		$[Mo(CO)_2\{P(CH_3)_2(C_6H_5)\}_3Cl_2]$ [269]
$[Cr(O_2)_2(CN)_3]^{3-}$ [263]		
$[Pb(\pi\text{-}C_6H_5)(AlCl_4)_3]\cdot C_6H_6$ [270]		
$[Fe(H_2O)(edta)]^{-}$ [271]		
$[M(DAPBH)(H_2O)(NO_3)]^{+}$ (M = Co, Ni) [271 b]		

[a] T = tropolonate; X = Cl, OH.

[b] DAPBH = 2,6-diacetylpyridinebis(benzoic acid hydrazone).

energy changes associated with reorganizations from 7-coordination to 8-coordination are very small, solvation of seemingly 7-coordinate species and change in geometry occur readily.

Certain macrocyclic (Sec. 7.3-2) and closely related ligands with molecular structures that bring five donor atoms (most commonly N, but also N in combination with O or S) into a planar array react with di- or tri-positive $3d$ ions and appropriate anions to give pentagonal bipyramidal complexes of the types $(M^{II}(L)X_2]$ $(X^- = NCS^-, ClO_4^-, NO_3^-, BF_4^-)$ and $[M^{III}(L)X_2]X$.[273, 274] Typical is the reaction of the ligand (L) 2,13-dimethyl-3,6,9,12,18-pentaazabicyclo[12.3.1]octadec-1(18),2,12,14,16-pentaene,

with Fe^{2+} or Fe^{3+} ion.[254, 255] Here the location of donor atoms has a controlling effect on geometry.

Only a limited number of species [e.g., NbF_7^{2-}, ZrF_7^{3-}, $V(CN)_7^{4-}$] contain nonchelating ligands. Isomerism, although possible in principle, has not been detected. A useful and unified stereochemical notation for mononuclear complexes in this coordination number, as well as coordination numbers 8 and 9, has been developed.[275]

7. Coordination number 8. This coordination number ranks next to six and four in numbers of characterized examples, largely as a consequence of relatively recent developments[253, 255, 275-279] which may well have been stimulated by the fundamental structural studies of J. L. Hoard and his coworkers in 1963.[280-282] Relatively few 8-coordinate complexes of the $3d$-transition metal ions have been described, but more are known for the $4d$ and $5d$ ions and, in particular, for the $4f$ (lanthanide) and $5f$ (actinide) ions. Inasmuch as the crowding of eight (or more) donor groups around a small cation is sterically difficult, ionic size is an important factor. The availability of automated x-ray diffractometers and of computer interpretations of diffraction data now allows the definitive determinations of molecular structures necessary to establish these higher coordination geometries.

Possible coordination polyhedra giving eight donor-atom positions are the cube (O_h point-group symmetry); the square, or Archimedean, antiprism (D_{4d}); the triangular face dodecahedron (D_{2d}); the rectangular face bicapped trigonal prism (C_{2v} or D_{3h}); and the hexagonal bipyramid (D_{6h}). These polyhedra are indicated in Fig. 7-12. The cube, although characteristic of certain ionic crystals

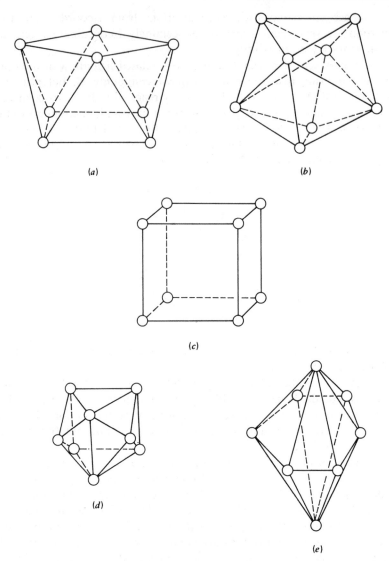

Figure 7-12 Polyhedra for C.N. = 8. (*a*) Square antiprism (D_{4d}); (*b*) dodecahedron (D_{2d}); (*c*) cube (O_h); (*d*) bicapped trigonal prism (C_{2v} or D_{3h}); (*e*) hexagonal bipyramid (D_{6h}).

(e.g., CsCl, NaCl, CaF$_2$ types) is not found in coordination complexes because its inefficiency in packing ligands around a metal ion would result in excessive ligand–ligand repulsions. The hexagonal bipyramid, with nonequivalent axial and equatorial positions, is restricted to uranyl (UO$_2^{2+}$) and related complexes (e.g., [UO$_2$(CO$_3$)$_3$]$^{4-}$,[283] [UO$_2$(NO$_3$)$_3$]$^{-}$,[284] [UO$_2$(C$_2$O$_4$)$_3$]$^{4-}$ [285]), where the linear O—M—O arrangement is axial. Bicapped trigonal prismatic geometry may exist[278] in a limited number of complexes (e.g., $[UF_8]^{4-}$,[286] $[Nd(acac)_3(OH_2)_2]$,[287] $[Ho(isonic)_3(OH_2)_2]$[288]),* but a number of apparent examples represent intermediates between, or distortions of, square antiprismatic or dodecahedral geometry.[278]

By a wide margin, known 8-coordinate geometries are either D_{2d} dodecahedral or D_{4d} square antiprismatic (Fig. 7-13). Each of these represents a distortion of cubic geometry to give equivalent donor sites but with reduced ligand–ligand repulsions (Fig. 7-14). The square antiprism may be regarded as formed from the cube by twisting one face of the latter through 45°, but distortion to produce the dodecahedron is not in a single plane. In projection, the latter can be considered, in terms of the donor sites, as two intersecting trapezoids.

The square-antiprismatic and the dodecahedral geometries may be regarded as limiting arrangements of about equal probability of existence.[289] By only minor rearrangements, each can convert into the other or into the bicapped trigonal-prismatic geometry.[278] Since the bond-angle and bond-distance parameters are not markedly different, decision as to the correct geometry in a given crystal is difficult, and only the most refined measurements can lead to correct geometry.† A few examples are summarized in Table 7-16.

All of the geometrical arrangements predict the existence of stereoisomers. That the isolation and characterization of isomers have not developed extensively may reflect both ready interconversions and the necessity for detailed x-ray crystallographic measurements for absolute identification. An 8-coordinate complex containing two bidentate ligands can, in principle, exist as either a *cis* isomer (both ligands on the same side of the metal atom) or a *trans* isomer (ligands on opposite sides of the metal atom). The relative stabilities of the two are unknown. The compounds $[W\{ON(CH_3)NO\}_2(CH_3)_4]$[299] and $[Ca(NO_3)_2(CH_3OH)_4]$[300] have *cis* arrangements. Examples of *trans* species include $[Zr(bpy)_2(NCS)_4]$ (square antiprism; bpy = 2,2'-bipyridine)[301] and $[Mo(diars)_2Cl_4]^+$ (dodecahedron).[302]

8. Coordination number 9. The packing of nine donors around a central atom or ion requires that the latter be comparatively large and in practice largely restricts this coordination number, except in the hydrido species $[TcH_9]^{2-}$ and $[ReH_9]^{2-}$, to the lanthanide and actinide ions.[253,278,303] Not uncommonly, systems of these types are polymeric. Both tricapped trigonal-prismatic (D_{3h}) and mono-capped square-antiprismatic (C_{4v}) idealized geometries have been found, but transition between the two is relatively easy since the calculated energy barrier for the transition is only about 0.1% of the total energy of either form.[278] These geometries are indicated for a specific species in Fig. 7-15 and noted for several complexes in Table 7-17. Stereoisomerism is possible but uncharacterized. In isostructural ethylenediaminetetraacetate complexes of

---

* Isonic = isonicotinate.

† Cases in point include α-$[Ce(acac)_4]$ and α-$[Th(acac)_4]$, which on close examination have neither square-antiprismatic nor dodecahedral geometry but, rather, bicapped trigonal-prismatic geometry [W. L. Steffen and R. C. Fay, *Inorg. Chem.*, **17**, 779 (1978)].

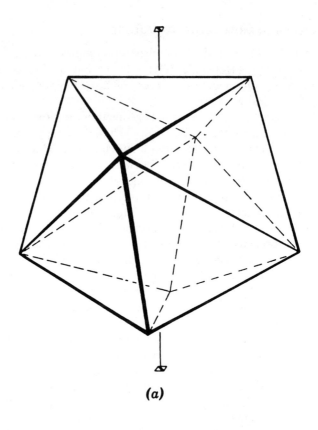

(a)

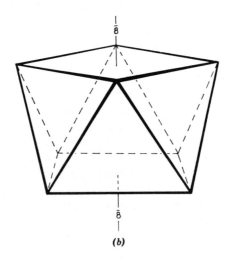

(b)

Figure 7-13 D_{2d} dodecahedron (*a*) and D_{4d} square antiprism (*b*). [J. L. Hoard and J. V. Silverton, *Inorg. Chem.*, **2**, 235 (1963), Figs. 1, 2. Reprinted with permission from the American Chemical Society.]

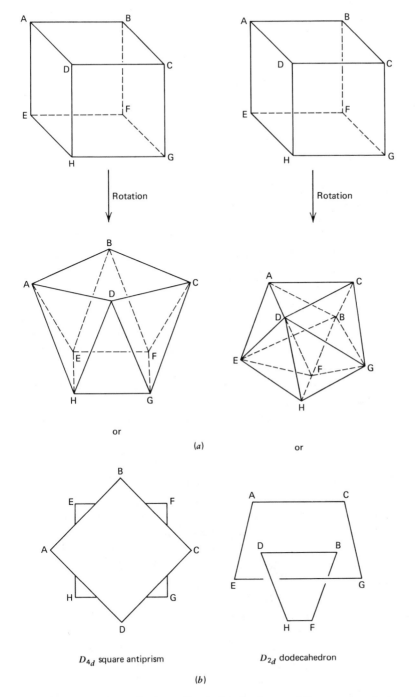

Rotation

Rotation

or

(a)

or

D_{4d} square antiprism

D_{2d} dodecahedron

(b)

Figure 7-14 Distortions of cube to D_{2d} dodecahedron and D_{4d} square antiprism. (a) In terms of geometry; (b) as projection on two-dimensional surface)

425

Table 7-16 Representative Examples of 8-Coordination

Square-Antiprismatic Geometry (D_{4d})		Dodecahedral Geometry (D_{2d})	
Formula[a]	Reference	Formula[a]	Reference
β-[Zr(acac)$_4$]	281	[Fe(napy)$_4$]$^{2+}$ (as ClO$_4^-$)	294
[TaF$_8$]$^{3-}$ (as Na$^+$ salt)	290	[Te{S$_2$CN(C$_2$H$_5$)$_2$}$_4$]	295
[U(NCS)$_8$]$^{4-}$ (as Cs$^+$ salt)	291	[Ti(NO$_3$)$_4$]	296
[W(CN)$_8$]$^{4-}$ (as acid)	292	[TiCl$_4$(diars)$_2$]	297
[La{OC(CHCCH$_3$)$_2$O}$_8$]$^{3+}$ as (ClO$_4^-$)	293	[Zr(C$_2$O$_4$)$_4$]$^{4-}$ (as Na$^+$ salt)	282
[Zr(dbm)$_4$]	298a	[Mo(CN)$_8$]$^{4-}$ (as K$^+$ salt)	298

[a] Abbreviations: acac (acetylacetonate), napy (1,8-naphthyridine), diars (o-phenylenebisdimethylarsine), dpm (dibenzoylmethanate).

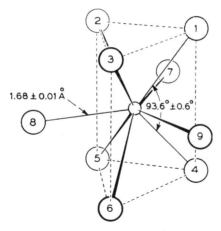

Figure 7-15 Geometry for a typical 9-coordinate species, $[ReH_9]^{2-}$. [S. C. Abrahams, A. P. Ginsberg, and K. Knox, *Inorg. Chem.*, **3** 558 (1964). Reprinted with permission from the American Chemical Society.]

compositions $[Ln(OH_2)_3\{(O_2CCH_2)_2NCH_2CH_2N(CH_2CO_2)_2\}]^-$ (Ln = La, Nd, Gd, Tb, Er), the coordination polyhedron (7 O and 2 N) (Fig. 7-16) is better defined in terms of the distorted D_{2d} dodecahedron with two water molecules in the same coordination site, rather than in terms of the listed 9-coordinate structures.[302]

   9. Coordination number 10. Requiring as it does both a large central atom or ion and a very compact ligand, this coordination number is restricted in complexes to the large lanthanide and actinide cations in association with small unidentate donor atoms or bidentate ligands with small "bites" (e.g., NO_3^-, CO_3^{2-}). Of the possible coordination polyhedra,[303] the bicapped square antiprism (D_{4d}) is the most stable based upon energy considerations.[278,316] The bicapped dodecahedron (D_2) and the expanded dodecahedron (C_{2v}) are closely related. These arrangements are shown in projection in Fig. 7-17. Transformations among these geometries have low-energy barriers. Some examples are summarized in Table 7-18, and the molecular structures of two typical species are shown in Fig. 7-18. The molecular structure of $[La(OH_2)_4\{(O_2CCH_2)_2-NCH_2CH_2N(CH_2CO_2)(CH_2CO_2H)\}]$, wherein there are two nitrogen-donor sites and six potential oxygen-donor sites, is like those of the $[Ln(OH_2)_2-\{(O_2CH_2)_2NCH_2CH_2N(CH_3CO_2)_2]^-$ ions except that three water molecules are close to one dodecahedral site.[315,317] Again, stereoisomerism among 10-coordinate species is expected, although not yet described.

   10. Coordination number 11. This coordination number is very uncommon, perhaps because no good idealized models for 11-atom polyhedra exist and polyhedra for 10- and 12-coordination are energetically more stable. In crystals of $Th(NO_3)_4 \cdot 5 H_2O$, the unit $[Th(NO_3)_4(OH_2)_3]$ can be distinguished in terms of Th—O bonding distances for the four bidentate NO_3^- ions and the three H_2O molecules,[318] which, if the NO_3^- groups are considered as single

Table 7-17 Representative Examples of 9-Coordination

Tricapped Trigonal-Prismatic Geometry (D_{3h})		Mono-Capped Square-Antiprismatic Geometry (C_{4v})	
Formula	Reference	Formula	Reference
$[ReH_9]^{3-}$, $[TeH_9]^{2-}$	304	$[Pr(NCS)_3(OH_2)_6]$	310
$[Nd(OH_2)_9]^{3+}$, as BrO_3^-	305	$[LaCl(OH_2)_7]_2^{4+}$	311
$[Ln(OH_2)_9]^{3+}$, as $C_2H_5SO_4^-$	306, 307	$[ThF_8]^{4-}$ (F bridged)	312
$[Ln(OH)_9]^{6-}$	308	$(LaOCl)_n$	313
Hydrous $Ln(OH)_3$ (Ln = La, Nd, Sm, Gd, Tb, Dy, Ho, Er, Y)	309	$Th(trop)_4(DMF)^a$	314

^{a} trop = tropolone; DMF = N,N-dimethylformamide.

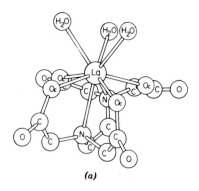

(a)

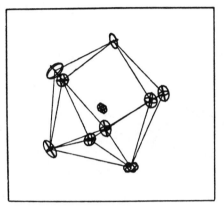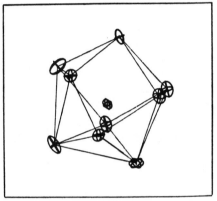

(b)

Figure 7-16 Geometries of (*a*) $[Ln(OH_2)_3(edta)]^-$ and (*b*) $[ThF_8]^{4-}$. [(*a*), J. L. Hoard, B. Lee, and M. D. Lind, *J. Am. Chem. Soc.*, **87**, 1612 (1965). Reprinted with permission from the American Chemical Society; (*b*), R. R. Ryan, R. A. Penneman, and A. Rosenzweig, *Acta Crystallogr.*, **25B**, 1958 (1969).]

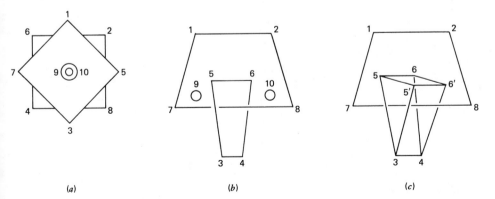

(a)   *(b)*   *(c)*

Figure 7-17 Projections of polyhedra for C.N. = 10. (*a*) Bicapped square antiprism, D_{4d}; (*b*) bicapped dodecahedron, D_2; (*c*) expanded dodecahedron, C_{2v}.

Table 7-18 Representative Examples of 10-Coordination

Bicapped Square-Antiprism (D_{4d})		Bicapped Dodecahedron (D_2)		Expanded Dodecahedron (C_{2v})	
Formula	Reference	Formula	Reference	Formula	Reference
$[Th(C_2O_4)_4]^{4-}$	323	$[La(NO_3)_3(bpy)_2]$	316	$[Ho(NO_3)_5]^{2-}$	324
$[La(OH_2)_2(C_{10}H_8O_6)_2]^-$	322[a]	$[Ce(NO_3)_5]^{2-}$ in $[(C_6H_5)_3(C_2H_5)P]^+$ salt	321	$[La(CO_3)_3O_4]$ in $La_2(CO_3)_3 \cdot 8H_2O^b$	325
		$[La(CO_3)_2O_6]$ in $La_2(CO_3)_3 \cdot 8H_2O^b$	325		

[a] $C_{10}H_8O_6^{2-}$ = benzene-1,2-dioxodiacetate ion. Complex ion in 4-hydrated sodium salt.
[b] Two geometries noted in this crystal.

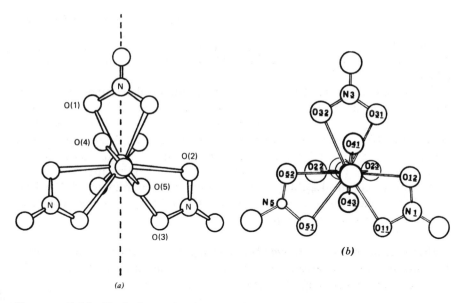

(a)

(b)

Figure 7-18 Typical molecular structures representing C.N. = 10. (a) [Ce(NO$_3$)$_5$]$^{2-}$; (b) [Ho(NO$_3$)$_5$]$^{2-}$. [(a), A. R. Al-Karaghouli and J. S. Wood, *J. Chem. Soc. Chem. Commun.*, **1970**, 135; (b) G. E. Toogood and C. Chieh, *Can. J. Chem.*, **53**, 831 (1975). Reproduced by permission of the National Research Council of Canada.

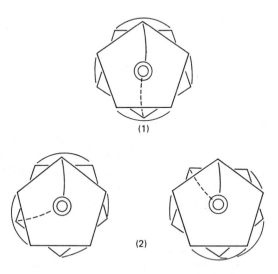

(1)

(2)

Figure 7-19 Projections of chelating arrangements in icosahedra for C.N. = 12. [M. G. B. Drew, *Coord. Chem. Rev.*, **24**, 179 (1977), Fig. 35.]

entities, approximates a mono-capped trigonal-prismatic (C_{2v}) geometry.[319] The dimeric complex molecule $[Th(NO_3)_3(OH_2)_3(OH)]_2$ has a somewhat different geometry.[320]

11. Coordination number 12. All experimentally determined geometries for this coordination number are based upon an icosahedral arrangement of donor atoms $(I_h$ symmetry). Ligands involved are chelating with relatively small "bites" (e.g., NO_3^-, SO_4^{2-}). In projection, these ligands can be arranged in two ways, noted as (1) and (2) in Fig. 7-19. Examples include: structure (1) $[Ce(NO_3)_6]^{2-}$,[326] $[Th(NO_3)_6]^{2-}$,[327] $[Pr_2(NO_3)_9]^{3-}$,[328] $[Pr(napy)_6]^{3+}$,[329] and one arrangement in $La_2(SO_4)_3 \cdot 9H_2O$ (other is 9-coordinate).[330,331] Two typical molecular structures are indicated in Fig. 7-20. Again, stereoisomerism, although not characterized, is possible.

***Optical Isomerism—Molecular Dissymmetry.**†** A molecule or ion so structured that two nonsuperposable, mirror-image forms can exist is said to be dissymmetric (e.g., the cis-$[Co(en)_2Cl_2]^+$ ion) as

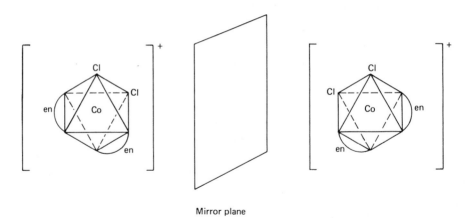

Mirror plane

Because of a comparable relationship between the right and left hands, the term *chiral* (Greek $\chi\epsilon\iota\rho$, hand) is also used. The nonsuperposable structures are termed *optical isomers, enantiomers,* or *enantiomorphs.* A chiral species has either no element of symmetry or only proper axes of symmetry (Sec. 6.4-1). The commonest examples are the 6-coordinate species cis-$[M(LL)_2L_2]^{n+}$ and $[M(LL)_3]^{n+}$, the point-group symmetries of which are C_2 and D_3, respectively. Isolation of optical isomers of these species is possible when the M—L bond is sufficiently covalent that rearrange-

---

*   Asymmetry and asymmetric as used in this book are in deference to long-standing practice. Strictly speaking, these terms mean "without symmetry" To allow for the existence of symmetry in certain species, the terms *dissymmetry* and *dissymmetric* are preferable.

†   For general surveys, see Ref. 175; S. Kirschner, in *Preparative Inorganic Reactions*, W. L. Jolly (Ed.), Vol. 1, Ch. 2, Interscience Publishers, New York (1964); and H. Brunner, *Angew. Chem., Int. Ed. Engl.*, **10**, 249 (1971).

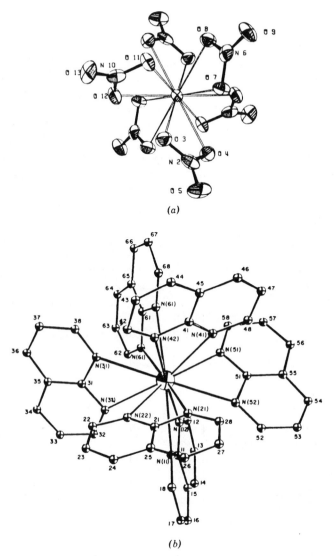

Figure 7-20 Typical molecular structures representing C.N. = 12. (a) $[Ce(NO_3)_6]^{2-}$; (b) $[Pr(napy)_6]^{3+}$. [(a), T. A. Beineke and J. Delgaudio, *Inorg. Chem.*, **7**, 715 (1968); (b) A. Clearfield, R. Gopal, and R. W. Olsen *Inorg. Chem.*, **16**, 911 (1977). Reprinted with permission from the American Chemical Society.]

ments between mirror-image structures are very slow (Sec. 11.2-4).

The center of dissymmetry in a complex species is the metal ion itself. Optical activity is most readily observed via the rotation of the plane of polarization of plane-polarized monochromatic radiation upon being passed through a solution in which one of the enantiomers is present in excess of the other. When the two enantiomers of a given species are present in equal concentration, no rotation occurs. Experimental data on optical isomers are restricted largely to coordination numbers 4 and 6.

    1. Coordination number 4. Optical isomerism in square-planar complexes is rare. The reported resolutions of the isobutylenediamine-*meso*-stilbenediamine complexes of platinum(II)[200] and palladium(II)[332] established the asymmetry in these species essential to this geometry.

    In tetrahedral geometry, no case of optical isomerism where four different unidentate ligands are present has been described. However, optical isomers containing no resolving reagent, have been obtained for a few species containing unsymmetrical chelating ligands: for example, the bis(benzoylpyruvato)-beryllium(II)[333] molecule, the bis(benzoylacetonato)beryllium(II)[334] molecule, and the bis(8-quinolino-5-sulfonic acid)zincate(II) ion.[335] The origin of dissymmetry is illustrated for the second of these species in Fig. 7-21.

    2. Coordination number 6. Previous discussions indicate that an octahedral species can exhibit asymmetry if at least two chelate rings are present.

    Molecular models or related diagrams show clearly that, although the 6-coordinate *trans*-diacidobis(bidentate) species are symmetrical, neither a plane nor an axis of symmetry characterizes the *cis* isomers, and two nonsuperposable, mirror-image isomeric forms should exist. This circumstance is readily illustrated for the *cis*-amminechlorobis(ethylenediamine)cobalt(III) ion, as

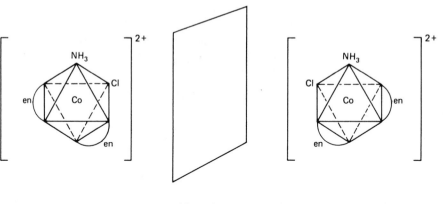

Mirror plane

That this conclusion is correct is shown by A. Werner's resolution of this ion (and its bromo analog) into *dextro*- and *levo*-rotatory forms by fractional crystallization of the diastereoisomeric *d*-α-bromocamphor-π-sulfonates.[336] This resolution was the first one reported for a substance predicted by Werner's theory to exhibit optical isomerism. In subsequent work, Werner himself resolved a number of other compounds of this type, and many other workers resolved other species, including the purely inorganic ion $[\text{Rh}\{(\text{HN})_2\text{SO}_2\}_2(\text{H}_2\text{O})_2]^-$.[337]

    Trisbidentate 6-coordinate species, $[\text{M(LL)}_3]^{n+}$, are also dissymmetric and capable of existing as mixtures of optical isomers, as

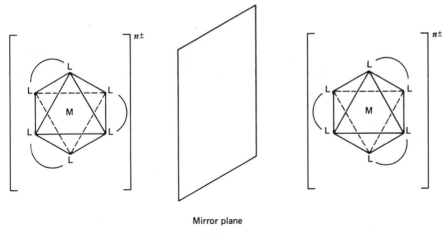

Mirror plane

It is of interest that although Werner's resolution of the *cis*-$[Co(en)_2(NH_3)Cl]^{2+}$ ion appeared to provide convincing proof of a significant structural portion of his theory, belief that optical isomerism could characterize only carbon-containing species was so prevalent that other investigators argued that the ethylenediamine groups themselves, albeit in some unexplained way, were responsible for the observed activity. Werner's contention that molecular dissymmetry relative to the metal ion exists was accepted only after his resolution[338] of the strictly inorganic complex ion $\left[Co\left\{ \begin{matrix} HO \\ HO \end{matrix} \!\!> Co(NH_3)_4 \right\}_3 \right]^{6+}$

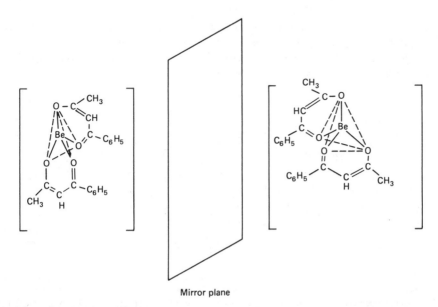

Mirror plane

Figure 7-21 Dissymetry in tetrahedral bis(benzoylacetonato)beryllium(II).

into forms of very large rotatory power (molecular rotation $= \pm 47,600°$).* In subsequent years, many other species of this type have been resolved: for example, $[M(en)_3]^{n+}$ $[M = Cr(III), Co(III), Rh(III), Ir(III), Pt(IV), Zn(II), Cd(II)]$; $[M(C_2O_4)_3]^{n-}$ $[M = Cr(III), Co(III), Rh(III), Ir(III), Ge(IV)^{339}]$; $[M(diket)_3]°$ $[diket = 1,3\text{-diketone}; M = Cr(III), Co(III), Y(III)]$;[340-342] $[M(phen)_3]^{2+}$ and $[M(bipy)_3]^{2+}$ $[phen = 1,10\text{-phenanthroline}; bipy = 2,2'\text{-bipyridine}; M = Fe, Ru, Os, Ni]$.[175]

Bis(terdentate), 6-coordinate species, $[M(LLL)_3]^{n+}$, can exist as three geometrical isomers,

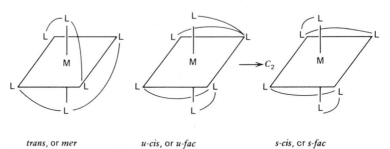

| *trans,* or *mer* | *u-cis,* or *u-fac* | *s-cis,* or *s-fac* |

the *u-* and *s-* designating unsymmetrical and symmetrical, respectively. Of these, the first two are chiral, but the third has a center and planes of symmetry and is not dissymmetric. Lack of symmetry in the *trans* form is possible only because of nonplanarity in the chelate ring structures. The *u-cis* has a C_2 axis is chiral. This situation is exemplified with the bis(diethylenetriamine)-cobalt(III) ion, $[Co(dien)_3]^{3+}$, all three geometrical isomers of which have been isolated and the *trans-* and *u-cis* resolved as optical isomers.[343]

Optical isomerism should characterize both *cis* isomers of the ion $[Co(teta)Cl_2]^+$ (teta = triethylenetetramine, a quadridentate ligand), but such isomerism could not be demonstrated.[344] Optical isomerism exists in sexidentate cobalt(III) complexes of the general type

$$\left[\begin{array}{c} \text{CH=N} \quad \text{Co} \quad \text{N=CH} \\ (CH_2)_x-S-(CH_2)_y-S-(CH_2)_z \end{array} \right]^+$$

where $x, y, z = 2,2,2.; 2,3,2; 3,2,3; 2,2,3;$ and $2,3,3$ and where NH may replace S if $x = y = z = 2$.[345-347] Each of these compounds has a resolvable green isomer and certain of them a resolvable brown isomer,

---

* Optical isomers are characterized in terms of either *specific rotation* or *molecular rotation*. The specific rotation, $[\alpha]_\lambda^t$ (at temperature t and wavelength λ), is defined by the relationship $[\alpha]_\lambda^t = r/cl$, where r is observed rotation in degrees, c is concentration in g ml^{-1}, and l is the cell length in dm. Specific rotation multiplied by g mole^{-1} gives molecular rotation.

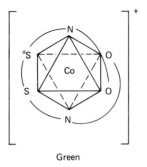

Green

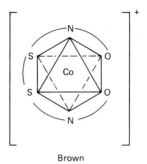

Brown

The cobalt(III) ethylenediaminetetraacetate ion, [Co(edta)]⁻, similarly contains a sexadentate ligand, and is chiral

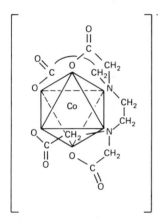

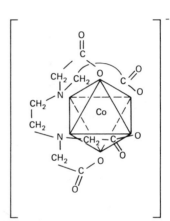

and has been resolved.[348, 349] Many other examples of optical isomerism have been reported for thermodynamically and/or kinetically (Chs. 9, 11) stable complexes containing multidentate ligands.

Optical isomerism also characterizes certain multinuclear complexes. For example, in 1913 Werner[350] separated the ion

$$\left[(en)_2 Co \overset{\displaystyle \overset{H_2}{N}}{\underset{\displaystyle \underset{O_2}{N}}{\diamondsuit}} Co(en)_2 \right]^{4+}$$

into three isomers by crystallization of the d-α-bromocamphor-π-sulfonates. Two of these were optically active, mirror-image isomers,

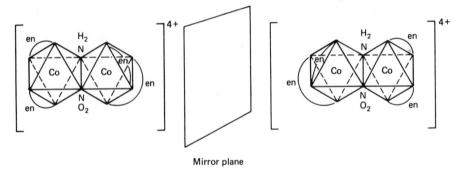

Mirror plane

The remaining optically inactive isomer was formulated as an internally compensated (*meso*) species, as

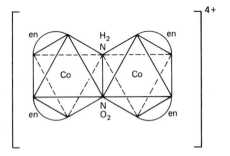

Both optically active isomers isomerized to the *meso* form when their aqueous solutions were heated.

Optical Isomerism—Ligand Dissymmetry. Dissymmetry of this type may be present in the coordinating ligand (e.g., in *d*- or *l*-propylenediamine or *d*- or *l*-tartrate ion) or may be induced as a result of coordination. The latter is of particular interest in indicating both the strength of the metal–ligand bond and in emphasizing again the directional characteristics of these bonds. Thus the anion

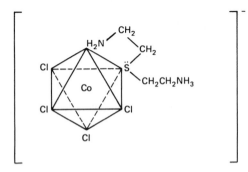

has been resolved because of the induced tetrahedral arrangement of bonds and nonbonded electron pair around the S atom,[351] and the square-planar ion has also been resolved because of the presence of a tetrahedral nitrogen atom bonded to four different groups.[352]

Optical Isomerism—Resolution Techniques. The resolution of inorganic isomers is effected by the same techniques used for organic isomers.[175, 353, 354] The following approaches have been useful:

1. **Selective crystallization of diastereomers.*** This technique, which is the most useful one, depends upon the fact that the diastereomeric salts obtained when the racemic mixture of the complex anion or cation in question is allowed to react with an appropriate optically active cation or anion, respectively, differ sufficiently in physical properties that they can be separated by fractional crystallization, precipitation, extraction, or adsorption (i.e., chromatographically). Useful cationic resolving agents are optically active protonated alkaloids (e.g., strychnine, brucine, cinchonine, quinine), cis-$[Co(en)_2(NO_2)_2]^+$, $[Co(en)_3]^{3+}$, and $[Ni(phen)_3]^{2+}$. Useful anionic resolving agents are optically active tartrate, antimonyl tartrate, α-bromocamphor-π-sulfonate, camphor-π-sulfonate, and $[Co(edta)]^-$ ions. Procedures involved in the resolution, by this technique, of the ions $[Co(edta)]^-$,[348,349] cis-$[Co(en)_2Cl_2]^+$,[355] $[Co(en)_3]^{3+}$,[356] $[Co(C_2O_4)_3]^{3-}$,[357] $[Co(en)_2(C_2O_4)]^+$,[358] cis-$[Co(en)_2(NH_3)Br]^+$,[359] $[Ge(C_2O_4)_3]^{2-}$,[339] $[M(phen)_3]^{2+}$ [360] are typical.

It is apparent, of course, that the procedure is useless for neutral complexes unless the ligand itself possesses a reactive functional group [e.g., the bis(8-quinolino-5-sulfonic acid)zincate(II) ion].[335] A special application involves sufficient alteration of the *activity* (Sec. 9.2) of the species in solution by addition of an optically active ion to effect selective precipitation of one enantiomer. Partial resolutions of the molecular species $[Co(acac)_3]$ and $[Cr(acac)_3]$ by addition of D-$[Co(en)_3]^{3+}$ to ethanolic solutions[340] are examples of this procedure of "configurational activity."

2. **Selective adsorption on optically active adsorbents.** This technique was first demonstrated by the partial resolution of the compound $[Co(dmg)_2(NH_3)Cl]$ (dmg = dimethylglyoxime) by stirring its saturated solution with powdered *l*-quartz, separating the solid, and examining the remaining solution for optical activity.[361] Although it is questionable whether the quartz

---

* Diastereomers are optically active isomers that are not enantiomers or mirror-image isomers. Thus D-$[Ni(phen)_3]$D-$[Co(C_2O_4)_3]^-$ and L-$[Ni(phen)_3]$D-$[Co(C_2O_4)_3]^-$ are diastereomers, as are D-$[Co(en)_2(d\text{-tart})]^+$ and L-$[Co(en)_2(d\text{-tart})]^-$ (tart = tartrate ion). Uppercase D and L refer to signs of rotation for complex ions at a stated wavelength; lowercase d and l refer to signs of rotation of ligands.

acts as a true adsorbent or a catalyst for racemization,[252] the procedure has been used to demonstrate optical isomerism in a number of substances[245, 334, 352] Column techniques, involving adsorption and selective elution with an appropriate solvent, have yielded better separations with starch[362-364] or d-lactose[341, 342, 365] as adsorbents. This procedure is particularly applicable to molecular species and to those that racemize too rapidly in solution to permit separation via diastereomers. Significant separations have been achieved with very long columns.[342] Similarly, selective adsorption from the vapor phase in a gas chromatographic column packed with an optically active substrate can effect resolution [e.g., of chromium(III)-hexafluoroacetylacetonate with powdered d-quartz].[366]

3. Selective extraction using optically active solvents. This technique, comparable in principle to selective adsorption, has been used only to a limited extent because of difficulty in preparing suitable solvent systems. However, using two immiscible liquid phases, one of which contained a substantial quantity of an optically active substance (e.g., d-isoamyl tartrate or l-fructose), partial resolution of the neutral tris-acetylacetonato complexes of chromium(III), cobalt(III), and rhodium(III) was effected.[366a, 366b]

4. Stereospecific reactions. Inorganic enantiomers often react selectively or at different rates with other optically active species (Sec. 11.2) and can thus be either concentrated or separated.[367-371] Thus although species of the type $[M(LL)_3]^{n+}$, where LL is an optically active ligand, should theoretically exist in the forms $Dddd$, $Dddl$, $Ddll$, $Dlll$, $Lddd$, $Lddl$, $Ldll$, and $Llll$, only the $Dlll$, $Lddd$, $Dddd$, and $Llll$ forms have been isolated.* A case in point is the selective formation of D-$[Co(d\text{-}pn)_3]^{3+}$ when D-$[Co(edta)]^-$ is treated with dl-pn (pn = 1,2-diaminopropane).[370]

5. Mechanical separation of racemic mixtures. Although situations comparable to L. Pasteur's separation of the enantiomers of sodium ammonium tartrate[371] are limited in inorganic systems, well-defined examples of separate crystallization of the antipodes are the salts $K_3[Co(C_2O_4)_3]$[372] and $Rb[Co(edta)]$.[373] Handpicking then gives the individual optical isomers. Seeding a solution of a racemate with a crystal of one of the enantiomers or of an isomorphous substance may cause selective crystallization of one isomer. Thus the D,L-$[Co(en)_2(C_2O_4)]^+$ ion has been so resolved by adding a crystal of the D-isomer.[374]

That a substance can be resolved is an indication that the asymmetry existing in the crystal is preserved in solution. Although it was believed originally that only covalently bonded species can be resolved, it is more probable that both the solution stability of the complex and the polarity of the solvent are factors of greater importance. Thus observations that the ions $[Al(C_2O_4)_3]^{3-}$, $[Ga(C_2O_4)_3]^{3-}$, and $[Fe(C_2O_4)_3]^{3-}$ cannot be resolved,[357, 375] whereas the

---

* In principle, numerous additional isomers could exist in terms of different orientations of the CH_3 group of one ligand relative to the CH_3 group of another ligand. Those listed are the most probable isomers.

ions $[Co(C_2O_4)_3]^{3-}$ and $[Cr(C_2O_4)_3]^{3-}$ can, is clearly related to thermodynamic stability in solution since these three species rapidly exchange with added oxalate ion containing carbon-14, whereas the cobalt(III) and chromium(III) analogs do not.[376] Furthermore, the substantially electrostatically bonded tris(acetylacetonato) chelates of certain lanthanide(III) ions can be resolved in a nonpolar benzene-petroleum ether medium but not in a more polar chloroform medium.[341]

Optical Isomerism—Conformational Analysis and Absolute Molecular Configurations. Another aspect of ligand dissymmetry, which is related to absolute molecular configurations, reflects conformational differences in saturated bidentate ligands. Thus, when bonded to a metal atom or ion, the ethylenediamine molecule, for example, assumes a *gauche* conformation (Fig. 7-22) in order to bring the two donor nitrogen atoms into correct position to interact.[370,377,378] The result is chirality, but resolution cannot be effected because of the absence of a sufficient energy barrier to prevent racemization (e.g., in the complex ion $[Co(en)(NH_3)_4]^{3+}$).[379] When viewed along the C—C axis, the two diagrams in Fig. 7-22 represent, respectively, a right-hand helix (δ) and a left-hand helix (λ).

When two or more such chelate rings are present in a given complex species, interactions may be expected to give certain conformations that are more stable thermodynamically than others. For a square-planar complex $[M(en)_2]^{n+}$, interactions (Fig. 7-23) should allow the conformations M$\delta\delta$, M$\lambda\lambda$, and M$\delta\lambda$, of which the first two are dissymmetric, and the third is an internally compensated *meso*, form. However, conformational analysis[370] shows that the M$\delta\delta$ and M$\lambda\lambda$ conformers should predominate because of unfavorable H—H repulsions in the *meso* form. For an octahedral tris chelate $[M(en)_3]^{n+}$, conformational analysis predicts M$\delta\delta\delta$, M$\delta\delta\lambda$, M$\delta\lambda\lambda$, and M$\lambda\lambda\lambda$, each of which will exist as D- and L-enantiomers because of the asymmetry of the $[M(en)_3]^{n+}$ structure.* Of these

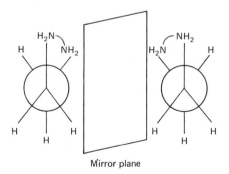

Mirror plane

Figure 7-22 Possible enantiometric conformations for *gauche* form of ethylenediamine (δ and λ).

* See also B. Lee, *Inorg. Chem.*, **11**, 1072 (1972).

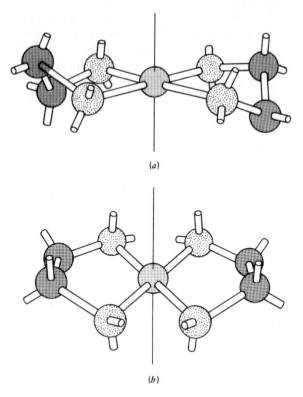

(a)

(b)

Figure 7-23 (a) $\lambda\lambda$ and (b) $\lambda\delta$ forms of square-planar $[M(en)_2]^{n+}$ complexes. [E. J. Corey and J. C. Bailar, *J. Am. Chem. Soc.*, **81**, 2620 (1959). Reprinted with permission from the American Chemical Society.]

eight isomers, only two are normally found. If the bidentate ligand is optically active [e.g., propylenediamine (1,2-diaminopropane)], conformational constraints result in $(-)$-propylenediamine bonding preferentially as the λ chelate and $(+)$-propylenediamine as the δ chelate.[370] Four isomers then result: D-M$\delta\delta\delta$, D-M$\lambda\lambda\lambda$, L-M$\delta\delta\delta$, and L-M$\lambda\lambda\lambda$ (Fig. 7-24).*

The assignment of absolute molecular configuration to an enantiomer by chemical means alone is difficult, but conformational analysis is helpful.[370] On the other hand, a number of absolute molecular configurations have been determined by x-ray crystallography[380, 381] (e.g., $(+)_{450}$-cis-$[Pt(en)_2Cl_2]^{2+}$ ion).[382] Relative configurations may be assigned on the basis of solubility data if the less soluble diastereomers formed by enantiomers of two ions with a given resolving counter ion are isomorphous and thus have related configurations.[383]

---

* D-enantiomers are referenced to D-$(+)_{589}[Co(en)_3]^{3+}$; L-enantiomers are their mirror images. In another system [T. S. Piper, *J. Am. Chem. Soc.*, **83**, 3908 (1961)], Λ and Δ enantiomers are referenced to left- and right-handed helices when viewed along the threefold axis. Thus for propylenediamine the D$\delta\delta\delta$ and $\Lambda\delta\delta\delta$ enantiomers are the same, as are the L$\delta\delta\delta$ and $\Delta\delta\delta\delta$ enantiomers.[357]

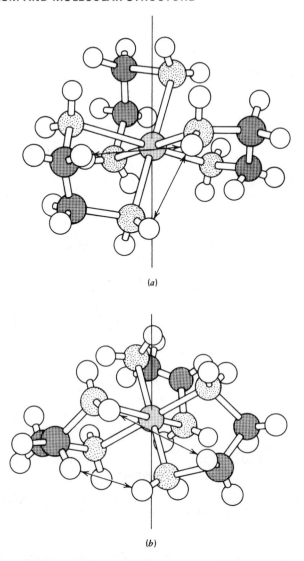

(a)

(b)

Figure 7-24 (a) $D(\Lambda)$ ($D\delta\delta\delta$ or $\Lambda\delta\delta\delta$) and $L(\Delta)$ ($L\delta\delta\delta$ or $\Delta\delta\delta\delta$) conformers of $[Co(pn)_3]^{3+}$ enantiomers. [E. J. Corey and J. C. Bailar, *J. Am. Chem. Soc.*, **81**, 2620 (1959). Reprinted with permission from the American Chemical Society.]

Similarly, absolute molecular configurations can be assigned on the bases of circular dichroism and optical rotatory dispersion data. The typical optical rotatory dispersion curves given in Fig. 7-25 indicate clearly sharp changes in specific rotation as absorption bands are approached — on occasion even change in sign at the wavelength of absorption. Abrupt reversal in sign of specific rotation is called the *Cotton effect*.[384] The Cotton effect may be either

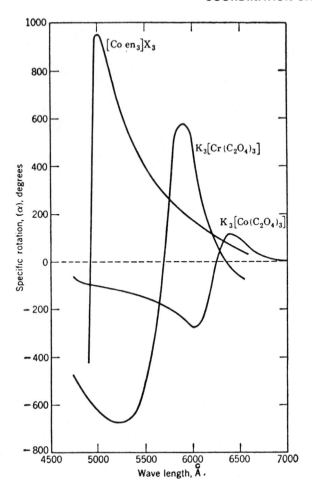

Figure 7-25 Typical optical rotatory dispersion curves. [T. Moeller, *Inorganic Chemistry*, Fig. 7.1, p. 268, John Wiley & Sons, New York (1952).]

positive or negative, as indicated in Fig. 7-26.[385] As a general rule, analogous complex species, showing for corresponding electronic transitions Cotton effects of the same sign, have the same optical configuration.

Many absolute configurations have been determined in this way.[385, 386] Typical are resolutions of racemic *cis*-[M(en)$_2$(N$_3$)$_2$](NO$_3$) (M = Cr, Co) with potassium antimonyl-(+)-tartrate to give the Λ-(+)-[M(en)$_2$(N$_3$)$_2$]$^+$ ions as the less soluble diastereomers and ultimate establishment of Λ-(+)-[Cr(en)$_2$-(N$_3$)(OH)]$^{2+}$, Λ-(+)-[Cr(en)$_2$(OH$_2$)(NCS)]$^{2+}$, Λ-(+)-[Cr(en)$_2$(OH$_2$)]$^+$, Λ-(+)-[Cr(en)$_2$Cl$_2$]$^+$, Λ-(+)-[Cr(en)$_2$(OH$_2$)Cl]$^{2+}$, Λ-(+)-[Co(en)$_2$(OH$_2$)(N$_3$)]$^{2+}$, and Λ-(+)-[Co(en)$_2$(OH$_2$)$_2$]$^{3+}$ in aqueous solution.[387]

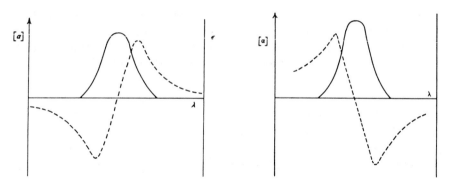

Figure 7-26 Positive and negative Cotton effects. [R. C. Gillard, *Progress in Inorganic Chemistry*, F. A. Cotton (Ed.), Vol. 7, Fig. 4, p. 275; Fig. 5, p. 276, Interscience Publishers, New York (1966).]

EXERCISES

7-1. Give an IUPAC-approved name for each of the following coordination entities: $[Cl_4ReReCl_4]^{2-}$, $[Pt(l\text{-pn})_2Cl_2]Cl_2$, $NH_4[Cr(SCN)_4(NH_3)_2]$, $K[PtCl_3(C_2H_4)]$, $[Nd(NCO)_3(NO_3)_3]^{3-}$, $K_3[Fe(CN)_5(CO)]$. $[Fe(OH)(H_2O)_5]^{2+}$, $\left[Co\left\{HO\!\!>\!\!HO\!\!>\!\!Co(NH_3)_4\right\}_3\right]^{6+}$, $K[CrOF_5]$, $[Co(en)_3][Fe(CN)_6]$.

7-2. Distinguish clearly between Type A and Type B donors and between Type A and Type B acceptors. Illustrate each with at least two examples. What regularities can be obeserved in reactions of these acceptors and these donors? Explain.

7-3. Can a noncoordinating anion exist? Explain your answer.

7-4. What factors make cryptands and sepulchrands strong ligands? Is the presence of d orbitals within an acceptor ion a factor of importance?

7-5. Give diagrammatic representations of all of the possible isomers of species of each of the following analytical compositions: $[Cr(NH_3)_3(H_2O)ClBr]^+$, $[Pt(en)_2(H_2O)Br]^{3+}$, $[Pd(C_5H_5N)(NH_3)(NH_2OH)(NO_2)]^+$, $[Co_2(NH_3)_6(OH)_2Cl_2]^{2+}$, $[Pd\{As(C_6H_5)_3\}_2(NCS)_2]$. Name each species. (*Note*: do not overlook linkage isomerism.)

7-6. Giving your reasons, predict the molecular structure of each of the following coordination entities: Nb_2Cl_{10}, $[Co_2(NH_3)_6(NH_2)(O_2)Cl_2]^{2+}$, $[Ni(CN)_5]^{3-}$, $[PuO_2(C_2O_4)_3]^{4-}$, $[Ce(NO_3)_6]^{2-}$. Which, if any, of these structures is potentially asymmetric?

7-7. Suggest a reason for the observation that although manganese pentacarbonyl, $Mn(CO)_5$, is dimeric, its hydrido derivative, $HMn(CO)_5$, is monomeric.

7-8. Prepare schematic drawings of possible isomeric structures for each of the following compositions, ignoring conformational isomers, and give for each the appropriate point group: $Cr(pn)_2Cl_3\{pn = H_2NCH_2CH(NH_2)CH_3\}$; $Co(NH_3)_3(\text{—}NO_2)_3$; $Pt(nta)Cl(H_2O)$ (nta = $N(CH_2CO_2)_3^{3-}$).

7-9. Which of the following octahedral compositions could exhibit optical isomerism: $[ML_2L'L''L'''L'''']$; $[ML_2L_2'(LL)]$; $(MLL'(LL)_2]$? (L, L', etc. = unidentate ligands; LL = symmetrical bidentate ligand; complexes may be molecules or ions, depending upon the ligands present.)

7-10. Give the most common oxidation state and the usual geometry of its complexes for four of the following elements: Ti, V, Cr, Re, Cu, Zn, P, Se.

7-11. A d-transition metal complex $[M(LL')_3]^-$, where LL' is an dissymmetric bidentate ligand, is isolated as two isomers, neither of which can be resolved into optical isomers. Depict possible molecular structures in terms of octahedral and trigonal prismatic geometries. Does the information given prove which of the two molecular geometries is correct? Explain.

7-12. For each of the following octahedral complexes, indicate the type of distortion you might expect to exist and give a reason for your choice: $trans$-$[Co(NH_3)_4Cl_2]^+$, $[Co(en)_3]Cl_3$.

7-13. Represent by diagrams the possible isomers for an octahedral complex containing six different unidentate ligands. Are any of these optical isomers? Explain.

7-14. Using the modified IUPAC Rules of Nomenclature (Appendix II), name each of the following species: $Fe(C_5H_5)_2$, $[Cr(NH_3)_5Cl]Cl_2$, $[Co(NH_3)_2(H_2O)_3(OH)]^{2+}$, $[Pt(NH_3)BrCl(NO_2)]^-$, $Na[Co(edta)]$, $(NH_4)_3[Fe(C_2O_4)_3]$, $[(C_2H_5)_4N]_3[Er(NCO)_3(NO_3)_3]$, $[Cr(C_6H_6)_2]$, $[Co\{Co(NH_3)_4(OH)_2\}_3]$, $[Pt(NH_3)_2(py)_2Cl_2]^{2+}$.

7-15. For each of the following pairs of isomers, indicate clearly how the two species can be distinguished from each other: cis-$[Cr(en)_2Cl_2]Cl$ and $trans$-$[Cr(en)_2Cl_2]Cl$; $[Cr(H_2O)_6]Cl_3$ and $[Cr(H_2O)_5Cl]Cl_2 \cdot H_2O$; $[Co(NH_3)_3(NO_2)_3]$ and $[Co(NH_3)_6][Co(NO_2)_6]$; $[Co(NH_3)_6][Cr(CN)_6]$ and $[Cr(NH_3)_6][Co(CN)_6]$; cis-$[Pt(NH_3)_2Cl_2]$ and $trans$-$[Pt(NH_3)_2Cl_2]$.

REFERENCES

1. W. Biltz, *Z. Anorg. Allg. Chem.*, **164**, 245 (1927); **223**, 321 (1935); **234**, 253 (1937).
2. P. Rây, *Z. Anorg. Allg. Chem.*, **174**, 189 (1928); *J. Indian Chem. Soc.*, **5**, 73 (1928).
3. R. W. Parry, *Chem. Rev.*, **46**, 507 (1950).
4. J. C. Bailar, Jr., and D. H. Busch, in *The Chemistry of the Coordination Compounds*, J. C. Bailar, Jr., and D. H. Busch (Eds.), Ch. 1, Reinhold Publishing Corp., New York (1956).
5. Citoyen Tassert, *Ann. Chim. Phys.*, **28**, 92 (1798).
6. J. C. Bailar, Jr., in *The Chemistry of the Coordination Compounds*, J. C. Bailar, Jr., and D. H.

Busch (Eds.), Ch. 2, Reinhold Publishing Corp., New York (1956). Early development.

7. G. B. Kauffman (Ed.), *Classics in Coordination Chemistry*, Parts I and II, Dover Publications, New York (1968, 1976). Translations, with detailed commentaries by G.B.K., of selected papers dating from 1828. Highly recommended as background material.

8. C. E. Claus, *Ann. Chem. Pharm.*, **98**, 317 (1856).

9. C. W. Blomstrand, *Chem. Ber.*, **4**, 40 (1871).

10. S. M. Jørgensen, *J. Prakt. Chem.*, [2], **27**, 433 (1883); [2], **41**, 429 (1890). *Z. Anorg. Chem.*, **19**, 109 (1899). The last paper describes modifications and extensions of the author's views and disputes Werner's proposals.

11. A. Werner, *Z. Anorg. Chem.*, **3**, 267 (1893).

12. A. Kekulé, *Ann. Chem.*, **106**, 129 (1858).

13. G. B. Kauffman, *Alfred Werner: Founder of Coordination Chemistry*, Springer-Verlag, New York (1966). Useful and interesting biography. Part I of Ref. 7 gives translations of six of Werner's most important papers.

14. A. Werner, *Neuere Anschauungen auf dem Gebiete der anorganischen Chemie*, 4th ed., p. 44, T. Viewig u. Sohn, Braunschweig (1920). Translation quoted from R. Schwarz. *The Chemistry of the Inorganic Complex Compounds*, p. 9, John Wiley & Sons, New York (1923) (as translated by L. W. Bass).

15. A. Werner, *Chem. Ber.*, **47**, 3087 (1914). English translation: Ref. 7, Part I, Paper 6.

16. A. Werner and A. Miolati, *Z. Phys. Chem.*, **12**, 35 (1893); **14**, 506 (1894); **21**, 225 (1896). English translations: Ref. 7, Part I, Papers 2 and 3.

17. E. Petersen, *Z. Phys. Chem.*, **22**, 410 (1897).

18. J. C. Bailar, Jr., Ref. 6, pp. 113–118.

19. N. V. Sidgwick, *J. Chem. Soc.*, **123**, 725 (1923).

20. T. M. Lowry, *J. Soc. Chem. Ind.*, **42**, 316 (1923).

21. N. V. Sidgwick, *The Electronic Theory of Valency*, Ch. 10, Clarendon Press, Oxford (1927).

22. A. A. Blanchard, *Chem. Rev.*, **21**, 3 (1937); **26**, 409 (1940).

23. A. Werner, Ref. 14, pp. 92–95.

24. W. P. Jorissen, H. Bassett, A. Damiens, F. Fichter, and H. Remy, *J. Am. Chem. Soc.*, **63**, 889 (1941).

25. W. C. Fernelius, *Chem. Eng. News*, **26**, 161 (1948).

26. W. C. Fernelius, E. M. Larsen, L. E. Marchi, and C. L. Rollinson, *Chem. Eng. News*, **26**, 520 (1948).

27. W. C. Fernelius, *Advances in Chemistry Series*, No. 8, pp. 9–37, American Chemical Society, Washington, D.C. (1953).

28. R. V. G. Ewens and H. Bassett, *Chem. Ind. (Lond.)*, **1949**, 131.

29. International Union of Pure and Applied Chemistry, *Nomenclature of Inorganic Chemistry*, Ch. 7, Butterworth, London (1959).

30. International Union of Pure and Applied Chemistry, *J. Am. Chem. Soc.*, **82**, 5523 (1960). Report cited in Ref. 29, with comments by U.S. authorities.

31. International Union of Pure and Applied Chemistry, *Nomenclature of Inorganic Chemistry*, 2nd ed., Ch. 7, Butterworth, London (1971); also *Pure Appl. Chem.*, **27**, 67 (1971).

32. W. C. Fernelius, K. Loening, and R. M. Adams, *J. Chem. Educ.*, **51**, 468, 603 (1974); **52**, 793 (1975).

33. T. Moeller, in *MIP International Review of Science, Inorganic Chemistry, Series One*, H. J. Emeléus (Consultant Ed.), Vol. 7, K. W. Bagnall (Ed.), pp. 275–298, Butterworth, London (1972).

34. J. H. Forsberg, *Coord. Chem. Rev.*, **10**, 195 (1973).

34a. T. Moeller and E. Schleitzer-Rust (Eds.), *Gmelin Handbuch der anorganischen Chemie*, System-No. 39, Part D1 (1980), Part D2 (1981), Part D3 (1982), Part D4 (1983), Springer-Verlag, Berlin.

35. A. D. Jones and G. R. Choppin, *Actinides Rev.*, **1**, 311 (1969).

36. B. G. F. Carleson and H. Irving, *J. Chem. Soc.*, **1954**, 4390.

37. S. Ahrland and R. Larsson, *Acta Chem. Scand.*, **8**, 354 (1954).

38. S. Ahrland, *Acta Chem. Scand.*, **10**, 723 (1956).

39. S. Ahrland and K. Rosengren, *Acta Chem. Scand.*, **10**, 727 (1956).

40. S. Ahrland and I. Grenthe, *Acta Chem. Scand.*, **11**, 1111 (1957).

41. S. Ahrland and B. Norén, *Acta Chem. Scand.*, **12**, 1595 (1958).

42. R. J. P. Williams, *Proc. Chem. Soc.*, **1960**, 20.

43. I. Leden and J. Chatt, *J. Chem. Soc.*, **1955**, 2936.

44. S. Ahrland, J. Chatt, and N. R. Davies, *Q. Rev. (Lond.)*, **12**, 265 (1958).

45. G. Schwarzenbach, in *Advances in Inorganic Chemistry and Radiochemistry*, H. J. Emeléus and A. G. Sharpe (Eds.), Vol. 3, pp. 257–285, Academic Press, New York (1961).

46. F. Basolo and R. G. Pearson, *Mechanisms of Inorganic Reactions: A Study of Metal Complexes in Solution*, 2nd ed., p. 22, John Wiley & Sons, New York (1967).

47. A. D. Allen and C. V. Senoff, *J. Chem. Soc. Chem. Commun.*, **1965**, 621.

48. A. D. Allen and F. Bottomley, *Acc. Chem. Res.*, **1**, 12 (1968).

49. R. Murray and D. C. Smith, *Coord. Chem. Rev.*, **3**, 429 (1968).

50. J. N. Armor, *Inorg. Chem.*, **17**, 213 (1978).

51. A. P. Ginsberg and W. E. Lindsell, *J. Am. Chem. Soc.*, **93**, 2082 (1971).

52. L. Klevan, J. Peone, Jr., and S. K. Madan, *J. Chem. Educ.*, **50**, 670 (1973).

53. L. D. Brown and K. N. Raymond, *Inorg. Chem.*, **14**, 2595 (1975).

54. D. A. Summerville, R. D. Jones, B. M. Hoffman, and F. Basolo, *J. Chem. Educ.*, **56**, 157 (1979).

55. R. A. Krause, *Inorg. Nucl. Chem. Lett.*, **7**, 973 (1971).

56. R. B. King, in *Progress in Inorganic Chemistry*, S. J. Lippard (Ed.), Vol. 15, pp. 287–473, Interscience Publishers, New York (1972).

57. R. D. Cramer, R. V. Lindsay, Jr., C. T. Prewitt, and U. G. Stolberg, *J. Am. Chem. Soc.*, **87**, 658 (1965).

58. M. Elder and D. Hall, *Inorg. Chem.*, **8**, 1268, 1273 (1969).

59. R. Kummer and W. A. G. Graham, *Inorg. Chem.*, **7**, 1208 (1968).

60. J. V. Quagliano, J. T. Summers, S. Kida, and L. M. Vallarino, *Inorg. Chem.*, **3**, 1557 (1964).

61. V. L. Goedken, L. M. Vallarino, and J. V. Quagliano, *Inorg. Chem.*, **10**, 2682 (1971).

62. M. R. Rosenthal, *J. Chem. Educ.*, **50**, 331 (1973).

63. C. C. Addison, N. Logan, S. C. Wallwork, and C. D. Garner, *Q. Rev. (Lond.)*, **25**, 289 (1971).

64. T. A. Beinke and J. Delgaudio, *Inorg. Chem.*, **7**, 715 (1968).

65. R. D. Larsen and G. H. Brown, *J. Phys. Chem.*, **68**, 3060 (1964).

66. G. F. Smith, V. R. Sullivan, and G. Frank, *Ind. Eng. Chem., Anal. Ed.*, **8**, 449 (1936).

67. G. F. Smith, *Cerate Oxidimetry*, G. Frederick Smith Chemical Company, Columbus, Ohio (1942).

68. B. J. Hathaway and A. E. Underhill, *J. Chem. Soc.*, **1961**, 3091.

69. B. J. Hathaway, D. G. Holah, and M. Hudson, *J. Chem. Soc.*, **1963**, 4586.

70. F. Madaule-Aubry and G. M. Brown, *Acta Crystallogr.*, **B24**, 745 (1968).

71. G. H. W. Milburn, M. R. Truter, and B. L. Vickery, *J. Chem. Soc. Chem. Commun.*, 1188 (1968).

72. W. C. Wolsey, *J. Chem. Educ.*, **50**, A335 (1973).

73. E. R. Birnbaum and S. Stratton, *Inorg. Chem.*, **12**, 379 (1973).

74. E. Lenz and R. K. Murmann, *Inorg. Chem.*, **7**, 1880 (1968).

75. S. H. Hunter, R. S. Nyholm, and G. A. Rodley, *Inorg. Chim. Acta*, **3**, 631 (1969).

76. H. G. Mayfield, Jr., and W. E. Bull, *J. Chem. Soc. A*, **1971**, 2279.

77. H. C. Clark and R. J. O'Brien, *Inorg. Chem.*, **2**, 1020 (1963).

78. N. F. Curtis, *J. Chem. Soc.*, **1965**, 924.

79. B. G. Segal and S. J. Lippard, *Inorg. Chem.*, **17**, 844 (1978).

80. M. J. Nolte, G. Gafner, and L. M. Haines, *J. Chem. Soc. Chem. Commun.*, 1406 (1969).

81. R. S. Schrock and J. A. Osborn, *Inorg. Chem.*, **9**, 2339 (1970).

82. L. J. Sauro and T. Moeller, *J. Inorg. Nucl. Chem.*, **30**, 953 (1968).

83. R. W. Bashioum, R. L. Dieck, and T. Moeller, *Inorg. Nucl. Chem. Lett.*, **9**, 773 (1973).

84. E. C. Baker, G. W. Halstead, and K. N. Raymond, *Struct. Bonding*, **25**, 23 (1976).

85. W. C. Zeise, *Overs. K. Dan. Vidensk. Selsk. Forh.*, **1825–1826**, 13; cited in Poggendorf's *Ann. Phys. Chem.*, **9**, 632 (1827); *Ann. Chem.*, **23**, 1 (1837).

86. K. Birnbaum, *Ann. Chem.*, **145**, 67 (1869).

87. B. E. Douglas, in *The Chemistry of the Coordination Compounds*, J. C. Bailar, Jr., and D. H. Busch (Eds.), Ch. 15, Reinhold Publishing Corp., New York (1956).

88. R. G. Guy and B. L. Shaw, in *Advances in Inorganic Chemistry and Radiochemistry*, H. J. Eméleus and A. G. Sharpe (Eds.), Vol. 4, pp. 78–131, Academic Press, New York (1962).

89. G. E. Coates, M. L. H. Green, P. Powell, and K. Wade, *Principles of Organometallic Chemistry*, Methuen, London (1968).

90. D. I. Hall, J. H. Ling, and R. S. Nyholm, *Struct. Bonding*, **15**, 3 (1973).

91. S. A. Miller, J. A. Tebboth, and J. F. Tremaine, *J. Chem. Soc.*, **1952**, 632.

92. T. J. Kealy and P. L. Pauson, *Nature*, **168**, 1039 (1951).

93. G. Wilkinson, M. Rosenblum, M. C. Whiting, and R. B. Woodward, *J. Am. Chem. Soc.*, **74**, 2125 (1952).

94. E. O. Fischer and W. Pfab, *Z. Naturforsch.*, **7b**, 377 (1952).

95. J. D. Dunitz, L. E. Orgel, and A. Rich, *Acta Crystallogr.*, **9**, 373 (1956).

96. G. Wilkinson and F. A. Cotton, in *Progress in Inorganic Chemistry*, F. A. Cotton (Ed.), Vol. 1, pp. 1–124, Interscience Publishers, New York (1959).

97. E. O. Fischer and H. P. Fritz, in *Advances in Inorganic Chemistry and Radiochemistry*, H. J. Eméleus and A. G. Sharpe (Eds.), Vol. 1, pp. 55–115, Academic Press, New York (1959).

98. F. A. Cotton, *Chem. Rev.*, **55**, 551 (1955).

99. H. H. Zeiss and M. Tsutsui, *J. Am. Chem. Soc.*, **79**, 3062 (1957).

100. E. O. Fischer and W. Hafner, *Z. Naturforsch.*, **10b**, 665 (1955); *Z. Anorg. Allg. Chem.*, **286**, 146 (1956).

101. F. Hein and H. Müller, *Chem. Ber.*, **89**, 2722 (1956).

102. E. Weiss and E. O. Fischer, *Z. Anorg. Allg. Chem.*, **286**, 142 (1956).

103. H. Diehl, *Chem. Rev.*, **21**, 39 (1937).

104. S. Chaberek and A. E. Martell, *Organic Sequestering Agents*, Chs. 2–4, John Wiley & Sons, New York (1959).

105. F. Basolo and R. G. Pearson, *Mechanisms of Inorganic Reactions: A Study of Metal Complexes in Solution*, 2nd ed., John Wiley & Sons, New York (1967).

106. J. Bjerrum, G. Schwarzenbach, and L. G. Sillén, *Stability Constants of Metal-Ion Complexes*, 2nd ed., Spec. Publ. No. 17, The Chemical Society, London (1964).

107. L. G. Sillén and A. E. Martell, *Stability Constants of Metal-Ion Complexes*, Suppl. No. 1, Spec. Publ. No. 25, The Chemical Society, London (1971).

108. R. G. Wilkins, *The Study of Kinetics and Mechanisms of Reactions of Transition Metal Ions*, Allyn and Bacon, New York (1974).

109. J. Kragten, *Atlas of Metal–Ligand Equilibria in Aqueous Solution*, Halsted Press, John Wiley & Sons, New York (1978).

110. H. Irving, R. J. P. Williams, D. J. Ferrett, and A. E. Williams, *J. Chem. Soc.*, **1954**, 3494.

111. L. J. Charpentier and T. Moeller, *J. Inorg. Nucl. Chem.*, **32**, 3575 (1970).

112. J. E. McDonald and T. Moeller, *J. Inorg. Nucl. Chem.*, **39**, 2287 (1977).

113. F. G. Mann, *J. Chem. Soc.*, **1926**, 2681.

114. G. Schwarzenbach and G. Anderegg, *Z. Anorg. Allg. Chem.*, **282**, 286 (1955).

115. G. Schwarzenbach, H. Senn, and G. Anderegg, *Helv. Chim. Acta*, **40**, 1186 (1957).

116. L. L. Merritt, Jr., *Rec. Chem. Prog.*, **10**, 59 (1949).

117. C. J. Pedersen, *J. Am. Chem. Soc.*, **89**, 7017 (1967).

118. C. J. Pedersen and H. K. Frensdorff, *Angew. Chem., Int. Ed. Engl.*, **11**, 16 (1972).

119. J. J. Christensen, D. J. Eatough, and R. M. Izatt. *Chem. Rev.*, **74**, 351 (1974).

119a. R. Izatt and J. J. Christensen (Eds.), *Synthetic Multidentate Macrocyclic Compounds*, Academic Press, New York (1978).

120. L. F. Lindoy, *Chem. Soc. Rev.*, **4**, 421 (1975).

120a. G. A. Melson, Ed., *Coordination Chemistry of Macrocyclic Compounds*, Plenum Press, New York (1979).

121. J. M. Lehn, *Struct. Bonding*, **16**, 1 (1973); *Pure Appl. Chem.*, **49**, 857 (1977). J. M. Lehn, B. Dietrich, and J. P. Sauvage, *Tetrahedron Lett.*, **1969**, 2885, 2889; *J. Am. Chem. Soc.*, **92**, 2916 (1970).

122. A. M. Sargeson, *Chem. Br.*, **15**, 23 (1979).

123. D. H. Busch, *Rec. Chem. Prog.*, **25**, 107 (1964).

124. D. Midgley, *Chem. Soc. Rev.*, **4**, 549 (1975).

125. H. K. Frensdorff, *J. Am. Chem. Soc.*, **93**, 600 (1971).

126. J. L. Dye, J. M. Ceraso, M. T. Lok, B. L. Barnett, and F. J. Tehan, *J. Am. Chem. Soc.*, **96**, 608 (1974); F. J. Tehan, B. L. Barnett, and J. L. Dye, *J. Am. Chem. Soc.*, **96**, 7203 (1974).

127. D. G. Adolphson, J. D. Corbett, and D. J. Merryman, *J. Am. Chem. Soc.*, **98**, 7234 (1976).

128. A. Cisar and J. D. Corbett, *Inorg. Chem.*, **16**, 632, 2482 (1977).

129. P. A. Edwards and J. D. Corbett, *Inorg. Chem.*, **16**, 903 (1977).

130. J. E. Parks, B. E. Wagner, and R. H. Holm, *J. Am. Chem. Soc.*, **92**, 3500 (1970); *Inorg. Chem.*, **10**, 2472 (1971).

131. E. Larsen, G. N. La Mar, B. E. Wagner, and R. H. Holm, *Inorg. Chem.*, **11**, 2652 (1972).

132. A. Werner, Ref. 14, pp. 327–386.

133. T. D. O'Brien, in *The Chemistry of the Coordination Compounds*, J. C. Bailar, Jr., and D. H. Busch (Eds.), Ch. 7, Reinhold Publishing Corp., New York (1956).

134. F. Basolo and R. G. Pearson, *Mechanisms of Inorganic Reactions: A Study of Metal Complexes in Solution*, 2nd ed., p. 13, John Wiley & Sons, New York (1967).

135. A. Werner and A. Gubser, *Chem. Ber.*, **34**, 1579 (1901).

136. N. Bjerrum, *Z. Phys., Chem.*, **59**, 336, 581 (1907).

137. A. Recoura, *C.R. Acad. Sci.*, **194**, 229 (1932).

138. R. E. Hamm and C. M. Shull, Jr., *J. Am. Chem. Soc.*, **73**, 1240 (1951).

139. A. V. Pamfilov and N. N. Gumenyuk, *J. Gen. Chem. (USSR)*, **23**, 1117 (1953).

140. D. Kiraly, K. Zalatnai, and M. T. Beck, *J. Inorg. Nucl. Chem.*, **11**, 170 (1959).

141. R. J. Angelici, *Synthesis and Technique in Inorganic Chemistry*, 2nd ed., pp. 63–70, W. B. Saunders, Philadelphia (1977).

142. J. H. Espenson and J. P. Birk, *Inorg. Chem.*, **4**, 527 (1965).

143. T. W. Swaddle and E. L. King, *Inorg. Chem.*, **4**, 532 (1965).

144. J. H. Espenson and S. G. Slocum, *Inorg. Chem.*, **6**, 906 (1967).

145. I. G. Dance and H. C. Freeman, *Inorg. Chem.*, **4**, 1555 (1965).

146. S. M. Jørgensen, *Z. Anorg. Chem.*, **5**, 147 (1893); **19**, 109 (1899).

147. A. Werner, *Chem. Ber.*, **40**, 765 (1907).

148. R. G. Pearson, P. M. Henry, J. G. Bergmann, and F. Basolo, *J. Am. Chem. Soc.*, **76**, 5920 (1954).

149. F. Basolo and G. S. Hamacher, *Inorg. Chem.*, **1**, 1 (1962). See also N. Sabbatini and V. Balzani, *Inorg. Chem.*, **10**, 209 (1971) for conflicting data on the platinum(IV) compounds.

150. F. Basolo and R. G. Pearson, *Mechanisms of Inorganic Reactions: A Study of Metal Complexes in Solution*, 2nd ed., pp. 291–300, John Wiley & Sons, New York (1967).

151. B. Adell, *Sven. Kem. Tidskr.*, **56**, 318 (1944); **57**, 260 (1945); *Z. Anorg. Chem.*, **252**, 272 (1944).

152. J. P. Faust and J. V. Quagliano, *J. Am. Chem. Soc.*, **76**, 5346 (1954).

153. R. B. Penland, T. J. Lane, and J. V. Quagliano, *J. Am. Chem. Soc.*, **78**, 887 (1956).

154. F. Basolo, J. L. Burmeister, and A. J. Poe, *J. Am. Chem. Soc.*, **85**, 1700 (1963).

155. I. Grenthe and E. Nordin, *Inorg. Chem.*, **18**, 1109, 1869 (1979).

156. D. A. Johnson and K. A. Pashman, *Inorg. Nucl. Chem. Lett.*, **11**, 23 (1975).

157. J. L. Burmeister and F. Basolo, *Inorg. Chem.*, **3**, 1587 (1964).

158. M. F. Farona and A. Wojcicki, *Inorg. Chem.*, **4**, 857 (1965).

159. A. Haim and N. Sutin, *J. Am. Chem. Soc.*, **87**, 4210 (1965); **88**, 434 (1966).

160. J. L. Burmeister, R. L. Hassel, and R. J. Phelen, *Inorg. Chem.*, **10**, 2032 (1972).

161. A. Diaz, M. Massacesi, G. Ponticelli, and G. Paschina, *J. Inorg. Nucl. Chem.*, **37**, 2469 (1975).

162. J. E. Huheey and S. O. Grim, *Inorg. Nucl. Chem. Lett.*, **10**, 973 (1974).

163. J. Halpern and S. Nakamura, *J. Am. Chem. Soc.*, **87**, 3002 (1965).

164. J. H. Espensen and J. P. Birk, *J. Am. Chem. Soc.*, **87**, 3280 (1965).

165. D. Gaswick and A. Haim, *J. Inorg. Nucl. Chem.*, **40**, 437 (1978).

166. D. E. Peters and R. T. M. Fraser, *J. Am. Chem. Soc.*, **87**, 2758 (1965).

167. F. Basolo, W. H. Baddley, and J. L. Burmeister, *Inorg. Chem.*, **3**, 1202 (1964).

168. F. Basolo, W. H. Baddley, and K. J. Weidenbaum, *J. Am. Chem. Soc.*, **88**, 1576 (1966).

169. P. Nicpon and D. W. Meek, *Inorg. Chem.*, **6**, 145 (1967).

169a. G. J. Palenik, W. L. Steffen, M. Mathew, M. Li, and D. W. Meek, *Inorg. Nucl. Chem. Lett.*, **10**, 125 (1974).

169b. F. A. Cotton, A. Davison, W. H. Ilsley, and H. S. Trop, *Inorg. Chem.*, **18**, 2719 (1979).

170. D. F. Shriver, S. A. Shriver, and S. E. Anderson, *Inorg. Chem.*, **4**, 725 (1965).

171a. D. Hall and A. A. Taggart, *J. Chem. Soc.*, **1965**, 1359; D. Dale and D. C. Hodgkin, *J. Chem. Soc.*, **1965**, 1364.

171b. R. D. Feltham, *Inorg. Chem.*, **3**, 1038 (1964).

172. M. C. Browning, J. R. Mellor, D. J. Morgan, S. A. J. Pratt, L. E. Sutton, and L. M. Venanzi, *J. Chem. Soc.*, **1962**, 693.

173. R. G. Hayter and F. S. Humiec, *J. Am. Chem. Soc.*, **84**, 2004 (1962); *Inorg. Chem.*, **4**, 1701 (1965).

174. B. T. Kilbourn, H. M. Powell, and J. A. C. Darbyshire, *Proc. Chem. Soc.*, **1963**, 207.

175. F. Basolo, in *The Chemistry of the Coordination Compounds*, J. C. Bailar, Jr., and D. H. Busch (Eds.), Ch. 8, Reinhold Publishing Corp., New York (1956). (C.N. = 6.)

176. B. P. Block, in *The Chemistry of the Coordination Compounds*, J. C. Bailar, Jr., and D. H. Busch (Eds.), Ch. 9, Reinhold Publishing Corp., New York (1956). (C.N. = 4.)

177. T. D. O'Brien, in *The Chemistry of the Coordination Compounds*, J. C. Bailar, Jr., and D. H. Busch (Eds.), Ch. 10, Reinhold Publishing Corp., New York (1956). (Less common C.N.'s.)

178. F. Basolo and R. G. Pearson, *Mechanisms of Inorganic Reactions: A Study of Metal Complexes in Solution*, 2nd ed., Ch. 4, John Wiley & Sons, New York (1967).

179. J. M. Fergusson, *Stereochemistry and Bonding in Inorganic Chemistry*, Prentice-Hall, Englewood Cliffs, N.J. (1974).

180. E. Cartmell and G. W. A. Fowles, *Valency and Molecular Structure*, 4th ed., Chs. 11, 12, Butterworth, London (1977).

181. G. E. Coates and N. D. Huck, *J. Chem. Soc.*, **1952**, 4501.

182. R. C. Cass, G. E. Coates, and R. G. Hayter, *J. Chem. Soc.*, **1955**, 4007.

183. R. H. Fenn, J. W. H. Oldham, and D. C. Phillips, *Nature*, **198**, 381 (1963).

184. J. A. Tiethof, J. K. Stalick, and D. W. Meek, *Inorg. Chem.*, **12**, 1170 (1973).

185. V. Albano, P. L. Bellon, and V. Scatturin, *J. Chem. Soc. Chem. Commun.*, **1966**, 507.

186. P. G. Eller and P. W. R. Corfield, *J. Chem. Soc. Chem. Commun.*, **1971**, 105.

187. I. S. Ahuja and K. S. Rao, *J. Inorg. Nucl. Chem.*, **37**, 586 (1975).

188. C. Brink and C. H. MacGillavry, *Acta Crystallogr.*, **2**, 158 (1949).

189. D. T. Cromer, *J. Phys. Chem.*, **61**, 1388 (1957).

190. P. G. Eller, D. C. Bradley, M. B. Hursthouse, and D. W. Meek, *J. Coord. Chem. Rev.*, **24**(1), 1 (1977).

191. D. P. Mellor, *Chem. Rev.*, **33**, 137 (1943).

192. E. L. Muetterties, *J. Am. Chem. Soc.*, **91**, 1636 (1969).

193. R. B. King, *J. Am. Chem. Soc.*, **91**, 7211 (1969).

194. Ref. 179, pp. 197–219.

195. J. Reiset, *C.R. Acad. Sci.*, **18**, 1103 (1844).

196. M. Peyrone, *Ann. Chem.*, **51**, 1 (1844).

197. A. D. Gel'man and E. Gorushkina, *C. R. (Dokl.) Acad. Sci. URSS*, **55**, 33 (1947).

198. A. D. Gel'man, E. F. Karandashova, and L. N. Essen, *Dokl. Akad. Nauk, SSSR*, **63**, 37 (1948).

199. I. I. Tscherniaev, *Ann. Inst. Platine*, **6**, 55 (1928).

200. W. H. Mills and T. H. H. Quibell, *J. Chem. Soc.*, **1935**, 839.

201. R. G. Dickinson, *J. Am. Chem. Soc.*, **44**, 2404 (1922).

202. E. L. Muetterties and R. A. Schunn, *Q. Rev. (Lond.)*, **20**, 245 (1966).

203. H. P. Klug and L. Alexander, *J. Am. Chem. Soc.*, **66**, 1056 (1944).

204. H. M. Powell and A. F. Wells, *J. Chem. Soc.*, **1935**, 359.

205. H. P. Klug, E. Kummer, and L. Alexander, *J. Am. Chem. Soc.*, **70**, 3064 (1948).

206. I. Lindquist, *Ark. Kemi Mineral. Geol.*, **24A**, 1 (1947).

207. M. Ciampolini and G. P. Speroni, *Inorg. Chem.*, **5**, 45 (1966).

208. T. V. Long, II, A. W. Herlinger, E. F. Epstein, and I. Bernal, *Inorg. Chem.*, **9**, 459 (1970).

209. D. W. Meek and J. A. Ibers, *Inorg. Chem.*, **9**, 465 (1970).

210. C. Furlani, *Coord. Chem. Rev.*, **3**, 141 (1968).

211. P. L. Orioli, *Coord. Chem. Rev.*, **6**, 285 (1971).

212. J. S. Wood, in *Progress in Inorganic Chemistry*, S. J. Lippard (Ed.), Vol. 16, pp. 227–486, Interscience Publishers, New York (1972).

213. R. Morassi, I. Bertini, and L. Sacconi, *Coord. Chem. Rev.*, **11**, 343 (1973).

214. M. Ciampolimi, *Struct. Bonding*, **6**, 52 (1969).

215. K. N. Raymond, D. W. Meek, and J. A. Ibers, *Inorg. Chem.*, **7**, 1111 (1968).

216. D. L. Stevenson and L. F. Dahl, *J. Am. Chem. Soc.*, **89**, 3424 (1967).

217. A. Terzis, K. N. Raymond, and T. G. Spiro, *Inorg. Chem.*, **9**, 2415 (1970).

218. M. E. Kimball and W. C. Kaska, *Inorg. Nucl. Chem. Lett.*, **7**, 119 (1971).

219. P. Meakin, J. P. Jesson, F. N. Tebbe, and E. L. Muetterties, *J. Am. Chem. Soc.*, **93**, 1797 (1971).

220. J. R. Shapley and J. A. Osborn, *Acc. Chem. Res.*, **6**, 305 (1973).

221. J. I. Musher, *J. Chem. Educ.*, **51**, 94 (1974).

222. K. N. Raymond, P. W. R. Corfield, and J. A. Ibers, *Inorg. Chem.*, **7**, 1362 (1968).

223. J. D. Stalick, P. W. R. Corfield, and D. W. Meek, *Inorg. Chem.*, **12**, 1668 (1973).

224. L. D. Brown and K. N. Raymond, *Inorg. Chem.*, **14**, 2590 (1975).

225. I. Bernal, A. Clearfield, and J. S. Ricca, Jr., *J. Cryst. Mol. Struct.*, **4**, 43 (1974).

226. J. D. Main Smith, *Chemistry and Atomic Structure*, p. 97, E. Bern, London (1924).

227. W. C. Fernelius and B. E. Bryant, *J. Am. Chem. Soc.*, **75**, 1735 (1953).

228. R. B. Trimble, Jr., *J. Chem. Educ.*, **31**, 176 (1954).

229. B. A. Kennedy, D. A. McQuarrie, and C. H. Brubaker, Jr., *Inorg. Chem.*, **3**, 265 (1964).

230. D. H. McDaniel, *Inorg. Chem.*, **11**, 2678 (1972).

231. J. C. Bailar, Jr., *J. Chem. Educ.*, **34**, 334, 626 (1957).

232. S. A. Mayper, *J. Chem. Educ.*, **34**, 623 (1957).

233. W. E. Bennett, *Inorg. Chem.*, **8**, 1325 (1969).

233a. Chung-Sun Chung, *J. Chem. Educ.*, **56**, 398 (1979).

234. R. Eisenberg and J. A. Ibers, *J. Am. Chem. Soc.*, **87**, 3776 (1965); *Inorg. Chem.*, **5**, 411 (1966).

234a. D. L. Kepert, *Inorg. Chem.*, **11**, 1561 (1972).

235. G. N. Schrauzer, in *Transition Metal Chemistry*, R. L. Carlin (Ed.), Vol. 4, pp. 299–335, Marcel Dekker, New York (1968).

236. J. A. McCleverty, in *Progress in Inorganic Chemistry*, F. A. Cotton (Ed.), Vol. 10, pp. 49–221, Interscience Publishers, New York (1968).

237. D. Coucouvanis, in *Progress in Inorganic Chemistry*, S. J. Lippard (Ed.), Vol. 11, pp. 233–371; Vol. 26, pp. 301–469, Interscience Publishers, New York (1970, 1979).

238. R. Eisenberg, in *Progress in Inorganic Chemistry*, S. J. Lippard (Ed.), Vol. 12, pp. 295–369, Interscience Publishers, New York (1970).

239. R. A. D. Wentworth, *Coord. Chem. Rev.*, **9**, 171 (1972/3).

240. J. R. Dilworth, J. Hyde, P. Lyford, P. Vella, K. Venkatasubramaman, and J. A. Zubeta, *Inorg. Chem.*, **18**, 268 (1979).

241. J. A. Bertrand, J. A. Kelley, and E. G. Vassian, *J. Am. Chem. Soc.*, **91**, 2394 (1969).

242. S. M. Jørgensen, *J. Prakt. Chem.*, **39**, 16 (1889); **41**, 448 (1890).

243. F. Basolo, *Chem. Rev.*, **52**, 459 (1953).

244. H. J. Emeléus and J. S. Anderson, *Modern Aspects of Inorganic Chemistry*, 3rd ed., Ch. 5, D. Van Nostrand, New York (1960).

245. J. C. Bailar, Jr., and D. F. Peppard, *J. Am. Chem. Soc.*, **62**, 105 (1940).

246. R. B. Hagel and L. F. Druding, *Inorg. Chem.*, **9**, 1496 (1970).

247. R. E. Connick and D. A. Fine, *J. Am. Chem. Soc.*, **83**, 3414 (1961).

248. E. E. Mercer and W. A. McAllister, *Inorg. Chem.*, **4**, 1414 (1965).

249. V. S. Orlova and I. I. Chernyaev, *Zh. Neorg. Khim.*, **11**, 1338 (1966).

250. V. S. Orlova and I. I. Chernyaev, *Zh. Neorg. Khim.*, **12**, 1274 (1967).

251. A. A. El-Awady, E. J. Bounsell, and C. S. Garner, *Inorg. Chem.*, **6**, 79 (1967).

252. B. D. Catsikis and M. L. Good, *Inorg. Chem.*, **8**, 1095 (1969).

253. E. L. Muetterties and C. M. Wright, *Q. Rev. (Lond.)*, **21**, 109 (1967).

254. R. K. Boggess and W. D. Wiegele, *J. Chem. Educ.*, **55**, 156 (1978).

255. S. J. Lippard, in *Progress in Inorganic Chemistry*, S. J. Lippard (Ed.), Vol. 21, pp. 91–103, Interscience Publishers, New York (1976). Seven- and eight-coordinate molybdenum complexes.

256. M. G. B. Drew, in *Progress in Inorganic Chemistry*, S. J. Lippard (Ed.), Vol. 23, pp. 63–210, Interscience Publishers, New York (1977).

257. D. Kepert, in *Progress in Inorganic Chemistry*, S. J. Lippard (Ed.), Vol. 25, pp. 41–144, Interscience Publishers, New York (1979).

257a. E. B. Dreyer, C. T. Lam, and S. J. Lippard, *Inorg. Chem.*, **18**, 1904 (1979).

258. E. L. Muetterties, *Inorg. Chem.*, **4**, 769 (1965).

259. W. H. Zachariasen, *Acta Crystallogr.*, **7**, 783 (1954).

260. H. J. Hurst and J. C. Taylor, *Acta Crystallogr.*, **B26**, 417, 2136 (1970).

261. R. A. Levenson and R. L. R. Towns, *Inorg. Chem.*, **13**, 105 (1974).

262. J. J. Park, D. M. Collins, and J. L. Hoard, *J. Am. Chem. Soc.*, **92**, 3636 (1970).

263. R. Stomberg, *Ark. Kemi*, **23**, 401 (1964).

264. J. L. Hoard, *J. Am. Chem. Soc.*, **61**, 1252 (1939).

265. D. L. Lewis and S. J. Lippard, *J. Am. Chem. Soc.*, **97**, 2697 (1975).

266. M. B. Williams and J. L. Hoard, *J. Am. Chem. Soc.*, **64**, 1139 (1942).

267. M. G. B. Drew and A. P. Walters, *J. Chem. Soc. Chem. Commun.*, **1972**, 457.

268. L. Manojlović-Muir, *Inorg. Nucl. Chem. Lett.*, **9**, 59 (1973).

269. A. Mawby and G. E. Pringle, *J. Inorg. Nucl. Chem.*, **34**, 517 (1972).

270. A. J. Gash, P. F. Rodesiler, and E. L. Amma, *Inorg. Chem.*, **13**, 2429 (1974).

271. J. J. Stezowski, R. Countryman, and J. L. Hoard, *Inorg. Chem.*, **12**, 1749 (1973).

271a. T. J. Giordano, G. J. Palenik, R. C. Palenik, and D. A. Sullivan, *Inorg. Chem.*, **18**, 2445 (1979).

272. P. Hoffmann, B. F. Beier, E. L. Muetterties, and A. R. Rossi, *Inorg. Chem.*, **16**, 511 (1977).

273. S. M. Nelson and D. H. Busch, *Inorg. Chem.*, **8**, 1859 (1969).

274. M. G. B. Drew, A. H. Othman, and S. M. Nelson, *J. Chem. Soc.* Dalton, **1976**, 1394.

275. M. F. Brown, B. R. Cook, and T. E. Sloan, *Inorg. Chem.*, **17**, 1563 (1978).

276. S. J. Lippard, in *Progress in Inorganic Chemistry*, F. A. Cotton (Ed.), Vol. 8, pp. 109–193, Interscience Publishers, New York (1967). (Review, C.N. = 8.)

277. D. G. Blight and D. L. Kepert, *Inorg. Chem.*, **11**, 1556 (1972).

278. M. G. B. Drew, *Coord. Chem. Rev.*, **24**, 179–275 (1977). (Review, C.N. = 8, 9, 10, 11, 12, 14.)

279. D. L. Kepert, in *Progress in Inorganic Chemistry*, S. J. Lippard (Ed.), Vol. 24, pp. 179–249, Interscience Publishers, New York (1978).

280. J. L. Hoard and J. V. Silverton, *Inorg. Chem.*, **2**, 235 (1963).

281. J. V. Silverton and J. L. Hoard, *Inorg. Chem.*, **2**, 243 (1963).

282. G. L. Glen, J. V. Silverton, and J. L. Hoard, *Inorg. Chem.*, **2**, 250 (1963).

283. R. Graziani, G. Bombieri, and E. Forsellini, *J. Chem. Soc. Dalton*, **1972**, 2059.

284. G. A. Barclay, T. M. Sabine, and J. C. Taylor, *Acta Crystallogr.*, **19**, 205 (1965).

285. N. W. Alcock, *J. Chem. Soc. Dalton*, **1973**, 1610.

286. G. Brunton, *J. Inorg. Nucl. Chem.*, **29**, 1631 (1967).

287. L. A. Aslanov, E. F. Korytnyi, and M. A. Porai-Koshits, *J. Struct. Chem.*, **12**, 600 (1971).

288. J. Jay, J. W. Moore, and M. D. Glick, *Inorg. Chem.*, **11**, 2818 (1972).

289. J. L. Hoard, G. L. Glen, and J. V. Silverton, *J. Am. Chem. Soc.*, **83**, 4293 (1961).

290. J. L. Hoard, W. J. Martin, M. E. Smith, and J. F. Whitney, *J. Am. Chem. Soc.*, **76**, 3820 (1954).

291. G. Bombieri, P. T. Moseley, and D. Brown, *J. Chem. Soc. Dalton*, **1975**, 1520.

292. S. S. Basson, L. D. C. Bok, and J. G. Leipoldt, *Acta Crystallogr.*, **B26**, 1209 (1970).

293. C. C. Bisi, A. D. Giusta, A. Coda, and V. Tazzoli, *Cryst. Struct. Commun.*, **3**, 381 (1974).

294. P. Singh, A. Clearfield, and I. Bernal, *J. Coord. Chem.*, **1**, 29 (1971).

295. S. Husebye and S. E. Svaeren, *Acta Chem. Scand.*, **27**, 763 (1973).

296. C. D. Garner and S. C. Wallwork, *J. Chem. Soc. A*, **1966**, 1496.

297. R. J. H. Clark, J. Lewis, R. S. Nyholm, P. Pauling, and G. B. Robertson, *Nature*, **192**, 222 (1961).

298. J. L. Hoard and H. H. Nordsieck, *J. Am. Chem. Soc.*, **61**, 2853 (1939).

298a. H. K. Chung, W. L. Steffen, and R. C. Fay, *Inorg. Chem.*, **18**, 2458 (1979).

299. S. R. Fletcher and A. C. Skapski, *J. Organomet. Chem.*, **59**, 299 (1973).

300. A. LeClaire, *Acta Crystallogr.*, **B30**, 2259 (1974).

301. E. J. Peterson, R. B. Von Dreele, and T. M. Brown, *Inorg. Chem.*, **15**, 309 (1976).

302. M. G. B. Drew, A. P. Wolters, and J. D. Wilkins, *Acta Crystallogr.*, **B31**, 324 (1974).

303. B. E. Robertson, *Inorg. Chem.*, **16**, 2735 (1977).

304. S. C. Abrahams, A. P. Ginsberg, and K. Knox, *Inorg. Chem.*, **3**, 558 (1964); A. P. Ginsberg, *Inorg. Chem.*, **3**, 567 (1964).

305. L. Helmholz, *J. Am. Chem. Soc.*, **61**, 1544 (1939).

306. D. R. Fitzwater and R. E. Rundle, *Z. Kristallogr., Kristallgeom., Kristallphys., Kristallchem.*, **112**, 362 (1959).

307. J. Albertsson and I. Elding, *Acta Crystallogr.*, **33B**, 1460 (1977).

308. K. Schubert and A. Seitz, *Z. Anorg. Chem.*, **254**, 116 (1947).

309. G. W. Beall, W. O. Milligan, and H. A. Wolcott, *J. Inorg. Nucl. Chem.*, **39**, 65 (1977).

310. P. I. Lazerev, L. A. Aslanov, and M. A. Porai-Koshits, *Koord. Khim.*, **1**, 979 (1975).

311. V. V. Bakakin, R. F. Klevstova, and L. P. Solov'eva, *J. Struct. Chem.*, **15**, 723 (1974).

312. R. R. Ryan, R. A. Penneman, and A. Rosenzweig, *Acta Crystallogr.*, **25b**, 1958 (1969).

313. D. H. Templeton and C. H. Dauben, *J. Am. Chem. Soc.*, **75**, 6069 (1953).

314. V. W. Day and J. L. Hoard, *J. Am. Chem. Soc.*, **92**, 3626 (1970).

315. J. L. Hoard, B. Lee, and M. D. Lind, *J. Am. Chem. Soc.*, **87**, 1612 (1965).

316. A. R. Al-Karaghouli and J. S. Wood, *Inorg. Chem.*, **11**, 2293 (1972).

317. M. D. Lind, B. K. Lee, and J. L. Hoard, *J. Am. Chem. Soc.*, **87**, 1611 (1965).

318. T. Ueki, A. Zalkin, and D. H. Templeton, *Acta Crystallogr.*, **20**, 836 (1966).

319. J. C. Taylor, M. H. Mueller, and R. L. Hitterman, *Acta Crystallogr.*, **20**, 842 (1966).

320. G. Johansson, *Acta Chem. Scand.*, **22**, 389 (1968).

321. A. R. Al-Karaghouli and J. S. Wood, *J. Chem. Soc. Chem. Commun.*, 135 (1970).

322. H. B. Kerfoot, G. R. Choppin, and T. J. Kistenmacher, *Inorg. Chem.*, **18**, 787 (1979).

323. M. N. Akhtar and A. J. Smith, *Acta Crystallogr.*, **B31**, 1361 (1975).

324. G. E. Toogood and C. Chieh, *Can. J. Chem.*, **53**, 831 (1975).

325. D. B. Shinn and H. A. Eick, *Inorg. Chem.*, **7**, 1340 (1968).

326. T. A. Beineke and J. Delgaudio, *Inorg. Chem.*, **7**, 715 (1968).

327. S. Ščavničar and B. Prodić, *Acta Crystallogr.*, **18**, 698 (1965).

328. W. T. Carnall, S. Siegel, J. R. Ferraro, B. Tani, and E. Gebert, *Inorg. Chem.*, **12**, 560 (1973).

329. A. Clearfield, R. Gopal, and R. W. Olsen, *Inorg. Chem.*, **16**, 911 (1977).

330. E. G. Sherry, *J. Solid State Chem.*, **19**, 271 (1976).

331. E. B. Hunt, Jr., R. E. Rundle, and A. J. Stosick, *Acta Crystallogr.*, **7**, 106 (1954).

332. A. G. Lidstone and W. H. Mills, *J. Chem. Soc.*, **1939**, 1754.

333. W. H. Mills and R. A. Gotts, *J. Chem. Soc.*, **1926**, 3121.

334. D. H. Busch and J. C. Bailar, Jr., *J. Am. Chem. Soc.*, **76**, 5352 (1954).

335. J. C. I. Liu and J. C. Bailar, Jr., *J. Am. Chem. Soc.*, **73**, 5432 (1951).

336. A. Werner, *Chem. Ber.*, **44**, 1887 (1911). English translation: Ref. 7, Part I, pp. 159–173.

337. F. G. Mann, *J. Chem. Soc.*, **1933**, 412.

338. A. Werner, *Chem. Ber.*, **47**, 3087 (1914). English translation: Ref. 7, Part I, pp. 177–184.

339. T. Moeller and N. C. Nielsen, *J. Am. Chem. Soc.*, **75**, 5106 (1953).

340. F. P. Dwyer and E. C. Gyarfas, *Nature*, **168**, 29 (1951).

341. T. Moeller, E. Gulyas, and R. H. Marshall, *J. Inorg. Nucl. Chem.*, **9**, 82 (1959).

342. J. P. Collman, R. P. Blair, R. L. Marshall, and L. Slade, *Inorg. Chem.*, **2**, 576 (1963).

343. F. R. Keene and G. H. Searle, *Inorg. Chem.*, **11**, 148 (1972).

344. F. Basolo, *J. Am. Chem. Soc.*, **70**, 2634 (1948).

345. J. Collins, F. P. Dwyer, and F. Lions, *J. Am. Chem. Soc.*, **74**, 3134 (1952).

346. F. P. Dwyer, N. S. Gill, E. C. Gyarfas, and F. Lions, *J. Am. Chem. Soc.*, **74**, 4188 (1952); **75**, 2443 (1953).

347. F. P. Dwyer and F. Lions, *J. Am. Chem. Soc.*, **72**, 1545 (1950).

348. F. P. Dwyer, E. C. Gyarfas, and D. P. Mellor, *J. Phys. Chem.*, **59**, 296 (1955).

349. F. P. Dwyer and F. L. Garvan, *Inorg. Synth.*, **6**, 192–194 (1960).

350. A. Werner, *Chem. Ber.*, **46**, 3674 (1913).

351. F. G. Mann, *J. Chem. Soc.*, **1930**, 1745.

352. J. R. Kuebler, Jr., and J. C. Bailar, Jr., *J. Am. Chem. Soc.*, **74**, 3535 (1952).

353. S. Kirschner, in *Preparative Inorganic Reactions*, W. L. Jolly (Ed.), Vol. 1, Ch. 2, Interscience Publishers, New York (1964).

354. P. H. Boyle, *Q. Rev. (Lond.)*, **25**, 323 (1971).

355. J. C. Bailar, Jr., *Inorg. Synth.*, **2**, 222 (1946).

356. D. H. Busch, *J. Am. Chem. Soc.*, **77**, 2747 (1955).

357. F. P. Dwyer and A. M. Sargeson, *J. Phys. Chem.*, **60**, 1331 (1956).

358. W. T. Jordan and L. R. Froebe, *Inorg. Synth.*, **18**, 96 (1978).

359. G. B. Kauffman and E. V. Lindley, Jr., *Inorg. Synth.*, **16**, 93 (1976).

360. W. W. Brandt, F. P. Dwyer, and E. C. Gyarfas, *Chem. Rev.*, **54**, 959 (1954).

361. R. Tsuchida, M. Kobayashi, and A. Nakamura, *J. Chem. Soc. Jpn.*, **56**, 1339 (1935); *Bull. Chem. Soc. Jpn.*, **11**, 38 (1936).

362. H. Krebs and R. Rasche, *Z. Anorg. Allg. Chem.*, **276**, 236 (1954).

363. H. Krebs, J. Diewold, and J. A. Wagner, *Angew. Chem.*, **67**, 705 (1955).

364. H. Krebs and J. Diewold, *Z. Anorg. Allg. Chem.*, **287**, 98 (1956).

365. T.-M. Hseu, D. F. Martin, and T. Moeller, *Inorg. Chem.*, **2**, 587 (1963).

366. R. E. Sievers, R. W. Mosier, and M. L. Morris, *Inorg. Chem.*, **1**, 966 (1962).

366a. N. S. Bowman, V. G'ceva, G. K. Schweitzer, and I. R. Supernaw, *Inorg. Nucl. Chem. Lett.*, **2**, 351 (1966).

366b. G. K. Schweitzer, I. R. Supernaw, and N. S. Bowman, *J. Inorg. Nucl. Chem.*, **30**, 1885 (1968).

367. H. B. Jonassen, J. C. Bailar, Jr., and E. H. Huffman, *J. Am. Chem. Soc.*, **70**, 756 (1948).

368. B. Das Sarma and J. C. Bailar, Jr., *J. Am. Chem. Soc.*, **77**, 5480 (1955).

369. S. Kirschner, Y.-K. Wei, and J. C. Bailar, Jr., *J. Am. Chem. Soc.*, **79**, 5877 (1957).

370. E. J. Corey and J. C. Bailar, Jr., *J. Am. Chem. Soc.*, **81**, 2620 (1959).

371. L. Pasteur, *Ann. Chim. Phys.*, **24**, 442 (1848).

372. F. M. Jaeger, *Rec. Trav. Chim.*, **38**, 171 (1919).

373. H. A. Weakliem and J. L. Hoard, *J. Am. Chem. Soc.*, **81**, 549 (1959).

374. A. Werner and J. Bosshart, *Chem. Ber.*, **47**, 2171 (1914).

375. T. Moeller and E. H. Grahn, *J. Inorg. Nucl. Chem.*, **5**, 53 (1957).

376. F. A. Long, *J. Am. Chem. Soc.*, **63**, 1353 (1941).

377. J. R. Gollogly, C. J. Hawkins, and J. K. Beattie, *Inorg. Chem.*, **10**, 317 (1971).

378. J. K. Beattie, *Acc. Chem. Res.*, **4**, 253 (1971).

379. Y. Shimura, *Bull. Chem. Soc. Jpn.*, **31**, 311 (1958).

380. J. M. Bijvoet, *Endeavor*, **14**, 71 (1955).

381. Y. Saito, K. Nakatsu, M. Shiro, and H. Kuroya, *Acta Crystallogr.*, **8**, 729 (1955).

382. C. F. Liu and J. A. Ibers, *Inorg. Chem.*, **9**, 773 (1970).

383. K. Garbett and R. D. Gillard, *J. Chem. Soc. A*, **1966**, 802.

384. A. Cotton, *C.R. Acad. Sci.*, **120**, 989, 1044 (1895).

385. R. D. Gillard, in *Progress in Inorganic Chemistry*, F. A. Cotton (Ed.), Vol. 7, pp. 215–276, Interscience Publishers, New York (1966).

386. C. J. Hawkins, *Absolute Configurations of Metal Complexes*, John Wiley & Sons, New York (1971).

387. I. J. Kindred and D. A. House, *J. Inorg. Nucl. Chem.*, **37**, 1359 (1975).

SUGGESTED SUPPLEMENTARY REFERENCES

J. C. Bailar, Jr., and D. H. Busch (Eds.), *The Chemistry of the Coordination Compounds*, Reinhold Publishing Corp., New York (1956). Detailed treatments of special topics. Older but still very useful.

F. Basolo and R. G. Pearson, *Mechanisms of Inorganic Reactions: A Study of Metal Complexes in Solution*, 2nd ed., John Wiley & Sons, New York (1967). Largely kinetics and mechanisms, but contains much other useful information.

J. M. Fergusson, *Stereochemistry and Bonding in Inorganic Chemistry*, Prentice-Hall, Englewood Cliffs, N.J. (1974). Brief but useful treatment.

M. and J. Hargittai, *The Molecular Geometries of Coordination Compounds in the Vapour Phase*, Elsevier, Amsterdam (1977). Unique treatment of gaseous species.

G. B. Kauffman, *Alfred Werner, Founder of Coordination Chemistry*, Springer-Verlag, New York (1966). Biography with many interesting comments on coordination chemistry and citations of all of Werner's publications.

G. B. Kauffman (Ed.), *Classics in Coordination Chemistry*, Parts I, II. Dover Publications, New York (1968, 1976). Translations of and commentaries on significant European papers dating from 1828.

G. B. Kauffman (Ed.), *Werner Centennial*, Vol. 62, Advances in Chemistry Series, American Chemical Society, Washington, D.C. (1967). Symposium papers in honor of the 100th birthday of Alfred Werner.

G. A. Melson (Ed.), *Coordination Chemistry of Macrocyclic Compounds*, Academic Press, New York (1979). Reactions of macrocyclic donors.

J. K. O'Loane, *Chem. Rev.*, **80**, 41 (1980). Detailed discussion of optical activity in small molecules, including asymmetry vs. dissymmetry.

A. Werner, *Neuere Anschauungen auf dem Gebiete der Anorganischen Chemie*, 4th ed., T. Viewig und Sohn, Braunschweig (1920). Werner's textbook, delineating many of his ideas and contributions.

R. G. Wilkins, *The Study of Kinetics and Mechanisms of Reactions of Transition Metal Ions*, Allyn and Bacon, Boston (1974).

CHAPTER EIGHT

COORDINATION CHEMISTRY, II BONDING — MODERN APPROACHES; STRUCTURE AND BONDING IN SPECIFIC TYPES OF COORDINATION ENTITIES

1. BONDING IN TRANSITION METAL COMPLEXES — MODERN APPROACHES*

The inadequacies of the simple Sidgwick extension of the Werner theory of bonding (Sec. 7.1-3) are emphasized by its inability to account even qualitatively for many of the observed properties of complex species as outlined in Chapter 7. Modern approaches involve the valence-bond, electrostatic, and molecular-orbital approximations discussed in general terms in Chapters 4 to 6. These are now considered in more detail as they apply, most specifically, to complexes of the d-transition metals. In general they are based upon magnetic, spectroscopic, and structural data.

Magnetic data (Sec. 2.4-4), relating as they do to unpaired electrons, aid in selection of proper theoretical interpretations and in distinction between alternative geometrical arrangements of donor atoms. Absorption spectra (Sec. 6.5-3), reflecting as they do electronic or molecular vibrations and transitions between energy states, aid in assigning correct formulas, in measuring relative thermodynamic stabilities, and in deducing reaction mechanisms (Chs. 9, 11); in assigning correct molecular structures, point-group symmetries, and absolute configurations; and thus in providing the bases for modern approaches to bonding.

Vibrational spectra are symmetry related and molecular-structure indicative. Electronic spectra (visible and ultraviolet regions) are due to electronic transitions and indicate the energy changes associated with those transitions. Absorption in the visible region is directly related to the observed color in the solid state or in solution.[1] Many d-transition metal ions in various oxidation states in a multitude of complexes are distinctively colored.

Broadly, electronic absorption spectra are characterized by three bands, namely in the 4500 to 5500 Å, 3200 to 4000 Å, and 1950 to 2500 Å regions. The first band has been associated with the transitions of nonbonding electrons of the metal ion; the second band with those of electrons involved in bonding the ligand to the metal ion; and the third band with those electrons that are

---

* A particularly pertinent, informative, and interesting historical account of the application of quantum mechanical principles to bonding in inorganic complexes appears in a series of papers by C. J. Ballhausen [*J. Chem. Educ.*, **56**, 215, 294, 357 (1979)].

influenced by the geometry of the species.[2] These associations are not necessarily exact. Displacements of the second band toward shorter wavelengths (hyposochromic shifts) among cobalt(III) ammines containing other coordinating groups have been associated with changes in thermodynamic stability according to the series[3]

$$NO_2^- > ONO^- > H_2O > NCS^- > OH^- > NO_3^- > Cl^- > CO_3^{2-} > Br^-$$

decreasing hyposochromic shift and decreasing thermodynamic stability

The significance of this early *spectrochemical series* is pointed out later (Sec. 8.1-2). The third band, although associated originally with *trans* geometry, also characterizes *cis* geometry, but at different wavelengths.[4] Electronic absorption spectra have been most useful in supporting and providing background for the crystal-field, ligand-field, and molecular-orbital approximations to bonding in *d*-transition metal complexes.[5-11]

1-1. Valence-, or Directed-, Bond Approximation

The valence-, or directed-, bond approximation (Sec. 5.1-1) was first extended to coordination compounds in 1931 by L. Pauling,[12,13] in terms of the fundamental concepts of orbital hybridization, double bonding between ligand and metal ion, and relationship of observed magnetic properties to bond type.

That appropriate linear combinations of hydrogenlike wave functions that are not geometrically equivalent may give sets of hybrid orbitals that are equivalent and geometrically oriented has been pointed out in earlier discussions (Secs. 5.1-1, 5.1-2). For complex species, the most important of these (Sec. 6.1-2) are sp (linear), sp^3 (tetrahedral), dsp^2 (square planar), dsp^3 (trigonal bipyramidal), and d^2sp^3 (octahedral). Hybrids of these types commonly result from presumed rearrangements to make orbitals available to electrons provided by the ligand, although on occasion such rearrangement may be unnecessary. In terms of the notation adopted earlier, typical examples are indicated in Table 8-1. It should be noted that hybrids involving only s and p orbitals similarly account for the observed geometries of complex species derived from the representative elements (e.g., tetrahedral $[BF_4]^-$ or $[BeF_4]^{2-}$). Hybrids of all of these types must involve orbitals lying along the x, y, and z axes (i.e., the d_{z^2}, $d_{x^2-y^2}$, p_x, p_y, and p_z orbitals, and s orbitals).

The unlikely accumulation of negative charge on the metal ion as a consequence of its acceptor properties may be precluded, in some instances at least, by the assumption of reverse donation of electrons to the ligand, or in other words by double bonding. This has a delocalizing effect and serves to spread negative charge over a number of atoms. The result is an overall stabilization of the system. Cases in point are the metal carbonyls and nitrosyls and the cyano complexes (Sec. 8.2-1), where observed shortening of metal–ligand bonds is in agreement with postulation.[1] Such situations can pertain only with ligands that themselves have vacant p or d orbitals. In more modern terminology, such orbitals can overlap with doubly occupied d_{xy}, d_{xz}, or d_{yz}

orbitals from the metal ion to give π bonds. It will be recalled that the position of these d orbitals between the coordinate axes favors this type of sidewise overlap. The tendency toward π bonding increases with increasing number of doubly occupied d orbitals in the metallic atom and is thus most characteristic of complexes of metals at the ends of the transition series. This type of overlap and increase in bond order is indicated in Fig. 8-1 for one Ni—C bond in the molecule $[Ni(CO)_4]$.

The rather general application of the spin-only simplification (Sec. 2.4-4) to complex species derived from transition metal ions of at least the $3d$ series permits decision as to the number of unpaired electrons in a typical substance and thus the assignment of a reasonable electronic configuration.[14] Thus, of the species summarized in Table 8-1, all except the $[CoCl_4]^{2-}$ and $[Cr(NH_3)_6]^{3+}$ ions are diamagnetic, in agreement with the completely spin-paired configurations recorded. The observed magnetic moments of these two ions (4.74 and 3.71 B.M., respectively) suggest the presence in each of three unpaired electrons, as indicated.

Many correlations have been made on the basis of magnetic data. Thus the marked thermodynamic stabilities of the iron(II), cobalt(III), and platinum(IV) species are associated with complete double occupancy of all $(n-1)d$ orbitals, whereas the ease of oxidation of octahedral cobalt(II) species and the ease of reduction of octahedral iron(III) species are associated with tendencies to achieve this type of arrangement through removal of a promoted unpaired electron or addition of an electron to complete pairing, respectively. A closer examination of magnetic data for coordination compounds derived from metals of the first transition series, however, reveals certain complications. If the metal atom has no more than three $3d$ electrons (e.g., d^0—Ti^{4+}, d^1—Ti^{3+}, d^2—Ti^{2+}, V^{3+}, or d^3—V^{2+}, Cr^{3+}), two vacant $3d$ orbitals are always available for hybridization with the $4s$ and $4p$ orbitals, and the magnetic properties of the

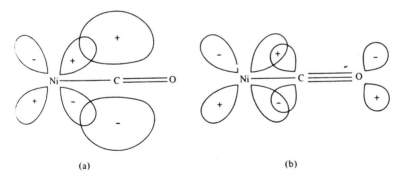

(a)   (b)

Figure 8-1 Formation of one π bond in the molecule $[Ni(CO)_4]$ by overlap of occupied d orbital of Ni atom with vacant orbital of a C atom. (*a*) Valence-bond: overlap of Ni d orbital with C p orbital; (*b*) molecular-orbital: overlap of Ni d orbital with π^* orbital of CO. [J. E. Huheey, *Inorganic Chemistry*, 2nd ed., Fig. 9, p. 347, Harper & Row, New York (1978).]

Table 8-1 Orbital Hybridization and Stereochemistry

Species	Electronic Distribution $(n-1)d$ / ns / np / nd	Spatial Arrangement of Bonds
Ag^+ $[Ag(CN)_2]^-$ — diamagnetic	sp	Linear
Co^{2+} $[CoCl_4]^{2-}$ — paramagnetic	sp^3	Tetrahedral
Ni^{2+} $[Ni(CN)_4]^{2-}$ — diamagnetic	dsp^2	Square planar
Pt^{2+} $[Pt(tetrars)Br]^+$ — diamagnetic	dsp^3	Trigonal bipyramidal

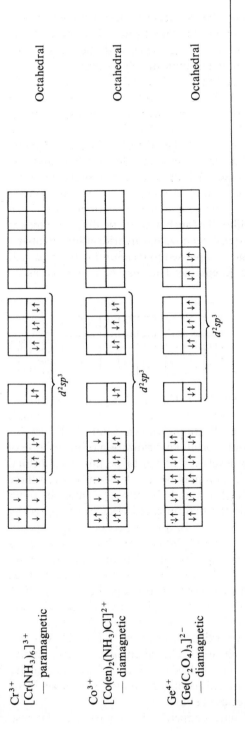

Cr^{3+}
[Cr(NH$_3$)$_6$]$^{3+}$
— paramagnetic

d^2sp^3

Octahedral

Co^{3+}
[Co(en)$_2$(NH$_3$)Cl]$^{2+}$
— diamagnetic

d^2sp^3

Octahedral

Ge^{4+}
[Ge(C$_2$O$_4$)$_3$]$^{2-}$
— diamagnetic

d^2sp^3

Octahedral

resulting octahedral complex species will be identical with those of the free ions. If more than three $3d$ electrons are present, the Hund principle requires single occupancy of as many of the $3d$ orbitals as possible, and the vacant pair necessary for octahedral hybridization is not immediately available.

For the d^4 (e.g., Cr^{2+}, Mn^{3+}), d^5 (e.g., Mn^{2+}, Fe^{3+}), and d^6 (e.g., Fe^{2+}, Co^{3+}) cases, this pair can result by pairing, to leave, respectively, two, one, and zero unpaired $3d$ electrons. Again, magnetic data agree with maximum spin pairing in many cases (especially the d^6), but with the d^4 and d^5 cases, there are significant deviations. Cases in point involve the $[Fe(CN)_6]^{3-}$ ion, the magnetic moment of which corresponds to a single unpaired $3d$ electron, and the $[FeF_6]^{3-}$ ion, the magnetic moment of which is the same as that of the free Fe^{3+} ion and corresponds to *five* unpaired electrons. Although this apparent anomaly was first ascribed to covalent bonding for the first species as opposed to ionic bonding for the second,[13] the unlikelihood of such a dramatic difference makes the assumption of an *outer orbital* (spin-free) covalent hybridization involving the available $4d$ orbitals for the second, as opposed to *inner orbital* (spin-paired) hybridization for the first,[15,16] more acceptable. This may be illustrated as

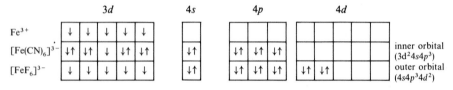

For the d^7 (e.g., Co^{2+}, Ni^{3+}), d^8 (e.g., Ni^{2+}, Cu^{3+}), d^9 (e.g., Cu^{2+}), and d^{10} (e.g., Ni^0, Cu^+, Zn^{2+}, Ga^{3+}, Ge^{4+}) cases, sufficient *inner* $3d$ orbitals cannot become available for octahedral hybridization. Known octahedral complex species must result then either from outer orbital hybridization or promotion of sufficient electrons to higher energy levels ($5s$ or $5p$) to permit inner orbital hybridization. On occasion, magnetic data may permit decision between these alternatives. Thus the species $[Ni(NH_3)_6]^{2+}$ and $[Ni(bipy)_3]^{2+}$ must be outer orbital hybrids ($4s4p^34d^2$) to provide the two unpaired electrons required by their observed paramagnetism ($\mu = 2.9$ B.M.), whereas the species $[Ni(diars)_3]^{2+}$ must be an inner orbital hybrid ($3d^24s4p^3$) with two promoted $5s$ electrons to account for its observed diamagnetic behavior. Although outer orbital species are commonly unstable with respect to oxidation (e.g., $[Co(NH_3)_6]^{2+}$) or dissociation into their components (e.g., $[Ga(C_2O_4)_3]^{3-}$), examples of exceptional stability (e.g., resistance of $[Ni(bipy)_3]^{2+}$ to oxidation and of this species and $[Ge(C_2O_4)_3]^{2-}$ to racemization) are known. An alternative, of course, is a lower coordination number with different hybridization. Thus the d^8 species can pair the two unpaired $3d$ electrons and give stable, diamagnetic square-planar $3d4s4p^2$ hybrids, and the d^{10} species can give diamagnetic tetrahedral $4s4p^3$ hybrids. This configuration is also characteristic of the paramagnetic d^7 cobalt(II) species (Table 8-1).

The d^9 configuration of 4-coordinate copper(II) is of particular interest, however. Since a single unpaired electron must be present regardless of whether the species is tetrahedral ($4s4p^3$) or square planar (either $3d4s4p^2$ with one $5s$ electron or $4s4p^24d$), magnetic data permit no decision as to configuration. Experimentally, however, these species are apparently square planar, as are the corresponding silver(II) species, and since they resist oxidation to copper(III) species, a promoted electron is unlikely. The planar outer-orbital hybrid is thus the more favorable.

It is apparent that although the valence-bond theory accounts for many of the aspects of complex-ion formation, a number of arbitrary assumptions are necessary to make it broadly applicable. Additional objections are (1) it fails to interpret quantitatively all magnetic data and any spectral data, (2) it gives no quantitative indication of the relative thermodynamic or solution stabilities of complex species and often gives incorrect qualitative indications, and (3) it does not predict exactly whether 4-coordinate complex species will be square planar or tetrahedral. The last of these is illustrated by nickel(II) species for which the hybridizations

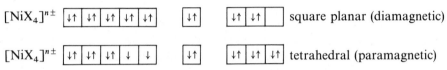

were regarded as probable and decision as to geometry was based only on magnetic moment. It is now known that the square-planar arrangement is preferred and that certain species that were assumed to be tetrahedral are actually octahedral (e.g., trimeric $[Ni(acac)_2]^{17}$) because of the greater stability of the latter arrangement.[18, 19] A few tetrahedral species (e.g., $[Ni\{P(C_6H_5)_3\}_2Br_2]$,[20] $[(C_6H_5)_3(CH_3)As]_2[NiBr_4]$,[21] and $[NiCl_4]^{2-}$ [22]) have been described, but their magnetic moments are more than one would expect on a spin-only basis.

1-2. Crystal-Field Approximation

Certain of the difficulties inherent in the covalent valence-bond approximation can be resolved by an electrostatic treatment. An early electrostatic approach, as developed in particular by A. E. Van Arkel and J. H. de Boer[23] and by F. J. Garrick[24] using the concepts of W. Kossel (Sec. 4.1) and K. Fajans (Sec. 5.1-8), permitted the evaluation of bond energies and enthalpies of formation in terms of the charges and radii of central ions and the charges, sizes, dipole moments, and polarizabilities of ligands. In addition, this approach showed that linear, tetrahedral, octahedral, and certain less common geometries are reasonable for various coordination numbers in that they allow maximum separations of ligands and thus minimize electrostatic ligand–ligand repulsions.[25, 26] Unfortunately, however, it could not account logically for square-planar geometry, for the thermal stabilities of species containing nonpolar ligands (e.g., CO), for differences in the thermodynamic stabilities of analogous complex species containing cations of the same charge and size [e.g., those of iron(III) and cobalt(III)], or for increased stabilities for cations of the same charge but larger size in proceeding from the first to the second or third d-transition series.

These problems arise because this simple approach was concerned with *overall* electrostatic interactions between metal ion and ligand, rather than with electrostatic interactions between spatially oriented d orbitals of the metal ion and the ligand. A correction essential to overcoming many of these difficulties lies in L. E. Orgel's extension[27] of the *crystal-field* approximation of H. Bethe[28] and J. H. Van Vleck[29] to the d-transition metal ions. In its original form, the crystal-field approach was concerned with the electrostatic effects of a surrounding array of anions or electric dipoles on the properties of a central cation in a crystalline solid. As applied to d-transition metal complexes, it was concerned with electrostatic interactions between surrounding ligands, treated as point charges, upon electrons in d orbitals of a central metal ion. It must be emphasized that *all interactions* in this approximation are completely electrostatic. More details than can be given in what follows are found in available summaries.[30-35]

Behavior of d Orbitals in Electrostatic Fields. The spatial orientations of the five d orbitals (Fig. 2-15) are indicated in Fig. 8-2 in relation to a circumscribed cube.* If six ligands are arranged *octahedrally* around the central metal ion M at the points (marked ●) where the three Cartesian axes emerge from the cube, it is apparent that the $d_{x^2-y^2}$ and d_{z^2} orbitals are exactly directed toward these ligands, whereas the d_{xy}, d_{yz}, and d_{xz} orbitals, directed toward the midpoints of the edges of the cube, lie between adjacent ligands. It may be expected, therefore, that electrons in the $d_{x^2-y^2}$ and d_{z^2} orbitals (commonly designated as d_γ or e_g orbitals) will be repelled extensively by negative donor sites in the ligands, whereas electrons in the d_{xy}, d_{yz}, and d_{xz} orbitals (designated as d_ε or t_{2g})† will not be as extensively repelled.‡ In terms of energy relationships, the degeneracy (equal energy) of d orbitals in an isolated gaseous atom in its ground state is removed by alteration of energy levels as indicated in Fig. 8-3.

In a *tetrahedral* arrangement of four negative ligand changes around a central metal atom or ion (points ▲, Fig. 8-2), exactly the opposite removal of degeneracy results because in this geometry the e_g orbitals are farther removed from the ligand positions than the t_{2g} orbitals and are those at lower relative energy. The relationship between octahedral and tetrahedral geometry is indicated in Fig. 8-4. In square-planar geometry, exact direction of the $d_{x^2-y^2}$ orbital toward negative ligand charges and direction of the d_{xy} orbital close to ligand positions (■ in Fig. 8-2) produces a still different splitting pattern

---

* A cubic representation of this type encompasses both octahedral and tetrahedral geometries, and, with reasonable distortions or variations, square-planar, tetragonal, square-pyramidal, and square-antiprismatic geometries as well. All of these represent cubic crystal fields.

† Doubly degenerate levels are designated e, triply degenerate levels as t. Subscript g indicates symmetric inversion through a center of symmetry; subscript 2 another symmetry property.

‡ Commonly referred to as the *splitting* of energy levels or orbitals.

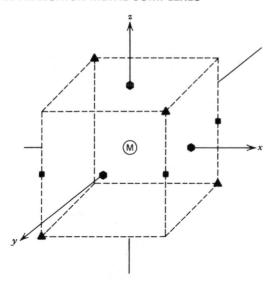

Figure 8-2 d Orbitals inside a circumscribed cube. ▲, tetrahedral; ■, square planar; ⬢, octahedral; ○, metal-ion center.

(Fig. 8-4). In trigonal bipyramidal geometry for 5-coordination, the d_{z^2} orbital is most affected (Fig. 8-5).

Energy Relationships and the Occupancy of Orbitals. It is apparent that because of preferential occupancy of lower-energy d orbitals the distribution of ligand charge about the central atom or ion is generally unsymmetrical, rather than symmetrical as assumed in early electrostatic concepts. The net result is a

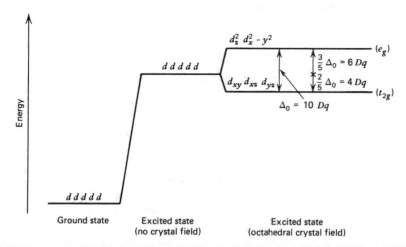

Figure 8-3 Removal of degeneracy of d orbitals in an octahedral crystal field.

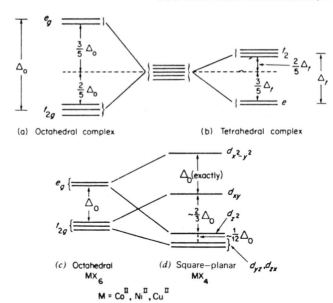

(a) Octahedral complex   (b) Tetrahedral complex

(c) Octahedral MX₆   (d) Square-planar MX₄

$M = Co^{II}, Ni^{II}, Cu^{II}$

Figure 8-4 Orbital-splitting energy-level diagrams for tetrahedral and square-planar geometries compared with octahedral geometry. [F. A. Cotton and G. Wilkinson, *Advanced Inorganic Chemistry*, 4th ed., Fig. 20.13, p. 641; Fig. 20.16, p. 643, John Wiley & Sons, New York (1980).]

gain in binding energy (i.e., the crystal-field stabilization energy). The energy separation between the lower and higher levels, indicated in Fig. 8-3 for octahedral geometry as Δ_0 (or as $10\,Dq$ in other treatments),* is of such magnitude that electronic transitions from the lower to the higher level can be effected by absorption of radiation in the visible region, thus accounting for the colors of many octahedral complexes.

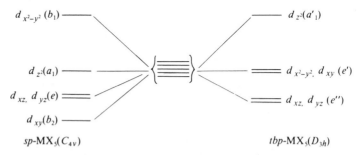

sp-MX₅(C_{4v}) tbp-MX₅(D_{3h})

Figure 8-5 Orbital-splitting energy-level diagram for 5-coordinate geometries (*sp*, square planar; *tbp*, trigonal bipyramidal). [F. A. Cotton and G. Wilkinson, *Advanced Inorganic Chemistry*, 4th ed., Fig. 20.14, p. 642, John Wiley & Sons, New York (1980).]

---

* It can be shown that in an octahedral field the six t_{2g} electrons must lie below the original degenerate level by 0.4 unit and the four e_g electrons above by 0.6 unit. If then the total energy separation between the two is expressed in nonfractional units of Dq, that separation becomes $10\,Dq$, an energy parameter.

The frequency, wave number, or wavelength of the absorbed energy can be used to determine the energy separation, or the crystal-field stabilization energy. Although this becomes quite complicated when several d electrons are involved, the principle can be illustrated quite simply using a d^1 case, namely the $[Ti(H_2O)_6]^{3+}$ ion, for which a single absorption band centering at 20,300 cm^{-1} (Fig. 8-6)[36] indicates the absorption of 58.0 kcal mole^{-1} in raising the $3d$ electron from the t_{2g} to the e_g state. Thus since 10 Dq for this case equals 20,300 cm^{-1} or 58.0 kcal mole^{-1}, Dq is then 2030 cm^{-1} or 5.8 kcal (24.2 kJ) mole^{-1}.

Consideration of these energy terms is important in determining the distributions of d electrons among orbitals. For an octahedral geometry, neither the d^1, d^2, or d^3 cases nor the d^8, d^9, or d^{10} cases allow choice between configurations, as indicated by the diagram

since both the Aufbau and Hund principles apply. For the d^4, d^5, d^6, and d^7 cases, however, additional d electrons beyond three may either pair with those already in the t_{2g} orbitals (low-spin or spin-paired) or occupy e_g orbitals singly before pairing begins (high-spin or spin-free), as indicated by the diagrams

The configuration that results for a specific complex species depends upon the relative magnitudes of the separation (Δ_0 or 10 Dq)* and pairing (P) energy terms. If the former is larger than the latter, the low-spin (or spin-paired) configuration is favored; if the reverse pertains, the high-spin (or spin-free) configuration is favored.† The pairing energy is not particularly sensitive to

---

* Theory indicates 10 $Dq \sim 5e\mu\bar{a}^4/r^6$, where e is the electronic charge, μ the charge or dipole moment of the six ligands, $\bar{a}^4$ the average of the fourth power of the distance of the electron from the nucleus, and r the metal–ligand distance.

† In valence-bond terminology, high-spin configurations characterize the outer-orbital hybridizations and low-spin configurations the inner-orbital hybridizations.

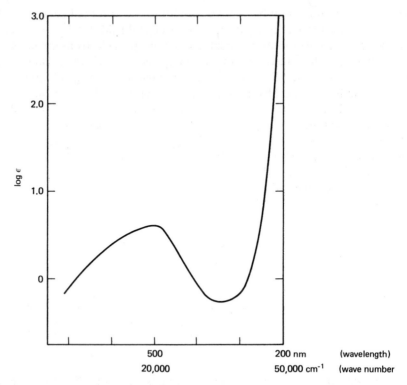

Figure 8-6 Absorption spectrum of 0.1 M [Ti(H$_2$O)$_6$]$^{3+}$ solution. [H. Hartman, H. L. Schläfer, and K. H. Hansen, *Z. Anorg. Allg. Chem.*, **284**, 153 (1956), redrawn from H. L. Schläfer and G. Glieman, *Basic Principles of Ligand Field Theory*, John Wiley & Sons, New York (1969).]

changes in the nature of the ligand and is of the same general magnitude for 3d, 4d, and 5d orbitals. The separation energy, however, is sensitive to changes in the ligand because it is affected by the imposed crystal field. Irrespective of the cation present, this energy *increases* in the order (*the spectrochemical series*):

$$I^- < Br^- < Cl^- < F^- < OH^- < H_2O \sim C_2O_4^{2-} < NCS^- < py \sim NH_3$$
$$< en < NO_2^- < o\text{-phen} \sim bipy < CN^-$$

$\overline{\text{weak-field ligands} \qquad \text{increase in } \Delta \text{ or } Dq \qquad \text{strong-field ligands}}\longrightarrow$

Obviously, *weak-field* ligands favor *high-spin* configurations, and *strong-field* ligands favor *low-spin* configurations. For 4d and 5d cations, respectively, the separation energy is larger by about 50% and 100% than for 3d cations. The effect is to cause resulting 4d and 5d complexes to be low-spin irrespective of the ligand present. For a given ligand, this energy term increases with increasing cationic charge because of decrease in the number of electrons to be accommodated in the d orbitals and increase in polarization of the ligand, but for complexes derived from the same ligand and cations of the same charge in a given transition series its observed values encompass only a limited range.

Certain of these considerations are illustrated by data in Tables 8-2 and 8-3.* For comparison, crystal-field stabilization energies for square-planar complexes are included in Table 8-2. The differences in energies for configurations d^4 through d^8 reflect differences in orbital splitting patterns (Fig. 8-3 vs. Fig. 8-4).

It is apparent from the data for octahedral complexes in Table 8-2 that overall completely spherical symmetry can be expected only for d^0, d^{10}, low-spin d^6 (i.e., Fe^{2+}, Co^{3+}), high-spin d^5 (i.e., Mn^{2+}, Fe^{3+}), and d^3 (i.e., V^{2+}, Cr^{3+}) cations. There is, therefore, a rather direct correlation between the acceptor ability of the cation, or the thermodynamic stabilities of the resulting complexes in solution (Sec. 11.1), and the crystal-field-determined occupancy of d orbitals. Although it might appear reasonable that at some position in the spectrochemical series a transition from high-spin to low-spin configuration should occur irrespective of the cation, such is not the case because of varying covalent contributions to bonding and energy differences as the number of d electrons changes. Thus although the spin-paired configurations are characteristic of manganese(III) and iron(III) only in association with the strongest-field ligands, that configuration characterizes cobalt(III) with all ligands except the fluoride ion.

High-spin to low-spin crossovers[37] in terms of the crystal-field strength of the ligand and pairing energy can be idealized graphically (e.g., for the d^6 case in Fig. 8-7). Complexes in the general vicinity of $10\,Dq$ may exist as equilibrium mixtures of the two spin states. Cases in point involve the iron(III) N,N-dialkyldithiocarbamates, $[Fe(S_2CNRR')_3]$.[38]

Applications. The applicability of the crystal-field approximation can be illustrated in terms of some complex species that cannot be treated logically by the valence-bond approximation.

1. **Iron(III) species.** That the magnetic moment of the $[Fe(CN)_6]^{3-}$ ion corresponds to the presence of *one* unpaired electrons whereas those of the $[Fe(H_2O)_6]^{3+}$ and $[FeF_6]^{3-}$ ions each correspond to the presence of *five* unpaired electrons is not a consequence of covalent vs. ionic bonding but rather a consequence of the crowding of the five $3d$ electrons in question into t_{2g} orbitals with maximum pairing by the strong-field CN^- ligand as opposed to the spreading of those electrons over both t_{2g} and e_g orbitals with maximum spin by the weaker-field H_2O and F^- ligands. The cyanide ion is more polarizable than either of the other ligands, a factor that increases its crystal field strength as well as promotes its tendency to form covalent bonds in the Pauling sense. Although the unmodified crystal-field approach cannot admit the double bonding proposed by Pauling for this ligand, the ligand-field extension (Sec. 8.1-3) does recognize its ability to form π bonds and thus enhance its ligand-field strength.

---

* By thermodynamic convention (Ch. 9), crystal-field stabilization energies should be expressed as negative quantities since they measure the stabilization of complexes. This practice is seldom followed. For octahedral geometry, calculation is effected as

weak field d^6 (i.e., $t_{2g}^4 e_g^2$): $\quad 4(-4\,Dq) + 2(+6\,Dq) = -4\,Dq$
strong field d^6 (i.e., t_{2g}^6): $\quad 6(-4\,Dq) = -24\,Dq$

Table 8-2 Crystal-Field Stabilization Energies for Octahedral and Square-Planar 3d Species

Case	Examples	Octahedral Geometry — Weak-Field, or High-Spin — Configuration (t_{2g} / e_g)	Dq (Crystal-Field Energy)	Octahedral Geometry — Strong-Field, or Low Spin — Configuration (t_{2g} / e_g)	Dq (Crystal-Field Energy[a])	Square-Planar Geometry — Weak-Field, or High-Spin Dq	Square-Planar Geometry — Strong-Field, or Low-Spin Dq
d^0	Ca^{2+}, Sc^{3+}, Ti^{4+}	t_{2g}: ( )( )( ) ; e_g: ( )( )	0	t_{2g}: ( )( )( ) ; e_g: ( )( )	0	0	0
d^1	Ti^{3+}, V^{4+}	t_{2g}: (↑)( )( ) ; e_g: ( )( )	4	t_{2g}: (↑)( )( ) ; e_g: ( )( )	4	5.14	5.14
d^2	Ti^{2+}, V^{3+}	t_{2g}: (↑)(↑)( ) ; e_g: ( )( )	8	t_{2g}: (↑)(↑)( ) ; e_g: ( )( )	8	10.28	10.28
d^3	V^{2+}, Cr^{3+}	t_{2g}: (↑)(↑)(↑) ; e_g: ( )( )	12	t_{2g}: (↑)(↑)(↑) ; e_g: ( )( )	12	14.56	14.56
d^4	Cr^{2+}, Mn^{3+}	t_{2g}: (↑)(↑)(↑) ; e_g: (↑)( )	6	t_{2g}: (↑↓)(↑)(↑) ; e_g: ( )( )	16	12.28	19.70
d^5	Mn^{2+}, Fe^{3+}	t_{2g}: (↑)(↑)(↑) ; e_g: (↑)(↑)	0	t_{2g}: (↑↓)(↑↓)(↑) ; e_g: ( )( )	20	0	24.84
d^6	Fe^{2+}, Co^{3+}	t_{2g}: (↑↓)(↑)(↑) ; e_g: (↑)(↑)	4	t_{2g}: (↑↓)(↑↓)(↑↓) ; e_g: ( )( )	24	5.14	29.12
d^7	Co^{2+}, Ni^{3+}	t_{2g}: (↑↓)(↑↓)(↑) ; e_g: (↑)(↑)	8	t_{2g}: (↑↓)(↑↓)(↑↓) ; e_g: (↑)( )	18	10.28	26.84
d^8	Ni^{2+}, Cu^{3+}	t_{2g}: (↑↓)(↑↓)(↑↓) ; e_g: (↑)(↑)	12	t_{2g}: (↑↓)(↑↓)(↑↓) ; e_g: (↑)(↑)	12	14.56	24.56
d^9	Cu^{2+}	t_{2g}: (↑↓)(↑↓)(↑↓) ; e_g: (↑↓)(↑)	6	t_{2g}: (↑↓)(↑↓)(↑↓) ; e_g: (↑↓)(↑)	6	12.28	12.28
d^{10}	Cu^+, Zn^{2+}, Ga^{3+}	t_{2g}: (↑↓)(↑↓)(↑↓) ; e_g: (↑↓)(↑↓)	0	t_{2g}: (↑↓)(↑↓)(↑↓) ; e_g: (↑↓)(↑↓)	0	0	0

Table 8-3 Comparison of Crystal-Field Splitting Energies (Δ_0) and Electron-Pairing Energies (P) for Octahedral 3d Species[a]

Configuration	Ion	P (cm^{-1})	Ligands	Δ_0 (cm^{-1})	Spin State Predicted	Spin State Observed
d^4	Cr^{2+}	23,500	6 H_2O	13,900	High	High
	Mn^{3+}	28,000	6 H_2O	21,000	High	High
d^5	Mn^{2+}	25,500	6 H_2O	7,800	High	High
	Fe^{3+}	30,000	6 H_2O	13,700	High	High
d^6	Fe^{2+}	17,600	6 H_2O	10,400	High	High
			6 CN^-	33,000	Low	Low
	Co^{3+}	21,000	6 F^-	13,000	High	High
			6 NH_3	23,000	Low	Low
d^7	Co^{2+}	22,500	6 H_2O	9,300	High	High

[a] F. A. Cotton and G. Wilkinson, *Advanced Inorganic Chemistry*, 4th ed., p. 646, Wiley-Interscience Publishers, New York (1980).

2. Nickel(II) species. The crystal-field stabilization, or separation, energies for this d^8 case for varying geometries are: octahedral, strong or weak field, 12 Dq; square-planar, weak field, 14.56 Dq; square-planar, strong field, 24.56 Dq; and tetrahedral, strong or weak field, 3.56 Dq.[34] Thus the tetrahedral geometry is energetically the least probable, and, for strong-field ligands, the square-planar geometry is the most probable. Experimental observations support these conclusions. For weaker-field ligands, the slightly larger stabilization energy for the square-planar configuration and the concentration of electronic charge

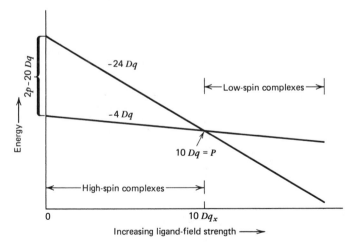

Figure 8-7 Relationship between crystal-field stabilization energy and pairing energy for d^6 complexes in terms of increasing crystal-field strength of ligand. [J. E. Huheey, *Inorganic Chemistry*, 2nd ed., Fig. 9.16, p. 359, Harper & Row, New York (1978).]

density in orbitals normal to the plane suggest that this arrangement should be favored over the octahedral. In practice, however, certain ligands (e.g., bpy, phen, NH_3, H_2O) do yield octahedral species. Since some of these complexes are paramagnetic (e.g., $[Ni(bpy)_3]^{2+}$) and others diamagnetic (e.g., $[Ni(diars)_3]^{2+}$), it is apparent that the e_g electrons can be differently arranged. Only with the weakest-field ligands (e.g., Cl^-, Br^-) have tetrahedral species been obtained. Magnetic data have been particularly useful in relating observed geometries to theory.[18]

The larger crystal-field stabilization energies of $4d^8$ and $5d^8$ cations (e.g., Pd^{2+}, Pt^{2+}, Au^{3+}) render all of their complexes square planar.

3. Copper(II) species.[39] Crystal-field stabilization energies for the d^9 species for either weak- or strong-field ligands are: square planar, 12.28 Dq; octahedral, 6 Dq; and tetrahedral, 1.78 Dq. The favored square-planar geometry does indeed characterize copper(II) complexes,[40] but a closer examination of structural data shows that the true coordination number is more commonly *six* than *four* and that the two additional ligands lie above and below the square plane but at longer distances than those in the plane. The geometry is thus tetragonal, or tetragonally distorted octahedral. In the tetragonal arrangement, the $d_{x^2-y^2}$ orbital is raised above and the d_{z^2} orbital dropped below their normal energies for an octahedral arrangement, and of the t_{2g} orbitals, the d_{xy} orbital is raised above the other two. The tetragonal configuration is not, in general, suggested by stoichiometry but is identified by crystal-structure determination. Thus crystals of anhydrous copper(II) chloride are built up of chains in which each copper atom is surrounded by a square of bridge-bonding chlorine atoms, but adjacent chains are so displaced that at somewhat longer Cu—Cl bond distances chlorine atoms lie above and below each copper atom.[41] Similarly, in crystals of the compound $[Cu(H_2O)_2Cl_2]$, the arrangement is[42,43]

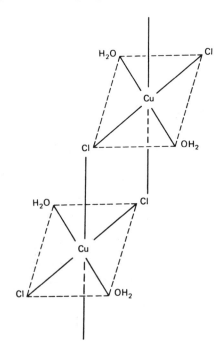

This rather unusual geometry and the somewhat anomalously large stabilities of these species are accounted for in terms of the *Jahn–Teller distortion*, or *effect*.[44,45] It is suggested that if degeneracy results because one d orbital is doubly occupied whereas another of equal energy is singly occupied, a change in geometry may resolve this degeneracy and yield a more stable state. With copper(II), alteration of the degenerate arrangement $d_{z^2}^2 d_{x^2-y^2}^1$ gives a tetragonal distribution of ligands. Similar, although less well defined Jahn–Teller distortions are found with chromium(II) and manganese(III) species (both d^4).

Cons and Pros. The unmodified crystal-field approximation is unrealistic since it is based upon the electrostatic interactions of point charges. Among ligands, not even ions are point charges. Furthermore, the overall properties of many complex species suggest, at the very least, covalent interactions between ligands and metal atoms or ions, which of course are not admissible in the crystal-field approach. On these bases, why then has so much space been devoted to this approach to bonding?

Broadly the answer to this question is that the fundamental ideas on energy and the splitting of orbitals provide a logical basis for concepts that admit covalent bonding. Furthermore, although unrealistic in terms of real species, the crystal-field approximation provided the first truly quantitative correlations among physical measurements, molecular structure, and bonding that gave experimentalists and theoreticians common ground for the researches that have made the chemistry of d-transition metal species the currently most popular and productive branch of inorganic chemistry. One need only to consult any issue of a journal dealing with inorganic chemistry, to note how many new journals covering coordination chemistry have appeared in the past few years, or to examine recently published monographs, reviews, or summaries of symposia to verify this statement. Imprecise though it may be, the crystal-field approximation was responsible for a renaissance of d-transition metal chemistry.

Among the useful contributions based upon the crystal-field approach are (1) predictions and correlations of stereochemistry and coordination number; (2) correlations between absorption spectra and energy states for complexes; (3) direct correlations between magnetic properties and molecular structures; (4) correlations between molecular structure and thermodynamic stability as indicated by redox potentials (Secs. 9.5, 11.3) and formation constants (Sec. 11.1); (5) postulation of mechanisms for reactions involving complex species (Sec. 11.2); (6) evaluation of coordination bond energies and correlation of these values with electronic configurations; (7) decisions as to appropriate reactions for syntheses (Sec. 11.3-3); (8) decisions as to site preferences in ionic crystals (Secs. 4.3, 10.3); and (9) substantial simplification of the otherwise confusing general chemistry of the d-transition metals.[34]

To supplement earlier discussions in this chapter and those in subsequent chapters alluded to above, a few comments on items (1) and (6) conclude this section. In Table 8-4, some early predictions by R. S. Nyholm[18,*] as to expected

---

* Relative to the late R. S. Nyholm, see R. J. P. Williams and J. D. Hale, *Struct. Bonding*, **15**, 1 (1973).

Table 8-4 Stereochemical Predictions Based on Crystal-Field Approximation[18]

Number of Nonbonding d Electrons	Number of Unpaired Electrons	Coordination Geometry	
		4-Coordinate	6-Coordinate
		HIGH SPIN	
0, 5, or 10	0 or 5	Regular tetrahedral	Regular octahedral
9 or 4	1	Square planar	Tetragonal
8 or 4	2	Distorted tetrahedral	Regular octahedral
7 or 2	3	Regular tetrahedral	Almost regular octahedral
6 or 1	4	Almost regular tetrahedral	Almost regular octahedral
		LOW SPIN	
1 or 2	—	—	—
3	1	Almost regular tetrahedral	—
4	0	Regular tetrahedral	Almost regular octahedral
5	1	Distorted tetrahedral	Almost regular octahedral
6	0	Distorted tetrahedral	Regular octahedral
7	1	Square planar	Tetragonal
8	0	Square planar	Tetragonal

476

geometries of 4- and 6-coordinate species are summarized. Many of these predicted geometries are realized in isolable species. Figure 8-8, indicating how by crystal-field corrections the rather erratic enthalpies of hydration of the gaseous di- and tri-positive $3d$ ions can be altered to an essentially smooth transition from d^0 through d^{10}, may be regarded as convincing evidence for validity of crystal-field ideas.[46,47]

1-3. Adjusted Crystal-Field, or Ligand-Field, Approximation

Although it was not exactly so developed, it is reasonable to consider the crystal-field approach as involving electrostatic interaction between a metal ion and its surrounding ligands with no overlap of orbitals, or sharing of electronic charge density, between the two. There is, of course, abundant evidence that overlap (i.e., covalency) does exist. Thus electron spin resonance spectra (Table 6-21)

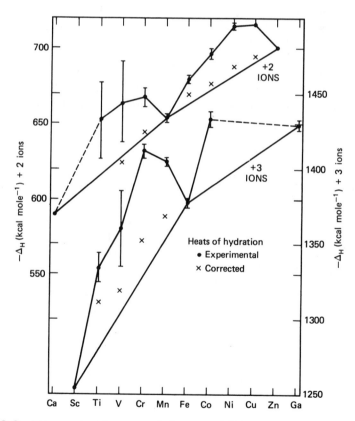

Figure 8-8 Measured and corrected, for crystal-field stabilization energies, enthalpies of hydration of $M^{2+}(g)$ and $M^{3+}(g)$ ions from $3d$ series. [D. S. McClure, in *Progress in Inorganic Chemistry*, F. A. Cotton (Ed.), Vol. 1, p. 418, Interscience Publishers, New York (1959).]

may show hyperfine structures which indicate that an unpaired electron in a metal-ion d orbital is fractionally in an orbital in a donor atom or ion (e.g., a $4d$ electron from Ir(IV) in the ion $[IrCl_6]^{2-}$ ca. 30% in orbitals of the Cl^- ions).[48] Similarly, in certain nuclear magnetic resonance spectra (Sec. 6.5-4) observed shifts of nuclear resonances of ligand atoms are explicable only in terms of transfer of unpaired electron spin density from metal-ion d orbitals to ligand-atom orbitals[49] (e.g., shift of resonance frequency of the ring-bonded proton in paramagnetic tris(acetylacetonato)vanadium(III) relative to its position in diamagnetic tris(acetylacetonato)aluminum(III)).

Indirect evidence of covalency is provided by the *nephelauxetic effect*.[6, 7, 50] Nephelauxetic (i.e., cloud expanding) refers to an effective increase in the size of orbitals as a consequence of the observation that electron–electron repulsion in a complex is less than in a free metal ion. Increase in orbital size appears to be a consequence of combination of metal-ion and ligand orbitals to form molecular orbitals. Ligands containing large atoms with d orbitals for π bonding show the largest nephelauxetic effects. Nephelauxetic series for ligands (h_X) and metal ions (k_M) are summarized in Table 8-5, with total effect in a complex $[MX_n]^{n\pm}$ being proportional to the product $h_X \cdot k_M$. For ligands, the series is largely independent of the metal ion and closely related to the spectrochemical series (Sec. 8.1-2).

Inasmuch as some degree of covalency is thus characteristic of many complexes, no theoretical approximation that fails to admit covalent interactions can be completely acceptable. Modification of the crystal-field

Table 8-5 Nephelauxetic Effects[51]

Ligand (X)		Metal Ion (M)	
Nature	h_X	Nature	k_M
F^-	0.8	Mn(II)	0.07
H_2O	1.0	V(II)	0.1
$HC(O)N(CH_3)_2$	1.2	Ni(II)	0.12
$OC(NH_2)_2$	1.2	Mo(III)	0.15
NH_3	1.4	Cr(III)	0.20
en	1.5	Fe(III)	0.24
$C_2O_4^{2-}$	1.5	Rh(III)	0.28
Cl^-	2.0	Ir(III)	0.28
CN^-	2.1	Tc(IV)	0.3
Br^-	2.3	Co(III)	0.33
N_3^-	2.4	Mn(IV)	0.5
I^-	2.7	Pt(IV)	0.6
dtp^{-a}	2.8	Pd(IV)	0.7
$dsep^{-b}$	3.0	Ni(IV)	0.8

[a] dtp^- = diethylthiophosphate ion.
[b] $dsep^-$ = diethylselenophosphate ion.

approach to allow small quantities of orbital overlap by considering all measurements of interelectronic interactions rather than equating them only to values for free ions gives the adjusted crystal-field, or ligand-field, approximation. Important interactions include spin-orbit coupling constants and interelectronic repulsions. Within these modifications, energy-level diagrams and magnetic behaviors are treated exactly the same as was done for crystal-field evaluations but with values smaller than free-ion values. This approach is useful for small degrees of covalency, but where covalent interactions are major, the molecular-orbital approximation is desirable.

1-4. Molecular-Orbital Approximation

The molecular-orbital approach (Sec. 5.1-1) was first applied to complexes by J. H. Van Vleck,[29] who pointed out that both the valence-bond and crystal-field approximations amount to simplified approaches to this more general approximation. Its subsequent development and comprehensive treatments of detail are found in the standard references.[8, 10, 33, 35, 52, 53] Only the pertinent aspects are summarized here.

As is true for simpler species, the molecular-orbital approximation assumes that as a consequence of overlaps of atomic orbitals, as permitted by symmetry and to the extent allowed by spatial orientation, molecular orbitals based upon linear combinations of these atomic orbitals result. Possible and impossible overlaps can be determined mathematically by application of group theory (Sec. 6.4), but for our purposes visualization of bonding via a more pictorial approach is adequate and more meaningful.

The notation used and the general methodology followed can be illustrated for octahedral species derived from the $3d$ elements, where orbitals at higher energies than the $4p$ can be neglected. Nine valence-shell orbitals for the metal ion must be considered: two e_g and three t_{2g} ($3d$), one a_{1g} ($4s$), and three t_{1u} ($4p$) orbitals.* Of these, the $d_{x^2-y^2}, d_{z^2}, 4s, 4p_x, 4p_y,$ and $4p_z$ orbitals have lobes directed along the metal–ligand bonds and are thus appropriate for formation of σ bonds. The $d_{xy}, d_{xz},$ and d_{yz} orbitals, however, are so oriented that they can form only π bonds. It can be assumed, at the moment, that each of the six ligand species possesses at least one σ orbital, which can be combined with one of the six orbitals from the metal ion that is suitably oriented for σ bonding. Thus each of these metal-ion orbitals overlaps with a ligand orbital to give a bonding and an antibonding σ molecular orbital. If π orbitals are present in the ligands, each of these must be combined with an appropriate metal-ion π orbital to give both bonding and antibonding π overlaps.

To illustrate, let us consider first an octahedral complex that involves no π bonding. For such a species, the six σ symmetry metal-ion orbitals, together with matching σ symmetry ligand orbitals, are indicated in Fig. 8-9. Symmetry-class

---

* Again, the symbols g and u indicate, respectively, centrosymmetric (g from *gerade*) and anticentrosymmetric (u from *ungerade*) orbitals.

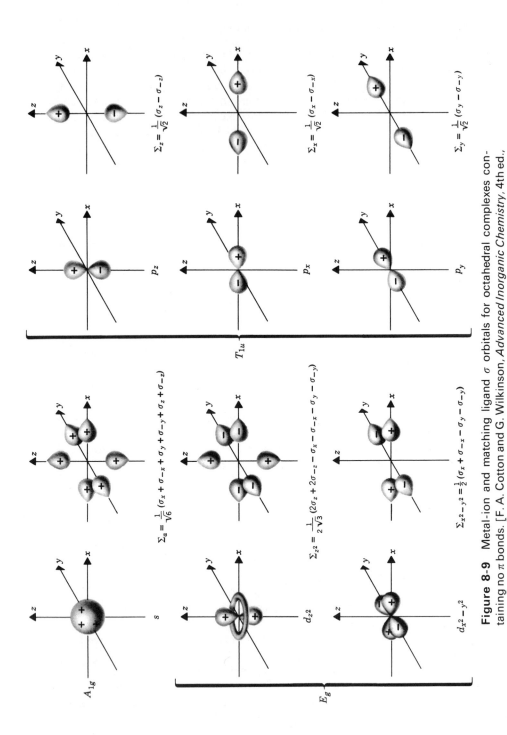

Figure 8-9 Metal-ion and matching ligand σ orbitals for octahedral complexes containing no π bonds. [F. A. Cotton and G. Wilkinson, *Advanced Inorganic Chemistry*, 4th ed.,

$$\Sigma_z = \frac{1}{\sqrt{2}}(\sigma_z - \sigma_{-z})$$

$$\Sigma_x = \frac{1}{\sqrt{2}}(\sigma_x - \sigma_{-x})$$

$$\Sigma_y = \frac{1}{\sqrt{2}}(\sigma_y - \sigma_{-y})$$

p_z

p_x

p_y

T_{1u}

$$\Sigma_a = \frac{1}{\sqrt{6}}(\sigma_x + \sigma_{-x} + \sigma_y + \sigma_{-y} + \sigma_z + \sigma_{-z})$$

$$\Sigma_{z^2} = \frac{1}{2\sqrt{3}}(2\sigma_z + 2\sigma_{-z} - \sigma_x - \sigma_{-x} - \sigma_y - \sigma_{-y})$$

$$\Sigma_{x^2 - y^2} = \frac{1}{2}(\sigma_x + \sigma_{-x} - \sigma_y - \sigma_{-y})$$

s

d_{z^2}

$d_{x^2 - y^2}$

A_{1g}

E_g

symbols representing metal-ion, ligand, and resulting molecular orbitals, namely A_{1g}, E_g, and T_{1u}, are included also. The resulting molecular orbitals, both bonding and antibonding, are then indicated as a general energy-level diagram (Fig. 8-10).

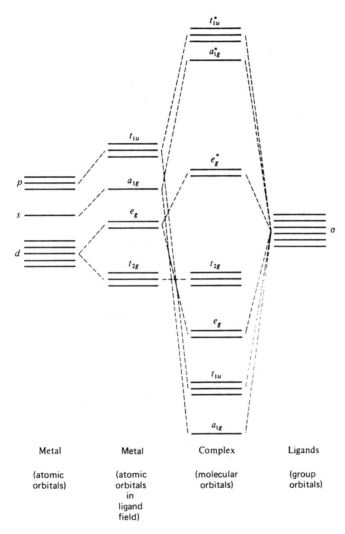

Metal	Metal	Complex	Ligands
(atomic orbitals)	(atomic orbitals in ligand field)	(molecular orbitals)	(group orbitals)

Figure 8-10 Qualitative molecular-orbital, energy-level diagram for $3d$ metal ion and six ligands, with no π bonding present. [J. E. Huheey, *Inorganic Chemistry*, 2nd ed., Fig. 9.48, p. 398 Harper & Row, New York (1978).]

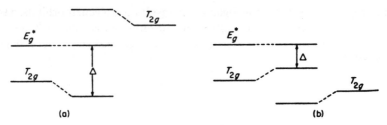

Figure 8-11 Effects of ligand π orbital interactions for cases where (a) they are at higher energy than metal-ion T_{2g} orbitals and (b) where the reverse is true. [F. A. Cotton and G. Wilkinson, *Advanced Inorganic Chemistry*, 4th ed., Fig. 20-6, p. 634, John Wiley & Sons, New York (1980).]

Let us next consider an octahedral complex derived from ligands that can also form π bonds by interaction with T_{2g} d orbitals. The value of Δ_0 can be affected differently depending upon whether the π orbitals from the ligand are of higher energy than the metal-ion T_{2g} orbitals or of lower energy, as indicated in Fig. 8-11. To illustrate the π-bonding situation, qualitative correlations for $[M(CO)_6]$ and $[M(CN)_6]^{n-}$ species are depicted in Fig. 8-12.

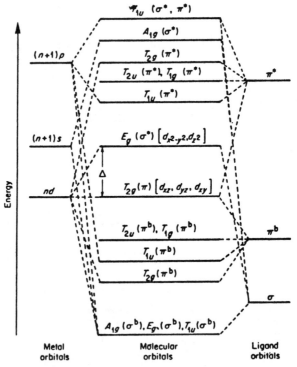

Figure 8-12 Qualitative molecular orbitals for $[M(CO)_6]$ and $[M(CN)_6]^{n-}$. [F. A. Cotton and G. Wilkinson, *Advanced Inorganic Chemistry*, 4th ed., Fig. 20-7, p. 635, John Wiley & Sons, New York (1980), adapted from H. B. Gray and N. A. Beach, *J. Am. Chem. Soc.*, **85**, 2922 (1963).]

Other geometries can be handled similarly. For example, representations applicable to σ-bonding cases only are included for tetrahedral and square-planar geometries in Fig. 8-13 and 8-14, respectively.

Molecular-orbital approximations are probably better in accord with experimentally determined or measured properties than the other bonding approximations. Inclusion of π bonding is a major contribution by the molecular-orbital approach. Thus the strong-field behavior of the cyanide ion is understandable when it is realized that an empty, antibonding d_π orbital can readily overlap with an orbital from the cation. Correspondingly, the strong-field behaviors of arsines and phosphines, as opposed to ammonia and the amines, are again related to unoccupied d_π orbitals. On the other hand, where the ligand π orbitals are stable and occupied (e.g., for the halide ions), some delocalization occurs, but metal-ion d electrons must occupy t_{2g}^* orbitals, giving a decrease in ligand-field stabilization energy that results in maximum numbers of unpaired spins. The importance of π bonding is most clearly emphasized, however, by the thermodynamic stabilities of complex species derived from carbon monoxide (Sec. 8.2-1), nitrogen(II) oxide (Sec. 8.2-1), or unsaturated hydrocarbons (Sec. 8.2-3).

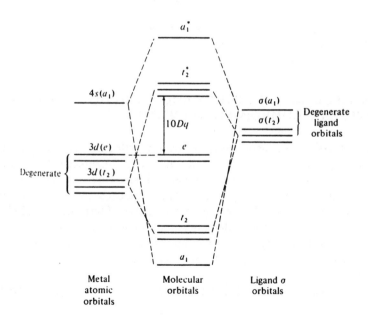

Figure 8-13 Qualitative molecular orbitals for tetrahedral complex with no π bonding. [J. E. Huheey, *Inorganic Chemistry*, 2nd ed., Fig. 9.58, p. 411, Harper & Row, New York (1978).]

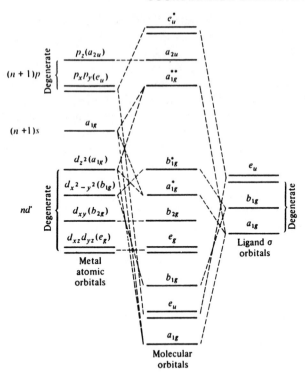

Figure 8-14 Qualitative molecular orbitals for a square-planar complex with no π bonding. [J. E. Huheey, *Inorganic Chemistry*, 2nd ed., Fig. 9.59, p. 412, Harper & Row, New York (1978).]

2. MOLECULAR STRUCTURE AND BONDING — SPECIFIC TYPES OF COORDINATION ENTITIES

Earlier discussions, in particular those dealing with donor species, have directed attention to several types of coordination species the molecular structures and bonding properties of which are particularly interesting and significant. Among these, and also some related species, are the metal carbonyls, nitrosyls, cyanides, and isocyanides; the zero-valent metal derivatives of terpositive periodic group VA compounds; the metal derivatives of olefins and related unsaturated hydrocarbons, including carbocyclic and arene donors; metal-cluster species; and the hetero- and isopoly acids and their anions.

2-1. Derivatives of 14-Electron Donor Molecules and Ions

Metal carbonyl, nitrosyl, and cyano complexes are derived from the iso-electronic 14-electron donor species CO, NO^+, and CN^-.[54] Organic iso-cyanides, or isonitriles ($R—N{=}C$), are also, if the R groups are discounted, pseudo 14-electron donor species[54] and are conveniently included with the others. These ligands all function as both σ donors and π acceptors, with σ-donor strength changing roughly as $CN^- > RNC > NO^+ \sim CO$ and π-acceptor strength changing roughly as $NO^+ > CO \gg RNC > CN^-$. Each of these donors is a strong-field ligand and thus tends to form low-spin, or spin-paired, complexes. All effect stabilization of low oxidation states, but not high oxidation states, in d-transition metal complexes. Only comparatively few related complexes derived from f-transition metals have been described.

Metal Carbonyls and Their Derivatives. Since the characterization of the volatile product of the reaction of elemental nickel with carbon monoxide as $Ni(CO)_4$ in 1890,[55] interest in the syntheses, molecular structures, bonding, chemical reactions, and applications of metal carbonyl compounds has continued to increase. For detailed information beyond that pertaining to molecular structures and bonding in the sections that follow and certain reactions as described in Chapter 10, available extensive reviews[56-60] can be consulted with profit.

Metal carbonyl compounds include the mononuclear carbonyls [e.g., $Ni(CO)_4$, $Fe(CO)_5$, $Cr(CO)_6$], the polynuclear carbonyls [e.g., $Fe_2(CO)_9$, $Rh_6(CO)_6$], the carbonyl hydrides [e.g., $H_2Fe(CO)_4$, $HCo(CO)_4$], the carbonyl halides [e.g., $Fe(CO)_4I_2$, $Ir(CO)\{P(C_6H_5)_3\}Cl$], carbonyls containing periodic group VIA to IVA donors [e.g., $Fe_2(CO)_6S_2$, $[Co(CO)_3(PR_3)]_2$, $Ni(CO)_2(C_5H_5)_3$], and the nitrosyl carbonyls [e.g., $Co(CO)_3(NO)$, $V(CO)_5(NO)$]. Many typical compounds of these types are listed in Tables 8-6 and 8-7.

The impossibility of discussing the molecular structures and the bonding patterns for all of these compound types is apparent. What follows, relating to uni- and multinuclear carbonyls, can be extended, in terms of the CO ligands, to other carbonyl species.

1. Molecular structures. Although the molecular structures of many of the metal carbonyls and their derivatives have been determined with precision by x-ray diffraction, significant data have also been obtained via electron diffraction; vibrational, mass, nuclear magnetic resonance, and Mössbauer spectroscopy; and dipole moment and magnetic susceptibility measurements.[57,59,61,62] It is significant that in every case each CO group is bonded to the metal atom through its carbon atom. A CO group may, however, either bond to a single metal atom [e.g., in the $Cr(CO)_6$ molecule] or act as a bridge by bonding simultaneously to two atoms [e.g., three of the CO groups in the $Fe_2(CO)_9$ molecule].

The molecular structures of typical mono- and polynuclear metal carbonyls are indicated in Fig. 8-15, together with some data on M—C bond distances. The mononuclear carbonyl molecules have simple and predictable geometries, namely, $Ni(CO)_4$ [63]—tetrahedral(T_d); $Fe(CO)_5$,[64] $Ru(CO)_5$,[65] and $Os(CO)_5$ [65]—trigonal bipyramidal (D_{3h}); and $V(CO)_6$,[66,67] $Cr(CO)_6$,[68,69] $Mo(CO)_6$,[69] and $W(CO)_6$ [69]—octahedral (O_h). In each of these molecular structures, the M—C—O arrangement is a collinear one.

Table 8-6 Uni- and Multinuclear Metal Carbonyls

Group VB	Group VIB	Group VIIB	Group VIIIB		
			Fe Family	Co Family	Ni Family
$V(CO)_6$ m.p. ca. 70°C (dec.) (deep green)	$Cr(CO)_6$ dec. 130°C sublimes (colorless)	$Mn_2(CO)_{10}$ m.p. 154°C (yellow)	$Fe(CO)_5$ m.p. −20.5°C; b.p. 103°C (yellow)	$Co_2(CO)_8$ m.p. 51°C (dec.) (orange)	$Ni(CO)_4$ m.p. −19.3°C; b.p. 42.1°C (colorless)
		$Mn_n(CO)_{4n}$	$Fe_2(CO)_9$ dec. 100°C (brown)	$Co_4(CO)_{12}$ (black)	
	$Mo(CO)_6$ sublimes (colorless)		$Fe_3(CO)_{12}$ dec. 140°C (deep green)	$Co_6(CO)_{16}$ m.p. 105°C (dec.) (black)	
		$Tc_2(CO)_{10}$ m.p. 159°C (colorless)	$Ru(CO)_5$ m.p. −22°C (colorless)	$Rh_2(CO)_8$ m.p. 76°C (dec.) (orange)	
		$Tc_4(CO)_{16}$ (yellow)	$Ru_3(CO)_{12}$ m.p. 150°C (dec.) (orange)	$Rh_4(CO)_{12}$ m.p. 150°C (dec.) (red)	
			$Ru_6(CO)_{18}$ m.p. 235°C (dec.) (red)	$Rh_6(CO)_{16}$ m.p. 220°C (dec.) (black)	

$W(CO)_6$
sublimes
(colorless)

$Re_2(CO)_{10}$
m.p. 177°C
(colorless)

$Re_n(CO)_{4n}$

$Os(CO)_5$
m.p. −15°C
(colorless)

$Os_3(CO)_{12}$
m.p. 224°C
(yellow)

$Ir_2(CO)_8$
solid
(yellow-green)

$Ir_4(CO)_{12}$
m.p. 210°C (dec.)
(yellow)

$Ir_6(CO)_{16}$
solid
(red)

Table 8-7 Derivatives of Metal Carbonyls

Type	Group VB	Group VIB	Group VIIB	Group VIIIB — Fe Family	Co Family	Ni Family	Group IB
Anions	$[V(CO)_6]^-$	$[M(CO)_5]^{2-}$ $[M_2(CO)_{10}]^{2-}$ $[M_3(CO)_{14}]^{2-}$ (M = Cr, Mo, W)	$[M(CO)_5]^-$ (M = Mn, Tc, Re) $[Mn_2(CO)_9]^{2-}$ $[Re_4(CO)_{16}]^{2-}$ $[FeMn(CO)_9]^-$ $[Fe_2Mn(CO)_{12}]^-$	$[Fe(CO)_4]^{2-}$ $[Fe_2(CO)_8]^{2-}$ $[Fe_3(CO)_{11}]^{2-}$ $[Fe_4(CO)_{13}]^{2-}$ $[HFe_3(CO)_{11}]^-$	$[Co(CO)_4]^-$ $[Co_6(CO)_{15}]^{2-}$ $[Co_6(CO)_{14}]^{4-}$ $[Co_3Fe(CO)_{12}]^{3-}$ $[Rh_6(CO)_{15}]^{2-}$ $[Rh_{12}(CO)_{30}]^{2-}$ $[Ir_4(CO)_{10}]^{2-}$	$[Ni_2(CO)_6]^{2-}$ $[Ni_3(CO)_8]^{2-}$ $[Ni_4(CO)_9]^{2-}$ $[Ni_5(CO)_9]^{2-}$	
Cations			$[Mn(CO)_6]^+$	$[Fe(CO)_6]^{2+}$ $[HFe(CO)_5]^+$	$[Co(CO)_5]^+$ (?)		
Hydrides			$HMn(CO)_5$ $H_2Mn_2(CO)_9$ $H_3Mn_3(CO)_{12}$ $HMnRe_2(CO)_{14}$ $HRe(CO)_5$ $H_3Re_3(CO)_{12}$ $HRe_3(CO)_{14}$	$H_2Fe(CO)_4$ $H_4Ru_4(CO)_{12}$ $H_2Ru_4(CO)_{13}$ $H_2Ru_6(CO)_{18}$ $H_2Os(CO)_4$ $H_4Os_4(CO)_{12}$ $H_4Os_4(CO)_{12}$	$HCo(CO)_4$		
Halides			$Mn(CO)_5Cl$ $[Mn(CO)_4Br]_2$ $[Re(CO)_3X]_n$ (X = Cl, Br)	$Fe(CO)_4I_2$ $Ru(CO)_4I_2$ $[Ru(CO)_3F_2]_4$ $[Ru(CO)_2Cl_2]_2$	$Co(CO)_4X$ (X = Cl, Br, I) $[Rh(CO)_2X]_2$ (X = Cl, Br, I) $[Rh(CO)_2X_2]_2$ (X = Cl, Br, I) $[Rh(CO)_2X_2]^-$ (X = Br, I) $[Rh(CO)X_5]^{2-}$ (X = Cl, Br) $Ir(CO)_3X$ (X = Cl, Br, I) $[Ir(CO)_2X_2]^-$ (X = Br, I) $[Ir_2(CO)_6Br_2]^-$	$Pd(CO)_2X_2$ (X = Cl, Br, I) $[Pd(CO)Cl_3]^-$ $[Pt(CO)_2X_2]$ (X = Cl, Br, I) $Pt_2(CO)_3X_4$ (X = Br, I) $[Pt(CO)X...]^-$	$[Cu(CO)Cl]_n$ $[Cu(CO)Cl]\cdot 2H_2O$ $Au(CO)Cl$

Compounds with group VIA donors	Fe₂(CO)₆S₂ Fe₂(CO)₉Z₂ (Z = S, Se, Te)	[Co₂(CO)₅S]₂ Co₃(CO)₉S	Ni(CO)₃{S(C₂H₅)₂}
Compounds with group VA donors	[Fe(CO)₃SR]₂ Fe(CO)₂(PR₃)₂I₂ Fe(CO)₄(PR₃) Fe(CO)₃(PR₃)₂	Co₃(CO)₇S₂ [Co(CO)₃SR]₂ [Co(CO)₃PR₃]₂ Co(CO)₃(PR₃)X (X = Cl, Br, I) [Co(CO)₃PR₃]⁻ Rh(CO)X₂(ZR₃)₂Cl Rh(CO)X(ZR₃)₂ (X = Cl, Br, I; Z = P, As, Sb; R = alkyl) trans-Ir(CO)H(PR₃)₂Clᵃ (R = C₆H₅)	Ni(CO)₃ZR₃ Ni(CO)₂(ZR₃)₂ Ni(CO)(ZR₃)₃ (Z = P, As, Sb) Ni(CO)₄₋ₙ(ZX₃)ₙ (X = Cl, Br, I; Z = P, As; n = 1–4) Pt(CO)₄₋ₙ{P(C₆H₅)₃}ₙ (n = 1–4) Pt₃(CO)₄{P(C₆H₅)₃}₃ Pt(CO)Cl₂{P(C₆H₅)₃}
Compounds with group IVA donors	Fe(CO)₂R(C₅H₅) (R = alkyl, aryl, vinyl) Fe(CO)₄(diene) (diene = butadiene) [Fe(CO)₂(C₂H₄)(C₅H₅)]⁺ Fe(CO)₃RX (R = π alkyl; X = Cl, Br, I) [Ru(CO)Cl₂(diene)]₂ (diene = norbornadiene) [Ru(CO)₂(C₅H₅)]⁻ Os₂(CO)₂(C₈H₈)	Co(CO)₄R (R = alkyl, aryl, acyl) Co₂(CO)₇(CH₂=CHR) [Co₂(CO)₆(HC≡CH)]₂ [Co(CO)₂(diene)]₂ Co(CO)₃(π-C₃H₅) [Co(CO)₃(C₅H₅)]₃ Rh(CO)(diene)Cl₂ Rh₂(CO)₃(C₅H₅)₂ [Rh(CO)₃(C₅H₅)]₃ Ir(CO)(diene)₂Cl	Ni(CO)Cl(π-C₃H₅) [Ni(CO)₂(C₅H₅)]₂ Ni(CO)₂(C₅H₅)₃ Ni(CO)I(C₅H₅) [Pt(CO)(C₅H₅)]₂ Pt(CO)I(C₅H₅)

ᵃ "Vaska's compound". See L. Vaska and J. W. Di Luzio, *J. Am. Chem. Soc.*, **84**, 679 (1962); L. Vaska, *Acc. Chem. Res.*, **1**, 335 (1968).

In the molecular structures of the polynuclear carbonyls, metal–metal cluster bonding, with or without bridging by CO groups, is present. A bridging CO group may bond either to two metal atoms [i.e., be doubly bridging, as in the $Fe_2(CO)_9$ or the $Co_4(CO)_{12}$ molecule] or to three metal atoms [i.e., be triply bridging, as in the $Rh_6(CO)_6$ or the $Fe_4(CO)_4(C_5H_5)_4$ molecule]. Every known example of CO bridging also involves a metal–metal bond.[62, 70] Although

doubly bridging CO groups are formulated as $-C\!\!\overset{\displaystyle O}{\underset{\|}{}}\!\!\smile$, it is improper to term them "ketonic bridges." More nearly properly, they are "edge-bridging" groups,[62] reflecting their function in bridging the edges of metal–metal polyhedra.

Vibrational spectroscopy is useful in distinguishing among these three types of CO groups. Thus the CO stretching vibration is noted in the following ranges: 1900 to 2150 cm^{-1} for terminal groups, 1750 to 1850 cm^{-1} for doubly bridging groups, and 1620 to 1730 cm^{-1} for triply bridging groups, all for neutral molecules. Although doubly bridging arrangements are usually symmetrical (i.e., equal M—C bond distances and equal M—C—O bond angles), a few unsymmetrical examples have been reported [e.g., in $Fe_3(CO)_{12}$[62]].

It is of interest that although the compounds $Fe_2(CO)_9$ and $Fe_3(CO)_{12}$ were characterized many years ago, precise data for their molecular structures were obtained only comparatively recently.[62] As normally obtained, crystals of the compound $Fe_2(CO)_9$ are obtained as extremely thin flakes that do not yield accurate x-ray diffraction data. A nearly equidimensional crystal obtained from a commercially prepared sample ultimately provided accurate data.[71] Similarly, problems caused by disorder in the orientation of molecules in crystals of $Fe_3(CO)_{12}$ were solved by better resolution techniques.[72]

Molecular structures of derivatives of the metal carbonyls, where determined, are closely related to those of the parent carbonyls. Typical structures are summarized in Fig. 8-16. Additional structures are summarized in cited reviews.[57, 59, 62]

2. Bonding. The ease of formation of the metal–carbon bond with refractory metals that are often resistant to oxidation combines with the general chemical inertness of the CO molecule, its relatively small basicity in the G. N. Lewis sense, and the retention of individuality by the CO groups in the carbonyls to obviate any simple explanation of bonding in these species.

It is significant that most of the metal carbonyls and many of their derivatives are diamagnetic and adhere to the effective atomic number rule of N. V. Sidgwick (Sec. 7.1-3), assuming that each CO group donates two electrons to the metal atom or atoms to which it is bonded (Table 8-8). The sole exception among the mononuclear carbonyls is the compound $V(CO)_6$, which is paramagnetic to the extent of one unpaired electron per molecule but which readily adds a single electron to form the diamagnetic anion $V(CO)_6^-$, and thus adhere to the rule. Adherence to the effective atomic number rule among the polynuclear carbonyls results if each metal-to-metal bond is assumed to add one

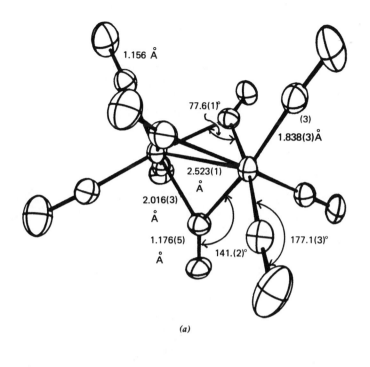

(a)

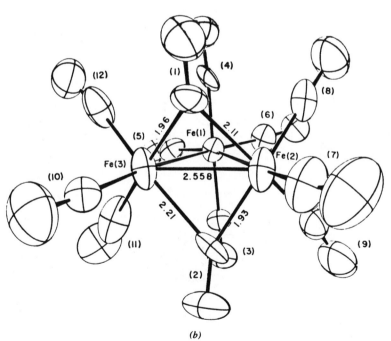

(b)

Figure 8-15

Figure 8-15

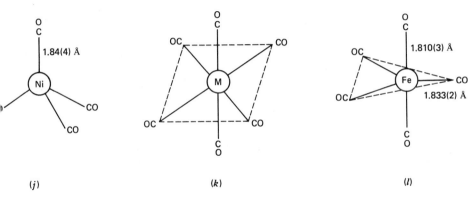

Figure 8-15 Molecular structures of mono- and polynuclear metal carbonyls. (*a*) $Fe_2(CO)_9$ (D_{3h}); (*b*) $Fe_3(CO)_{12}$ (C_{2v}); (*c*) $M_2(CO)_{10}$ (M = Mn, Tc, Re) (D_{4d}); (*d*) $Co_2(CO)_8$ (C_{2v}); (*e*) (D_{3d}); (*f*) $M_3(CO)_{12}$ (M = Ru, Os) (D_{3h}); (*g*) $Ir_4(CO)_{12}$ (T_d); (*h*) $M_4(CO)_{12}$ (M = Co, Rh) (C_{3v}); (*i*) $Rh_6(CO)_{16}$ (T_d); (*j*) $Ni(CO)_4$ (T_d); (*k*) $M(CO)_6$ (M = Cr, Mo, W) (O_h); M—C = 1.92(4) (M = Cr), = 2.06(2) (M = Mo), = 2.06(4) (M = W). (*l*) $Fe(CO)_5$. [(a), (b), F. A. Cotton, in *Progress in Inorganic Chemistry*, S. J. Lippard (Ed.), Vol. 21, pp. 1–28, Interscience, New York (1975); (*c*)–(*i*), F. A. Cotton and G. Wilkinson, *Advanced Inorganic Chemistry*, 4th ed., Fig. 25-1, pp. 1052–1053, John Wiley & Sons, New York (1980), Modified.]

electron to each of the bonded atoms (Table 8-8). Metal–metal bonding is implicit in structural data (Fig. 8-15).

Adherence to the effective atomic number rule is equivalent to adherence to the *18-electron rule*: that is, for a *d*-transition metal complex the sum of the outer electrons in the free cation, one electron contributed in each covalent bond to that ion (including a metal–metal bond), and two electrons contributed by each coordinated group (e.g., CO, one double bond) is 18.[73-75] Certain carbonyl species (e.g., $[Ir(CO)\{P(C_6H_5)_3\}_2Cl]$) obey an analogous 16-electron rule.[75]

In molecular-orbital terminology, the 18-electron rule is a generalization relating to the usage of orbitals. Thus for an $(n-1)d\,ns\,np$ combination, a complex species will contain either a σ-nonbonding or a σ-bonding–antibonding electron pair. The rule is obeyed whenever the σ-bonding and the σ-nonbonding orbitals are completely occupied and the σ-antibonding orbitals are all empty. The rule is restrictive to strong-field ligands, of which the CO molecule is an example. The σ-nonbonding orbitals must be stabilized by π bonding, as, for example, in the $Cr(CO)_6$ molecule.

This generalization is particularly useful (1) in suggesting analogies involving chemically dissimilar species such as $[Co(CO)_4]^-$ and $[Cr(CO)_3(C_5H_5)]^-$, (2) in rationalizing and to a degree predicting the molecular structures of complicated species, and (3) in raising questions about the bonding in and chemical reactivity of deviating species within a general class to which the rule is generally applicable. Thus the $Fe(C_5H_5)_2$ molecule, which obeys the 18-electron rule, is

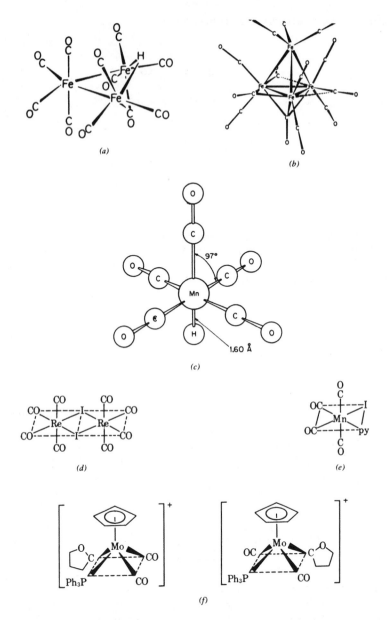

Figure 8-16 Molecular structures of derivatives of metal carbonyls. (*a*) [HFe₃(CO)₁₁]⁻; (*b*) [Fe₄(CO)₁₃]²⁻; (*c*) HMn(CO)₅; (*d*) [Re(CO)₄I]₂; (*e*) Mn(CO)₄(py)I (py = pyridine); (*f*) [(η⁵ — C₅H₅) Mo(CO)₂(PPh₃) (C₂H₄O)]⁺ isomers (Ph = C₆H₅). [F. A. Cotton and G. Wilkinson, *Advanced Inorganic Chemistry*, 4th ed., Figs. 25.4, 25.6, 25–XVII, 25–XIX, 25–13, John Wiley & Sons, New York (1980).]

Table 8-8 Metal Carbonyls and the Effective Atomic Number (18-Electron) Rule

Compound	Atomic Number of Metal (M), Z	CO Groups per Atom M		Metal–Metal Bonds per Atom M	Electrons per Atom M			Total Electrons per Atom M or Effective Atomic Number
		Terminal	Bridging		Z	From CO	M—M Bonds	
Ni(CO)$_4$	28	4	0	0	28	8		36 (Kr)
Fe(CO)$_5$	26	5	0	0	26	10		36 (Kr)
Cr(CO)$_6$	24	6	0	0	24	12		36 (Kr)
V(CO)$_6$	23	6	0	0	23	12		35
Ru(CO)$_5$	44	5	0	0	44	10		54 (Xe)
Mo(CO)$_6$	42	6	0	0	42	12		54 (Xe)
Os(CO)$_5$	76	5	0	0	76	10		86 (Rn)
W(CO)$_6$	74	6	0	0	74	12		86 (Rn)
Co$_2$(CO)$_8$	27	4	0	1	27	8	1	36 (Kr)
		3	2	1	27	8	1	36 (Kr)
Fe$_2$(CO)$_9$	26	3	3	1	26	9	1	36 (Kr)
Mn$_2$(CO)$_{10}$	25	5	0	1	25	10	1	36 (Kr)
Ru$_3$(CO)$_{12}$	44	4	0	2	44	8	2	54 (Xe)
Tc$_2$(CO)$_{10}$	43	5	0	1	43	10	1	54 (Xe)
Ir$_4$(CO)$_{12}$	77	3	0	3	77	6	3	86 (Rn)
Re$_2$(CO)$_{10}$	75	5	0	1	75	10	1	86 (Rn)

comparatively unreactive, whereas the $Ni(C_5H_5)_2$ molecule, which contains two additional electrons, is very reactive. The 18-electron rule suggests that as many of the electrons in the valence shell as possible should participate in bonding. There are, however, many exceptions to this generalization.

Terminal M—CO bonds are formulated in valence-bond terminology, as

$$M \leftarrow C \equiv O \qquad or \qquad \overset{-}{M} - \overset{+}{C} \equiv O:$$

To account for the observed multiple-bond character of the metal–carbon monoxide interaction, formulation in terms of a resonance hybrid, as

$$\overset{-}{M} - \overset{+}{C} \equiv O: \leftrightarrow M = C = \overset{..}{O}:$$

is suggested. More usefully and more in keeping with energy considerations, the bond can be described as a combination of σ and π interactions. An $M \leftarrow C$ σ bond forms by donation of the unshared electron pair on the C atom into a metal orbital of σ symmetry directed to the center of the π system of the ligand; simultaneously a filled d_π or hybrid dp_π orbital of the metal atom overlaps with an empty $2p_\pi^*$ (antibonding) orbital of the carbon atom. The result is a *synergic* effect which leads to enhanced M—C bond strength and an M—C bond of essential electroneutrality. This model[76, 77] is shown diagrammatically in Fig. 8-17. A qualitative molecular-orbital, energy-level diagram (Fig. 8-12)[78] shows the complete analogy between the isoelectronic species $[M(CO)_6]$ and $[M(CN)_6]^{n-}$.

Doubly and triply bridging CO groups also bond by synergic "forward" donations of electronic charge density from the carbon atom to the metal atom and "back" donations of electronic charge density from the metal atom to the carbon atom. These are shown diagrammatically in Fig. 8-18.

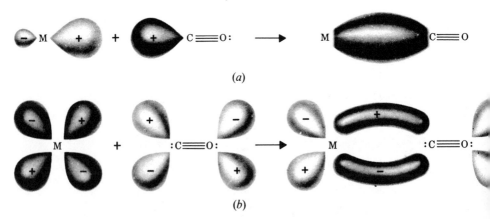

(a)

(b)

Figure 8-17 (a) σ and (b) π metal–carbon monoxide interactions. [F. A. Cotton and G. Wilkinson, *Advanced Inorganic Chemistry*, 4th ed., Fig. 3-6, p. 83, John Wiley & Sons, New York (1980).]

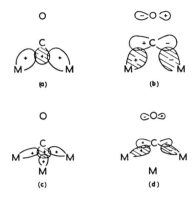

Figure 8-18 σ and π overlaps in doubly and triply bridging M—CO bonds. (*a*) Forward donation to two metal atoms; (*b*) back bonding from two metal atoms; (*c*) forward donation to three metal atoms; (*d*) back donation from three metal atoms (one member of doubly degenerate set). [P. S. Braterman, *Struct. Bonding*, **10**, 68 (1972), Fig. 5.]

For more extensive treatments of bonding in metal carbonyl species, various summaries can be consulted.[59, 61]

Metal Nitrosyls and Their Derivatives. Although a number of types of *d*-transition metal compounds and ions that contain the NO (nitrosyl) group have been described (Table 8-9), only a single metal nitrosyl, namely $Co(NO)_3$, has been clearly identified. Early characterization of various nitrosyl carbonyls (Table 8-9) and establishment of similarities to the metal carbonyls resulted in development of a nitrosyl chemistry somewhat parallel to carbonyl chemistry.[54, 79-88] Metal nitrosyl derivatives also include halides and pseudo-halides; amines, phosphines, arsines, and stibines; π-cyclopentadienyl species; and organo compounds (Table 8-9). Characteristics beyond molecular structure and bonding are also discussed in the cited references.

1. **Molecular structures.** Typical molecular structures for various types of metal nitrosyl compounds are indicated in Fig. 8-19. The NO group is invariably bonded to the metal atom through its nitrogen atom. Terminal NO groups may be either linear (e.g., in $Co(CO)_3(NO)$, $[Cr(NO)(CN)_5]^{3-}$) or bent (e.g., in $[Ir(NO)(CO)\{P(C_6H_5)_3\}_2Cl]^+$,[89] black $[Co(NO)(NH_3)_5]^{2+}$). Both linear and bent NO groups may appear in the same species (e.g., $[Ru(NO)_2\{P(C_6H_5)_3\}_2Cl]^+$). Furthermore, small deviations from absolute linearity (161 to 175° vs. 180°) may characterize so-called "linear" MNO arrangements. Bent MNO arrangements have bond angles in the region 120 to 140°. Bridging NO groups are less common than bridging CO groups. Both doubly and triply bridging NO groups can be distinguished [e.g., three of the former and one of the latter in the compound $(C_5H_5)_3Mn_3(NO)_4$].[90] The NO stretching vibration in the infrared region allows distinction among the various types of NO groups,[91] viz., linear terminal NO (1800 to 1900 cm^{-1}), bent terminal NO (1525 to 1690 cm^{-1}), bridging NO (1300 to 1550 cm^{-1}). Triply bridging NO groups absorb at smaller wave numbers than do doubly bridging NO groups. Further details on stereochemistry are available in a comprehensive review.[92]

2. **Bonding.** The molecular orbital description of the NO(g) molecule is comparable with that of the CO(g) molecule (Fig. 5-14) except for the presence of

Table 8-9 Metal Nitrosyls and Typical Derivatives of Metal Nitrosyls

Type	Group VB	Group VIB	Group VIIB	Group VIIIB			Group IB
				Fe Family	Co Family	Ni Family	
Nitrosyls				$Fe(NO)_4$ (?) dec. 0°C (black) $Ru(NO)_4$ or $Ru(NO)_5$ (?) (red)	$Co(NO)_3$ (brown)		
Nitrosyl carbonyls	$V(NO)(CO)_5$ (red-violet)		$Mn(NO)_3(CO)$ m.p. 27°C (dark green) $Mn(NO)(CO)_4$ m.p. −1.5°C (red) $Mn_2(NO)_2(CO)_7$ (purple)	$Fe(NO)_2(CO)_2$ m.p. 18.4°C (red) $[Fe(NO)(CO)_3]^-$ (red)	$Co(NO)(CO)_3$ m.p. −11°C (red)		
Nitrosyl halides or pseudohalides	$V(NO)Cl_4$ (purple) $V_2(NO)Cl_2$ (purple) $V_2(NO)_5Cl_8$ (purple) $[V(NO)_3Cl_2]_n$ (brown) $[V(NO)(CN)_5]^{5-}$ (as K⁺ salt)	$[Cr(NO)(CN)_5]^{4-}$ (as K⁺ salt) (blue) $[Cr(NO)(CN)_5]^{3-}$ (as K⁺ salt) (green) $[Mo(NO)_2Cl_2]_n$ (dark green)	$[Mn(NO)(CN)_5]^{3-}$ (as K⁺ salt) (purple) $[Re(NO)Cl_5]^{2-}$	$[Fe(NO)_2X_2]_2$ (X = Cl, Br, I) (black-brown) $Fe(NO)_3X$ (X = Cl, Br, I) (black-brown) $[Fe(NO)(CN)_5]^{2-}$ (as Na⁺ salt) (red-brown) $[Ru(NO)(CN)_5]^{2-}$ (as K⁺ salt)	$[Co(NO)_2X]_2$ (X = Cl, Br, I, NCS) (black-brown) $[Co(NO)(CN)_5]^{3-}$ (as K⁺ salt) (yellow) $[Rh(NO)_2X]_n$ (X = Cl, Br, I)	$[Ni(NO)X]_n$ (X = Cl, Br, I) (dark green-black) $[Ni(NO)X_2]_n$ (X = NO₂, Cl, Br) [blue (NO₂), green] $[Ni(NO)(CN)_3]^{2-}$ (as K⁺ salt) (red) $[Pd(NO)Cl]_n$ (yellow)	$Cu(NO)X_2$ (X = Cl, Br) (violet)

[Mo(NO)-
(CN)₅]⁴⁻
(as K⁺ salt)
(purple)

[Re(NO)-
(CN)₇]³⁻
(as Ag⁺ salt)
(mauve)

Pd(NO₂)₂Cl₂
(brown)

[Pd(NO)X₅]²⁻
(X = Cl, CN)
(green)

[Pt(NO)XX′₄]²⁻
(X = X′ = Cl as
  Na⁺, K⁺ salts;
 X = X′ = CN⁻
  as K⁺ salt;
 (X = Cl, X′ = NO₂
  as K⁺ salt)
 (X = NO₃, X′ = NO₂
  as K⁺, Cs⁺ salts)
 [Pt(NO)(NO₂)₃Cl₂]⁻
 (as K⁺ salt)

[W(NO)₂Cl₂]ₙ
(green)

[Os(NO)Cl₅]²⁻
(red as K⁺, Cs⁺ salts)

[Ir(NO)Br₅]⁻
(golden brown as
 K⁺ salt)

Ni(NO){(C₆H₅)₃Z}₂X
(Z = P, As, Sb;
 X = Br, I)

[Ni(NO)(C₆H₁₁)₃PX]₂
(X = Br, I; cis-
 and trans-)

[Ni(NO)(C₆H₅O)₃PX]₂
(X = Br, I)

Ni(NO){(C₆H₅)₃P}₂X
(X = Cl, Br, I, OCN,
 NO₃, NO₂, OH, OCH₃)

Co(NO)₂{(C₆H₅)₃Z}X
(Z = P, As, Sb;
 X = Cl, Br, I)

Co(NO)₂(C₅H₁₁N)Cl

[Co(NO)₂(phen)]ClO₄

Co(NO)(R₃,P)X₂
(R = C₆H₅, C₂H₅;
 X = Cl, Br, I)

Rh(NO){(C₆H₅)₃Z}₂Cl₂
(Z = P, As, Sb)

Fe(NO)₂{(C₆H₅)₃Z}X
(Z = P, As, Sb;
 X = Cl, Br, I)

Fe(NO)₂(C₅H₁₁N)Br

Nitrosyl
amines,
phosphines,
arsines,
stibines

Mn(NO)₂
{(C₆H₅)₃-
P}₂X
(X = Br, I)

Mn(NO)₂-
{(C₆H₅)₃-
Z}₂I
(Z = P, As, Sb)

Mn(NO)₂-
{(C₆H₅)₃-
Z}₂X
(Z = P, As, Sb;
 X = Cl, Br, I)

Mn(NO)₂-
{(C₆H₅)₃-
P}₂H

Re(NO)₂-
{(C₆H₅)₃-
P}₂X₂
(X = Cl, Br,
 I, NO₃)

Table 8-9 (*Continued*)

Type	Group VB	Group VIB	Group VIIB	Group VIIIB			Group IB
				Fe Family	Co Family	Ni Family	
Nitrosyl-π-cyclo-pentadienyl compounds		$Cr(C_5H_5)-(NO)_2X$ ($X = Cl$, Br, I, NCS, CH_3, C_2H_5, C_6H_5) (green; 1-black) $[Cr(C_5H_5)-(NO)_2]_2$ m.p. 158–9°C (red-purple) $Cr(C_5H_5)-(NO)(CO)_2$ m.p. 67–68°C (orange) $[Cr(C_5H_5)-(NO)_2(CO)]^+$ (as PF_6^- salt) (green)	$[Mn(C_5H_5)-(NO)(CO)]_2$ m.p. 200°C (dec.) (purple-brown) $Mn_6(C_5H_5)_6-(NO)_8$ m.p. 220°C (dec.) (black) $Mn(C_5H_5)_3-(NO)_3$ m.p. 100°C (dec.) (red-black)		$Rh(NO)_2(C_5H_5N)_2Cl$	$[Pt(NO)(CN)_5]^{2-}$ (as K^+ salt) (green)	
Organo nitrosyl compounds				$Fe(C_3H_5)(NO)(CO)_2$ oil (red) $Fe(C_3H_4)(CO_2CH_3)-(NO)(CO)_2$ oil (red)	$Co(C_2H_5)-(NO)_2\{P(C_6H_5)_3\}$ m.p. 65°C (brown-violet)		

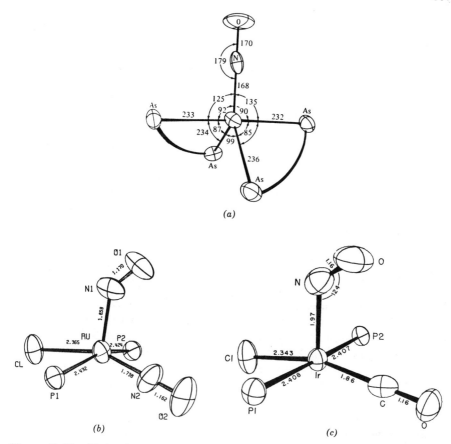

Figure 8-19 Molecular structures of typical metal nitrosyl derivatives. (a) $[Co(NO)(diars)_2]^{2+}$; (b) $[RuCl(NO)_2(P\{C_6H_5\}_3)_2]^+$; (c) $[IrCl(CO)(NO)\{P(C_6H_5)_3\}_2]^+$. [(a), J. H. Enemark and R. D. Feltham, *Proc. Natl. Acad. Sci. USA*, **69**, 3524 (1974); (b), C. G. Pierpont, D. G. Van Derveer, W. Durland, and R. Eisenberg, *J. Am. Chem. Soc.*, **92**, 4760 (1970); (c), D. J. Hodgsen, N. C. Payne, J. A. McGinnety, R. G. Pearson, and J. A. Ibers, *J. Am. Chem. Soc.*, **90**, 4486 (1968). (b) and (c) reprinted with permission from the American Chemical Society.]

an additional electron in a π^* orbital. The general similarities between carbonyl and nitrosyl compounds already noted are thus reasonable, but with the added differences that this electron is responsible for bent terminal geometry in the NO ligand and for the useful concept of regarding the NO molecule as a three-electron donor to a metal atom or ion. Regularities that result in the stoichiometries of comparable compounds are:

a. Nitrosyl carbonyl compounds isoelectronic with carbonyl compounds $M(CO)_n$ have compositions $M'(CO)_{n-1}(NO)$, $M''(CO)_{n-2}(NO)_2$, $M'''(CO)_{n-3}(NO)_3$, and so on, where M', M'', M''', and so on, have atomic numbers that are 1, 2, 3, and so on, less than that of M. Examples include the series $Ni(CO)_4$, $Co(CO)_3(NO)$, $Fe(CO)_2(NO)_2$, $Mn(CO)(NO)_3$ and $Fe(CO)_5$, $Mn(CO)_4(NO)$.

b. Two NO groups replace three CO groups [e.g., in the pairs $Fe(CO)_5$–$Fe(CO)_2(NO)_2$ and $Mn(CO)_4(NO)$–$Mn(CO)(NO)_3$].

It is significant that many of the complex nitrosyl species also obey the effective atomic number, or 18-electron, rule.

One can envision the NO molecule as coordinating to a metal atom or ion in three different ways: (1) by transfer of the odd electron to the metal atom or ion, followed by donation of a nonbonded electron pair in the resulting NO^+ ion; (2) transfer of an electron from the metal atom or ion to the NO molecule to form the NO^- ion, which then bonds like a halide ion; and (3) donation of an electron pair by the N atom of the NO molecule, with retention of the unpaired electron by the ligand. Most usefully, the NO molecule is regarded as a three-electron donor, when present as either a terminal or a bridging ligand.

Rather than attempt to describe the bonding mechanism in terms of prior formation of an NO^+ or NO^- ion, it is reasonable to assume pairing of the π^* electron of the NO molecule with a d_π electron of the metal atom or ion. The NO molecule forms particularly strong π bonds. The NO^+ ion description can result in assignment of questionable oxidation numbers to metals [e.g., -1 in $Co(CO)_3(NO)$ or -3 in $Mn(CO)(NO)_3$]. However, the NO^+ ion does exist in certain saltlike nitrosonium compounds such as $NO^+ClO_4^-(s)$ and $NO^+HSO_4^-(s)$. The NO^- ion description is useful in preserving reasonable oxidation numbers in certain species [e.g., cobalt(III) in the red $[Co(NH_3)_5(NO)]^{2+}$ ion]. However, detailed investigation has shown this ion to be dimeric and to contain the bridging hyponitrite ion, $N_2O_2^{2-}$.[93]

Metal Cyano, Isocyano, and Isonitrile Complexes. As indicated by the examples cited in Table 8–10, the CN^- group forms a variety of complexes with many of the d-transition metal and post-d-transition metal ions in numerous oxidation states. Stabilization of lower oxidation states is particularly apparent. The ambidentate character of the CN^- ion (Sec. 7.4-5) allows both the cyano (M—CN) and isocyano (M—NC) molecular arrangements, although the former appears to be much the more common one. Most of the species listed in Table 8–10 are so formulated, but only limited numbers of these formulations are supported by definitive structure-indicating data. Cyano, or isocyano, groups are terminal groups in most individual molecular structures. Linear bridging groups in the arrangement

$$M—C{\equiv}N—M$$

are found in crystals of metal cyanides and cyano complexes [e.g., $Zn(CN)_2$ and $KFeFe(CN)_6$ (Fe in both $+2$ and $+3$ oxidation states)].

The organic isocyanides, or isonitriles, R—NC, are common carbon donors toward d-transition metal ions (Table 8–11) and yield complexes that are remarkably similar to the carbonyls in many of their properties.

For additional information, comprehensive summaries[54,94-97] can be consulted.

1. Molecular structures. Molecular and ionic species summarized in Table 8–10 exemplify all coordination numbers from 2 through 8. Common experimentally determined molecular geometries and specific examples are given for cyano complexes in Table 8–12. These structures emphasize the ubiquitous M—C linkage.

Of particular interest from the structural point of view are the partially oxidized tetracyanoplatinate compounds, $M_2[Pt(CN)_4]X_n \cdot y H_2O$ (M = alkali metal or NH_4^+, X = monoanion, $n < 1$, y may = O), which behave as one-dimensional electrical conductors.[98-102] Crystals of these compounds contain nearly square-planar $Pt(CN)_4$ groups stacked along the c axis to form linear Pt–Pt chains with short Pt–Pt separations ascribable to overlap of d_z^2 orbitals. Conductivity occurs in the direction of these overlapping orbitals. Many of the crystal structures have been determined by neutron diffraction (Sec. 6.5-1).

Definitive data for the molecular structures of organo isocyanide (or isonit-rile) species are much less numerous. Isosteric species obtained by the partial or complete substitution of CNR groups for CO groups in carbonyls appear to be isostructural with the parent carbonyls. This conclusion was supported by early infrared data[126] and was later verified by crystal-structure determinations. Thus the cation in the compound of empirical formula $Co(CNCH_3)_5(ClO_4)_2$ has been shown to be dimeric with the $Co_2(CNCH_3)_{10}^{4+}$ group isostructural with the isosteric $Mn_2(CO)_{10}$ molecule.[127]

Single-crystal, x-ray diffraction measurements have indicated a slightly distorted trigonal bipyramidal geometry for the $[Co(CNCH_3)_5]^+$ cation in its perchlorate salt.[128] The blue $[Rh_2(bridge)_4]^{2+}$ ion (bridge = 1,3-diisocyano-propane) in concentrated aqueous hydrochloric acid solution yields molecular hydrogen and the yellow ion $[Rh_2(bridge)_4Cl_2]^{2+}$ upon irradiation at 546 nm.[129] X-ray diffraction studies on a single crystal of $[Rh_2(bridge)_4Cl_2]Cl_2 \cdot 8 H_2O$ showed the cation to have the "windmill-like", or "paddle-wheel," molecular geometry indicated in Fig. 8–20.[130]

2. Bonding. The discussion of bonding in the section on metal carbonyls is also applicable to these complexes. The position of the CN^- ion in the spectro-chemical series (Sec. 8.1-2) and the resulting large nephelauxetic (Sec. 8.1-3) and *trans*-directing effects are consistent with π bonding in the M—CN linkage. The stabilization of low oxidation states by the CN^- ion is indicative of its accept-ance of electronic charge density into π^* orbitals in terms of the Dewar–Chatt–Duncanson picture (Fig. 8–17),[76,77] but the negative charge of the ion probably decreases this tendency somewhat, thereby rendering the CN^- ion less effective than the CO molecule in effecting this stabilization. The CNR molecules appear to be stronger σ donors than the CO molecule, but infrared data suggest that their abilities to accept π electronic charge density are comparable with that of the CO molecule.[126-128,131]

Table 8-10 Some Metal Cyano and Isocyano Species

Group IVB	Group VB	Group VIB	Group VIIB	Group VIIIB			Group IB	Group IIB
				Fe Family	Co Family	Ni Family		
III $[Ti(CN)_6]^{3-}$ (as K^+ salt) (blue) (?)	0 $[V(CN)_5(NO)]^{5-}$ (as K^+ salt) (pink) (?)	0 $[Cr(CN)_6]^{6-}$ (as K^+ salt) (green)	I $[Mn(CN)_6]^{5-}$ (as Na^+, K^+ salts) (colorless)	I $[Fe(CN)_5]^{4-}$ (as Na^+, K^+, etc., salts)	-1 $[Co(CN)_3(NO)]^{3-}$ (as K^+ salt) (dark violet)	0 $[Ni(CN)_4]^{4-}$ (as K^+ salt) (copper colored)	I $[Cu(CN)_2]^-$ (as K^+ salt) (colorless)	II $[Zn(CN)_4]^{2-}$ (as K^+ salt) (colorless)
III $[Ti(CN)_8]^{5-}$ (as K^+ salt) (green) (?)	0 $[V(CN)_7]^{7-}$ (as K^+ salt) (chocolate-brown) (?)	0 $[Cr(CN)_m(CO)_{6-m}]^{n-}$ (n = 4, 3, 2, 1)	I $[Mn(CN)_n-(CO)_{6-n}]^{1-n}$ (n = 1-4)	I $[Fe(CN)_5(NO)]^{3-}$ (as Na^+ salt) (ochre yellow)	-1 $[Co(CN)(CO)_2(NO)]^+$ (as K^+ salt)	0 $[Ni(CN)_2(CO)]^{2-}$ (as K^+ salt)	I $[Cu(CN)_3]^{2-}$	II $[Zn(CN)_3]^-$
	I $[V(CN)_5(NO)]^{3-}$ (as K^+ salt) (orange) (?)	I $[Cr(CN)_4]^{3-}$ (as K^+ salt) (brown)	I $[Mn(CN)_5(NO)]^{3-}$ (as K^+ salt) (purple)	II $[Fe(CN)_6]^{4-}$ (yellow)	0 $[Co_2(CN)_8]^{8-}$ (as K^+ salt) (brown-violet)	I $[Ni_2(CN)_6]^{4-}$ (as K^+ salt) (red)	I $[Cu(CN)_4]^{3-}$	II $[Zn(CN)_4]^{2-}$ (colorless)
	II $[V(CN)_6]^{4-}$ (as K^+ salt) (yellow-brown)	I $[Cr(CN)_5(NO)]^{3-}$ (as K^+ salt) (olive green)	II $[Mn(CN)_6]^{4-}$ (as Na^+, K^+) (K^+-blue; Na^+-yellow)	II $[Fe(CN)_5(NO)]^{2-}$ (red)	I $[Co(CN)_5]^{4-}$	II $[Ni(CN)_4]^{2-}$ (yellow)		
	III $[V(CN)_7]^{4-}$ (as K^+ salt) (scarlet)	II $[Cr(CN)_6]^{4-}$ (red)	III $[Mn(CN)_6]^{3-}$ (as K^+ salt) (red)	II $[Fe(CN)_5(L)]^{3-}$ (L = H_2O, NH_3, CO, etc.)	I $[Co(CN)_3(CO)_2]^{2-}$ (as $Fe(phen)_3^{2+}$ salt) (red)	II $[Ni(CN)_5]^{3-}$ (red)		
	IV $[VO(CN)_5]^{3-}$ (green)	III $[Cr(CN)_6]^{3-}$ (pale yellow)	IV $[Mn(CN)_6]^{2-}$ (as K^+ salt) (red)	III $[Fe(CN)_6]^{3-}$ (as Na^+, K^+ salts) (red)	II $[Co(CN)_5]^{3-}$ (olive green)	II $[Ni(CN)_2(PR_3)_2]$		

IV [Nb(CN)₈]⁴⁻ → $[Nb(CN)_8]^{4-}$ (as K⁺ salt) (orange)

$[Nb(CN)Cl_4]$ (as ether adduct) (red-brown)

$[Cr(CN)_n\text{-}(H_2O)_{6-n}]^{n-3}$

0 $[Mo(CN)(CO)_5]^-$ (as Na⁺ salt) (colorless)

0 $[Mo(CN)_3(CO)_3]^{3-}$ (as K⁺ salt) (colorless)

0 $[Mo(CN)_5(NO)]^{4-}$ (as K⁺ salt) (deep violet)

II $[Mo(CN)_6]^{4-}$ (as K⁺ salt) (green to black)

II $[Mo(CN)_2\text{-}(\pi\text{-}C_5H_5)\text{-}(CO)_2]^-$ (as K⁺ salt)

III $[Mo(CN)_7]^{4-}$ (as K⁺ salt) (greenish black)

III $[Mo(CN)_6]^{3-}$ (as K⁺ salt) (greenish brown)

IV $[Mo(CN)_8]^{4-}$ (as K⁺ salt) (golden yellow)

I $[Tc(CN)_6]^{5-}$ (as K⁺ salt) (green)

IV $[Tc(CN)_4(OH)_3]^{3-}$ (as Tl⁺ salt) (dark brown)

III $[Co(CN)_6]^{3-}$ (pale yellow)

I $[Ru(CN)_5(NC)]^{5-}$ (as K⁺ salt)

II $[Ru(CN)_6]^{4-}$ (as K⁺ salt) (colorless)

II $[Ru(CN)_2(bipy)_2]$ (orange-red)

II $[Ru(CN)_5(NO)]^{2-}$ (as K⁺ salt) (red-brown)

III $[Ru(CN)_6]^{3-}$ (?)

0 $[Rh_2(CN)_8]^{8-}$ (as K⁺ salt)

I $[Rh(CN)_4]^{3-}$ (as $[Co(en)_3]^{3+}$ salt)

II $[Rh(CN)_6]^{4-}$

II $[Rh(CN)_6]^{3-}$ (as K⁺ salt) (pale yellow)

$[Rh(py)_2\text{-}(CN)_2Cl]\cdot 2H_2O$ (pale yellow)

0 $[Pd(CN)_4]^{4-}$ (as K⁺ salt) (yellowish white)

II $[Pd(CN)_4]^{2-}$ (as K⁺ salt) (colorless)

II $[Pd(CN)_2\text{-}P(C_6H_5)_3)_2]$ (colorless)

IV $[Pd(CN)_6]^{2-}$ (as K⁺ salt)

$[Ag(CN)_2]^-$ (as K⁺ salt) (colorless)

$[Ag(CN)_n]$ (n = 2, 3, 4) (colorless)

$[Cd(CN)_4]^{2-}$ (as K⁺ salt) (colorless)

$[Cd(CN)_3]^-$ (colorless)

Table 8-10 (Continued)

Group IVB	Group VB	Group VIB	Group VIIB	Group VIIIB			Group IB	Group IIB
				Fe Family	Co Family	Ni Family		
		0 $[W(CN)_n(CO)_m]^{n-}$ $(n=1-4; m=6-n)$ (as Na$^+$, K$^+$ salts)	I $[Re(CN)_6]^{5-}$ (as K$^+$ salt) (blue-green)	II $[Os(CN)_6]^{4-}$ (as K$^+$ salt) (colorless)	I $[Ir(CN)(CO)\{P(C_6H_5)_3\}]$ (yellow)	0 $[Pt(CN)_4]^{4-}$ (as K$^+$ salt) (colorless)	$[Au(CN)_2]^{-}$ (as K$^+$ salt) (colorless)	$[Hg(CN)_4]^{2-}$ (as K$^+$ salt) (colorless)
		II $[W(CN)_6]^{4-}$ (as K$^+$ salt) (black or green)	II $[Re(CN)_5(H_2O)]^{3-}$ (as Na$^+$, K$^+$ salts) (brown)	II $[Os(CN)_5(NO)]^{2-}$ (as K$^+$ salt) (red)	II $[Ir(CN)_5]^{3-}$ (trapped in KCl crystal)	II $[Pt(CN)_4]^{2-}$ (as K$^+$ salt) (pale yellow)	III $[Au(CN)_4]^{-}$ (as K$^+$ salt) (colorless)	$[HgCN]^+$ (as BF$_4^-$ or AsF$_6^-$ salt) (colorless)
		II $[W(CN)_n(\pi\text{-}C_5H_5)(CO)_m]^{(1-n)}$ $(n=1, 2;$ $m=4-n)$	III $[Re(CN)_6]^{3-}$ (as K$^+$ salt) (green)	III $[Os(CN)_6]^{3-}$ [as N(C$_4$H$_9$)$_4$ salt] (yellow)	III $[Ir(CN)_6]^{3-}$ (as K$^+$ salt) (pale yellow)	II $[Pt(CN)_4]^{2-}$ (as Mg^{2+} salt) (dark red)	III $[Au(CN)_3X]^{-}$ (X = Cl, Br, I) (as K$^+$ salts)	$[Hg(CN)_2]$ (s, aq) (colorless)
		III $[W(CN)_6]^{3-}$ (as K$^+$ salt) (brown)	III $[Re(CN)_3(OH)_3]^{3-}$ (as K$^+$ salt) (blue)	VI $[Os(CN)_4O_2]^{2-}$ (as K$^+$ salt) (orange-red)	III $[Ir(CN)_5(H_2O)]^{2-}$	*cis*- and *trans*- $[Pt(CN)_2$- $(NH_3)_2]$	*trans*- $[Au(CN)_2X_2]^{-}$ (X = Cl, Br) (as K$^+$ salts)	$[Hg(CN)_3]^-$ (colorless)
		IV $[W(CN)_8]^{4-}$ (as K$^+$ salt) (orange)	IV $[Re(CN)_4O_2]^{4-}$ (as K$^+$ salt) (grey-black)		III $[Ir(CN)(NCS)(CO)$- $\{P(C_6H_5)_3\}_2Cl]$	IV $[Pt(CN)_6]^{2-}$ (as K$^+$ salt) (colorless)	$[R_2Au(CN)]_4$ (R = alkyl)	$[Hg(CN)_2X]^{-}$ (X = Cl, Br, I) (as M^I salts)

IV
trans-
[W(CN)$_4$O$_2$]$^{4-}$
(as K$^+$ salt)
(yellow-brown)

V
[W(CN)$_8$]$^{3-}$
(as K$^+$ salt)
(pale yellow)

V
[Re(CN)$_8$]$^{3-}$
(as K$^+$ salt)
(brown)

V
[Re(CN)$_4$O$_2$]$^{3-}$
(as K$^+$ salt)
(orange)

−I
[Re(CN)$_8$]$^{2-}$
[as As(C$_6$H$_5$)$_4$]$^+$
salt]
(purple)

IV
[Pt(CN)$_4$(NH$_3$)$_2$]

IV
trans-[Pt(CN)$_4$-
X$_2$]$^{2-}$
(X = Cl, Br, I)
(as K$^+$ salts)
(yellow—Cl, Br;
brown—I)

Table 8-11 Some Isocyanide (*Isonitrile*) Complexes

Group IVB	Group VB	Group VIB	Group VIIB	Group VIIIB			Group IB	Group IIB
				Fe Family	Co Family	Ni Family		
		$\overset{0}{\mathrm{Cr}}(CNC_6H_5)_6$ m.p. 178.5°C (red)	I [Mn(CNCH$_3$)$_6$]ClO$_4$ m.p. 161°C (dec.) (white)	$\overset{0}{\mathrm{Fe}}(CNCH_3)(CO)_4$ m.p. 31.5°C (pale yellow)	$\overset{0}{\mathrm{Co}}(CNC_7H_7)_2(NO)(CO)$ m.p. 156–159°C (dec.) (orange-red)	$\overset{0}{\mathrm{Ni}}(CNCH_3)_3(CO)$ m.p. 110°C (dec.) (pale yellow)	I Cu(CNC$_6$H$_5$)Cl (white)	Zn(CNC$_5$H$_{11}$)$_2$(CN)$_2$ (golden yellow)
		$\overset{0}{\mathrm{Cr}}(CNC_7H_7)_6$ m.p. 152.8°C (red)	I [Mn(CNC$_6$H$_5$)$_6$]I m.p. 216°C (yellow)	$\overset{0}{\mathrm{Fe}}(CNC_2H_5)_2(CO)_3$ m.p. 65.6–66°C	I Co(CNR)$_5$X (X = I, ClO$_4$, NO$_3$; R = CH$_3$, C$_6$H$_5$) (yellow)	$\overset{0}{\mathrm{Ni}}(CNC_6H_5)_4$ m.p. 202–204°C (dec.) (canary yellow)	I Cu(CNC$_4$H$_9$)CN (colorless)	Zn(CNC$_7$H$_7$)$_2$Cl$_2$ (white)
		$\overset{0}{\mathrm{Cr}}(CNC_7H_4Cl)_6$ m.p. 156°C (orange-red)	I [Mn(CNR)$_6$]$^+$ (with various anions; R = alkyl, aryl)	II Fe(CNCH$_3$)(CO)$_3$I$_2$ m.p. 115°C (dec.) (brown-red)	II [Co(CNCH$_3$)$_4$]X$_2$ (X = CoCl$_4$, Cl, Br) (green or violet)	$\overset{0}{\mathrm{Ni}}(CNC_{10}H_7)_4$ (yellow-red)	I Cu(CNC$_7$H$_7$)$_4$ClO$_4$ (colorless)	Zn(CNC$_7$H$_7$)$_2$I$_2$ (pale yellow)
			II [Mn(CNC$_7$H$_7$)$_6$][CdBr$_4$] m.p. 145°C (blue)	II Fe(CNC$_5$H$_{11}$)$_2$Cl$_2$ (golden yellow)	III [Co(CNC$_6$H$_5$)$_4$Br$_2$]-Br (dark green)			
				II [Fe(CNCH$_3$)$_6$]$^{2+}$ (with various anions) (colorless)	III [Co(CNC$_7$H$_7$)$_5$I]-(ClO$_4$)$_2$ (red-violet)			
		$\overset{0}{\mathrm{Mo}}(CNC_7H_7O)(CO)_5$ m.p. 105–106°C (colorless)	—	II Ru(CNR)$_4$Cl$_2$ (R = CH$_3$, C$_2$H$_5$, C$_7$H$_7$) (yellow)	I [Rh(CNR)$_4$]ClO$_4$ (R = C$_6$H$_5$, C$_7$H$_7$, C$_7$H$_7$O) m.p. 172–200°C (yellow or green)	Pd(CNR)$_2$ (R = C$_6$H$_5$, C$_7$H$_7$, C$_7$H$_7$O) m.p. 150–190°C (dec.) (dark brown)	I Ag(CNCH$_3$)(CN) m.p. 75–76°C (colorless)	II Cd(CNC$_7$H$_7$)$_2$Cl$_2$ (white)

Mo(CNC₇H₇)₆

$\overset{0}{\text{Mo}}(CNC_7H_7)_6$
(orange)

$\overset{0}{\text{Mo}}(CNC_6H_4Cl)_6$
(yellow-orange)

$\overset{0}{\text{W}}(CNC_6H_5)_6$
m.p. 120–130°C
(dark red)

$\overset{0}{\text{W}}(CNC_7H_6Cl)_6$
m.p. 205–207°C
(red)

Re(CNC₇H₇)-
(CO)₄Cl
(dec. 200°C)

II
Ru(CNC₇H₇)₄-
Cl₂I₂
(red)

II
[Rh(CNC₆H₅)₈I₂]-
X₂
(X = I, ClO₄)
(dark)

III
Rh(CNC₆H₅)₄Cl₂]X
(X = Cl, ClO₄)
blue-violet, green

II
Pd(CNR)₂X₂
(R = C₆H₅, C₇H₇,
C₇H₇O, C₆H₄Cl;
X = Cl, Br, I)
(yellow to orange)

II
Pt(CNR)₂X₂
(R = CH₃, C₂H₅,
C₄H₉, C₆H₅;
X = CN, Cl, Br, I)
(colorless, yellow)

II
[Pt(CNR)₄]-
[PtCl₄]
(R = C₄H₉, C₆H₅)
(red to violet)

I
Ag(CNC₅H₁₁)(CN)
(colorless)

I
[Ag(CNC₇H₇)₄]-
ClO₄
m.p. 101°C
(white)

I
Au(CNR)Cl
(R = C₆H₅, C₇H₇)
(white)

[Au(CNR)₄]X
(R = C₆H₅, C₇H₇,
C₇H₇O; X = Cl,
B(C₆H₅)₄)
m.p. 175–200°C
(white)

III
Au(CNR)Cl₃
(R = C₇H₇,
C₇H₇O)
(yellow)

II
Hg(CNC₇H₇)₂Cl₂
(colorless)

Table 8-12 Coordination Numbers and Molecular Geometries of Cyano Complexes

Coordination Number	Molecular Geometry	Examples
2	Linear	$[Ag(CN)_2]^-$,[103] $[Au(CN)_2]^-$,[104]
4	Tetrahedral (T_d)	$[Zn(CN)_4]^{2-}$,[105] $[Cd(CN)_4]^{2-}$,[105] $[Hg(CN)_4]^{2-}$,[106,107]
4	Square planar (D_{4h})	$[Ni(CN)_4]^{2-}$,[108,109] $[Pd(CN)_4]^{2-}$,[110] $[Pt(CN)_4]^{2-}$,[111] $[Au(CN)_4]^-$,[112] $[Au(CN)_2Cl_2]^-$[113]
6	Octahedral (O)	$[VO(CN)_5]^{3-}$,[114] $[Cr(NO)(CN)_5]^{3-}$,[115] $[Cr(CN)_6]^{3-}$,[116] $[Mo(NO)(CN)_5]^{4-}$,[117] $[Mn(NO)(CN)_5]^{3-}$,[118] $[Mn(CN)_6]^{4-}$,[119] $[Mn(CN)_6]^{3-}$,[120] $[Fe(CN)_6]^{4-}$,[119] $[Fe(CN)_6]^{3-}$[121]
7	Pentagonal bipyramidal (D_{5h})	$[V(CN)_7]^{4-}$[122]
8	Dodecahedral (D_{2d})	$[Mo(CN)_8]^{4-}$,[123] $[Mo(CN)_8]^{3-}$[124]
8	Square antiprismatic (D_{4d})	$[W(CN)_8]^{4-}$[125]

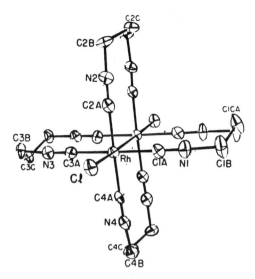

Figure 8-20 Molecular structure of $[Rh_2(bridge)_4Cl_2]^{2+}$ ion. [K. R. Mann, R. A. Bell, and H. B. Gray, *Inorg. Chem.*, **18**, 2671 (1979), Fig. 2. Reprinted with permission from the American Chemical Society.]

2-2. Zero-Valent Metal Derivatives of Terpositive Group VB Donors

Coordination to *d*-transition species by molecules such as PR_3, PX_3, AsR_3, and AsX_3 (R = organo group, X = halogen) has been noted in earlier discussions. Among the more interesting species of this type are those containing metals in the zero-valent state, some examples of which are summarized in Table 8–13. Although most of these compounds are of the type $M^0(ZY_3)_n(CO)_m$, a limited number have the composition $M^0(ZY_3)_n$ (Z = P, As, Sb, or Bi; Y = halide, alkyl, aryl, alkoxo, or aroxo or some combination of these). These compounds are structurally comparable with corresponding metal carbonyl compounds.[132-134]

Ligand molecules of this type are relatively strong Lewis bases and thus give complexes that do not involve *d* orbitals with acceptor species such as BX_3 and BR_3. However, empty d_π orbitals in the structures of the group VB atoms in these ligands can act as π acceptors in the back-bonding sense. An interaction of this type between a filled d_{xy} or d_{yz} orbital of a transition metal atom and an empty *d* orbital of the donor atom, as indicated in Fig. 8–21, can then form a bond. Based upon decreases in the wave number of the stretching vibration of the CO group in compounds of the types $M(ZY_3)_n(CO)_m$ and $M(ZY_3)_n(CO)_o(NO)_p$, π-acceptor strengths among ligands of this type follow generally the order[135-137]

$$CO \approx PF_3 > PCl_3 \approx AsCl_3 \approx SbCl_3 > PCl_2(OR) > PCl_2R > PCl(OR)_2$$
$$> PClR_2 \sim P(OR)_3 > PR_3 \approx AsR_3 \approx SbR_3$$

Table 8-13 Some Metal Complexes of ZY₃ Ligands[a]

Group VB	Group VIB	Group VIIB	Fe Family	Co Family	Ni Family	Group IB	Group IIB
				Group VIIIB			
$\overset{0}{V}(CO)_4\{P(C_6H_5)_3\}_2$	$\overset{0}{Cr}(CO)_5\{Z(C_6H_5)_3\}$ (Z = P, As, Sb)	$\overset{0}{Mn}(CO)_4\{P(C_6H_5)_3\}$	$\overset{0}{Fe}(CO)_4\{P(C_6H_5)_3\}$	$[Co(CO)_3\{P(C_6H_5)_3\}]_2$	$\overset{0}{Ni}(CO)_2(PR_3)_2$ (R = C₂H₅, C₄H₉, C₆H₅)	—	—
$\overset{-1}{[V}(CO)_5\{P(C_6H_5)_3\}]^-$	$\overset{0}{Cr}(CO)_4\{Z(C_6H_5)_3\}_2$ (Z = P, As, Sb)	$Mn_2(CO)_8\{P(C_6H_5)_2\text{-}CH_2P(C_6H_5)_2\}$	$Fe(CO)_3\{P(C_6H_5)_3\}_2$	$[Co(CO)_3\{As(C_6H_5)_3\}]_2$	$Ni(CO)_3\{Z(C_6H_5)_3\}$ (Z = P, As, Sb)		
$\overset{0}{V}(CO)_4(NO)\{P(C_6H_5)_3\}$	$\overset{0}{Cr}(CO)_2\{(C_6H_5)_2\text{-}PC_2H_4P(C_6H_5)_2\}_2$		$\overset{0}{Fe}(CO)_3\{SbCl_3\}_2$	$[Co(CO)_3\{Sb(C_6H_5)_3\}]_2$	$Ni(PX_3)_4$ (X = F, Cl, Br, C₂H₅)		
$\overset{0}{V}\{P(CH_3)_2C_2H_4P\text{-}(CH_3)_2\}_3$	$\overset{0}{Mo}(CO)_5\{As(C_6H_5)_3\}$	—	—	—	$Pd\{Z(C_6H_5)_3\}_4$ (Z = P, As)	—	—
	$Mo(CO)_4\{P(C_2H_5)_3\}_2$				$Pd(PF_3)_4$		
	$Mo(CO)_2\{(C_6H_5)_2As\text{-}C_2H_4As(C_6H_5)_2\}$				$Pd\{(CH_3)_2PC_2H_4P\text{-}(CH_3)_2\}_4$		
—	$\overset{0}{W}(CO)_5\{P(C_6H_5)_3\}$	$[Re(CO)_4\{P(C_6H_5)_3\}]_2$	—	—	$\overset{0}{Pt}(PF_3)_4$	—	—
	$\overset{0}{W}(CO)_4\{As(C_6H_5)_3\}_2$				$Pt\{(C_6H_5)_2PC_2H_4P\text{-}(C_6H_5)_2\}_2$		

[a] (Z = P, As, Sb, Bi; Y = R, halogen).

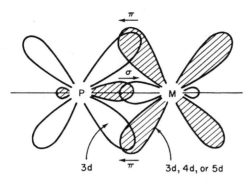

Figure 8-21   Back-donation π bonding in a d-transition metal—PZ_3 species. [W. Levason and C. A. McAuliffe, *Advances in Inorganic Chemistry and Radiochemistry*, H. J. Eméleus and A. G. Sharpe (Eds.), Vol. 14, p. 173, Academic Press, New York (1972).]

That the phosphorus(III) fluoride molecule in particular, and the various chloride molecules are strong complexing agents must be associated with their strengths as π acceptors since the relatively large electronegativities of the halogen atoms materially reduce the donor strengths of these molecules. That the PF_3 molecule is at least as strong a π acceptor as the CO molecule is indicated by photoelectron spectroscopic measurements.[138,139] Compounds of the type $M_n(PF_3)_m$ are strict analogs of the corresponding $M_n(CO)_m$ compounds. Species such as $\overset{0}{Pd}(PF_3)_4$, $\overset{0}{Pt}(PF_3)_4$, and $\overset{-I}{HCo}(PF_3)_4$ [140] emphasize the ability of this ligand to stabilize low oxidation states. The doubly bridging PF_2 group is found in $(F_3P)_3M(PF_2)_2M(PF_3)_3$, where $M = Co$ or Fe.[141,142] Geometrical isomers have been characterized in a number of cases (e.g., *cis*- and *trans*-$[Mo(CO)_4\{P(C_6H_5)_3\}_2]$, *cis*- and *trans*-$[W(CO)_4\{P(C_6H_5)_3\}_2]$).

2-3. Derivatives of Olefins, Carbocyclic Compounds, Arenes, and Related Species

The development of this area of d-transition metal chemistry has been, and continues to be, so extensive (Sec. 7.3-2) and so thoroughly reviewed[143-147],* that to tabulate representative compounds, as has been done in prior sections of this discussion, is virtually impossible. The importance of these organometallic compounds lies not only in their molecular structures and bonding patterns but also in their unique properties as catalysts.† The organometallic compounds discussed in this section are distinguished from those derived from σ-bonding organic anions in their derivation from chain (e.g., $H_2C{=}CH_2$, $H_2C{=}CHCH{=}CH_2$) or cyclic (e.g., $C_5H_5^-$, C_6H_6) π donors. There are, of course,

---

*   Chapter 7, Refs. 87–90, 96, 97.

†   Chapter 7, Ref. 89.

certain overlaps with previously discussed species, in particular the metal carbonyls and their derivatives.

Substantial adherence of species of these types to the 16-electron or the 18-electron rule has prompted two postulates relating to the compounds and their reactions[75]:

1. A diamagnetic organometallic complex of a *d*-transition metal may exist in significant quantity only if the valence shell of the metal contains 16 or 18 electrons.

2. Organometallic reactions, catalytic reactions included, proceed by elementary steps that involve only intermediates having 16 or 18 valence electrons.

In arriving at the number of valence electrons, each σ-bonded CO group donates one electron pair, each π-bonded olefinic linkage one electron pair, each η^5-$C_5H_5^-$ ion three electron pairs,* each η^1-$C_5H_5^-$ ion one electron pair, and each C_6H_6 molecule three electron pairs.

1. Molecular structures. The molecular geometry of the anion of Zeise's salt (Sec. 7.3-2), $[PtCl_3(C_2H_4)]^-$, as determined by neutron diffraction measurements,[149] is indicated in Fig. 8-22. Agreement with the results of an earlier x-ray diffraction study[150] is excellent. The PtCl$_3$ group is very nearly planar, with the

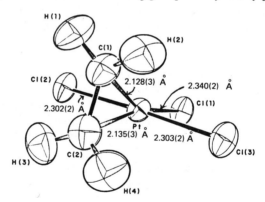

Figure 8-22 Molecular geometry of $[Pt(C_2H_4)Cl_3]^-$ ion. Angles: Cl(1)—Pt—Cl(2), 90.05(8)°; Cl(2)—Pt—Cl(3), 177.65(9)°; Cl(1)—Pt—Cl(3) 90.43(8)°; C(1)—Pt—C(2) 37.6(1)°. [R. A. Love, T. F. Koetzle, G. J. B. Williams, L. C. Andrews, and R. Bau, *Inorg. Chem.*, **14**, 2653 (1975), Fig. 1. Reprinted with permission from the American Chemical Society.]

---

* The η^n or h^n (*hapto* — Greek ηαπτιν, to fasten) notation[148] indicates the number of atoms of the ligand species that are attached, or bonded, to a metal atom or ion in a complex ion or molecule. Thus the notation $Cr(\eta^6$-$C_6H_6)_2$ indicates that all six carbon atoms in each benzene molecule are bonded to the chromium atom, and the notation $Fe(CO)_3(\eta^4$-$C_4H_4)$ indicates that all four carbon atoms in the butadiene-1,3 molecule are bonded to the iron atom. These two compounds are named, respectively, bis(hexahaptobenzene)chromium and tricarbonyltetrahaptobutadiene-1,3-iron. A variant is to replace *hapto* by *eta*. An η^5 species would then be termed either *penta-hapto* or *eta-five*.

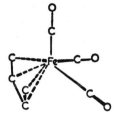

Figure 8-23 Molecular geometry of $[Fe(CO)_3(C_4H_9)]$. [O. S. Mills and G. Robinson, *Proc. Chem. Soc.*, 421 (1960).]

Pt atom only 0.03 Å out of plane. The C=C bond is displaced by only 5.9° from a normal to this plane, with the central point between the two carbon atoms 0.22 Å out of the $PtCl_3$ plane. Thus the geometry around the Pt atom is square planar, the midpoint of the C=C bond occupying one corner of the plane. Other molecular structures involving mono-olefin ligands are similar in indicating that the two carbon atoms in an olefinic bond are essentially equidistant from the metal atom and suggesting interaction involving the π-electronic charge-density cloud in the double bond. The molecular geometry of the butadiene-1,3 compound $[Fe(CO)_3(C_4H_4)]$,[151] shown diagrammatically in Fig. 8-23, indicates chelating interaction involving the conjugated double bonds of the ligand molecule, with three carbon atoms from the CO groups and the two terminal carbon atoms from the C_4H_4 molecule lying in a roughly square-pyramidal arrangement around the iron atom. Chelates formed by other conjugated, noncyclic polyolefins are structurally comparable, with each double bond functioning as a two-electron donor.

Cyclic polyenes, or carbocyclic compounds, give a variety of molecular arrangements, ranging from those in which one or more of the C=C bonds interact with a metal atom or ion, as noted above, to those in which the cyclic arrangement interacts as a unit with all carbon atoms then being equidistant from the metal atom or ion and bonded to it. The latter situation is particularly true of cyclopentadiene and cyclooctatetraene complexes (Table 7-6).

Extreme cases are illustrated for two cyclopentadiene derivatives of cobalt(II) in Fig. 8-24. The molecular geometry of $[Co(\eta^5\text{-}C_5H_5)_2]$, cobaltocene, involves two coplanar delocalized C_5 rings lying on each side of the cobalt(II) ion, with all C—Co bond distances equal and each ring acting as a six-electron donor group.[152] Reaction with C_2F_4 gives a bridged compound in the molecular

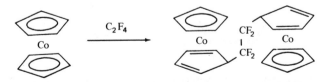

Figure 8-24 Change in structure and bonding on reaction of $Co(\eta^5-C_5H_5)_2$ with C_2F_4. [J. E. Huheey, *Inorganic Chemistry*, 2nd ed., portion of Fig. 13.68, p. 548, Harper & Row, New York (1978).]

structure of which each of the cobalt atoms is similarly bonded to an η^5-C_5H_5 ring but is also bonded through two conjugated double bonds to an η^4-C_5H_5, four-electron donor group. Similarly, the cyclooctatetraenide ion, $C_8H_8^{2-}$, with the static configuration

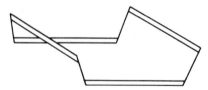

can bond in an η^8-C_8H_8 fashion, as in $[Th(C_8H_8)_2]$ (Fig. 8-25), through one double bond to a metal atom, as in $[Ru(C_8H_8)(CO)_3]$, or through two *trans* double bonds to two metal atoms, as in $[Ru(C_8H_8)(CO)_3]_2$.

Of the carbocyclic ligands, cyclopentadienide ion (almost universally abbreviated as Cp^-) has been particularly extensively studied from the structural point of view.[153] Ionic derivatives (e.g., MC_5H_5, where M = Na, K, Rb, Cs; $M(C_5H_5)_2$, where M = Mg, Sr, Ba, Eu, Yb; $M(C_5H_5)_3$, where M = lanthanide, actinide) are properly termed cyclopentadienides, whereas covalent derivatives are properly referred to as *cyclopentadienyl* compounds. Not uncommonly, the latter term is applied to both types.

Monohapto C_5H_5 rings are occasionally found where other ligands are bonded to the metal atoms but rarely where they are the only ligands. Compounds of this type include $Hg(\eta^1$-$C_5H_5)_2$ and $Al(\eta^1$-$C_5H_5)_3$. Dihapto compounds include $Ti(\eta^5$-$C_5H_5)(\eta^2$-$C_5H_5)(s)$ and $Al(\eta^2$-$C_5H_5)(CH_3)_2(g)$. Tri- and tetrahapto species have not been clearly established. Pentahapto arrangements are particularly common, especially in the compounds $M(\eta^5$-$C_5H_5)_2$ (M = V, Cr, Mn, Fe, Co, Ni, Mo, W, Ru, Os), the *metallocenes* (after *ferrocene*, the first one to be identified).

Properties of the $3d$ metallocenes are summarized in Table 8-14. The molecular structures of these "sandwich" compounds are characterized by parallel C_5 rings that can be oriented in either of two ways,

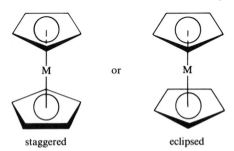

staggered   eclipsed

The staggered geometry is characteristic of solid compounds where M = Fe, Co, Mg; the eclipsed geometry of solid $Ru(C_5H_5)_2$ and gaseous compounds where M = V, Cr, Mn, Fe, Co, Ni, Ru, Mg. Bridging η^5-η^5 rings are present in solid $M(C_5H_5)$ where M = In, Tl. In these compounds, long chains of bridged groups are present. Polyhapto, tilted rings intermediate between tri- and penta-hapto

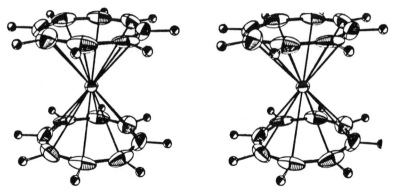

Figure 8-25 Molecular structure of $Th(C_8H_8)_2$. [A. Avdeef, K. N. Raymond, K. O. Hodgson, and A. Zalkin, Fig. 2, *Inorg. Chem.*, **11**, 1083 (1972), Fig. 2. Reprinted with permission from the American Chemical Society.]

extremes have been detected in crystalline $Mo(C_5H_5)_3(NO)$ and $M(C_5H_5)_4$ (M = Ti, Zr, Hf).[153]

Structural data for complexes of benzene and other arenes are much more limited. The molecular structure of dibenzenechromium(0),[154,155] which is typical of these "sandwich" species, can be represented as

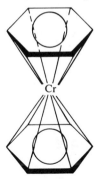

with each C_6H_6 molecule bonded in an η^6 fashion.

Complexes derived from allenes and acetylenes (or alkynes) represent both related and somewhat different molecular structures.[143] Allene complexes, based upon the structural unit $\diagup C=CH-CH_2-$, may contain η^1 σ-bonding ligands [e.g., in $(CH_2=CH-CH_2-)Mn(CO)_5$] or a combination of η^1 σ-bonding and η^3 π-bonding ligands [e.g., in $(CH_2\cdots CH\cdots CH_2)Mn(CO)_4$]. Structurally, the tri-hapto arrangement can be indicated as

Table 8-14 Some Properties of 3d Metallocenes and Related Species

Formula	Color	Melting Point (°C)	Unpaired Electrons	Synthesis[a]	Chemical Properties
$(\eta^5\text{-}C_5H_5)_2TiCl_2$	Bright red	290	0	$(C_5H_5)MgBr + TiCl_4$	Slightly soluble H_2O, yielding $(\eta^5\text{-}C_5H_5)TiOH^+$; reduced to $(\eta^5\text{-}C_5H_5)_2TiCl$
$(\eta^5\text{-}C_5H_5)_2V$	Purple	167–168	3	$(C_5H_5)MgBr(excess) + VCl_4$	
$(\eta^5\text{-}C_5H_5)_2Cr$	Scarlet	172–173	2		Air-sensitive; $HCl(aq) \rightarrow [\eta^5\text{-}C_5H_5CrCl]^+$ (?)
$(\eta^5\text{-}C_5H_5)_2Mn$	Amber	172–173	5	$NaC_5H_5 + MnBr_2$	
$(\eta^5\text{-}C_5H_5)_2Fe$	Orange	173–174	0	$C_5H_6 + base + FeCl_2$	
$(\eta^5\text{-}C_5H_5)_2Co$	Purple	173–174	1	$C_5H_6 + (C_2H_5)_2NH + CoX_2$	
$(\eta^5\text{-}C_5H_5)_2Co^+$	Yellow (aqueous)	—	0		Gives many salts, strongly basic hydroxide; not decomposed below ca. 400°C
$(\eta^5\text{-}C_5H)_2Ni$	Green	173–174 (decomp.)	2	$C_5H_6 + (C_2H_5)_2NH + NiX_2$	Reasonably air-stable; oxidized to $[\eta^5\text{-}C_5H_5)_2Ni]^+$

[a] Other procedures available in most cases.

518

with delocalization of electronic charge density among the three carbon atoms and the central carbon atom lying closer to the metal atom than the other carbon atoms.[156,157] The C—C—C bond angle in complex species is commonly in the range 140 to 160°.

Complexes derived from acetylene and other alkyne molecules are structurally both similar to and different from those derived from ethylene and comparable alkene molecules.[143] Thus in molecules of an analog of Zeise's salt (see Fig. 8-22), $[PtCl_2(t-C_4H_9C≡Ct-C_4H_9)(CH_3C_6H_4NH_2)]$, the C≡C group is positioned vertically at one corner of the square-planar arrangement of ligands around the Pt(II) center.[158] In the molecule $(\mu-C_6H_5C≡CC_6H_5)$-$[(CO)_3Co—Ni(\eta^5-C_5H_5)]$, the alkyne molecule is perpendicularly bonded to the Co—Ni bond.[159] However, in the molecule $\{(C_6H_5)_3P\}_2Pt(C_6H_5C≡CC_6H_5)$, the alkyne molecule lies almost exactly in the PtP_2 plane, with each carbon atom in the —C≡C— moiety occupying a coordination position in a square-planar geometry around the Pt atom.[160] In other cases,[161,162] the alkyne molecules appear as bridging groups in molecular structures, as indicated in Fig. 8-26.

2. Bonding. Bonding involving d-transition metal species and alkene or alkyne molecules is comparable to that existing among the carbonyls, nitrosyls, and cyano complexes (Sec. 8.2-1) and is conveniently explained in terms of the Dewar–Chatt–Duncanson treatment (Sec. 8.2-1). For example, in the ethylene

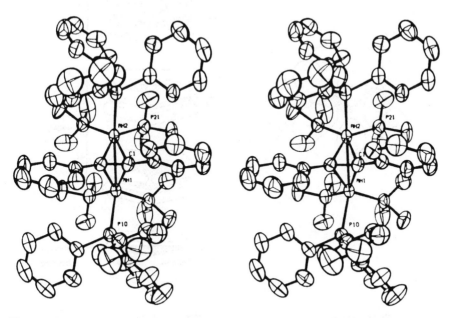

Figure 8-26 Molecular structure of $[(F_3P)_2 (C_6H_5)_3 PRh (C_6H_5C≡CC_6H_5)Rh-P(C_6H_5)_3(PF_3)_2(C_2H_5O)]$. [M. A. Bennett, R. N. Johnson, G. B. Robertson, T. W. Turney, and P. O. Whimp, *Inorg. Chem.*, **15**, 97 (1976). Reprinted with permission from the American Chemical Society.]

molecule, a bonding (π) orbital holds two electrons, and an antibonding (π^*) orbital is empty. The former is oriented with respect to a metal–ligand bond in essentially the same way as a σ orbital of a more conventional ligand and can thus provide an electron pair to a coordinate bond. Inasmuch as even d^{10} species form comparatively stable ethylene adducts, it is apparent that empty metal orbitals are unnecessary for bonding. It is thus reasonable to conclude that the metal ion must simultaneously donate an electron pair to the empty π^* orbital of the ethylene molecule. Since this π^* orbital has the correct symmetry to interact with d_{xz} orbital from the d species, bonding occurs. Alkyne molecules, because of the presence of doubly degenerate orbitals involving both the p_y and p_z orbitals of the carbon atoms, can thus form two metal–ligand bonds at right angles to each other and thereby behave as bridging ligands.

Carbocyclic species and arene molecules are characterized by molecular orbitals spread over several carbon atoms and generated by delocalization of π-

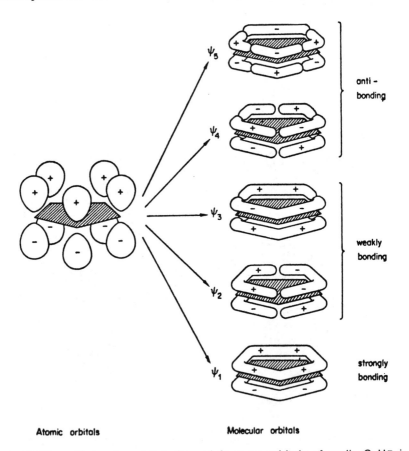

Atomic orbitals   Molecular orbitals

Figure 8-27 π-Molecular orbitals formed from $p\pi$ orbitals of cyclic $C_5H_5^-$ ion. [F. A. Cotton and G. Wilkinson, *Advanced Inorganic Chemistry*, 3rd ed., Fig. 23-9, p. 741, Interscience Publishers, New York (1972).]

type p orbitals of the carbon atoms. Of the cyclic species mentioned, only the C_6H_6 molecule has just sufficient electrons to complete the necessary molecular orbital. With the cyclopentadiene, methylcyclopentadiene, indene, and fluorene molecules, one electron must be added to each to give the necessary delocalization to form the molecular orbital. The cyclooctatetraene molecule requires the addition of two electrons and then π interaction as the dinegative ion $C_8H_8^{2-}$. The cycloheptatriene molecule, on the other hand, requires the loss of one electron, and subsequent π interaction as the cation $C_7H_8^+$. Characteristic development of molecular orbitals for the $C_5H_5^-$ species is shown diagrammatically in Fig. 8-27. Adherence to the 18-electron rule is common.

As noted above, sandwich-type molecular structures are characteristic of compounds formed between the cyclic ligands and many d species. Two compounds of particular importance and substantial thermal stability are
$$\quad\quad\quad\quad II \quad\quad\quad\quad\quad\quad\quad\quad\quad IV$$
ferrocene, $Fe(C_5H_5)_2$, and uranocene, $U(C_8H_8)_2$. A significant similarity in terms of molecular orbitals exists between these two compounds.[163] In the ferrocene molecule, the highest occupied molecular orbital of the $C_5H_5^-$ ion has e_{1g} symmetry (in the D_{5d} point group), as do also the d_{xy} and d_{yz} orbitals of the Fe^{2+} ion. Each bonding component then has a single nodal plane. In the uranocene molecule, the highest occupied molecular orbital of the $C_8H_8^-$ ion has e_{2u} symmetry (in the D_{8h} point group), as do also the f_{xyz} and $f_{z(x^2-y^2)}$ orbitals of the U^{4+} ion. Each bonding component has a single nodal plane in the ferrocene structure and two nodal planes in the uranocene structure. These relationships and similarities are shown in Fig. 8-28.[163]

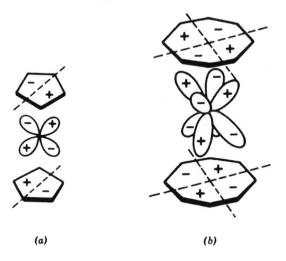

(a)   (b)

Figure 8-28 Orbital interactions in ferrocene and uranocene molecules. (a) $e_{1g}-d_{xz}$, d_{yz} interaction as in ferrocene; (b) proposed $e_{2u}-f_{xyz}$, $f_{z(x^2-y^2)}$ interaction in uranocene. [A. Streitwieser, Jr., U. Müller-Westerhoff, G. Sonnichsen, F. Mares, D. G. Morrell, and C. A. Harmon, *J. Am. Chem. Soc.*, **98**, 8644 (1973). Reprinted with permission from the American Chemical Society.]

Table 8-15 Typical Metal-Cluster Species (2- through 6-Atom Clusters)

Classification (Metal Atom Array)	Group IVB	Group VB	Group VIB	Group VIIB	Group VIIIB			Group IB
					Fe Family	Co Family	Ni Family	
2-Atom (linear)	$[(\pi\text{-}C_5H_5)_2Ti\text{-}Al(C_2H_5)_2]_2$	$V_2\{2,6(CH_3O)_2C_6H_3\}_4$ $[(\pi\text{-}C_5H_5)V\{S(CH_3)_2\}]_2$ $Nb_2Br_6(SC_4H_8)_3$ $Ta_2Br_6(SC_4H_8)_3$	$Cr_2(O_2CR)_4$ (R = CH$_3$, C(CH$_3$)$_3$, CF$_3$) $[Cr_2(CH_3)_8]^{4-}$ $[Cr_2(C_4H_8)_4]^{4-}$ $[Mo_2Cl_8]^{4-}$ $[Mo_2Cl_9]^{3-}$ $Mo_2(O_2CCH_3)_4$ $[W_2Cl_9]^{3-}$ $W_2[CH_2Si(CH_3)_3]_6$ $W_2(C_8H_8)_3$	$Mn_2(CO)_{10}$ $[Mn_2(CO)_9]^-$ $Mn_2(NO)_2(CO)_4$ $Tc_2(CO)_{10}$ $[Tc_2Cl_8]^{2-}$ $[Re_2X_8]^{2-}$ (X = Cl, Br, NCS) $Re_2Cl_4(O_2CCH_3)_2$ $(R_4N)_3Re_2(NCS)_{10}$ (R = n-C$_4$H$_9$)	$Fe_2(CO)_9$ $[Fe_2(CO)_8]^{2-}$ $Fe_2(NO)_4X_4$ (X = Cl, Br, I) $[Ru_2(O_2CR)_4]^+$ (R = CH$_3$, C$_3$H$_7$)	$Co_2(CO)_8$ $Co_2(NO)_4X_2$ (X = Cl, Br, I, NCS) $Rh_2(O_2CCH_3)_4\cdot\{P(C_6H_5)_3\}_2$ $[Rh_2(aq)]^{4+}$	$[Ni_2(CO)_6]^{2-}$ $[Ni_2Cl_8]^{4-}$ $[Pd_3(CNR)_6]^{2+}$ $Pd_2(dpm)_2Cl_2{}^a$ $[Pt_2(CNR)_6]^{2+}$ $Pt_2(dpm)_2Cl_2{}^a$	$Cu_2(O_2CCH_3)_4$ $[Cu_2Cl_8]^{4-}$
3-Atom (triangle)	—	Nb_3X_8 (X = Cl, Br, I) $[Ta_3\{C_6(CH_3)_6\}_3Cl_2]^+$	II $M_2Mo_3O_8$ $[Mo_3Cl_9]^{3-}$ $[W_3X_{12}]^{3+}$ (X = Cl, Br)	$Mn_3(C_5H_5)_3(NO)_4$ $Mn_3H_3(CO)_{12}$ $[Re_3Cl_{12}]^{3-}$ Re_3Cl_9	$Fe_3(CO)_{12}$ $[Fe_3(CO)_{11}]^{2-}$ $Ru_3(CO)_{12}$ $Os_3(CO)_{12}$	$Co_3(CO)_9S$ $[Co_3(CO)_{10}]^-$ $[Rh(CO)_3(C_5H_5)]_3$	$Ni_3(C_5H_5)_3(CO)_2$ $Ni_3(C_5H_5)_3S_2$ $Pd_3(CO)_3\{P(C_6H_5)_3\}_4$ $Pt_3(CO)_3\{P(C_6H_5)_3\}_4$	$Pd_3(CNR)_6$
4-Atom (tetrahedron)	—	—	—	—	$[Fe_4(CO)_{13}]^{2-}$ $Fe_4(CO)_4(C_5H_5)_4$	$Co_4(CO)_{12}$ $Co_4H_4(C_5H_5)_4$ $Co_4(CO)_{10}S_2$	$Pt_3(CNR)_6$ $Ni_4(CNR)_7$ $Ni_4(CO)_6(PR_3)_4$	—

Structure							
	—	—	[Re$_4$H$_6$(CO)$_{12}$]$^{2-}$	Ru$_4$H$_4$(CO)$_{12}$ Ru$_3$FeH$_2$(CO)$_{13}$ Os$_4$H$_4$(CO)$_{12}$	Rh$_4$(CO)$_{12}$ Rh$_2$Fe$_2$(C$_5$H$_5$)$_2$(CO)$_8$ Ir$_4$(CO)$_{12}$ Ir$_4$H$_2$(CO)$_{11}$ Ir$_4$H$_2$(CO)$_8$ {P(CH$_3$)(C$_6$H$_5$)$_2$}$_4$	Pt$_4$(CO)$_5$(PR$_3$)$_4$	—
5-Atom (trigonal bipyramid)	—	[Mo$_2$Ni$_3$(CO)$_{16}$]$^{2-}$ [W$_2$Ni$_3$(CO)$_{16}$]$^{2-}$	—	[Ru$_5$Cl$_{12}$]$^{2-}$ (?) Os$_5$(CO)$_{16}$	—	[Ni$_5$(CO)$_{12}$]$^{2-}$	—
6-Atom (octahedron)	Zr$_6$I$_{12}$ [Zr$_6$Cl$_{12}$]$^{3+}$ Nb$_6$F$_{12}$ [Nb$_6$X$_{12}$]$^{2+}$ [Ta$_6$X$_{12}$]$^{2+}$ (X = Cl, Br, I)	Mo$_6$Cl$_{14}^{2-}$ Mo$_6$Cl$_{12}$(PR$_3$)$_2$	—	[Fe$_6$C(CO)$_{16}$]$^{2-}$ Ru$_6$C(CO)$_{17}$ Ru$_6$H$_2$(CO)$_{18}$ [Os$_6$(CO)$_{18}$]$^{2-}$ Os$_6$H$_2$(CO)$_{18}$	Co$_6$(CO)$_{16}$ Co$_6$C(CO)$_{15}$ Rh$_6$(CO)$_{16}$ [Rh$_6$C(CO)$_{15}$]$^{2-}$ [Ir$_6$(CO)$_{15}$]$^{2-}$	Pt$_3$Sn$_2$Cl$_6$(COD)$_2^+$ [Ni$_6$(CO)$_{12}$]$^{2-}$ Pt$_6$(CO)$_{12}^{2-}$	Cu$_6$H$_6$(PR$_3$)$_6$ [Au$_6$(PR$_3$)$_6$]$^{2+}$

^{a} dpm = (C$_6$H$_5$)$_2$PC$_2$H$_4$P(C$_6$H$_5$)$_2$.

Table 8-16 Typical Metal-Cluster Species[182] (Large-Atom Clusters)

Classification	Examples	Polyhedron Defined by Metal-Atom Framework
7-atom	$[Rh_7(CO)_{16}]^{3-}$, $[Rh_7(CO)_{15}]^{2-}$, $Os_7(CO)_{21}$	C_{3v}-capped octahedron
	Pb_7^{3-}, Sb_7^{3-}	C_{3v}-capped trigonal prism
8-atom	$Os_8(CO)_{23}$	
	$[Co_8C(CO)_{18}]^{2-}$	Tetragonal antiprism[a]
	$[Au_8\mathbf{Au}(PR_3)_8]^{3+}$	D_{2h}
	Bi_8^{2+}	Square antiprism
9-atom	Bi_9^{5+}	D_{3h}-tricapped trigonal prism
	Sn_9^{4-}, Pb_9^{4-}	C_{4v}-capped square antiprism
10-atom	$Au_{10}\mathbf{Au}(PR_3)_7(SCN)_3$	Defect icosahedron
12-atom	$[Rh_{12}\mathbf{RhH}_3(C)_{24}]^{2-}$	D_{3h}-hexagonal close-packed
13-atom	$[Rh_{13}\mathbf{Rh}(CO)_{25}]^{4-}$	Body-centered cubic
14-atom	$[Rh_{14}\mathbf{RhC}_2(CO)_{28}]^{-}$	Tetracapped pentagonal prism
15-atom	$[Pt_{15}(CO)_{30}]^{2-}$	Stacked trigonal prisms
16-atom	$[Rh_{16}\mathbf{RhS}_2(CO)_{32}]^{3-}$	Four staggered prisms
17-atom	$[Pt_{17}\mathbf{Pt}_2(CO)_{12}]^{4-}$	Three staggered pentagons bi-end-capped
18-atom	$[Pt_{18}(CO)_{36}]^{2-}$	Stacked trigonal prisms
30-atom	$[Pt_{30}(CO)_{60}]^{2-}$	Stacked trigonal prisms

[a] Atoms indicated in boldface type lie within the indicated clusters.

It has been noted above that the cyclopentadienide ion does not always bond in the η^5- fashion and that more than one eta pattern can be found in a single molecular entity. Another example of the latter is found in the molecular structure of the compound $Sc(C_5H_5)_3$, where each Sc^{3+} ion is η^5-bonded to two terminal $C_5H_5^-$ groups and η^1-bonded to one bridging $C_5H_5^-$ group.[164]

Metal-Cluster Species. Prior to the mid-1960s, metal–metal bonds were largely limited to the metals and their alloys and certain solids with metallic properties (Sec. 4.2-4). Ionic or molecular species, the compositions and properties of which strongly suggested the presence of metal–metal bonds (e.g., the Hg_2^{2+} ion;[165, 166] the $[Mo_6Cl_8]^{4+}$ ion;[167] the $[M_6Cl_{12}]^{2+}$ ions, M = Nb, Ta;[168] the polynuclear carbonyls such as $Fe_2(CO)_9$[169] and $Mn_2(CO)_{10}$[170]) were limited in number and were considered to be unusual. By 1964, scores of these species had been identified,[171] and subsequently, the literature has expanded very substantially.[172-185] These are the metal-cluster species, defined by F. A. Cotton as "those containing a finite group of metal atoms which are held together entirely, mainly, or at least to a significant extent, by bonds directly between the metal atoms even though some non-metal atoms may be associated with the cluster."[171, 186] Typical metal-cluster species are summarized in Tables 8-15 and 8-16.

Very broadly, metal-cluster species have been classified as (1) the polynuclear carbonyls and related species containing nitrosyl groups, organo π donors, and other π acceptor acids (Sec. 9.6-4), and (2) the lower halides and certain oxides.[171] To these classes may be added cluster ions such as Pb_n^{4-}, Bi_5^{3+}, and Bi_4^+, which are discussed in Secs. 10.1 and 10.2. Metal-cluster species of the second broad class are most characteristic of the 4d and 5d metals, namely Zr, Nb, Mo, Tc, Ru, Rh, Pd and Hf, Ta, W, Re, Os, Ir, and Pt. These are the elements characterized by large enthalpies of atomization (Table 4-13), or large M—M

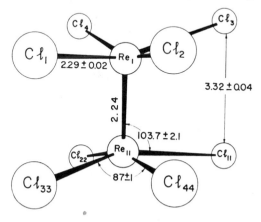

Figure 8-29 Molecular structure of $[Re_2Cl_8]^{2-}$ ion. [F. A. Cotton and C. B. Harris, *Inorg. Chem.*, **4**, 330 (1965). Reprinted with permission from the American Chemical Society.]

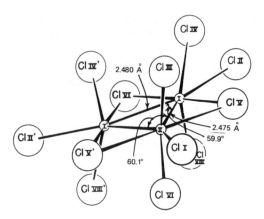

Figure 8-30 Molecular structure of $[Re_3Cl_{12}]^{3-}$ ion. ⓘⓘ'ⓘ = Re atoms. [J. A. Bertrand, F. A. Cotton, and W. Dollase, *Inorg. Chem.*, **2**, 1166 (1963). Reprinted with permission from the American Chemical Society.]

bond energies in the solid state.[187, 188] In all the cluster species, the metals are in low formal states of oxidation. Inasmuch as covalent interactions involving d orbitals are apparently involved, metal–metal interactions should be maximized by d orbitals large enough to overlap effectively, a condition that should result as atomic size increases and formal oxidation number decreases.

    1. Molecular structures. Each molecular structure is best defined in terms of the geometrical arrangement of the bonded metal atoms (Table 8-15) or the polyhedron defined by those metal atoms (Table 8-16). Experimentally determined molecular structures are largely limited to small metal clusters. Typical of these are some of the structures depicted in Fig. 8-15, 8-16, and 8-20 and those given in Figs. 8-29 to 8-33. Two broad structural classes are distinguishable: (1) that in which only two-centered bonds, single or multiple, are present, and (2) that in which three or more metal atoms appear in polygonal or polyhedral arrays.

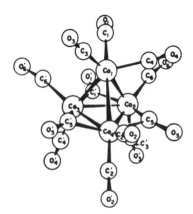

Figure 8-31 Molecular structure of $Co_4(CO)_{12}$. [C. H. Wei and L. F. Dahl, *J. Am. Chem. Soc.*, **91**, 1351 (1969). Reprinted with permission from the American Chemical Society.]

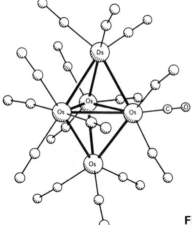

Figure 8-32 Molecular structure of $Os_5(CO)16$.

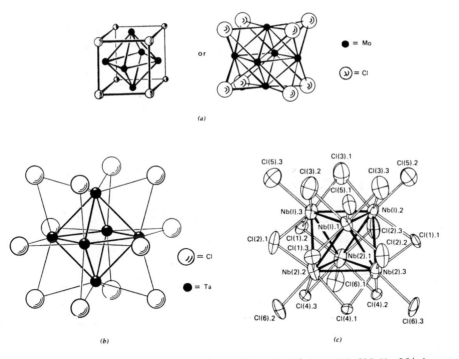

Figure 8-33 Molecular structures of (a) $[Mo_6Cl_8]^{4+}$ ion, (b) $[M_6X_{12}]^{2+}$ ions (M = Nb, Ta), and (c) $[(Nb_6Cl_{12})Cl_6]^{3-}$ ion. [(a), F. A. Cotton, *Acct. Chem. Res.*, **2**, 240 (1969); (b), P. A. Vaughn, J. H. Sturdivant, and L. Pauling, *J. Am. Chem. Soc.*, **72**, 5477 (1950); (c), F. W. Koknat and R. E. McCarley, *Inorg. Chem.*, **13**, 295 (1974). Reprinted with permission from the American Chemical Society.]

Table 8-17 M—M Internuclear Distances

Type of Cluster	M	Cluster Species	M—M Distance (Å) Cluster	Parent Metal Observed[a]	Parent Metal Calculated[b]
2-Atom	Ta	$Ta_2Br_6 \cdot 3\,THF$	2.710	2.86	2.94
	Cr	$Cr_2(O_2CCH_3)_4 \cdot 2\,H_2O$	2.386	2.498	2.58
		$[Cr_2(C_4H_8)_4]^{4-}$	1.976		
		$Cr_2\{CH_3NC(C_6H_5)NCH_3\}_4$	1.843		
	Mo	$Mo_2(O_2CCH_3)_4$	2.096	2.725	2.80
		$Mo_2[(CH_3)_2P(CH_2CH_3)]_4$	2.082		
		$[Mo_2Cl_8]^{2-}$	2.139		
	Tc	$[Tc_2Cl_8]^{3-}$	2.117	2.703	2.70
	Re	$Re_2(O_2CC_6H_5)_2I_4$	2.198	2.741	2.74
		$[Re_2Cl_8]^{2-}$	2.24		
		$[Re_2Br_8]^{2-}$	2.23		
	Fe	$Fe_2(CO)_9$	2.523	2.482	2.52
	Ru	$Ru_2(O_2CC_3H_7)_4Cl$	2.281	2.650	2.68
	Rh	$Rh_2(O_2CCH_3)_4\{P(C_6H_5)_3\}_2$	2.449	2.690	2.68
3-Atom	Re	Re_3Cl_9	2.48	2.741	2.74
		$[Re_3Cl_{12}]^{3-}$	2.47		
	Fe	$Fe_3(CO)_{12}$	2.558, 2.677, 2.683	2.482	2.52
	Os	$Os_2(CO)_{12}$	2.87	2.675	2.70

4-Atom	Fe	[Fe(C$_5$H$_5$)(CO)]$_4$	2.52	2.482	2.52
	Co	Co$_4$(CO)$_{12}$	2.49	2.506	2.50
	Rh	Rh$_4$(CO)$_{12}$	2.73	2.690	2.68
6-Atom	Ta	[Ta$_6$Cl$_{12}$(OH$_2$)$_6$]$^{2+}$	2.90	2.86	2.94
	Mo	[Mo$_6$Cl$_{14}$]$^{2-}$	2.65	2.725	2.80
		[Mo$_6$Cl$_{12}$]Cl$_2$	2.61		
		[Mo$_6$Cl$_8$]Cl$_4$	2.62		
	Co	[Co$_6$(CO)$_{14}$]$^{4-}$	2.50	2.506	2.50
	Ru	Ru$_6$H$_2$(CO)$_{18}$	2.85, 2.95	2.650	2.68

[a] As tabulated in Ref. 182.
[b] As given in Ref. 182. Calculated for 12-coordination.

Observed internuclear distances (Table 8-17) approximate, and are commonly shorter than, those in the parent metals themselves. Metal atom–metal atom interactions beyond those characteristic of the pure metals are indicated. As cluster size increases, the structure of the parent metal is more and more closely approached, and parallels between metal clusters and metal surfaces as regards chemisorption and potential catalysis can be distinguished.[182, 188a]

Fluxional behavior,[189] that is, possession of more than one molecular configuration of minimum energy but of chemical equivalence, is characteristic of certain cluster species. This type of behavior is typically indicated by nuclear magnetic resonance measurements since the time scale for such measurements is consistent with the time involved in the rapid interconversion between idealized geometries of small energy difference without appreciable distortion.[190] Thus the complex $Fe_2(CO)_4(\eta^5\text{-}C_5H_5)_2$, which exists at $-70°C$ in *cis* and *trans* forms (Fig. 8-34), each with its own separate proton resonance signal and a rate of conversion of ca. 8×10^{-12} sec^{-1}, gives only a single, sharply defined proton resonance signal intermediate between these two at room temperature, where the rate of conversion is ca. 4×10^{-3} sec^{-1}.[191]

2. Bonding. Bond orders of 1, 2, 3, and 4 have been assigned to metal–metal interactions in various cluster species. Substantial adherence to the 18-electron rule, except with six-atom clusters, is often useful in suggesting both molecular geometry and bond order (e.g., for the 3*d* metal carbonyls as summarized in Table 8-18).

Metal–metal bonds are assumed to exist where their contributions are sufficient to provide a significant portion of the overall observed enthalpy of formation.[192] They are considered, often in terms of diamagnetic character that must involve spin pairing between metal atoms, to be covalent in nature. That metal–metal interactions are not necessarily the same in species of the same compositional type can be illustrated by the ions $[M_2Cl_9]^{3-}$ (M = Cr, W) and the molecules $M_2(O_2CR)_4$ (M = Cu, Mo). The two $M_2Cl_9^{3-}$ ions have similar confacial molecular structures (Fig. 8-35), but the Cr—Cr separation (3.12 Å) is substantially larger than the W—W separation (2.41 Å) even though the observed internuclear distances in the metals (Cr—Cr, 2.498 Å; W—W, 2.741 Å) are in the reverse order. In the $W_2Cl_9^{3-}$ ion, shortening of the W—W distance suggests covalent overlap of metal orbitals and metal–metal bonding.

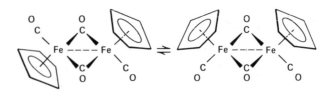

Figure 8-34 Molecular structures of $[(\eta^5\text{-}C_5H_5)_2 Fe_2(CO)_4]$. [A. F. Purcell and J. C. Kotz, *Inorganic Chemistry*, 17–122, p. 958, W. B. Saunders, Philadelphia (1977). Reprinted with permission of Holt, Rinehart and Winston.]

Table 8-18 Examples of the 18-Electron Rule

				Outer-Shell Electrons			Shared per M Atom in M—M Bonds[b]	M—M Bond Order
Compound	Molecular Structure	From $M_n{}^a$	From $(CO)_m$	Total	Per M Atom			
$Cr(CO)_6$	Octahedron	6	12	18	18		0	0
$Mn_2(CO)_{10}$	Two octahedra, sharing one corner	14	20	34	17		1	1
$Fe_3(CO)_{12}$	Fe_3 triangle	24	24	48	16		2	2
$Co_4(CO)_{12}$	Co_4 tetrahedron	36	24	60	15		3	3
$Ni(CO)_4$	Tetrahedron	10	8	18	18		0	0

[a] $4s + 3d$ electrons, M^0.
[b] To satisfy 18-electron rule.

531

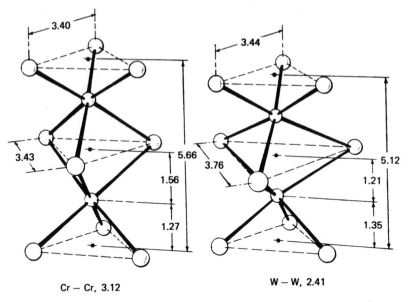

Cr — Cr, 3.12   W — W, 2.41

Figure 8-35 Molecular structures of $[M_2Cl_9]^{3-}$ ions (distances in Å). [F. A. Cotton, *Rev. Pure Appl. Chem.*, **17**, 25 (1967).]

In the $Cr_2Cl_9^{3-}$ ion, covalent interaction, if present at all, must be very weak. The $M_2(O_2CR)_4$ species have "paddle-wheel" molecular structures of the type

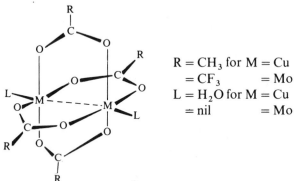

$$R = CH_3 \text{ for } M = Cu$$
$$= CF_3 \qquad = Mo$$
$$L = H_2O \text{ for } M = Cu$$
$$= nil \qquad = Mo$$

with the Cu—Cu separation (2.65 Å) larger than the Mo—Mo separation (2.109 Å). Again, M—M covalent bonding is weak, if any, in the Cu species but strong in the Mo species.

The $[Re_2Cl_8]^{2-}$ ion, the molecular structure of which is depicted in Fig. 8-29, was the first to be characterized in terms of a quadruple metal–metal bond.[193] Each Re^{3+} ion (configuration $5d^4$) can be considered to use its $d_{x^2-y^2}$ orbitals to bond to two Cl^- ions, leaving its d_{z^2}, d_{xy}, d_{xz}, and d_{yz} orbitals for Re—Re bonding. Overlap of the d_{z^2} orbitals of the two Re atoms gives a σ bond, overlap of the d_{xz} and of the d_{yz} orbitals gives two π bonds, and overlap of the d_{xy} orbitals of the two Re atoms gives a δ bond. The result is a remarkably strong bond (bond energy = 115 to 132 kcal mole^{-1}). These interactions are indicated in Fig. 8-36. This concept is in accord with the observed eclipsed arrangement of the Cl atoms in the $Re_2Cl_8^{2-}$ ion (Fig. 8-29) since, as shown in Fig. 8-36, only this arrangement allows overlap of the d_{xy} orbitals.

This treatment of bonding, which is general for many metal-cluster species, can be expressed in molecular-orbital terms.[178,180,194-196] Alternatively, L. Pauling[197,198] has described metal–metal bonds in terms of hybrid orbitals and the quadruple bond as involving also four bent bonds. This approach is also in accord with observed properties and estimates of bond lengths.

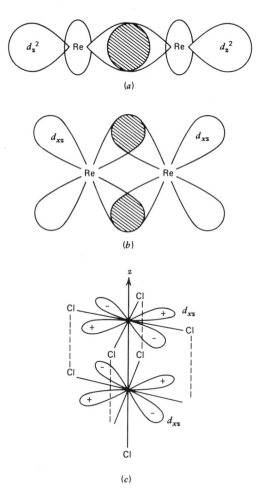

(a)

(b)

(c)

Figure 8-36 Bonding in $[Re_2Cl_8]^{2-}$ ion. (a) (σ bond); (b) (π bond) (second π bond in xy plane); (c) (δ bond) (Re d_{xz} orbitals). [J. E. Huheey, *Inorganic Chemistry*, 2nd ed., Fig. 14.44, p. 640, Harper & Row, New York (1978), modified. Reprinted with permission from the American Chemical Society. In part from F. A. Cotton *Acc. Chem. Res.*, **2**, 240 (1969). Adapted by permission.]

2-4. Iso- and Heteropoly Acids and Anions

There exist numerous oxo acids and/or oxo anions the compositions of which are suggestive of derivation, in each example, from more than one mole of the acid anhydride and formation by dehydration condensation (Sec. 10.4-1). Each of these *poly acids* and *poly anions* can be based either on a single type of anhydride (i.e., *isopoly*) or on more than one type of anhydride (i.e., *heteropoly*). Isopoly species are characteristic of both certain nonmetals (e.g., S, P, As) and certain *d*-transition metals (especially V, Mo, W). Heteropoly species are characteristic of certain *d*-transition metals (especially V, Mo, W) in combination with other elements (e.g., I, Te, P, As, Si, Ge, Sn, B, Al, Ga, Ce, Zr, Hf, Cr, Mn, Fe, Co, Ni, Cu).

Structurally, each of these species exemplifies bridging via oxygen atoms. The molecular structures of isopoly species derived from the nonmetallic elements are most commonly based upon corner shared EO_4 tetraheda (E = S, P, As) in linear or cyclic arrangements, as indicated in Fig. 8-37. The isopoly species derived from *d*-transition metals exemplify a variety of molecular structures, most of which are based upon MO_4 tetrahedra and MO_6 octahedra (Fig. 8-38). The molecular structures of almost all heteropoly species involve combinations of oxygen-shared tetrahedra and octahedra (Fig. 8-39).

The iso- and heteropoly acids and anions derived from the *d*-transition metals, although differing in many ways from the more classical complexes, qualify as complexes in terms of discussions in earlier sections of this chapter. Many aspects of molecular structure and bonding have been described in various available summaries.[199-203] Representative anionic species are listed in Table 8-19.

It has been particularly difficult in this area of chemistry to correlate the compositions and potential molecular structures shown to exist in aqueous solutions by various physical methods with the compositions and molecular structures of anions in isolable crystalline salts. In solution, pH-dependent

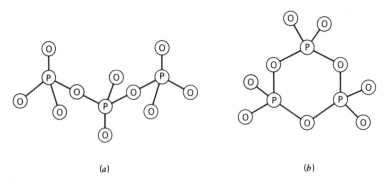

(a)   (b)

Figure 8-37   Molecular structures of (a) $P_3O_{10}^{5-}$ and (b) $P_3O_9^{3-}$ ions.

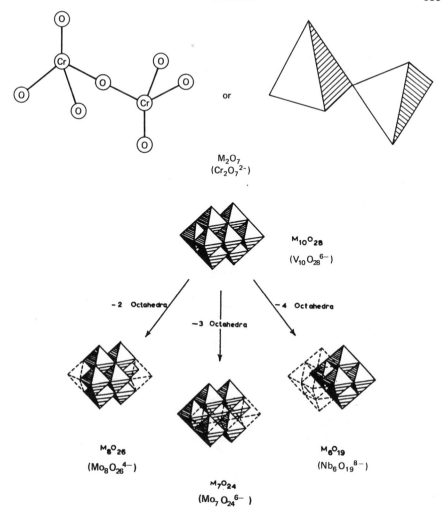

M_2O_7
$(Cr_2O_7^{2-})$

or

$M_{10}O_{28}$
$(V_{10}O_{28}^{6-})$

−2 Octahedra

−3 Octahedra

−4 Octahedra

M_8O_{26}
$(Mo_8O_{26}^{4-})$

M_7O_{24}
$(Mo_7O_{24}^{6-})$

M_6O_{19}
$(Nb_6O_{19}^{8-})$

Figure 8-38 Molecular structures of $Cr_2O_7^{2-}$, $V_{10}O_{28}^{6-}$, $Mo_8O_{26}^{4-}$, $Mo_7O_{24}^{6-}$, and $Nb_6O_{19}^{8-}$ ions. [D. L. Kepert, *Inorg. Chem.*, **8**, 1556 (1969); except $Cr_2O_7^{2-}$.] Reprinted with permission from the American Chemical Society.

aggregation–deaggregation equilibria

$$\text{monoanions} \underset{\text{increasing pH}}{\overset{\text{decreasing pH}}{\rightleftharpoons}} \text{polyanions} + n\,H_2O$$

establish pH-concentration regions within which the various poly species can exist.

These regions are delineated for the vanadate(V) system in Fig. 8-40. In order of *decreasing* pH, the specific skeleton equilibria that are involved are:[201]

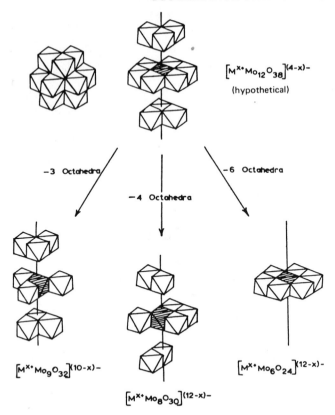

$$\left[M^{x+}Mo_{12}O_{38}\right]^{(4-x)-}$$

(hypothetical)

−3 Octahedra   −4 Octahedra   −6 Octahedra

$$\left[M^{x+}Mo_{9}O_{32}\right]^{(10-x)-}$$ $$\left[M^{x+}Mo_{6}O_{24}\right]^{(12-x)-}$$

$$\left[M^{x+}Mo_{8}O_{30}\right]^{(12-x)-}$$

Figure 8-39 Molecular structures of heteropoly anions. [D. L. Kepert, *Inorg. Chem.*, **8**, 1556 (1969). Reprinted with permission from the American Chemical Society.]

$$VO_4^{3-} \underset{-H^+}{\overset{+H^+}{\rightleftharpoons}} HVO_4^{2-}$$
$$\updownarrow$$
$$V_2O_7^{4-}$$

$$HVO_4^{2-} \underset{-H^+}{\overset{+H^+}{\rightleftharpoons}} H_2VO_4^- \text{ (or } VO_3^-)$$
$$\updownarrow$$
$$V_3O_9^{3-} \text{ and } V_4O_{12}^{4-}$$

$$H_2VO_4^- \text{ (or } VO_3^-) \underset{-H^+}{\overset{+H^+}{\rightleftharpoons}} HVO_3$$
$$\updownarrow \qquad\qquad \updownarrow$$
$$V_{10}O_{28}^{6-}, HV_{10}O_{28}^{5-}, H_2V_{10}O_{28}^{4-}$$
$$\updownarrow \qquad +H^+ \quad \updownarrow$$
$$HVO_3 \underset{-H^+}{\overset{}{\rightleftharpoons}} VO_2^+$$

Molybdate(VI) is present in solutions as the monoanion, MoO_4^{2-}, above pH 8, but acidification leads to the equilibria

$$7MoO_4^{2-} + 8H^+ \rightleftharpoons [Mo_7O_{24}]^{6-} + 4H_2O$$

$$8MoO_4^{2-} + 12H^+ \rightleftharpoons [Mo_8O_{26}]^{4-} + 6H_2O$$

Table 8-19 Typical Isopoly and Heteropoly Derivatives of Group VB and VIB Metals

Isopoly Species

Parent Atom	Formula	Solution	Salt
V(V)	$[V_3O_9]^{3-}$	+	
	$[V_4O_{12}]^{4-}$	+	
	$[V_{10}O_{28}]^{6-}$	+	+
Nb, Ta(V) (= M)	$[M_6O_{19}]^{8-}$	+	+
Cr(VI)	$[Cr_2O_7]^{2-}$	+	+
Mo(VI)	$[Mo_2O_7]^{2-}$		+
	$[Mo_7O_{24}]^{6-}$	+	+
	$[Mo_8O_{26}]^{4-}$	+	+
W(VI)	$[HW_6O_{21}]^{5-}$	+	
	$[HW_{12}O_{40}]^{6-}$	+	+
	$[HW_{12}O_{40}]^{7-}$	+	+
	$[W_{12}O_{41}]^{10-}$	+	+
	$[W_{12}O_{42}]^{12-}$		+
	$[W_{12}O_{46}]^{10-}$		+

Heteropoly Species

Group	Parent Atom	Hetero Atom	Hetero Atom : Parent Atom	Formula	Solution	Solid
	V(V)–Mo(VI)	V(V)	1 : 12	$[VMo_9V_3O_{40}]^{6-}$	+	+
1	Mo(VI)	P(V)	1 : 12	$[PMo_{12}O_{40}]^{3-}$		+
		As(V)	1 : 12	$[AsMo_{12}O_{40}]^{3-}$		+
		Si(IV)	1 : 12	$[SiMo_{12}O_{40}]^{4-}$		+
		Ti(IV)	1 : 12	$[TiMo_{12}O_{40}]^{4-}$		+
	W(VI)	P(V)	1 : 12	$[PW_{12}O_{40}]^{3-}$		+
		Si(IV)	1 : 12	$[SiW_{12}O_{40}]^{4-}$		+
		B(III)	1 : 12	$[BW_{12}O_{40}]^{5-}$		+
		Ga(III)	2 : 10	$[Ga_2W_{10}O_{36}]^{6-}$		+
		As(V)	2 : 18	$[As_2W_{18}O_{62}]^{6-}$		+
2	Mo(VI)	Te(VI)	1 : 6	$[TeMo_6O_{24}]^{6-}$		+
		Ga(III)	1 : 6	$[GaMo_6O_{24}]^{9-}$		+
		Cr(III)	1 : 6	$[CrMo_6O_{24}]^{9-}$		+
		Rh(III)	1 : 6	$[RhMo_6O_{24}]^{9-}$		+
3	Mo(VI)	Mn(IV)	9 : 32	$[MnMo_9O_{32}]^{6-}$		+
		Ni(IV)	9 : 32	$[NiMo_9O_{32}]^{6-}$		+
		Ce(IV)	1 : 12	$[CeMo_{12}O_{42}]^{8-}$		+
	W(VI)	Ce(IV)	1 : 8	$[CeW_8O_{28}]^{4-}$		+
	W(VI)	As(III)	1 : 18	$[H_2AsW_{18}O_{60}]^{7-}$		+[a]

[a] An $[AsW_9O_{33}]$ unit bonded through 6 oxygen atoms to an $[H_2W_9O_{33}]$ unit [Y. Jeannin and J. Martin-Frère, *Inorg. Chem.*, **18**, 3010 (1979)].

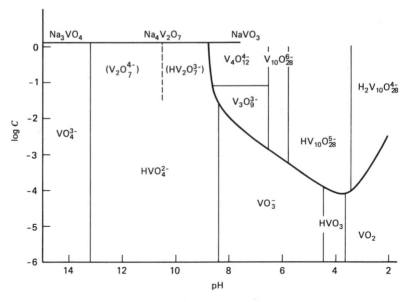

Figure 8-40 Isopolyvanadate formation—log $C_{v\,\text{total}}$ vs. pH. [D. L. Kepert, in *Comprehensive Inorganic Chemistry*, A. F. Trotman-Dickenson (Executive Ed.), Vol. 4, Fig. 5, p. 626, Pergamon Press, Oxford (1973).]

The solution chemistry of the isopolytungstates(VI) is both the most extensively investigated and the least clearly delineated,[199, 200] but a number of species derived from the groups $W_6O_{21}^{6-}$, $W_{12}O_{40}^{10-}$, and $W_{12}O_{42}^{12-}$ have been identified.

The heteropoly species form in solution or precipitate when mixtures containing mono-anions are acidified. The classic example is bright yellow 12-molybdophosphoric acid,

$$12\text{MoO}_4^{2-}(\text{aq}) + \text{PO}_4^{3-}(\text{aq}) + 27\,\text{H}^+(\text{aq}) \rightleftharpoons \text{H}_3[\text{PMo}_{12}\text{O}_{40}](\text{s}) + 12\,\text{H}_2\text{O}(\text{l})$$

first reported in 1826 by J. J. Berzelius and subsequently used for both the qualitative detection and the quantitative determination of orthophosphate(V) ion. Most of the known heteropoly species are derived from molybdenum and tungsten. Fewer, and much less stable, heteropolyvanates, as well as a limited number of heteropolyniobates and heteropolytantalates, are known. More than 30 different hetero atoms appear in these species. Most common of the hetero species are P(V), As(V), Ge(IV), Si(IV), and B(III); others include I(VII), Se(VI), Te(VI), Sb(V), Sn(IV), Al(III), Ga(III), all $3d$ elements from Ti(IV) through Zn(II), Nb(V), Pt(IV), Ce(IV), and Th(IV). Some of the elements noted also appear in other oxidation states.

1. Molecular structures. Known molecular structures are consistent with the view that they represent specific fragments of metal oxide crystal lattices.[202, 204] The metal atoms in oxides and polyanions occupy positions in close-packed arrays of the oxide ions. In metal oxides, these arrays appear as infinite chains, sheets, or three-dimensional lattices. In polyanions, similar arrays are limited in size and appear as discrete units, commonly of high symmetry. In heteropoly anions, MO_6 octahedra are usually linked to EO_4 tetrahedra or EO_6 octahedra, where E represents the hetero atom. In a limited number of cases (e.g., E = Ce, Th), EO_{12} icosahedra are present. The molecular structures depicted in Figs. 8-38 and 8-39 are consistent with these interpretations.

Early in the proton-induced polymerizations of the ions VO_4^{3-}, MoO_4^{2-}, and WO_4^{2-}, tetrahedral coordination is supplanted by octahedral coordination. Reduced electrostatic repulsions in the larger octahedra then allow edge sharing, and occasionally face sharing, of oxygen atoms as well as apex sharing, giving small clusters of MO_6 octahedra. This process is ultimately terminated by increasing electrostatic repulsions. The smaller the central cation (Table 4-4), the larger the number of octahedra per distinguishable cluster, as indicated by the series $V_{10}O_{28}^{6-}$, $Mo_8O_{26}^{4-}$, $Mo_7O_{24}^{6-}$, $HW_6O_{21}^{5-}$, $Nb_6O_{19}^{8-}$, $Ta_6O_{19}^{8-}$. These relationships are shown in Fig. 8-41.[205]

Three fundamental structural groups, depending upon the coordination number of the hetero atom, can be distinguished among the heteropoly anions[205]:

Group 1. Tetrahedrally coordinated heteroatoms, particularly in 1 : 12[199] and 2 : 18[206] heteropoly molybdates and tungstates and their reduced, substituted, or partially hydrolyzed derivatives.* The fundamental structural units are two or three edge-shared MO_6 octahedra.

Group 2. Octahedrally coordinated hetero atoms, particularly in the 1 : 6 anions $[Te^{VI}Mo_6O_{24}]^{6-}$[207] and $[E^{III}Mo_6O_{24}]^{9-}$ (E^{III} = Al, Ga, Cr, Fe, Co, Rh)[208] and in the 1 : 9 anions $[E^{IV}Mo_9O_{32}]^{6-}$ (E = Mn, Ni).[209] These molecular structures can be considered as fragments of the molecular structures of the symmetrical 1 : 12 anions $(E^{x+}Mo_{12}O_{38})^{(4-x)-}$, as indicated in Fig. 8-39.

Group 3. Icosahedrally coordinated heteroatoms, specifically in the 1 : 12 anion $[Ce^{IV}Mo_{12}O_{42}]^{8-}$, in which six pairs of face-shared MoO_6 octahedra surround the Ce^{4+} ion.[210]

Figure 8-41 Keggin structure for the species $[E^{x+}M_{12}O_{40}]^{(8-x)-}$. [F. A. Cotton and G. Wilkinson, *Advanced Inorganic Chemistry*, 4th ed., Fig. 22–C–6, p. 858, John Wiley & Sons, New York (1980).]

---

* Designations such as 1 : 12 and 2 : 18 indicate hetero atom to parent atom mole ratios.

It is significant to note that these concepts of molecular structures date to L. Pauling's conclusion from early x-ray diffraction data that in the structures of the 1 : 12 heteroacids a central EO_4 tetrahedron must be surrounded by 12 MoO_6 or WO_6 octahedra bonded to each other by the mutual sharing of three oxygen atoms.[211] However, this arrangement would require accommodation of protons by unshared oxygen atoms to give stable neutral groups of compositions $M_{12}O_{18}(OH)_{36}$ and thereby exclude any acid containing less than 18 molecules of water per molecular group, even though anhydrous species of the type $H_{8-n}[EMo_{12}O_{40}]$ (E = P, Si, Ti) had been isolated. That problem was resolved by J. F. Keggin's proposal[212, 213] of a central EO_4 tetrahedron each oxygen atom of which is shared with three MO_6 octahedra, with each octahedron in turn sharing four other oxygen atoms with other octahedra. The Keggin structure (Fig. 8-41) is characteristic of the anions $(E^{x+}M_{12}O_{40})^{(8-x)-}$ (M = Mo, W),[214, 215] as well as other related species such as the salt $(NH_4)_{6.5}Na[H_{1.5}\overset{II}{Mn}O_6GaO_4W_{11}O_{30}]$.[216]

2. Bonding. Bonding between polyhedra in these species is of the usual covalent type and requires no special treatment.[217]

EXERCISES

8-1. Give a valence-bond electronic configuration for each of the following complex species and suggest a hybrid that accounts for the indicated molecular geometry: linear $[Ag(CN_2)]^-$, square-planar $[AuCl_4]^-$, octahedral $[Fe(CN)_6]^{3-}$, octahedral $[Ti(H_2O)_6]^{3+}$, tetrahedral $[GaCl_4]^-$.

8-2. Determine the effective atomic number of the central atom in each of the following complex species: $[Au(CN)_2]^-$, $[Ni(CO)_4]$, $[HfF_7]^{3-}$, $[Mn(C_5H_5)_3(NO)_3]$, $[Tc(CN)_6]^{5-}$, $[Fe_2(CO)_9]$, $[Fe(CO)_4\{P(C_6H_5)_3\}]$, $[PtCl_3(C_2H_4)]^-$, $[Pr(C_5H_5)_3]$, $[Re_3Cl_9]$.

8-3. By reference to appropriate energy-level diagrams, indicate the occupancy of orbitals for each of the following: d^4, octahedral, low spin; d^6, tetrahedral, high spin; d^7, octahedral, high spin; d^5, octahedral, low spin; d^8, octahedral, tetragonal distortion.

8-4. Indicate the electronic configurations that could be responsible for Jahn–Teller distortions in tetrahedral and octahedral complex species in nonexcited energy states. To what extents are these influenced by high-spin and low-spin configurations?

8-5. Why are the qualitative molecular-orbital schemes for $[M(CO)_6]$ and $[M(CN)_6]^{n-}$ (Fig. 8-12) the same?

8-6. Check the original literature and indicate in detail the justification for stronger covalent interaction in the $W_2Cl_9^{3-}$ ion than in the $Cr_2Cl_9^{3-}$ ion.

8-7. Indicate clearly the essence of (a) the 16-electron rule, (b) the 18-electron rule. Cite examples of reactions the courses of which can be related to transfers between these two arrangements.

8-8. On the basis of the observation that the complex $[Cr_2(CH_3COO)_4(H_2O)_2]$ is diamagnetic, what conclusions can be drawn relative to the natures of the Cr—O bonds? Explain.

8-9. Why are numerous tetrahedral cobalt(II) complexes thermodynamically stable whereas the corresponding complexes of nickel(II) are not?

8-10. The following effective magnetic moments (μ_{eff}) have been recorded: $K_3[Fe(C_2O_4)_3]\cdot 3H_2O$, 5.85–5.95; $K_3[Fe(CN)_6]$, 2.3–2.4. Discuss metal-ligand bonding in these two species in terms of these magnetic data.

8-11. What is your understanding of a 14-electron donor species? Cite a number of similarities in molecular structure and bonding derived from different chemical species of this type. Account for these similarities.

8-12. What advantages does the Keggin molecular structure for 12-molybdo or 12-tungsto heteropoly anions have over previously proposed molecular structures? Consult cited references before answering.

8-13. Consult Reference 203 and A. R. Middleton, *J. Chem. Educ.*, **10**, 726 (1933) and summarize the essence of the Miolati–Rosenheim–Copeaux suggestions as to molecular structures of heteropoly and isopoly acids and anions. Why are these considerations invalid?

8-14. Cite specific examples of η^{1-5}-C_5H_5 organometallic complex species and interpret these notations in terms of molecular structure and bonding.

8-15. Summarize data and argument that support the view that $[Fe(C_5H_5)_2]$ and $[U(C_8H_8)_2]$ are comparable species.

REFERENCES

1. B. M. Loeffler and R. G. Burns, *Am. Sci.*, **64**, 636 (1976).
2. R. C. Brasted and W. E. Cooley, in *The Chemistry of the Coordination Compounds*, J. C. Bailar, Jr., and D. H. Busch (Eds.), Ch. 18, Reinhold Publishing Corp., New York (1956).
3. R. Tsuchida, *Bull. Soc. Chem. Jpn.*, **13**, 388, 436, 471 (1938).
4. F. Basolo, *J. Am. Chem. Soc.*, **72**, 4393 (1950).
5. C. J. Ballhausen, *Introduction to Ligand Field Theory*, McGraw-Hill Book Company, New York (1962).
6. Chr. Klixbüll Jørgensen, *Orbitals in Atoms and Molecules*, Academic Press, New York (1962).
7. Chr. Klixbüll Jørgensen, *Absorption Spectra and Chemical Bonding in Complexes*, Academic Press, New York (1962).
8. F. Basolo and R. G. Pearson, *Mechanisms of Inorganic Reactions: A Study of Complexes in Solution*, 2nd ed., Ch. 2, John Wiley & Sons, New York (1967).
9. R. Mason, *Chem. Soc. Rev.*, **1**, 431 (1972).
10. E. Cartmell and G. W. A. Fowles, *Valency and Molecular Structure*, 4th ed., Chs. 11, 12, Butterworth, London (1977).
11. J. N. Murrell, S. F. A. Kettle, and J. M. Tedder, *The Chemical Bond*, Ch. 12, John Wiley & Sons, New York (1978).
12. L. Pauling, *J. Am. Chem. Soc.*, **53**, 1367 (1931).
13. L. Pauling, *The Nature of the Chemical Bond*, 3rd ed., Chs. 5, 9, Cornell University Press, Ithaca, N.Y. (1960).

14. R. S. Nyholm, *Q. Rev. (Lond.)*, **7**, 377 (1953).

15. D. P. Craig, A. Maccoll, R. S. Nyholm, L. E. Orgel, and L. E. Sutton, *J. Chem. Soc.*, **1954**, 332.

16. H. Taube, *Chem. Rev.*, **50**, 69 (1952).

17. G. J. Bullen, R. Mason, and P. Pauling, *Nature*, **189**, 291 (1961).

18. R. S. Nyholm, *Proc. Chem. Soc.*, **1961**, 273.

19. W. Manch and W. C. Fernelius, *J. Chem. Educ.*, **38**, 192 (1961).

20. L. M. Venanzi, *J. Chem. Soc.*, **1958**, 719.

21. N. S. Gill, R. S. Nyholm, and P. Pauling, *Nature*, **182**, 168 (1958).

22. D. M. L. Goodgame, M. Goodgame, and F. A. Cotton, *J. Am. Chem. Soc.*, **83**, 4161 (1961).

23. A. E. Van Arkel and J. H. de Boer, *Rec. Trav. Chim.*, **47**, 593 (1928).

24. F. J. Garrick, *Philos. Mag.*, [7], **9**, 131 (1930); **10**, 71, 76 (1930); **11**, 741 (1931); **14**, 914 (1932).

25. R. J. Gillespie and R. S. Nyholm, *Q. Rev. (Lond.)*, **11**, 339 (1957).

26. R. J. Gillespie and R. S. Nyholm, in *Progress in Stereochemistry*, W. Klyne and P. B. D. de la Mare (Eds.), Vol. 2, pp. 261–305, Academic Press, New York (1958).

27. L. E. Orgel, *J. Chem. Soc.*, **1952**, 4756; *J. Chem. Phys.*, **23**, 1819 (1955).

28. H. Bethe, *Ann. Phys. (Leipzig)*, [5], **3**, 133 (1929).

29. J. H. Van Vleck, *Phys. Rev.*, **41**, 208 (1932); *J. Chem. Phys.*, **3**, 803, 807 (1935).

30. L. E. Orgel, in *Quelques problèmes de chimie minérale*, R. Stoops (Ed.), Dixième Conseil de Chimie, Institut International de Chimie Solvay, Bruxelles (1956).

31. J. S. Griffith and L. E. Orgel, *Q. Rev. (Lond.)*, **11**, 381 (1957).

32. W. Moffitt and C. J. Ballhausen, *Ann. Rev. Phys. Chem.*, **7**, 107 (1956).

33. L. E. Orgel, *An Introduction to Transition-Metal Chemistry: Ligand-Field Theory*, Ch. 2, John Wiley & Sons, New York (1960).

34. F. Basolo and R. G. Pearson, Ref. 8, pp. 65–89.

35. F. A. Cotton and G. Wilkinson, *Advanced Inorganic Chemistry: A Comprehensive Text*, 4th ed., Ch. 20, Wiley-Interscience Publishers, New York (1980).

36. H. Hartman, H. L. Schläfer, and K. H. Hansen, *Z. Anorg. Allg. Chem.*, **284**, 153 (1956).

37. R. L. Martin and A. H. White, in *Transition Metal Chemistry*, R. L. Carlin (Ed.), Vol. 4, pp. 113–198, Marcel Dekker, New York (1968).

38. L. Cambi, L. Szegö, and A. Cagnasso, *Atti Accad. Naz. Lincei*, **15**, 266, 329 (1932).

39. L. E. Orgel and J. D. Dunitz, *Nature*, **179**, 462 (1957).

40. E. G. Cox and K. C. Webster, *J. Chem. Soc.*, **1935**, 731.

41. A. F. Wells, *J. Chem. Soc.*, **1947**, 1670.

42. D. Harker, *Z. Kristallogr., Kristallgeom, Kristallphys., Kristallchem.*, **93**, 136 (1936).

43. C. H. MacGillavry and J. M. Bijvoet, *Z. Kristallogr., Kristallgeom., Kristallphys., Kristallchem.*, **94**, 231 (1936).

44. H. A. Jahn and E. Teller, *Proc. R. Soc. Lond. Ser. A*, **161**, 220 (1937).

45. J. D. Dunitz and L. E. Orgel, in *Advances in Inorganic Chemistry and Radiochemistry*, H. J. Emeléus and A. G. Sharpe (Eds.), Vol. 2, pp. 1–60, Academic Press, New York (1960).

46. O. G. Holmes and D. S. McClure, *J. Chem. Phys.*, **26**, 1686 (1957).

47. P. George and D. S. McClure, in *Progress in Inorganic Chemistry*, F. A. Cotton (Ed.), Vol. 1, pp. 381–463, Interscience Publishers, New York (1959). An extensive and comprehensive summary dealing with the effects of inner orbital splitting on the thermodynamic properties of d-transition metal ions and compounds.

48. J. Owen, *Disc. Faraday Soc.*, **19**, 127 (1955).

49. R. S. Drago, *Physical Methods in Chemistry*, Ch. 12, W. B. Saunders Company, Philadelphia (1977).

50. Chr. Klixbüll Jørgensen, in *Progress in Inorganic Chemistry*, F. A. Cotton (Ed.), Vol. 4, pp. 73–124, Interscience Publishers, New York (1962).

51. Chr. Klixbüll Jørgensen, *Oxidation Numbers and Oxidation States*, p. 106, Springer-Verlag, New York (1969).

52. R. F. Fenske, in *Progress in Inorganic Chemistry*, S. J. Lippard (Ed.), Vol. 21, pp. 179–208, Interscience Publishers, New York (1976).

53. J. E. Huheey, *Inorganic Chemistry*, 2nd ed., pp. 390–412, Harper & Row, New York (1978).

54. W. P. Griffith, in *Comprehensive Inorganic Chemistry*, A. F. Trotman-Dickenson (Executive Ed.), Vol. 4, pp. 105–195, Pergamon Press, Oxford (1973).

55. L. Mond, C. Langer, and F. Quincke, *J. Chem. Soc.*, **57**, 749 (1890).

56. F. Calderazzo, R. Ercoli, and G. Natta, *Organic Syntheses via Metal Carbonyls*, Vol. 1, Ch. 1, John Wiley & Sons, New York (1968).

57. R. D. Johnston, in *MTP International Review of Science, Inorganic Chemistry, Series One*, Vol. 6, pp. 1–41, Butterworth, London (1972).

58. H. J. Emeléus and A. G. Sharpe, *Modern Aspects of Inorganic Chemistry*, 4th ed., Ch. 20, John Wiley & Sons, New York (1973).

59. E. W. Abel and F. G. A. Stone, *Q. Rev. (Lond.)*, **24**, 498 (1970).

60. W. Hieber, W. Beck, and G. Braun, *Angew. Chem., Int. Ed. Engl.*, **0**, 65 (1961).

61. P. S. Braterman, *Struct. Bonding*, **10**, 57 (1972); **26**, 1 (1976).

62. F. A. Cotton, in *Progress in Inorganic Chemistry*, S. J. Lippard (Ed.), Vol. 21, pp. 1–28, Interscience Publishers, New York (1975).

63. J. Ladell, B. Post, and I. Fankuchen, *Acta Crystallogr.*, **5**, 795 (1952). (X-ray.)

64. J. Donohue and A. Caron, *Acta Crystallogr.*, **17**, 663 (1964). (X-ray.)

65. F. Calderazzo and F. L'Epplattenier, *Inorg. Chem.*, **6**, 1220 (1967). (Infrared.)

66. G. Natta, R. Ercoli, F. Calderazzo, A. Alberola, P. Corradini, and G. Allegra, *Atti Acad. Naz. Lincei, Rend. Classe Sci. Fis. Mat. Nat.*, (8), **27**, 107 (1959). (X-ray.)

67. W. Beck and R. E. Nitschmann, *Z. Naturforsch.*, **17b**, 577 (1962).

68. W. Rüdorff and U. Hofmann, *Z. Phys. Chem. B*, **28**, 351 (1935).

69. L. O. Brockway, R. V. G. Ewens, and M. W. Lister, *Trans. Faraday Soc.*, **34**, 1350 (1938).

70. F. A. Cotton and D. L. Hunter, *Inorg. Chem.*, **13**, 2044 (1974).

71. F. A. Cotton and J. M. Troup, *J. Chem. Soc. Dalton*, **1974**, 800.

72. F. A. Cotton and J. M. Troup, *J. Am. Chem. Soc.*, **96**, 4155 (1974).

73. N. V. Sidgwick, *The Electronic Theory of Valency*, p. 163, Clarendon Press, Oxford (1927).

74. A. A. Blanchard, *Chem. Rev.*, **21**, 7 (1937).

75. C. A. Tolman, *Chem. Soc. Rev.*, **1**, 337 (1972).

76. M. J. S. Dewar, *Bull. Soc. Chim. Fr.*, **18**, C79 (1951).

77. J. Chatt and L. A. Duncanson, *J. Chem. Soc.*, **1953**, 2939.

78. H. B. Gray and N. A. Beach, *J. Am. Chem. Soc.*, **85**, 2922 (1963).

79. C. C. Addison and J. Lewis, *Q. Rev. (Lond.)*, **9**, 115 (1955).

80. F. Seel, *Angew Chem.*, **68**, 272 (1956).

81. J. Lewis, *Sci. Prog. (Lond.)*, **47**, 506 (1959).

82. B. F. G. Johnson and J. A. McCleverty, in *Progress in Inorganic Chemistry*, Vol. 7, F. A. Cotton (Ed.), pp. 277–359, Interscience Publishers, New York (1966).

83. W. P. Griffith, *Adv. Organomet. Chem.*, **7**, 211 (1968).

84. J. H. Enemark and R. D. Feltham, *Coord. Chem. Rev.*, **13**, 339 (1974).

85. R. Hoffmann, M. M. L. Chen, M. Elian, A. R. Rossi, and D. M. P. Mingos, *Inorg. Chem.*, **13**, 2666 (1974).

86. K. G. Caulton, *Coord. Chem. Rev.*, **14**, 317 (1975).

87. R. Eisenberg and C. D. Meyer, *Acc. Chem. Res.*, **8**, 26 (1975).

88. F. Bottomley, *Coord. Chem. Rev.*, **26**, 7 (1978).

89. D. J. Hodgson, N. C. Payne, J. A. McGinnerty, R. G. Pearson, and J. A. Ibers, *J. Am. Chem. Soc.*, **90**, 4486 (1968).

90. R. C. Elder, F. A. Cotton, and R. A. Schunn, *J. Am. Chem. Soc.*, **89**, 3645 (1967).

91. J. Lewis, R. J. Irving, and G. Wilkinson, *J. Inorg. Nucl. Chem.*, **7**, 32 (1958); W. P. Griffith, J. Lewis, and G. Wilkinson, *J. Inorg. Nucl. Chem.*, **7**, 38 (1958).

92. B. A. Frenz and J. A. Ibers, *MTP International Review of Science, Physical Chemistry, Series One*, Vol. 11, pp. 33–72, Butterworth, London (1973).

93. R. D. Feltham, *Inorg. Chem.*, **3**, 1038 (1964).

94. B. M. Chadwick and A. G. Sharpe, in *Advances in Inorganic Chemistry and Radiochemistry*, H. J. Emeléus and A. G. Sharpe (Eds.), Vol. 8, pp. 83–176, Academic Press, New York (1966). Cyanides, cyano complexes.

95. A. G. Sharpe, *The Chemistry of Cyano Complexes of the Transition Metals*, Academic Press, New York (1976).

96. L. Malatesta, in *Progress in Inorganic Chemistry*, F. A. Cotton (Ed.), Vol. 1, pp. 283–379, Interscience, New York (1959). Isocyanide complexes.

97. L. M. Malatesta and F. Bonati, *Isocyanide Complexes of Metals*, Interscience, New York (1969).

98. J. M. Williams, P. L. Johnson, A. J. Schultz, and C. C. Coffee, *Inorg. Chem.*, **17**, 834 (1978). $Rb_2[Pt(CN)_4]Cl_{0.3} \cdot 3.0 H_2O$.

99. P. L. Johnson, A. J. Schultz, A. E. Underhill, D. M. Watkins, D. J. Wood, and J. M. Williams, *Inorg. Chem.*, **17**, 839 (1978). $(NH_4)_2[Pt(CN)_4]Cl_{0.3} \cdot 3 H_2O$.

100. R. K. Brown and J. M. Williams, *Inorg. Chem.*, **17**, 2607 (1978). $Cs_2[Pt(CN)_4]Cl_{0.30}$.

101. R. K. Brown, D. A. Vidusek, and J. M. Williams, *Inorg. Chem.*, **18**, 801 (1979). $Cs_2[Pt(CN)_4](N_3)_{0.25} \cdot 0.5 H_2O$.

102. R. K. Brown and J. M. Williams, *Inorg. Chem.*, **18**, 1922 (1979). $Rb_2[Pt(CN)_4]Cl_{0.3} \cdot 3.0 H_2O$.

103. J. L. Hoard, *Z. Kristallogr., Kristallgeom., Kristallphys., Kristallchem.*, **84**, 231 (1933).

104. A. Rosenzweig and D. T. Comer, *Acta Crystallogr.*, **12**, 709 (1959).

105. R. G. Dickinson, *J. Am. Chem. Soc.*, **44**, 774 (1922).

106. J. Hvoslef, *Acta Chem. Scand.*, **12**, 1568 (1958).

107. R. C. Seccombe and C. H. L. Kennard, *J. Organomet. Chem.*, **18**, 243 (1969).

108. E. M. Holt and K. J. Watson, *Acta Chem. Scand.*, **23**, 14 (1969).

109. F. K. Larsen, R. G. Hazell, S. E. Rasmussen, *Acta Chem. Scand.*, **23**, 61 (1969).

110. L. Dupont, *Acta Crystallogr.*, **B26**, 964 (1970).

111. L. Dupont, *Bull. Soc. R. Sci. Liège*, **36**, 471 (1967).

112. C. Bertinotti and A. Bertinotti, *Acta Crystallogr.*, **B26**, 422 (1970).

113. C. Bertinotti and A. Bertinotti, *C. R. Acad. Sci.*, **273**, 33 (1971).

114. S. Jagner and N.-G. Vannerberg, *Acta Chem. Scand.*, **27**, 3482 (1973).

115. N.-G. Vannerberg, *Acta Chem. Scand.*, **20**, 1571 (1966).

116. S. Jagner, E. Ljungström, and N.-G. Vannerberg, *Acta Chem. Scand.*, **A28**, 623 (1974).

117. D. H. Svedung and N.-G. Vannerberg, *Acta Chem. Scand.*, **22**, 1551 (1968).

118. A. Tullberg and N.-G. Vannerberg, *Acta Chem. Scand.*, **21**, 1462 (1967).

119. A. Tullberg and N.-G. Vannerberg, *Acta Chem. Scand.*, **25**, 343 (1971); **A28**, 551 (1974).

120. B. I. Swanson and R. R. Ryan, *Inorg. Chem.*, **12**, 283 (1973).

121. N.-G. Vannerberg, *Acta. Chem. Scand.*, **26**, 2863 (1972).

122. R. A. Levenson and R. L. R. Towns, *Inorg. Chem.*, **13**, 105 (1974).

123. J. L. Hoard, T. A. Hamor, and M. D. Glick, *J. Am. Chem. Soc.*, **90**, 3177 (1968).

124. B. J. Corden, J. A. Cunningham, and R. Eisenberg, *Inorg. Chem.*, **9**, 356 (1970).

125. S. S. Basson, L. D. C. Bok, and J. G. Leipoldt, *Acta Crystallogr.*, **B26**, 1209 (1970).

126. F. A. Cotton and F. Zingales, *J. Am. Chem. Soc.*, **83**, 351 (1961).

127. F. A. Cotton, T. G. Dunne, and J. S. Wood, *Inorg. Chem.*, **3**, 1495 (1964).

128. F. A. Cotton, T. G. Dunne, and J. S. Wood, *Inorg. Chem.*, **4**, 318 (1965).

129. K. R. Mann, N. S. Lewis, V. M. Mickowski, D. K. Erwin, G. S. Hammond, and H. B. Gray, *J. Am. Chem. Soc.*, **99**, 5525 (1977).

130. K. R. Mann, R. A. Bell, and H. B. Gray, *Inorg. Chem.*, **18**, 267 (1979).

131. R. C. Taylor and W. D. Horrocks, Jr., *Inorg. Chem.*, **3**, 584 (1964).

132. G. Booth, in *Advances in Inorganic Chemistry and Radiochemistry*, H. J. Eméleus and A. G. Sharpe (Eds.), Vol. 6, pp. 1–69, Academic Press, New York (1964).

133. W. Levason and C. A. McAuliffe, in *Advances in Inorganic Chemistry and Radiochemistry*, H. J. Eméleus and A. G. Sharpe (Eds.), Vol. 9, pp. 173–253, Academic Press, New York (1972).

134. A. Stelzer, in *Topics in Phosphorus Chemistry*, E. J. Griffith and M. Grayson (Eds.), Vol. 9, pp. 1–229, John Wiley & Sons, New York (1977).

135. W. D. Horrocks, Jr., and R. C. Taylor, *Inorg. Chem.*, **2**, 723 (1963).

136. F. A. Cotton, *Inorg. Chem.*, **3**, 702 (1964).

137. W. Strohmeier and F. J. Müller, *Chem. Ber.*, **100**, 2812 (1967).

138. J. C. Green, D. I. King, and J. H. D. Eland, *J. Chem. Soc. Chem. Commun.*, **1970**, 1121.

139. J. H. Hillier, V. R. Saunders, M. J. Ware, R. J. Bassett, D. R. Lloyd, and N. Lynaugh, *J. Chem. Soc. Chem. Commun.*, **1970**, 1316.

140. T. Kruck, *Angew. Chem., Int. Ed. Engl.*, **6**, 53 (1967).

141. T. Kruck and W. Lang, *Angew. Chem., Int. Ed. Engl.*, **6**, 454 (1967).

142. P. L. Timms, *J. Chem. Soc. Chem. Commun.*, **1969**, 1033.

143. R. G. Guy and B. L. Shaw, in *Advances in Inorganic Chemistry and Radiochemistry*, H. J. Eméleus and A. G. Sharpe (Eds.), Vol. 4, pp. 77–131, Academic Press, New York (1962). Olefin, acetylene, π-allylic complexes.

144. H. W. Quinn and J. H. Tsai, in *Advances in Inorganic Chemistry and Radiochemistry*, H. J. Eméleus and A. G. Sharpe (Eds.), Vol. 12, pp. 217–373, Academic Press, New York (1969). Olefin complexes.

145. M. R. Churchill, in *Progress in Inorganic Chemistry*, Vol. 11, S. J. Lippard (Ed.), pp. 53–98, Interscience, New York (1970). Azulene complexes.

146. D. I. Hall, J. H. Ling, and R. S. Nyholm, *Struct. Bonding*, **15**, 3 (1973). Olefin complexes.

147. J. S. Thayer, *Adv. Organometal. Chem.*, **13**, 1 (1975). History.

148. F. A. Cotton, *J. Am. Chem. Soc.*, **90**, 6230 (1968).

149. R. A. Love, T. F. Koetzle, G. J. B. Williams, L. C. Andrews, and R. Bau, *Inorg. Chem.*, **14**, 2653 (1975).

150. J. A. J. Jarvis, B. T. Kilbourn, and P. G. Owston, *Acta Crystallogr.*, **B26**, 876 (1970); **B27**, 366 (1971).

151. O. S. Mills and G. Robinson, *Proc. R. Soc. Lond.*, 421 (1960).

152. W. Bünder and E. Weiss, *J. Organometal. Chem.*, **92**, 65 (1975).

153. E. Maslowsky, Jr., *J. Chem. Educ.*, **55**, 276 (1978).

154. F. A. Cotton, W. A. Dollase, and J. S. Wood, *J. Am. Chem. Soc.*, **85**, 1543 (1963).

155. G. Albrecht, E. Förster, D. Sippel, F. Eichhorn, and E. Kurras, *Z. Chem.*, **8**, 311 (1968).

156. T. Kashiwagi, N. Yasuoka, N. Kasai, and M. Kukudo, *J. Chem. Soc. Chem. Commun.*, **1969**, 317. [RhI(allene){P(C_6H_5)$_3$}$_2$.]

157. P. Racanelli, G. Pantini, A. Immirizi, G. Allegra, and L. Porri, *J. Chem. Soc. Chem. Commun.*, **1969**, 361. [Rh$_2$(acac)$_2$(allene)(CO)$_2$.]

158. G. R. Davis, W. Hewertson, R. H. B. Mais, P. G. Owston, and G. Patel, *J. Chem. Soc. A*, **1970**, 1873.

159. B. H. Freeland, J. E. Hux, N. C. Payne, and K. G. Tyers, *Inorg. Chem.*, **19**, 693 (1980).

160. J. O. Glanville, J. M. Stewart, and S. O. Grim, *J. Organometal. Chem.*, **7**, P9 (1967).

161. W. G. Sly, *J. Am. Chem. Soc.*, **81**, 18 (1959). [(OC)$_3$Co(C$_6$H$_5$C≡CC$_6$H$_5$)Co(CO)$_3$.] (Co–Co bonding also present.)

162. M. A. Bennett, R. N. Johnson, G. B. Robertson, T. W. Turney, and P. O. Whimp, *Inorg. Chem.*, **15**, 97 (1976).

163. A. Streitwieser, Jr., and U. Müller-Westerhoff, *J. Am. Chem. Soc.*, **90**, 7364 (1968); A. Streitwieser, Jr., U. Müller-Westerhoff, G. Sonnichsen, F. Mares, D. G. Morrell, K. O. Hodgson, and C. A. Harmon, *J. Am. Chem. Soc.*, **95**, 8644 (1973).

164. J. L. Atwood and K. D. Smith, *J. Am. Chem. Soc.*, **95**, 1488 (1973).

165. R. J. Havighurst, *Am. J. Sci.*, **10**, 15 (1925); *J. Am. Chem. Soc.*, **48**, 2113 (1926).

166. L. A. Woodward, *Philos. Mag.*, [7], **18**, 823 (1934).

167. P. A. Vaughan, *Proc. Natl. Acad. Sci. USA*, **36**, 461 (1950).

168. P. A. Vaughan, J. H. Sturdivant, and L. Pauling, *J. Am. Chem. Soc.*, **72**, 5477 (1950).

169. H. M. Powell and R. V. G. Ewens, *J. Chem. Soc.*, **1939**, 286.

170. L. F. Dahl, E. E. Ishishi, and R. E. Rundle, *J. Chem. Phys.*, **26**, 1750 (1957).

171. F. A. Cotton, *Q. Rev. (Lond.)*, **20**, 389 (1966). Survey.

172. M. C. Baird, in *Progress in Inorganic Chemistry*, Vol. 9, F. A. Cotton (Ed.), pp. 1–159, Interscience Publishers, New York (1968). Survey.

173. R. B. King, in *Progress in Inorganic Chemistry*, Vol. 15, S. J. Lippard (Ed.), pp. 287–473, Interscience Publishers, New York (1972). Survey.

174. D. L. Kepert and K. Vrieze, in *Comprehensive Inorganic Chemistry*, A. F. Trotman-Dickenson (Executive Ed.), Vol. 4, pp. 197–354, Pergamon Press, Oxford (1973). Survey.

175. T. J. Meyer, in *Progress in Inorganic Chemistry*, Vol. 19, S. J. Lippard (Ed.), pp. 1–50, Interscience Publishers, New York (1975). Oxidation–reduction and related reactions.

176. F. A. Cotton, *Chem. Soc. Rev.*, **4**, 27 (1975). Quadruple bonds.

177. K. Wade, in *Advances in Inorganic Chemistry and Radiochemistry*, H. J. Emeléus and A. G. Sharpe (Eds.), Vol. 18, pp. 1–66, Academic Press, New York (1976). Structure, bonding.

178. H. Vahrenkamp. *Struct. Bonding*, **32**, 1 (1977). Structure, bonding.

179. F. A. Cotton, *Acc. Chem. Res.*, **11**, 225 (1978). Quadruple bonds — interpretation.

180. W. Trogler and H. B. Gray, *Acc. Chem. Res.*, **11**, 232 (1978). Quadruple bonds — spectra, photochemistry.

181. E. Band and E. L. Muetterties, *Chem. Rev.*, **78**, 639 (1978). Mechanisms of rearrangements.

182. E. L. Muetterties, T. N. Rhodin, E. Band, C. F. Brucker, and W. R. Pretzer, *Chem. Rev.*, **79**, 91 (1979). Clusters and surfaces.

183. A. P. Humphries and H. D. Kaesz, in *Progress in Inorganic Chemistry*, Vol. 25, S. J. Lippard (Ed.), pp. 145–221, Interscience Publishers, New York (1979). Hydrido clusters.

184. B. F. G. Johnson (Ed.), *Transition Metal Clusters*, John Wiley & Sons, New York (1979).

185. T. G. Spiro, in *Progress in Inorganic Chemistry*, Vol. 11, S. J. Lippard (Ed.), pp. 1–51, Interscience Publishers, New York (1970). Vibrational spectra.

186. F. A. Cotton, *Inorg. Chem.*, **3**, 1217 (1964).

187. J. Lewis and R. S. Nyholm, *Sci. Prog. (Lond.)*, **52**, 557 (1964).

188. H. Schäfer and H. G. Schnering, *Angew. Chem.*, **76**, 833 (1964).

188a. R. Rawls, *Chem. Eng. News*, Apr. 28, 1980, p. 27. Review of industrial catalytic reactions of carbon monoxide with hydrogen.

189. F. A. Cotton, *Chem. Br.*, **4**, 345 (1968).

190. E. L. Muetterties, *Inorg. Chem.*, **4**, 769 (1965).

191. J. G. Bullitt, F. A. Cotton, and T. J. Marks, *J. Am. Chem. Soc.*, **92**, 2155 (1970).

192. J. A. Bertrand, F. A. Cotton, and W. A. Dollase, *Inorg. Chem.*, **2**, 1166 (1963).

193. F. A. Cotton and C. B. Harris, *Inorg. Chem.*, **4**, 330 (1965).

194. F. A. Cotton and T. E. Haas, *Inorg. Chem.*, **3**, 10 (1964).

195. F. A. Cotton, *Inorg. Chem.*, **4**, 334 (1965).

196. R. G. Woolley, *Inorg. Chem.*, **18**, 2945 (1979).

197. L. Pauling, *The Nature of the Chemical Bond*, 3rd ed., pp. 436–442, Cornell University Press, Ithaca, N.Y. (1960).

198. L. Pauling, *Proc. Natl. Acad. Sci. USA*, **72**, 3799 (1975).

199. D. L. Kepert, in *Progress in Inorganic Chemistry*, F. A. Cotton (Ed.), Vol. 4, pp. 199–274, Interscience Publishers, New York (1962). Isopolytungstates.

200. M. T. Pope and B. W. Dale, *Q. Rev. (Lond.)*, **22**, 527–548 (1968). Isopolyvanadates, -niobates, -tantalates.

201. D. L. Kepert, in *Comprehensive Inorganic Chemistry*, A. F. Trotman-Dickenson (Executive Ed.), Vol. 4, pp. 607–671, Pergamon Press, Oxford (1973). Isopoly- and heteropoly-anions.

202. T. J. R. Weakley, *Struct. Bonding*, **18**, 131–176 (1974). Heteropolymolybdates, -tungstates.

203. A. Rosenheim, in Abegg's *Handbuch der anorganischen Chemie*, Vol. 4, Pt. *T*, ii, pp. 977–1065, Verlag von S. Hirzel, Leipzig (1921), A comprehensive and historically significant summation of the early structural systematization, now unacceptable, of Miolati, Rosenheim, and Copaux.

204. L. C. W. Baker, in *Advances in the Chemistry of the Coordination Compounds*, S. Kirschner (Ed.), pp. 604–612, Macmillan, New York (1961).

205. D. L. Kepert, *Inorg. Chem.*, **8**, 1556 (1969).

206. B. Dawson, *Acta Crystallogr.*, **6**, 113 (1953).

207. H. T. Evans, Jr., *J. Am. Chem. Soc.*, **70**, 1291 (1948); **90**, 3275 (1968).

208. L. C. W. Baker, G. Foster, W. Tan, F. Scholnick, and T. P. McCutcheon, *J. Am. Chem. Soc.*, **77**, 2136 (1955).

209. J. L. T. Waugh, D. P. Schoemaker, and L. Pauling, *Acta Crystallogr.*, **7**, 438 (1954).

210. D. D. Dexter and J. V. Silverton, *J. Am. Chem. Soc.*, **90**, 3589 (1968).

211. L. Pauling, *J. Am. Chem. Soc.*, **51**, 2868 (1929).

212. J. F. Keggin, *Proc. R. Soc. Lond. Ser. A*, **144**, 75 (1934).

213. J. W. Illingworth and J. F. Keggin, *J. Chem. Soc.*, **1935**, 575.

214. H. T. Evans, Jr., in *Perspectives in Structural Chemistry*, J. D. Dunitz and J. A. Ibers (Eds.), Vol. 4, pp. 1–59, John Wiley & Sons, New York (1971).

215. R. W. G. Wyckoff, *Crystal Structures*, 2nd ed., Vol. 3, pp. 887–893, Interscience Publishers, New York (1965).

216. O. W. Rollins and C. R. Skolds, *J. Inorg. Nucl. Chem.*, **42**, 371 (1980).

217. M. R. V. Sahyun, *J. Chem. Educ.*, **57**, 239 (1980). Simple molecular-orbital treatment.

SUGGESTED SUPPLEMENTARY REFERENCES

J. C. Bailar, Jr., and D. H. Busch (Eds.), *The Chemistry of the Coordination Compounds*, Reinhold Publishing Corp., New York (1956). Detailed treatments of special topics. Older but still very useful.

C. J. Ballhausen, *Introduction to Ligand Field Theory*, McGraw-Hill Book Company, New York (1962). Detailed supplement to chapter discussions.

F. Basolo and R. G. Pearson, *Mechanisms of Inorganic Reactions: A Study of Complexes in Solution*, 2nd ed., Chs. 2, 7. John Wiley & Sons, New York (1967). Structure and bonding.

R. L. Carlin and A. J. Van Duyneveldt, *Magnetic Properties of Transition Metal Compounds*, Springer-Verlag, New York (1977). Detailed treatment.

B. N. Figgis, *Introduction to Ligand Fields*, Interscience Publishers, New York (1966). General.

W. P. Griffith, in *Comprehensive Inorganic Chemistry*, A. F. Trotman-Dickinson (Exec. Ed.), Vol. 4, pp. 105–195, Pergamon Press, Oxford (1973). Carbonyl, nitrosyl, cyano, and isonitrile complexes.

B. F. G. Johnson (Ed.), *Transition Metal Clusters*, John Wiley & Sons, New York (1979). Detailed summary of structure, bonding, ligand mobility, catalysis, reactions, etc.

D. L. Kepert, in *Comprehensive Inorganic Chemistry*, A. F. Trotman-Dickenson (Exec. Ed.), Vol. 4, pp. 607–671, Pergamon Press, Oxford (1973). Isopoly- and heteropoly anions — detailed summary.

D. L. Kepert and K. Vrieze, in *Comprehensive Inorganic Chemistry*, A. F. Trotman-Dickenson (Exec. Ed.), Vol. 4, pp. 197–354, Pergamon Press, Oxford (1973). Survey of metal–metal bonded complex species.

E. L. Muetterties, T. N. Rhodin, E. Band, C. F. Brucker, and W. R. Pretzer, *Chem. Rev.*, **79**, 91 (1979). Metal cluster species and metal surfaces.

K. Nakamoto, *Infrared and Raman Spectra of Inorganic and Coordination Compounds*, 3rd ed., John Wiley & Sons, New York (1978). Summation and interpretation of vibrational spectroscopic data, especially for coordination entities.

L. E. Orgel, *An Introduction to Transition-Metal Chemistry: Ligand-Field Theory*, John Wiley & Sons, New York (1960). Older but still a very useful introduction.

A. G. Sharpe, *The Chemistry of Cyano Complexes of the Transition Metals*, Academic Press, New York (1976). Detailed summation of preparation, properties.

A. Stelzer, in *Topics in Phosphorus Chemistry*, E. J. Griffith and M. Grayson (Eds.), Vol. 9, pp. 1–229, John Wiley & Sons, New York (1977). Metal complexes of terpositive phosphorus ligands.

A. E. Underhill and D. M. Watkins, *Chem. Soc. Rev.*, **9**, 429 (1980). Review dealing with one-dimensional metallic complexes.

G. Wilkinson (Ed.), *Comprehensive Organometallic Chemistry*, Pergamon Press, Maxwell House, Elmsford, N.Y. (1981—, 6 volumes).

CHAPTER NINE

INORGANIC CHEMICAL REACTIONS, I PRINCIPLES; REDOX AND ACID–BASE REACTIONS

A chemical reaction may be defined as a process in which a chemical bond (or bonds) is (or are) formed or broken or both formed and broken, with the concomitant conversion of a pure substance (or substances) into another pure substance (or other pure substances) with different physical and chemical properties. Inasmuch as chemical bonds are believed to be the consequences of appropriate combinations and distributions of electronic charge densities (Chs. 4–8), it may be assumed that a chemical reaction is a consequence of a redistribution of electronic charge density among energetically more favorable states. Such redistributions of electronic charge density may take place in any of a number of ways, thus allowing us to classify chemical reactions under several broad headings. In inorganic chemistry, it is useful to discuss chemical reactions under the general classifications of oxidation–reduction (or redox) reactions and metathetical (or acid–base) reactions. This classification can then be implemented by discussions of reactions that occur in nonaqueous media and in the solid state, of reactions that are peculiar to coordination species, and of polymerization reactions.

The chemical reactions of both organic and inorganic substances are fundamentally the same and involve the same principles. Those of organic substances involve primarily the formation and/or rupture of covalent bonds. Those of inorganic substances involve ionic bonds as well. In inorganic chemistry, there is the further complication of differences induced by the wide variety of elements and ions that must be treated. As a consequence, it is essentially impossible and not at all profitable to provide the detailed classifications of inorganic chemical reactions in the manner that is so useful for organic chemical reactions.[1]

1. CHEMICAL REACTIVITY

The formation and/or rupture of chemical bonds in a specific reaction require at least the following:

1. Collisions among particles at the atomic, molecular, or ionic level.
2. A significant difference in energy to allow redistribution of electronic charge density.
3. A suitable pathway, or mechanism, at the atomic, molecular, or ionic level, to allow this redistribution to occur.

Each chemical reaction is fundamentally reversible (i.e., capable of proceeding in either direction) given an appropriate set of conditions. Each reaction thus proceeds, under a specific combination of conditions, until a state of dynamic equilibrium results. This state of equilibrium is also a function of the reacting species. The point of equilibrium, or the percentage of conversion of reactant(s) into product(s), may vary from about 0% to about 100%. The ca. 0% and ca. 100% cases tempt one to think of such reactions as being irreversible. Thus the observations described by the equations

$$Mg(s) + H_2O(g) \rightarrow MgO(s) + H_2(g)$$

$$MgO(s) + H_2(g) \rightarrow \text{no observable change}$$

$$CuO(s) + H_2(g) \rightarrow Cu(s) + H_2O(g)$$

$$Cu(s) + H_2O(g) \rightarrow \text{no observable change}$$

suggest that in each instance the first reaction is irreversible. Actually, in each instance an equilibrium does exist, and reversibility is evident at very high temperatures.

No chemical reaction can occur unless the energy change associated with it is favorable. The energy factors related to equilibrium reactions are treated in terms of *chemical thermodynamics*, that is, "the discipline which enables us to apply our relevant experience with regard to energy and spontaneity to any process which involves a change in a macroscopic property of a system between two states of rest."[2] Chemical thermodynamics allows us to establish whether, under a specific set of conditions, a given chemical reaction is possible. Chemical thermodynamics provides answers to such questions as:

1. Where, in terms of quantities of reactants and products in a given chemical system under specified conditions, does the point of equilibrium lie?
2. How is the point of equilibrium affected by changes in temperature, pressure, or activity?
3. What changes in energy are associated with a reaction under various sets of conditions?

Chemical thermodynamics can only indicate whether or not a particular reaction can take place under the conditions specified. It can give no information as to whether or not a reaction that is energetically feasible will take place within an observable period of time. Thus chemical thermodynamics predicts that at 25°C elemental hydrogen (1 atm) will reduce copper(II) ion (unit activity) to elemental copper in acidic, aqueous solution and that under acidic conditions cerium(IV) ion (unit activity) will be quantitatively reduced by water. Yet one can bubble hydrogen gas through copper(II) salt solutions for prolonged periods of time without observing reduction, and acidic cerium(IV) salt solutions are useful volumetric oxidants, the titers of which remain unchanged over substantial periods of time. Such predicted reactions are not observed experimentally because of unfavorable kinetics, which in turn is a consequence of unfavorable reaction mechanisms.

Chemical kinetics allows us to measure the rate of a thermodynamically feasible chemical reaction and in terms of the dependence of rate upon the activities* (or concentrations) of the reacting substances to postulate a reasonable *pathway*, or *mechanism*, by way of which the reaction occurs. Chemical kinetics provides answers to such questions as:

1. How is the rate of the reaction affected by concentrations of individual reactants or by temperature?
2. Does the reaction proceed in a single step or by a series of steps, the summation of which is the observed total change?
3. What logical mechanism or mechanisms can be postulated for the reaction?

Thus a detailed understanding of a particular chemical reaction requires consideration of each of the following:

1. The nature and quantities of the products obtained — factual and measurable.
2. The energy change — factual and either measurable or calculable.
3. The reaction rate — factual and measurable.
4. The reaction mechanism — interpretable only. Inasmuch as more than one mechanistic interpretation may be consistent with the observed kinetics, the apparently most logical choice must be made. That choice *may* or *may not* be the correct one.

---

* Activity (a) is a thermodynamic concentration, which for solutions usually has the units of molality (m) and is related to molality as $a = \gamma m$, where γ is an *activity coefficient*. Activities may be expressed in the units of molarity (M). Gases at 1 atm and pure liquids and solids, by convention, are at unit activity.

2. SOME USEFUL CHEMICAL THERMODYNAMICS

It is neither pertinent nor proper that space in a book of this type be devoted to a detailed discussion of chemical thermodynamics. Rather, it is useful to adopt a more operational point of view and to limit our discussion to useful applications. More detailed background from courses in physical chemistry or from innumerable references[2-5a] can be reviewed or consulted as necessary. It is important to reemphasize that chemical thermodynamics deals with the characteristics of systems at equilibrium, that it has nothing to do with time, that it is independent of any assumptions as to molecular structure and reaction mechanism, that it relates exactly data from a variety of measurements, and that it is really concerned only with the initial and final states of a chemical reaction. Chemical thermodynamics allows for the compact storage of a great abundance of information about chemical substances in terms of tables of so-called thermodynamic functions. From this information one can reconstruct the information from which the data resulted and predict whether or not previously uninvestigated reactions can take place under the conditions specified.

2-1. Thermodynamic Functions

The thermodynamic functions include, for a given chemical reaction, the change in *enthalpy*, the change in *entropy*, the change in *free energy*, the *equilibrium constant*, and, for redox reactions, the *electrode potential*. Tabulated values (Appendix IV for electrode potentials) are most commonly given for so-called *standard-state conditions* of *unit activity* [i.e., for a gas, an ideal gas at 1 atm partial pressure; for a solid or liquid, the pure substance at 1 atm external pressure; for a solute, an ideal solution at $1\ m$ or $1\ M$ concentration, or unit activity].[5b] The temperature is most commonly $25°C\ (=298.14°K)$, and for pure solids values are often given for the most stable modifications at $298.14°K$ and 1 atm.

Standard-state conditions are indicated by a superscripted zero in symbols for these functions. The functions may be reviewed briefly as follows:

1. Standard change in *enthalpy*, ΔH^0. The change in enthalpy is the change in *heat* or *thermal* energy for the reaction in question. The common unit is kcal $mole^{-1}$ (or in SI units, kJ $mole^{-1}$, where $1\ cal = 4.184\ J$), or if written in association with a balanced equation, kcal or kJ. By convention, a negative sign is used if heat energy is released (exothermic process), a positive sign if heat energy is absorbed (endothermic process). The ΔH^0 value for a particular reaction measures the changes in bond energies or strengths between products and reactants. The ΔH^0 value for a spontaneous reaction is usually, *but not always*, negative.

2. Standard change in *entropy*, ΔS^0. Operationally, the change in entropy reflects the change in randomness in the reaction in the sense that to produce order energy must be absorbed whereas an increase in disorder, or randomness, releases energy. Thus for the reaction described by

$$[Ni(H_2O)_6]^{2+}(aq) + 6\,NH_3(aq) \rightarrow [Ni(NH_3)_6]^{2+}(aq) + 6\,H_2O(l)$$

the total number of moles (i.e., fundamental particles) is unchanged, the degree of randomness remains the same, and ΔS^0 is expected to be zero. However, for the change described as

$$[Ni(H_2O)_6]^{2+}(aq) + 3\,en(aq) \rightarrow [Ni(en)_3]^{2+}(aq) + 6\,H_2O(l)$$

(en = ethylenediamine, $H_2NC_2H_4NH_2$), the total number of moles is increased, the degree of randomness increases, and ΔS^0 is expected to be larger than zero. The common unit of entropy is cal deg.$^{-1}$ mole^{-1} (i.e., an entropy unit) or cal deg.$^{-1}$ if associated with a balanced equation. By convention a positive sign is used if entropy increases; a negative sign if it decreases. For a spontaneous process, the change in entropy is normally positive. Multiplication of ΔS^0 by T yields an energy quantity, as shown by the unit conversion

$$T\,\Delta S^0 = (^\circ K)(cal)(mole)^{-1}(^\circ K)^{-1} = cal\,mole^{-1}$$

Each pure substance may be assumed to have an entropy of zero at $0^\circ K$ and a positive entropy at any other temperature.

   3. Standard change in *Gibbs free energy*, **ΔG^0.**   The change in Gibbs free energy* for a chemical reaction is a measure of the energy available from that reaction to do external work. The common units are kcal mole^{-1} or kJ mole^{-1}, or kcal or kJ if associated with a balanced equation. By convention, a negative sign is used if the reaction releases energy that can do work. *A reaction for which ΔG^0 is negative is spontaneous. It cannot occur as formulated and under the conditions specified if ΔG^0 is positive.* The change in Gibbs free energy is related to the changes in enthalpy and entropy as

$$\Delta G^0 = \Delta H^0 - T\,\Delta S^0 \tag{9-1}$$

Spontaneity of reaction is thus determined by both ΔH^0 and ΔS^0, which quantities may act either in concert or in opposition.

   4. Thermodynamic *equilibrium constant* **(K).**   The thermodynamic equilibrium constant for a chemical reaction represented by the equation

$$vA + wB \rightleftharpoons xC + yD$$

is expressed as

$$K = \frac{(a_C)^x(a_D)^y}{(a_A)^v(a_B)^w} \tag{9-2}$$

or as

$$K = \frac{(\gamma_C m_C)^x(\gamma_D m_D)^y}{(\gamma_A m_A)^v(\gamma_B m_B)^w} \tag{9-3}$$

---

*   For J. W. Gibbs, a pioneer in chemical thermodynamics. Other types of free energy are used also. Chemists usually drop the term "Gibbs."

using appropriate activity coefficients. Thermodynamic relationships apply rigorously only when activities are used. Commonly used equilibrium constants for reactions in solution are expressed in terms of *molarities*. To minimize deviations from truly thermodynamic conditions, they are determined either for very dilute solutions, where molarities and molalities approach each other closely and interionic attraction effects are minimized, or for solutions of fixed and constant *ionic strength* (I), where activity coefficients are constant. The ionic strength is calculated as

$$I = \tfrac{1}{2}(m_1 z_1^2 + m_2 z_2^2 + m_3 z_3^2 + \cdots)$$ (9-4)

where the subscripts refer to the different ions, and z_1, z_2, ... are the numerical charges borne by those ions. The units in Eq. 9-2 or 9-3 are mole(kg H_2O)$^{-1}$ or mole(liter solution)$^{-1}$ raised to the power indicated by combining appropriately the values of v, w, x, y for the reaction in question. A long-standing convention is to omit units for equilibrium constants.

Many reactions, in particular those involved in the formation of complex species, proceed by successive steps, each of which can be described by its own equilibrium constant. Thus, neglecting solvating water molecules, the formation of the tetraamminecopper(II) ion in aqueous solution may be described as

$$Cu^{2+} + NH_3 \rightleftharpoons [Cu(NH_3)]^{2+} \quad K_1 = \frac{(a_{Cu(NH_3)^{2+}})}{(a_{Cu^{2+}})(a_{NH_3})}$$

$$[Cu(NH_3)]^{2+} + NH_3 \rightleftharpoons [Cu(NH_3)_2]^{2+} \quad K_2 = \frac{(a_{Cu(NH_3)_2^{2+}})}{(a_{Cu(NH_3)^{2+}})(a_{NH_3})}$$

$$[Cu(NH_3)_2]^{2+} + NH_3 \rightleftharpoons [Cu(NH_3)_3]^{2+} \quad K_3 = \frac{(a_{Cu(NH_3)_3^{2+}})}{(a_{Cu(NH_3)_2^{2+}})(a_{NH_3})}$$

$$[Cu(NH_3)_3]^{2+} + NH_3 \rightleftharpoons [Cu(NH_3)_4]^{2+} \quad K_4 = \frac{(a_{Cu(NH_3)_4^{2+}})}{(a_{Cu(NH_3)_3^{2+}})(a_{NH_3})}$$

The overall reaction is then formulated as the summation of these steps, as

$$Cu^{2+} + 4 NH_3 \rightleftharpoons Cu(NH_3)_4^{2+} \quad \beta_4 = \frac{(a_{Cu(NH_3)_4^{2+}})}{(a_{Cu^{2+}})(a_{NH_3})^4} = K_1 K_2 K_3 K_4$$

5. Standard electrode, or reduction, potential, E_{298}^0 or E^0. The standard electrode potential is the potential assigned to a redox half-reaction occurring in aqueous solution and formulated as*

$$oxidant(a = 1) + n e^- \rightleftharpoons reductant \ (a = 1)$$

using an arbitrarily selected reference potential of $E^0 = 0.000$ V for the hydrogen(I)–hydrogen(0) couple

---

* Rigorously, unit *mean* ionic activity, not unit *individual* ionic activities, is meant. See O. Robbins, Jr., *J. Chem. Educ.*, **48**, 737 (1971).

$$2\,H^+(aq, a = 1) + 2\,e^- \rightleftharpoons H_2(g, 1\text{ atm})$$

By convention, the electrode potential is written with a *positive* sign if the oxidant gains electrons more readily than the H^+ ion ($a = 1$)* and is thus reducible by molecular hydrogen (at 1 atm). Similarly, the electrode potential is written with a negative sign if the reductant loses electrons more readily than molecular hydrogen (at 1 atm) and is thus oxidizable by the H^+ ion ($a = 1$). The numerical value of the electrode potential is thus a measure of the strength of the oxidant or the reductant in aqueous systems relative to the H(I)–H(0) couple. Reversal of the equation for a specific half-reaction reverses the sign of the potential for that reaction. Standard potentials associated with half-reactions formulated as

$$\text{reductant }(a = 1) \rightleftharpoons \text{oxidant }(a = 1) + n\,e^-$$

are termed *oxidation potentials*.† In this book we use electrode (reduction) potentials as standards of reference. Many useful electrode potentials are tabulated for specific equilibria in Appendix IV.[5-7a] Applications and related topics are discussed later in this chapter (Sec. 9.5).

2-2. Relationships Involving Thermodynamic Functions

These thermodynamic functions are usefully summarized and interrelated as‡

$$\Delta G^0 = \Delta H^0 - T\,\Delta S^0 = -RT\ln K = -n\,\Delta E^0 \mathscr{F} \tag{9-6}$$

---

* In aqueous systems, the proton is properly symbolized H_3O^+ (Sec. 5.2-5) since its hydration energy is substantial (ca. -103 kcal mole^{-1}). The notation H^+ is used only for simplification. See P. A. Giguère, *J. Chem. Educ.*, **56**, 571 (1979), "The Great Fallacy of the H^+ Ion."

† The sign convention used for oxidation potentials, first suggested by G. N. Lewis and M. Randall[5] and subsequently developed in detail by W. M. Latimer,[6] has been termed the *American convention* because of its widespread adoption and usage in the United States. A better terminology might be the *thermodynamic convention*,[6a] since this convention assigns to the electromotive force of a half-reaction a sign such that

$$\Delta G^0 = -nE^0 \mathscr{F} \tag{9-5}$$

where n is the number of equivalents ($=$ the number of electrons in the equation for the half-reaction and $\mathscr{F}$ is the Faraday constant ($=23.066$ kcal volt^{-1} equiv^{-1}). This convention is often used in physical chemistry. In the so-called *European convention*, which is used in particular in analytical chemistry and in electrochemistry, the sign associated with a particular potential is invariant and is determined by the electrostatic charge of the metal electrode involved in the couple in question. Continuing confusion and argument relative to the two systems have arisen largely from the failure of an author to indicate at the outset the convention being used. In an attempt to resolve this problem, it was recommended at the 17th Conference of the International Union of Pure and Applied Chemistry (IUPAC) at Stockholm in 1953 that the term *electrode potential* be used only in describing the equation for a half-reaction written as a *reduction process* and with the same sign as the electrostatic charge on the metal electrode in that half-reaction. Use of the term *oxidation potential* violates the Stockholm recommendation only in associating the term *potential* with a quantity opposite in sign from that electrostatic charge. Continuing arguments should be readily answerable by statement of the convention being used. Further details are available in many excellent references.[8-11]

‡ In Eq. (9-6) and in subsequent discussions, ΔE^0 is the change in potential for a redox reaction obtained by combining appropriate electrode potentials.

where R is the ideal gas-law constant (1.987 cal mole^{-1} deg^{-1}), and the other terms are as previously noted.

The functions ΔH^0 and ΔG^0 are strictly additive for combinations of reactions, in accordance with the law of Hess.* Two examples, using accepted ΔH^0 and ΔG^0 values of formation, may be cited to illustrate this point.

1. Calculate ΔH^0 for the reaction described as

$$Ag^+(1\ m) + NaCl(s) \rightarrow AgCl(s) + Na^+(1\ m)$$

$$\Delta H^0_{reaction} = \{[\text{moles AgCl} \times \Delta H^0_{AgCl}] + [\text{moles Na}^+ \times \Delta H^0_{Na^+}]\}$$
$$- \{[\text{moles Ag}^+ \times \Delta H^0_{Ag^+}] + [\text{moles NaCl} \times \Delta H^0_{NaCl}]\}$$

$$= \left\{ \left[1 \text{ mole AgCl} \times \left(\frac{-30.36 \text{ kcal}}{\text{mole AgCl}} \right) \right] \right.$$
$$\left. + \left[1 \text{ mole Na}^+ \times \left(\frac{-57.28 \text{ kcal}}{\text{mole Na}^+} \right) \right] \right\}$$
$$- \left\{ \left[1 \text{ mole Ag}^+ \times \left(\frac{25.31 \text{ kcal}}{\text{mole Ag}^+} \right) \right] \right.$$
$$\left. + \left[1 \text{ mole NaCl} \times \left(\frac{-98.23 \text{ kcal}}{\text{mole NaCl}} \right) \right] \right\}$$

$$= -14.72 \text{ kcal}$$

This calculation reflects the assumed formation of each substance from its elements and a summation of these steps.

2. Calculate ΔG^0 and K (at 25°C) for the reaction described as

$$Cl_2(g, 1 \text{ atm}) + 3F_2(g, 1 \text{ atm}) \rightleftharpoons 2ClF_3(g, 1 \text{ atm})$$

$$\Delta G^0_{reaction} = \{[\text{moles ClF}_3 \times \Delta G^0_{ClF_3}]$$
$$- \{[\text{moles Cl}_2 \times \Delta G^0_{Cl_2}] + [\text{moles F}_2 \times \Delta G^0_{F_2}]\}$$

$$= \left\{ \left[2 \text{ moles ClF}_3 \times \left(\frac{-29.0 \text{ kcal}}{\text{mole ClF}_3} \right) \right] \right\}$$
$$- \left\{ \left[1 \text{ mole Cl}_2 \times \left(\frac{0 \text{ kcal}}{\text{mole Cl}_2} \right) \right] \right.$$
$$\left. + \left[3 \text{ moles F}_2 \times \left(\frac{0 \text{ kcal}}{\text{mole F}_2} \right) \right] \right\}$$

$$= -58.0 \text{ kcal}$$

---

* In 1840, G. H. Hess noted that the heat change associated with a specific chemical reaction is the same regardless of the intermediate steps involved. Since $T\Delta S^0$ is also an energy term, the law applies equally well to ΔG^0.

$$\ln K = 2.303 \log K = -\frac{\Delta G^0}{RT}$$

$$\log K = -\frac{-58.0 \times 10^3 \text{ cal}}{(1.987 \text{ cal deg}^{-1})(298 \text{ deg})(2.303)}$$

$$= 42.5$$

or

$$K = 3.1 \times 10^{42}$$

Both the large negative value of ΔG^0 and the large numerical value of K indicate the thermodynamic ease with which chlorine(III) fluoride can be obtained from its elements.

Applications of these fundamental thermodynamic principles are illustrated both in subsequent sections of this chapter and in later chapters.

3. SOME USEFUL CHEMICAL KINETICS

The practical limitations of chemical thermodynamics can be illustrated in terms of the familiar reaction between molecular hydrogen and molecular oxygen summarized as

$$H_2(g) + \tfrac{1}{2}O_2(g) \rightarrow 2 H_2O(l \text{ or } g) \qquad \Delta G^0 = -56.687 \text{ kcal for } H_2O(l)$$

$$= -54.634 \text{ kcal for } H_2O(g)$$

Thermodynamically, this reaction is highly favorable at 25°C. In practice, however, the reaction is infinitely slow at 25°C, and mixtures of molecular hydrogen and oxygen can be preserved indefinitely at room temperature in the absence of a catalyst. The introduction of a suitable catalyst or the addition of a sufficient amount of energy causes the reaction to proceed so rapidly that it becomes explosive in nature.

The apparent anomaly is explained in terms of a kinetically unfavorable pathway involving simultaneous participation by both molecular species. Chemical kinetics provides a rationale for accounting for this and many similar situations. Again, our purpose is to provide in this book only an operationally useful acquaintance with chemical kinetics. More detailed background from courses in physical chemistry or from various references[12-16] can be reviewed or consulted as necessary.

3-1. Preliminary Concepts

Modern approaches are based upon two broad concepts: the *collision concept* and the *activated complex concept*.

It is generally agreed that a chemical reaction results only after collisions of fundamental particles of the reacting substance(s). Many such collisions are not effective. Effective collisions require both proper orientation of the colliding particles and adequate energy imparted by these particles. Any conditions that enhance either or both of these requirements result in a reaction of increased rate.

It is believed also that the chemical change from reactants to products occurs via the formation and subsequent decomposition of an intermediate or transition-state substance called an *activated complex*. The significance of the activated complex on an energy basis is shown qualitatively in Fig. 9-1, where an internal energy coordinate and a coordinate representing the progress of the reaction (reaction coordinate) are related.

If the reactants lie at energy E_r, with respect to a zero-energy reference, and the product at a lower energy E_p, the difference is the observed energy released in the reaction

$$\Delta E_{react} = E_r - E_p \tag{9-7}$$

It is assumed, however, that collision of particles produces a metastable intermediate or activated complex at higher energy E_{comp}, or E_c. The energy of activation of the reactants is then given as

$$\Delta E_{act} = E_c - E_r \tag{9-8}$$

Decomposition of this activated complex to the products releases energy $\Delta E'_{act}$ as

$$\Delta E'_{act} = E_c - E_p \tag{9-9}$$

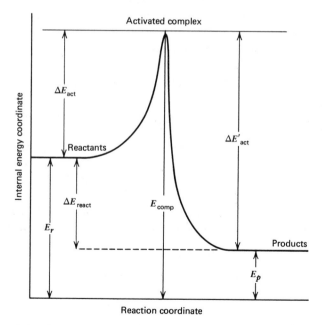

Figure 9-1 Concept of the activated complex in a chemical reaction.

whence

$$\Delta E_{react} = \Delta E'_{act} - \Delta E_{act} = E_r - E_p \qquad (9\text{-}10)$$

In a real situation, a smooth change from reactants to products via the activated complex is expected, as indicated by the curve in Fig. 9-1. For the reverse reaction, the energy of activation is E'_{act}.

An activated complex is seldom an isolable entity and commonly only a hopefully logical postulation. In any event, it must be a species the components of which are held together by relatively weak bonds.

3-2. The Rate Constant

The rate of a chemical reaction can be expressed as the ratio of the total quantity of a reactant that has disappeared, or of a product that has formed, to the total time elapsed. Quantities are usually expressed as moles or moles liter^{-1}; so the units of reaction rate are moles sec^{-1} or moles liter^{-1} sec^{-1}, raised to an appropriate power. Rates of chemical reactions are affected by at least the following:

1. **Contact between reacting species.** The frequency of effective collisions increases with subdivision of the reactants, that is, for a specific reaction on going from solid to liquid or solution, or from solid to gas.

2. **Presence of a *catalyst*.** The only function of a catalyst is to increase* the rate of a chemical reaction by providing an alternative pathway, or mechanism, of decreased energy of activation. A catalyst cannot initiate a reaction that is thermodynamically impossible. Nor can a catalyst displace the point of equilibrium for a particular system, for it affects equally the rates of both the forward and the reverse reactions.

3. **Temperature.** Inasmuch as increase in temperature increases the kinetic energies of fundamental particles, increase in temperature increases both the number and the intensity of effective collisions, and thus increases the rate of a chemical reaction.

4. **Concentration.** Quite reasonably, an increase in the concentration of a reacting substance (or substances) increases the frequency of fundamental particle collisions and thereby the rate of a chemical reaction.

Concentration and temperature are significant and controllable factors. Experimentally, for a chemical reaction at a fixed temperature formulated as

$$v\,A + w\,B \rightarrow x\,C + y\,D$$

the reaction rate (R) is given by the expression

$$R = k[A]^{n_A}[B]^{n_B} \qquad (9\text{-}11)$$

---

*   Conversely, a reaction rate is decreased by an *inhibitor*.

where the square brackets denote concentrations, n_A and n_B are small whole numbers that may or may not be equal, respectively, to v and w in the chemical equation, and k is the specific rate constant with the usual dimension sec^{-1}. Credit for the initial statement of this type is usually given to C. M. Guldberg and P. Waage, who about 1867 related reaction rates to the product of the "masses" of the reactants, each raised to some definite power. Presently, "masses" are treated as pressures, activities, molalities, or molarities. For each chemical reaction, then, a rate law of this type can be written. Usually, such a law is expressed as a differential equation dealing with a decrease (negative sign) in concentration of a particular reactant, for example, for the reaction formulated

$$2\,N_2O_5(g, \text{ or in } CCl_4) \rightarrow 2\,N_2O_4(g, \text{ or in } CCl_4) + O_2(g)$$

$$R = -\frac{d[N_2O_5]}{dt} = k[N_2O_5] \tag{9-12}$$

(coefficient and exponent not equal), or for the reaction formulated as

$$2\,NO(g) + O_2(g) \rightarrow 2\,NO_2(g)$$

$$R = -\frac{d[NO]}{dt} = k[NO]^2[O_2] \tag{9-13}$$

(coefficients and exponents equal). The effect of temperature on the specific rate constant, and thus upon the reaction rate, can be expressed in integrated form as (S. Arrhenius, 1889)

$$k = Ae^{-E_a/RT} \tag{9-14}$$

where A is a frequency factor and E_a is the energy of activation. The latter may be determined from the slope of the straight line that results for a given reaction from plotting $\log k$ vs. $1/T$.

3-3. Order of Reaction

The overall order of a given chemical reaction is equal to the sum of the exponents n_A, n_B, and so on (Eq. 9-11). The order with respect to a given substance is given by the power to which the concentration of that species is raised in the defining rate equation. Thus from Eq. 9-13, the reaction in question has an overall order of 3, is second order in NO, and is first order in O_2. The thermal decomposition of N_2O_5 (Eq. 9-12) is first order in N_2O_5 even though the defining equation is balanced with two moles of the compound. First-, second-, and third-order reactions are common. A zero-order reaction, where the rate is given by

$$-\frac{d[A]}{dt} = k \tag{9-15}$$

and is unaffected by concentration, has a rate dependence on such items as quantity of catalyst present or absorption of radiation.

Order with respect to a given reactant is determined by measuring reaction rates at constant temperature as a function of varying concentrations of that reactant, while the concentrations of other reactants are maintained constant. Appropriate mathematical treatments show[12-16] the following correlations: zero-order, linear plot of [A] vs. time (t), with $k = -$slope; first-order, linear plot of log [A] vs. t, with $k = (-2.303)$(slope); second-order, linear plot of 1/[A] vs. t, with $k =$ slope; and so on.

Equation 9-12 is a specific example of the general first-order rate constant expression

$$-\frac{d[A]}{dt} = k[A] \qquad (9\text{-}16)$$

Integration of Eq. 9-16 between the limits of time zero (t_0) and any time t, where the concentrations are $[A]_0$ and $[A]$, and conversion to ordinary logarithms gives ultimately

$$\log [A] = \frac{-kt}{2.303} + \log [A_0] \qquad (9\text{-}17)$$

from which the graphical method of establishing order and evaluating k, as noted in the preceding paragraph, emerges. It is of considerable practical interest to characterize a first-order reaction in terms of the time necessary for one-half of the initial material to react (i.e., the half-life). If in Eq. 9-17 $t = t_{1/2}$ and $[A] = ([A]_0)/2$, substitution and rearrangement give

$$t_{1/2} = \frac{2.303}{k} \ \log \ \left(\frac{[A]_0}{[A]_0/2}\right) = \frac{0.693}{k} \qquad (9\text{-}18)$$

Once k has been evaluated for a particular change, the half-life can be calculated. This approach has been particularly useful in characterizing spontaneous nuclear decay processes (Sec. 2.2), which are first-order changes. For each of these processes, the rate constant k is called the *decay constant* and symbolized λ. Nuclides are described in terms of their λ and $t_{1/2}$ values.

3-4. Molecularity of Reaction

As indicated earlier, chemical reactions in general proceed in a series of steps, each of which for a particular reaction is referred to as an *elementary reaction*. In discussing the combination of such steps into the mechanism of the reaction, the *molecularity* of each step (i.e., the number of reacting fundamental particles) is important. For an elementary reaction, the molecularity (uni-, bi-, ter-, etc. molecular) and the order (first-, second-, third-order) are identical, but the terms are not synonymous as regards the overall rate law for the entire reaction.

A unimolecular elementary reaction is exemplified by the equation

$$N_2O_5 \rightarrow NO_2 + NO_3$$

and a bimolecular elementary reaction by the equation

$$NO_2 + NO_3 \rightarrow NO + O_2 + NO_2$$

Inasmuch as the probabilities of three-body and larger multibody collisions are very small, ter- and larger molecularities are very uncommon.

3-5. Reaction Mechanism

Measurement of the order of a reaction with respect to each reactant has as its ultimate objective interpretation in terms of a reasonable mechanism. Thus (a) the thermal decomposition of dinitrogen(V) pentaoxide, which is kinetically a first-order process (Eq. 9-12), is believed to proceed in steps, as

$$N_2O_5 \rightleftharpoons NO_2 + NO_3 \qquad \text{(fast)}$$
$$NO_2 + NO_3 \rightarrow NO + O_2 + NO_2 \qquad \text{(slow)}$$
$$NO + NO_3 \rightarrow 2\,NO_2 \qquad \text{(fast)}$$

(b) the chloride ion-catalyzed reduction, represented by the equation

$$2\,Fe^{3+}(aq) + Sn^{2+}(aq) \rightarrow 2\,Fe^{2+}(aq) + Sn^{4+}(aq)$$

the rate expression for which is

$$-\frac{d[Fe^{3+}]}{dt} = k[Fe^{3+}][Sn^{2+}] \qquad (9\text{-}19)$$

is considered to proceed in steps, as

$$Sn^{2+} + 3\,Cl^- \rightleftharpoons SnCl_3^- \qquad \text{(fast)}$$
$$Fe^{3+} + SnCl_3^- \rightleftharpoons Fe^{2+} + SnCl_3 \qquad \text{(fast)}$$
$$Fe^{3+} + SnCl_3 \rightarrow Fe^{2+} + SnCl_3^+ \qquad \text{(slow)}$$

and (c) the base-catalyzed hydrolyses of the acidopentaamminecobalt(III) ions, represented by the overall equation

$$[Co(NH_3)_5X]^{2+}(aq) + OH^-(aq) \rightarrow [Co(NH_3)_5OH]^{2+}(aq) + X^-(aq)$$

where X = Cl, Br, SCN, and so on, the rate of which is first-order in the complex ion and first-order in hydroxide ion, is assigned the mechanism represented as

$$[Co(NH_3)_5X]^{2+} + OH^- \rightleftharpoons [Co(NH_3)_4(NH_2)X]^+ + H_2O \qquad \text{(fast)}$$
$$[Co(NH_3)_4(NH_2)X]^+ \rightarrow [Co(NH_3)_4(NH_2)]^{2+} + X^- \qquad \text{(slow)}$$
$$[Co(NH_3)_4(NH_2)]^{2+} + H_2O \rightarrow [Co(NH_3)_5(OH)]^{2+} \qquad \text{(fast)}$$

In each case, the slow step is the rate-controlling step.

Gas-phase, redox, and ligand-exchange reactions have been studied particularly effectively from the kinetic point of view. Many of these reactions are considered in later discussions. It must be emphasized, however, that usually the mechanism proposed for a particular chemical reaction is one that seems most logical in terms of the kinetic data at hand and is not necessarily supported by other evidence. More than one mechanism may agree with the experimental kinetic data for a specific reaction.

4. CLASSIFICATION OF INORGANIC CHEMICAL REACTIONS

Chemical reactions of inorganic substances may be classified either in terms of the nature of the redistributions of electronic charge densities that occur or in terms of the natures of the species involved or formed and/or the conditions under which these reactions proceed.

The first classification, which is the more fundamental and which embraces the second, amounts to:

1. *Redox reactions*, that is, reactions that involve changes in oxidation states and result from the transfer of electronic charge density from one species to another. Typical examples can be formulated as

$$Fe(s) + 2\,H^+(aq) \rightleftharpoons Fe^{2+}(aq) + H_2(g)$$

$$24\,CuS(s) + 64\,H^+(aq) + 16\,NO_3^-(aq) \rightleftharpoons 24\,Cu^{2+}(aq) + 16\,NO(g)$$
$$+ 32\,H_2O(l) + 3\,S_8(s)$$

$$4\,NH_3(g) + 5\,O_2(g) \rightleftharpoons 4\,NO(g) + 6\,H_2O(g)$$

$$8\,Al(s) + 3\,NO_3^-(aq) + 5\,OH^-(aq) + 18\,H_2O(l) \rightleftharpoons 8\,[Al(OH)_4]^-(aq)$$
$$+ 3\,NH_3(g)$$

2. *Metathetical reactions*, that is, reactions that involve no changes in oxidation states and result in "exchanges of partners" as a consequence of changes in electronic charge distribution. Typical examples can be formulated as

$$ZnS(s) + 2\,H^+(aq) \rightleftharpoons Zn^{2+}(aq) + H_2S(g)$$

$$2\,Pb^{2+}(aq) + Cr_2O_7^{2-}(aq) + H_2O(l) \rightleftharpoons 2\,PbCrO_4(s) + 2\,H^+(aq)$$

$$[Cu(H_2O)_4]^{2+}(aq) + 4\,NH_3(aq) \rightleftharpoons [Cu(NH_3)_4]^{2+}(aq) + 4\,H_2O(l)$$

By one definition or other (Sec. 9.6-1), each such reaction may be called an acid–base reaction.

Reactions of these two types are considered in subsequent sections of this chapter.

The second classification could include reactions based upon physical states (solids, liquids, gases, solutions), reactions of polymerization and depolymerization, and certain reactions of coordination entities. Some of these types of reactions are treated in detail in Chapters 10 and 11.

5. REDOX REACTIONS

Although a redox reaction is easily definable in terms of electron transfer from one chemical species to another, with the term *oxidation* being associated with electron loss and *reduction* with electron gain, neither the mechanism by which such transfer occurs nor the assignment of specific oxidation states in a given chemical reaction is necessarily simple or evident. Mechanisms can best be handled later in this discussion (Sec. 9.5-3), but knowledge of oxidation states and their changes is of paramount initial importance.

There is no ambiguity in assigning oxidation numbers or states to simple monatomic ions (e.g., $+1$ for Na^+, Tl^+; $+2$ for Ca^{2+}, Eu^{2+}; -1 for F^-, At^-; -2 for O^{2-}, S^{2-}; etc.).* However, assignments to individual elements in polyatomic molecules or ions normally require prior knowledge of constancy for particular elements (e.g., -2 for oxygen, $+1$ for hydrogen) and algebraic summation (e.g., $+7$ for Cl in ClO_4^-). Fortunately, the ion–electron equation† for the half-reaction involving the change in question often provides, in combination with zero values for the free elements and values for monatomic species, unambiguous assignments. This approach is illustrated in Table 9-1.

Continuing reference should be made to half-reactions and their standard potentials as summarized for acidic and alkaline aqueous solutions in Appendix IV. The magnitudes of the potential values are definitive indicators of comparative oxidizing or reducing strengths under these specified conditions — the larger the standard electrode potential in the *positive* direction, the stronger the oxidant; the larger the standard electrode potential in the *negative* direction, the stronger the reductant. Comparisons on this basis must be restricted to conditions of unit activity and 25°C.

Table 9-1   Oxidation States from Ion–Electron Equations

Ion–Electron Equation	Electron Change	Indicated Oxidation State
$Al(OH)_3 + 3e^- \rightleftharpoons Al + 3OH^-$	$3e^-$ per Al	Al(III) in $Al(OH)_3$
$SO_4^{2-} + 8H^+ + 6e^- \rightleftharpoons \frac{1}{8}S_8 + 4H_2O$	$6e^-$ per S	S(VI) in SO_4^{2-}
$ClO_3^- + 3H_2O + 6e^- \rightleftharpoons Cl^- + 6OH^-$	$6e^-$ per Cl	Cl(V) in ClO_3^-
$PuO_2^{2+} + 4H^+ + 2e^- \rightleftharpoons Pu^{4+} + 2H_2O$	$2e^-$ per Pu	Pu(VI) in PuO_2^{2+}
$\frac{1}{2}N_2 + 4H^+ + 3e^- \rightleftharpoons NH_4^+$	$3e^-$ per N	N($-$III) in NH_4^+ or NH_3
$N_2H_5^+ + 3H^+ + 2e^- \rightleftharpoons 2NH_4^+$	$1e^-$ per N	N($-$II) in $N_2H_5^+$ or N_2H_4
$NO_3^- + 6H_2O + 8e^- \rightleftharpoons NH_3 + 9OH^-$	$8e^-$ per N	N(V) in NO_3^-

* Our convention is to use Arabic numbers for oxidation numbers and Roman numbers for oxidation states. The Arabic zero (0) is used for both.

† Although in each such equation the electron is simply symbolized as e^-, it is, of course, hydrated in every aqueous system [see E. J. Hart, *Science*, **146**, 19 (1964); D. C. Walker, *Q. Rev. (Lond.)*, **21**, 79 (1967)].

5-1. Factors Affecting the Numerical Magnitudes of Electrode Potentials

These factors include the inherent chemical natures of the substances in question and such controllable variables as concentration and temperature.

Inherent Chemical Nature. As indicated by electrode-potential data, the relative reducing strength of a metal or the relative oxidizing strength of a nonmetal is associated qualitatively with electronegativity (Sec. 2.4-3). Electronegativity is itself the consequence of a combination of other factors, some of which are of particular importance.

The standard electrode potential associated with a particular $M^{n+}(aq)-M(s)$ couple under acidic conditions indicates the change in standard free energy for a process described as

$$M^{n+}(aq, a = 1) + ne^{-}(aq) \rightleftharpoons M(s)$$

Correspondingly, the standard electrode potential associated with a particular simple $A_2(g)-A^{n-}(aq)*$ couple under acidic conditions indicates the change in standard free energy for a process

$$A_2(g, 1 \text{ atm}) + ne^{-}(aq) \rightleftharpoons 2 A^{n-}(aq, a = 1)$$

For each of these processes, the total change in free energy is, by Hess's law, the sum of the changes in free energy for the hypothetical steps that lead to the overall change. Unfortunately, the absence of free-energy data for all such steps precludes the exact calculation of standard potential values. However, enthalpy data are available, are qualitatively proportional to free-energy data, and allow some useful predictions and conclusions.

Application of the generalized Born–Fajans–Haber cyclic approach (Sec. 4.3-4) to the couple $M(s)-M^{n+}(aq)$ yields the formulation

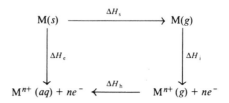

where ΔH_e, ΔH_s, ΔH_i, and ΔH_h represent, respectively, the enthalpy changes associated with electrode potential, sublimation of the metal, ionization of the gaseous, metal, and hydration of the gaseous ion and electron. From Hess's law,

$$\Delta H_e = \Delta H_s + \Delta H_i + \Delta H_h \tag{9-20}$$

ΔH_s and ΔH_i being endothermic terms (positive), ΔH_h exothermic (negative), and

---

* Gaseous diatomic molecules are assumed. Greater molecular complexity introduces no additional considerations, but the existence of the nonmetal as a liquid or a solid necessitates the inclusion of an endothermic enthalpy term to effect vaporization.

ΔH_e thus either exothermic or endothermic, depending on the others. Inasmuch as ΔH_i and ΔH_h are usually large with respect to ΔH_s, their relative magnitudes are significant in determining ΔH_e and indicating E^0.

Equation 9-20 indicates that decreased reducing strength, or increased nobility, for a metal is favored by large values of its energy of sublimation and ionization and a small value of the energy of hydration of its gaseous ion. Inasmuch as sublimation energy is somewhat dependent upon boiling point (Table 4-13), metals of high boiling point (e.g., Au, Pt metals) are expected (and found) to be quite noble. To illustrate, let us compare the formation of K^+ ($E^0 = -2.925$ V for $K^+ + e^- \rightleftharpoons K$) with the formation of Ag^+ ($E^0 = +0.7991$ V for $Ag^+ + e^- \rightleftharpoons Ag$). The stepwise reactions and their enthalpy changes (in kcal mole^{-1}) are

	K	Ag
$M(s) \rightleftharpoons M(g)$	21.5	67
$M(g) \rightleftharpoons M^+(g) + e^-$	100.0	174.7
$M^+(g) \rightleftharpoons M^+(aq)$	-76	-111
$M(s) \rightleftharpoons M^+(aq) + e^-$	45.5	130.7

Thus potassium is the more readily oxidized because the much smaller energies for its sublimation and ionization counterbalance the smaller energy released in the hydration of the relatively large $K^+(g)$ ion.

Of particular interest are variations in E^0 for the alkali metal ion–alkali metal atom and the seemingly anomalously large negative value for lithium (Appendix IV). Helpful stepwise data to aid in accounting for these variations in ease of oxidation are (again in kcal mole^{-1})

	Li	Na	K	Rb	Cs
$M(s) \rightleftharpoons M(g)$	37.1	26.0	21.5	20.5	18.8
$M(g) \rightleftharpoons M^+(g) + e^-$	124.3	118.5	100.0	96.3	89.8
$M^+(g) \rightleftharpoons M^+(aq)$	-121	-95	-76	-69	-62
$M(s) \rightleftharpoons M^+(aq) + e^-$	40.4	49.5	45.5	47.8	46.6

Although these data do not correlate completely with E^0 values, they do indicate that the reduced ionization and sublimation energies of potassium, rubidium, and cesium make these metals stronger reducing agents than sodium. The large negative electrode potential for lithium is due to the fact that the energy released in the hydration of the lithium ion counterbalances effectively the larger quantities of energy needed to effect its sublimation and ionization. In contact with aqueous systems, lithium is the strongest reducing agent among the alkali metals (omitting francium). In the absence of water or other strongly solvating medium (Sec. 10.1-1), however, it is not.

A comparable cyclic treatment can be applied to the formation of a simple anion from a nonmetal molecule, as

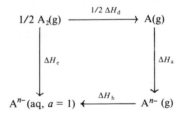

where ΔH_e has the same connotation as before, ΔH_d is the enthalpy change for dissociation into atoms (positive), ΔH_a is the enthalpy change for formation of gaseous anions or the electron affinity (positive or negative), and ΔH_h is again a hydrational enthalpy change.

An application to the halogens shows that the decrease in oxidizing strength with increasing atomic number is due primarily to a substantial decrease in energy of hydration as size of the halide ion increases, with some contribution from the electron affinity, changes in the enthalpies of dissociation being insignificant by comparison. For nonmetals that in the solid state are composed of polyatomic, and often giant, molecules, combination of the energies of vaporization and dissociation may be large enough to determine the standard electrode potentials.

Concentrations of Species Involved. It is seldom, of course, that a redox system can be studied under conditions of unit activity. Correction of the standard potential values for alterations in activity (in the concentration and pressure senses) is effected by use of the Nernst relationship

$$E = E^0 - \frac{RT}{n\mathscr{F}} \ln Q \tag{9-21}$$

where Q has the same form as the equilibrium constant K but incorporates the actual activities of products and reactants for the reaction,* and not the equilibrium values, and E is the potential for that case. Insertion of numerical constants and conversion to common logarithms then gives at $T = 298.14°K$,

$$E = E^0 - \frac{0.05916}{n} \log Q \tag{9-22}$$

The effects of changes in activity† may be illustrated by some examples.† Thus if the activity of the hydrogen ion is reduced to 10^{-7} m (i.e., the value for pure

---

*   For the electrode-potential convention, Q and K for a couple amount to activity–product ratios with reductants in the numerator and oxidants in the denominator.

†   Again molalities (or molarities) are commonly used for practicality. Although absolute numerical accuracy is sacrificed, the ultimate conclusions are the same. Potentials calculated in terms of concentration changes are sometimes called *formal electrode potentials*.

water, pH = 7) and the pressure of molecular hydrogen is maintained at 1 atm, the potential of the H(0)–H(I) couple is altered as

$$E = E^0 - \frac{0.05916}{n} \log \left(\frac{p_{H_2}}{a_{H^+}^2} \right) = 0 - \frac{0.05916}{2} \log \left[\frac{1}{(10^{-7})^2} \right] = -0.414 \text{ V}$$

Still further decrease in the activity of the hydrogen ion as the alkaline range is entered results in still larger negative potentials. This is, of course, a fundamental reason for the fact that couples involving H^+ or OH^- ions have potential values which are more negative under alkaline conditions than under acidic conditions (Appendix IV). To emphasize this point, let us consider the half-reaction described as

$$MnO_4^-(aq) + 8 H^+(aq) + 5 e^- \rightleftharpoons Mn^{2+}(aq) + 4 H_2O(l) \qquad E^0 = +1.51 \text{ V}$$

for which

$$E = E^0 - \frac{0.05916}{5} \log \left(\frac{a_{Mn^{2+}}}{a_{MnO_4} \times a_{H^+}^8} \right)$$

In pure water with $a_{Mn^{2+}} = a_{MnO_4^-} = 1$ m, calculation gives $E = +1.42$ V. Oxidation of Mn^{2+} to MnO_4^- is thus somewhat easier to effect in pure water than in 1 m acid. The effect of the activity of the H^+ ion is the controlling factor because of the high power to which it is raised.

pH and Potential-pH Diagrams. Although change in activity (or concentration) of any species in a redox system alters the reduction potential, change in pH is the most significant single factor and the one most apt to characterize reactions in aqueous solutions. In every aqueous system, two pH-dependent redox couples involving water and its ions are important:

1. The H(I)–H(0) couple, formulated for acidic and alkaline systems, respectively, as

$$2 H^+(aq) + 2 e^- \rightleftharpoons H_2(g) \qquad\qquad E^0 = 0.000 \text{ V}$$

and

$$2 H_2O(l) + 2 e^- \rightleftharpoons H_2(g) + 2 OH^-(aq) \qquad E^0 = -0.828 \text{ V}$$

2. The O(0)–O(−II) couple, formulated for acidic and alkaline systems, respectively, as

$$O_2(g) + 4 H^+(aq) + 4 e^- \rightleftharpoons 2 H_2O(l) \qquad E^0 = 1.229 \text{ V}$$

and

$$O_2(g) + 2 H_2O(l) + 4 e^- \rightleftharpoons 4 OH^-(aq) \qquad E^0 = 0.401 \text{ V}$$

The effect of change in pH upon potentials associated with hydrogen or oxygen evolution (p_{H_2} or $p_{O_2} = 1$ atm) from aqueous solutions is shown in Fig. 9-2.[17]

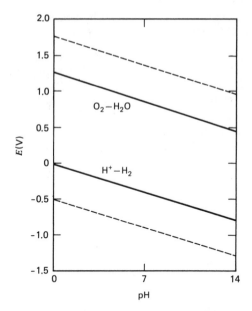

Figure 9-2 Effect of pH on electrode potentials involving the H_2-H_2O-O_2 system. [Redrawn from W. L. Jolly, *J. Chem. Educ.*, **43**, 198 (1966), Fig. 1.]

Linearity between potential and pH arises from application of the Nernst relationship to yield the two equations

$$E^0_{\text{H(I)}\rightarrow\text{H(0)}} = -0.05916 \log a_{\text{H}^+} = -0.059 \text{ pH} \qquad (9\text{-}23)$$

and

$$E^0_{\text{O(0)}\rightarrow\text{O(-II)}} = +1.229 - 0.05916 \log a_{\text{H}^+} = +1.229 - 0.059 \text{ pH} \quad (9\text{-}24)$$

The regions of evolution of molecular hydrogen, evolution of molecular oxygen, and stability of water or OH^- ion with respect to oxidation or reduction are indicated. For kinetic reasons, they are better defined by the dashed lines at ca. ± 0.5 V (overvoltage) than by the calculated solid lines. This difference should remind one again that thermodynamic prediction has its limitations.

It is apparent that the potential associated with any couple in which water and its ions are involved is pH dependent. If the other species present remain unchanged as the pH is altered, the potential varies uniformly and only with pH. The Cl(VII)–Cl(0) couple is a case in point, since the defining reactions may be formulated as

$$\text{ClO}_4^- + 8\,\text{H}^+ + 7e^- \rightleftharpoons \tfrac{1}{2}\text{Cl}_2 + 4\text{H}_2\text{O} \qquad \text{(acidic)}$$

and

$$\text{ClO}_4^- + 4\,\text{H}_2\text{O} + 7e^- \rightleftharpoons \tfrac{1}{2}\text{Cl}_2 + 8\,\text{OH}^- \qquad \text{(alkaline)}$$

Where there is a change in species, the potential is correspondingly altered.

Typical examples involve metal ions that form difficultly soluble hydroxides. Thus for the Fe(II)–Fe(0) couple, we have

$$Fe^{2+} + 2e^- \rightleftharpoons Fe \quad \text{(acidic)} \qquad E^0 = -0.440 \text{ V}$$

compared with

$$Fe(OH)_2 + 2e^- \rightleftharpoons Fe + 2OH^- \quad \text{(alkaline)} \qquad E^0 = -0.877 \text{ V}$$

If, however, the hydroxide is soluble, the potential is unchanged as the pH changes, for example,

$$\overset{+}{Cs} + e^- \rightleftharpoons Cs \quad \text{(acidic or alkaline)} \qquad E^0 = -2.923 \text{ V}$$

These situations are indicated in Fig. 9-3.

Potential-pH diagrams are particularly useful in defining the regions of stable existence of various species when several oxidation states can be distinguished for a given element and in allowing one to predict changes among them.[18, 19] In the construction of such diagrams, three equilibrium relationships between E^0 and pH are utilized:

1. Equilibrium independent of pH (i.e., oxidant $+ ne^- \rightleftharpoons$ reductant). A case in point is the Cs(I)–Cs(0) couple noted in the preceding paragraph. As

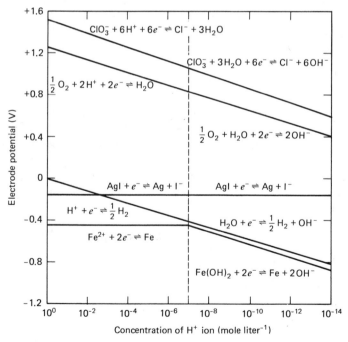

Figure 9-3 Some typical relationships between electrode potential and pH. [Redrawn from T. Moeller, *Inorganic Chemistry*, Fig. 8.1, p. 293, John Wiley & Sons, New York (1952).]

indicated in Fig. 9-4, this variation is shown as a horizontal line of a plot of E (as ordinate) vs. pH (as abscissa).

2. Equilibrium independent of E, that is, a nonredox reaction such as might be exemplified by

$$Fe^{2+} + 2\,OH^- \rightleftharpoons Fe(OH)_2(s)$$

As indicated in Fig. 9-4, this variation is shown as a vertical line on the E vs. pH plot.

3. Equilibrium dependent on both E^0 and pH (i.e., a redox reaction that involves water and/or its ions) such as might be exemplified by

$$Fe(OH)_2(s) + 2\,H^+ + 2\,e^- \rightleftharpoons Fe(s) + 2\,H_2O$$

This variation appears as an angular line of negative slope in the E vs. pH plot.

A typical potential-pH diagram for the familiar $Fe-H_2O-H^+$ system is given in Fig. 9-5. Regions within which the indicated species have stable existence are bounded by solid lines. The dashed lines have the same connotation as those in Fig. 9-2. Equilibria denoted by the various boundary lines are apparent within Fig. 9-5. The tetraoxoferrate(VI) species, FeO_4^{2-}, is not included, since it forms only above $E^0 = -2.20$ V, and little is known about its area of existence.

The Pourbaix references should be consulted for additional details. Variations in these diagrams are periodic in character, and a detailed treatment from this point of view can be consulted with profit.[20]

5-2. Applications of Electrode-Potential Data

Redox systems can be treated quantitatively via electrode-potential data. Consideration of several types of applications through some specific examples provides useful background for studying the detailed chemistries of the elements and their compounds.[17]

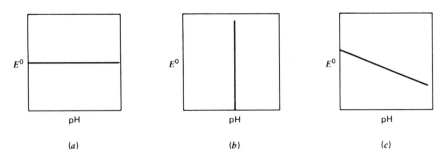

Figure 9-4 Variations of E^0 with pH in potential-pH diagrams. Equilibrium (a) independent of pH—E^0 constant, pH variable; (b) independent of E^0—pH constant, E^0 variable; (c) dependent on both E^0 and pH—pH variable, E^0 variable.

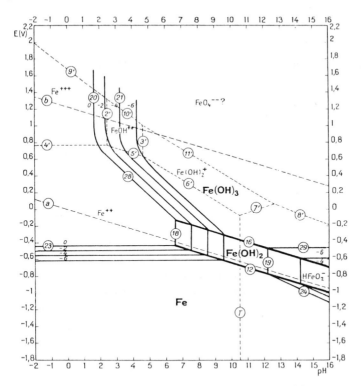

Figure 9-5 Potential-pH diagram for the system $Fe-H^+-H_2O-OH^-$. [M. Pourbaix, *Atlas of Electrochemical Equilibria in Aqueous Solutions*, Fig. 5, p. 313, Pergamon Press, Oxford (1966).]

Combinations of Half-Reactions. By appropriate combinations of equations and potentials for known half-reactions, potentials for other half-reactions and completed reactions can be calculated. It must be remembered that a reduction potential is an intensive property that depends only upon the redox process in question and not upon doubling, tripling, and so on, the coefficients in the defining equation. Thus the couple $O(0)-O(-II)$ in acidic solution is characterized by $E^0 = +1.229$ V for all the following formulations:

$$O_2(g) + 4 H^+(aq) + 4 e^- \rightleftharpoons 2 H_2O(l)$$

$$\tfrac{1}{2} O_2(g) + 2 H^+(aq) + 2 e^- \rightleftharpoons H_2O(l)$$

$$5 O_2(g) + 20 H^+(aq) + 20 e^- \rightleftharpoons 10 H_2O(l)$$

Furthermore, when equations for half-reactions are combined to give equations for other half-reactions, the free-energy changes calculated as in Eq. 9-6, not the electrode potentials, must be combined.

To illustrate, let us calculate the E^0 value for the unlisted half-reaction described by the equation

$$MnO_2(s) + 4 H^+(aq) + e^- \rightleftharpoons Mn^{3+}(aq) + 2 H_2O(l)$$

Reference to Appendix IV gives us the following pertinent information:

(1) $\quad MnO_2(s) + 4 H^+(aq) + 2 e^- \rightleftharpoons Mn^{2+}(aq) + 2 H_2O(l)$

$$E^0 = +1.23 \text{ V}$$

$$\Delta G^0 = -nE^0\mathscr{F} = -2.46\mathscr{F}$$

(2) $\quad Mn^{3+}(aq) + e^- \qquad\qquad \rightleftharpoons Mn^{2+}(aq) \qquad E^0 = +1.51 \text{ V}$

$$\Delta G^0 = -nE^0\mathscr{F} = -1.51\mathscr{F}$$

Subtraction of (2) and (1) gives the equation for the desired half-reaction, for which

$$\Delta G^0 = -nE^0\mathscr{F} = (-2.46\mathscr{F}) - (-1.51\mathscr{F}) = -0.95\mathscr{F}$$

Since $n = 1$, the calculated value of $E^0 = +0.95$ V. It is obviously unnecessary to evaluate ΔG^0 completely since the Faraday constant cancels. From the calculated value of E^0 for the Mn(IV)–Mn(III) couple, $\Delta G^0 = -21.9$ kcal and $K = $ ca. 10^{+16}, thus indicating the thermodynamic ease of effecting this reduction.

Combinations of half-reactions to give completed reactions allow direct addition of E^0 values.* Thus the completed reaction formulated for acidic solutions as

$$2 Al(s) + 3 Cu^{2+}(aq) \rightleftharpoons 3 Cu(s) + 2 Al^{3+}(aq) \qquad \Delta E^0 = 1.999 \text{ V}$$

follows from combination of the two half-reactions

$$Al^{3+}(aq) + 3 e^- \rightleftharpoons Al(s) \qquad E^0 = -1.662 \text{ V}$$

and

$$Cu^{2+}(aq) + 2 e^- \rightleftharpoons Cu(s) \qquad E^0 = +0.337 \text{ V}$$

This theoretical potential could be approximated but not exactly achieved in a real galvanic cell based upon these reactions.

Electrode-Potential and Oxidation-State Diagrams. The use of electrode-potential data in permitting one to predict the products that can be formed in reactions involving elements that exhibit several oxidation states and in making direct comparisons among a variety of species is facilitated by electrode-potential diagrams. To illustrate, data for manganese in its various oxidation states (Appendix IV) can be summarized as:

Acidic media

---

* The symbol ΔE or ΔE^0 is used for the potential of a completed reaction, as obtained by combinations of E or E^0 values for half-reactions.

Alkaline media

$$\text{MnO}_4^- \xrightarrow{0.564} \text{MnO}_4^{2-} \xrightarrow{0.60} \text{MnO}_2 \xrightarrow{\quad} \text{Mn(OH)}_3 \xrightarrow{0.15} \text{Mn(OH)}_2 \xrightarrow{-1.55} \text{Mn}$$
$$\underset{0.588}{\underline{\qquad\qquad\qquad}} \qquad \underset{-0.05}{\underline{\qquad\qquad\qquad}}$$

Similarly, data for chlorine can be summarized as:

Acidic media

$$\text{ClO}_4^- \xrightarrow{1.19} \text{ClO}_3^- \xrightarrow{\quad} \text{ClO}_2 \xrightarrow{1.275} \text{HOClO} \xrightarrow{1.645} \text{HOCl} \xrightarrow{1.63} \text{Cl}_2 \xrightarrow{1.3595} \text{Cl}^-$$
$$\underset{1.21}{\underline{\qquad\qquad\qquad\qquad}}$$

Alkaline media

$$\text{ClO}_4^- \xrightarrow{0.36} \text{ClO}_3^- \xrightarrow{\quad} \text{ClO}_2 \xrightarrow{1.16} \text{ClO}_2^- \xrightarrow{0.66} \text{ClO}^- \xrightarrow{\quad} \text{Cl}_2 \xrightarrow{1.3595} \text{Cl}^-$$
$$\underset{0.33}{\underline{\qquad\qquad\qquad}} \qquad \underset{0.89}{\underline{\qquad\qquad\qquad}}$$

With reference to data involving water, it is evident that thermodynamically the conversion Mn → Mn(II) can be effected at any pH, whereas the conversion Mn(II)–Mn(III) by elemental oxygen can be effected only under alkaline conditions. Oxidation of Mn(IV) to Mn(VII) or reduction of Mn(VII) to Mn(IV) is direct under acidic conditions but can readily go by way of Mn(VI) under alkaline conditions. Under acidic conditions Cl($-$I) is more readily oxidized to Cl(0) than to Cl(I), whereas under alkaline conditions the reverse is true. Reduction of Cl(VII) is easier to effect under acidic than under alkaline conditions, as is also true of the other positive oxidation states of chlorine. Under acidic conditions, Mn(VII) should oxidize Cl($-$I) to Cl(0) but Mn(IV) should not. In arriving at these and other possible conclusions, it must be remembered that conditions of unit activity are specified and that only thermodynamic considerations are offered. Observed reactions may differ from prediction, for example, the formation of elemental chlorine when $12\,M$ HCl solution is warmed with manganese(IV) oxide.

The chemistries of elements that have several oxidation states are often characterized by *disproportionation*[21] reactions, that is, reactions in which the same element is both oxidized and reduced. Cases in point are illustrated by the equations

$$\text{Hg}_2^{2+}(\text{aq}) \rightleftharpoons \text{Hg}(l) + \text{Hg}^{2+}(\text{aq})$$

$$\text{Cl}_2(g) + \text{H}_2\text{O}(l) \rightleftharpoons \text{H}^+(\text{aq}) + \text{Cl}^-(\text{aq}) + \text{HOCl}(\text{aq})$$

In terms of an electrode-potential diagram, disproportionation may be expected if a given oxidation state is a stronger oxidizing agent (more positive E^0 value) than the next higher state. Thus the diagrams in the preceding paragraph indicate that the species that are definitely unstable with respect to disproportionation are Mn^{3+}, MnO_4^{2-}, Cl_2 (alkaline media), and HOClO. Similarly, the data

$$O_2 \overset{0.68}{\text{——}} H_2O_2 \overset{1.77}{\text{——}} H_2O$$
$$\underset{1.23}{\lfloor\text{——————————}\rfloor}$$

indicate that hydrogen peroxide undergoes disproportionation under acidic conditions. Hydrogen peroxide is a common intermediate in oxidation reactions effected by molecular oxygen, and its formation affects the rates of such reactions. Thus possible reactions using reductants with E^0 values more positive than 0.68 V proceed slowly (e.g., $Br_3^- - Br^-$, $E^0 = 1.065$ V) but those involving reductants with less positive E^0 values may proceed more rapidly (e.g., $I_3^- - I^-$, $E^0 = 0.536$ V).[17]

The lack of additivity of electrode potentials may render electrode-potential diagrams misleading. It has been suggested, with considerable merit, that diagrams relating oxidation states to free energies[22] or to volt equivalents[23] (a volt equivalent being defined as the product of oxidation state and potential measured relative to the element in the free state, including the usual signs) replace simple electrode-potential diagrams. Typical diagrams describing the oxidation states of manganese and chlorine, respectively, are included as Figs. 9-6

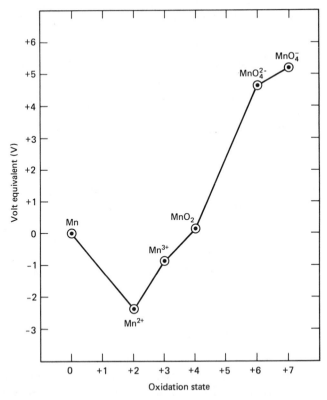

Figure 9-6 Volt-equivalent oxidation-state diagram for manganese in acidic medium.

and 9-7, with oxidation state as abscissa and volt equivalent as ordinate.*

If the free-energy change is expressed in electron volts, the reduction potential is then the *positive slope* of the line joining the two oxidation states in question. The tendency of a reaction to proceed toward the product of lower free energy is measured by the steepness of such a line. In terms of the sign convention used, a positive slope indicates that the oxidant in the couple will convert elemental hydrogen to hydrogen ion (acidic media), whereas a negative slope indicates that the reverse reaction will occur.

Comparisons of diagrams of this type allow one to predict whether or not reactions between species derived from more than one element can be expected. Thus comparison of Fig. 9-6 and 9-7 shows that the slope for the $Cl^- - Cl_2$ couple (-1.3595) is less than that for the $Mn^{2+} - Mn^{3+}$ ($+1.51$), $Mn^{2+} - MnO_4^-$ ($+1.51$), $MnO_2 - MnO_4^-$ ($+1.695$), and $MnO_2 - MnO_4^{2-}$ ($+2.26$) couples, indicating that under acidic conditions the species Mn^{3+}, MnO_4^{2-}, and MnO_4^- are all

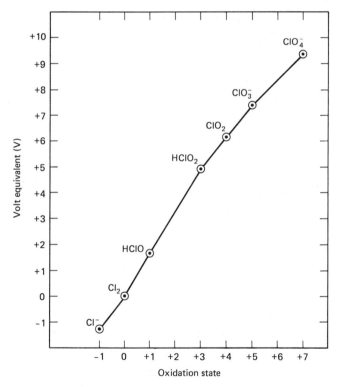

Figure 9-7 Volt-equivalent-oxidation-state diagram for chlorine in acidic medium.

* For additional examples, consult in particular Ref. 23 (Vols. 1, 2); M. E. Muller and M. O. Naumann, *Free Energy-Oxidation State Diagrams* (*Frost Diagrams*), City College of San Francisco, San Francisco, Calif. (1970); R. T. Myers, *J. Chem. Educ.*, **51**, 444 (1974); and A. Friedel and R. Murray, *J. Chem. Educ.*, **54**, 485 (1977).

capable of oxidizing chloride ion to elemental chlorine. A disproportionation reaction may be expected whenever the free energy or volt equivalent of the species in question lies above the straight line joining the adjacent higher and lower oxidation states. Thus in Fig. 9-6, both Mn^{3+} and MnO_4^{2-} ions are in such positions, and both are known to disproportionate in acidic aqueous solution as

$$2\,Mn^{3+}(aq) + 2\,H_2O(l) \rightarrow Mn^{2+}(aq) + MnO_2(s) + 4\,H^+(aq)$$

$$3\,MnO_4^{2-}(aq) + 4\,H^+(aq) \rightarrow MnO_2(s) + 2\,MnO_4^-(aq) + 2\,H_2O(l)$$

Further applications appear in subsequent discussions.

Stabilization of Oxidation States. In Chapters 7 and 8 many of the coordination species described contained metal ions in apparently unusual oxidation states, and it was noted that certain oxidation states can be stabilized against oxidation or reduction by coordination involving specific ligands. Of particular importance is the fact that most of the known cobalt complexes contain cobalt(III) even though the couple

$$Co^{3+}(aq) + e^- \rightleftharpoons Co^{2+}(aq) \qquad E^0 = 1.808\ V$$

indicates that the aquated cobalt(III) ion will oxidize many coordinating species, including water. Yet in the presence of cyanide ion, for example, even water will oxidize cobalt(II), with the formation of the exceedingly stable species $[Co(CN)_6]^{3-}$. Similarly, although the Mn^{3+} ion will oxidize water readily, the $[Mn(CN)_6]^{3-}$ ion resists reduction. Neither copper(II) iodide nor copper(II) hypophosphite can be prepared because of reduction to copper(I or 0) by the anion. Yet bis(ethylenediamine)copper(II) iodide or hypophosphite is readily obtainable and resistant to internal reduction. Although simple copper(I) compounds containing oxidizing anions such as nitrate cannot be prepared in aqueous media, coordination of copper(I) to acetonitrile stabilizes these species. The aquated silver(II) ion is one of our more potent oxidizing agents

$$Ag^{2+}(aq) + e^- \rightleftharpoons Ag^+(aq) \qquad E^0 = 1.980\ V$$

but many comparatively stable salts containing the ion $[Ag(py)_4]^{2+}$ (py = pyridine) can be prepared without difficulty.

Many examples of elements in seemingly unusual oxidation states are known among difficultly soluble compounds. Thus cobalt(III) hydroxide is readily obtained by air oxidation of the cobalt(II) compound. Manganese(III) orthophosphate is precipitated after oxidation of aqueous manganese(II) by nitric acid, even though nitric acid is ordinarily incapable of oxidizing manganese(II). Similarly, many metals are much more readily oxidized in the presence of anions that yield difficultly soluble products than in their absence.

Many other examples of thermodynamic stabilization of oxidation states involving coordination or precipitation could be cited.[24] Although the effects of coordination have been the more strongly emphasized,[25-27a] the two phenomena are both related to alterations in the activities of the species involved and are interpretable in these terms. Although electrode-potential data are available for only a limited number of systems, those summarized in Tables 9-2, 9-3, and 9-4 may be regarded as typical.

For each of the series in these tables, comparisons should be made with the couple involving only the simple aquated ions. These aquated ions are, of course, aquo complex species, but their exact chemical compositions are commonly either unknown or variable, and their thermodynamic stabilities relative to other complexes or difficultly soluble compounds are comparatively small. The data exemplify what has been suggested, namely that if a complex of large thermodynamic stability (i.e., large formation constant) or a compound of limited solubility (i.e., small solubility product constant) can be formed, oxidation or reduction to yield that species will be favored. Thus oxidation of iron(0) to iron(II) is most readily effected when the iron(II) ion is strongly bonded in a very stable complex ion (e.g., $[Fe(CN)_6]^{4-}$) or difficultly soluble compound (e.g., FeS). The same trends are discernible among the other examples cited in these tables. Ease of oxidation parallels, of course, increase in magnitude of E^0 in the negative direction. Although the data cited in general suggest how the higher of two oxidation states can be stabilized, it is apparent from data for the Fe(III)–Fe(II) couples that by proper choice of ligand either state may be stabilized. Thus, by comparison with the $Fe^{3+}(aq) \rightleftharpoons Fe^{2+}(aq)$ couple, the bipositive state is stabilized by bipyridyl, 1,10-phenanthroline, and 5-nitro-1,10-phenanthroline, whereas the terpositive state is stabilized by the other ligands listed.

Although it is unlikely that these and other instances of the stabilization of oxidation states can be accounted for in any simple way, some insight can be gained by considering activity (or concentration) effects in terms of a variant of Eq. 9-22. Neglecting all species except the oxidant (ox) and the reductant (red) and assuming that both species are present in the reaction in the same stoichiometry ($= y$), one can rewrite this equation as

$$E = E^0 - \frac{0.059}{n} \log \left(\frac{a_{red}}{a_{ox}} \right)^y \tag{9-25}$$

where E^0 refers to the aquated couple in acidic medium. Thus alteration of the activity of either the oxidant or the reductant without corresponding alteration of the activity of the other changes the observed potential for the system. It is unlikely that the activities of both oxidant and reductant will change the same in a real system. Alteration of the standard potential in a negative direction may be expected if the activity of the oxidant is reduced below that of the reductant, and vice versa. Since the thermodynamic stabilities of complexes or the solubilities of compounds of a particular metal ion in two states of oxidation are not the same the potentials for the simple aquated systems are altered accordingly.

It must not be implied from this discussion that oxidation and/or reduction in all systems involving complex species or precipitates proceeds via dissociation to aquated ions, electron transfer through these ions, and subsequent re-

Table 9-2 Stabilization of Oxidation States of Iron and Cobalt

Couple	Equation for Half-Reaction[a]	E^0 (V)
Fe(II)–Fe(0)	$Fe^{2+}(aq) + 2e^- \rightleftharpoons Fe(s) + x\,H_2O$	-0.4402
	$FeCO_3(s) + 2e^- \rightleftharpoons Fe(s) + CO_3^{2-}(aq)$	-0.756
	$Fe(OH)_2(s) + 2e^- \rightleftharpoons Fe(s) + 2\,OH^-(aq)$	-0.877
	$FeS(s) + 2e^- \rightleftharpoons Fe(s) + S^{2-}(aq)$	-0.95
	$[Fe(CN)_6]^{4-}(aq) + 2e^- \rightleftharpoons Fe(s) + 6\,CN^-(aq)$	ca. -1.5
Fe(III)–Fe(II)	$Fe^{3+}(aq) + e^- \rightleftharpoons Fe^{2+}(aq) + x\,H_2O$	$+0.771$
	$[Fe(NO_2\text{-}phen)_3]^{3+}(aq) + e^- \rightleftharpoons [Fe(NO_2\text{-}phen)_3]^{2+}(aq)$	$+1.25$
	$[Fe(phen)_3]^{3+}(aq) + e^- \rightleftharpoons [Fe(phen)_3]^{2+}(aq)$	$+1.14$
	$[Fe(bpy)_3]^{3+}(aq) + e^- \rightleftharpoons [Fe(bpy)_3]^{2+}(aq)$	ca. $+1.1$
	$[FeF_6]^{3-}(aq) + e^- \rightleftharpoons Fe^{2+}(aq) + 6\,F^-(aq)$	$> +0.4$
	$[Fe(CN)_6]^{3-}(aq) + e^- \rightleftharpoons [Fe(CN)_6]^{4-}(aq)$	$+0.36$
	$Fe(OH)_3(s) + e^- \rightleftharpoons Fe(OH)_2(s) + OH^-(aq)$	-0.56
	$Fe_2S_3(s) + 2e^- \rightleftharpoons 2\,FeS(s) + S^{2-}(aq)$	-0.715
Fe(VI)–Fe(III)	$FeO_4^{2-}(aq)\ 8\,H^+(aq) + 3e^- \rightleftharpoons Fe^{3+}(aq) + 4\,H_2O(l)$	$+2.20$
	$FeO_4^{2-}(aq) + 3\,H_2O(l) + 3e^- \rightleftharpoons Fe(OH)_3(s) + 5\,OH^-(aq)$	$+0.72$
Co(II)–Co(0)	$Co^{2+}(aq) + 2e^- \rightleftharpoons Co(s) + x\,H_2O(l)$	-0.277
	$[Co(NH_3)_6]^{2+}(aq) + 2e^- \rightleftharpoons Co(s) + 6\,NH_3(aq)$	-0.42
	$CoCO_3(s) + 2e^- \rightleftharpoons Co(s) + CO_3^{2-}(aq)$	-0.64
	$Co(OH)_2(s) + 2e^- \rightleftharpoons Co(s) + 2\,OH^-(aq)$	-0.73
	$CoS(s,\beta) + 2e^- \rightleftharpoons Co(s) + S^{2-}(aq)$	-1.07
Co(III)–Co(II)	$Co^{3+}(aq) + e^- \rightleftharpoons Co^{2+}(aq)$	$+1.808$
	$Co(OH)_3(s) + e^- \rightleftharpoons Co(OH)_2(s) + OH^-(aq)$	$+0.17$
	$[Co(NH_3)_6]^{3+}(aq) + e^- \rightleftharpoons [Co(NH_3)_6]^{2+}(aq)$	$+0.108$
	$[Co(CN)_6]^{3-}(aq) + e^- \rightleftharpoons [Co(CN)_5]^{3-}(aq) + CN^-(aq)$	-0.83

[a] bpy = bipyridyl; phen = 1,10-phenanthroline; NO_2-phen = 5-nitro-1,10-phenanthroline.

Table 9-3 Stabilization of Oxidation States of Copper

Couple	Equation for Half-Reaction	E^0 (V)
Cu(I)–Cu(0)	$Cu^+(aq) + e^- \rightleftharpoons Cu(s) + x\,H_2O$	+0.521
	$[CuCl_2]^-(aq) + e^- \rightleftharpoons Cu(s) + 2\,Cl^-(aq)$	+0.19
	$CuCl(s) + e^- \rightleftharpoons Cu(s) + Cl^-(aq)$	+0.137
	$[CuBr_2]^-(aq) + e^- \rightleftharpoons Cu(s) + 2\,Br^-(aq)$	+0.05
	$CuBr(s) + e^- \rightleftharpoons Cu(s) + Br^-(aq)$	+0.033
	$[CuI_2]^-(aq) + e^- \rightleftharpoons Cu(s) + 2\,I^-(aq)$	0.00
	$[Cu(NH_3)_2]^+(aq) + e^- \rightleftharpoons Cu(s) + 2\,NH_3(aq)$	-0.12
	$CuI(s) + e^- \rightleftharpoons Cu(s) + I^-(aq)$	-0.1852
	$CuSCN(s) + e^- \rightleftharpoons Cu(s) + SCN^-(aq)$	-0.27
	$Cu_2O(s) + H_2O + 2\,e^- \rightleftharpoons 2\,Cu(s) + 2\,OH^-(aq)$	-0.358
	$[Cu(CN)_2]^-(aq) + e^- \rightleftharpoons Cu(s) + 2\,CN^-(aq)$	-0.429
	$Cu_2S(s) + 2\,e^- \rightleftharpoons 2\,Cu(s) + S^{2-}(aq)$	-0.89
Cu(II)–Cu(0)	$Cu^{2+}(aq) + 2\,e^- \rightleftharpoons Cu(s) + x\,H_2O$	+0.337
	$CuCO_3(s) + 2\,e^- \rightleftharpoons Cu(s) + CO_3^{2-}(aq)$	+0.053
	$[Cu(NH_3)_4]^{2+}(aq) + 2\,e^- \rightleftharpoons Cu(s) + 4\,NH_3(aq)$	-0.05
	$Cu(OH)_2(s) + 2\,e^- \rightleftharpoons Cu(s) + 2\,OH^-(aq)$	-0.224
	$CuS(s) + 2\,e^- \rightleftharpoons Cu(s) + S^{2-}(aq)$	-0.74
Cu(II)–Cu(I)	$Cu^{2+}(aq) + e^- \rightleftharpoons Cu^+(aq)$	+0.153
	$Cu^{2+}(aq) + 2\,CN^-(aq) + e^- \rightleftharpoons [Cu(CN)_2]^-(aq)$	+1.103
	$Cu^{2+}(aq) + I^-(aq) + e^- \rightleftharpoons CuI(s) + x\,H_2O$	+0.86
	$Cu^{2+}(aq) + Br^-(aq) + e^- \rightleftharpoons CuBr(s) + x\,H_2O$	+0.640
	$Cu^{2+}(aq) + Cl^-(aq) + e^- \rightleftharpoons CuCl(s) + x\,H_2O$	+0.538
	$[Cu(NH_3)_4]^{2+}(aq) + e^- \rightleftharpoons [Cu(NH_3)_2]^+(aq) + 2\,NH_3(aq)$	-0.01
	$2\,Cu(OH)_2(s) + 2\,e^- \rightleftharpoons Cu_2O(s) + 2\,OH^-(aq) + H_2O$	-0.080
	$2\,CuS(s) + 2\,e^- \rightleftharpoons Cu_2S(s) + S^{2-}(aq)$	-0.54

Table 9-4 Stabilization of Oxidation States of Silver

Couple	Equation for Half-Reaction	E^0 (V)
Ag(I)–Ag(0)	$Ag^+(aq) + e^- \rightleftharpoons Ag(s) + x\,H_2O$	+0.7991
	$Ag_2SO_4(s) + 2e^- \rightleftharpoons 2\,Ag(s) + SO_4^{2-}(aq)$	+0.654
	$AgC_2H_3O_2(s) + e^- \rightleftharpoons Ag(s) + C_2H_3O_2^-(aq)$	+0.643
	$AgNO_2(s) + e^- \rightleftharpoons Ag(s) + NO_2^-(aq)$	+0.564
	$AgBrO_3(s) + e^- \rightleftharpoons Ag(s) + BrO_3^-(aq)$	+0.546
	$Ag_2MoO_4(s) + 2e^- \rightleftharpoons 2\,Ag(s) + MoO_4^{2-}(aq)$	+0.486
	$Ag_2CO_3(s) + 2e^- \rightleftharpoons 2\,Ag(s) + CO_3^{2-}(aq)$	+0.47
	$Ag_2CrO_4(s) + 2e^- \rightleftharpoons 2\,Ag(s) + CrO_4^{2-}(aq)$	+0.464
	$AgNCO(s) + e^- \rightleftharpoons Ag(s) + OCN^-(aq)$	+0.41
	$[Ag(NH_3)_2]^+(aq) + e^- \rightleftharpoons Ag(s) + 2\,NH_3(aq)$	+0.373
	$AgIO_3(s) + e^- \rightleftharpoons Ag(s) + IO_3^-(aq)$	+0.354
	$Ag_2O(s) + H_2O + 2e^- \rightleftharpoons 2\,Ag(s) + 2\,OH^-(aq)$	+0.345
	$[Ag(SO_3)_2]^{3-}(aq) + e^- \rightleftharpoons Ag(s) + 2\,SO_3^{2-}(aq)$	+0.295
	$AgCl(s) + e^- \rightleftharpoons Ag(s) + Cl^-(aq)$	+0.2222
	$Ag_4[Fe(CN)_6](s) + 4e^- \rightleftharpoons 4\,Ag(s) + Fe(CN)_6^{4-}(aq)$	+0.194
	$AgSCN(s) + e^- \rightleftharpoons Ag(s) + SCN^-(aq)$	+0.09
	$AgBr(s) + e^- \rightleftharpoons Ag(s) + Br^-(aq)$	+0.0713
	$[Ag(S_2O_3)_2]^{3-}(aq) + e^- \rightleftharpoons Ag(s) + 2\,S_2O_3^{2-}(aq)$	+0.017
	$AgCN(s) + e^- \rightleftharpoons Ag(s) + CN^-(aq)$	-0.017
	$AgI(s) + e^- \rightleftharpoons Ag(s) + I^-(aq)$	-0.1518
	$[Ag(CN)_2]^-(aq) + e^- \rightleftharpoons Ag(s) + 2\,CN^-(aq)$	-0.31
	$Ag_2S(s) + 2e^- \rightleftharpoons 2\,Ag(s) + S^{2-}(aq)$	-0.66
Ag(II)–Ag(I)	$Ag^{2+}(aq) + e^- \rightleftharpoons Ag^+(aq)$	+1.980
Ag(III)–Ag(II)	$2\,AgO(s) + H_2O + 2e^- \rightleftharpoons Ag_2O(s) + 2\,OH^-(aq)$	+0.607
	$Ag_2O_3(s) + H_2O + 2e^- \rightleftharpoons 2\,AgO(s) + 2\,OH^-(aq)$	+0.739

association with the complexing or precipitating species. It does follow that the larger the reduction in the activity of an equilibrium aquated species the more profound the alteration in the characteristic electrode potential.*

5-3. Mechanisms of Redox Reactions

In both theoretical and practical senses, knowledge that a particular redox reaction is thermodynamically possible should be supplemented with knowledge as to its kinetic feasibility and a reasonable postulation, if possible, of its mechanism. With the development of sophisticated instrumental techniques for determining the rates of even extremely rapid reactions, mechanistic interpretations have become extensive,[15, 28-31] particularly for redox systems involving complex species, including oxoanions, in aqueous solution.

Although one can envision a transfer of electrons from one species to another in the gaseous state as proceeding without particular difficulty when appropriate collisions occur, the effects of chemical interactions involving water and other ligands and of ionic interactions in solution render reaction kinetics less favorable and kinetic data for such systems less readily interpretable. It is convenient to consider reactions under the mechanistic headings of oxygen-atom transfer, outer-sphere interactions, and inner-sphere interactions.

Oxygen-Atom Transfer Reactions. Cases in point are reactions such as those indicated by the equations

$$ClO_3^-(aq) + 6 X^-(aq) + 6 H^+(aq) \rightarrow Cl^-(aq) + 3 X_2(g) + 3 H_2O(l)$$

$$ClO^-(aq) + NO_2^-(aq) \rightarrow Cl^-(aq) + NO_3^-(aq)$$

(where $X = Cl, Br, I$) and the reactions of XO^-, XO_2^-, XO_3^-, ClO_2, or Cl_2O with SO_3^{2-} to yield X^- and SO_4^{2-} (where $X = Cl, Br$), which have been studied using ^{18}O as a tracer.[32, 33] In every instance, oxygen-atom transfer occurs. Thus in the ClO^-–NO_2^- reaction, using ^{18}O-labeled ClO^- ion, all of the extra oxygen atoms that distinguish NO_3^- from NO_2^- come from the ClO^- ion. Such reactions are commonly faster under acidic conditions, where protonation of an oxygen atom in an oxo-anion may so reduce the negative charge on that oxygen atom as to allow the more rapid loss of an OH^- group or an H_2O molecule from the transition state.

Outer-Sphere Reactions.[34] In reactions of this type, the coordination spheres of the species in question are not altered. The simplest outer-sphere reactions are between species derived from the same metal ion that differ only in charge (e.g., $[Fe(CN)_6]^{4-}$ and $[Fe(CN)_6]^{3-}$, $[Os(bpy)_3]^{2+}$ and $[Os(bpy)_3]^{3+}$, and $[IrCl_6]^{2-}$ and $[IrCl_6]^{3-}$). Somewhat more complex are reactions involving two different metal ions (e.g., $[Os(bpy)_3]^{2+}$ and $[Mo(CN)_8]^{3-}$, and $[Fe(CN)_6]^{4-}$

---

* For a more detailed discussion of the effects of complexing groups, see Ref. 11 and D. D. Pervin, *Rev. Pure Appl. Chem. (Aust.)*, **9**, 257 (1959).

and $[Mo(CN)_8]^{3-}$). In general, these reactions are quite slow and are kinetically second-order. Support for the view that the electron transfer occurs through an extended, or outer-orbital, activated complex in which the first coordination shell remains unaltered (i.e., a tunneling process) is provided by the observations that (1) the rate-law expressions are in agreement with the existence of an activated complex containing all the ligands in the first coordination shells of both metal ions, and (2) electron transfer is more rapid than the exchange of ligands between the complex ions. To account for the catalytic effect of moderately large cations with loosely bonded hydration spheres (e.g., Cs^+) on such exchanges as MnO_4^{2-}–MnO_4^- and $[Fe(CN)_6]^{4-}$–$[Fe(CN)_6]^{3-}$, outer-sphere bridged activated complexes (e.g., $[O_3MnO\cdots Cs\cdots OMnO_3]^-$) are proposed.

Inner-Sphere Reactions.[34]   In reactions of this type, electron transfer is believed to occur via an inner-sphere or ligand-bridged intermediate wherein the two metal ions are joined by a common ligand which was originally the sole property of one of those metal ions. This intermediate then decomposes to the products. A classic example[35] is the reaction between the nonlabile, low-spin species $[Co(NH_3)_5Cl]^{2+}$ and the labile, high-spin species $[Cr(H_2O)_6]^{2+}$ in acidic solution. This reaction gives the nonlabile, low-spin species $[Cr(H_2O)_5Cl]^{2+}$ and the labile, high-spin species $[Co(H_2O)_6]^{2+}$, as

$$[Co(NH_3)_5Cl]^{2+} + [Cr(H_2O)_6]^{2+} + 5H_3O^+ \rightarrow [Co(H_2O)_6]^{2+}$$

$$+ [Cr(H_2O)_5Cl]^{2+} + 5NH_4^+$$

A mechanism is probably best postulated in terms of the formation of the chlorine-bridged intermediate

$$[(H_3N)_5Co^{III}—Cl—Cr^{II}(H_2O)_5]^{4+}$$

electron transfer via the bridging chloride group, subsequent decomposition via removal of the now labile NH_3 ligands as NH_4^+ ions, and substitution of water molecules to give the products. This mechanism is supported by the observation that if the reaction is carried out in the presence of $^{36}Cl^-$ ion, no ^{36}Cl appears in the $[Cr(H_2O)_5Cl]^{2+}$ ion that results.

Although the inner-sphere, bridging mechanism is probably quite common, it has been demonstrated unequivocally in only a limited number of cases. Minimally, this mechanism requires that the reactant complex of one metal ion and the product complex of the other be inert to substitution. Other reactions that are believed to proceed via this mechanism include those of $[Co(CN)_5]^{3-}$ with $[Co(NH_3)_5X]^{2+}$ (X = Cl, N_3, SCN, OH); of $[Cr(H_2O)_6]^{2+}$ with $[IrCl_6]^{2-}$; and of $[Co(CN)_5]^{3-}$ with $[Fe(CN)_6]^{3-}$. Only rarely does an intermediate have a sufficiently long half-life to allow for its isolation, but some have been identified by spectroscopic means.

6. ACID–BASE REACTIONS

Few subjects in chemistry have, over a long period of time, excited more interest and led to more controversy than acids and bases. Although the terms *acid* and *base* are so common in our general parlance that they are usually accepted without question or definition, careful consideration indicates that they are applied to a bewildering array of substances that seem to be unrelated to each other and that they have a variety of meanings. Much of the problem lies in the primarily historical development of the subject, wherein the definition of an acid progressed from behavioristic points of view (R. Boyle, 1680), through characterization in terms of specific elements (oxygen by A. L. Lavoisier, 1787, H. Lux, 1939; hydrogen by H. Davy, 1811, J. von Liebig, 1838, S. Arrhenius, 1887, J. N. Brønsted and T. M. Lowry, 1923), to definition in terms of bonding (G. N. Lewis, 1923; M. Usanovich, 1929; and others).[36-40] Modern definitions date to the *water-ion* approach of Arrhenius, namely that *in aqueous solution* an acid is a hydrogen-containing substance that yields hydrogen (now hydronium or oxonium) ions, and a base is a hydroxo compound that yields hydroxide ion.

It must be emphasized that, contrary to what characterizes much of the literature, the proposals advanced by Arhenius and others who followed him chronologically are *definitions* of the terms acid and base and are not *theories*. It was not until G. N. Lewis[41] emphasized the need for experimental justification of definitions in terms of so-called "phenomenological criteria" that acid–base behavior was approached in a truly fundamental way. The criteria offered by Lewis are:

1. When an acid and a base can combine, the process of combination, or neutralization, is a rapid one.

2. An acid or a base will replace a weaker acid or base from its compounds.

3. Acids and bases may be titrated against one another by the use of substances, usually colored, known as indicators.

4. Both acids and bases play an extremely important part in promoting chemical processes through their action as catalysts.

These criteria should be used in the evaluation of any approach to acid–base behavior.

6-1. Semimodern and Modern Definitions

Water-Ion Definitions. Because of its overall familiarity and widespread usage in more elementary presentations, it is unnecessary to treat this approach in more detail than a brief delineation of its advantages and disadvantages. The former are indicated by the ease with which the water-ion approach handles aqueous acid–base chemistry. Included are the common reactions of neutralization, the quantitative evaluation of acid strengths in terms of K_a and pK_a

values, the development and general application of the pH scale, the constancy of the enthalpy of neutralization of strong acids and strong bases, and the correlation of many catalytic properties with pH.

The disadvantages are essentially limitations. Thus acid–base behavior is restricted to aqueous solutions, although comparable reactions can take place in both protonic (e.g., liquid ammonia, liquid hydrogen fluoride) and nonprotonic [e.g., liquid sulfur(IV) oxide, liquid dinitrogen(IV) tetraoxide] nonaqueous solvents or even in the absence of solvent [e.g., the reaction of $BaO(s)$ with SO_3(l or g) to give $BaSO_4(s)$, which is strictly comparable with the reaction of $Ba(OH)_2(aq)$ with $H_2SO_4(aq)$ to give $BaSO_4(s)$]. Furthermore, the aqueous OH^- ion is the only allowable base, although its reactions differ only in degree, but not in kind, from those of other aqueous anions (e.g., CN^-, F^-, CO_3^{2-}) or even molecular species (e.g., NH_3, amines).

Protonic Definitions.   The definitions are commonly referred to as the Brønsted–Lowry definitions in terms of publications that appeared nearly simultaneously.[42,43,*] Acids and bases are interrelated by the equation

$$acid(HB) \rightleftharpoons base(B) + proton(H^+)$$

with an *acid* being defined as any hydrogen-containing molecule or ion that is capable of *releasing* a proton (i.e., a *proton donor*) and a *base* as any molecule or ion that is capable of *combining with* a proton (i.e., a *proton acceptor*). Thus acids and bases are independent of solvent and bear no relationship to salts. Neutralization has no meaning as such, and acid–base reactions are equilibrium reactions based upon proton transfer from one base to another in terms of a general equilibrium equation

$$\underset{(acid_1)}{HB_1} + \underset{(base_2)}{B_2} \rightleftharpoons \underset{(acid_2)}{HB_2} + \underset{(base_1)}{B_1}$$

The base formed by the loss of a proton from an acid is termed the *conjugate base* of that acid (e.g., B_1 and B_2 are the conjugate bases of acids HB_1 and HB_2, respectively). The term *conjugate acid* has a comparable connotation.

An obvious advantage of the protonic approach is that it extends the Arrhenius approach to all protonic systems, irrespective of physical state or of the presence or absence of solvent. Thus the equations that follow all describe acid–base reactions

$$NH_4^+ + NH_2^- \rightleftharpoons NH_3 + NH_3 \quad \text{(in liquid ammonia)}$$

---

* The absolute accuracy of the assignment of names can be questioned. Brønsted's original publication is comprehensive and clearly defines an acid and a base. Lowry's publication deals with the "uniqueness of hydrogen" but does not define an acid or a base. However, Lowry recognized the importance of the proton and its transfer in both aqueous and nonaqueous media and the extent of proton transfer as a measure of base strength. Statements such as "the real function of a base is that of an acceptor of hydrogen nuclei" and "one of the simplest cases of neutralization is the acceptance of a hydrogen ion by a molecule of ammonia" suggest that even if Lowry did not publish definitions, he was a pioneer in developing the protonic approach. Often only Brønsted's name is used. Uniformly, however, the very substantial contributions of Brønsted's colleague, N. Bjerrum, are ignored.

$$HCl(g) + NH_3(g) \rightleftharpoons NH_4^+Cl^-(s)$$

$$2\,NH_4NO_3(l) + CaO(s) \overset{\Delta}{\longrightarrow} Ca(NO_3)_2(s) + 2\,NH_3(g) + H_2O(g)$$
$$\underset{\text{(acid)}}{} \qquad \underset{H^+}{\overset{}{}} \underset{\text{(base)}}{}$$

In every case the stronger acid and the stronger base of each conjugate pair react the more extensively, and the equilibrium is displaced accordingly.

The establishment of a uniform scale of acid strength (in terms of K_a or pK_a values) in aqueous systems involves quantitative evaluation of equilibria of the type

$$HB_1 + H_2O \rightleftharpoons H_3O^+ + B_1$$

where the solvent water is the reference base. Typical values are summarized in Table 9-5.

Acids may, of course, be molecular (e.g., HCl, H_2SO_4, HNO_3, H_2O), cationic [e.g., H_3O^+, NH_4^+, $Cr(H_2O)_6^{3+}$], or anionic (e.g., HCO_3^-, HSO_4^-, $H_2PO_4^-$). Similarly, all anions are bases, but there are also molecular (e.g., NH_3, H_2O, amines) and cationic [e.g., $Al(H_2O)_5(OH)^{2+}$, $Cu(H_2O)_3(OH)^+$] bases. That some substances may behave as either acids or bases is a function of the relative strength of the other substance present. Thus the water molecule may either accept or donate protons as

$$HNO_3 + H_2O \rightleftharpoons H_3O^+ + NO_3^-$$

$$\underset{\text{(acid)}}{H_2O} + \underset{\text{(base)}}{NH_3} \rightleftharpoons \underset{\text{(acid)}}{NH_4^+} + \underset{\text{(base)}}{OH^-}$$

since it is a stronger base than the NO_3^- ion and a weaker acid than the NH_4^+ ion.

Protonic solvents may be classified in terms of their relative abilities to donate or accept protons, as in Table 9-6. The effect of solvent type upon the acid–base character of a given solute may be profound. Thus urea behaves as an acid when dissolved in the strongly basic solvent anhydrous liquid ammonia,

$$(H_2N)_2CO + NH_3(l) \rightleftharpoons NH_4^+ + (H_2N)(HN)CO^-$$

but as a base when dissolved in the strongly acidic solvent anhydrous formic acid

$$HCOOH + (H_2N)_2CO \rightleftharpoons (H_3N)(H_2N)CO^+ + HCOO^-$$

It follows that the basic strength of a solute increases as the acid strength of the solvent increases and that the acidic strength of the solute increases as the basic strength of the solvent increases. Thus the choice of an appropriate solvent permits the quantitative titration of many acids and bases that are too weak to be determined titrametrically in aqueous solution.[44]

The immediate and essentially quantitative conversion by water of any acid that is stronger than the H_3O^+ ion* to that ion limits actual measurements of

* More properly $H^+(aq)$ or $H(H_2O)_n^+$ since such species as $H_5O_2^+$, $H_9O_4^+$, and so on, are also known [see H. L. Clever, J. Chem. Educ., **40**, 637 (1963); P. A. Giguère, J. Chem. Educ., **56**, 571 (1979)].

Table 9-5 Relative Strengths of Protonic Acids in Aqueous Solution[a]

Acid	Conjugate Base	pK_a	Conditions[b] Temperature (°C)	Ionic Strength (M)
HSO_3CF_3	$SO_3CF_3^-$	Negative		
$HClO_4$	ClO_4^-	Negative		
$HClO_3$	ClO_3^-	Negative		
$HBrO_3$	BrO_3^-	Negative		
HNO_3	NO_3^-	Negative		
HI	I^-	Negative		
HBr	Br^-	Negative		
HCl	Cl^-	Negative		
H_2SO_4	HSO_4^-	Negative		
H_2SeO_4	$HSeO_4^-$	Negative		
$H_5P_3O_{10}$	$H_4P_3O_{10}^-$	Negative		
H_3O^+	H_2O	-1.74		
H_2CrO_4	$HCrO_4^-$	-0.2		
$H_2S_2O_3$	$HS_2O_3^-$	0.6		
HIO_3	IO_3^-	0.77		
$H_4P_2O_7$	$H_3P_2O_7^-$	0.8		
$HNCS$	NCS^-	0.9		
HSO_3NH_2	$SO_3NH_2^-$	0.988		
$H_4P_3O_{10}^-$	$H_3P_3O_{10}^{2-}$	1.0	25	0.1
H_3PO_2	$H_2PO_2^-$	1.23		
H_3PO_3	$H_2PO_3^-$	1.5		
H_5IO_6	$H_4IO_6^- + IO_4^-$	1.58		
$HS_2O_3^-$	$S_2O_3^{2-}$	1.6		
$HSeO_4^-$	SeO_4^{2-}	1.70		
H_2SO_3	HSO_3^-	1.91		

HClO$_2$	ClO$_2^-$	1.95		
HSO$_4^-$	SO$_4^{2-}$	1.99		
HP$_3$O$_9^{2-}$	P$_3$O$_9^{3-}$	2.05		
H$_3$PO$_4$	H$_2$PO$_4^-$	2.148		
H$_3$P$_2$O$_7^-$	H$_2$P$_2$O$_7^{2-}$	2.2		
[Fe(H$_2$O)$_6$]$^{3+}$	[Fe(H$_2$O)$_5$(OH)]$^{2+}$	2.2	18	?
H$_3$AsO$_4$	H$_2$AsO$_4^-$	2.24		
H$_2$SeO$_3$	HSeO$_3^-$	2.27	25	1.0
H$_3$P$_3$O$_{10}^{2-}$	H$_2$P$_3$O$_{10}^{3-}$	2.5		
H$_2$Te	HTe$^-$	2.64	25	?
HNO$_2$	NO$_2^-$	3.15		
HF	F$^-$	3.17		
HOCN	OCN$^-$	3.48		
H$_2$Se	HSe$^-$	3.89		
[Cr(H$_2$O)$_6$]$^{3+}$	[Cr(H$_2$O)$_5$(OH)]$^{2+}$	4	25	?
HN$_3$	N$_3^-$	4.65		
CH$_3$CO$_2$H	CH$_3$CO$_2^-$	4.74	25	?
[Al(H$_2$O)$_6$]$^{3+}$	[Al(H$_2$O)$_5$(OH)]$^{2+}$	4.89	18	?
NH$_2$OH	NHOH$^-$	5.96	20	0.1
H$_2$S$_4$	HS$_4^-$	6.3		
H$_2$CO$_3$	HCO$_3^-$	6.352		
HCrO$_4^-$	CrO$_4^{2-}$	6.51		
H$_2$P$_2$O$_7^{2-}$	HP$_2$O$_7^{3-}$	6.70		
H$_2$PO$_3^-$	HPO$_3^{2-}$	6.79		
H$_2$AsO$_4^-$	HAsO$_4^{2-}$	6.96		
H$_2$S	HS$^-$	7.02		
HSO$_3^-$	SO$_3^{2-}$	7.18		
H$_2$N$_2$O$_2$	HN$_2$O$_2^-$	7.18		
H$_2$PO$_4^-$	HPO$_4^{2-}$	7.199		
HClO	ClO$^-$	7.53		
H$_6$TeO$_6$	H$_5$TeO$_6^-$	7.66		

589

Table 9-5 (*Continued*)

Acid	Conjugate Base	pK_a	Conditions[b] Temperature (°C)	Ionic Strength (M)
$HSeO_3^-$	SeO_3^{2-}	7.78	25	1.0
$N_2H_5^+$	N_2H_4	7.98	25	?
$[Cu(H_2O)_4]^{2+}$	$[Cu(H_2O)_3(OH)]^+$	8.00	25	1.0
$HSeO_3^-$	SeO_3^{2-}	8.05	25	0
H_2MoO_4	MoO_4^{2-}	8.24	20	0
$HBrO$	BrO^-	8.63	10–60	0
HCN	CN^-	9.21		
HBO_3^{2-}	BO_3^{3-}	9.236		
NH_4^+	NH_3	9.244		
$HP_3O_{10}^{4-}$	$P_3O_{10}^{5-}$	9.25		
$HP_2O_7^{3-}$	$P_2O_7^{4-}$	9.40		
$H_2P_3O_{10}^{3-}$	$HP_3O_{10}^{4-}$	9.54		
HCO_3^-	CO_3^{2-}	10.329		
HIO	IO^-	10.64		
$H_5TeO_6^-$	$H_4TeO_6^{2-}$	11.0		
H_2WO_4	WO_4^{2-}	11.30	25	1.0
$HAsO_4^{2-}$	AsO_4^{3-}	11.50		
$HN_2O_2^-$	$N_2O_2^{2-}$	11.54		
H_2O_2	HO_2^-	11.65		
HPO_4^{2-}	PO_4^{3-}	12.35		
$[Ca(H_2O)_x]^{2+}$	$[Ca(H_2O)_{x-1}(OH)]^+$	13	25	?
HS^-	S^{2-}	13.9		
H_2O	OH^-	13.997		
HSe^-	Se^{2-}	15.0	22	0
OH^-	O^{2-}	36	25	?

[a] Most values from R. M. Smith and A. E. Martell, *Critical Stability Constants*, Vol. 4, Plenum Press, New York (1976).

Table 9-6   Classification of Protonic Solvents

Solvent Type	General Tendency	Examples
Acidic	Donate protons	HF, H_2SO_4, HSO_3F, HCOOH, CH_3COOH, C_6H_5OH
Amphiprotic	Donate or accept protons	H_2O, alcohols
Basic	Accept protons	NH_3, NH_2OH, N_2H_4, amines
Aprotic	Neither donate nor accept protons	C_6H_6, $C_2H_4Cl_2$, $CHCl_3$

acid strength in aqueous solutions to acids weaker than H_3O^+ (Table 9-5). All acids for which pK_a values are more negative than for the H_3O^+ ion are indistinguishable in strength from each other in aqueous solution. Similarly, the acidic properties of water are such that all bases stronger than the OH^- ion are converted by water to that ion. Water thus has a leveling effect on acid and base strengths and is referred to as a *leveling solvent*. Solvents that have little or no tendency to donate or accept protons allow one to differentiate among the strengths of strong acids and strong bases and are referred to as *differentiating solvents*. The strengths of protonic acids in general are considered in further detail in a subsequent section of this chapter.

The disadvantages of the protonic definitions lie largely in their inability to:

1. Accommodate formally similar reactions that occur in nonprotonic systems: for example, the $BaO(s) + SO_3$(l or g) reaction or the titration in anhydrous liquid bromine(III) fluoride,[45] formulated as (Table 9-8)

$$BrF_2SbF_6 + AgBrF_4 \xrightarrow{\text{BrF}_3(\text{l})} AgSbF_6 + 2\,BrF_3$$

2. Relate acid–base behavior to neutralization.

Furthermore, the proton is considered to be a unique species. Although the small size of the proton results in a very large concentration of positive charge, the only difference from other cations is in degree as affected by this charge concentration.

Oxide-Ion Definitions.   The lack of applicability of protonic definitions to reactions of oxides at elevated temperatures, particularly in ceramic systems, led Lux[46] to define an *acid* as an *oxide-ion acceptor* and a *base* as an *oxide-ion donor*. These definitions are indicated and illustrated by the equations

$$base \rightleftharpoons acid + O^{2-}$$

$$CaO(s) \rightleftharpoons Ca^{2+} + O^{2-}$$

$$SO_4^{2-} \rightleftharpoons SO_3(\text{l or g}) + O^{2-}$$

in terms of which the overall reaction described as

$$CaO(s) + SO_3(l \text{ or } g) \rightleftharpoons Ca^{2+}SO_4^{2-}(s)$$

is a neutralization process. Flood and his coworkers,[47,48] in applying the Lux definitions to a number of high-temperature systems, emphasize the importance of ionization energy and distinguish three types of behavior:

1. Atoms of large ionization energy frequently participate in processes involving change in coordination number, for example,

$$\underset{(\text{C.N.}=3)}{SO_3^{2-}} + O^{2-} \rightleftharpoons \underset{(\text{C.N.}=4)}{SO_4^{2-}}$$

2. Atoms of intermediate ionization energy commonly undergo no change in coordination number, for example,

$$\underset{(\text{C.N.}=3)}{(B_2O_3)_n(s)} + 3n\,O^{2-} \rightleftharpoons \underset{(\text{C.N.}=3)}{2n\,BO_3^{3-}(s)}$$

    but conversion of the acid to the base involves the destruction of a highly polymeric structure.

3. Atoms of small ionization energy usually yield an oxide phase, for example,

$$Mg^{2+} + O^{2-} \rightleftharpoons MgO$$

Qualitative evaluation of equilibrium constants defined as

$$K = \frac{a_{\text{acid}}a_{O^{2-}}}{a_{\text{base}}} \tag{9-26}$$

using oxide-ion activities determined either potentiometrically or in terms of indicator equilibria such as

$$2\,\underset{\text{yellow}}{CrO_4^{2-}} \rightleftharpoons 2\,\underset{\text{green}}{Cr^{3+}} + 5\,O^{2-} + \tfrac{3}{2}O_2$$

show acid strength to change as

$$PO_3^- > BO_2^- \geqslant SiO_2 > TiO_3^{2-}$$

The limitations and applicability of these definitions are apparent. However, the syntheses of various anhydrous chlorides (e.g., $TiCl_4$, $SnCl_4$, $AsCl_3$, $SbCl_5$) by reaction of oxides with the molten $AlCl_3$–$NaCl$ eutectic (Sec. 10.2-2) are readily explained by reaction of $AlCl_3$, as a Lux–Flood acid, to give Al_2O_3.[48a]

Ionotropic Definitions. The proton transfer and oxide ion transfer definitions are specific cases of what one might refer to in general as ion-transfer, or *ionotropic*[49,50] definitions of acids and bases. In order to develop the definitions, let us consider reactions in anhydrous liquid hydrogen fluoride (Sec. 10.1-5), the autoionization of which may result from either proton transfer

$$\overset{\displaystyle H^+}{\overset{\displaystyle\frown}{HF} + \overset{\displaystyle}{HF}} \rightleftharpoons H_2F^+ + F^- \text{ (or } HF_2^-)$$
(base) (acid)

or fluoride ion transfer

$$\overset{\displaystyle F^-}{\overset{\displaystyle\frown}{HF} + \overset{\displaystyle}{HF}} \rightleftharpoons H^+ \text{ (or } H_2F^+) + HF_2^-$$
(base) (acid)

the net result being the same if solvation of the ions is considered. Experimentally, the concentration of the H_2F^+ ion can be increased by the addition of anhydrous $HClO_4$ or anhydrous SbF_5, as

$$\overset{\displaystyle H^+}{\overset{\displaystyle\frown}{HClO_4} + \overset{\displaystyle}{HF}} \rightleftharpoons H_2F^+ + ClO_4^-$$

$$\overset{\displaystyle F^-}{\overset{\displaystyle\frown}{SbF_5} + \overset{\displaystyle}{2\,HF}} \rightleftharpoons H_2F^+ + SbF_6^-$$

If the formation of the H_2F^+ ion is accepted as indicating the presence of an acid, as the protonic definition would require, then an acid in this solvent may be either a *proton donor* or an *anion acceptor* and a base either a *proton acceptor* or an *anion donor*.

More broadly, then, acid–base behavior is a manifestation of ion transfer in an appropriate solvent. Substances may be either *cationotropic* or *anionotropic*, depending on the type of ion transferred. In a cationotropic system, *acids* are defined as *cation donors* and *bases* as *cation acceptors*. In an anionotropic system, *acids* are defined as *anion acceptors* and *bases* as *anion donors*. Cationotropic systems appear to be limited to protonic systems, but many other anionotropic systems can be distinguished. For a given solvent, one can distinguish *solvo-acids* and *solvo-bases*, as indicated by the following examples:

1. In water (proton transfer)

$$HCl \quad + \quad H_2O \quad \rightleftharpoons H_3O^+ + Cl^-$$
solvo-acid
(proton donor)

$$H_2O \quad + \quad NH_3 \quad \rightleftharpoons NH_4^+ + OH^-$$
solvo-base
(proton acceptor)

2. In bromine(III) fluoride (fluoride-ion transfer)

$$VF_5 \quad + \quad BrF_3 \quad \rightleftharpoons BrF_2^+ + VF_6^-$$
solvo-acid
(fluoride acceptor)

$$BrF_3 \quad + \quad \underset{\substack{\text{solvo-base} \\ \text{(fluoride donor)}}}{CsF} \quad \rightleftharpoons Cs^+ + BrF_4^-$$

The general aspects of ionotropism for a number of solvents are summarized in Table 9-7.

The ionotropic approach is advantageous in relating proton transfer and oxide ion transfer to each other, in fitting these definitions into a broadened one, and in providing a logical bridge to the solvent-system, Lewis, and Usanovich definitions that follow. It is, of course, strictly limited in direct applicability to solvents that are capable of autoionization.

Solvent-System Definitions. It is impossible to ascribe the definitions used in this approach to any one person or even to a specific group of persons, although the fundamental ideas may well be traceable to E. C. Franklin, who recognized the acidic properties of ammonium salts and the basic properties of metal amides, imides, and nitrides, when dissolved in anhydrous liquid ammonia, and incorporated this information into his *ammono system* of acids, bases, and salts (Sec. 10.1-3).[51,52] Similar considerations, often primarily by analogy, showed that acid–base behavior can characterize a variety of other anhydrous protonic solvents (e.g., N_2H_4, NH_2OH, CH_3COOH, HCN, H_2SO_4, HSO_3F) and nonprotonic solvents (e.g., SO_2, $SeOCl_2$, BrF_3, IF_5, ICl, I_2, $AsCl_3$, $POCl_3$, N_2O_4) that undergo measurable autoionization (Ch. 10). A number of these are indicated in Table 9-7, but it must be understood that no claim can be made that the solvated or unsolvated formulations given are completely accurate.

These and related considerations have led to several definitions for acids and bases,[53–55] which may be combined to admit as an *acid* any substance that yields, either by direct dissociation or by interaction with the solvent, the *cation characteristic of that solvent*, and as a *base* any substance that yields, either by direct dissociation or by interaction with the solvent, the *anion characteristic of the solvent*. Within these definitions, it is possible to distinguish both donor and acceptor acids and bases as[56]

1. *Donor acid* — a substance that can split off solvent cations or can unite with solvent molecules to form cations.
2. *Acceptor acid* — a substance that can combine with solvent anions.
3. *Donor base* — a substance that can split off solvent anions or can unite with solvent molecules to form anions.
4. *Acceptor base* — a substance that can combine with solvent cations.

It is apparent that acids may either be derived from the solvent or bear no relationship to the solvent except in terms of the cation. Thus hydrazoic acid (HN_3) is a derivative of ammonia and a true ammono acid whereas nitric acid (HNO_3) is not a derivative of ammonia and thus an aquo acid. Yet both of these compounds behave as acids when dissolved in liquid ammonia. The same is true

Table 9-7 Ionotropism in Various Solvents

Type of Solvent	Solvent	$Acid_1$	+	Auto-ionization Process[a] $Base_2 \rightleftharpoons Acid_2$	+	$Base_1$	Ion Transferred
Cationotropic							
Prototropic	H_2O	H_2O	+ H_2O	H_3O^+	+ OH^-		H^+
	NH_3	NH_3	+ NH_3	NH_4^+	+ NH_2^-		H^+
	$HC_2H_3O_2$	$HC_2H_3O_2$	+ $HC_2H_3O_2$	$H_2C_2H_3O_2^+$	+ $C_2H_3O_2^-$		H^+
	NH_2OH	NH_2OH	+ NH_2OH	NH_3OH^+	+ $NHOH^-$		H^+
Anionotropic							
Fluoridotropic	BrF_3	BrF_3	+ BrF_3	BrF_2^+	+ BrF_4^-		F^-
	IF_5	IF_5	+ IF_5	IF_4^+	+ IF_6^-		F^-
Chloridotropic	$COCl_2$	$COCl_2$	+ $COCl_2$	$COCl^+$	+ $COCl_3^-$		Cl^-
	$POCl_3$	$POCl_3$	+ $POCl_3$	$POCl_2^+$	+ $POCl_4^-$		Cl^-
	ICl	ICl	+ ICl	I^+	+ ICl_2^-		Cl^-
Bromidotropic	IBr	IBr	+ IBr	I^+	+ IBr_2^-		Br^-
	$HgBr_2(l)$	$HgBr_2$	+ $HgBr_2$	$HgBr^+$	+ $HgBr_3^-$		Br^-
Iodidotropic	$I_2(l)$	I_2	+ I_2	I^+	+ I_3^-		I^-
Oxidotropic	SO_2	SO_2	+ SO_2	SO^{2+}	+ SO_3^{2-}		O^{2-}
	N_2O_4	N_2O_4	+ N_2O_4	$2 NO^+$	+ $2 NO_3^-$		O^{2-}

[a] Solvation of ions generally ignored.

of bases but to a lesser degree. A salt may be regarded as any substance that gives in a particular solvent solutions with larger electrical conductivity than the pure solvent and yields at least one ion different from those characteristic of the solvent. Neutralization amounts to a combination of the solvent anion with the solvent cation to produce the solvent molecule and does not emphasize, although it involves, salt formation. Some examples are included in Table 9-8.

Protonic systems are handled without problem, but many of the applications to nonprotonic systems have been based only on analogy.[57]

Although liquid sulfur(IV) oxide was apparently suggested as a medium for acid–base reactions by Cady and Elsey,[55] much of the earlier interpretative work was carried out by G. Jander and his associates.[57,58] The conductimetric titration of cesium sulfite with thionyl chloride in liquid sulfur(IV) oxide to a 1 : 1 end point (Table 9-8) was interpreted, by analogy, in terms of combination of the solvent ions SO^{2+} and SO_3^{2-}. On this basis, and again by analogy to aqueous or liquid ammonia chemistry, the experimental observations that treatment of a solution of aluminum chloride in liquid sulfur(IV) oxide with tetramethylammonium sulfite precipitates aluminum sulfite, which dissolves when an excess of the sulfite is added and then reprecipitates when thionyl chloride is added, were interpreted as a case of amphoterism as

$$2\,Al^{3+} + 3\,SO_3^{2-} \rightarrow Al_2(SO_3)_3(s)$$
$$\text{(base)}$$

$$Al_2(SO_3)_3(s) + 3\,SO_3^{2-} \rightarrow 2\,[Al(SO_3)_3]^{3-}$$
$$\text{(base)}$$

$$2\,[Al(SO_3)_3]^{3-} + 3\,SO^{2+} \rightarrow Al_2(SO_3)_3(s) + 6\,SO_2(l)$$
$$\text{(acid)}$$

However, exchange studies using isotopically labeled substances[59-64] raise questions about these interpretations. Thus the fact that neither sulfur nor oxygen exchanges no more than extremely slowly between liquid sulfur(IV) oxide and either thionyl chloride or thionyl bromide suggests that the SO^{2+} ion is an unlikely species. The facts that ionic halides catalyze the exchange of sulfur between thionyl bromide and sulfur(IV) oxide and that chlorine exchanges rapidly between tetramethylammonium chloride and thionyl chloride in liquid sulfur(IV) oxide suggest strongly that the acidic species in these solutions are the $SOCl^+$ and $SOBr^+$ ions. That tetramethylammonium pyrosulfite, $[(CH_3)_4N]_2S_2O_5$, exchanges sulfur rapidly with liquid sulfur(IV) oxide does support, however, the presence of the SO_3^{2-} as a base. Oxide ion transfer is thus a more reasonable interpretation of acid–base behavior.[60,64]

On the other hand, the observation of rapid exchange of chlorine between a dissolved chloride and liquid $SeOCl_2$[63] is in agreement with the autoionization equilibrium

$$SeOCl_2(l) \rightleftharpoons SeOCl^+ + Cl^-$$

proposed by G. B. L. Smith[65] to account for his finding that in this solvent the

Table 9-8 Neutralization Reactions in Various Solvents

Solvent	Acid Ion	Base Ion	Typical Neutralization Equation				
			Acid	+ Base	→ Salt	+	Solvent
H_2O	$H_3O^+(H^+)$	OH^-	HBr	KOH	KBr		H_2O
NH_3	$NH_4^+(H^+)$	NH_2^-	NH_4Cl	KNH_2	KCl		$2\,NH_3$
NH_2OH	$NH_3OH^+(H^+)$	$NHOH^-$	$(NH_3OH)Cl$	NaNHOH	NaCl		$2\,NH_2OH$
CH_3COOH	$CH_3CO_2H_2^+(H^+)$	$CH_3CO_2^-$	$HClO_4$	$Na(CH_3CO_2)$	$NaClO_4$		CH_3COOH
SO_2	SO^{2+}	SO_3^{2-}	$SOCl_2$	Cs_2SO_3	$2\,CsCl$		$2\,SO_2$
$SeOCl_2$	$SeOCl^+$	Cl^-	$(SeOCl)_2SnCl_6$	$2\,KCl$	K_2SnCl_6		$2\,SeOCl_2$
N_2O_4	NO^+	NO_3^-	NOCl	$AgNO_3$	AgCl		N_2O_4
BrF_3	BrF_2^+	BrF_4^-	$(BrF_2)(SbF_6)$	$Ag(BrF_4)$	$AgSbF_6$		$2\,BrF_3$

neutralization of the acid tin(IV) chloride with bases such as potassium or calcium chloride can be followed conductimetrically.

The dissociation of liquid N_2O_4 as

$$N_2O_4(l) \rightleftharpoons \underset{\text{(acid)}}{NO^+} + \underset{\text{(base)}}{NO_3^-}$$

is well supported chemically by observations[66-70] that nitrosyl chloride (NOCl) is neutralized in this solvent by silver nitrate; that solutions of nitrosyl chloride when treated with active metals (e.g., Zn, Fe, Sn) liberate nitrogen(II) oxide and form either insoluble chlorides or chloro complexes, $(NO)_x[MCl_y]$; that solutions of diethylammonium nitrate in this solvent react with elemental zinc to liberate nitrogen(II) oxide and form $[(C_2H_5)_2NH_2]_2[Zn(NO_3)_4]$; and that strongly basic metal oxides are converted to anhydrous nitrates by reactions with the solvent that can be formulated in general as[69, 70]

$$O^{2-} + 2\,NO^+NO_3^-(l) \rightarrow \underset{\text{(i.e., } N_2O_3)}{(NO)_2O} + 2\,NO_3^-$$

Similarly, the autodissociation of liquid BrF_3

$$2\,BrF_3(l) \rightleftharpoons BrF_2^+ + BrF_4^-$$

is chemically supported by the large specific conductance of the pure liquid (8.0×10^{-3} ohm^{-1} cm^{-1} at $25°C$),[71] the increase in conductance by addition of compounds such as BrF_2SbF_6, $(BrF_2)_2SnF_6$, $KBrF_4$, and $AgBrF_4$,[71] and the neutralization of BrF_2SbF_6 with $AgBrF_4$, as shown in Table 9-8.[45]

The observation that in anhydrous liquid $POCl_3$ tetramethylammonium chloride and iron(III) chloride react to yield $[(CH_3)_4N][FeCl_4]$ is accounted for in terms of chloride-ion transfer (Table 9-7)[50] and the solvent-systems approach as

$$(CH_3)_4NCl \xrightarrow{\text{POCl}_3(l)} (CH_3)_4N^+ + Cl^-$$

$$FeCl_3 + POCl_3 \rightleftharpoons POCl_2^+ + FeCl_4^-$$

$$\underset{\text{(acid)}}{POCl_2^+} + \underset{\text{(base)}}{Cl^-} \rightleftharpoons POCl_3$$

Inasmuch as exactly the same overall reaction occurs in liquid triethylphosphate, $PO(OC_2H_5)_3$, a nonchloride solvent of comparable molecular structure and dielectric constant, it has been suggested that the reaction is better explained in both media as

$$FeCl_3 + Cl^- \rightleftharpoons FeCl_4^-$$

(i.e., an electron-pair donor–acceptor reaction).[72]

Thus it is apparent that although analogy is often useful and frequently predictive of overall reactions, correct explanations of observed acid–base behavior in nonprotonic media must be based upon experimental data. The solvent-systems definitions have been very important in elucidating the chemistries of many nonaqueous systems, but alternative explanations are available.[73,73a]

Electron-Pair Donor–Acceptor Definitions. The fundamentals of these definitions were first stated by G. N. Lewis in 1923,[74] but it was not until 1938 that they were emphasized.[41] More recently, the approach has been extensively reviewed.[73,75,76] An *acid* is defined as any molecule, radical, or ion in which an atom, because of the presence of an incomplete electronic grouping, can *accept* one or more *electron pairs* (i.e., as an *electron-pair acceptor*). Correspondingly, a *base* is defined as a species that has within its molecular structure an atom(s) that is(are) capable of *donating* an *electron pair(s)* (i.e., as an *electron-pair donor*). Acids and bases thus exemplify the electron-pair acceptors and donors suggested by N. V. Sidgwick[77] to account for the formation of complex ions and coordination compounds (Sec. 7.1-3). *Neutralization* then amounts to the formation of a coordinate covalent bond, with the resulting product undergoing rearrangement, dissociation, or no change, as dictated by the requirements of thermodynamic and kinetic stability.

It is apparent from the examples summarized in Table 9-9 that acids or bases defined by any and all of the approaches previously discussed are similarly defined by the Lewis approach. In addition, every case of coordination that can be described in terms of electron-pair interaction is an example of acid–base interaction. Significantly, in terms of the Lewis definitions, acid–base behavior is

Table 9-9   Acids, Bases, and Neutralization Products in Terms of the Lewis Definitions

Acid	Base	Neutralization Product(s)
HNO_3	H_2O	$H_2O: \dashrightarrow HNO_3 \rightarrow H_3O^+ + NO_3^-$
H_2O	NH_3	$H_3N: \dashrightarrow HOH \rightarrow NH_4^+ + OH^-$
$NH_4NO_3(l)$	CaO	$:\ddot{O}:^{2-} \dashrightarrow (HNH_3)^+ \Big\}$ $\left.\begin{array}{c} \downarrow \\ (HNH_3)^+ \end{array}\right\} \rightarrow H_2O + 2NH_3$
NH_4^+	NH_2^-	$H_2\ddot{N}:^- \dashrightarrow (HNH_3)^+ \rightarrow NH_3 + NH_3$
H_3O^+	OH^-	$HO^-: \dashrightarrow HOH_2^+ \rightarrow H_2O + H_2O$
SO_3	O^{2-}	$:\ddot{O}:^{2-} \dashrightarrow SO_3 \rightarrow SO_4^{2-}$
$HClO_4$	HF	$HF: \dashrightarrow HClO_4 \rightarrow H_2F^+ + ClO_4^-$
BrF_3	CsF	$CsF: \dashrightarrow BrF_3 \rightarrow Cs^+ + BrF_4^-$
$FeCl_3$	Cl^-	$:\ddot{Cl}:^- \dashrightarrow FeCl_3 \rightarrow FeCl_4^-$
H_2O	CN^-	$NC:^- \dashrightarrow HOH \rightarrow HCN + OH^-$
Cu^{2+}	NH_3	$(H_3N: \dashrightarrow)_4Cu^{2+} \rightarrow Cu(NH_3)_4^{2+}$
BF_3	$(C_2H_5)_2O$	$(C_2H_5)_2O: \dashrightarrow BF_3 \rightarrow (C_2H_5)_2O:BF_3$
Co^{3+}	Cl^-, NH_3	$(H_3N: \dashrightarrow)_5Co^{3+} \dashleftarrow :\ddot{Cl}:^- \rightarrow Co(NH_3)_5Cl^{2+}$

made independent of any one element, of any specific combination of elements, of the presence or absence of ions, or of the presence or absence of a solvent.

That the Lewis definitions represent interpretations of experimental data rather than conceptual deduction is evident in terms of the Lewis "phenomenological criteria."[41] From the many examples and illustrations that have been given,[75, 78, 79] the following may be cited as both representative and typical:

1. **Rapidity of neutralization.** Well-known reactions such as those of oxonium and hydroxide ions, gaseous hydrogen chloride and ammonia, nickel(II) ion with aqueous ammonia, and boron(III) chloride with triethylamine are so rapid as to support Lewis's view that the energy of activation of a neutralization process must be very small.

2. **Displacement of a weaker acid or base.** The hydrated proton displaces the copper(II) ion from the tetraamminecopper(II) ion because it is a stronger acid than the copper(II) ion.

3. **Titration, using indicators.** Methyl violet serves as an indicator for titrations of the acid HCl with the base NH_3 in aqueous medium, of the acid BCl_3 with the base phthalic anhydride in chlorobenzene, and of the acid Al_2Cl_6 with the base pyridine in chloroform, the color change in each case being from yellow in the presence of excess acid to violet in the presence of excess base.

4. **Catalysis.** The well-known catalytic reactions summarized in Table 9-10 are systematized in terms of the Lewis definitions, which allow the formation of metastable intermediates via electron-pair donation or acceptance.[75]

The Lewis definitions are obviously highly advantageous and widely applicable. Unfortunately, they admit the common protonic acids only through the formalism indicated in Table 9-9 where the hydrogen bond in the intermediate is equated to electron-pair donation to the already bonded hydrogen atom. Perhaps a more serious disadvantage is the absence of a uniform scale of acid or base strengths. Instead, these are made variable by dependence upon the reaction or method used for their evaluation. One technique of measurement is to equate acid or base strength with enthalpy of reaction using a reference acid or base, the enthalpies being evaluated calorimetrically from the temperature coefficients of equilibrium constants or from alterations in infrared spectra.[80-82] In terms of enthalpy changes, the strengths of Lewis acids and bases have been related as[83]

$$-\Delta H = E_A E_B + C_A C_B \qquad (9\text{-}27)$$

where E_A and E_B are parameters of the acid and the base, respectively, that are related to the susceptibilities of the species to interact electrostatically, and C_A and C_B are additional parameters related to the susceptibility of the species to interact covalently. The references cited, including Ref. 76, should be consulted for details and numerical results. The further disadvantage that salt formation

Table 9-10   Acids and Bases as Catalysts in the Lewis Sense

Reaction Type	Typical Catalysts
Acid-catalyzed reactions	
Alkylation (Friedel–Crafts)	BF_3, $AlCl_3$, $GaCl_3$, H_2SO_4, P_4O_{10}, HF
Acylation (Friedel–Crafts)	$AlCl_3$, $ZnCl_2$, HF
Esterification of aldehydes (Cannizzaro)	$Al(OR)_3$
Condensation of carbonyls (semicarbazone)	HX
Halogenation	$FeBr_3$, $AgClO_4$, HCl, $SbCl_5$, $ZnCl_2$, $SnCl_4$
Hydrolysis	Hg^{2+}
Sulfonation, nitration	BF_3, HF
Cyclic ketone formation	$AlCl_3$, $SnCl_4$, HF, H_2SO_4, P_4O_{10}
Base-catalyzed reactions	
Self-condensation	
Aldol	$CH_3CO_2^-$, CO_3^{2-}, pyridine, amines
Aldolization	OH^-
Claisen	$OC_2H_5^-$, $(C_6H_5)_3C^-$
Condensation with carbonyls	
Perkin	$CH_3CO_2^-$, CO_3^{2-}, SO_3^{2-}, PO_4^{3-}, $(C_2H_5)_3N$, C_5H_5N
Knoevenagel	NH_3
Cannizzaro	OH^-
Benzoin condensation	CN^-

may or may not be related to acid–base behavior also characterizes some of the other definitions.*

The Usanovich Definitions.   In an apparent attempt to include all recognized aspects of acid–base chemistry, M. Usanovich[84,85] defined an *acid* as any substance that *forms salts with bases via neutralization, gives up cations,* or *combines with anions* or *electrons.* Similarly, he defined a *base* as any substance that *neutralizes an acid, gives up anions* or *electrons,* or *combines with cations.* Salt formation is an important consideration, as indicated in Table 9-11.

All substances definable as acids and bases in terms of the other approaches that have been discussed are correctly classified by the Usanovich definitions. In addition, and contrary to our practice in this chapter, redox reactions are also included as acid–base reactions. Although it is advantageous to make this

---

* For very extensive and detailed treatments of the Lewis definitions the following references are recommended: W. B. Jensen, *Chem. Rev.*, **78**, 1 (1978), with 156 literature citations; W. B. Jensen, *The Lewis Acid–Base Concepts: An Overview*, John Wiley & Sons, New York (1980); D. P. N. Satchell and R. S. Satchell, *Q. Rev. (Lond.)*, **25**, 171 (1971).

Table 9-11   Acid–Base Relationships in the Usanovich Sense

Acid	+ Base	→	Salt	Justification
SO_3	Na_2O		$Na_2^+SO_4^{2-}$	Na_2O yields O^{2-} ion
				SO_3 combines with O^{2-} ion
$Fe(CN)_2$	KCN		$K_4^+[Fe(CN)_6]^{4-}$	KCN yields CN^- ion
				$Fe(CN)_2$ combines with CN^- ion
As_2S_5	$(NH_4)_2S$		$(NH_4)_3^+[AsS_4]^{3-}$	$(NH_4)_2S$ donates S^{2-} ion
				As_2S_5 adds S^{2-} ion
Cl_2	K		K^+Cl^-	K loses an electron
				Cl gains an electron

distinction, it is not always clearly possible to do so. Thus the formation of pyridine-*N*-oxide by oxidation of pyridine

$$\bigcirc N + 1/2\,O_2 \longrightarrow \bigcirc NO$$

may also be formulated as a combination of the base pyridine with atomic oxygen as the acid[86]

$$\bigcirc N{:} + \overset{..}{\underset{..}{O}}{:} \longrightarrow \bigcirc N{:}\overset{..}{\underset{..}{O}}{:}$$

(base)   (acid)

The Usanovich definitions stress primarily the positive and negative traits of acids and bases and, to a lesser extent, coordinate unsaturation. The acidic function is determined by the presence of coordinately unsaturated positive particles; the basic function by the presence of coordinately saturated negative particles. Although in general acidity is favored by highly charged positive particles (e.g., Al_2O_3 vs. Na_2O) and basicity by highly charged negative particles (e.g., Na_2O vs. NaCl) an exact balancing of both ionic size and charge is essential to explain a number of observed trends (e.g., increased basicity in the series BeO–RaO). The Usanovich definition is particularly advantageous in correctly classifying together all examples of acidic and basic behavior, but it suffers from its extreme generality of literally bringing together under one heading all the reactions of inorganic chemistry. Many inorganic chemists do now, and will continue to, prefer to consider various reactions from other points of view.

6-2.   Acid–Base Definitions in Summary

Several definitions of acids and bases have been presented and discussed in the general order of increasing comprehensiveness. It is only natural for one to ask whether these definitions represent conflicting points of view and what one should really know about approaches to acid–base behavior. Of course, each set of definitions is correct within its own framework, and contrary to unfortunate presentations in the literature, there is no conflict among these definitions. Proper breadth dictates that one should be conversant with each point of view and should apply whatever approach seems to be best suited to the problem he is considering. No single set of definitions is necessarily the most useful for all circumstances. Each has its own realm of specialized applicability. Knowledge and utilization of the fundamentals of all the definitions are essential.

Inasmuch as the general ideas of *donating a positive species* or *accepting a negative species* characterize all definitions of an *acid* and of *donating a negative species* or *accepting a positive species* characterize all definitions of *bases*, one can combine all the acid–base definitions into an all-encompassing statement. Thus *acidity* refers to the "*positive character of a chemical species which is decreased by reaction with a base*," and *basicity* refers to the "*negative character of a chemical species which is decreased by reaction with an acid*."[86]

6-3. Trends in the Strengths of Protonic Acids in Aqueous Media

The acids listed in Table 9-5 are of three general types: hydro acids (e.g., HCl, H_2S, HS^-), oxo acids (e.g., $HClO_4$, HSO_4^-, $H_2PO_4^-$), and solvated cations [e.g., $Al(H_2O)_x^{3+}$, NH_4^+]. Within each of these types significant trends in strength are discernable in terms of pK_a values.

Among the neutral hydro acids that characterize the elements of a particular periodic family, there is a general increase in acid strength with increasing formula weight or covalent radius of the element other than hydrogen. This trend appears, at first glance, to violate what might be expected in terms of decrease in electronegativity of an atom of that element and decrease in polarity of its bond to a hydrogen atom (Sec. 5.1-8). Thus in the series $H_2O–H_2S–H_2Se–H_2Te$, the last compound is both the strongest acid and the least polar substance. However, the strength of a protonic acid is measured by the ease with which the proton can be removed from the atom to which it is bonded. The weakening of this bond due to increased covalent radius of the nonmetal atom more than compensates for decrease in polarity and results in increased acidity.

For the hydrogen halides (HX), this situation can be illustrated via application of the Born–Fajans–Haber cyclic treatment[87, 87a] as

$$
\begin{array}{ccc}
HX(g) & \xrightarrow{\Delta H_{diss}} & H(g) \quad + X(g) \\
\downarrow{\scriptstyle \Delta H} & \quad\downarrow{\scriptstyle \Delta H_{ion}} & \quad\downarrow{\scriptstyle \Delta H_{aff}} \\
H^+(aq) + X^-(aq) & \xleftarrow{\Delta H_{hyd}} H^+(g) & + X^-(g)
\end{array}
$$

where the ΔH of electrolytic dissociation ($\sim pK_a$) is determined by summation of the enthalpy changes of dissociation of HX(g) (ΔH_{diss}), of ionization of H(g) (ΔH_{ion}) and X(g) (ΔH_{aff}), and of hydration of $H^+(g)$ and $X^-(g)$ (ΔH_{hyd}). In Pauling's terminology,[88] partial ionic character so stabilizes the covalent bond that the anion-forming tendencies of the most electronegative atoms are overcome. Thus in aqueous solution HF is a relatively weak acid ($pK_a = 3.17$), whereas HCl, HBr, and HI are all so strong (pK_a = negative) that differences among them cannot be distinguished. On the other hand, in any horizontal series of the periodic table, acid strength increases with the electronegativity of the nonmetal atom (e.g., $H_4C < H_3N < H_2O < HF$). Here changes in covalent radii are too small to be effective.*

Among the oxo acids, strength is determined by both the covalent radius and the electronegativity of the central atom in the molecular structure, although the latter appears to be the more important. Within each molecular structure, the grouping (E = any element)

$$: \overset{..}{\underset{..}{E}} : \overset{..}{\underset{..}{O}} : H$$

appears. The smaller and the more electronegative atom E is, the more strongly electrons in its outermost shell are attracted, and the more polar the O—H bond becomes. When atom E is small and highly electronegative, the proton is readily removed, and the compound is acidic, but as the size of atom E increases and its electronegativity decreases, the grouping becomes amphiprotic and ultimately basic (OH^- ion forms). According to Gallais,[89] basic character is dominant when the electronegativity of E (Pauling scale) is less than 1.7; acidic character when it exceeds 1.7. The presence of additional oxygen atoms bonded to atom E will, of course, increase acidity.

These principles are in agreement with the following observations:

1. ClOH is acidic; NaOH is strongly basic.
2. Acidity increases markedly in the series $HOCl–HOClO–HOClO_2–HOClO_3$.
3. H_2SeO_3 is a stronger acid than H_2SO_3.
4. Among the O—H compounds of manganese, we have $Mn(OH)_2$—basic, MnO_2—amphiprotic, $MnO(OH)_3$—acidic, $MnO_2(OH)_2$—strongly acidic, $MnO_3(OH)$—very strongly acidic.

Many years ago, L. Pauling[90] pointed out that the strengths of oxo acids in aqueous solution can be approximated in terms of two empirical rules:

1. For a given polyprotic acid, the successive constants $pK_{a_1}, pK_{a_2}, pK_{a_3}$ stand in the ratio 0 : 5 : 10 (see Table 9-5).

---

*   To eliminate the effects of solvation, comparisons can be made in terms of *proton affinities*, that is, changes in enthalpy for reactions described by $B(g) + H^+(g) \rightarrow HB^+$, which values reflect qualitatively expectations based upon inductive effects and charges. See J. E. Huheey, Ref. 86, pp. 271–272.

2. The magnitude of K_{a_1} is dependent upon the value of m in the formulation $EO_m(OH)_n$. If $m = O$ [e.g., $B(OH)_3$], the acid is very weak ($pK_a \geqslant 7$); if $m = 1$ [e.g., $ClO(OH)$], the acid is weak ($pK_a \simeq 2$); if $m = 2$ [e.g., $ClO_2(OH)$], the acid is strong ($pK_a \simeq -3$); and if $m = 3$ [e.g., $ClO_3(OH)$], the acid is very strong ($pK_a \simeq -8$).

These rules are invaluable as approximations. Pauling's second rule is in close agreement with J. E. Ricci's generalization[91]

$$pK_a = 8.0 - 9.0a + 4.0m \qquad (9\text{-}28)$$

where a is the formal charge (Sec. 5.1) on atom E and m is the Pauling number. Values calculated from Eq. 9-28 agree closely with those listed in Table 9-5.

For oxo acids in the molecular structures of which hydrogen atoms are bonded to central nonmetal atoms, the formal charge borne by the central atom may serve as a better indicator of strength in aqueous solution than the oxidation number. The mono acids of phosphorus are cases in point, as indicated in the tabulation

molecular formulation	H_3PO_2	H_3PO_3	H_3PO_4
electronic formulation	$\begin{matrix} H \\ H:P:O:H \\ :O: \end{matrix}$	$\begin{matrix} H \\ :O:P:O:H \\ :O: \\ H \end{matrix}$	$\begin{matrix} :O: \\ H:O:P:O:H \\ :O: \\ H \end{matrix}$
oxidation number of P	+1	+3	+5
formal charge of P	1+	1+	1+
pK_a	1.23	1.5	2.148

Thus near constancy in pK_a values parallels constancy in formal charge.

It is of interest that the polyphosphoric acids are substantially stronger than monophosphoric acids (pK_a = negative for $H_5P_3O_{10}$; 0.8 for $H_4P_2O_7$; 2.148 for H_3PO_4). Isopolyacids (Sec. 10.4-1) of other elements are too extensively hydrolyzed in aqueous solution to establish this observation as a general one, but Gillespie's calculations[92] for the sulfuric acids ($pK_a = -12$ or -13 for $H_2S_2O_7$; -2 or -3 for H_2SO_4) suggest that it may be. Indeed in anhydrous sulfuric acid medium, disulfuric acid (pK_a = ca. 2) is much stronger than perchloric acid (pK_a = 4). Enhanced strength of the poly oxo acids may be attributed to the large inductive effect of the acyl-type group that replaces a hydrogen atom in the parent oxo acid.

Acid strength of hydrated cations reflects displacement of equilibria of the type

$$[M(H_2O)_x]^{n+}(aq) + H_2O(l) \rightleftharpoons [M(H_2O)_{x-1}(OH)]^{n-1}(aq) + H_3O^+(aq)$$

The smaller the cation M^{n+} and the larger its charge ($n+$), the more polar the

O—H bond in a solvating water molecule, and the more extensively a proton is released to the solvent. Thus (Table 9-5), acid strength decreases as

$$[Fe(H_2O)_x]^{3+} > [Cr(H_2O)_x]^{3+} > [Al(H_2O)_x]^{3+} > Zn[(H_2O)_x]^{2+}$$
$$> Ca[(H_2O)_x]^{2+}$$

6-4. Hard and Soft Acids and Bases — The HSAB Classification

Some of the problems relating to the quantitative evaluation of acid strength and base strength in terms of the Lewis definitions (Sec. 9.6-3)[93] have been approached through the formation of coordination species. Thus comparisons of the thermodynamic stabilities of complexes, with respect to dissociation into their components, have revealed the following regularities:*

1. *The Irving–Williams stability series,*[94, 94a] based upon the observation that for a variety of ligands stabilities of complexes derived from dipositive ions, as indicated by formation constants (Table 11-1) for aqueous systems, vary as

    $$Ba^{2+} < Sr^{2+} < Ca^{2+} < Mg^{2+} < Mn^{2+} < Fe^{2+} < Co^{2+} < Ni^{2+}$$
    $$< Cu^{2+} > Zn^{2+}$$

    These variations are in general accord with changing cationic radius and changing ligand-field interactions (Sec. 8.1-3). Comparable trends are less regular for terpositive ions.

2. *The A–B acceptor classification,*[95–97a] based upon the observation that ligands containing certain donor atoms form their most stable complexes with cations such as Al^{3+}, Ln^{3+}, Co^{3+}, and Th^{4+} and their least stable complexes with cations such as Cu^+, Hg^{2+}, and Pt^{2+}, whereas for ligands containing other donor groups essentially the reverse is true (Sec. 7.3). On this basis cations have been classified as

    (a) *Type A cations*, which in their most stable oxidation states give complexes the thermodynamic stabilities of which *decrease* as the donor atom changes in the following series†:

    $$R_3N \ggg R_3P > R_3As > R_3Sb$$
    $$R_2O \ggg R_2S > R_2Se > R_2Te$$
    $$F^- > Cl^- > Br^- > I^-$$

    Type A cations include the noble gas (d^0) ions of the periodic group IA (Li^+–Cs^+), IIA (Be^{2+}–Ra^{2+}), IIIA (Al^{3+}–Tl^{3+}), and IIIB (Sc^{3+}–Ac^{3+}, Ln^{3+}) families, the actinide ions, and the lighter *d*-transition metal ions

---

*   See also Chapter 7, pp. 374–376.
†   R = H, alkyl, etc.

in higher oxidation states (e.g., Ti^{4+}, Cr^{3+}, Fe^{3+}, Co^{3+}). Ligands that bond preferentially to these cations are termed *Type A ligands*.

(b)   *Type B cations*, which give complexes the thermodynamic stabilities of which change in roughly the *reverse order* as the donor atom changes, as

$$R_3N \lll R_3P > R_3As > R_3Sb$$

$$R_2O \ll < R_2S < R_2Se \sim R_2Te$$

$$F^- \; < \; Cl^- \; < Br^- \; < I^-$$

Type B cations include transition metal ions in lower oxidation states (e.g., Pd^{2+}, Pt^{2+}, Cu^+, Ag^+, Hg_2^{2+}) and in general the heavier *d*-transition metal ions. Ligands that bond preferentially to these cations are termed *Type B ligands*.

More recently, R. G. Pearson[98][100] has broadened these somewhat empirical classifications into an overall approach to Lewis acid–base interactions based upon so-called *hard* and *soft* acids and bases. *Hard bases* are those substances that hold their electrons tightly as a consequence of the large electronegativities, low polarizabilities, and difficulty of oxidation of their donor atoms. *Soft bases* are then those substances that hold their electrons more loosely as a consequence of the decreased electronegativities, enhanced polarizabilities, and enhanced ease of oxidation of their donor atoms. *Hard acids* are those acids that react preferentially with *hard bases*, whereas *soft acids* are those that react preferentially with *soft bases*. The general principle of the HSAB approach is indicated by the preceding sentence, namely that hard reacts best with hard and soft reacts best with soft. To accommodate those substances that are not clearly hard or soft, Pearson also uses the designation *borderline*.*

Tables 9-12 and 9-13 classify, respectively, bases and acids under these headings. It is apparent that in general the hard acids include the Type A cations and the soft acids the Type B cations.†

Hardness and softness are not synonymous with the strengths of acids and bases. Thus although both the OH^- ion and the F^- ion are hard bases, the hydroxide ion is ca. 10^{13} times stronger as a base than the fluoride ion. A sufficiently strong acid (or base) can displace a weaker acid (or base) in violation of the hard–hard, soft–soft principle. A case in point is represented by the equation

$$\underset{\text{soft}}{SO_3^{2-}} + \underset{\text{hard–hard}}{HF} \rightleftharpoons \underset{\text{hard–soft}}{HSO_3^-} + \underset{\text{hard}}{F^-} \qquad K = \text{ca. } 10^4$$

where the hard, relatively weak base F^- is displaced by the softer, stronger base

---

*   Pearson notes[98] that the descriptive terms "hard" and "soft" were suggested by D. H. Busch.

†   R. T. Myers [*Inorg. Chem.*, **13**, 2040 (1974)] suggests that there is no relationship between Type A and Type B behavior and polarizability for the metal–ion acceptors which are the most important hard acids.

Table 9-12   Pearson Classification of Lewis Bases[a]

Hard	Borderline	Soft
F^-, Cl^-	Br^-	I^-, H^-, R^-
H_2O, OH^-, O^{2-}	SO_3^{2-}, NO_2^-	CN^-, SCN^-, $S_2O_3^{2-}$
ROH, R_2O, OR^-	N_2, N_3^-	RSH, R_2S, SR^-
ClO_4^-, SO_4^{2-}, NO_3^-, PO_4^{3-}, $CH_3CO_2^-$	C_5H_5N	R_3P, $(RO)_3P$, R_3As
	$C_6H_5NH_2$	CO, RNC
NH_3, RNH_2, N_2H_4		C_2H_4, C_6H_6

[a] R = alkyl group.

SO_3^{2-}. However, in an appropriate competitive reaction, both strength and hardness are involved, and the hard–hard soft–soft principle applies, as

$$CH_3HgF + HSO_3^- \rightleftharpoons CH_3HgSO_3^- + HF \qquad K = ca.\ 10^3$$

soft–hard   hard–soft   soft–soft   hard–hard

Thus in applying the HSAB concept, both inherent strength and hardness-softness must be considered.

Unlike strength, hardness or softness is not an inherent property of a donor or acceptor atom. Rather it is affected by the substituents on that atom. Thus boron(III) is a hard acceptor in the compounds BF_3 and $B(OR)_3$, a borderline acceptor in the compound $B(CH_3)_3$, and a soft acceptor in the radical BH_3*
(Table 9-13). The hard–hard soft–soft principle is exemplified by the hydride ion
(H^-) exchange

$$BH_3F^- + BF_3H^- \rightleftharpoons BF_4^- + BH_4^-$$

soft–hard   hard–soft   hard–hard   soft–soft

The mutual stabilizing effect of the grouping of soft ligands or of hard ligands about a central atom has been termed *symbiosis*.[101]

Applications   A number of experimental observations are in accord with expectations and predictions in terms of the HSAB principle. Thus an element in a high positive oxidation state exemplifies hardness and is stabilized by surrounding its "ions" with hard bases such as O^{2-}, OH^-, or F^- ions (e.g., PtF_6, H_4XeO_6, ClO_4^-). Conversely, an element in a low oxidation state (soft acid) is stabilized by bonding its "ions" to soft bases such as H^-, CO, CN^-, R_3P, R_3As, RNC, C_2H_4, C_6H_6 [e.g., $Ni(CO)_4$, $Cu(CN)_2^-$, $Pd[\sigma\text{-}C_6H_4\{P(C_2H_5)_2\}_2]$, ReH_9^{2-}]. Symbiosis is evident in these examples. Bulk transition metals, which are soft acids, selectively chemisorb soft bases such as olefins and carbon monoxide in their

---

* Since BH_3 has no independent existence, it is referred to as a radical. Species such as H_3BCO and B_2H_6 are known.

Table 9-13 Pearson Classification of Lewis Acids[a]

Hard	Borderline	Soft
H^+, Li^+, Na^+, K^+, (Rb^+, Cs^+)	SO_2	Cl, Br, I, I_2, Br_2
Be^{2+}, Mg^{2+}, Ca^{2+}, Sr^{2+}, (Ba^{2+})	NO^+, Sb^{3+}, Bi^{3+}	I^+, Br^+, ICN, etc.
Al^{3+}, Ga^{3+}, In^{3+}	R_3C^+, $C_6H_5^+$, Sn^{2+}, Pb^{2+}	O, HO^+, RO^+, RO, RO_2
Sc^{3+}, Y^{3+}, Ln^{3+}, (An^{3+})[b]	$B(CH_3)_3$, GaH_3	RS^+, RSe^+, RTe^+, Te^{4+}
Ce^{4+}, Th^{4+}, U^{4+}, Pu^{4+}, (An^{4+})	Zn^{2+}	N
Ti^{4+}, Zr^{4+}, Hf^{4+}	Cu^{2+}	CH_2; carbenes; trinitrobenzene, etc.;
VO^{2+}, MoO^{3+}, WO^{4+}, UO_2^{2+}	Fe^{2+}, Co^{2+}, Ni^{2+}	chloranil, quinones, etc.;
$Cl(III), Cl(VII), I(V), I(VII), N(III), As(III)$	Rh^{3+}, Os^{2+}, Ir^{3+}	tetracyanoethylene, etc.
SO_3, CO_2		BH_3, $Ga(CH_3)_3$, $GaCl_3$, GaI_3,
		$InCl_3$, Tl^+, Tl^{3+}, $Tl(CH_3)_3$
RSO_2^+, $ROSO_2^+$, RPO_2^+, $ROPO_2^+$, RCO^+, RNC		Cd^{2+}, Hg_2^{2+}, Hg^{2+}, CH_3Hg^+
$(CH_3)_2Sn^{2+}$, CH_3Sn^{3+}, $Al(CH_3)_3$, $Be(CH_3)_2$		Cu^+, Ag^+, Au^+
		Pd^{2+}, Pt^{2+}, Pt^{4+}
BF_3, $B(OR)_3$, $AlCl_3$, AlH_3		$Co(CN)_5^{2-}$
HX (hydrogen-bonding molecules)		Metals, atoms or bulk

[a] R = alkyl group; ionic charges sometimes real, sometimes formal.

[b] Ln = any lanthanide element (i.e., La–Lu); An = actinide element (i.e., Ac–Lr).

roles as heterogeneous catalysts. Very soft bases containing elements such as P, As, Sb, Se, or Te are so strongly sorbed by these bulk metals as to act as poisons in catalysis, but harder bases containing oxygen and nitrogen donor centers do not act as poisons. The use of acidic clays in the catalytic cracking of petroleum depends upon hard acid sites such as H^+ or Al^{3+}, which sites are poisoned by hard bases (e.g., amines, oxo anions). Water, a *hard solvent* in terms of both acidic and basic functions, is an excellent solvent for hard acids, hard bases, and hard complexes. The various alcohols increase in softness as the size of the alkyl group increases and also increase in a parallel fashion in their ability to dissolve *soft solutes*. The ease with which a metal is oxidized in aqueous solution increases with the enthalpy of hydration of its cations, which in turn is maximized by hard–hard interactions. The least readily oxidized metals (e.g., Au, Pt metals, Ag, Hg) all give cations that are soft acids. Ease of oxidation of these metals is enhanced by adding soft bases to water (e.g., CN^- ion to form cyanocomplexes). Other examples can be discerned in discussions of the chemical properties of the elements and their compounds.

Theoretical Bases The fundamentally phenomenological HSAB principle has been approached theoretically very largely in terms of bonding (Chs. 4, 6, 7). No simple or completely satisfying theoretical explanation has yet resulted. It is useful to regard hard interactions as reflecting electrostatic bonding and soft interactions as reflecting covalent bonding. Soft interactions involving π bonding are undoubtedly important, in particular with d-transition metal ions. Both London and van der Waals forces have been suggested as important. Electronegativity is significant in defining hard and soft bases, but strict application of electronegativity differences to calculation of enthalpy changes for exchanges leading to hard–hard and soft–soft products often yields results that are grossly in error.[99] For more detailed treatment, one may consult with profit publications by Pearson[98-100] and Huheey.[102,103]

In summary, we may with profit quote from Pearson as follows:[98]

> Whatever the explanations, it appears that the "principle of hard and soft acids and bases" does describe a wide range of chemical phenomena in a qualitative way, if not quantitative. It can be used as an aid in correlating and storing large amounts of data and in predicting results. It is not infallible, since many apparent discrepancies and exceptions exist. The exceptions are usually an indication that some special factor exists in these examples. In such cases the principle can still be of value by calling attention to the need for further consideration.

EXERCISES

9-1. Criticize, *pro* and *con*, the statement "All reactions involving inorganic chemical species are acid–base reactions."

9-2. Summarize the *pros* and *cons* of considering a specific chemical reaction from both the thermodynamic and the kinetic points of view. Citing examples may be of assistance.

9-3. Write a balanced ion–electron equation to describe each of the following conversions in aqueous solution under the conditions described: Mn(VII)–Mn(II), alkaline; Mn(VII)–Mn(II), acidic; I(VII)–I(–I), acidic; P(III)–P(I), alkaline; Pu(VI)–Pu(IV), acidic; $S_4O_6^{2-}$–$S_2O_3^{2-}$, acidic; Nb(V)–Nb(0), acidic; Hg(II)–Hg(I), I^- present, acidic; Sn(IV)–Sn(II), alkaline; Ho(III)–Ho(0), alkaline.

9-4. Discuss briefly the *pros* and *cons* of the *oxidation potential* and the *reduction potential* conventions. Consult additional references as necessary.

9-5. Define each of the following terms succinctly: activated complex, reaction rate constant, reaction order vs. molecularity, Usanovich acid, hard acid, soft base, ionotropic, solvo acid, proton affinity, Type B anion.

9-6. Indicate clearly why the $E°$ value for the Co(III)–Co(II) reduction couple varies with the ligand or the anion present.

9-7. What principles can be applied to support variation of protonic acid strength in aqueous media in terms of Eq. 9–28?

9-8. Why is it preferable to speak of definitions of acids and bases rather than of theories of acids and bases?

9-9. Given the electrode-potential diagrams for manganese for acidic and alkaline aqueous systems, under which condition is manganese(III) expected to disproportionate more readily? Explain.

9-10. Distinguish clearly between "inner-sphere" and "outer-sphere" reactions as probable mechanisms.

9-11. For each of the following combinations of reagents and stated conditions, write a balanced equation for the probable chemical reaction that occurs and indicate clearly why it is an acid–base reaction: $La_2O_3(s)$ $+ N_2O_4(l) +$ heat; $BF_3(g) +$ tetrahydrofuran (l); $CH_3CO_2K + HClO_4$ in $CH_3CO_2H(l)$; $InCl_3(aq) + NH_3(aq)$; $SnS_2(s) + NH_3(aq) + CH_3C(S)NH_2(aq)$; $Fe(s) + Cl_2(g) +$ heat; $[Al(H_2O)_6]^{3+}(aq)$; $Hg(NO_3)_2(aq) + HCl(aq)$; $[Cu(H_2O)_4]^{2+} + C_5H_5N(pyridine) + NCS^-$; $KNH_2 + NH_4NO_3$ in $NH_3(l)$.

9-12. Discuss concisely the relationships between Type A and Type B cations and hard and soft acids.

9-13. Why are hard–hard and soft–soft combinations preferred to hard–soft or soft–hard combinations?

9-14. Arrange the following substances in the order of increasing acidity in aqueous solution and give reasons for your arrangement: HCl, $HClO_4$, HI, CH_4, $H_2PO_4^-$, $H_3SO_4^+$, $CH_3CO_2H_2^+$, NH_3, HSO_4^-, $C_5H_5NH^+$.

9-15. What is meant by the leveling of a series of bases? Illustrate by describing the reaction of each of the following bases with $NH_3(l)$: O^{2-}, OH^-, NH_2^-, $CH_3CO_2^-$.

REFERENCES

1. For implementation of this point of view, the reader is referred to the series *Organic Reactions*. Wiley-Interscience, New York (1942–   ). As of the publication date of this textbook, 26 volumes have appeared in this series.

2. H. F. Franzen and B. C. Gerstein, *Rudimentary Chemical Thermodynamics*, p. 3, D. C. Heath, Lexington, Mass. (1971).

3. R. A. Alberty and F. Daniels, *Physical Chemistry*, 5th ed., Chs. 1–7, John Wiley & Sons, New York (1979).

4. A. W. Adamson, *A Textbook of Physical Chemistry*, 2nd ed., Chs. 4–7, 10, Academic Press, New York (1979).

5. G. N. Lewis and M. Randall, *Thermodynamics*, 2nd ed. (revised by K. S. Pitzer and L. Brewer), McGraw-Hill Book Company, New York (1961).

5a. D. A. Johnson, *Some Thermodynamic Aspects of Inorganic Chemistry*, Cambridge University Press, Cambridge (1968).

5b. H. Carmichael, *J. Chem. Educ.*, **53**, 695 (1976).

6. W. M. Latimer, *The Oxidation States of the Elements and Their Potentials in Aqueous Solutions*, 2nd ed., Prentice-Hall, New York (1952). A scientific classic.

6a. Ref. 5, Ch. 24.

7. A. J. de Bethune and N. A. S. Loud, *Standard Aqueous Electrode Potentials and Temperature Coefficients at 25°C*, Clifford A. Hampel, Skokie, Ill. (1964).

7a. G. Milazzo and S. Caroli, *Tables of Standard Electrode Potentials*, Wiley-Interscience, New York (1978).

8. A. J. de Bethune and T. S. Licht, *J. Chem. Educ.*, **34**, 433 (1957).

9. F. C. Anson, *J. Chem. Educ.*, **36**, 394 (1959).

10. M. C. Day, Jr., and J. Selbin, *Theoretical Inorganic Chemistry*, 2nd ed., pp. 342–343, Reinhold Publishing Corp., New York (1969).

11. H. A. Laitinen and W. E. Harris, *Chemical Analysis: An Advanced Text and Reference*, Ch. 12, McGraw-Hill Book Company, New York (1975).

12. R. A. Alberty and F. Daniels, *Physical Chemistry*, 5th ed., Chs. 14–18, John Wiley & Sons, New York (1979).

13. A. W. Adamson, *A Textbook of Physical Chemistry*, 2nd ed., Chs. 14, 15, Academic Press, New York (1979).

14. S. W. Benson, *The Foundations of Chemical Kinetics*, McGraw-Hill Book Company, New York (1960).

15. F. Basolo and R. G. Pearson, *Mechanisms of Inorganic Reactions*, 2nd ed., John Wiley & Sons, New York (1967).

16. H. Eyring and E. M. Eyring, *Modern Chemical Kinetics*, Reinhold Publishing Corp., New York (1963).

17. W. L. Jolly, *J. Chem. Educ.*, **43**, 198 (1966). Article written in terms of oxidation potentials.

18. P. Delahay, M. Pourbaix, and P. van Rysselberghe, *J. Chem. Educ.*, **27**, 683 (1950).

19. M. Pourbaix, *Atlas of Electrochemical Equilibria in Aqueous Solution*, Pergamon Press, Oxford (1966).

20. J. A. Campbell and R. A. Whiteker, *J. Chem. Educ.*, **46**, 90 (1969).

21. H. N. Wilson and J. G. M. Bremner, *Q. Rev. (Lond.)*, **2**, 1 (1948). A review article.

22. A. A. Frost, *J. Am. Chem. Soc.*, **73**, 2680 (1951).

23. C. S. G. Phillips and R. J. P. Williams, *Inorganic Chemistry*, Vol. 1, pp. 314–321, Oxford University Press, New York (1965).

24. J. Kleinberg, *Unfamiliar Oxidation States and Their Stabilization*, University of Kansas Press, Lawrence, Kans. (1950); *J. Chem. Educ.*, **27**, 32 (1950); **29**, 324 (1952); **32**, 73 (1956).

25. J. V. Quagliano and R. L. Rebertus, in *The Chemistry of the Coordination Compounds*, J. C. Bailar, Jr., and D. H. Busch (eds.), Ch. 11, Rheinhold Publishing Corp., New York (1956).

26. M. M. Jones, *Elementary Coordination Chemistry*, pp. 226–231, Prentice-Hall, Englewood Cliffs, N.J. (1964).

27. H. J. Emeléus and A. G. Sharpe, *Modern Aspects of Inorganic Chemistry*, 4th ed., pp. 529–532, John Wiley & Sons, New York (1973).

27a. V. Gutmann, *Struct. Bonding*, **15**, 141 (1973). A detailed discussion of changes in redox properties as affected by coordination.

28. H. Taube, in *Advances in Inorganic Chemistry and Radiochemistry*, H. J. Emeléus and A. G. Sharpe (Eds.), Vol. 1, pp. 1–53, Academic Press, New York (1959).

29. S. W. Benson, Ref. 14, Ch. 16.

30. J. Halpern, *Q. Rev. (Lond.)*, **15**, 207 (1961).

31. J. O. Edwards (Ed.), *Progress in Inorganic Chemistry*, Vol. 13: *Inorganic Reaction Mechanisms*, Wiley-Interscience, New York (1970).

32. J. Halpern and H. Taube, *J. Am. Chem. Soc.*, **74**, 375 (1952).

33. M. Anbar and H. Taube, *J. Am. Chem. Soc.*, **80**, 1073 (1958).

34. H. Taube, *Chem. Rev.*, **50**, 69 (1952). A fundamental article dealing with substitution reactions of complexes in solution and discussing the role of outer and inner orbitals. See Sec. 11.3-1 for a detailed discussion.

35. H. Taube, H. Myers, and R. Rich, *J. Am. Chem. Soc.*, **75**, 4118 (1953).

36. P. Walden, *Salts, Acids, and Bases*, McGraw-Hill Book Company, New York (1929).

37. R. P. Bell, *The Proton in Chemistry*, 2nd ed., Cornell University Press, Ithaca, N.Y. (1969).

38. L. F. Audrieth, *Acids, Bases, and Non-Aqueous Solvents*, 23rd Annual Priestley Lectures, Pennsylvania State University, University Lithoprinters, Ypsilanti, Mich. (1949).

39. L. F. Audrieth and J. Kleinberg, *Non-Aqueous Solvents: Applications as Media for Chemical Reactions*, Ch. 2, John Wiley & Sons, New York (1953).

40. A. J. Ihde, *The Development of Modern Chemistry*, pp. 547–549, Harper & Row, New York (1964).

41. G. N. Lewis, *J. Franklin Inst.*, **226**, 293 (1938). A classical discussion.

42. J. N. Brønsted, *Rec. Trav. Chim.*, **42**, 718 (1923).

43. T. M. Lowry, *Chem. Ind. (Lond.)*, **42**, 43 (1923); *Trans. Faraday Soc.*, **20**, 13 (1924).

44. J. A. Riddick, *Anal. Chem.*, **26**, 77 (1954).

45. A. A. Woolf and H. J. Emeléus, *J. Chem. Soc.*, **1949**, 2865.

46. H. Lux, *Z. Elektrochem.*, **45**, 303 (1939).

47. H. Flood and T. Förland, *Acta Chem. Scand.*, **1**, 592, 781 (1947).

48. H. Flood, T. Förland, and B. Roald, *Acta Chem. Scand.*, **1**, 790 (1947).

48a. C. S. Sherer, *J. Inorg. Nucl. Chem.*, **34**, 1615 (1972).

49. V. Gutmann and I. Lindqvist, *Z. Phys. Chem.*, **203**, 250 (1954).

50. V. Gutmann, *Sven. Kem. Tidskr.*, **68**, 1 (1956); *J. Phys. Chem.*, **63**, 378 (1959).

51. E. C. Franklin, *J. Am. Chem. Soc.*, **27**, 820 (1905); *Am. Chem. J.*, **47**, 285 (1912).

52. E. C. Franklin, *The Nitrogen System of Compounds*, Reinhold Publishing Corp., New York (1935).

53. E. C. Franklin, *J. Am. Chem. Soc.*, **46**, 2137 (1924).

54. A. F. O. Germann, *Science*, **61**, 71 (1925); *J. Am. Chem. Soc.*, **47**, 2461 (1925).

55. H. P. Cady and H. M. Elsey, *Science*, **56**, 27 (1922); *J. Chem. Educ.*, **5**, 1425 (1928).

56. L. Ebert and N. Konopik, *Oesterr. Chem.-Ztg.*, **50**, 184 (1949).

57. G. Jander, *Die Chemie in wasserähnlichen Lösungsmitteln*, Springer-Verlag, Berlin (1949).

58. G. Jander and K. Wickert, *Z. Phys. Chem.*, **A178**, 57 (1936).

59. R. Muxart, *C. R. Acad. Sci.*, **231**, 1489 (1950).

60. R. E. Johnson, T. H. Norris, and J. L. Huston, *J. Am. Chem. Soc.*, **73**, 3052 (1951).

61. E. C. M. Grigg and I. Lander, *Trans. Faraday Soc.*, **46**, 1039 (1950).

62. R. H. Herber, T. H. Norris, and J. L. Huston, *J. Am. Chem. Soc.*, **76**, 2015 (1954).

63. B. J. Masters, N. D. Potter, D. R. Asher, and T. H. Norris, *J. Am. Chem. Soc.*, **78**, 4252 (1956).

64. T. H. Norris, *J. Phys. Chem.*, **63**, 383 (1959).

65. G. B. L. Smith, *Chem. Rev.*, **23**, 165 (1938).

66. C. C. Addison and R. Thompson, *Nature*, **162**, 369 (1948); *J. Chem. Soc.*, **1949**, 211, 218 (Suppl. Issue No. 1).

67. C. C. Addison and J. Lewis, *J. Chem. Soc.*, **1951**, 2833, 2843; **1953**, 1869, 1874.

68. C. C. Addison and C. P. Conduit, *J. Chem. Soc.*, **1952**, 1390, 1399.

69. J. R. Ferraro and G. Gibson, *J. Am. Chem. Soc.*, **75**, 5747 (1953).

70. T. Moeller, V. D. Aftandilian, and G. W. Cullen, *Inorg. Synth.*, **5**, 37 (1957).

71. A. A. Banks, H. J. Emeléus, and A. A. Woolf, *J. Chem. Soc.*, **1949**, 2861.

72. D. W. Meek and R. S. Drago, *J. Am. Chem. Soc.*, **83**, 4322 (1961).

73. R. S. Drago and K. F. Purcell, in *Progress in Inorganic Chemistry*, F. A. Cotton (Ed.), Vol. 6, pp. 271–322, Interscience Publishers, New York (1964).

73a. J. E. Huheey, *J. Inorg. Nucl. Chem.*, **24**, 1011 (1962). Resolution of a "semantic tangle" on the POCl₃ system.

74. G. N. Lewis, *Valence and the Structure of Atoms and Molecules*, The Chemical Catalog Co., New York (1923).

75. W. F. Luder and S. Zuffanti, *The Electronic Theory of Acids and Bases*, 2nd rev. ed., Dover Publications, New York (1961).

76. R. S. Drago and N. A. Matwiyoff, *Acids and Bases*, D. C. Heath, Boston (1968).

77. N. V. Sidgwick, *The Electronic Theory of Valency*, Oxford University Press, New York (1927).

78. W. F. Luder, W. S. McGuire, and S. Zuffanti, *J. Chem. Educ.*, **20**, 344 (1943).

79. W. F. Luder, *J. Chem. Educ.*, **22**, 301 (1945).

80. N. J. Rose and R. S. Drago, *J. Am. Chem. Soc.*, **81**, 6138 (1959).

81. C. D. Schmulbach and R. S. Drago, *J. Am. Chem. Soc.*, **82**, 4484 (1960).

82. T. D. Epley and R. S. Drago, *J. Am. Chem. Soc.*, **89**, 5770 (1967).

83. R. S. Drago and B. B. Wayland, *J. Am. Chem. Soc.*, **87**, 3571 (1965).

84. M. Usanovich, *J. Gen. Chem. (USSR)*, **9**, 182 (1939); *Izvest. Akad. Nauk SSR*, [**101**], *Ser. Khim.*, No. 4, 97 (1951).

85. H. Gehlen, *Z. Phys. Chem.*, **203**, 125 (1954).

86. J. E. Huheey, *Inorganic Chemistry: Principles of Structure and Reactivity*, 2nd ed., pp. 263–264, Harper & Row, New York (1978).

87. J. C. McCoubrey, *Trans. Faraday Soc.*, **51**, 743 (1955).

87a. S. D. Lessley and R. O. Ragsdale, *J. Chem. Educ.*, **53**, 19 (1976).

88. L. Pauling, *J. Chem. Educ.*, **33**, 16 (1956); **53**, 762 (1976). But see also Ref. 87a R. T. Myers, and *Inorg. Chem.*, **16**, 2671 (1977).

89. F. Gallais, *Bull. Soc. Chim. Fr.*, [5], **14**, 425 (1947).

90. L. Pauling, *General Chemistry*, 1st ed., p. 394, W. H. Freeman, San Francisco (1947).

91. J. E. Ricci, *J. Am. Chem. Soc.*, **70**, 109 (1948).

92. R. J. Gillespie, *J. Chem. Soc.*, **1950**, 2537.

93. D. P. N. Satchell and R. S. Satchell, *Q. Rev.* (*Lond.*), **25**, 171 (1971).

94. H. M. N. H. Irving and R. J. P. Williams, *Nature*, **162**, 746 (1948); *J. Chem. Soc.*, **1953**, 3192.

94a. R. J. P. Williams and J. D. Hale, *Struct. Bonding*, **1**, 250 (1966).

95. G. Schwarzenbach, *Experientia Suppl.*, **5**, 162 (1956).

96. S. Ahrland, J. Chatt, and N. R. Davies, *Q. Rev.* (*Lond.*), **12**, 265 (1958).

97. G. Schwarzenbach, in *Advances in Inorganic Chemistry and Radiochemistry*, H. J. Emeléus and A. G. Sharpe (Eds.), Vol. 3, pp. 257–285, Academic Press, New York (1961).

97a. S. Ahrland, *Struct. Bonding*, **1**, 207 (1966); **15**, 167 (1973).

98. R. G. Pearson, *J. Am. Chem. Soc.*, **85**, 3533 (1963); *J. Chem. Educ.*, **45**, 581, 643 (1968).

99. R. G. Pearson, in *Survey of Progress in Chemistry*, A. F. Scott (Ed.), Vol. 5, pp. 1–52, Academic Press, New York (1969).

100. R. G. Pearson (Ed.), *Hard and Soft Acids and Bases*, Dowden, Hutchinson, & Ross, Stroudsburg, Pa. (1973).

101. Chr. Klixbüll Jørgensen, *Inorg. Chem.*, **3**, 1201 (1964).

102. J. E. Huheey and R. S. Evans, *J. Chem. Soc. Chem. Commun.*, **1969**, 968; *J. Inorg. Nucl. Chem.*, **32**, 383 (1970).

103. R. S. Evans and J. E. Huheey, *J. Inorg. Nucl. Chem.*, **32**, 373, 777 (1970).

SUGGESTED SUPPLEMENTARY REFERENCES

A. W. Adamson, *A Textbook of Physical Chemistry*, 2nd ed., Chs. 4–7, 11, 13–15, Academic Press, New York (1979). Established readable textbook.

R. A. Alberty and F. Daniels, *Physical Chemistry*, 5th ed., Chs. 1–7, 14–18, John Wiley & Sons, New York (1979). Established readable textbook.

L. F. Audrieth and J. Kleinberg, *Non-aqueous Solvents: Applications as Media for Chemical Reactions*, John Wiley & Sons, New York (1953). Acid–base definitions and acid–base reactions in nonaqueous media.

F. Basolo and R. G. Pearson, *Mechanisms of Inorganic Reactions: A Study of Complexes in Solution*, 2nd ed., John Wiley & Sons, New York (1967). Applications of kinetic data to proposals of mechanisms for reactions involving complex species.

S. W. Benson, *The Foundations of Chemical Kinetics*, McGraw-Hill Book Company, New York (1960). In-depth treatment of subject.

H. Eyring, S. H. Lin, and S. M. Lin, *Basic Chemical Kinetics*, John Wiley & Sons, New York (1980). Combination of experiment with theory.

A. A. Frost and R. G. Pearson, *Kinetics and Mechanism: A Study of Homogeneous Reactions*, 2nd ed., John Wiley & Sons, New York (1961). Relationships between kinetic data and reaction mechanisms.

D. A. Johnson, *Some Thermodynamic Aspects of Inorganic Chemistry*, Cambridge University Press, Cambridge (1968). Significant applications to inorganic chemistry.

W. M. Latimer, *The Oxidation States of the Elements and Their Potentials in Aqueous Solution*, 2nd ed., Prentice-Hall, New York (1952). Classic treatment of reaction potentials.

G. N. Lewis and M. Randall, *Thermodynamics*, 2nd ed. (revised by K. S. Pitzer and L. Brewer), McGraw-Hill Book Company, New York (1961). Classic reference.

R. G. Pearson, Ed., *Hard and Soft Acids and Bases*, Dowden, Hutchinson & Ross, Stroudsburg, Pa. (1973). Details of hard and soft definitions, with applications.

CHAPTER TEN

INORGANIC CHEMICAL REACTIONS, II REACTIONS UNDER VARIOUS CONDITIONS

Having discussed the general principles applicable to the chemical reactions of inorganic substances, broadly classified these reactions in terms of involvement of electrons, and cited a variety of specific examples, we may now examine reactions in terms of the environmental conditions under which they take place. In this chapter, attention is directed to chemical reactions in nonaqueous and fused salt media, reactions involving solids at elevated temperatures, and inorganic polymerization reactions. In Chapter 11, reactions involving coordination entities are considered in detail.

1.   REACTIONS IN NONAQUEOUS MEDIA

The ability of water to dissolve substances of a wide variety of chemical types, molecular structures, and chemical compositions has often led to the assumption that this compound is a unique solvent. Historically, such properties of water as its availability and nonantagonistic behavior toward human beings, its convenience in handling, its comparatively long and extremely convenient liquid range, and its ability to form hydrates (i.e., solvates) have undoubtedly combined with its versatility as a solvent to direct major attention to it rather than to other solvents. The Arrhenius limitation of electrolytic dissociation to aqueous solutions undoubtedly did much to enhance this attention. Indeed, in 1893, no less a scientist than W. Ostwald was convinced that insofar as its ability to form electrolytic solutions or to bring about ionization was concerned, water stood in a class by itself, a position "which is not even approximately simulated by any other substance."[1] Although it must be admitted that subsequent researches have failed to reveal another solvent that closely approaches water as a solvent, they have demonstrated that differences among solvents are differences of *degree* rather than of *kind* and that every characteristic of aqueous solutions is duplicated for solutions in other solvents. Realization that water is not a unique solvent has helped to broaden the fundamental understanding of solution chemistry that is so essential to the systematization of both inorganic and physical chemistry.

From a historical point of view, investigations of solvents and reactions therein encompass three broad periods. In the five- to seven-year interval subsequent to 1892, attention was devoted primarily to organic compounds (e.g., alcohols, ester, ketones) as electrolytic solvents for inorganic substances. The results obtained were generally qualitative since the electrolytes were usually selected because they were soluble rather than because they were suited to meaningful conductance measurements. Furthermore, the reported data usually embraced such limited ranges of concentration that only approximate extrapolations to infinite dilution could be made. It is reasonable to conclude also that standards and achievement of purity were markedly different from those currently acceptable.

A second period, beginning with the pioneering studies of solutions in liquid ammonia initiated in 1897 at the University of Kansas by H. P. Cady[2] and extended there and later at Stanford and Brown Universities, respectively, by E. C. Franklin and C. A. Kraus[3] and others and with the extensive investigations of liquid sulfur(IV) oxide solutions by P. Walden,[4] introduced the third period, namely that of the detailed chemical and physicochemical studies of the twentieth century. During this period, particularly rapid development of theoretical interpretation has paralleled increasingly sophisticated experimentation.

That interest in nonaqueous reaction media continues to increase is indicated by extensive publication of both original research and summary reviews. Of the latter, many older[5-9] and more recent[10-17] ones can be consulted with profit. Among the former, Walden's monograph[6] is truly a classic and is undeniably the basis for much modern effort. A number of these accounts contain fascinating blends of history with data and theoretical interpretation that will add to the reader's knowledge of and interest in this area.

Inasmuch as it is impossible to divorce a discussion of nonaqueous solvents from considerations of acid–base behavior, many of the topics included in Chapter 8 provide useful background for those that follow. Emphasis in this chapter is primarily on the properties of solvents that make them useful as media for chemical reactions that cannot be carried out in aqueous systems and on those reactions themselves.

1-1. Classifications of Nonaqueous Solvents

Solvents may be classified in many ways. Perhaps the most obvious classification, that based upon chemical constitution, is the least valuable since it can do no more than bring together substances with only generally similar solvent characteristics. Knowledge derived in this way is commonly more qualitative than quantitative. More fundamental and useful classifications follow.

Electrolytic and Nonelectrolytic Solvents. The electrolytic characteristics of a solvent are related to the dipole moment of its molecules (Sec. 5.1-8), association and dissociation involving its molecules, and its dielectric constant.*

Solvents, the molecules of which are polar (Sec. 5.1-8), are commonly ionizing media, either as a consequence of the strong forces of attraction existing between these molecules and the ions present in a crystalline salt or as a consequence of the formation of ions by interaction between these molecules and the polar molecules of a covalently bonded solute (e.g., a molecular protonic acid). The polarity of a solvent molecule is indicated both by its dipole moment (Table 10-1) and by the magnitudes of its enthalpies of fusion and vaporization (Table 10-1). The latter are measures of intermolecular forces and thus degree of molecular association. The degree of association is measured by the entropy change of vaporization, ΔS_v, or Trouton's constant, defined as

$$\Delta S_v = \text{Trouton's constant} = \frac{\Delta H_v}{T_b} \tag{10-2}$$

where T_b is the normal boiling point in °K. For an associated liquid, $\Delta S_v = $ or > 21.5 cal deg^{-1} mole^{-1}.

As indicated by Eq. 10-1, the force of attraction between two charged particles in a solute decreases as the dielectric constant of the solvent increases. Changes in the apparent degrees of electrolytic dissociation (α) of several related solutes in solvents with varying dielectric constants, as indicated in Table 10-2, are typical. Our best ionizing solvents are those having large dielectric constants (Table 10-1). Substances with small dielectric constants, such as the hydrocarbons and

---

* The dielectric constant (ε) of a medium is defined in terms of the equation

$$F = \frac{q_1 q_2}{C \varepsilon r^2} \tag{10-1}$$

where F is the force of attraction acting on each of two charges q_1 and q_2, r the distance between the two charges, and C a constant. The dielectric constant of a vacuum is unity. For a given medium, the magnitude of its dielectric constant increases with temperature.

Table 10-1 Numerical Properties of Water and Certain "Waterlike" Solvents

Property	H_2O	NH_3	NH_2OH	N_2H_4	HF	H_2SO_4	HSO_3F	CH_3COOH	HCN	HNO_3	SO_2	BrF_3
Formula weight (amu)	18.015	17.032	33.029	32.045	20.006	98.073	100.065	60.052	27.026	63.01	64.060	136.899
Liquid density (g ml^{-1})	0.958 (100°C)	0.683 (−33.35°C)	1.204 (33°C)	1.0114 (15°C)	0.991 (19.5°C)	1.8269 (25°C)	1.726 (25°C)	1.049 (20°C)	0.681 (25.6°C)	1.504 (25°C)	1.46 (−10°C)	2.8030 (25°C)
Molecular volume, (ml mole^{-1})	18.80	24.94	27.43	31.68	20.19	53.68	57.98	57.25	39.69	41.89	43.88	48.84
Melting point (°C)	0	−77.7	33	1.8	−89.37	10.371	−88.98	16.6	−13.35	−41.62	−75.52	8.9
Normal boiling point (°C)	100	−33.35	58 (22 mm)	113.5	19.51	ca. 270 (dec.)	162.7	118.1	25.6	82.6	−10.08	125.8
Critical temperature (°C)	374.1	132.4		380	230.2			321.6	183.5		157.50	
Critical pressure (atm)	217.7	112.3		145				57.2	55		77.8	
Enthalpy of fusion (kcal mole^{-1})	1.435	1.35		3.2	1.094	2.56		2.684	2.009	2.503	1.769	2.87
Enthalpy of vaporization (kcal mole^{-1})	9.719	5.58		10.7(25°C)	7.24			5.81	6.027	9.426	5.96	10.2
Entropy of vaporization (cal deg^{-1} mole^{-1})	25.0	23.3		27.7	24.7			14.9	20.2	26.5	22.7	25.7
Specific conductance (ohm^{-1} cm^{-1})	6×10^{-8} (25°C)	1×10^{-11} (−33.4°C)		2.3–2.8×10^{-6} (25°C)	1.4×10^{-5} (−15°C)	1.04×10^{-2} (25°C)	1.085×10^{-4} (25°C)	1.2×10^{-8} (25°C)	4.5×10^{-7} (0°C)	3.72×10^{-2} (25°C)	4×10^{-8} (−10°C)	8.0×10^{-3} (25°C)
Dielectric constant	81.1 (18°C)	22 (−34°C)		51.7 (25°C)	83.6 (0°C)	101 (25°C)		9.7 (18°C)	123.5 (15°C)	50±10 (14°C)	17.4 (−19°C)	
Viscosity (centipoise)	0.959 (25°C)	0.254 (−33.5°C)		0.9 (25°C)	0.256 (0°C)	24.54 (25°C)	1.56 (25°C)	1.040 (30°C)	0.201 (20.2°C)	7.46 (25°C)	0.4285 (−10°C)	2.22 (25°C)
Molal cryoscopic constant (°C mole kg^{-1})	1.859	0.97					3.93	3.9		2.68	3.01	
Molal ebullioscopic constant (°C mole kg^{-1})	0.51	0.34						3.07	0.85		1.48	
Dipole moment of gas (Debye)	1.85	1.47		1.83	1.9				2.8	2.16	1.61	1.19

620

Table 10-2 Effect of Dielectric Constant on Electrolytic Behavior

Solvent	Dielectric Constant	α at 2000 liters Dilution		
		$(C_2H_5)_4NPi^a$	$(C_2H_5)_2NH_2Pi^a$	$(C_2H_5)_2NH_2Cl$
H_2O	81.1 (18°C)	0.99	0.97	0.97
CH_3OH	35.4 (13°C)	0.95	—	0.92
C_2H_5OH	25.4 (25°C)	0.89	—	0.88
CH_3CN	36.4 (20°C)	0.95	0.90	0.31
$(CH_3)_2CO$	20.7 (20°C)	0.88	0.76	0.10
C_5H_5N	12.4 (21°C)	0.74	0.53	0.04
$C_2H_4Cl_2$	10.9 (0°C)	0.47	0.005	ca. 0.0

[a] Pi = picrate ion [i.e., $C_6H_2(NO_2)_3O^-$].

their halogenated derivatives, are poor electrolytic solvents, Decreasing the dielectric constant in a series of solvents is accompanied by alteration in the nature of the solute from relatively free solvated ions to at least ion pairs.[18, 19]

Electrolytic behavior is most characteristic in those solvents that both undergo self-ionization (specific conductance data in Table 10-1) and solvate ionic species (e.g., H_2O, HF, NH_3, HCN, H_2SO_4, SO_2). However, of two solvents of these types having comparable properties, the one with the smaller viscosity is commonly the better solvent. Presumably this is a consequence of the larger mobilities of ionic species. Thus liquid ammonia, with a relatively small viscosity, is a better ionizing solvent than its dielectric constant would indicate.

The problems of interpreting acid–base behavior in nonprotonic solvents (Sec. 9.6-1) also characterize the interpretation of ionization phenomena in these media. The literature records an extensive and sometimes bitter controversy between V. Gutmann and R. S. Drago, with the former championing auto-ionization in the broad solvent-system context (Sec. 9.6-1) and the latter espousing ionization as a consequence of coordination involving solvent molecules.[20–23] To a substantial degree, the differences may be more semantic than otherwise since the two approaches are not mutually exclusive but represent, rather, different points of view as regards dissociation phenomena.[24] A significant point in rationalizing the differing viewpoints is that for a given solvent its solvocation and solvoanion are, respectively, the strongest acid and the strongest base that can exist in that medium without being leveled.[23]

Molten salts, which are of course completely ionic, are the ultimate in electrolytic solvents (Sec. 10.2).

Leveling and Differentiating Solvents. The concept of leveling and differentiating solvents in acid–base behavior can be extended to electrolytic solvents in general. Leveling solvents (i.e., those in which most soluble electrolytes appear to be extensively ionized) are those the molecules of which are the most highly polar (e.g., HF, H_2O, NH_3). Differentiating solvents (i.e., solvents in which dissolved electrolytes vary widely in strength) are those the molecules of which are less polar (e.g., certain amines, halogenated hydrocarbons).

Acidic and Basic Solvents. Solvents are classifiable in this way in terms of a number of the definitions of acids and bases (Chapter 9), although this type of classification is most useful in the Brønsted–Lowry sense. Solvents that are neither acidic nor basic are termed *inert* solvents.[25] Inert solvents include benzene, chlorobenzene, *n*-hexane, cyclohexane, carbon tetrachloride, and carbon disulfide. Such solvents have small dielectric constants and are nonelectrolytic in behavior.

Parent Solvents. This classification reflects the concept that certain solvents may be considered to be parents of systems of compounds (Sec. 10.1-3). Classification is based upon *formally equivalent* groups of atoms derived from solvent molecules, as indicated in Table 10-3. Compounds derived from these formally equivalent groups, when dissolved in their parent solvents, are thus expected to react similarly. Subsequent discussions support the correctness of the expectation.

Those nonaqueous solvents that have "waterlike" properties[8] have been studied most extensively. The physical constants of such solvents are often closely comparable (Table 10-1). Discussions that follow should be implemented by references to Table 10-1.

1-2. Solubility in Polar Solvents

Nonpolar solutes are quite generally difficultly soluble in polar solvents. Polar covalent and ionic solutes owe their solubilities to electrostatic interactions between solvent molecules and fundamental solute particles. If the molecules of

Table 10-3 Formally Equivalent Groups among Solvents

Solvent	Equivalent Ions		
H_2O	H_3O^+	OH^-	O^{2-}
HF	H_2F^+	F^- or HF_2^-	—
NH_3	NH_4^+	NH_2^-	NH^{2-} or N^{3-}
N_2H_4	$N_2H_5^+$	$N_2H_3^-$	$N_2H_2^{2-}$
NH_2OH	NH_3OH^+	$NHOH^-$	NOH^{2-}
$(CH_3CO)_2O$	CH_3CO^+	$CH_3CO_2^-$	—
H_2SO_4	$H_3SO_4^+$	HSO_4^-	SO_4^{2-}
SO_2	SO^{2+} (?)	—	SO_3^{2-}
N_2O_4	NO^+	NO_3^-	—
BrF_3	BrF_2^+	BrF_4^-	—
$AsCl_3$	$AsCl_2^+$	$AsCl_4^-$	

both the solute and the solvent are sufficiently polar, bond rupture may occur with the formation of ions, as for example,

$$HBr(g) + NH_3(l) \rightleftharpoons NH_4^+(solv) + Br^-(solv)$$

$$\underset{solvent}{}$$

$$SbF_5(l) + 2\ HF(l) \rightleftharpoons H_2F^+(solv) + SbF_6^-(solv)$$

$$\underset{solvent}{}$$

where (solv) indicates an indefinite degree of solvation. An ionic solute dissolves by removal of existing ions from its crystal lattice, but dissolution occurs only if the energy released in solvating the ions exceeds the lattice energy (Sec. 4.3-4) of the crystal. Although such processes are dependent upon free-energy changes, necessary free-energy data are commonly lacking, and solubility predictions must be based upon proportional enthalpy data. Where such data are available, useful information can be obtained by application of enthalpy cycles of the Born–Fajans–Haber type (Sec. 4.3-4). For a uni-univalent salt M^+X^-, the cycle, for standard-state conditions, becomes

$$M^+X^-(s) \xleftarrow{\Delta H^0_{lat}} M^+(g)\ +\ X^-(g)$$

with vertical arrows $\Delta H^0_{solv(+)}$ and $\Delta H^0_{solv(-)}$, and

$$\Delta H^0_{soln} \longrightarrow M^+(solv) + X^-(solv)$$

where ΔH^0_{soln} is the enthalpy of solution (exothermic or endothermic), ΔH_{lat} is the lattice energy (exothermic), and $\Delta H^0_{solv(+)}$ and $\Delta H^0_{solv(-)}$ are, respectively, the enthalpies of solvation of the gaseous cation and the gaseous anion (both exothermic). The cycle then leads to

$$\Delta H^0_{soln} = \Delta H^0_{solv(+)} + \Delta H^0_{solv(-)} - \Delta H^0_{lat} \tag{10-3}$$

Unlike aqueous systems for which ionic hydration enthalpies are readily available,[26] few data exist for nonaqueous solvents.

Dissolution of an ionic compound is enhanced by the weakening of interionic attractions in the crystal when a solvent of large dielectric constant is used, as suggested by Eq. 10-1. Solvents of this type (e.g., H_2O, HF, H_2SO_4) also dissolve a larger variety of ionic compounds than those with smaller dielectric constants. For a given case, the dielectric constant and the enthalpy of solvation can be related in terms of the Born expression

$$\Delta H^0_{solv} = \frac{q^2}{2r}\left(1 - \frac{1}{\varepsilon}\right) \tag{10-4}$$

with enthalpy of solvation increasing as the dielectric constant increases. Not uncommonly, the dielectric constant can be the major factor in determining solubility.

Inasmuch as both the enthalpy of solvation and the lattice energy increase as cationic and anionic radii decrease, it is difficult to relate the solubilities of salts in polar solvents directly to ionic sizes. However, differences in the magnitudes of the energy quantities are such that solubility usually increases as cationic or anionic radius increases. With increase in either cationic or anionic charge, however, the lattice energy increases much more rapidly than the enthalpy of solvation, and a parallel decrease in solubility in a given solvent is noted. The electronic atmospheres of the ions are also important because of the polarizing effects of cations on anions (Sec. 5.1-8). If the anions are more readily polarized than the solvent molecules, the lattice energy increases more than the enthalpy of solvation, and solubility decreases. If the solvent molecules are the more readily polarized, solubility increases. For a given periodic group, cations with underlying 18-electron arrangements are more strongly polarizing than cations of the same charge with underlying 8-electron arrangements and thus give generally less soluble salts with a common anion. Specific examples noted in subsequent discussions and appearing in tables of solubility data support these predicted trends.

Discussions of specific solvents that follow are limited to selected basic solvents (NH_3, derivatives of NH_3), acidic solvents (HF, H_2SO_4, HSO_3F, superacid systems), nonprotonic solvents (SO_2, BrF_3, N_2O_4), and molten salts. The general references already cited[5-17] provide data for many other systems.

1-3. Liquid Ammonia as a Basic Solvent[*]

Probably because of its particularly pronounced waterlike characteristics (Table 10-1), ammonia has been investigated more extensively than any other single nonaqueous solvent.[12, 27-32b] Like water, ammonia is associated, but because the N$\cdots$H—N bond is weaker than the O$\cdots$H—O bond (Sec. 5.2-2), those properties that reflect association are less pronounced with ammonia. Ammonia is a poorer electrolytic solvent than water (smaller ε). Although its reduced viscosity does enhance its electrolytic properties, it is probable that ion-pairing characterizes solutions of all salts in ammonia.[33, 34] Indeed, solutions of uni-univalent electrolytes in ammonia are comparable in this respect with solutions of bi-bivalent electrolytes in water.[34] There are also structural differences between the two liquids that account for the low mobility of the proton in ammonia as compared with its very high mobility in water.[34, 35] On the other hand, thermodynamic treatments of solutions in these two solvents are comparable.[33, 36]

It should be noted that although experimental studies of solutions in ammonia are usually carried out at its boiling point ($-33.35°C$) or below, the ease with which ammonia can be liquefied by application of pressure at room temperature allows investigations to be carried out at this more convenient temperature.

---

[*] Inasmuch as all of the specific solvents discussed are liquids, the adjective "liquid" is generally omitted.

Table 10-4 Representative Solubilities in Liquid Ammonia[a]

Solute	Solubility (g solute/ 100 g NH_3)	Temperature (°C)	Solute	Solubility (g solute/ 100 g NH_3)	Temperature (°C)
Li	11.3	−33	$LiNO_3$	243.66	25
Na	24.6	−33	$NaNO_3$	97.60	25
K	49.0	−33	KNO_3	10.4	25
S	21	30	NH_4NO_3	390.0	25
NaF	0.35	25	$Ca(NO_3)_2$	80.22	25
NaCl	3.02	25	$Sr(NO_3)_2$	87.08	25
NaBr	137.95	25	$Ba(NO_3)_2$	97.22	25
NaI	161.90	25	$AgNO_3$	86.04	25
KCl	0.04	25	NaSCN	205.50	25
KBr	13.50	25	NH_4SCN	312.0	25
KI	182.0	25	KOCN	1.70	25
NH_4Cl	102.5	25	Li_2SO_4	0	25
NH_4Br	237.9	25	Na_2SO_4	0	25
NH_4I	368.5	25	K_2SO_4	0	25
$BaCl_2$	0	25	$KClO_3$	2.52	25
MnI_2	0.02	25	$KBrO_3$	0.002	25
AgCl	0.83	25	KIO_3	0	25
AgBr	5.92	25	NH_4ClO_4	137.93	25
AgI	206.84	25	$NH_4C_2H_3O_2$	253.16	25
ZnI_2	0.10	25	NH_4HCO_3	0	25
$NaNH_2$	0.004	25	$(NH_4)_2CO_3$	0	25
KNH_2	3.6	25	$(NH_4)_2HPO_4$	0	25
K_2CO_3	0	25	$(NH_4)_2S$	120.0	25
H_3BO_3	1.92	25	ZnO	0	25

[a] Data taken largely from H. Hunt, *J. Am. Chem. Soc.*, **54**, 3509 (1932); H. Hunt and L. Boncyk, *J. Am. Chem. Soc.*, **55**, 3528 (1933).

Solubilities in Ammonia. Inasmuch as the solubilities of specific substances in ammonia are markedly different from those in water[3] and inasmuch as the nature of a chemical reaction is often dependent upon solubilities,[9] a detailed summation is in order (Table 10-4).

Undeniably, a major difference from water is the ability of ammonia to dissolve, without immediate oxidation, free metals that are outstandingly powerful reducing agents. Thus the periodic group IA metals dissolve readily to yield uniformly blue dilute solutions* from which the metals, except lithium, can be recovered by evaporation of the solvent. When a 20% solution of lithium is evaporated, a bronze-colored electrically conducting, low melting (89°K), low

* First noted by W. Weyl [*Ann. Phys. (Leipzig)*, **121**, 601 (1864)].

density (0.5 g cm^{-3}) solid of composition Li(NH$_3$)$_4$ is obtained.[37] Calcium, strontium, barium, and the inherently divalent lanthanide metals europium and ytterbium[38] also dissolve readily, but upon evaporation of the solvent these solutions deposit initially relatively unstable body-centered cubic ammines of composition M(NH$_3$)$_6$.

Powder neutron diffraction and ^{3}H nuclear magnetic resonance investigations using crystalline Ca(ND$_3$)$_6$, as typical of the metal hexaammines, have shown that the calcium atoms are surrounded exactly octahedrally by nitrogen atoms at the rather long Ca—N bond distance of ca. 3 Å.[38a–38c] Interestingly, each ND$_3$ molecule is nearly planar, unlike normal trigonal pyramidal ND$_3$ molecules, with one N—D bond distance in each molecule relatively short (ca. 0.9 Å) and the other two N—D bond distances comparatively long (ca. 1.4 Å). The pseudotrigonal axes are not coincident with the Ca—N bond but lie at an angle of about 13° to it. Each ND$_3$ molecule has fourfold rotational disorder, giving a total of 4^6 possible orientations of ND$_3$ molecules in each complex species. One of these is depicted in Fig. 10-1. No hydrogen bonds are apparent, and nearly all nonbonded contacts are larger than the van der Waals

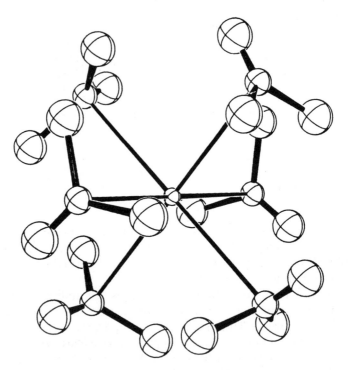

Figure 10-1 One possible molecular structure of the complex species Ca(ND$_3$)$_6$ in metallic solid Ca(ND$_3$)$_6$. [W. S. Glaunsinger, T. R. White, R. B. Von Dreele, D. A. Gordon, R. F. Marzke, A. L. Bowman, and J. L. Yarnell, *Nature*, **271**, 5644 (1978, Fig. 1.]

distance. Isomorphism with all other metal hexaammines is indicated by newly identical unit cells, lattice parameters, magnetic resonance data, and chemical behavior.

Magnesium dissolves slightly, as does aluminum.[39] The same is apparently true, but to a still lesser degree, of beryllium, zinc, gallium, manganese, lanthanum, and cerium. Several other solvents, including water,[40,41] ethers,[42,43] hexamethylphosphoramide, $OP[N(CH_3)_2]_3$, and some amines (Sec. 10.1-4), dissolve metals to give strikingly similar solutions, but in many instances oxidation by the solvent is so rapid that such solutions have only transient existence.

Nonmetals such as iodine, sulfur, selenium, and phosphorus are somewhat soluble in ammonia, apparently all because of reaction with the solvent. Thus sulfur dissolves to form ammonium sulfide and tetrasulfurtetranitride, S_4N_4.

The solubilities of inorganic salts differ markedly from those in water. Salts that are the most extensively ammonated (i.e., solvated) are the most soluble. The nature of the anion appears to be more significant in ammonia than in water, but the nature of the cation is commonly of lesser importance, except that ammonium salts are quite generally soluble irrespective of the anion present. Certain ammonium salts, notably the nitrate, thiocyanate, and acetate are so soluble and yield solutions over which the vapor pressure is so small that they deliquesce in an atmosphere of ammonia (i.e., are ammonodeliquescent) even at 0°C. The equilibrium pressure of ammonia over a saturated solution of ammonium nitrate in ammonia is so small that this solution (Divers solution) can exist even at room temperature.

For a given cation, solubilities of its halides increase in general from fluoride to iodide. Nearly all fluorides are difficultly soluble, the only readily soluble chlorides are those of ammonium and beryllium, but even the iodides of silver(I), mercury(II), and lead(II) are very soluble. Soluble inorganic salts usually contain such anions as I^-, ClO_4^-, NO_3^-, NCS^-, CN^-, and NO_2^-, whereas salts derived from the anions CO_3^{2-}, $C_2O_4^{2-}$, SO_4^{2-}, SO_3^{2-}, S^{2-}, AsO_3^{3-}, PO_4^{3-}, OH^-, and O^{2-} are difficultly soluble.[29] Hydroxylamine and hydrazine are very soluble.

The solubilities of organic substances in ammonia depend upon the functional groups present and parallel closely those in the lower alcohols. Amines, amides, nitro compounds, and heterocyclic compounds are usually soluble. Like water, ammonia acts as a dispersion medium for such high polymers as cellulose and the proteins.

The direct interpretation of observed solubilities is complicated by the simultaneous influence of large dielectric constant, large dipole moment, high basicity, tendency to form hydrogen bonds, large dispersion forces, and comparatively small molecular volume. Application of simple theory is difficult, but some observed phenomena can be accounted for in terms of single influential factors.[29,44,45] Thus the solubilities of ionic compounds are closely dependent on the dielectric constant. That ammonia is both a strong base and a good hydrogen-bonding agent enhances its ability to dissolve acidic substances. With water, dipole attractions for other species predominate; with ammonia, dispersion and dipole interactions are nearly equal. Hence, ammonia is not as good a solvent as water for dipolar species but better than water for nonpolar (especially organic) substances and for polarizable species containing many electrons (e.g., iodine compounds).

In terms of Hildebrand's treatment,[44,45] change in solubility in a series such as

$$NaF < NaCl < NaBr < NaI$$

is associated with decreases in both melting point and lattice energy. The extremely large solubility of sodium iodide is related also to the formation of $NaI \cdot 4 NH_3$ as an equilibrium solid phase. Decrease in solubility from lithium halides to cesium halides parallels decrease in tendency for ammoniated cations to form. Dissolution of the free metals appears to be favored by the almost negligible self-ionization of anhydrous ammonia, the resistance of ammonia to reduction by solvated electrons, and the ability of ammonia molecules to solvate metal ions. The large variation in solubility of the alkali metal halides results from changes in factors that favor dissolution (e.g., solvation energies of gaseous cations and the electron and favorable changes in entropy) and those that oppose dissolution (e.g., enthalpies of sublimation and ionization energies).

Solutions of Metals in Ammonia.   Of the many published summaries that describe these systems, several are outstanding for their completeness and detailed coverage of the original literature.[12, 27-32, 46-56]

As background for discussion of theoretical implications, it is useful to note a number of the observed characteristics of these systems. The alkali metals dissolve without significant thermal changes. In the absence of certain catalysts (e.g., Fe, Fe_2O_3, Pt) or radiation of wavelength 2150 to 2550 Å, all of which favor amide formation in terms of the general equation

$$M(s) + NH_3(l) \rightarrow MNH_2(solv) + \tfrac{1}{2} H_2(g)$$

the dissolution process involves no observable chemical change. If concentrated ($> 1.0\,M$), the resulting solutions are bronze-colored and metallic in appearance and other properties. If either moderately concentrated (1.0 to 0.5 M) or dilute ($< 0.5\,M$), they are uniformly blue, irrespective of the metal used. In intermediate concentration ranges, separation into a bronze-colored liquid phase floating on a less concentrated blue liquid phase occurs with all the alkali metals except cesium.[57] Transition from metallic character, as indicated by very large equivalent conductance, to electrolytic character, as indicated by ultimate increase in conductance with dilution, occurs in the general 0.05 M range (Fig. 10-2). Solutions of other strongly reducing metals are comparable. It is of interest, in this connection, that blue solutions containing tetraalkylammonium groups are obtained by cathodic reduction of solutions of tetraalkylammonium chlorides in ammonia.[58, 59] These solutions resemble those of the alkali metals in other properties.[60-62]

The dissolution process is accompanied by a substantial increase in volume,

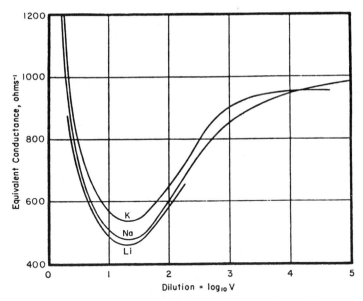

Figure 10-2 Effect of concentration of conductance for solutions of lithium, sodium, and potassium in liquid ammonia at $-33.5°C$. [W. L. Jolly, *Progress in Inorganic Chemistry*, F. A. Cotton (Ed.), Vol. 1, Fig. 6, p. 255, Interscience Publishers, New York (1959)]

or decrease in density. With sodium and potassium, this volume change is largest at a concentration of about 3 M. With lithium, it increases with dilution, but with cesium it decreases with dilution.[63, 64] These solutions are strongly paramagnetic, with molar susceptibilities decreasing with increasing concentration.[65] In general, magnetic data are consistent with the assumption of nearly complete electron pairing at concentrations $>0.1\,M$ and no pairing at concentrations $<0.001\,M$. An absorption band at ca. 6500 Å has been associated with electron pairing; another characteristic, intense absorption band at ca. 15,000 Å has been ascribed to unpaired electrons,[66] although these proposals are not supported by other data. The blue color is due to the short wavelength tail of the band at 15,000 Å.

Chemically, solutions of metals in ammonia are very strong reducing agents and react very much like the free metals themselves. The potentialities of reduction reactions are suggested by the electrode-potential data summarized in Table 10-5.[33] Comparisons with data for aqueous systems (Appendix IV) should be made.

The products of a variety of reduction reactions involving inorganic substances are summarized in Table 10-6.[28, 29, 37, 67] Many organic compounds are also usefully reduced by solutions of the strongly electropositive metals in ammonia.[68-70] In every instance, the reduction process appears to involve comparatively free, possibly ammoniated, electrons, $e^-(NH_3)_x$. The resistance of

Table 10-5 Standard Electrode Potentials in Liquid Ammonia

Couple	Equation	$E°$ (V)
I. ACIDIC SOLUTIONS		
Li(I)–Li(0)	$Li^+ + e^- \rightleftharpoons Li$	-2.34
Sr(II)–Sr(0)	$Sr^{2+} + 2e^- \rightleftharpoons Sr$	-2.3
Ba(II)–Ba(0)	$Ba^{2+} + 2e^- \rightleftharpoons Ba$	-2.2
Ca(II)–Ca(0)	$Ca^{2+} + 2e^- \rightleftharpoons Ca$	-2.17
Cs(I)–Cs(0)	$Cs^+ + e^- \rightleftharpoons Cs$	-2.08
Rb(I)–Rb(0)	$Rb^+ + e^- \rightleftharpoons Rb$	-2.06
K(I)–K(0)	$K^+ + e^- \rightleftharpoons K$	-2.04
$NH_3-e(NH_3)_x^-$	$x\,NH_3 + e^- \rightleftharpoons e(NH_3)_x^-$	-1.95
Na(I)–Na(0)	$Na^+ + e^- \rightleftharpoons Na$	-1.89
Mg(II)–Mg(0)	$Mg^{2+} + 2e^- \rightleftharpoons Mg$	-1.74
Mn(II)–Mn(0)	$Mn^{2+} + 2e^- \rightleftharpoons Mn$	-0.56
Zn(II)–Zn(0)	$Zn^{2+} + 2e^- \rightleftharpoons Zn$	-0.54
Cd(II)–Cd(0)	$Cd^{2+} + 2e^- \rightleftharpoons Cd$	(-0.2)
Ni(II)–Ni(0)	$Ni^{2+} + 2e^- \rightleftharpoons Ni$	(-0.1)
Co(II)–Co(0)	$Co^{2+} + 2e^- \rightleftharpoons Co$	(0.0)
Fe(II)–Fe(0)	$Fe^{2+} + 2e^- \rightleftharpoons Fe$	(0.0)
H(I)–H(0)	$NH_4^+ + e^- \rightleftharpoons \frac{1}{2}H_2 + NH_3$	0.0
N(0)–N($-$III)	$\frac{1}{2}N_2 + 3\,NH_4^+ + 3e^- \rightleftharpoons 4\,NH_3$	$+0.04$
Co(III)–Co(0)	$Co^{3+} + 3e^- \rightleftharpoons Co$	$(+0.2)$
Tl(I)–Tl(0)	$Tl^+ + e^- \rightleftharpoons Tl$	$+0.25$
Pb(II)–Pb(0)	$Pb^{2+} + 2e^- \rightleftharpoons Pb$	$+0.28$
Cu(I)–Cu(0)	$Cu^+ + e^- \rightleftharpoons Cu$	$+0.36$
Cu(II)–Cu(0)	$Cu^{2+} + 2e^- \rightleftharpoons Cu$	$+0.40$
Hg(II)–Hg(0)	$Hg^{2+} + 2e^- \rightleftharpoons Hg$	$+0.67$
N(I)–N($-\frac{1}{3}$)	$3\,N_2O + 6\,NH_4^+ + 8e^- \rightleftharpoons 2\,N_3^- + 3\,H_2O + 6\,NH_3$	$(+0.7)$
Ag(I)–Ag(0)	$Ag^+ + e^- \rightleftharpoons Ag$	$+0.76$
N(V)–N(0)	$NO_3^- + 6\,NH_4^+ + 5e^- \rightleftharpoons \frac{1}{2}N_2 + 3\,H_2O + 6\,NH_3$	$+1.17$
I(0)–I($-$I)	$\frac{1}{2}I_2(s) + e^- \rightleftharpoons I^-$	$+1.26$
O(0)–O($-$II)	$\frac{1}{2}O_2 + 2\,NH_4^+ + 2e^- \rightleftharpoons H_2O + 2\,NH_3$	$+1.28$
Br(0)–Br($-$I)	$\frac{1}{2}Br_2(l) + 2e^- \rightleftharpoons Br^-$	$+1.73$
Cl(0)–Cl($-$I)	$\frac{1}{2}Cl_2(g) + e^- \rightleftharpoons Cl^-$	$+1.91$
F(0)–F($-$I)	$\frac{1}{2}F_2(g) + e^- \rightleftharpoons F^-$	$+3.50$
II. ALKALINE SOLUTIONS		
Ca(II)–Ca(0)	$Ca(NH_2)_2 + 2e^- \rightleftharpoons Ca + 2\,NH_2^-$	-2.83
Li(I)–Li(0)	$LiNH_2 + e^- \rightleftharpoons Li + NH_2^-$	-2.70
Cs(I)–Cs(0)	$Cs^+ + e^- \rightleftharpoons Cs$	-2.08
Rb(I)–Rb(0)	$Rb^+ + e^- \rightleftharpoons Rb$	-2.06
K(I)–K(0)	$K^+ + e^- \rightleftharpoons K$	-2.04
Na(I)–Na(0)	$NaNH_2 + e^- \rightleftharpoons Na + NH_2^-$	-2.02
$NH_3-e(NH_3)_x^-$	$x\,NH_3 + e^- \rightleftharpoons e(NH_3)_x^-$	-1.95
Zn(II)–Zn(0)	$Zn(NH_2)_4^{2-} + 2e^- \rightleftharpoons Zn + 4\,NH_2^-$	(-1.8)
Mn(II)–Mn(0)	$Mn(NH_2)_4^{2-} + 2e^- \rightleftharpoons Mn + 4\,NH_2^-$	(-1.7)

Table 10-5 (*Continued*)

Couple	Equation	$E°$ (V)
H(I)–H(0)	$NH_3 + e^- \rightleftharpoons \frac{1}{2}H_2 + NH_2^-$	-1.59
N(0)–N(–III)	$\frac{1}{2}N_2 + 2NH_3 + 3e^- \rightleftharpoons 3NH_2^-$	-1.55
Pb(II)–Pb(0)	$Pb(NH_2)_3^- + 2e^- \rightleftharpoons Pb + 3NH_2^-$	(-1.4)
Cd(II)–Cd(0)	$Cd(NH_2)_4^{2-} + 2e^- \rightleftharpoons Cd + 4NH_2^-$	(-1.4)
Cu(I)–Cu(0)	$Cu(NH_2)_3^{2-} + e^- \rightleftharpoons Cu + 3NH_2^-$	(-1.3)
Ni(II)–Ni(0)	$Ni(NH_2)_4^{2-} + 2e^- \rightleftharpoons Ni + 4NH_2^-$	(-1.3)
Tl(I)–Tl(0)	$Tl(NH_2)_2^- + e^- \rightleftharpoons Tl + 2NH_2^-$	(-1.3)
Hg(II)–Hg(0)	$Hg_3N_2 + 4NH_3 + 6e^- \rightleftharpoons 3Hg + 6NH_2^-$	(-1.1)
Ag(I)–Ag(0)	$Ag(NH_2)_2^- + e^- \rightleftharpoons Ag + 2NH_2^-$	(-1.0)
N(V)–N(0)	$NO_3^- + 3NH_3 + 5e^- \rightleftharpoons \frac{1}{2}N_2 + 3OH^- + 3NH_2^-$	-0.14
O(0)–O(–II)	$\frac{1}{2}O_2 + NH_3 + 2e^- \rightleftharpoons OH^- + NH_2^-$	(-0.06)
N(I)–N($-\frac{1}{3}$)	$3N_2O + 3NH_3 + 8e^- \rightleftharpoons 2N_3^- + 3OH^- + 3NH_2^-$	(0.0)
I(0)–I(–I)	$\frac{1}{2}I_2(s) + e^- \rightleftharpoons I^-$	$+1.26$
Br(0)–Br(–I)	$\frac{1}{2}Br_2(l) + e^- \rightleftharpoons Br^-$	$+1.73$
Cl(0)–Cl(–I)	$\frac{1}{2}Cl_2(g) + e^- \rightleftharpoons Cl^-$	$+1.91$
F(0)–F(–I)	$\frac{1}{2}F_2(g) + e^- \rightleftharpoons F^-$	$+3.50$

ammonia itself to reduction provides a medium in which highly reduced states are themselves thermodynamically stable.

Some systematization of these reactions results if one considers that each reaction is initiated by one of the following:[28]

1. Electron addition without bond cleavage,

$$e^-(NH_3)_x + X \rightarrow Y^-$$

for example, such conversions as $MnO_4^- \rightarrow MnO_4^{2-}$, $[Ni(CN)_4]^{2-} \rightarrow [Ni(CN)_4]^{4-}$, $NO_2^- \rightarrow NO_2^{2-}$.

2. Bond cleavage by addition of one electron,

$$e^-(NH_3)_x + X—Y \rightarrow X\cdot + .Y^-$$

for example, $NH_3 \rightarrow NH_2^- + \frac{1}{2}H_2$, $NH_4^+ \rightarrow NH_3 + \frac{1}{2}H_2$, $(C_6H_5)_3SnBr \rightarrow (C_6H_5)_3Sn\cdot + Br^-$

3. Bond cleavage by addition of two electrons,

$$2e^-(NH_3)_x + X—Y \rightarrow X^- + Y^-$$

for example, $Ge_2H_6 \rightarrow 2GeH_3^-$, $C_6H_5NHNH_2 \rightarrow C_6H_5NH^- + NH_2^-$

Kraus's early conclusion[49] that cations and mobile electrons present in a highly electropositive metal are separated by reversible solvation,

$$M^+e^-(s) + (x+y)NH_3 \rightleftharpoons M^+(NH_3)_y + e^-(NH_3)_x$$

Table 10-6 Typical Reduction Reactions Produced by Metals in Liquid Ammonia

Substance	Metal	Products	Substance	Metal	Products
NH_4Cl	Ca	H_2, $CaCl_2$	$BF_3 \cdot NH_3$	Li	$(H_2N)_2BNHBNH$
$(NH_4)_2S$	Li	H_2, Li_2S		Na	$(H_2N)_2BNHBFNH_2$
$NH_4C_2H_3O_2$	Na	H_2, $NaC_2H_3O_2$		K	H_2NBF_2
			$AlCl_3$	K	Al
O_2	Li	Li_2O, Li_2O_2	$GaBr_3$	Na	Ga
	Na	Na_2O_2, $NaOH$, $NaNH_2$, $NaNO_2$, NaO_2	InI_3	K	In
S	Li	Li_2S, Li_2S_x	ZnI_2	Na	$NaZn_4$, $NaZn_{12}$
	Na	Na_2S, Na_2S_x	CdI_2	Na	$NaCd_{5-7}$
	Cs	Cs_2S_x	HgI_2	K	Hg, KHg_2
Se	Na	Na_2Se, Na_2Se_x	CuI	Na	Cu
	K	K_2Se, K_2Se_x	CuO	K	Cu_2O
Te	Na	Na_2Te, Na_2Te_x	AgI	K	Ag
			Ag_2O	K	Ag
P	Li	LiP			
	Na	$Na_3P_2H_3$, $Na_3P \cdot 3NH_3$	MnI_2	Na	Mn
As	Na	Na_3As	$NiBr_2$	K	Ni, $Ni(NH_2)_2$
Sb	Li	$Li_3Sb \cdot NH_3$			
			$(C_6H_5)_4Ge$	Na	$(C_6H_5)_3GeNa$

Starting material	Cation	Product	Starting material	Cation	Product
Ge	Na	Na_4Ge_x	AgCN	K	Ag
Pb	Na	$NaPb$, $NaPb_2$, Na_4Pb_9	$Hg(CN)_2$	Ca	Ca_3Hg_2
Hg	Na	$NaHg_8$	$NaNO_2$	Na	Na_2NO_2
N_2O	K	N_2, KOH, KNH_2, KN_3	$NaNO_3$	Na	Na_2NO_2
NO	Na	$NaNO$	$Co(NO_3)_2$	Na	Co
NO_2	Ba	BaN_2O_2	$K_3[Mn(CN)_6]$	K	$K_5[Mn(CN)_6] \cdot K_6[Mn(CN)_6]$
Na_4P_2	Na (+NH_4^+)	$NaPH_2$	$K_3[Co(CN)_6]$	K	$K_4[Co(CN)_4]$
BiI_3	Na	Na_3Bi, Na_3Bi_x	$K_2[Ni(CN)_4]$	K	$K_4[Ni_2(CN)_6]$
GeH_4	Na	$NaGeH_3$	$[Pt(NH_3)_4]Br_2$	K	$[Pt(NH_3)_4]$
GeS	Na	Ge, Na_4Ge_x	$[Fe(CO)_5]$	Na	$Na_2[Fe(CO)_4]$
PbI_2	Na	Na_4Pb_9	$[Mn_2(CO)_{10}]$	K	$K[Mn(CO)_5]$

is consistent with observed conductance phenomena and the transition from metallic solutions to ionic solutions on dilution, but it does not account for all the other properties discussed.

Of the various theoretical approaches that have been advanced,[50, 51, 71] two appear to be most useful: the *expanded metal theory*[72] and the *cavity theory*.[73-77] The former approach considers the valence electrons of the metal as occupying *expanded orbitals* around the ammoniated cation. The neutral species in solution is then an ammoniated metal ion with valence electrons localized on the protons of the solvating ammonia molecules. This species then dissociates reversibly like an ion pair, and two such species may dimerize by sharing electrons of opposed spin. This interpretation is consistent with x-ray data that indicate slight clustering in solution and with magnetic data. The second approach considers that the solvated electrons of Kraus are electrons that occupy cavities produced by oriented ammonia molecules, with their protons directed inward. Dilution gives, in order, electron pairs in these cavities and single electrons. It is the trapped electrons that are believed to be responsible for the blue color. The radii of these cavities have been estimated to be ca. 3.2 Å,[73-75, 78] but other data suggest that they are much smaller. The cavity concept is in accord with the bulk of the experimental data for dilute solutions.[79-81] In more concentrated solutions, clusters of ammoniated metal ions bonded by electrons are present,[30, 50] whereas the metallic-appearing solutions have metallic structures (Sec. 4.4).[30] Support for a *composite model* is given by absorption, ion-transport, and activity-coefficient data for solutions of sodium. The presence of "free" electrons is also supported by polarographic data.[60] Further details may be obtained from extensive published summations.[54, 81-82]

Nonredox Reactions in Ammonia. Reactions of this type may be categorized as (1) ammonation, (2) ammonolysis, (3) acid–base metathetical, and (4) non-acid–base metathetical reactions.

Ammonation reactions involve the direct addition of ammonia molecules to Lewis-acid species and yield ammine-type complexes, for example, $CrCl_3 \rightarrow [Cr(NH_3)_6]Cl_3$; $BF_3 \rightarrow [F_3B(NH_3)]$; $MCl_3 \rightarrow [Cl_3M(NH_3)]$, where M = Al, Ga, In; $SiF_4 \rightarrow [F_4Si(NH_3)_2]$. The resulting ammonates are comparable with hydrates in many of their property types, although they are generally more stable to subsequent ammonolysis at elevated temperatures than hydrates are to hydrolysis. Thus the diammonate of silicon(IV) fluoride is stable at 300°C, and the monoammonates of aluminum, gallium, and indium trichlorides can be volatilized without change in composition. The term ammonation is applied also to a real or hypothetical process in which an aquo compound is converted to its ammono analog.

Ammonolysis reactions are methathetical reactions in which ammonia is a reactant. Like hydrolysis reactions, they are protolytic equilibrium processes and are thus related to acid–base behavior. Here they are considered separately because they involve ammonia directly rather than solutes that react as acids or bases when dissolved in ammonia. An ammonolysis reaction is ordinarily less extensive than an hydrolysis reaction, as may be expected from the smaller ion-product of ammonia (1.9×10^{-33} at $-50°C$, vs. 1×10^{-14} for water at $25°C$). Ammonolysis reactions have been reviewed extensively.[9,32,83] A few typical inorganic reactions are summarized in Table 10-7. For other examples, in particular of the organic type, the indicated references should be consulted. The catalytic effect of an ammonium salt on the ammonolysis of an ester is analogous to the effect of an aqueous acid on the corresponding hydrolysis reaction. As may be expected, ammonolysis is particularly extensive in the presence of amide ion.[32]

Acid–base metathetical reactions are commonly reactions between the ammonium ion (acid) and either soluble or insoluble amides, imides, or nitrides (bases). Amphoteric behavior is noted, as for example in the dissolution of zinc amide, $Zn(NH_2)_2$, in a solution of potassium amide in ammonia to yield the ion $[Zn(NH_2)_4]^{2-}$, which is an ammono analog of the tetrahydroxozincate(II) ion, $[Zn(OH)_4]^{2-}$. Similarly, ammonobasic compounds (e.g., $PbNH_2I$ or $HgNH_2Cl$) are obtained readily. Additionally, many ammono and aquoammono derivatives of nonmetals such as sulfur, phosphorus, and carbon undergo acid–base reactions in ammonia.

Non-acid–base metathetical reactions are often of particular importance in synthesis but do not involve ammonia as a reactant. They depend upon favorable solubility relationships, the enhanced thermodynamic stabilities that result at low temperatures, the almost negligible oxidizing strength of the solvent, and reduced tendencies toward solvolysis. Differences in solubility allow reactions that do not take place in aqueous medium, for example, the precipitation of ammonated barium bromide by reaction of silver(I) bromide with barium nitrate; the formation of ammonium nitrite by removal of sodium chloride that precipitates when solutions of sodium nitrite and ammonium chloride are mixed;[84] the recovery of methylphosphine, CH_3PH_2, from the filtrate remaining after removal of precipitated sodium chloride in the reaction of $NaPH_2$ with methyl chloride;[85] and the precipitation of peroxides of Mg, Cd, and Zn from the decomposition of superoxides prepared by reaction of metal nitrates with sodium or potassium superoxide.[86] Decrease in ease of oxidation or solvolysis is useful in the preparation of cyclopentadiene compounds (Sec. 8.2-3) by the route

$$[M^{II}(MH_3)_6](NCS)_2 \xrightarrow{M^I C_5 H_5} [M^{II}(NH_3)_6](C_5H_5)_2 \xrightarrow{\text{heat}} [M^{II}(C_5H_5)_2]$$

where M^{II} = Mn, Fe, Co, or Ni and M^I = Li, Na, or K,[87,88] or in the preparation of cyanocarbonyl compounds such as $K_2[Ni(CN)_3(CO)]$ or $K_2[Ni(CN)_2(CO)_2]$ from cyano complexes and carbon monoxide.[89]

The Nitrogen System of Compounds. The concept of parent solvents and the existence of formally equivalent groups based upon these solvents (Table 10-3) suggest that the familiar oxygen system of compounds, based upon water, can be paralleled by other systems based upon other parent solvents. The best known

Table 10-7 Ammonolysis Reactions Involving Typical Inorganic Substances

Substance	Ammonolysis Product	Substance	Ammonolysis Product
Cl_2	NH_2Cl, N_2H_4	BCl_3, BBr_3	$B(NH_2)_3$
I_2	$NH_4I + NI_3 \cdot n\,NH_3$	BI_3	$B_2(NH)_3$
		$AlBr_3$	$Al(NH_2)_3$, AlN
S_2Cl_2	S_4N_4, S_7NH	$GaBr_3$	$Ga(NH_2)_3$
SO_2Cl_2	$SO_2(NH_2)_2$, $NH_4N(SO_2NH_2)_2$		
$TeBr_4$	Te_3N_4	Hg_2Cl_2	$HgNH_2Cl + Hg$
		$HgCl_2$	$HgNH_2Cl$
NH_2Cl	N_2H_4	$HgBr_2$	Hg_2NBr
$NOCl$	$NONH_2$		
PCl_3	$P(NH_2)_3$	Na_2O	$NaNH_2$
PCl_5	$[P(NH)(NH_2)]_x$	NaH	$NaNH_2$
$POCl_3$	$PO(NH_2)_3$, $(PON)_x$		
$(PNCl_2)_x$	$[PN(NH_2)_2]_x$	$TiCl_4$	$Ti(NH_2)_3Cl$
$SbCl_3$	SbN	$ZrBr_4$	$Zr(NH_2)Br_3 + Zr(NH_2)_2Br_2$
		VCl_3	$V(NH_2)Cl_2 \cdot 4\,NH_3$
CCl_4	$C(NH)(NH_2)_2$	VCl_4	$V(NH_2)Cl_3 + V(NH_2)_2Cl_2$
$SiCl_4$	$Si(NH_2)_4$, $Si(NH)(NH_2)_2$, $Si(NH_2)_2$	$VOCl_3$	$VO(NH_3)_3$
		$NbCl_5$	$Nb(NH_2)_2Cl_3 \cdot x\,NH_3$
$SiHCl_3$	$[SiH(NH)]_2NH$	$TaCl_5$	$Ta(NH_2)_2Cl_3 \cdot x\,NH_3$
SiH_2Cl_2	$[SiH_2(NH)]_x$	$MoBr_3$	$Mo(NH_2)_2Br \cdot NH_3$
SiH_3Cl	$(SiH_3)_3N$	$MoCl_5$	$Mo(NH)NH_2$
Si_2Cl_6	$[Si(NH)(NH_2)]_2$		
$GeCl_4$	$Ge(NH_2)_2$		
$SnCl_4$	$2\,Sn(NH_2)_3Cl \cdot NH_4Cl$		
SnI_4	$Sn(NH_2)_2I_2$		

of these is the nitrogen system, as developed in particular by E. C. Franklin[7] and extended by other investigators, notably W. C. Fernelius and L. F. Audrieth. In general, substitution of nitrogen-containing groups for equivalent oxygen-containing groups (Table 10-3) gives the nitrogen-system analogs of the oxygen-system substances.

Direct analogies between the two systems are made more apparent by the summation given in Table 10-8,[7] wherein compounds on a given horizontal line are not only formally analogous but also closely similar chemically. Where all oxygen functions have been substituted by nitrogen functions, the resulting species are referred to as *ammono* substances; whereas if both oxygen and nitrogen functions appear, the substances are *aquo-ammono* species. When aqueous solutions of aquo compounds are compared with ammoniacal solutions of ammono analogs, chemical similarities are particularly striking.

Furthermore, analogies may be seen in the comparisons of derivatives of carbonic acid[7] and sulfuric acid,[90] as summarized in formalized hydration–dehydration, ammonation–deammonation, and hydrolysis–ammonolysis schemes in Tables 10-9 and 10-10, respectively. These tabulations are intended only to show interrelationships among the compounds included. It must not be implied that the compounds included are necessarily preparable from each other by the schemes indicated. It may be pointed out also that the reduction of the aquo acid HNO_3 has a striking parallel in the reduction of the ammono acid HN_3 (hydrazoic acid), as

$$HNO_3 \xrightarrow{e^-} HNO_2 \xrightarrow{e^-} (HO)_2NH \xrightarrow{e^-} HONH_2 \xrightarrow{e^-} NH_3$$

nitric   nitrous   dihydroxylimine   hydroxylamine   ammonia
acid   acid

$$HNNN \xrightarrow{e^-} H_2NN{=}NH \xrightarrow{e^-} (H_2N)_2NH \xrightarrow{e^-} H_2NNH_2 \xrightarrow{e^-} NH_3$$

hydrazoic   triazene   triazane   hydrazine   ammonia
acid

The nitrogen-system concept is particularly helpful in placing many seemingly unrelated oxygen and nitrogen compounds into a unified picture in either inorganic or organic chemistry. One is warned, however, that these analogies can be extended too far.

1-4. Derivatives of Ammonia as Basic Solvents

The only other basic solvents of significance are derivatives of ammonia: for example, hydroxylamine, hydrazine, amines and polyamines, and acid amides.[9]

Hydroxylamine. The incomplete numerical data given in Table 10-1 suggest that hydroxylamine, in accordance with its aquoammono composition (Table 10-8), should be comparable with both water and ammonia as a solvent. Early observations[91,92] indicated rather striking resemblances to water in

Table 10-8 Comparisons between Oxygen-System and Nitrogen-System Compounds

General Classes	Aquo Compounds	Aquo-Ammono Compounds	Ammono Compounds
Inorganic compounds			
Metal hydroxides, oxides vs. amides, imides, nitrides	$LiOH$		$LiNH_2$
	CaO		Ca_3N_2
	$Al_2O_3 \cdot x\,H_2O$		$AlN \cdot y\,NH_3$
Acids, acid oxides vs. acids, acid nitrides	$HONO_2$		$HNNN$
	$HONO$		H_2NNNH
	$(HO)_3PO$		$(H_2N)_2PN$
	$HOPO_2$		$HNPN$
	$HOCl$		H_2NCl
	$(HO)_2SO_2$	$HOSO_2NH_2$	
		$(H_2N)_2SO_2$	
Salts vs. salts	Cl_2O		Cl_3N
	CO_2		$(C_3N_4)_x$
	KNO_3		KN_3
	$HgOHCl$		$HgNH_2Cl$
	$K_2[Zn(OH)_4]$		$(K_2[Zn(NH_2)_4])$
	K_2CO_3		K_2CN_2
Miscellaneous	$HOOH$	$HONH_2$	H_2NNH_2
Organic compounds			
Alcohols vs. amines (primary)	CH_3OH		CH_3NH_2
Phenols vs. aromatic amines (primary)	C_6H_5OH		$C_6H_5NH_2$
Ethers vs. amines (secondary, tertiary)	$(C_2H_5)_2O$		$(C_2H_5)_2NH$
Carboxylic acids vs. acid amides, amidines	CH_3COOH	CH_3CONH_2	$CH_3C(NH)NH_2$
Acid anhydrides vs. nitriles	$(CH_3CO)_2O$		CH_3CN
Salts vs. salts	CH_3COOK		$CH_3C(NH)NHK$

Table 10-9 Some Nitrogen Derivatives of Carbonic Acid

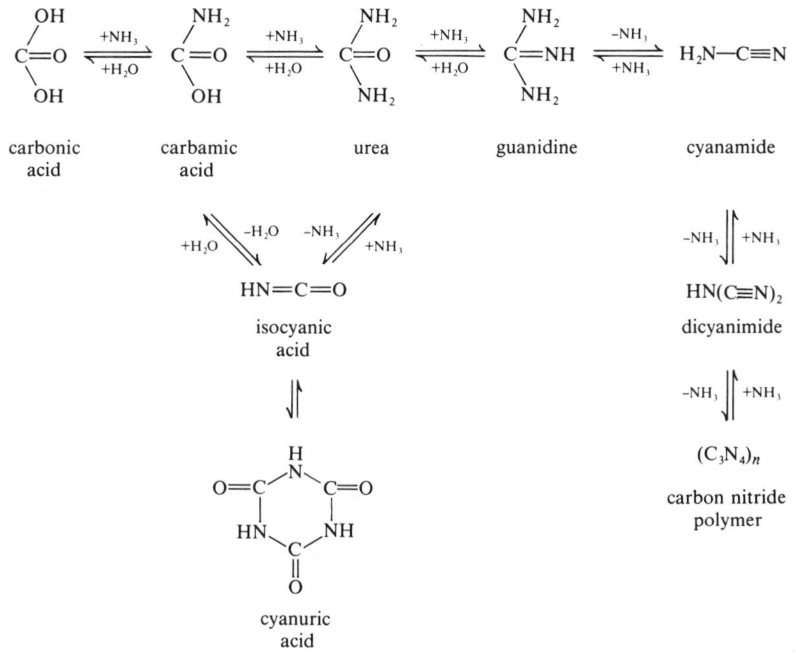

carbonic acid   carbamic acid   urea   guanidine   cyanamide

HN=C=O
isocyanic acid

HN(C≡N)$_2$
dicyanimide

cyanuric acid

(C$_3$N$_4$)$_n$
carbon nitride polymer

solvent character, solvation, amphoterism, and reactions with active metals and recorded sizable solubilities for certain compounds [e.g., KI, KBr, KCN, NaOH, Ba(OH)$_2$, NH$_3$] and lesser solubilities for others [e.g., NaCl, KCl, NaNO$_3$, Ba(NO$_3$)$_2$]. As a practical solvent, anhydrous hydroxylamine has the disadvantages of melting at 32°C, decomposing explosively at higher temperatures, being very hygroscopic, and reacting with both dioxygen and carbon dioxide.[9] It is a strong base in both the Brønsted–Lowry and Lewis senses. Although its solvent properties have not been explored extensively, it can be regarded as the parent solvent of species such as HONHSO$_3$H, HON(NO)OH, HONNOH, and HON(NO)H.[9] Fulminic acid (HONC), known largely in terms of dangerously explosive heavy-metal salts, is related to hydroxylamine in the same way that hydrocyanic acid is related to ammonia.

Hydrazine. Anhydrous hydrazine is an associated liquid with a convenient liquid range and a large dielectric constant (Table 10-1). Like hydroxylamine, it reacts with atmospheric dioxygen and carbon dioxide. It functions as an electrolytic solvent for a variety of salts[9, 93–98] but not for covalent compounds. It is also a strong base and thus an excellent solvating and complexing agent toward metal ions. Solubilities in hydrazine resemble those in ammonia. The alkali metals dissolve by reduction of hydrogen and the formation of hydrazides

Table 10-10 Some Nitrogen Derivatives of Sulfuric Acid

$$\begin{array}{c} OH \\ / \\ SO_2 \\ \backslash \\ OH \end{array} \underset{+H_2O}{\overset{+NH_3}{\rightleftharpoons}} \begin{array}{c} NH_2 \\ / \\ SO_2 \\ \backslash \\ OH \end{array} \underset{+H_2O}{\overset{+NH_3}{\rightleftharpoons}} \begin{array}{c} NH_2 \\ / \\ SO_2 \\ \backslash \\ NH_2 \end{array} \underset{+NH_3}{\overset{-NH_3}{\rightleftharpoons}} \left(\begin{array}{c} NH \\ \| \\ SO_2 \end{array} \right)^a$$

sulfuric acid	amidosulfuric acid (sulfamic acid)	diamidosulfuric acid (sulfamide)	imidosulfuric acid (sulfimide)

+H₂SO₄ ⇅ +H₂O / −H₂O   +H₂NSO₃H ⇅ +NH₃ / −NH₃   +H₂NSO₃NH₂ ⇅ +NH₃ / −NH₃

$$\begin{array}{c} SO_2OH \\ / \\ O \\ \backslash \\ SO_2OH \end{array} \underset{+H_2O}{\overset{+NH_3}{\rightleftharpoons}} \begin{array}{c} SO_2OH \\ / \\ HN \\ \backslash \\ SO_2OH \end{array} \qquad \begin{array}{c} SO_2NH_2 \\ / \\ HN \\ \backslash \\ SO_2NH_2 \end{array} \qquad \begin{array}{c} H \\ N \\ O_2S \diagup \ \diagdown SO_2 \\ | \quad\quad | \\ HN \diagdown \ \diagup NH \\ S \\ O_2 \end{array}$$

disulfuric acid	imidodisulfuric acid	imidodisulfuric diamide (imidodisulfamide)	(trisulfimide)

or

$$\begin{array}{c} H \\ N \\ O_2S \diagup \ \diagdown SO_2 \\ | \quad\quad\quad | \\ HN \quad\quad NH \\ | \quad\quad\quad | \\ O_2S \diagdown \ \diagup SO_2 \\ N \\ H \end{array}$$

(tetrasulfimide)

+H₂SO₄ ⇅ +H₂O / −H₂O   +H₂NSO₃H ⇅ +NH₃ / −NH₃   +H₂NSO₂NH₂ ⇅ +NH₃ / −NH₃

$$\begin{array}{c} SO_2OH \\ / \\ O \\ \backslash \\ SO_2 \\ / \\ O \\ \backslash \\ SO_2OH \end{array} \underset{+H_2O}{\overset{+NH_3}{\rightleftharpoons}} \begin{array}{c} SO_2OH \\ / \\ HN \\ \backslash \\ SO_2 \\ / \\ HN \\ \backslash \\ SO_2NH_2 \end{array} \underset{+H_2O}{\overset{+NH_3}{\rightleftharpoons}} \begin{array}{c} SO_2NH_2 \\ / \\ HN \\ \backslash \\ SO_2 \\ / \\ HN \\ \backslash \\ SO_2NH_2 \end{array}$$

trisulfuric acid	diimidotrisulfuric acid	diimidotrisulfuric diamide (diimidotrisulfamide)

^{a} Not isolable, but a reasonable intermediate.

(MN₂H₃). Compounds of unreactive metals (e.g., Cu, Hg, Ag, Bi, Po, Pt, Pd) are reduced by the solvent to the free metals. Compounds containing active halide or ester groups are solvolyzed (e.g., $KSO_3F \rightarrow KSO_3N_2H_3$, $RCOOR' \rightarrow \rightarrow RCON_2H_3$). Hydrazinium chloride and sulfate, as acids in this solvent, are neutralized by soluble metal hydrazides and dissolve active metals with evolution of hydrogen.

Amines and Polyamines. The lower aliphatic amines have several of the solvent characteristics of ammonia, but as the sizes, and numbers, of aliphatic groups increase, these compounds become more organic in character and lose their ability to dissolve ionic substances.[9] Thus although methylamine is a generally poorer electrolytic solvent than ammonia, it does dissolve a variety of salts and saltlike compounds, in particular nitrates, halides, and lithium salts, but the higher aliphatic amines are relatively poor solvents for these substances. Solubility also decreases markedly from primary to secondary to tertiary amines. The alkali metals dissolve readily in anhydrous methylamine, are less soluble in ethylamine, and are essentially insoluble in other primary amines and in all secondary and tertiary amines.[99] That methylamine is more basic than ammonia renders solutions of the alkali metals in this solvent less prone to oxidation and more convenient to handle. Many of the properties of solutions of metals in methylamine parallel those of solutions in ammonia.[81, 100-102]

Aromatic amines are generally poorer electrolytic solvents than their aliphatic analogs. Of the heterocyclic amines, pyridine has been the most extensively investigated.[9, 103] The ease with which pyridine forms solvates apparently overbalances its unfavorable dielectric constant (12.5 at 20°C) and allows this amine to dissolve a variety of salts.[104, 105] Conductance data[106, 107] and metathetical reactions in this medium[9] indicate the presence of at least ion pairs, if not solvated ions, in solutions of electrolytes. A number of metals (e.g., Li, Na, K, Mg, Ca, Ba, Ag, Cu, Zn, Fe, Pb) have been electrodeposited from solutions of their salts in pyridine.[108]

Anhydrous ethylenediamine (m.p. 11.0°C; b.p. 116.2°C; ΔH_{vap}, 112 kcal mole^{-1} at 20°C; dielectric constant, 12.9 at 25°C; specific conductance, 9×10^{-8} ohm^{-1} cm^{-1}; viscosity, 1.725 centipoise at 25°C) is an excellent electrolytic solvent,[9, 109-111] as indicated by solubility data for salts,[110] conductance data,[109, 112-114] polarographic data,[115, 116] and data for the electrodeposition of metals, particularly the readily oxidizable metals.[109, 112, 113] Anhydrous ethylenediamine is very hygroscopic and strongly basic, and it behaves as a powerful bidentate chelating agent. The ethanolamines are also good electrolytic solvents,[117] but their large viscosities make them difficult to handle experimentally.

Solutions of the alkali metals in amines, and ethers, and hexamethylphosphoramide differ from those in ammonia in exhibiting, except for lithium, two types of absorption bands, one of which is metal-independent and assignable to the solvated electron and the other, the so-called V-band in the infrared, which is metal dependent. The latter is associated with a diamagnetic alkali metal anion M$^-$.* Studies of this type of anion have been particularly extensive with anhydrous ethylenediamine as solvent. Thus treatment of solutions of a sodium salt with cesium in this medium yielded Na$^-$ ions,[118] as did subjecting solutions of sodium bromide or iodide to an electron pulse from a linear accelerator.[119] In each instance, solvated electrons were shown to react with Na$^+$ ions to form Na$^-$ ions.[56]

Enthalpy-cycle calculations for the formation of a salt Na$^+$Na$^-$(s) yielded a value of $+13.7$ kcal (57.3 kJ) mole^{-1},[120, 121] suggesting that if the Na$^+$ ion could be strongly enough bonded by a complexing ligand that is itself resistant to reduction, an isolable salt might result. Reaction of sodium in ethylenediamine

---

* Reference 56 summarizes data from various sources confirming the presence of M$^-$ ions.

with the cryptand $C_{2,2,2}$ (Table 7-9) yielded the isolable salt $Na(C_{2,2,2})^+Na^-$ as a crystalline compound.[120, 122] Elemental sodium dissolves only very slightly (ca. 10^{-6} M) in ethylamine, but when $C_{2,2,2}$ is added solubility increases dramatically (up to 0.2 M), presumably as

$$2\,Na(s) + C \rightarrow (NaC)^+ + Na^-$$

When cooled to $-15°C$ or below, these solutions deposited shiny, gold-colored crystals, which after washing with ether and handled in dry, inert atmospheres, analyzed to $Na_2C_{18}H_{36}O_6N_2$ (i.e., Na_2C). A detailed crystal structure determination[120] agreed completely with the assignment $(NaC)^+Na^-$ with alternate layers of $(NaC)^+$ and Na^- in a hexagonally close-packed arrangement, as shown in Fig. 10-3. Other comparable compounds have been prepared also.*

It is of interest also that by reaction in anhydrous ethylenediamine, 2,2,2-crypt has been used to stabilize alkali metal cations in the formation of crystalline salts containing homopolyatomic anions such as Te_3^{2-},[124] Sb_7^{3-},[125] Bi_4^{2-},[126] Sn_5^{2-},[127] and Pb_5^{2-}.[127] Preparative procedures involve treatment of substances such as $K_2Te + Te$, K_5Bi_4, and $NaPb_{1.7-2}$ with ethylenediamine and the cryptand, followed by crystallization of the resulting highly colored solutions. Comparable simple polyanionic lead compounds (e.g., Na_4Pb_9 and Na_4Pb_7) are obtainable from and are soluble in ammonia.[128]

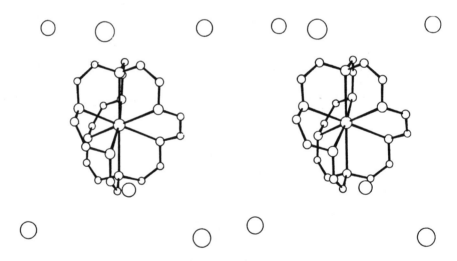

Figure 10-3 Stereoscopic projection of molecular structure of $Na(C_{2,2,2})^+$ and Na^- in $Na(C_{2,2,2})^+Na^-$ crystals. [F. J. Tehan, B. L. Barnett, and J. L. Dye, *J. Am. Chem. Soc.*, **96**, 7203 (1974), Fig. 1. Reprinted with permission from the American Chemical Society.]

---

* In an extension of these studies, by dissolving lithium, potassium, or cesium in liquid ammonia in the presence of a cryptand to trap the M^+ ion, J. L. Dye has obtained a true solution of solvated electrons (i.e., an *electride* system different from an alkalide, or M^-, system). Such a system will dissolve other metals (e.g., Au to give ammonated Au^- ions).[123]

Acid Amides. The low-molecular-weight acid amides and N-substituted acid amides are of particular interest because of their large dielectric constants, convenient liquid ranges, and rather sizable specific conductances (Table 10-11)[129] and of their excellent solvent properties.[9,129] The latter were first emphasized for formamide by P. Walden,[130] who also pointed out its waterlike characteristics and behavior as an electrolytic solvent. More recent investigations have yielded quantitative data on solubilities,[129,131] on electrode potentials (e.g., -2.872 V for K^+, K; -0.759 V for Zn^{2+}, Zn; -0.193 V for Pb^{2+}, Pb),[132] conductances,[133] and transference phenomena.[134] Fused acetamide also resembles water as a solvent,[135,136] giving solutions of salts that are excellent conductors and lend themselves to metathetical and electrolysis reactions. Potentiometric titrations in this medium are comparable with those in water, and autoprotolysis of the solvent to $CH_3CONH_3^+$ and CH_3CONH^- ions ($K = 3.2 \times 10^{-11}$ at 94°C) indicates that acetamide is both a parent solvent and a medium for acid–base reactions.[136]

The N-methyl and N,N-dimethyl acid amides are also interesting electrolytic solvents, with dielectric constants that decrease in the series N-methyl > nonmethyl > N,N-dimethyl acid amide (Table 10-11). Very large dielectric constants also characterize N-ethylacetamide (129.0), N-n-propylacetamide (117.8), N-n-butylacetamide (100.3), N-ethylpropionamide (126.8), N-n-propylpropionamide (118.1), and N-methylbutyramide (107.0).[99] N-Methylacetamide,[137] N-methylpropionamide,[138,139] and N-methylbutyramide[138] have been studied extensively as electrolytic solvents. N,N-Dimethylformamide (DMF) is an excellent solvent for a wide variety of inorganic and organic substances. Many crystalline solvates are known [e.g., $BF_3 \cdot DMF$,[140] and $LnI_3 \cdot 8\,DMF$ (Ln = La, Pr, Nd, Sm, Gd)].[141] Conductance,[141,142] polarographic,[143] and absorption spectral[144] data all emphasize similarities between solutions of electrolytes in water and dimethylformamide.

The excellent summaries by Audrieth and Kleinberg[9] and Vaughn[129] should be consulted for additional data.

1-5. Hydrogen Halides as Acidic Solvents

Comparison of the numerical properties of the higher hydrogen halides (Table 10-12) with those of hydrogen fluoride (Table 10-1) suggests that in the liquid state they should be much poorer electrolytic solvents. Although this suggestion is quite generally correct, extensive work during the last two decades shows that these substances do function as acidic electrolytic solvents and as media in which many reactions can be carried out.[145,146]

The greater versatility of anhydrous hydrogen fluoride and its more extensive use as a solvent require that major emphasis be placed upon it in this section. However, several aspects of solutions in the higher hydrogen halides can be discussed to advantage. After the first definitive investigations that established that these compounds are ionizing solvents,[147] little attention was given to the

Table 10-11 Numerical Properties of Certain Acid Amides

Property	$HCONH_2$	$HCONH(CH_3)$	$HCON(CH_3)_2$	CH_3CONH_2	$CH_3CONH(CH_3)$	$CH_3CON(CH_3)_2$
Molecular weight (amu)	45.041	59.068	73.094	59.068	73.094	87.121
Density (g ml^{-1})	1.1296	0.9988	0.9445	1.159	0.9503	0.9366
	(25°C)	(25°C)	(25°C)	(20°C)	(30°C)	(25°C)
Molecular volume (ml mole^{-1})	39.87	59.14	77.39	50.96	76.92	93.02
	(25°C)	(25°C)	(25°C)	(20°C)	(30°C)	(25°C)
Melting point (°C)	2.55		-61	81.1	29.8	-20
Boiling point (°C)	193	111.2	153	222	206	166–167
Enthalpy of vaporization (kcal mole^{-1})			11.37		14.2 (115–205°C)	10.36
Specific conductance (ohm^{-1} cm^{-1})	3×10^{-5} (20°C)	8×10^{-7} (20°C)		1.5×10^{-5}	1×10^{-7} (40°C)	1×10^{-7}
Dielectric constant	109.5	182.4	36.71	60.6	178.9	37.78
	(25°C)	(25°C)	(25°C)	(94°C)	(30°C)	(25°C)
Viscosity (centipoise)	3.764	1.65	0.796	1.32	3.885	0.919
	(20°C)	(25°C)	(25°C)	(105°C)	(30°C)	(25°C)
Dipole moment (D)	3.37		3.86	3.89	3.5	3.79

Table 10-12 Numerical Constants for HCl, HBr, and HI

Property	HCl	HBr	HI
Formula weight (amu)	36.461	80.912	127.912
Density (g ml^{-1})	1.187	2.603	2.85
	($-114°C$)	($-84°C$)	($-47°C$)
Molecular volume (ml mole^{-1})	30.72	31.08	44.88
	($-114°C$)	($-84°C$)	($-47°C$)
Melting point (°C)	-114.25	-86.92	-50.85
Boiling point (°C at 1 atm)	-85.09	-66.78	-35.41
Critical temperature (°C)	51.3	91.0	150.5
Critical pressure (atm)	80.5	84	80.8
Enthalpy of fusion (kcal mole^{-1})	0.476	0.600	0.686
Enthalpy of vaporization (kcal mole^{-1})	3.860	4.210	4.724
Entropy of vaporization (cal mole^{-1} deg^{-1})	20.5	20.4	20.0
Specific conductance (ohm^{-1} cm^{-1})	3.5×10^{-9}	1.4×10^{-10}	8.5×10^{-10}
	($-85°C$)	($83.6°C$)	($-45°C$)
Dielectric constant of liquid ε	9.28	7.0	3.39
	($-95°C$)	($-85°C$)	($50°C$)
Viscosity (centipoise)	0.51	0.83	1.35
	($-95°C$)	($-67°C$)	($-35.4°C$)

area until 1958, when the first hydrogen dihalide compounds were investigated.[148] Self-ionization occurs as

$$3\,HX \rightleftharpoons H_2X^+ + HX_2^-$$

In the collinear $H-X\cdots H^+$ ions, energies of the hydrogen bonds are 14.2 (X = Cl), 12.8 (X = Br), and 12.4 (X = I) kcal mole^{-1} vs. 58 ± 5 kcal mole^{-1} (X = F).[146] Crystalline salts, stable at 25°C or above and containing HX_2^- ions, are formed with ions such as R_4N^+ (R = CH_3, C_2H_5, n-C_4H_9, C_6H_5), $(C_6H_5)_4As^+$, and $(C_5H_7O_2)_3Si^+$.

Many organic compounds of the Lewis base type dissolve extensively in hydrogen chloride to give conducting solutions. Inorganic compounds are usually somewhat less soluble, but some covalent chlorides (e.g., BCl_3, $SiCl_4$, PCl_3) dissolve extensively without enhancing the conductance of the solvent. Chemical reactions with the solvent impart larger conductances to solutions of substances like PCl_5, $(C_6H_5)_3PO$, $(C_6H_5)_3P$, and $(C_6H_5)_3As$. Typical reactions in hydrogen chloride are summarized by the equations[146, 149]

$$2\,(CH_3)_4NCl + B_2Cl_4 \rightarrow [(CH_3)_4N]_2B_2Cl_6$$

$$(C_6H_5)_3CCl + HCl \rightarrow [(C_6H_5)_3C]^+ + [HCl_2]^-$$

$$(C_6H_5)_3P + BCl_3 + HCl \rightarrow [(C_6H_5)_3PH]^+ + [BCl_4]^-$$

$$PCl_5(s) + HCl \rightarrow [PCl_4]^+ + [HCl_2]^-$$

$$HSO_3Cl + HCl \rightarrow [H_2Cl]^+ + [SO_3Cl]^-$$

$$2\,AsF_5 + 4\,HCl \rightarrow [AsCl_4]^+ [AsF_6]^- + 4\,HF$$

The acid–base, the solvolysis, and some oxidation reactions of the compounds $(C_6H_5)_3M$ (M = N, P, As, Sb, Bi) in hydrogen chloride have been investigated.[150, 151]

In hydrogen bromide, solubility and conductance are limited. Thus the compounds $(C_6H_5)_3M$ (M = N, P, As, Sb, Bi) in ca. 0.1 to 0.2 M solution gave conductance values in the range 7 to 39 cm^2 ohm^{-1} mole^{-1}.[152] Hydrogen chloride showed no acidic properties, but boron tribromide behaved as a solvo acid giving $(C_6H_5)_3NH^+BBr_4^-$, $(C_6H_5)_3PH^+BBr_4^-$, and $(C_6H_5)_3As \cdot BBr_3$.[152] Elemental bromine proved to be the strongest oxidizing agent existent in this solvent, forming, for example, $(C_6H_5)_3PBr^+$ and HBr from $(C_6H_5)_3PH^+$ and Br_3^- from $Br(HBr)_x^-$. Tetrabutylammonium chloride dissolved readily, yielding $(C_4H_9)_4N^+HBr_2^-$ after volatilization of the solvent. Triphenylphosphine yielded $(C_6H_5)_3PH^+Br^-$, but triphenylarsine dissolved and was recovered unchanged. Triphenylstibine yielded $(C_6H_5)_xSbBr_{3-x}$, and triphenylbismuthine was converted completely to $BiBr_3$.

In hydrogen iodide, difficultly soluble compounds (e.g., CsI, CaI_2, AgI, CdI_2, PbI_2, and SnI_4), soluble compounds that give conducting solutions (e.g., BI_3, TlI, GeI_4, PI_3, HgI_2), and soluble compounds that undergo solvolysis or reduction (e.g., $GeCl_4 \rightarrow GeI_4$, $SO_3 \rightarrow H_2S$, $BCl_3 \rightarrow BI_3$) could be distinguished.[153]

In spite of its chemical and physiological reactivity and the experimental difficulties associated with its handling, hydrogen fluoride is one of our most important strongly acidic electrolytic solvents.[154-156] The data in Table 10-13 indicate the versatility of this substance as a solvent.[157,158] Sizable solubility among inorganic compounds is limited essentially to the fluorides. Among organic compounds, those that dissolve extensively do so as a consequence of reaction with hydrogen fluoride.[159] Use of hydrogen fluoride as a medium for organic reactions depends more on its ability to act as an acid catalyst (Table 9-10) than on its solvent power. In terms of the Hammett acidity function (Sec. 10.1-9), anhydrous hydrogen fluoride is one of the most acidic of all pure solvents.

H. Fredenhagen's original summation[160] of dissolution behaviors is still very useful:

1. Normal dissociation of the solute into ions, for example, metal fluorides and some sulfates, perchlorates, and fluoro salts.
2. Prior addition of hydrogen fluoride to the solute, followed by electrolytic dissociation, for example, $SbF_5 \rightarrow SbF_5 \cdot 2HF \rightarrow H_2F^+ + SbF_6^-$; $XeF_6 \rightarrow XeF_6 \cdot HF \rightarrow XeF_5^+ + HF_2^-$.
3. Liberation of a protonic acid, for example, $MX \rightarrow HX$, where M = alkali metal, X = Cl, Br, I, CN.
4. Reaction that alters the solute completely, for example, $SO_4^{2-} \rightarrow SO_3F^-$, $ClO_3^- \rightarrow ClO_2 + O_2 + H_2O$.

It is convenient, however, to supplement this summation by the following classification:[156]

1. Metal fluorides. Soluble ionic fluorides (Table 10-13) enhance conductance by adding metal ions and increasing the F^- and HF_2^- ion concentrations. Conductance behavior indicates essentially complete ionization in dilute solutions. Such solutions are comparable to aqueous solutions of soluble ionic hydroxides—the fluoride in reacting as a base in the one case, the hydroxide in the other.

2. Nonmetal, nonacidic fluorides. The halogen and certain noble gas fluorides dissolve readily in hydrogen fluoride by transfer of fluoride ion to the solvent molecule: for example,

$$ClF_3 + HF \rightleftharpoons ClF_2^+ + HF_2^-$$

$$XeF_6 + HF \rightleftharpoons XeF_5^+ + HF_2^-$$

Conductivity measurements can be used to follow the neutralization of such solutions, as

$$(BrF_2^+ HF_2^-) + (H_2F^+ SbF_6^-) \xrightarrow{\text{(HF)}} (BrF_2^+ SbF_6^-) + 3 HF$$
$$\underset{\text{acid}}{}$$

or

Table 10-13 Solubilities of Some Inorganic Substances in Hydrogen Fluoride

Very Soluble[a]	Slightly Soluble[a]	Difficultly Soluble[a]	Soluble with Reaction	Reaction Product Insoluble	Insoluble and Unreactive
H_2O	BeF_2(0.015, 11°)	AlF_3(<0.002, 11°)	$KCN(\rightarrow HCN)$	$AlCl_3$	$ZnCl_2$
LiF(10.3, 12°)	MgF_2(0.025, 12°)	FeF_2(0.006, 12°)	$NaN_3(\rightarrow HN_3)$	$FeCl_2$	$SnCl_2$
NaF(30.1, 11°)	CaF_2(0.817, 12°)	FeF_3(0.008, 12°)	$K_2SiF_6(\rightarrow SiF_4)$	$MnCl_2$	$NiCl_2$
KF(36.5, 8°)					
RbF(110.0, 20°)	ZnF_2(0.024, 14°)	ZrF_4(0.009, 12°)	$KClO_3(\rightarrow ClO_2, O_2)$	$CeCl_3$	$CdCl_2$
CsF(199.0, 10°)	CdF_2(0.201, 14°)	ThF_4(<0.006, 18°)	$KBrO_3(\rightarrow Br_2)$	MgO	$CuCl_2$
NH_4F(32.6, 17°)	HgF_2(0.54, 12°)	CuF_2	Alkali and	CaO	$AgCl$
AgF(83.2, 19°)	Hg_2F_2(0.87, 12°)	HCl	alkaline earth	Al_2O_3	$AgBr$
TlF(580.0, 12°)	TlF_3(0.081, 12°)	HBr	metal chlorides,	CuO	AgI
SrF_2(14.83, 12°)	SbF_3(0.536, 12°)	HI	bromides, iodides		HgI_2
BaF_2(5.60, 12°)	CoF_3(0.257, 12°)	SiF_4	$(\rightarrow HX)$		HgO
PbF_2(2.62, 12°)	CeF_4(0.10, 12°)	$Co(NO_3)_2$			PbO_2
NbF_5(6.8, 25°)	$CaSO_4$	$Cu(NO_3)_2$			MnO_2
TaF_5(15.2, 25°)	$KClO_4$	$Pb(NO_3)_3$			SnO_2
SbF_5(∞, 25°)	H_2S	$Bi(NO_3)_3$			Cr_2O_3
AsF_5					
XeF_6	CO	Ag_2SO_4			WO_3
SnF_4					
BF_3					
$Hg(CN)_2$	CO_2	$CuSO_4$			
KNO_3		$ZnSO_4$			
$NaNO_3$		$CdSO_4$			
$AgNO_3$					
K_2SO_4					
Na_2SO_4					

[a] In parentheses: solubility in g/100 g HF, temperature. See Ref. 158.

648

$$(BrF_2^+HF_2^-) + (Cs^+BrF_4^-) \xrightarrow{\text{(HF)}} (Cs^+HF_2^-) + 2\,BrF_3$$
$$\text{base}$$

Ionization in these solutions is always limited.

3. **Proton acceptors.** In this type of reaction, proton transfer from solvent molecules to solute species occurs when the latter is water or a soluble alcohol, aldehyde, ketone, ether, carboxylic acid or anhydride, carbohydrate, amine, or aromatic hydrocarbon. The large proton availability in hydrogen fluoride allows bonding of the proton to aromatic species through delocalized π-electron clouds. It is important to note that acetic acid (a weak acid in aqueous solution) is much more extensively ionized in hydrogen fluoride than trifluoroacetic acid (a very strong acid in aqueous systems). The base strength of the solute species controls this type of reaction.

4. **Fluoride-ion acceptors.** Only a few protonic substances (e.g., $HClO_4$, HSO_3F) are stronger acids than hydrogen fluoride and thus capable of acting as acids in this solvent. The most common acids are covalent fluorides such as BF_3, AsF_5, and SbF_5.[161-163] The compounds NbF_5, TaF_5, and SeF_4 are much weaker acids, and the compounds PF_5, SiF_4, and TiF_4 do not exhibit acidic properties in this solvent. The reactions of antimony(V) and arsenic(V) fluorides are typical, as[164,165]

$$SbF_5(l) + 2\,HF(l) \rightleftharpoons H_2F^+ + SbF_6^-$$

$$2\,AsF_5(g) + 2\,HF(l) \rightleftharpoons H_2F^+ + As_2F_{11}^-$$

5. **Nonfluoride salts.** Solvolysis in hydrogen fluoride is so extensive that except for the perchlorate ion oxoanions are converted to other species. Thus potassium nitrate reacts as[166]

$$K^+NO_3^-(s) + 6\,HF(l) \rightarrow K^+ + NO_2^+ + H_3O^+ + 3\,HF_2^-$$

alkali metal sulfates form fluorosulfates that dissolve as such, chlorates yield chlorine(IV) oxide, bromates are reduced to elemental bromine, and potassium perrhenate yields ReO_3F.[167] On the other hand, compounds such as KBF_4, $NaPF_6$, $Ba(PF_6)_2$, $NaAsF_6$, and $NaSbF_6$,[162,168] to whatever extents they dissolve, do so without destruction of the anions.

6. **Dissolution without ionization.** Trifluoroacetic acid and sulfur(VI) oxide appear to dissolve without significantly altering the acidity of hydrogen fluoride, and thus they function as diluents. Covalent hexafluorides might be expected to dissolve without transfer of fluoride ions.[169] Although uranium(VI) fluoride does not dissolve extensively, that portion which is in solution gives no evidence of chemical interaction or ionization.[170] Xenon(VI) fluoride, however, dissolves extensively to give conducting solutions and undergoes rapid fluorine exchange with the solvent, suggesting reaction as noted above.[171,172]

Solvolytic reactions allow the syntheses of numerous species not otherwise readily obtainable (e.g., fluorophosphoric acids and their salts[173] and fluoro-sulfates).[174] Salts containing readily hydrolyzable fluoroanions are preparable by treatment of suitable bases or metals with solutions of certain covalent fluorides in hydrogen fluoride.[161] A case in point is the synthesis of $[XeF_3]^+[SbF_6]^-$ by reaction of xenon(IV) fluoride with antimony(V) fluoride in anhydrous hydrogen fluoride.[175] Other examples are included in subsequent discussions. Anhydrous hydrogen fluoride is also useful for stabilizing, and thus allowing the investigations of, strongly oxidizing substances (e.g., $[N_2F]^+[AsF_6]^-$ and $[N_2F_3]^+[AsF_6]^-$,[176] $[NF_4]^+[AsF_6]^-$,[177] and $[NF_4]_2^+[GeF_6]^{2-}$,[178] and $[XeOF_4]^-$).[179] Some standard electrode potentials for reactions in anhydrous hydrogen fluoride at 273°K have been reported[180] [e.g., -0.29 V for $CdF_2(s)$–$Cd(s)$, 0.00 V for $HF(l)$–$H_2(g)$, $+0.94$ V for $AgF(in\ HF)$–$Ag(s)$, and $+2.708$ V for $F_2(g)$–$F^-(in\ HF)$].

1-6. Anhydrous Sulfuric Acid as an Acidic Solvent

Early studies by A. Hantzsch on anhydrous sulfuric acid as a solvent and reaction medium[181] have been extended from a physicochemical point of view by many investigators, notably L. P. Hammett[182] and R. J. Gillespie[183] and their coworkers. Of the numerous reviews that have been published, several can be consulted with particular profit.[184-187]

Numerical constants for anhydrous sulfuric acid are summarized in Table 10-1. The substantial self-ionization indicated by the large specific conductance is confirmed by equilibrium-constant data as given in Table 10-14.[185]

Although this solvent is primarily acidic in behavior, it is also clearly amphiprotic. The HSO_4^- ion is a strong base in anhydrous sulfuric acid. All bases that are inherently more strongly basic than this ion are completely solvolyzed, as indicated by the following equations:

$$H_2O + H_2SO_4 \rightarrow HSO_4^- + H_3O^+$$

$$SO_4^{2-} + H_2SO_4 \rightarrow 2\,HSO_4^-$$

$$ClO_4^- + H_2SO_4 \rightarrow HSO_4^- + HClO_4$$

$$HIO_3 + 2\,H_2SO_4 \rightarrow HSO_4^- + IO_2HSO_4 + H_3O^+$$

$$SeO_2 + H_2SO_4 \rightarrow HSO_4^- + HSeO_2^+$$

$$N_2O_3 + 3\,H_2SO_4 \rightarrow 3\,HSO_4^- + 2\,NO^+ + H_3O^+$$

$$NO_3^- + H_2SO_4 \rightarrow HSO_4^- + HNO_3$$

$$HNO_3 + 2\,H_2SO_4 \rightarrow 2\,HSO_4^- + NO_2^+ + H_3O^+$$

$$N_2O_5 + 3\,H_2SO_4 \rightarrow 3\,HSO_4^- + 2\,NO_2^+ + H_3O^+$$

$$PO_4^{3-} + 3\,H_2SO_4 \rightarrow 3\,HSO_4^- + H_3PO_4$$

$$H_3PO_4 + H_2SO_4 \rightarrow HSO_4^- + H_4PO_4^+$$

$$C_2H_5OH + 2\,H_2SO_4 \rightarrow HSO_4^- + C_2H_5HSO_4 + H_3O^+$$

The very strong acids in aqueous solution (e.g., $HClO_4$, $H_2S_2O_7$, $H_2S_3O_{10}$,

Table 10-14 Self-Ionization Processes in Anhydrous Sulfuric Acid

Equilibrium Expression	Equilibrium Constant Expression[a]	K^b (at 25°C)
$2\,H_2SO_4 \rightleftharpoons H_3SO_4^+ + HSO_4^-$	$K_{ap} = [H_3SO_4^+][HSO_4^-]$	2.7×10^{-4}
$2\,H_2SO_4 \rightleftharpoons H_3O^+ + HS_2O_7^-$	$K_{id} = [H_3O^+][HS_2O_7^-]$	5.1×10^{-5}
$2\,H_2SO_4 \rightleftharpoons H_2O + H_2S_2O_7$		
$H_2SO_4 + H_2S_2O_7 \rightleftharpoons H_3SO_4^+ + HS_2O_7^-$	$K_{H_2S_2O_7} = [H_3SO_4^+][HS_2O_7^-]/[H_2S_2O_7]$	1.4×10^{-2}
$H_2SO_4 + H_2O \rightleftharpoons H_3O^+ + HSO_4^-$	$K_{H_2O} = [H_3O^+][HSO_4^-]/[H_2O]$	1

[a] K_{ap} = autoprotolysis constant; K_{id} = ionic self-dehydration constant. Numerical values reflect *molal* concentrations.

[b] Comparable values for D_2SO_4 are: $K_{ap} = 4.6 \times 10^{-5}$, $K_{id} = 5.5 \times 10^{-5}$, $K_{D_2S_2O_7} = 2.8 \times 10^{-3}$, $K_{D_2O} = 0.2$.

HSO_3F) are weak acids in anhydrous sulfuric acid. The only acid of even moderate strength thus far described in the literature is tetra(hydrogensulfato)-boric acid, $HB(HSO_4)_4$, which is formed in solution by the reaction of boric acid (or its anhydride) with disulfuric acid, or oleum,[188] as

$$H_3BO_3 + 3 H_2S_2O_7 \rightarrow H_3SO_4^+ + B(HSO_4)_4^- + H_2SO_4$$

This acid can be titrated conductimetrically with the strong base $KHSO_4$ to form the salt $K^+[B(HSO_4)_4]^-$.[188]

Self-ionization is sufficiently suppressed by added HSO_4^- ion that the complicating effects of interionic forces and ion association are eliminated, and solutions in anhydrous sulfuric acid behave ideally with respect to freezing points, conductances, and solubilities. Anhydrous sulfuric acid is a much poorer solvent for nonelectrolytes than for electrolytes, although dissolution is complicated by both solvolysis and the large viscosity of the solvent. Soluble salts include the alkali metal sulfates, nitrates, and arsenates, and thiocyanates; silver nitrate and sulfate; and calcium sulfate, phosphate, and fluoride.[189] Most other sulfates are either difficultly soluble or insoluble. Chlorides, bromides, iodides, and so on, are solvolyzed.

The convenient freezing point (10.371°C) and large freezing point depression constant (6.12°C kg mole^{-1}) combine with properties previously discussed to make anhydrous sulfuric acid particularly useful as a medium for cryoscopic measurements. Such studies, pioneered in 1950 by R. J. Gillespie,[183,190] have provided much of the information necessary to understanding this solvent system.

Disulfuric acid, $H_2S_2O_7$, is closely comparable as a strongly acidic solvent that gives solutions amenable to study by cryoscopic and conductimetric techniques.[191,192] It dissolves elemental sulfur, giving both colorless (S_4^{2+} ion) and blue (S_8^{2+} ion) conducting solutions;[193] converts lead(IV) and tin(IV) acetates to the species $H_2[M(HSO_4)_6]$, which are weak acids in this medium;[193] forms the ions $I(HSO_4)_2^+$ and $I(SO_3F)_2^+$ with iodine(III) acetate and fluorosulfate, respectively, and blue I^{2+} with iodine(I) acetate, bromide, cyanide, or nitrate;[194] with the appropriate anhydrous chlorides yields the acids $H_2[Sn(HSO_4)_6]$, $H_2[Ti(HSO_4)_6]$, $H[B(HSO_4)_4]$, and $H[SbCl_2(SO_3Cl)_4]$; and with antimony(V) fluoride gives the acid $H[SbF_2(SO_3F)_4]$.[195] The first two acids are weak in this medium; the others strong.

1-7. Fluorosulfuric Acid as an Acidic Solvent

Initial studies on the fluorosulfuric acid solvent system by A. A. Woolf[196] have been extended very largely by R. J. Gillespie and his coworkers.[197–202] Extensive reviews can be consulted for more nearly complete details.[203–205]

Inasmuch as fluorosulfuric acid is definitely acidic (albeit weakly so) in anhydrous sulfuric acid, it is a more acidic solvent than sulfuric acid. A comparison of its numerical properties with those of sulfuric acid (Table 10-1) indicates significant advantages in much lower freezing point and viscosity.

Fluorosulfuric acid is readily purified by distillation, and when free from hydrogen fluoride, it does not attack glass and is thus conveniently handled in conventional glass apparatus. Moisture must, of course, be excluded. As suggested by its smaller conductance, autoprotolysis,

$$2\,HSO_3F \rightleftharpoons H_2SO_3F^+ + SO_3F^-$$

is much less extensive than with sulfuric acid. The equilibrium concentration of each ion is ca. 2×10^{-4} mole kg^{-1}. Thus autoprotolysis does not interfere significantly with interpretations of conductance and cryoscopic data for dilute solutions.

Bases in this solvent are those species that yield SO_3F^- ion, either by direct dissolution (e.g., KSO_3F and other soluble fluorosulfate salts) or by reaction with the solvent (e.g., amines, $C_6H_5NO_2$, ClO_4^-). Acids yield the $H_2SO_3F^+$ ion, but few, if any species are stronger proton donors and thus capable of displacing an equilibrium

$$HA + HSO_3F \rightleftharpoons H_2SO_3F^+ + A^-$$

toward the $H_2SO_3F^+$ ion. On the other hand, antimony(V) fluoride gives, in anhydrous fluorosulfuric acid, a conducting solution that can be titrated conductimetrically with a solution of potassium fluorosulfate to a 1 : 1 end point.[198] This observation, plus ^{19}F nuclear magnetic resonance data showing the presence of such species as $SbF_5(SO_3F)^-$ and $Sb_2F_{10}(SO_3F)^-$, indicate that at least the initial reaction can be formulated as

$$SbF_5 + 2\,HSO_3F \rightleftharpoons SbF_5(SO_3F)^- + H_2SO_3F^+$$

Furthermore, the addition of sulfur(VI) oxide to the HSO_3F–SbF_5 solutions results in reactions that may be formulated as

$$n\,SO_3 + SbF_5 \rightleftharpoons SbF_{5-n}(SO_3F)_n \quad (n = 1, 2, 3)$$

$$SbF_{5-n}(SO_3F)_n + HSO_3F \rightleftharpoons H[SbF_{5-n}(SO_3F)_{n+1}]$$

Each of the protonic species formed is an acid in the solvent, and $H[SbF_2(SO_3F)_4]$ is a strong acid,

$$H[SbF_2(SO_3F)_4] + HSO_3F \rightleftharpoons [SbF_2(SO_3F)_4]^- + H_2SO_3F^+$$

Fluorosulfuric acid and solutions of antimony(V) fluoride in this solvent are examples of superacids (Sec. 10.1-9). Gold(III) fluorosulfate reacts, in fluorosulfuric acid medium, as an excellent SO_3F^- acceptor, yielding salts $(Cat)^+[Au(SO_3F)_4]^-$, $(Cat)^+ = Li^+$, K^+, Cs^+, NO^+, ClO_2^+, Br_3^+, Br_5^+, $[Br(SO_3F)_2]^+$, and $[I(SO_3F)_2]^+$.[205a]

As a strongly acidic solvent, fluorosulfuric acid is an excellent medium for carrying out some unusual oxidation reactions (Sec. 10.1-9). It is also a very good fluorinating agent toward many oxides and oxoacids and their salts (Table 10-15).[203,206] The presence of polyfluorosulfuric acids, $H(SO_3)_nF$ ($n = 2$, 3, ..., 7), in solutions of SO_3 in HSO_3F but not in solutions of SbF_5 in HSO_3F, is indicated by ^{19}F nuclear magnetic resonance data.[207]

Table 10-15   Reactions of Inorganic Species with Fluorosulfuric Acid

Reactant	Product
KCl or KF	KSO_3F
$CaCl_2$ or CaF_2	$Ca(SO_3F)_2$
$KClO_4$	ClO_3F
$BaSeO_4$	SeO_2F_2
$BaTeO_4$	$Te(OH)F_5$
BaH_4TeO_6	$TeF_5(SO_3F)$
N_2O_3	$NOSO_3F$
N_2O_5	NO_2SO_3F
NOCl	$NOSO_3F$
P_4O_{10}	POF_3
As_4O_6	AsF_3
As_4O_{10}	$AsF_2(SO_3F)_3$, $AsF_3(SO_3F)_2$, AsF_5
$Bi(OH)_3$	BiF_3
SiO_2	SiF_4
CrO_3, K_2CrO_4, $K_2Cr_2O_7$	CrO_2F_2
$KMnO_4$	MnO_3F

1-8. Some Other Acidic Protonic Solvents

Space does not allow systematic discussion of all other solvents of this type that have been investigated. Consultation of the references cited can provide extensive and useful data for the following substances: hydrogen sulfide,[208] hydrogen cyanide,[209] nitric acid,[210] acetic acid,[9, 211] other carboxylic acids,[211] trifluoroacetic acid,[212] methanesulfonic acid,[213] and trifluoromethanesulfonic acid.[214]

1-9. Protonic Superacid Solvent Systems

The preceding discussions of the anhydrous hydrogen fluoride, sulfuric acid, and fluorosulfuric acid solvent systems have suggested acidities in these media that substantially exceed those achievable in aqueous systems. Because the H_3O^+ ion is the strongest acid that can exist in aqueous systems, all intrinsically stronger acids (e.g., HNO_3, H_2SO_4, HSO_3F, HSO_3CF_3, $HClO_4$) are leveled to its strength by complete reaction with water. In some acidic nonaqueous solvents where leveling processes cannot occur, acid strengths 10^{10} or more times those achievable in water are obtainable.

In dilute aqueous solution, acidity is conveniently measured in terms of pH, that is,

$$pH = -\log [H_3O^+] = -\log [H^+] \tag{10-5}$$

but in sulfuric acid solutions of increasing concentration, the value of $[H_3O^+]$ passes through a maximum even though acidity, as measured by the tendency of the system to react with other bases, continues to increase. Under these conditions, acidity is better measured in terms of displacement of an equilibrium

$$H^+ + B \rightleftharpoons HB^+$$

where B is a suitable indicator (e.g., picramide). For this type of equilibrium, the Hammett acidity function,[215-217] H_0, is defined as

$$H_0 = pK_{BH^+} - \log\left(\frac{[BH^+]}{[B]}\right) \qquad (10\text{-}6)$$

where K_{BH^+} is the ionization constant of the conjugate acid BH^+ and $[BH^+]/[B]$ is an ionization ratio, the values of $[BH^+]$ and $[B]$ ordinarily being determined spectrophotometrically.[218] Thus the more negative the value of H_0 the larger the acidity, or strength, of the acid that transfers its proton to the indicator base.

In dilute aqueous solutions of a particular protonic acid, pH and H_0 are equal.[219] However, as Fig. 10-4 clearly shows, decreasing the water content increases the acid strength markedly and for anhydrous sulfuric acid and hydrogen fluoride gives H_0 values of ca. -12 and -11, respectively, (i.e., acid

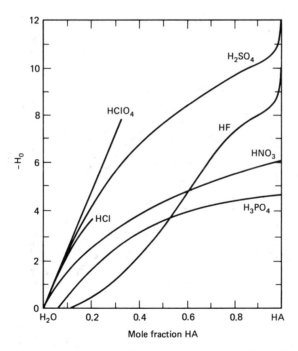

Figure 10-4 Hammett acidity functions for various protonic acid–water systems. [R. J. Gillespie, *Endeavor*, **32**, 3 (1973), Fig. 1. Published by Pergamon Press, Oxford.]

strengths 10^{12} and 10^{11} times those in 1 M aqueous solutions). With the possible exceptions of the $HSO_3CF_3-H_2O^{220}$ and $HClO_4-H_2O$, for which data are incomplete, it appears unlikely that acidities larger than that of anhydrous $H_2SO_4(H_0 = -12)$ can be achieved in other than nonaqueous systems. Those systems with H_0 values larger than -12 are called *superacid systems.*[219, 219a]

As the data in Fig. 10-5 indicate, the systems $H_2SO_4-HB(HSO_4)_4$, $H_2SO_4-SO_3$ (or $H_2SO_4-H_2S_nO_{3n+1}$), $H_2SO_4-HSO_3F$, and $H_2SO_4-HSO_3Cl$ are superacids. Although fluorosulfuric acid is a relatively weak acid in sulfuric acid, there is a rapid increase in H_0 as its concentration increases, until a value of -15.1 is attained for 100% HSO_3F. As previously indicated,[202] the addition of strong Lewis acids (AsF_5, SbF_5, SO_3) enhances the acidity of this substance very markedly. These effects are shown in Fig. 10-6, with maximum measured $H_0 = -19.3$ recorded for a 7 mole % solution of $SbF_5\cdot3SO_3$ in HSO_3F.[221] A 1:1 mole ratio combination of HSO_3F and SbF_5, has been termed "magic acid,"[221] although Fig. 10-5 would suggest that this combination does not represent unique strength. The fact that the compounds SbF_5 and AsF_5, in particular, and NbF_5, TaF_5, and SeF_4, to lesser degrees, give conducting solutions in anhydrous hydrogen fluoride indicates probable minus H_0 values exceeding 11 and possible superacid behavior.

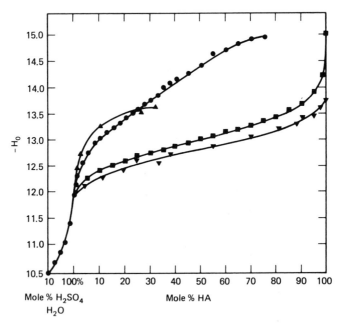

Figure 10-5 Hammett acidity functions for various H_2SO_4-added acid systems. ▲ $H_2SO_4-HB(HSO_4)_4$; ● $H_2SO_4-SO_3$; ■ $H_2SO_4-HSO_3F$; ▼ $H_2SO_4-HSO_3Cl$. [R. J. Gillespie, *Endeavor*, **32**, 3 (1973), Fig. 2. Published by Pergamon Press, Oxford.]

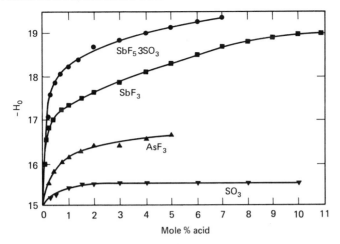

Figure 10-6 Hammett acidity functions for various HSO_3F-added acid systems. [R. J. Gillespie, *Endeavor*, **32**, 3 (1973), Fig. 3. Published by Pergamon Press, Oxford.]

Reactions characteristic of superacid solutions include the following:[219]

1. Protonation of very weak bases such as acetamide, acetic acid, carbonic acid, aromatic species, and biological substances. Aromatic species yield carbonium ions such as p-$HC_6H_5F^+$. The formation of carbonium ions, such as $(C_6H_5)_3C^+$ from $(C_6H_5)_3COH$ in 100% H_2SO_4 or of $(CH_3)_3C^+$ from $(CH_3)_3COH$ in HSO_3F–SbF_5–SO_3, presumably involves protonation, followed by dissociation to R_3C^+ and H_2O.[221] Similarly, the formation of NO_2^+ ion from NO_3^- or N_2O_5 in 100% H_2SO_4 probably involves initial protonation. There are also evidences of protonation of the species Cl_2, Br_2, CO_2, and Xe in SbF_5–HF solutions.[222,223]

2. Formation of halogen and interhalogen cations by oxidation of molecular halogens or interhalogens.[203,219,224,225] Thus the blue I_2^+ ion is readily obtained from elemental iodine by reaction with 65% oleum or with peroxodisulfuryldifluoride ($S_2O_6F_2$) in fluorosulfuric acid.[226] With excess iodine, the latter reagent also forms the ions I_3^+ and I_5^+.[227] Similarly, in the same $S_2O_6F_2$–HSO_3F system, elemental bromine is converted to Br_3^+, $BrSO_3F$, and $Br(SO_3F)_3$,[227] whereas in the system SbF_5–$3SO_3$–HSO_3F, bromine yields the Br_2^+ ion[227] and chlorine(I) fluoride an oxofluoride ion, $OClF^+$ or O_2ClF^+.[228]

3. Formation of sulfur, selenium, and tellurium cations by oxidation of the elements. Elemental sulfur, upon dissolution in oleum, is oxidized successively to S_{16}^{2+} (yellowish orange), S_8^{2+} (blue), and S_4^{2+} (yellow), whereas only the first two species are obtained in the $S_2O_6F_2$–HSO_3F system.[229,229a] Similarly, green Se_8^{2+} and orange Se_4^{2+} are obtainable from elemental selenium and oleum or $S_2O_6F_2$–HSO_3F and red Te_4^{2+} from elemental tellurium and anhydrous H_2SO_4 or HSO_3F.[229,229a]

Additional data on polyhomoatomic cations are presented elsewhere in this volume (Sec. 10.2-2).

1-10. Nonprotonic Solvents

Inasmuch as the major investigative efforts involving nonprotonic solvents have been directed toward acid–base relationships, as delineated in Chapter 9, only a limited amount of information is included here, and for only three representative solvents of this type.

Sulfur(IV)Oxide. Detailed discussions of this solvent system may be consulted.[230-233] Liquid sulfur(IV) oxide is somewhat associated, has a moderately large dielectric constant and a low viscosity, is not difficult to handle (normal b.p. $-10.08°C$), and has roughly the same specific conductance as water (Table 10-1). It functions as a ionizing solvent, but a system based upon self-dissociation to SO^{2+} and SO_3^{2-} ions is not justifiable (Sec. 9.6-1).

Covalently bonded substances are quite generally more soluble than ionically bonded substances. Thus Br_2, ICl, $SOCl_2$, $SOBr_2$, BCl_3, CS_2, $POCl_3$, PCl_3, $AsCl_3$, $SbCl_3$, and Al_2Cl_6 are either completely miscible or very soluble; CCl_4, $SiCl_4$, and other liquid group IVA tetrahalides are miscible above critical miscibility temperatures; and many amines, ethers, esters, alcohols, organic sulfides, mercaptans, carboxylic acids, and aromatic hydrocarbons and their halo and nitro derivatives dissolve readily. Aliphatic hydrocarbons have limited solubilities, as does water.

The solubilities of a number of salts are summarized in Table 10-16.[8,234] Many salts are soluble to the extent of 10^{-2} to 10^{-3} mole liter^{-1}. The anion present is commonly more significant than the cation, with iodides, thiocyanates, and acetates being more soluble than corresponding chlorides or fluorides. Oxides, hydroxides, sulfides, sulfates, and chlorates have small solubilities. So also do sulfites, except for the alkali-metal, thallium(I), ammonium, and tetramethylammonium compounds. A number of substances that might be expected to dissolve in the molecular state yield conducting solutions [e.g., iodine, bromine, certain interhalogens (IBr, ICl, ICl_3); some nonmetal halides and oxohalides (S_2Br_2, PBr_3, $SbCl_5$, $SOCl_2$, $SOBr_2$); a variety of carbinols, ketones, tertiary alkyl halides, and acyl halides; and sulfur(VI) oxide]. These substances are termed *ionogens*.

Soluble ionic compounds (*inophores*) yield conducting solutions as a consequence of the separation of ions induced by their solvation. For a given anion, conductance increases with increasing cationic radius, as

$$Na^+ < NH_4^+ < K^+ < Rb^+ < (CH_3)_4N^+ < (C_6H_5)_3C^+$$

For anions, where for a given cation conductance changes as

$$SCN^- < ClO_4^- < Cl^- < Br^- < SbCl_6^-$$

and there appears to be no size dependence.

Table 10-16 Some Representative Solubilities in Liquid Sulfur(IV) Oxide at 0°C

Solubility in millimoles/1000 g SO_2[a]

Cation	SO_3^{2-}	SO_4^{2-}	F^-	Cl^-	Br^-	I^-	SCN^-	CN^-	ClO_4^-	$CH_3CO_2^-$
Li^+	1.37	1.55	23.0	2.82	6.0	1490.0				3.48
Na^+		i	6.9	i	1.36	1000.0	80.5	3.67		8.90
K^+	1.58	i	3.1	5.5	40.0	2490.0	502.0	2.62		0.61
Rb^+	1.27			27.2	s	s				
NH_4^+	2.67	5.07		1.67	6.0	580.0	6160.0	0.522	2.14	141.0
Tl^+	4.96	0.417	i	0.292	0.60	1.81	0.915	1.42	0.43	285.0
Ag^+	i	i	i	20.07	0.159	0.68	0.845			1.02
Be^{2+}				5.8						
Mg^{2+}				1.47	1.3	0.50				
Ba^{2+}	i		i	i	i	18.15	i			
Cu^{2+}				i		1.17				
Zn^{2+}				11.75		3.45	40.4			
Hg^{2+}		0.338		3.8	2.06	0.265	0.632	0.556		2.98
Pb^{2+}		i	2.16	0.69	0.328	0.195	0.371	0.386		2.46
Co^{2+}				1.00		12.2	i			
Ni^{2+}				i			i			
Al^{3+}				vs	0.60	5.64				0.08
Sb^{3+}			0.56	575.0	21.8	0.26				
Bi^{3+}				0.60	3.44					

[a] i = insoluble, s = soluble, vs = very soluble.

No simple explanation accounts for the dissociation of all ionogens in this solvent. Conductimetric titrations of covalent chlorides such as NOCl, CH_3COCl, and C_6H_5COCl against $SbCl_5$ in this solvent, show end points at $1 : 1$ mole ratios, suggesting adduct formation followed by dissociation, as[235, 236]

$$ORCl + SbCl_5 \rightleftharpoons ORCl \rightarrow SbCl_5 \rightleftharpoons OR^+ + SbCl_6^-$$

but the fact that crystal structure determinations for some of these adducts show coordination through oxygen atoms suggests as a generally more reasonable process[237]

$$2\,SbCl_5 + 2\,ORCl \rightleftharpoons 2\,(ClRO \rightarrow SbCl_5) \rightleftharpoons [(ClRO)_2SbCl_4]^+ + SbCl_6^-$$

Ionization and conductance phenomena have been treated extensively by N. N. Lichtin.[238]

Reactions in sulfur dioxide as a solvent can be classified as:

1. Solvate formation. A number of solvates of alkali metal, alkaline earth metal, ammonium, and substituted ammonium halides, thiocyanates, and sulfates have been isolated. Mole ratios of SO_2 to salt are commonly from 1 to 4; the standard enthalpies of formation are ca. 10 kcal mole^{-1}; the products are often red or yellow; and appreciable pressures of sulfur(IV) oxide are apparent at room temperature. Monoamines commonly form $1 : 1$ adducts; diamines adducts with two moles SO_2 per one mole diamine. Adducts are also formed with π-donor species such as olefins and aromatic compounds. Sulfur(IV) oxide apparently acts as a donor species to strong Lewis acids, forming adducts such as $AsF_5 \cdot SO_2$ [239] and $2\,TiCl_4 \cdot SO_2$.[240]

2. Solvolysis. Typical solvolysis reactions are summarized by the equations

$$Zn(C_2H_5)_2 + SO_2(l) \longrightarrow ZnO + (C_2H_5)_2SO$$

$$PCl_5 + SO_2(l) \longrightarrow POCl_3 + SOCl_2$$

$$CCl_4 + SO_2(l) \xrightarrow{\ Al_2Cl_6\ } COCl_2 + SOCl_2$$

$$COCl_2 + SO_2(l) \xrightarrow{\ Al_2Cl_6\ } CO_2 + SOCl_2$$

$$WCl_6 + SO_2(l) \longrightarrow WOCl_4 + SOCl_2$$

Early suggestions that alkali-metal bromides and iodides are solvolyzed in redox processes to give sulfates, halogens, and S_2Br_2 or S_8 have been disputed[238] and the observed reactions ascribed to dissolved oxygen.

3. Metathetical reactions. Reactions in liquid sulfur(IV) oxide useful for synthesis are described by the equations

$$KI + (CH_3)_3NHCl \rightarrow KCl(s) + (CH_3)_3NHI$$

$$KI + HCl(g) \rightarrow KCl(s) + HI(g)$$

$$NOCl + KSCN \rightarrow KCl(s) + NOSCN$$

$$SOCl_2 + 2\,NH_4SCN \rightarrow 2\,NH_4Cl(s) + SO(SCN)_2$$

$$SOCl_2 + 2\,KBr \rightarrow 2\,KCl(s) + SOBr_2$$

The so-called "neutralization" reactions described by Jander (Sec. 9.6-1) are probably metathetical reactions that do not involve the SO^{2+} ion, for example,

$$K_2SO_3 + SOCl_2 \rightarrow 2\,KCl + 2\,SO_2$$

4. Redox reactions. Liquid sulfur(IV) oxide does not itself normally participate in redox processes, but it does serve as a medium for such reactions. Soluble iodides are oxidized to elemental iodine by reagents such as iron(III) chloride, antimony(V) chloride, or nitrosyl halides or ethoxide (i.e., ethylnitrite).[206] Conversely, elemental iodine can be reduced to iodide ion by soluble sulfites.[241] Treatment of silver thiocyanate with elemental bromine at low temperatures yields thiocyanogen, $(SCN)_2$, which then gives the polypseudohalide $K(SCN)_3$ with potassium thiocyanate.[242]

The hydrogen electrode is quite stable and reproducible in this medium. The calomel electrode is nonreproducible, and the oxygen electrode is irreversible.[243] A few standard potentials have been measured.[243] Electrolytic oxidations and reductions have been studied extensively but with conflicting results.[230,244] Cathodic products appear to be mixtures of sulfites, pyrosulfites, thionates, and so on, but not elemental sulfur.[245] Electrolysis of a mixture of hydrogen bromide and water in sulfur(IV) oxide yields bromine at the anode and water and hydrogen at the cathode.[246] The cathodic products are considered as classic evidence for the existence of the H_3O^+ ion.

5. Complexation reactions. That sulfur(IV) oxide can act as a ligand was first demonstrated by the syntheses of the complexes $[Ru(NH_3)_4(SO_2)(X)]X$ and $[Ru(NH_3)_5(SO_2)]X_2$.[247] In liquid sulfur(IV) oxide, the carbonyl $Fe_2(CO)_9$ is converted to red $Fe_2(CO)_8(SO_2)$,[248] $(\pi\text{-}C_5H_5)Mn(CO)_3$ to $(\pi\text{-}C_5H_5)Mn(CO)_2(SO_2)$,[249] and $Ir(Cl)(CO)[P(C_6H_5)_3]_2$ to $Ir(Cl)(CO)(SO_2)[P(C_6H_5)_3]_2$.[250]

6. Inert medium. Advantages of sulfur(IV) oxide as an inert reaction medium are its low boiling point and thus ease of removal, its low melting point and thus use at low temperatures where side reactions are minimized, its chemical inertness at low temperatures, and its substantial solvating power for both ionic and covalent compounds.

Inorganic reactions of interest include (a) preparation of anhydrous solutions of HBr, as[251]

$$Br_2(l) + 2\,H_2O + SO_2(l) \rightarrow 2\,HBr(in\ SO_2) + H_2SO_4(s)$$

(b) preparation of the mixed anhydrides $N_2O_3 \cdot 3SO_3$, $N_2O_4 \cdot 3SO_3$, and $N_2O_5 \cdot 4SO_3$ by reactions of nitrogen oxides with sulfur(VI) oxide;[252] and (c) preparation of chloroselenic acid, $HSeO_3Cl$ by reaction of SeO_3 with HCl.[253] Liquid sulfur(IV) oxide proves to be an excellent medium for Friedel–Crafts, sulfonation with SO_3, olefin polymerization, carbonium ion, and protonation reactions (using HSO_3F–SbF_5–SO_2 solutions).[232]

Dinitrogen(IV) Tetraoxide. Liquid N_2O_4 has only a reasonably convenient liquid range (m.p. $-12.3°C$; b.p. $21.3°C$), a small specific conductance (1.3×10^{-12} ohm^{-1} cm^{-1} at $17°C$), and a small dielectric constant (2.42 at $18°C$), all properties that militate against its potential as an electrolytic solvent. However, it is well established that both NO^+ (nitrosonium) and NO_3^- ions can exist in this solvent. Furthermore, its behaviors as a medium for acid–base reactions (Sec. 9.6-1) and as a solvolyzing agent, for example,

$$[(C_2H_5)_2NH_2]^+Cl^- + N_2O_4(l) \rightarrow NOCl + [(C_2H_5)_2NH_2]^+NO_3^-$$

suggest that electrolytic dissociation can occur, probably when strongly polarizing groups are present. The characteristics of this solvent system have been reviewed extensively by C. C. Addison.[254]

Liquid dinitrogen(IV) tetraoxide, either pure or containing dissolved nitrosylchloride, converts elemental zinc into the solvate $Zn(NO_3)_2 \cdot 2 N_2O_4$, with the liberation of nitrogen(II) oxide;[255] reacts with zinc in the presence of diethylnitrosoamine to give the salt $[\{(C_2H_5)_2NNO\}_2NO]_2^+[Zn(NO_3)_4]^{2-}$ and liberate nitrogen(II) oxide;[256] dissolves zinc in the presence of alkylammonium nitrates to give the salts $[RNH_3]_2^+[Zn(NO_3)_4]^{2-}$ and liberate nitrogen(II) oxide;[257] and converts basic metal oxides to solvated or ansolvous nitrates at elevated temperatures (Sec. 9.6-1). Each of these reactions is consistent with a presumed composition $(NO)^+(NO_3)^-$ or with dissociation into NO^+ and NO_3^- ions. The compound N_2O_4 behaves only as a weak electrolyte in ionizing solvents in which it is chemically stable.[258]

Bromine(III) Fluoride. The convenient liquid range, large specific conductance, and large entropy of vaporization of bromine(III) fluoride (Table 10-1) favor its behavior as an electrolytic solvent, but its ability to fluorinate nearly every substance that it contacts limits its potential as a solvent[259] and limits solubility in it to inorganic fluorides.

Soluble compounds are of two types: (1) alkali metal, silver(I), and barium fluorides; and (2) fluorides of high-valent nonmetals and metals (e.g., BF_3, SiF_4, GeF_4, SnF_4, PF_5, AsF_5, SbF_5, TiF_4, VF_5, NbF_5, TaF_5, RuF_5, PtF_4, and AuF_3). Detailed studies[260] suggest that dissolution of fluorides of the first group proceeds by fluoride ion acceptance by solvent molecules, for example,

$$KF(s) + BrF_3(l) \rightarrow K^+ + [BrF_4]^-$$

whereas fluorides of the second group dissolve by fluoride ion donation by solvent molecules, for example,

$$SbF_5(l) + BrF_3(l) \rightarrow [BrF_2]^+ + [SbF_6]^-$$

The fact that any member of the first group will react in this solvent with any member of the second group to yield a salt of a fluoro complex anion can be rationalized in terms of a neutralization process, that is,

$$BrF_2^+ + BrF_4^- \rightleftharpoons 2\,BrF_3(l)$$
$$\text{acid} \quad \text{base} \quad \text{solvent}$$

Although the self-dissociation of liquid bromine(III) fluoride,

$$2\,BrF_3(l) \rightleftharpoons BrF_2^+ + BrF_4^- \quad K = c_{BrF_2^+} \times c_{BrF_4^-} = \text{ca. } 4 \times 10^{-4}$$

is well established, as is also the existence of the BrF_4^- ion in salts, independent absolute characterization of the BrF_2^+ ion is lacking[259, 261] and is based more on analogy to x-ray evidence for the ICl_2^+ ion in crystals of the compounds $IAlCl_6$ and $ISbCl_8$.[262] For solutions in anhydrous hydrogen fluoride, however, Raman, infrared, and conductance measurements indicate for the equilibrium

$$BrF_3(l) + HF(l) \rightleftharpoons BrF_2^+ + HF_2^-$$

a maximum BrF_2^+ ion concentration of 1.2 m for a 30 mole % BrF_3 solution.[263]

Bromine(III) fluoride is an excellent medium for the syntheses of fluorides or salts containing fluoro anions, either by direct reactions with other salts or by indirect reactions of the acid–base type.[264–267] Solvolysis is, of course, very extensive. High positive oxidation states [e.g., Pd(IV) in K_2PdF_6, Cr(V) in $KCrOF_4$] are stabilized in this medium.

1-11. Other Electrolytic Solvents

The impracticality of discussing more systems in detail suggests the utility of providing only references to extensive treatments of other nonaqueous solvents. Among those that have been investigated most extensively are: tetramethyl urea;[268] cyclic carbonates;[269] sulfolane;[270, 271] acyl halides;[272] covalent halides and oxohalides,[273–276] especially phosphorus(V) oxotrichloride (pp. 598–599)[20, 21] and selenium(IV) oxodichloride;[277] various halogens and interhalogens;[259, 278] and trifluoroacetic acid.[279]

1-12. Electrochemical Reactions in Nonaqueous Media

Although most of the reactions already considered in this chapter have been of the nonredox type, it has been emphasized that lower oxidation states are thermodynamically better stabilized in basic solvents and higher oxidation states in acidic solvents, and a number of redox reactions have been at least mentioned. Electrochemical oxidation and reduction offer potentially greater versatility in appropriate nonaqueous systems than their chemical counterparts in aqueous systems, and proper choice of solvent avoids the problems of preferential hydrogen or oxygen discharge, hydrolytic decomposition, and limitations on ionic behavior inherent in aqueous systems.[280]

Although a wide variety of inorganic substances, ranging from simple salts of unusual composition [e.g., $Sb(ClO_4)_3$, $Bi(CN)_3$] through organometallic compounds to many compounds of the nonmetals, can be prepared from appropriate nonaqueous solutions by electrolytic processes,[281] limitations in space suggest that the discussion be restricted to the electrodeposition of metals.

The use of strongly basic nonaqueous solvents as reaction media is a reasonable approach to the elimination of preferential hydrogen discharge and to the electrodeposition of metals that are too strongly reducing to exist in contact with water. The use of strongly acidic nonaqueous solvents is a reasonable approach to the elimination of solvolytic destruction of electrolyte or electrolysis product.

Many of the aspects of successful and unsuccessful electrodeposition of individual elements from specific solvents have been reviewed.[281-284] It is apparent that such reactions are most probable with solvents of large dielectric constant, large entropy of vaporization, and low viscosity and with solutes exhibiting at least reasonable conductance in these media.[285] The ionic character of the solute is commonly used as a qualitative measure of the last criterion. That successful electrodeposition is not necessarily dependent upon an optimum combination of these factors, however, has been pointed out by A. Brenner,[283] who notes that diethyl ether ($\varepsilon = 4.33$ at 20°C, $\Delta S_{vapn} = 22$ cal mole^{-1} deg^{-1} at 32.22°C, $\eta = 0.2332$ centipoise at 20°C) is an excellent solvent for the electrodeposition of aluminum, beryllium, and magnesium from solutes that are extensively covalent in the pure state and that ionic solutes do not necessarily give conducting solutions in nonaqueous solvents. Brenner's general conclusion that the formation of a loose ionic complex between solute particles and solvent molecules is essential to conductance and is of prime importance in determining the success or failure of electrodeposition emphasizes the importance of solvation and/or solvolysis in these systems.

Although it is tempting to conclude that any metal can be electrodeposited, given a suitable combination of solute and solvent, it is discouragingly evident that relatively few metals that cannot be deposited from aqueous systems have been successfully electrodeposited from nonaqueous media. Even in aqueous solution where electrode potentials appear to be favorable (e.g., with Ge, Mo, W), electrodeposition is not always achieved because of unfavorable kinetics or reduction mechanisms. The same is true for nonaqueous media. The nature of the cathode is at least as important in nonaqueous media as it is in aqueous medium. Alloy formation (e.g., an amalgam) often permits the deposition of a metal that cannot otherwise be deposited.

The successful electrodeposition of the alkali, alkaline earth, and lanthanide metals from ammonia or amines (Sec. 10.1-3, 10.1-4) indicates the utility of basic solvents for strongly reducing metals. Deposition of arsenic and antimony from anhydrous acetic acid[286, 287] or acetic anhydride[288] illustrates the application of acidic solvents. Cathodic formation of lanthanide–metal amalgams from ethanol[289] or N,N-dimethyl formamide[290] suggests the influence of the cathode. Of interest also are the electrodeposition of aluminum from anhydrous solutions of aluminum chloride and ethylpyridinium bromide in benzene or toluene[291] or from anhydrous solutions of aluminum chloride or of lithium aluminum hydride in diethyl ether;[292, 293] of titanium and zirconium alloys,[294] beryllium and beryllium alloys,[295] or magnesium alloys[296] from comparable ethereal systems; and of germanium from solutions of its tetrachloride in propylene glycol.[297] These few examples illustrate the potentialities of nonaqueous solvents in this aspect of electrochemistry.

2. REACTIONS IN MOLTEN SALT MEDIA

The first applications of reactions involving molten (or fused) salts were those using fluxes to produce slags in metallurgical operations, giving glazes on pottery, and yielding glasses. Such reactions date back into antiquity. Modern attention may well date to Davy's and Bunsen's isolation of the alkali metals by electrolytic decomposition of their molten salts and to the Hall–Héroult process for isolating aluminum by the electrolytic reduction of its oxide in a molten fluoride solvent. Although a variety of industrial applications were effected in the interim, it was not until the 1950s that extensive fundamental research was initiated, largely as a consequence of the need for highly purified refractory metals and for homogeneous molten fuel-solvent materials for power and breeder nuclear reactors. Developments, both fundamental and practical, have been extensive. The brief treatment presented here should be supplemented by reference to detailed summaries.[298–304b]

Molten salts are frequently better and more versatile solvents than water and the other liquids discussed in earlier sections of this chapter. They dissolve substances of widely diverse properties (e.g., water, metals, oxides, other salts, nonmetals, and many covalently bonded substances). Solvent properties and reaction capabilities are affected by the wide range of temperatures available (<100 to $>1000°C$). Media that differ widely in oxidizing–reducing or acidic–basic properties are available. In molten salts, the range in stability of oxidation–reduction systems is substantial, reaction rates are large because of higher temperatures, and yields in chemical reactions in these media are commonly very high (ca. 100%).

The physical properties of molten salts[305] suggest that they are suitable media for a variety of fundamentally ionic reactions. Support of this suggestion is found in a review of well-known industrial processes for the electrolytic preparation of metals and certain nonmetals;[298, 304b–306] by the establishment of a scale of half-reaction potentials (e.g., for LiCl–KCl melts);[307] by investigations of acid–base equilibria in molten media;[304, 308–310] by the study of solvolytic reactions;[311] by establishment of catalysis in these media;[312] and by the use of molten salts as media for syntheses.[313]

2-1. Physical Properties and Classification of Molten-Salt Solvents

The physical properties of molten salts are those associated with both nonionic and ionic liquids.[303, 304] Thus, as shown in Table 10-17, certain of the properties of molten sodium and potassium chlorides are not materially different from those of liquid argon, water, or benzene, even though the temperatures are widely different. The parallelism between molten potassium chloride and liquid argon is particularly striking and may reflect the fact that the species K^+, Cl^-, and Ar are isoelectronic. A major difference is that the presence of ions makes molten salts much better electrical conductors. Thus molten sodium chloride has a specific conductance ca. 10^8 times that of liquid water, and molten potassium chloride at $800°C$ has a specific conductance ca. 22 times that of 1 M aqueous potassium chloride at $20°C$. Molten salts, as a class, are the best ionic conductors known. The strong cohesive forces resulting from electrostatic attractions between oppositely charged ions impart to molten salts long liquid ranges (e.g., ca. $600°C$ for sodium chloride vs. $100°C$ for water). These strong cohesive forces lead to ordered structures based upon ion pairs and complex ions.

On a structural basis, molten salts can be classified roughly in terms of primarily ionic interactions and of significant covalent interactions. Substances of the first type, as exemplified by the alkali metal halides, exhibit short-range order (coordination number ca. 4) based upon anion–cation interactions, are excellent electrolytes, and behave normally cryoscopically. Substances of the second type, as exemplified by mercury(II) bromide,[314] are electrical conductors because of autoionization, as

$$2\ HgBr_2(l) \rightleftharpoons HgBr^+ + HgBr_3^-$$

For these substances, the concentrations of ions present can be altered by the dissolution of other substances. Thus addition of mercury(II) sulfate to molten mercury(II) bromide increases the concentration of the $HgBr^+$ ion (acts as an acid),

$$HgSO_4(s) + HgBr_2(l) \rightarrow 2\ HgBr^+ + SO_4^{2-}$$

and addition of potassium bromide increases the concentration of the $HgBr_3^-$ ion (acts as a base),

$$KBr(s) + HgBr_2(l) \rightarrow K^+ + HgBr_3^-$$

2-2. Chemical Reactions in Molten Salts [300, 302, 304, 304a, 315]

Observed chemical reactions are of several types: for example, (1) those in which the molten salts are reacting species, (2) those that depend upon the molten salts as solvents only, and (3) those for which the molten salts are catalysts. We shall use, however, a more convenient classification in terms of (1) reactions in halide

Table 10-17 Comparison of Some Physical Properties

Property	NaCl (850°C)	KCl (800°C)	H$_2$O (25°C)	C$_6$H$_6$ (25°C)	Ar (−187°C)
Density (g ml^{-1})	1.539	1.510	1.00	0.87	1.407
Viscosity (millipoise)	12.5	11.0	8.95	6.14	2.8
Surface tension (dyne cm^{-1})	111.8	96.8	72	28.2	13.2
Refractive index	1.408	1.385	1.332	1.501	
Thermal conductivity (10^{4}°K) (J sec^{-1} cm^{-1} °C^{-1})		ca. 5–10			12
Heat capacity per atom or ion (10^{16}C$_v$) [erg (°C)$^{-1}$]		3.9			3.2
Self-diffusion coefficient (10^{4} D) (cm^2 sec^{-1})		1.48 (K$^+$) 1.34 (Cl$^-$)			0.16 (Ar)

melts, (2) reactions in other melts, (3) solutions of metals, (4) practical electrochemical reactions, and (5) high-temperature acid–base reactions.

It is important to realize that many reactions either are of the acid-base type or depend upon the acidity or basicity of the molten-salt medium. No uniform overall scale of acid and base strengths has been established for molten-salt systems, although nephelauxetic effects (Sec. 8.1-3) in the outer s and p orbitals of the probe ions Tl^+, Pb^{2+}, and Bi^{3+} noted in ultraviolet spectra have been used to establish scales of optical Lewis basicity for oxide systems and extended to LiCl–KCl(41% KCl) and NaCl–AlCl$_3$(67% AlCl$_3$) melts.[316] In a chloride melt, the chloride ion is a Lewis base, and the larger its concentration, the more basic the system is. Chloride ion concentration may be expressed as pCl$^-$, analogous to pH in an aqueous system. In acidic sodium tetrachloroalumi-nate(III) melts at 175°C, the ions O^{2-}, S^{2-}, Se^{2-}, Te^{2-} $(=X^{2-})$ react as

$$3\,Al_2Cl_7^- + X^{2-} \rightleftharpoons AlXCl + 5\,AlCl_4^-$$

but in basic melts they react as

$$2\,Al_2Cl_7^- + X^{2-} \rightleftharpoons AlXCl_2^- + 3\,AlCl_4^-$$

with base strength varying as Te^{2-}, $Se^{2-} < S^{2-} < O^{2-}$.[317]

Reactions in Halide Melts. The molten LiCl–KCl eutectic (450°C) has been extensively investigated.[307] Typical are investigations of the reduction of the CrO_4^{2-} ion, for which an initial step results in the formation of relatively stable Li_3CrO_4.[318] Disproportionation of this compound is catalyzed by divalent cations, acting as acids.[319] Calcium and cadmium ions yield CrO_4^{2-} and $LiCrO_2$(s), but with increasing acidity of the cation (e.g., Ni^{2+}, Zn^{2+}, Mg^{2+}) $Li_xM_y^{II}CrO_4$ $(x + 2y = 5)$ results.[319]

Complications with simple alkali metal halide melts resulting from both high melting points and large halide-ion activities (concentrations) are reduced by the addition of zinc or aluminum halides. Typically for the NaCl–AlCl$_3$ system, the equilibria described as

$$AlCl_3 + Cl^- \rightleftharpoons [AlCl_4]^-$$

$$2\,AlCl_3 + Cl^- \rightleftharpoons [Al_2Cl_7]^-$$

are strongly displaced toward the chlorocomplexes, thereby allowing the chloride ion concentration (i.e., the basicity of the system) to be controlled.* These effects are indicated in Figs. 10-7 and 10-8. Melting points are in the range 100 to 200°C, and nuclear magnetic resonance data[320] suggest essentially ionic $M^+Al_2Cl_7^-$ as the nature of the melt.

Acidic systems of this type have proved ideal for the formation of homopoly-atomic cations such as Cd_2^{2+},[321] Bi_5^{3+} and Bi_4^+,[322] S_2^{2+} and S_4^{2+},[323] and Te_4^{2+},

---

*   A potentiometric investigation of the aluminum(III) chloride-1-butylpyridinium chloride system at 30° gave $K = <3.8 \times 10^{-13}$ for the equilibrium [319a]

$$2\,AlCl_4^- \rightleftharpoons Al_2Cl_7^- + Cl^-$$

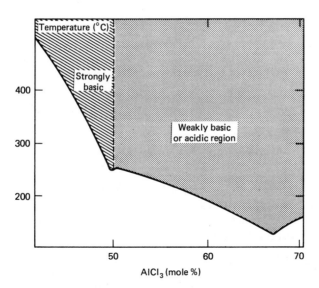

Figure 10-7 Schematic liquidus phase diagram for an alkali metal chloride–aluminum(III) chloride system. [J. H. R. Clarke and G. J. Hills, *Chem. Br.*, **9**, 12 (1973), Fig. 5.]

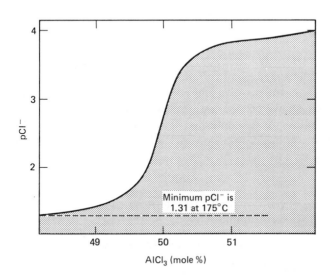

Figure 10-8 Changes in chloride-ion activity with composition in molten NaCl–AlCl$_3$ system at 175°C. [J. H. R. Clarke and G. J. Hills, *Chem. Br.* **9**, 12 (1973)., Fig. 6.]

Te_6^{2+}, and Te_8^{2+}.[324-326] Similarly, the compound $Bi^+(Bi_9)^{5+}[HfCl_6]_3^{2-}$ was isolated.[327] In each of these instances, reaction of the free element with its anhydrous chloride and anhydrous aluminum chloride, or hafnium tetra-chloride, at elevated temperatures was used. Ions of this general type[328] are subject to disproportionation by solvolysis involving chloride ion, as, for example,

$$Cd_2^{2+} + 4\,Cl^- \rightleftharpoons [CdCl_4]^{2-} + Cd(s)$$

Stabilization is thus effected by reduction of the activity of the chloride ion through formation of chloroaluminate (or chlorohafnate) ion in a resulting system of low basicity. The transparency of chloroaluminate and chlorozincate solutions in the ultraviolet region of the spectrum has allowed spectrophoto-metric characterization of equilibria such as[329]

$$2\,Bi(l) + Bi^{3+} \rightleftharpoons 3\,Bi^+ \qquad \text{(in NaCl–ZnCl}_2\text{)}$$

$$4\,Bi(l) + Bi^{3+} \rightleftharpoons [Bi_5]^{3+} \qquad \text{(in NaCl–AlCl}_3\text{)}$$

$$6\,Bi^+ \rightleftharpoons [Bi_5]^{3+} + Bi^{3+} \qquad \text{(in NaCl–ZnCl}_2\text{)}$$

Somewhat different is the reaction of S_8 with fused aluminum(III) chloride at 175°C in a sealed ampoule to yield white $SCl_3^+ AlCl_4^-$, as characterized by x-ray powder patterns and Raman and infrared data.[329a]

Large halide-ion concentrations also favor the formation and allow the investigation of halocomplex species, that, because of competition by solvent molecules, cannot be adequately stabilized in aqueous systems (e.g., $[CoCl_4]^{2-}$, $[CrCl_6]^{3-}$, $[TiCl_6]^{3-}$, $[NiCl_4]^{2-}$, $[NiCl_3]^-$).

Reactions in Other Molten Salts. Molten alkali metal acetates, nitrites, nitrates, and thiocyanates and certain of their eutectic mixtures have desirably low melting points and dissolve many ionic compounds. However, only nitrates and thiocyanates have been investigated extensively.

Molten nitrates are distinguished from molten halides in being effective oxidizing agents. It has been proposed,[330,331] that the oxidant is the nitronium ion (NO_2^+), formed as

$$NO_3^- \rightleftharpoons NO_2^+ + O^{2-}$$

the activity, or concentration, of this ion being enhanced by the addition of a strong Lewis acid, for example, $S_2O_7^{2-}$,

$$NO_3^- + S_2O_7^{2-} \rightleftharpoons NO_2^+ + 2SO_4^{2-}$$

and reduced by reaction with nitrate ion,

$$NO_2^+ + NO_3^- \rightarrow 2\,NO_2(g) + \tfrac{1}{2}O_2(g)$$

Much of the chemistry associated with molten nitrate systems appears to be explicable in these terms.[331] However, it has been suggested[332] also that the $S_2O_7^{2-}$ ion reacts with nitrate ion as

$$2 NO_3^- + S_2O_7^{2-} \rightarrow 2 NO_2(g) + \tfrac{1}{2}O_2(g) + 2 SO_4^{2-}$$

and that the oxide ion can be present only in very small concentration as a consequence of peroxide or superoxide formation,[333]

$$NO_3^- + O^{2-} \rightarrow NO_2^- + O_2^{2-}$$

$$2 NO_3^- + O_2^{2-} \rightarrow 2 NO_2^- + 2 O_2^-$$

Furthermore, the position of the equilibrium

$$NO_2^- + O_2^{2-} \rightleftharpoons NO_3^- + O^{2-}$$

is apparently important in determining the nature of basic species in both fused nitrates and nitrites.[334]

The $LiNO_3$–KNO_3 eutectic (38 mole % $LiNO_3$, m.p. 132°C) has been investigated extensively.[335-337a] A number of metals (e.g., Na, Mg, Ca, Hg) are oxidized to insoluble oxides in the pure nitrate melt but yield simple cations (e.g., Mg^{2+}, Ca^{2+}, Mn^{2+}, Fe^{3+}, Co^{2+}) when the melts are rendered acidic with $K_2S_2O_7$ or $K_2Cr_2O_7$, with evolution of $NO_2(g)$, $N_2O(g)$, and $N_2(g)$.[335] Reaction of manganese(IV) oxide with the eutectic melt at 160°C gave as products Mn^{2+} ion, $NO_2(g)$, and $O_2(g)$, consistent with the presence of NO_2^- ions.[336] Mercury(I) and (II) sulfates are converted initially in this pure eutectic, or the eutectic containing Na_2O_2, Na_2O, or NaOH, to green $HgSO_4 \cdot 2\,HgO$ and ultimately to brown HgO.[336] Zirconium(IV) sulfate acts as a Lux–Flood acid (Sec. 9.6-1), yielding first a basic nitrate and then $ZrO_2(s)$, with evolution of $NO_2(g)$ and $O_2(g)$.[337] For other details, a review by D. H. Kerridge[338] can be consulted.

The alkali metal thiocyanates and their eutectic mixtures melt in convenient temperature ranges (ca. 120 to 320°C) to liquids of moderate density (1.6 to 1.7 g ml^{-1}) and good conductivity (0.67 and 0.16 ohm^{-1} cm^{-1} for NaSCN and KSCN at their melting points).[339] Thermal decomposition above ca. 300°C yields blue solutions, containing presumably the S_2^{2-} ion,[340] which evolve nitrogen above 400°C. Electrolytic reduction of molten NaSCN–KSCN occurs at -1.75 V and oxidation to parathiocyanogen, $(SCN)_x(s)$, at $+0.25$ V (both vs. the Ag/AgCl electrode). Dissolution in molten thiocyanates is often accompanied by reduction (e.g., $I_2 \rightarrow I^-$, $K_2Cr_2O_7 \rightarrow Cr_2O_3$, $FeCl_3 \rightarrow Fe^{2+}$, $Cu^{2+} \rightarrow Cu^+$) or complexation (e.g., $Ca(SCN)_2 \rightarrow [Ca(SCN)_4]^{2-}$) by the thiocyanate ion.[339] The electronic spectra of iron(II), cobalt(II), and nickel(II) are consistent with N-bonded, or isothiocyanato, complexes of octahedral, tetrahedral, and octahedral geometry, respectively.[341] Molybdate(VI) ion and MoO_3 were reduced at 200°C, largely to insoluble molybdenum(IV and III) sulfides; rhodium(III) chloride and silver(I) oxide formed the corresponding sulfides; and silver(I) nitrate gave a mixture of the sulfide and cyanide.[342]

Solutions of Metals. Molten salts are commonly good solvents for metals.[300,302,303,343] Ordinarily, a given metal will dissolve only in one of its own molten salts, although displacement solubility is also observed. Thus elemental potassium dissolves in molten sodium chloride as a consequence of

$$K(l) + NaCl(l) \rightarrow KCl(l) + Na(l)$$

both products being soluble. Dissolution of a metal gives an intense color, for example, deep blue for sodium in sodium chloride, red for cadmium in cadmium halides. Solubilities vary widely, from complete miscibility of cesium in molten cesium halides to analytically nearly complete immiscibility of lead or tin in its molten halides. Two classes of solutions can be distinguished: those solutions in which solute–solvent interactions are negligible (e.g., alkali metal–alkali metal halide; alkaline earth metal–alkaline earth metal halide) and those solutions in which solute–solvent interactions are substantial (e.g., zinc family metals–zinc family halides; $Bi-BiCl_3$).

Solutions of the first class are excellent electrical conductors. Thus a solution of potassium in molten potassium chloride, for which the mole fraction of potassium is 0.42, has a conductance ca. 400 times that of the molten salt. Extrapolations to infinite dilution indicate that conduction is electronic, resulting from oxidation of the metal atoms, as

$$K \rightleftharpoons K^+ + e^-$$

These solutions are called *metallic solutions*. It has been suggested that the trapping of electrons to give M_2 species becomes increasingly important at higher metal concentrations. The proportion of diatomic species in molten alkali metal halides decreases as Li > Na > K > Rb > Cs, the reverse of solubility changes.

Solutions of the second class, or *nonmetallic solutions*, have metallic luster, but their conductances differ only insignificantly from those of the pure molten salts. Electrons are thus strongly held and unavailable for conduction. The d-transition, post-d-transition, and f-transition metal halide-metal solutions are examples. The formation of lower halides is characteristic, with stabilization effected by removal of halide ions as halocomplex species (e.g., Cd_2Cl_2 as $Cd_2^{2+}[AlCl_4]_2^-$, Bi_5Cl_3 as $Bi_5^{3+}[AlCl_4]_3^-$). The lanthanide metals (Ln) dissolve extensively in their molten halides, but the resulting systems are varied and somewhat more complex,[343-345] with in some instances reduction of LnX_3 to LnX_2 (e.g., $PrCl_2$, LaI_2) accompanying dissolution. Not infrequently, solidification results in disproportionation to the original metal and its salt.

Practical Electrochemistry.[304b]   Technically, a number of the more reactive metals are recovered by electrodeposition from molten salts (e.g., Li, Na, Mg from fused chlorides; Al from oxide dissolved in molten fluorides). A primary cell, described as

$$\text{steel} | \text{Li(l)} | \text{LiCl(l) at } 900°\text{K} | Cl_2(g) | \text{porous graphite}$$

is an example of a lightweight combination that delivers the high voltage at large power density (2.8 V at 30 W cm^{-2}) that is desirable for electrically powered small vehicles.[346]

Table 10-18 Some High-Temperature, Molten-Salt, Acid–Base Reactions in the Lewis Sense

Base[a]	Acid	Neutralization Products	Applications
O^{2-} (from M_2O, MOH, M_2CO_3, M_2SO_4)	SiO_2	SiO_4^{4-}	Manufacture of glass, cement, ceramic products; slag formation
	B_2O_3	BO_3^{3-}	
	Al_2O_3	AlO_3^{3-}	
O^{2-} (from M_2O)	$B_4O_7^{2-}$	BO_3^{3-}	Borax bead tests
	$(PO_3)_n^{n-}$	PO_4^{3-}	Phosphate bead tests
O^{2-} (from M_2O, etc.)	$S_2O_7^{2-}$	SO_4^{2-}	Opening refractory minerals
	HSO_4^{-}	SO_4^{2-}	
S^{2-} (from M_2S)	FeS	FeS_2^{2-}	Orford process for Ni–Fe separation
	Cu_2S	CuS^{-}	
F^{-} (from MF)	BeF_2	BeF_4^{2-}	Electrolytic melts, syntheses of complex ions
	AlF_3 or Al_2O_3	AlF_6^{3-}	
	TaF_5	TaF_7^{2-}	

[a] M = alkali metal (Na, K) or, in some instances, alkaline earth metal (especially Ca).

High-Temperature Acid–Base Reactions.   In the Brønsted–Lowry sense, fused "onium" salts are expected to react as acids.[347] It is observed, for example, that molten ammonium nitrate converts many metal oxides, hydroxides, and carbonates to nitrates and that ammonium chloride at elevated temperatures is effective in soldering fluxes, in the dehydration of hydrated metal chlorides that might otherwise undergo hydrolysis on heating, and in the preparation of anhydrous, nonvolatile chlorides (e.g., $LnCl_3$) by reaction with basic oxides. Many other reactions, some involving fusion and other reactions between solids, are conveniently classified as acid–base reactions in the Lewis sense. A few practical examples chosen from the fields of ceramics and metallurgy are summarized in Table 10-18.[347]

Although these reactions are formulated in simple terms, many are undoubtedly very complex. Thus an acidic oxide like SiO_2 or B_2O_3 exists as a giant molecule (Sec. 5.1-4), which must undergo depolymerization before reacting as indicated. Strong anion bases (e.g., O^{2-} ion) at elevated temperatures effect depolymerization in the same sense that isopolyanions are depolymerized in aqueous solution by OH^- ion (Sec. 10.4-1). Metaphosphate reactions proceed similarly, since no metaphosphate ion simpler than the cyclic trimer, $P_3O_9^{3-}$, can be distinguished (Sec. 6.2-1). Many other comparable examples could be cited.

3. REACTIONS INVOLVING SOLIDS

Detailed interest in the nature and applications of inorganic reactions involving solids has increased so substantially in the last few years that journal and monograph coverage and symposia are now quite common, and solid-state chemistry has become a significant area in inorganic chemistry.* To a large degree, this interest has been prompted by unique electronic properties which are dependent upon crystal structures, but a necessary component involves reactions in the solid state. The very brief treatment presented here can be supplemented by consultation of more thorough discussions.[348–356]

Reactions involving solids are broadly classified as those occurring between or within solids and those occurring between solids and either liquids or gases. Reactions of the first class are the ones to which solid-state chemistry is usually restricted. Many reactions of the second class have been considered in earlier sections of this book, and the ones we consider here, because of their pertinence to those of the first class, are chemical transport reactions from solid through gas to solid.

3-1. Reactions Involving Only Solids

Earlier discussions on crystal structures and the general properties of solids indicate clearly that, unlike reactions in the gaseous state or the liquid state,

---

*   Various aspects of solid-state chemistry, in particular as they apply to instructional presentations, are covered by various authors in *J. Chem. Educ.*, **57**, 531–590 (1980).

reactions in the solid state are controlled by the spatial coordinates of systems in question and depend upon the movement of boundaries between solid phases.[350] Two fundamental processes are involved: (1) the chemical reaction itself, involving bond-breaking and bond-making operations that result in the production of nuclei of a new phase, and (2) transport of material to the reaction zone to effect the growth of these nuclei.

The rate of the latter process depends upon factors that are of little importance in reactions between fundamental particles in the gaseous or liquid phase, that is, the mean length of the diffusion path of the grain of solid in question (ca. μm) and the small rate of diffusion within a solid (ca. 10^{-6} that in a liquid or gas at the same temperature). In general, diffusion becomes significant enough to allow measurable reaction only when the temperature is about two-thirds the fusion temperature of the lower-melting solid (Tammann's rule). Lattice defects or imperfections (Sec. 4.3-5) serve as centers at which chemical reactions occur,[351] and the concentrations of these imperfections increase with increasing temperature. Such regions of microheterogeneity[357] promote the formation of new defects and the development of new structural entities. Equilibrium considerations are thus important.[348, 351]

Many reactions of this type proceed as a consequence of the diffusion of ions. For example, the reaction described as

$$2\,CoO(s) + TiO_2(s) \rightarrow Co_2TiO_4(s)$$

apparently occurs as a consequence of diffusion of Co^{2+} and Ti^{4+} ions across a solid–solid boundary, leading to reactions that can be formulated as[358]

$$Ti^{4+} + 4\,CoO(s) \rightarrow Co_2TiO_4 + 2\,Co^{2+}$$

$$2\,TiO_2(s) + 2\,Co^{2+} \rightarrow Co_2TiO_4 + Ti^{4+}$$

Within a given compound, however, the anion and the cation seldom have the same mobility. Thus the Ag^+ ion moves more rapidly than the S^{2-} ion, and in the tarnishing of elemental silver by sulfur, silver ions move through a film of Ag_2S on the outer surface to contact additional sulfide ions. In general, the rates of diffusion of metal ions dictate the rates of formation of spinels and other similar species from metal oxides.

Many different types of inorganic substances can be synthesized by solid-state reactions.[359, 360]

3-2. Chemical Transport Reactions

A chemical transport reaction is one in which a condensed phase reacts with a gaseous phase to form only gas-phase reaction products, which themselves react at a different location in the system to give a condensed phase as a final product.[361-363] Each condensed phase is normally a solid. In the simplest sense, a reversible process occurs and can be formulated as

$$x\,A(s) + y\,B(g) \underset{T_2}{\overset{T_1}{\rightleftharpoons}} z\,C(g) \qquad T_2 > T_1$$

solid A having negligible vapor pressure at the lower temperature T_1. An early example was R. Bunsen's explanation in 1852 that hydrogen chloride in volcanic gases effected transport of iron(III) oxide as

$$Fe_2O_3(s) + 6\,HCl(g) \rightleftharpoons 2\,FeCl_3(g) + 3\,H_2O(g)$$

An excellent example of a commercial application dating to the 1890s is the Mond–Langer process[364] for the purification of elemental nickel via the formation of the gaseous tetracarbonyl at 45 to 50°C and its transport to another zone where it decomposes at 180 to 200°C:

$$Ni(s) + 4\,CO(g) \; \underset{\text{180 200 C}}{\overset{\text{45 50 C}}{\rightleftharpoons}} \; Ni(CO)_4(g)$$

Inasmuch as the extent to which transport occurs depends upon the movement of molecules in the gaseous phase, the difference in partial pressures (ΔP) of the gaseous species between the initial and final positions must determine whether or not transport occurs. Both the change in enthalpy (ΔH) and the change in entropy (ΔS) affect ΔP and the equilibrium constant. The following general rules are significant:

1.   No measurable transport will take place if the equilibrium constant is excessively large.
2.   For transport to occur, ΔH cannot equal zero.
3.   Transport to higher-temperature zones is favored for exothermic reactions ($\Delta H < 0$), to lower-temperature zones for endothermic reactions ($\Delta H > O$).
4.   If $\Delta S = 0$, a specific temperature difference between the hot and cold zones will allow maximum transport.
5.   If ΔS is small, transport can occur in either direction, depending on the magnitude of ΔH; if ΔS is large, transport is possible only if ΔH and ΔS have the same sign.
6.   Maximum transport increases with increasing ΔH when ΔS changes accordingly.

Additional details, including the design of equipment and the interpretation of results, can be obtained from cited summaries.[361-363]

Applications to inorganic systems can be classified and exemplified as:

1.   **Purification of elements.** Carbon monoxide is apparently restricted to nickel as a transporting agent, but the halogens are more versatile. Iodine is particularly useful because of the lower decomposition temperatures of the anhydrous iodides. The iodide process of A. E. van Arkel and J. H. de Boer,[365, 366] illustrated for zirconium as

$$Zr(s) + 2\,I_2(g) \; \underset{T_2}{\overset{T_1}{\rightleftharpoons}} \; ZrI_4(g) \qquad T_1 = 280°C, \; T_2 = 1450°C$$

wherein decomposition of the gaseous iodide occurs on a glowing filament, yields highly purified, ductile, crystalline refractory metals (e.g., Ti, Zr, Hf, V, Nb, Ta, Cr, Th, U).[367] An interesting application is the use of iodine in quartz-iodine incandescent bulbs to obviate problems resulting from vaporization of tungsten filaments via formation of volatile tungsten iodides that then redeposit tungsten by decomposition on the filaments.[368] Luminous efficiency, operational temperature, and bulb efficiency are thereby increased. Other examples involve transport of elements with hydrogen chloride (e.g., Fe, Co, Ni, Cu), gaseous sodium chloride (e.g., Be, Ti, V, U), gaseous aluminum chloride (e.g., Si), oxygen (e.g., Ir as IrO_3, Pt via PtO_2), and oxides (e.g., C using CO_2, Mo using MoO_3).

   2. **Syntheses of compounds.** Syntheses are effected by combining chemical transport with other processes by either (a) chemical conversion, followed by removal of the product by transport, or (b) transport of one reactant to a second reactant with which it is not in direct contact.[361] Typical examples of the first type of synthesis include pure, crystalline IrO_2 from elemental Ir, as

$$Ir(s) + \tfrac{3}{2}O_2(g) \rightarrow IrO_3(g)$$

$$IrO_3(g) \rightarrow IrO_2(s) + \tfrac{1}{2}O_2(g)$$

and pure, crystalline VOCl from V_2O_3 and VCl_3, as

$$VCl_3(s) + V_2O_3(s) \rightarrow 3\,VOCl(s)$$

$$VOCl(s) + 2\,VCl_4(g) \rightarrow VOCl_3(g) + 2\,VCl_3(g)$$

the VCl_4 being formed by the disproportionation of VCl_3. One example of the second type is the formation of Ca_2SnO_4 and $SrSnO_3$ from the component oxides in the presence of H_2 or CO, as

$$2\,CaO(s) + SnO_2(s) \rightarrow Ca_2SnO_4(s)$$

$$SrO(s) + SnO_2(s) \rightarrow SrSnO_3(s)$$

wherein SnO_2 is transported via SnO(g) formed as

$$SnO_2(s) + H_2(g) \rightleftharpoons SnO(g) + H_2O(g)$$

or

$$SnO_2(s) + CO(g) \rightleftharpoons SnO(g) + CO_2(g)$$

Other examples are the syntheses of ferrites (MFe_2O_4, where M = Mn, Co, Ni) from the oxides in the presence of HCl(g) as the transporting agent,

$$Fe_2O_3(s) + 6\,HCl(g) \rightleftharpoons 2\,FeCl_3(g) + 3\,H_2O(g)$$

$$MO(s) + 2\,HCl(g) \rightleftharpoons MCl_2(g) + H_2O(g)$$

$$2\,FeCl_3(g) + MCl_2(g) + 4\,H_2O(g) \rightleftharpoons MFe_2O_4(s) + 8\,HCl(g)$$

   3. **Preparation of single crystals.** Single crystals can be grown as small entities for x-ray diffraction studies, as whiskers or filaments, or as sizable crystals, depending on conditions, by transport reactions, as illustrated by the typical equations

$$AlOCl(s) + NbCl_5(g) \rightleftharpoons \tfrac{1}{2} Al_2Cl_6(g) + NbOCl_3(g) \qquad T_1 = 380°C, \; T_2 = 400°C$$

$$Si(s) + SiCl_4(g) \rightleftharpoons 2\,SiCl_2(g) \qquad\qquad\qquad = 900°C \qquad = 1100°C$$

$$ZnS(s) + I_2(g) \rightleftharpoons ZnI_2(g) + \tfrac{1}{2} S_2(g) \qquad\qquad = 700°C \qquad = 1000°C$$

Doped crystals, defect structures, and epitaxial growth can be obtained similarly.

4. INORGANIC POLYMERIZATION REACTIONS

Polymerization reactions are so characteristic of organic systems that one may be tempted to conclude that they are of little importance in inorganic systems. However, the recognized examples of polymerization in inorganic chemistry exceed in type and variety, and perhaps even in number, those in organic chemistry.[369-376] Structural relationships, as delineated in Chapter 6, lead to the classifications as summarized in Table 10-19, together with specific examples. The discussion here is restricted to reactions that yield various polymeric

Table 10-19   Structural Classification of Inorganic Polymers

Type	Examples
Linear polymers	Fibrous sulfur
	Sulfur nitride (polythiazyl)
	Selenium
	Sulfur(VI) oxide (asbestos type)
	Selenium(IV) oxide
	Linear polyphosphazenes
	Polymetaphosphates
	Cyamelide
	Silicon(IV) sulfide
	Silicates (extended linear anions)
	Polymetaborates
Two-dimensional polymers	Black phosphorus
	Graphite and its derivatives
	Calcium silicide, siloxene
	Silicates (layer lattices)
	Boron nitride (hexagonal)
Three-dimensional polymers	Diamond
	Silicon
	Silicon carbide
	Silicon(IV) oxide
	Silicates (three-dimensional anions)
	Boron
	Boron nitride (cubic)

substances. These reactions are conveniently classified as condensation, addition, and coordination reactions.

4-1. Condensation Polymerization Reactions

These reactions are essentially desolvation reactions since they proceed via the elimination of small molecules (e.g., H_2O, NH_3) of useful solvents. As such, they take place either in solution or at elevated temperatures. Ionic aggregation, desolvation, and solvolysis–desolvation are important types of these reactions.

Cationic Aggregation.[369,372]   Although the hydrous precipitate formed on treatment of an aqueous solution containing a high-valent metal cation with hydroxide ion is usually formulated as a simple oxide or hydroxide, the precipitation process involves the conversion of aquo complexes, that may be partially hydrolyzed or multinuclear,[377] either into basic salts as intermediates or into end products. An initial Brønsted–Lowry equilibrium

$$M(H_2O)_n^{m+} + H_2O \rightleftharpoons M(H_2O)_{n-1}(OH)^{m-1} + H_3O^+$$

may be followed, in sequence, by the bonding of partially hydrolyzed species through hydroxo bridges (*olation*) and the intramolecular and intermolecular loss of water molecules to form more thermodynamically stable oxo bridges (*oxolation*).[378] Such processes are favored by increasing pH since they depend upon continuing removal of hydrated protons. The properties of partially hydrolyzed systems and of the colloidal hydrous oxides and hydroxides are consistent with this concept of cationic polymerization.[369,372,379]

There are evidences of comparable aggregation in solutions of amides of high-valent metal ions in liquid ammonia,[380] of lanthanide acetates in glacial acetic acid,[381] and of metal alkoxides in anhydrous alcohols.[382] It is probable that cationic aggregation is involved in the conversion of hydrated salts into basic salts at elevated temperatures and in the absence of solvents.

Anionic Aggregation.[369]   By contrast, anionic aggregation processes are acid-induced desolvation reactions. In aqueous solution, each such reaction apparently involves the protonation of an oxoanion, followed by the elimination of a water molecule by interanionic condensation. For the simple case of chromium(VI), the initial reactions may be formulated

$$CrO_4^{2-} + H^+ \rightleftharpoons HCrO_4^-$$

$$HCrO_4^- + HCrO_4^- \rightleftharpoons Cr_2O_7^{2-} + H_2O$$

$$Cr_2O_7^{2-} + H^+ \rightleftharpoons HCr_2O_7^-$$

$$HCr_2O_7^- + HCrO_4^- \rightleftharpoons Cr_3O_{10}^{2-} + H_2O$$

As the pH continues to decrease, this type of process continues until complete aggregation to $CrO_3(s)$ results. Comparable isopoly anion formation (Sec. 8.2-3) characterizes certain other d-transition metal oxo anions (e.g., vanadate, molybdate, tungstate). The reactions are reversible, with depolymerization accompanying increase in pH.

Anionic aggregation with decreasing pH is also characteristic of aqueous solutions of silicates, except that the only distinguishable polymeric species is the disilicate ion, $Si_2O_7^{6-}$ (at pH 13.6 to 10.9), and that the addition of hydrogen-bonding reagents that do not themselves alter the pH significantly markedly inhibits polymerization in these systems.[383]

Comparable proton-dependent anionic aggregation processes are also characteristic of phosphate, sulfate, molybdate, and tungstate systems at elevated temperatures. Thus when sodium hydrogen orthophosphates (Na_2HPO_4, NaH_2PO_4) are heated, aggregation to yield di-, tri-, and higher polyphosphates increases with increasing mole ratio of proton to orthophosphate,[384, 385]

$$2\,HPO_4^{2-} \rightarrow P_2O_7^{4-} + H_2O \qquad H^+/PO_4^{3-} = 1.00$$

(that is, $2\,PO_4^{3-} + 2\,H^+ \rightarrow P_2O_7^{4-} + H_2O$)

$$H_2PO_4^- + 2\,HPO_4^{2-} \rightarrow P_3O_{10}^{5-} + 2\,H_2O \qquad\qquad = 1.33$$

(that is, $3\,PO_4^{3-} + 4\,H^+ \rightarrow P_3O_{10}^{5-} + 2\,H_2O$)

$$n\,H_2PO_4^- \rightarrow (PO_3)_n^{n-} + n\,H_2O \qquad\qquad = 2.00$$

(that is, $n\,PO_4^{3-} + 2n\,H^+ \rightarrow P_nO_{3n}^{n-} + n\,H_2O$)

Similarly, the conversion of HPO_4^{2-} to $H_2P_2O_7^{2-}$ at elevated temperatures is favored by the presence of HSO_4^- ion.[386]

Desolvation at Elevated Temperature. Reactions of this type are closely related to those just discussed, the clearest well-defined examples being those that yield polymeric oxides by *dehydration* of oxo acids or hydrous hydroxides. The formation of the various polyphosphoric acids and ultimately polymeric phosphorus(V) oxide[387] or of metal oxides is typical. *Deammonation* reactions may be *molecular* [e.g., conversion of $Si(NH_2)_4$ through intermediates to polymeric $(Si_3N_4)_x$],[388] *cationic* (as noted above),[380] or *anionic* [e.g., $n(NaO)P(O)(NH_2)_2 \rightarrow [(NaO)P(O)(NH)]_n + n\,NH_3$].[388] *Dehydrogenation* reactions, although similar in type but not involving a solvent, yield polymers in the phosphino-,[389] arseno-,[390] and sulfo-[391] borine systems. Similarly, *demethanation* reactions lead to polymers containing aluminum[392] and gallium[393] [e.g., $(CH_3)_2AlN(CH_3)_2)_2$ and $(CH_3GaO)_n$].

Solvolysis–Desolvation Reactions. This type of reaction depends upon the initial formation of a monomer by partial solvolysis of a solvent-sensitive compound (e.g., a halide) and the subsequent reaction of this partially solvolyzed substance with unreacted solvent-sensitive compound to yield a polymer. A necessary condition is that the solvolyzing reagent contain more than a single replaceable hydrogen atom per molecule. Thus partial hydrolysis of phosphorus(V) chloride yields, among other products, dichlorophosphoric acid, $HOP(O)Cl_2$, which then yields polymeric $(PO_2Cl)_n$;[394] partial hydrolysis of titanium(IV) ethoxide gives $Ti_{3(n+1)}O_{4n}(OC_2H_5)_{4(n+3)}$;[382] partial ammonolysis of phosphorus(V) chloride, with ammonium halides to control the quantity of ammonia available, yields polymeric halophosphazenes, $(NPCl_2)_n$ and $(NPBr_2)_n$ ($n = 3$ or larger);[395 397] and partial thiosolvolysis gives polysulfanes, R_2S_n.[398] Many other examples are known.

4-2. Addition Polymerization Reactions

In a reaction of this type, a monomer becomes a repeating unit in the ultimate polymeric molecular structure. Other species that are present in the polymerization reaction may act as initiating or terminating (end) groups. Reactions of this type lead to hydrogen-bonded species, to the polymeric allotropic forms of the elements, and to a limited number of molecular species.

Hydrogen-Bond Formation. Among purely inorganic systems, this type of polymerization is most evident in the aggregation of molecules in polar solvents. The existence of thermally stable systems is precluded by the relatively small energy of the hydrogen bond (Sec. 5.2-5).

Polymeric Forms of the Nonmetallic Elements. The most important reactions of this type are those in which the discrete species S_8 [399] and P_4 [400] are converted into high polymers. With sulfur, the initial process must involve opening the S_8 ring. Ring cleavage can be effected thermally or by the addition of nucleophilic (i.e., electron donating) reagents such as triphenylphosphine;[401] cyanide ion;[402] sulfite ion;[403] ions such as S^{2-}, SH^-, HSO_3^-, and OH^-;[404] and tertiary amines.[405] In each instance, the formation of a donor–acceptor adduct,

$$donor^+ : \ddot{S} - S_6 - \ddot{S} :^-$$

is an initial reaction step.

With phosphorus, the process may follow one of several pathways. Thus δ-white phosphorus, the crystal structure of which is a cubic array of P_4 molecules, results merely from the collection of P_4 molecules into an appropriate lattice, but the two-dimensional network in orthorhombic black phosphorus (Fig. 5-22) suggests that both bond rupture and rearrangement are essential to the formation of this allotropic modification. The changes are effected by heating under high pressure[406] or with mercury.[407] Red phosphorus, which has been described in a variety of crystalline forms, appears structurally to amount to an array of phosphorus atoms into which other atoms can be incorporated. It is obtained when white phosphorus is heated, either alone or in the presence of light or catalysts such as iodine, sulfur, or phosphorus(III) bromide. Again, bond rupture and rearrangement are involved.[408]

Polymerization of Molecular Units. Addition polymerization of this type is characteristic of sulfur(VI) oxide, the phosphazenes, cyanogen, thiocyanogen, and sulfur nitride.[409] It is unlikely that a monomeric unit characterizes any of these species, except possibly in the gaseous state. The simplest polymeric units are linear dimers (cyanogen, thiocyanogen), cyclic trimers [sulfur(VI) oxide, phosphazenes], and a cyclic tetramer (tetrasulfur tetranitride). Thus the formation of a high polymer from an oligomer in any of these systems must involve an initial activation process that can then result in either direct addition or ring cleavage.

Cyanogen, $(CN)_2$, and thiocyanogen, $(SCN)_2$, polymerize spontaneously at ca. 500°C and room temperature, respectively, to difficultly soluble solids, the molecular structures of which are very poorly characterized. Little is known about these polymerization processes. Both the alpha and beta modifications of sulfur(VI) oxide (Sec. 6.2-1) are asbestoslike solids that consist of long, staggered chains of SO_3 units,

the difference between the two being that in the alpha modification the chains are combined into layers. Conversion of the cyclic trimer into the β-form is catalyzed by water, which may involve an initial nucleophilic attack, followed by opening of the ring structure to form the species

$$H\overset{+}{\underset{\underset{H}{|}}{O}}-SO_2-O-SO_2-O-SO_2-\overset{-}{O}$$

which in turn opens another trimeric ring, with this process continuing until a high polymer results.

Polymerization of oligomeric chlorocyclophosphazenes, $(NPCl_2)_{3,4}$ (Sec. 6.2-1), at temperatures in the range 250 to 350°C,[410] or at lower temperatures in the presence of certain catalysts (e.g., ethers, carboxylic acids, ketones, alcohols, metals),[411] can yield macrocyclic, linear, or cross-linked high polymers. It has been suggested that the mechanism of the polymerization reaction involves either (1) a free-radical chain process or (2) an ionic process.[412] The latter process is consistent with the bulk of the experimental data, for example, the most effective catalysts are those substances that rupture P—Cl covalent bonds to yield Cl^- ions. For the cyclic trimer as a starting material, the process can be formulated in simple terms as[413] where~~indicates an $—N{=}PCl_2—$ chain.* Branching or cross-linking can be accounted for equally well in expanding this proposed mechanism. In the absence of an organic substance that can react initially to form hydrogen chloride, polymerization is slow even at 211°C. In the presence of an excess of phosphorus(V) chloride, the final high polymer may be end-grouped as

$$Cl_4P—(NPCl_2)\mathbin{\rlap{\,\sim\!\sim\!\sim}}$$

Polythiazyl, $(SN)_n$, is obtained by the slow polymerization of disulfur dinitride, S_2N_2, at room temperature, followed by heating at 75°C.[414] Disulfur dinitride is itself obtained from tetrasulfur tetranitride, S_4N_4, by high-temperature sublimation through silver wool. Crystals of polythiazyl consist of fibers, each made up of essentially planar chains of alternating N and S atoms. Oriented thin films can be made on a variety of substrates. Polythiazyl exhibits metallic properties (e.g., luster, electronic conduction) — with superconduction at 0.3°K.[415-419] Slow oxidation and/or hydrolysis occurs on exposure to air.[420] Again, the polymerization reaction must involve a bond breaking–bond making process.

4-3. Coordination Polymerization Reactions

Any polymer that contains coordinately bonded metal ions, irrespective of the arrangement of those ions or the manner in which they were introduced, can be termed a coordination polymer. Four general types of coordination polymers, and thus four general approaches to their preparation, can be distinguished,[421,422] as:

1. Polymerization occurs when a metal ion and a bifunctional ligand combine, as illustrated by the general equation†

Linear polymers are most likely to result when a 4-coordinate metal ion reacts with a bis(bidentate) ligand or when a 6-coordinate metal ion reacts with a bis(terdentate) ligand. Cases in point include high-molecular-weight substances derived from copper(II) or nickel(II) ions and dithiooxamides[421] and probably based upon the repeating unit

---

* Use of alternating double and single bonds in the cyclic formulations is a convenient formalism. Delocalization of electronic charge density via a d_π-p_π approximation has been discussed earlier (Sec. 6.2-1).

† The notation used designates a chelating group as ⌣ with linking structures, usually organic (e.g., $H_2NCH_2CH_2NH_2$). Chelating groups bond more strongly than nonchelating groups and also better saturate the coordination numbers of the metal ions to minimize cross-linking.

$$\left(\begin{array}{c} \overset{R}{\underset{|}{}} \\ \diagdown\text{S--C}{=}\text{N} \diagup \\ \text{M} \\ \diagup\text{N}{=}\text{C--S}\diagdown \\ \underset{|}{R} \end{array}\right)_n$$

M = Cu, Ni

R = H, CH$_3$, C$_6$H$_5$CH$_2$

2. Polymerization occurs when monomeric units already containing metal ions react through functional groups not associated with those metal ions, as

$$n\text{--X--}\bigcirc\bigcirc\text{--X--} + n\text{--Y--Y--} \longrightarrow$$

$$\left[\text{--X--}\bigcirc\bigcirc\text{--X--Y--Y--}\right]_n$$

A possible linear polymer might be

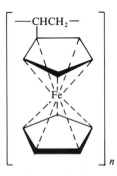

formed by the reaction of a diacid complex of a dipositive metal ion with a diol, HO(CH$_2$)$_x$OH. Typical polymers formed by reactions of this type include the π-cyclopentadienyliron(II) derivatives[423]

$$\left[\begin{array}{c} \text{--CHCH}_2\text{--} \\ \text{Fe} \end{array}\right]_n$$

3. Polymerization occurs by addition as a result of ring opening in a cyclic monomer containing a coordinated metal ion, as indicated by the general equation

Examples of this type of reaction are limited in number.[422]

4. Polymerization occurs by reaction of a preformed polymer having donor atom sites with a metal ion, as indicated by the general equation

a reported example is the product obtained by the addition of Cu^{2+} or Zn^{2+} ions to a polymer of graphic structure[424, 425]

All of the polymerization reactions noted depend upon polyfunctional ligands. Some typical polyfunctional ligands are summarized in Table 10-20.

An important class of coordination polymers not related specifically to any of the systems just considered is that of the heteropolyacids and their salts. Each of these species is the product of aggregation of two types of anions (e.g., molybdate with phosphate, tungstate with silicate) and is formally polymeric. However, for each a structurally discrete array of atoms can be distinguished, not an infinite orderly array such as characterizes the coordination polymers

Table 10-20 Typical Polyfunctional Ligands

Ligand	Type Formula[a]
Bis-(β-diketones)	
Bis-(o-hydroxyaldehydes)	
Bis-(α-amino acids)	
Bis(salicylaldehyde)-diimines	
Bis-(8-quinolinols)	
Bis-(nitrosophenols)	
Bis-(glyoximes)	
Bis-(thiocarbamates)	
Bis-(xanthates)	

[a] X represents an aliphatic or aromatic bridge.

Table 10-21 Cationic Cluster Species

Periodic Group	Formula	Molecular Geometry	Typical Synthesis	References
IIA	Cd_2^{2+}	Square antiprism (D_{4d})	$Cd + CdX_2 + Al_2X_6$ (X = Cl, Br, I)	321
VA	Bi_4^+	Trigonal bipyramid (D_{3h})	$Bi + BiCl_3 + Al_2Cl_6$	322
	Bi_5^{3+}	Tricapped trigonal prism (C_{3h})	$Bi + ZF_5$ (Z = P, As, Sb) in SO_2	322, 426
	$Bi^+(Bi_9)^{5+}$		$Bi + BiCl_3 + HfCl_4$	327
VIA	S_2^{2+}		anodic oxidation S_8 in $AlCl_4^-$	229a, 323, 328
	S_4^{2+}	Square plane	anodic oxidation S_8 in $AlCl_4^-$	229a, 323, 328
	Se_4^{2+}	Square plane	$Se + S_2O_6F_2$ in HSO_3F	229a
	Te_4^{2+}	Square plane	$Te + S_2O_6F_2$ in HSO_3F	229a, 324, 326
	$Te_2Se_2^{2+}$	Square plane	$Se–Te + SbF_5$ in SO_2	229a
	Te_6^{2+}		$Te + TeCl_4$ in $AlCl_4^-$	326
	$Te_3S_3^{2+}$	Six-membered ring, boat conformer	$Te_4(AsF_6)_2 + S_8(AsF_6)_2$ in SO_2	229a, 427
	$Te_2Se_4^{2+}$	Six-membered ring, boat conformer	$Te + Se + SbF_5$ in SO_2	229a, 427
	Te_6^{4+}	Trigonal prism	$Te + AsF_5$ in AsF_3 or SO_2	229a, 428
	S_8^{2+}	Eight-membered ring, exo-endo	$S_8 + AsF_5$ in SO_2	229a
	Se_8^{2+}	Eight-membered ring, exo-endo	$Se + SeCl_4 + Al_2Cl_6$	229a, 429
	Te_8^{2+}		$Te + TeCl_4$ in $AlCl_4^-$	328
	Se_{10}^{2+}	Six- and eight-membered rings fused		229a
	$Te_2Se_8^{2+}$	Six- and eight-membered rings fused	$Te + Se_8(AsF_6)_2$ in SO_2	429
	S_{19}^{2+}	Two seven-membered rings joined by five-atom chain	$S_8 + H_2S_2O_7$ or $S_8 + SbF_5$ in SO_2	229a

noted above and in the isopolyacids and their salts. The heteropoly species have been discussed as coordination entities in Sec. 8.2-3.

Homopolyatomic cations and anions are also describable as low-molecular-weight coordination polymers since they do indeed result from metal–metal or nonmetal–nonmetal coordinate interaction. Metal–metal interactions and the molecular structures of the resulting metal clusters have been discussed in Sec. 8.2-3. Comparable clusters of the main-group elements, as noted earlier, can be either cationic or anionic, but most of the well described species are cationic.[229a] The more important cations are listed in Table 10-21. As noted in earlier sections of this chapter, these ions are usually obtained from solvents or molten salt systems of low basicity as salts with counterions of low basicity (e.g., SO_3F^-, $HS_2O_7^-$, SbF_6^-, $Sb_2F_{11}^-$, AsF_6^-, $AlCl_4^-$, $Al_2Cl_7^-$, $HfCl_6^{2-}$).

EXERCISES

10-1. Liquid ammonia is a leveling solvent for the polar hydrides HCl and HF, but water is a differentiating solvent for these hydrides. Explain.

10-2. The reaction in $N_2O_4(l)$ described by the equation $NOClO_4 + KNO_3 \rightarrow KClO_4 + N_2O_4$ is a neutralization reaction. Explain.

10-3. Show clearly that there is no conflict between the following statements: (1) In a given solvent at a given total concentration, the acid that is the most extensively ionized is the strongest acid. (2) An acid will give its most acidic solution in the solvent in which it is ionized to the smallest extent.

10-4. Under what conditions are H_0 and pH identical? Under what conditions are H_0 and pH different? Explain.

10-5. Arrange the following substances in the order of increasing acidity, first in $H_2O(l)$, next in $NH_3(l)$, and finally in $H_2SO_4(l)$: CH_4, NH_3, H_2O, HF, HI, H_3PO_2, $S_2O_7^{2-}$. Assume reasonable solubility in each solvent. Give reasons for your arrangements.

10-6. Classify each of the following as a Lewis base or a Lewis acid, giving a reason for each choice: BH_4^-, $C_3H_7^-$, $B(OCH_3)_3$, PCl_3, $Hg(NO_3)_2$.

10-7. Using Tables 10-9 and 10-10 as models, develop a comparable aquo-ammono scheme for H_3PO_4. Limit your scheme to species containing no more than *four* phosphorus atoms. Using suitable auxiliary references, name these species.

10-8. Cite three specific reactions that can be carried out in a strongly acidic solvent but not in water. Cite three additional reactions that can be carried out in a strongly basic solvent but not in water. Explain.

10-9. Indicate clearly why molten aluminum(III) chloride is classified as a acidic solvent. Select a molten salt that is a basic solvent and account for your choice.

10-10. Distinguish clearly among condensation, addition, and coordination polymerization processes in *inorganic* chemistry.

10-11. Point out the significant differences between inorganic anionic and cationic aggregation reactions in aqueous systems and explain the significance of pH.

10-12. What factors militate against the formation of linear high polymers by reactions involving polyfunctional monomers? Suggest one or more possible solutions to this problem.

10-13. Summarize the special conditions that favor the formation of cationic cluster species. Indicate clearly why these conditions are essential.

10-14. Criticize, *pro* and *con*, the classification of the summary equation

$$SOCl_2 + Cs_2SO_3 \xrightarrow{SO_2(l)} 2CsCl + 2SO_2$$

as a neutralization reaction.

10-15. Why do reactions in the gas or liquid phase commonly yield stoichiometric products, whereas those in the solid phase commonly yield nonstoichiometric products? Illustrate with several definitive examples.

REFERENCES

1. W. Ostwald, *Lehrbuch der allgemeinen Chemie*, Vol. 2, Pt. 1, p. 705, Verlag von W. Engelmann, Leipzig (1893).

2. H. P. Cady, *J. Phys. Chem.*, **1**, 707 (1897).

3. E. C. Franklin and C. A. Kraus, *Am. J. Chem.*, **21**, 8 (1899).

4. P. Walden, *Chem. Ber.*, **32**, 2862 (1899).

5. C. A. Kraus, *The Properties of Electrically Conducting Systems*, Chemical Catalog Co., New York (1922).

6. P. Walden, *Elektrochemie nichtwässriger Lösungen*, Verlag J. A. Barth, Leipzig (1924).

7. E. C. Franklin, *The Nitrogen System of Compounds*, Reinhold Publishing Corp., New York (1935).

8. G. Jander, *Die Chemie in wasserähnlichen Lösungsmitteln*, Springer-Verlag, Berlin (1949).

9. L. F. Audrieth and J. Kleinberg, *Non-Aqueous Solvents: Applications as Media for Chemical Reactions*, John Wiley & Sons, New York (1953).

10. T. C. Waddington (Ed.), *Non-Aqueous Solvent Systems*, Academic Press, New York (1965). Ten chapters, each on a different system.

11. J. J. Lagowski (Ed.), *The Chemistry of Non-Aqueous Solvents*, Vol. 1: *Principles and Techniques* (1966); Vol. 2: *Acidic and Basic Solvents* (1967); Vol. 3: *Inert, Aprotic, and Acidic Solvents* (1970); Vol. 4: *Solution Phenomena and Aprotic Solvents* (1976); Vol. 5A: *Principles and Basic Solvents* (1978); Vol. 5B: *Acidic and Aprotic Solvents* (1978), Academic Press, New York. Collection of comprehensive reviews, each covering a specific area.

12. W. L. Jolly (Ed.), *Metal-Ammonia Solutions*, Dowden, Hutchinson, & Ross, Stroudsburg, Pa. (1972). Reprints of 63 selected papers in this area.

13. G. Jander, H. Spandau, and C. C. Addison (Eds.), *Chemistry in Nonaqueous Ionizing Solvents*, Vol. 1–, Pergamon Press, Oxford, (1967–   ).

14. V. Gutmann (Ed.), *Pure Applied Chem.*, **41**(3), 275–393 (1975). Plenary lectures of 4th International Conference on Non-Aqueous Solutions, Vienna (1974).

15. J. B. Gill (Ed.), *Pure Applied Chem.*, **49**(1), 1–124 (1977). Plenary and section lectures of 5th International Conference on Non-Aqueous Solutions, Leeds (1976).

16. G. J. Janz and R. P. T. Tomkins, *Non-Aqueous Electrolytes Handbook*, Vols. 1, 2, Academic Press, New York (1972, 1973).

17. C. Charlot and B. Trémillon (translated by P. J. J. Harvey), *Chemical Reactions in Solvents and Melts*, Pergamon Press, Elmsford, N.Y. (1969). Comprehensive, extensively referenced.

18. C. A. Kraus, *J. Phys. Chem.*, **60**, 129 (1956).

19. K. H. Stern and A. E. Martell, *J. Am. Chem. Soc.*, **77**, 1983 (1955).

20. V. Gutmann, *Q. Rev. (Lond.)*, **10**, 451 (1956); *Chem. Br.*, **7**, 102 (1971).

21. V. Gutmann, *Coordination Chemistry in Non-Aqueous Solutions*, Springer-Verlag, New York (1968).

22. R. S. Drago and K. F. Purcell, in *Progress in Inorganic Chemistry*, F. A. Cotton (Ed.), Vol. 6, pp. 271–322, Interscience, New York (1964).

23. R. S. Drago, *Chem. Br.*, **3**, 516 (1967).

24. J. E. Huheey, *J. Inorg. Nucl. Chem.*, **24**, 1011 (1962).

25. M. M. Davis, Ref. 11, Vol. 3, pp. 1–135.

26. D. W. Smith, *J. Chem. Educ.*, **54**, 540 (1977).

27. L. F. Audrieth and J. Kleinberg, Ref. 9, Chs. 3–6.

28. W. L. Jolly and C. J. Hallada, Ref. 10, Ch. 1.

29. J. J. Lagowski and G. A. Moczygemba, Ref. 11, Vol. 2, Ch. 7.

30. W. L. Jolly, in *Progress in Inorganic Chemistry*, F. A. Cotton (Ed.), Vol. 1, pp. 235–281, Interscience Publishers, New York (1959).

31. G. W. A. Fowles, in *Developments in Inorganic Nitrogen Chemistry*, C. B. Colburn (Ed.), Vol. 1, pp. 522–576, Elsevier, New York (1966).

32. G. W. A. Fowles and D. Nicholls, *Q. Rev. (Lond.)*, **16**, 19 (1962).

32a. J. J. Lagowski, *J. Chem. Educ.*, **55**, 752 (1978).

32b. D. Nicholls, *Inorganic Chemistry in Liquid Ammonia*, Elsevier, Amsterdam (1979).

33. W. L. Jolly, *J. Chem. Educ.*, **33**, 512 (1956).

34. J. B. Gill, *J. Chem. Educ.*, **47**, 619 (1970).

35. J. B. Gill, *J. Chem. Soc.*, **1965**, 5730.

36. W. L. Jolly, *Chem. Rev.*, **50**, 351 (1952).

37. N. Mammano and M. J. Sienko, *J. Am. Chem. Soc.*, **90**, 6322 (1968).

38. J. C. Warf and W. L. Korst, *J. Phys. Chem.*, **60**, 1590 (1956).

38a. W. S. Glaunsinger, T. R. White, R. B. Von Dreele, D. A. Gordon, R. F. Marzke, A. L. Bowman, and J. L. Yarnell, *Nature*, **271**, 414 (1978).

38b. W. S. Glaunsinger, *J. Phys. Chem.*, **84**, 1163 (1980).

38c. R. B. Von Dreele, W. S. Glaunsinger, P. Chieux, and P. Damay, *J. Phys. Chem.*, **84**, 1172 (1980).

39. A. W. Davidson, J. Kleinberg, W. E. Bennett, and A. D. McElroy, *J. Am. Chem. Soc.*, **71**, 377 (1949); A. D. McElroy, J. Kleinberg, and A. W. Davidson, *J. Am. Chem. Soc.*, **72**, 5178 (1950).

40. H. J. Wolthorn and W. C. Fernelius, *J. Am. Chem. Soc.*, **56**, 551 (1934).

41. J. Jortner and G. Stein, *Nature*, **175**, 893 (1955).

42. J. L. Down, J. Lewis, B. Moore, and G. Wilkinson, *J. Chem. Soc.*, **1959**, 3767.

43. J. L. Dye, M. G. DeBacker, and V. A. Nicely, *J. Am. Chem. Soc.*, **92**, 5226 (1970).

44. J. H. Hildebrand, *J. Chem. Educ.*, **25**, 74 (1948).

45. J. H. Hildebrand and R. L. Scott, *Regular Solutions*, Prentice-Hall, Englewood Cliffs, N.J. (1962).

46. W. C. Johnson and W. C. Fernelius, *J. Chem. Educ.*, **5**, 664, 828 (1928); **6**, 20, 441 (1929); **7**, 981 (1930).

47. W. C. Johnson and A. W. Meyer, *Chem. Rev.*, **8**, 273 (1931).

48. W. C. Fernelius and G. W. Watt, *Chem. Rev.*, **20**, 195 (1937).

49. C. A. Kraus, *J. Chem. Educ.*, **30**, 83 (1953).

50. M. C. R. Symons, *Q. Rev. (Lond.)*, **13**, 99 (1959); *Chem. Soc. Rev.*, **5**, 337 (1977).

51. E. C. Evers, *J. Chem. Educ.*, **38**, 590 (1961).

52. G. Lepoutre and M. J. Sienko (Eds.), *Metal–Ammonia Solutions*, Colloque Weyl I, W. A. Benjamin, New York (1964).

53. J. J. Lagowski and M. J. Sienko (Eds.), *Metal–Ammonia Solutions*, Colloque Weyl II, Butterworth, London (1970).

54. J. Jortner and N. R. Kestner (Eds.), *Electrons in Fluids — The Nature of Metal–Ammonia Solutions*, Colloque Weyl III, Springer-Verlag, Berlin (1973).

55. Numerous authors, *Electrons in Fluids — The Nature of Metal–Ammonia Solutions*, Colloque Weyl IV, Springer-Verlag, Berlin (1973); *J. Phys. Chem.*, **79**, 2789–3079 (1975).

56. J. L. Dye, *J. Chem. Educ.*, **54**, 332 (1977).

57. K. S. Pitzer, *J. Am. Chem. Soc.*, **80**, 5046 (1958).

58. H. H. Schlubach, *Chem. Ber.*, **53B**, 1689 (1920).

59. H. S. Forbes and C. E. Newton, *J. Am. Chem. Soc.*, **48**, 2278 (1926).

60. H. A. Laitinen and C. J. Nyman, *J. Am. Chem. Soc.*, **70**, 3002 (1948).

61. W. L. Jolly, *J. Am. Chem. Soc.*, **77**, 4958 (1955).

62. D. A. Hazlehurst, A. K. Holliday, and G. Pass, *J. Chem. Soc.*, **1956**, 4653.

63. W. C. Johnson, A. W. Meyer, and R. D. Martens, *J. Am. Chem. Soc.*, **72**, 1842 (1950).

64. P. Marshall and H. Hunt, *J. Am. Chem. Soc.*, **77**, 5016 (1955).

65. C. A. Hutchison, Jr., and R. C. Pastor, *J. Chem. Phys.*, **21**, 1959 (1953).

66. G. W. A. Fowles, W. R. McGregor, and M. C. R. Symons, *J. Chem. Soc.*, **1957**, 3329.

67. G. W. Watt, *Chem. Rev.*, **46**, 289 (1950).

68. A. J. Birch, *Q. Rev. (Lond.)*, **4**, 69 (1950).

69. A. J. Birch and H. Smith, *Q. Rev. (Lond.)*, **12**, 17 (1958).

70. G. W. Watt, *Chem. Rev.*, **46**, 317 (1950).

71. M. D. Newton, *J. Phys. Chem.*, **79**, 2795 (1975).

72. E. Becker, R. H. Lindquist, and B. J. Alder, *J. Chem. Phys.*, **25**, 971 (1956).

73. R. A. Ogg, Jr., *J. Chem. Phys.*, **13**, 533 (1945); **14**, 114, 295, 399 (1946); *Phys. Rev.*, **69**, 243, 544, 668 (1946).

74. W. N. Lipscomb, *J. Chem. Phys.*, **21**, 52 (1953).

75. J. Jortner, *J. Chem. Phys.*, **30**, 839 (1959).

76. D. A. Copeland, N. R. Kestner, and J. Jortner, *J. Chem. Phys.*, **53**, 1189 (1970).

77. N. R. Kestner and J. Jortner, *J. Phys. Chem.*, **77**, 1040 (1973).

78. R. A. Stairs, *J. Chem. Phys.*, **27**, 1431 (1957).

79. R. C. Douthit and J. L. Dye, *J. Am. Chem. Soc.*, **82**, 4472 (1960).

80. J. L. Dye, R. F. Sackuer, and G. E. Smith, *J. Am. Chem. Soc.*, **82**, 4797, 4803 (1960).

81. J. C. Thompson, Ref. 11, Vol. 2, pp. 265–317.

82. E. J. Hart (Ed.), *Solvated Electron*, Vol. 50, Advances in Chemistry Series, American Chemical Society, Washington, D.C. (1965).

83. W. C. Fernelius and G. B. Bowman, *Chem. Rev.*, **26**, 3 (1940).

84. L. Larbouillet-Linemann, *C. R. Acad. Sci.*, **238**, 902 (1954).

85. R. I. Wagner and A. B. Burg, *J. Am. Chem. Soc.*, **75**, 3869 (1953).

86. D. L. Schechter and J. Kleinberg, *J. Am. Chem. Soc.*, **76**, 3297 (1954).

87. E. O. Fischer and W. Hafner, *Z. Naturforsch.*, **8b**, 444 (1953).

88. E. O. Fischer and R. Jira, *Z. Naturforsch.*, **8b**, 217, 327 (1953).

89. R. Nast and H. Roos, *Z. Anorg. Allg. Chem.*, **272**, 242 (1953).

90. L. F. Audrieth, M. Sveda, H. H. Sisler, and M. J. Butler, *Chem. Rev.*, **26**, 49 (1940).

91. V. Köhlschütter and K. A. Hofmann, *Ann. Chem.*, **307**, 314 (1899).

92. C. A. Lobry de Bruyn, *Rec. Trav. Chim.*, **11**, 18 (1892).

93. L. F. Audrieth and B. A. Ogg, *The Chemistry of Hydrazine*, Ch. 10, John Wiley & Sons, New York (1951).

94. T. W. B. Welsh and H. J. Broderson, *J. Am. Chem. Soc.*, **37**, 816, 825 (1915).

95. P. Walden and H. Hilgert, *Z. Phys. Chem.*, **A165**, 241 (1933).

96. L. J. Vieland and R. P. Seward, *J. Phys. Chem.*, **59**, 466 (1955).

97. C. C. Clark, *Hydrazine*, Mathieson Chemical Corp., Baltimore, Md. (1953).

98. D. Bauer and P. Gaillochet, Ref. 11, Vol. 5A, pp. 251–275.

99. C. A. Kraus, *Chem. Rev.*, **8**, 251 (1931); **26**, 95 (1940).

100. E. C. Evers, A. E. Young, and A. J. Pauson, *J. Am. Chem. Soc.*, **79**, 5118 (1957).

101. D. S. Berns, E. C. Evers, and P. W. Frank, *J. Am. Chem. Soc.*, **82**, 310 (1960).

102. A. J. Pauson and E. C. Evers, *J. Am. Chem. Soc.*, **82**, 4468 (1960).

103. J.-M. Nigretto and M. Jozefowicz, Ref. 11, Vol. 5A, pp. 179–250.

104. A. Naumann, *Chem. Ber.*, **37**, 4609 (1904).

105. C. F. Nelson, *J. Am. Chem. Soc.*, **35**, 658 (1913).

106. P. Walden, L. F. Audrieth, and E. J. Birr, *Z. Phys. Chem.*, **A160**, 337 (1932).

107. D. S. Burgess and C. A. Kraus, *J. Am. Chem. Soc.*, **70**, 706 (1948).

108. L. F. Audrieth and H. W. Nelson, *Chem. Rev.*, **8**, 335 (1931).

109. G. L. Putnam and K. A. Kobe, *Trans. Electrochem. Soc.*, **74**, 609 (1938).

110. H. S. Isbin and K. A. Kobe, *J. Am. Chem. Soc.*, **67**, 464 (1945).

111. S. Bruckenstein and L. K. Mukherjee, *J. Phys. Chem.*, **64**, 1610 (1960).

112. T. Moeller and P. A. Zimmerman, *Science*, **120**, 539 (1954); *J. Am. Chem. Soc.*, **75**, 3950 (1953).

113. T. Moeller and G. W. Cullen, *J. Inorg. Nucl. Chem.*, **10**, 148 (1959).

114. J. Peacock, F. C. Schmidt, R. E. Davis, and W. B. Schaap, *J. Am. Chem. Soc.*, **77**, 5829 (1955).

115. W. B. Schaap, A. E. Messner, and F. C. Schmidt, *J. Am. Chem. Soc.*, **77**, 2683 (1955).

116. V. Gutmann and G. Schröber, *Monatsh. Chem.*, **88**, 206 (1957); **89**, 649 (1958).

117. R. W. Brewster, F. C. Schmidt, and W. B. Schaap, *J. Am. Chem. Soc.*, **81**, 5532 (1959).

118. M. G. DeBacker and J. L. Dye, *J. Phys. Chem.*, **75**, 3092 (1971).

119. J. L. Dye, M. G. DeBacker, J. A. Eyre, and L. M. Dorfman, *J. Phys. Chem.*, **76**, 839 (1972).

120. F. J. Tehan, B. L. Barnett, and J. L. Dye, *J. Am. Chem. Soc.*, **96**, 7203 (1974).

121. J. L. Dye, C. W. Andrews, and S. E. Mathews, *J. Phys. Chem.*, **79**, 3065 (1975).

122. J. L. Dye, J. M. Ceraso, M. T. Lok, B. L. Barnett, and F. J. Tehan, *J. Am. Chem. Soc.*, **96**, 608 (1974).

123. J. L. Fox, *Chem. Eng. News*, Dec. 10, 1979, pp. 32–35. Interesting account of investigations by J. L. Dye, J. J. Lagowski, M. J. Sienko, and W. J. Glaunsinger on metal–ammonia systems.

124. A. Cisar and J. D. Corbett, *Inorg. Chem.*, **16**, 632 (1977).

125. D. G. Adolphson, J. D. Corbett, and D. J. Merryman, *J. Am. Chem. Soc.*, **98**, 7234 (1976).

126. A. Cisar and J. D. Corbett, *Inorg. Chem.*, **16**, 2482 (1977).

127. P. A. Edwards and J. D. Corbett, *Inorg. Chem.*, **16**, 903 (1977).

128. E. Zintl and H. Kaiser, *Z. Anorg. Allg. Chem.*, **211**, 113 (1933).

129. J. W. Vaughn, Ref. 11, Vol. 2, pp. 191–264.

130. P. Walden, *Z. Phys. Chem.*, **46**, 103 (1903); **54**, 129 (1905); **55**, 207, 683 (1905); **59**, 385 (1907); **75**, 555 (1910). Classic papers on organic solvent systems.

131. E. Colton and R. E. Brooker, *J. Phys. Chem.*, **62**, 1595 (1958).

132. T. Pavlapoulas and H. Strehlow, *Z. Phys. Chem., Neue Folge*, **2**, 89 (1954).

133. L. R. Dawson, T. M. Newell, and W. J. McCreary, *J. Am. Chem. Soc.*, **76**, 6024 (1954).

134. L. R. Dawson and C. Berger, *J. Am. Chem. Soc.*, **79**, 4269 (1957).

135. O. F. Stafford, *J. Am. Chem. Soc.*, **55**, 3987 (1933).

136. G. Jander and G. Winkler, *J. Inorg. Nucl. Chem.*, **9**, 24, 32, 39 (1959).

137. L. R. Dawson, G. R. Lester, and P. G. Sears, *J. Am. Chem. Soc.*, **80**, 4233 (1958).

138. L. R. Dawson, R. H. Graves, and P. G. Sears, *J. Am. Chem. Soc.*, **79**, 298 (1957).

139. T. B. Hoover, *Pure Appl. Chem.*, **37**(4), 579 (1974).

140. E. L. Muetterties and E. G. Rochow, *J. Am. Chem. Soc.*, **75**, 490 (1953).

141. T. Moeller and V. Galasyn, *J. Inorg. Nucl. Chem.*, **12**, 259 (1959).

142. A. B. Thomas and E. G. Rochow, *J. Am. Chem. Soc.*, **79**, 1843 (1957).

143. G. H. Brown and R. Al-Urfali, *J. Am. Chem. Soc.*, **80**, 2133 (1958).

144. R. T. Pflaum and A. I. Popov, *Anal. Chim. Acta*, **13**, 165 (1955).

145. M. E. Peach and T. C. Waddington, Ref. 10, pp. 83–115.

146. F. Klanberg, Ref. 11, Vol. 2, pp. 1–41.

147. E. H. Archibald and D. McIntosh, *Proc. R. Soc. Lond.*, **73**, 454 (1904); D. McIntosh and B. D. Steele, *Proc. R. Soc. Lond.*, **73**, 450 (1904); B. D. Steele, D. McIntosh, and E. H. Archibald, *Z. Phys. Chem.*, **55**, 128 (1906).

148. T. C. Waddington, *J. Chem. Soc.*, **1958**, 1708.

149. D. A. Symon and T. C. Waddington, *J. Chem. Soc. Dalton*, 1879 (1973).

150. M. E. Peach and T. C. Waddington, *J. Chem. Soc.*, **1961**, 1238.

151. J. A. Salthouse and T. C. Waddington, *J. Chem. Soc. A*, **1967**, 1096.

152. M. E. Peach, *J. Inorg. Nucl. Chem.*, **39**, 565 (1977).

153. F. Klanberg and H. W. Kohlschütter, *Z. Naturforsch.*, **16b**, 69 (1961).

154. L. F. Audrieth and J. Kleinberg, Ref. 9, Ch. 10.

155. H. H. Hyman and J. J. Katz, Ref. 10, pp. 47–81.

156. M. Kilpatrick and J. G. Jones, Ref. 11, Vol. 2, pp. 43–98a.

157. A. W. Jache and G. H. Cady, *J. Phys. Chem.*, **56**, 1106 (1952).

158. J. H. Simons, *Chem. Rev.*, **8**, 213 (1931).

159. L. Quarterman, H. H. Hyman, and J. J. Katz, *J. Phys. Chem.*, **65**, 90 (1961).

160. H. Fredenhagen, *Z. Anorg. Allg. Chem.*, **242**, 23 (1939).

161. A. F. Clifford, H. C. Beachell, and W. M. Jack, *J. Inorg. Nucl. Chem.*, **5**, 57 (1957).

162. A. F. Clifford and A. G. Morris, *J. Inorg. Nucl. Chem.*, **5**, 71 (1957).

163. H. H. Hyman, L. A. Quaterman, M. Kilpatrick, and J. J. Katz, *J. Phys. Chem.*, **65**, 123 (1961).

164. R. J. Gillespie and K. C. Moss, *J. Chem. Soc. A*, **1966**, 1170.

165. P. A. W. Dean, R. J. Gillespie, R. Hulme, and D. A. Humphreys, *J. Chem. Soc. A*, **1971**, 341.

166. F. P. Del Greco and J. W. Gryder, *J. Phys. Chem.*, **65**, 922 (1961).

167. H. Selig and U. El-Gud, *J. Inorg. Nucl. Chem.*, **35**, 3517 (1973).

168. A. F. Clifford and S. Kongpricha, *J. Inorg. Nucl. Chem.*, **5**, 76 (1957).

169. E. L. Muetterties and W. D. Phillips, *J. Am. Chem. Soc.*, **81**, 1084 (1959).

170. G. P. Rutledge, R. L. Jarry, and W. Davis, Jr., *J. Phys. Chem.*, **57**, 541 (1953).

171. J. C. Hindman and A. Svirmickas, in *Noble-Gas Compounds*, H. H. Hyman (Ed.), pp. 251–262, University of Chicago Press, Chicago (1963).

172. H. H. Hyman and L. A. Quarterman, in *Noble-Gas Compounds*, H. H. Hyman (Ed.), p. 275, University of Chicago Press, Chicago (1963).

173. M. M. Woyski, *Inorg. Synth.*, **3**, 111 (1950).

174. M. M. Woyski, *J. Am. Chem. Soc.*, **72**, 919 (1950).

175. R. J. Gillespie, B. Landa, and G. J. Schrobilgen, *Inorg. Chem.*, **15**, 1256 (1976).

176. D. Moy and A. R. Young, *J. Am. Chem. Soc.*, **87**, 1889 (1965); *Inorg. Chem.*, **6**, 178 (1967).

177. J. P. Guertin, K. O. Christie, and A. E. Pavlath, *Inorg. Chem.*, **5**, 1921 (1966).

178. K. O. Christie, C. J. Schack, and R. D. Wilson, *Inorg. Chem.*, **15**, 1275 (1976).

179. H. H. Selig, L. A. Quarterman, and H. H. Hyman, *J. Inorg. Nucl. Chem.*, **28**, 2063 (1966).

180. A. F. Clifford, W. D. Pardieck, and M. W. Wadley, *J. Phys. Chem.*, **70**, 3241 (1966).

181. A. Hantzsch, *Z. Phys. Chem.*, **61**, 257 (1907); **62**, 626 (1908); **65**, 41 (1908); **68**, 204 (1909). *Chem. Ber.*, **B63**, 1782 (1930).

182. L. P. Hammett and A. J. Deyrup, *J. Am. Chem. Soc.*, **55**, 1900 (1933); H. P. Treffers and L. P. Hammett, *J. Am. Chem. Soc.*, **59**, 1708 (1937).

183. R. J. Gillespie, E. D. Hughes, and C. K. Ingold, *J. Chem. Soc.*, **1950**, 2473; R. J. Gillespie and J. B. Senior, *Inorg. Chem.*, **3**, 440, 972 (1964).

184. L. F. Audrieth and J. Kleinberg, Ref. 9, Ch. 9.

185. R. J. Gillespie and E. A. Robinson, in *Advances in Organic Chemistry and Radiochemistry*, H. J. Emeléus and A. G. Sharpe (Eds.), Vol. 1, pp. 385–423, Academic Press, New York (1959).

186. R. J. Gillespie and E. A. Robinson, Ref. 10, pp. 117–210.

187. W. H. Lee, Ref. 11, Vol. 2, pp. 99–150.

188. R. H. Flowers, R. J. Gillespie, and E. A. Robinson, *Can. J. Chem.*, **38**, 1363 (1960).

189. A. W. Davidson, *J. Am. Chem. Soc.*, **47**, 968 (1925).

190. R. J. Gillespie, *J. Chem. Soc.*, **1950**, 2493, 2516, 2537, 2542, 2997.

191. R. J. Gillespie and K. C. Malhotra, *J. Chem. Soc. A*, **1967**, 1994; **1968**, 1933.

192. R. J. Gillespie, J. Graham, E. D. Hughes, C. K. Ingold, and E. R. A. Peeling, *J. Chem. Soc.*, **1950**, 2504.

193. R. C. Paul, J. K. Puri, and K. C. Malhotra, *Inorg. Nucl. Chem. Lett.*, **7**, 729 (1971); *J. Inorg. Nucl. Chem.*, **35**, 403 (1973).

194. A. Bali and K. C. Malhotra, *J. Inorg. Nucl. Chem.*, **38**, 411 (1976).

195. R. C. Paul, J. K. Puri, V. P. Kapila, and K. C. Malhotra, *J. Inorg. Nucl. Chem.*, **34**, 2141 (1972).

196. A. A. Woolf, *J. Chem. Soc.*, **1955**, 433.

197. J. Barr, R. J. Gillespie, and R. C. Thompson, *Inorg. Chem.*, **3**, 1149 (1964).

198. R. C. Thompson, J. Barr, R. J. Gillespie, J. B. Milne, and R. A. Rothenbury, *Inorg. Chem.*, **4**, 1641 (1965).

199. R. J. Gillespie, J. B. Milne, and R. C. Thompson, *Inorg. Chem.*, **5**, 468 (1966).

200. R. J. Gillespie, J. B. Milne, and J. B. Senior, *Inorg. Chem.*, **5**, 1233 (1966).

201. R. J. Gillespie and J. B. Milne, *Inorg. Chem.*, **5**, 1236 (1966).

202. R. J. Gillespie, K. Ouchi, and G. P. Pez, *Inorg. Chem.*, **8**, 63 (1969).

203. R. J. Gillespie, *Acc. Chem. Res.*, **1**, 202 (1968).

204. A. W. Jache, in *Advances in Inorganic Chemistry and Radiochemistry*, H. J. Emeléus and A. G. Sharpe (Eds.), Vol. 16, pp. 177–200, Academic Press, New York (1974).

205. S. Natarajan and A. W. Jache, Ref. 11, Vol. 5B, pp. 53–155.

205a. K. C. Lee and F. Aubke, *Inorg. Chem.*, **18**, 389 (1979); **19**, 119 (1980).

206. A. Engelbrecht, *Angew. Chem., Int. Ed. Engl.*, **4**, 641 (1965).

207. P. A. W. Dean and R. J. Gillespie, *J. Am. Chem. Soc.*, **92**, 2362 (1970).

208. F. Fehér, Ref. 11, Vol. 3, pp. 219–240.

209. G. Jander and B. Grüttner, *Chem. Ber.*, **81**, 102, 107, 114 (1948).

210. W. H. Lee, Ref. 11, Vol. 2, pp. 151–189.

211. A. I. Popov, Ref. 11, Vol. 3, 241–337, 339–379.

212. J. B. Milne, Ref. 11, Vol. 5B, pp. 1–52.

213. R. C. Paul, K. K. Paul, and K. C. Malhotra, *J. Chem. Soc. A*, **1970**, 2712.

214. J. R. Dalziel and F. Aubke, *Inorg. Chem.*, **12**, 2707 (1973).

215. L. P. Hammett and A. J. Deyrup, *J. Am. Chem. Soc.*, **54**, 2721 (1932).

216. L. P. Hammett, *Physical Organic Chemistry*, Ch. 9, McGraw-Hill Book Company, New York (1940).

217. M. A. Paul and F. A. Long, *Chem. Rev.*, **57**, 1 (1957).

218. R. J. Gillespie and T. E. Peel, in *Advances in Physical Organic Chemistry*, V. Gold (Ed.), Vol. 9, pp. 1–24, Academic Press, New York (1971).

219. R. J. Gillespie, *Endeavor*, **32**, 3 (1973).

219a. G. A. Olah, G. K. Surya Prakash, and J. Sommer, *Science*, **206**, 13 (1979). General Discussion plus applications in organic syntheses.

220. T. Gramstad, *Tidsskr. Kemi, Bergves., Metall.*, **19**, 62 (1959).

221. R. J. Gillespie and T. E. Peel, *J. Am. Chem. Soc.*, **95**, 5173 (1937).

222. G. A. Olah, *Chem. Br.*, **8**, 281 (1972).

223. G. A. Olah and J. Shen, *J. Am. Chem. Soc.*, **95**, 3582 (1973).

224. R. J. Gillespie and M. J. Morton, *Q. Rev. (Lond.)*, **25**, 553 (1971).

225. R. J. Gillespie and M. J. Morton, in *MTP International Review of Science, Inorganic Chemistry, Series One*, Vol. 3, V. Gutmann (Ed.), pp. 199–213, Butterworth, London (1972).

226. R. J. Gillespie and J. B. Milne, *Inorg. Chem.*, **5**, 1577 (1966).

227. R. J. Gillespie and M. J. Morton, *Inorg. Chem.*, **11**, 586 (1972).

228. R. J. Gillespie and M. J. Morton, *Inorg. Chem.*, **11**, 591 (1972).

229. R. J. Gillespie and J. Passmore, in *Advances in Inorganic Chemistry and Radiochemistry*, H. J. Emeléus and A. G. Sharpe (Eds.), Vol. 17, pp. 49–87, Academic Press, New York (1975).

229a. R. J. Gillespie, *Chem. Soc. Rev.*, **8**, 315 (1979).

230. L. F. Audrieth and J. Kleinberg, *Ref. 9*, Ch. 11.

231. T. C. Waddington, Ref. 10, pp. 253–284.

232. D. F. Burow, Ref. 11, Vol. 3, pp. 137–185.

233. W. Karcher and H. Hecht, in *Chemistry in Non-Aqueous Ionizing Solvents*, G. Jander, H. Spandau, and C. C. Addison (Eds.), Vol. III, Pt. 2, p. 79, Pergamon (Macmillan), New York (1967).

234. *Gmelin Handbuch der anorganischen Chemie*, System-No. 9, Vol. Bl, pp. 310–314, Verlag Chemie, Weinheim (1953).

235. F. Seel, *Z. Anorg. Allg. Chem.*, **250**, 331 (1943); **252**, 24 (1943).

236. F. Seel and H. Bauer, *Z. Naturforsch.*, **2b**, 397 (1947).

237. I. Lindqvist, *Inorganic Adduct Molecules of Oxo-Compounds*, pp. 109–111, Academic Press, New York (1963).

238. N. N. Lichtin, *Prog. Phys. Org. Chem.*, **1**, 75 (1963).

239. L. Kolditz and W. Rehak, *Z. Anorg. Allg. Chem.*, **342**, 32 (1966).

240. P. A. Bond and W. E. Belton, *J. Am. Chem. Soc.*, **67**, 1691 (1945).

241. F. Seel, A. K. Bocz, and J. Nograli, *Z. Anorg. Allg. Chem.*, **264**, 298 (1951).

242. F. Seel and E. Müller, *Chem. Ber.*, **88**, 1747 (1955).

243. K. Cruse, *Z. Elektrochem.*, **46**, 571 (1940).

244. P. J. Elving and J. M. Markowitz, *J. Chem. Educ.*, **37**, 75 (1960).

245. H. P. Cady and R. Taft, *J. Phys. Chem.*, **29**, 1075 (1925).

246. L. S. Bagster and G. Cooling, *J. Chem. Soc.*, **117**, 693 (1920).

247. K. Gleu, W. Breuel, and K. Rehm, *Z. Anorg. Allg. Chem.*, **235**, 201, 211 (1938).

248. E. H. Braye and W. Hübel, *Angew. Chem.*, **75**, 345 (1963).

249. W. Strohmeier and J. F. Guttenberger, *Chem. Ber.*, **97**, 1871 (1964).

250. L. Vaska and S. S. Bath, *J. Am. Chem. Soc.*, **88**, 1333 (1966).

251. J. Ross, J. H. Percy, R. L. Brandt, A. I. Gebhart, J. Mitchell, and S. Yolles, *Ind. Eng. Chem.*, **34**, 924 (1942).

252. G. H. Weinreich, *Bull. Soc. Chim. Fr.*, **1963**, 2820.

253. M. Schmidt, R. Bornmann, and I. Wilhelm, *Angew. Chem. Int. Ed. Engl.*, **2**, 691 (1963).

254. C. C. Addison, *Angew. Chem.*, **72**, 193 (1960).

255. C. C. Addison, J. Lewis, and R. Thompson, *J. Chem. Soc.*, **1951**, 2829, 2833.

256. C. C. Addison and C. D. Condiut, *J. Chem. Soc.*, **1952**, 1399.

257. C. C. Addison, N. Hodge, and R. Thompson, *J. Chem. Soc.*, **1954**, 1143.

258. C. C. Addison, N. Hodge, and J. Lewis, *J. Chem. Soc.*, **1953**, 2631.

259. A. G. Sharpe, Ref. 10, Ch. 7.

260. I. Sheft, H. H. Hyman, and J. J. Katz, *J. Am. Chem. Soc.*, **75**, 5221 (1953).

261. A. A. Woolf, in *Advances in Inorganic Chemistry and Radiochemistry*, H. J. Emeléus and A. G. Sharpe (Eds.), Vol. 9, pp. 217–314, Academic Press, New York (1966). Heterocations of all types.

262. C. G. Vonk and E. H. Wiebenga, *Acta Crystallogr.*, **12**, 859 (1959).

263. T. Surles, H. H. Hyman, L. A. Quarterman, and A. I. Popov, *Inorg. Chem.*, **10**, 611 (1971).

264. A. G. Sharpe and H. J. Emeléus, *J. Chem. Soc.*, **1948**, 2135.

265. A. G. Sharpe, *J. Chem. Soc.*, **1950**, 3444; **1953**, 197.

266. A. A. Woolf, *J. Chem. Soc.*, **1950**, 1053, 3678.

267. P. Bony, *Ann. Chim.*, [13], **4**, 853 (1959).

268. B. J. Barker and J. A. Caruso, Ref. 11, Vol. 4, pp. 110–128.

269. W. H. Lee, Ref. 11, Vol. 4, pp. 167–245.

270. R. L. Benoit and R. Pichet, *J. Electroanal. Chem. Interfacial Electrochem.*, **43**, 59 (1973).

271. J. Martinmaa, Ref. 11, Vol. 4, pp. 247–287.

272. R. C. Paul and S. S. Sandhu, Ref. 11, Vol. 3, pp. 187–217.

273. D. S. Payne, Ref. 10, pp. 301–352.

274. V. Gutmann, in *Halogen Chemistry*, V. Gutman (Ed.), Vol. 2, pp. 399–438, Academic Press, New York (1967).

275. E. C. Baughan, Ref. 11, Vol. 4, pp. 130–165.

276. R. C. Paul and G. Singh, Ref. 11, Vol. 5B, pp. 197–268.

277. G. B. L. Smith, *Chem. Rev.*, **23**, 165 (1938).

278. D. Martin, R. Rousseau, and J.-M. Weulersee, Ref. 11, Vol. 5B, pp. 157–195.

REFERENCES

279. J. B. Milne, Ref. 11, Vol. 5B, pp. 1–52.

280. M. Rumeau, Ref. 11, Vol. 4, pp. 76–107.

281. B. L. Laube and C. D. Schmulbach, in *Progress in Inorganic Chemistry*, S. J. Lippard (Ed.), Vol. 14, pp. 65–118, Wiley-Interscience, New York (1971). Comprehensive review; 451 references.

282. L. F. Audrieth and H. W. Nelson, *Chem. Rev.*, **8**, 335 (1931).

283. A. Brenner, *J. Electrochem. Soc.*, **103**, 652 (1956); *Rec. Chem. Prog.*, **16**, 241 (1957).

284. V. Gutmann and G. Schröber, *Oesterr. Chem.-Ztg.*, **59**, 321 (1958).

285. J. H. Roberts, Ref. 11, Vol. 4, pp. 1–17. Conductivities.

286. C. W. Stillwell and L. F. Audrieth, *J. Am. Chem. Soc.*, **54**, 472 (1932).

287. H. Schmidt, *Z. Anorg. Allg. Chem.*, **270**, 188 (1952).

288. H. Schmidt, I. Wittkopf, and G. Jander, *Z. Anorg. Chem.*, **256**, 113 (1948).

289. E. E. Jukkola, with L. F. Audrieth and B S. Hopkins, *J. Am. Chem. Soc.*, **56**, 303 (1934).

290. T. Moeller and V. Galasyn, *J. Inorg. Nucl. Chem.*, **12**, 259 (1959).

291. T. P. Wier and F. H. Hurley (to Rice Institute), U.S. Patent 2, 446, 349, August 3, 1948.

292. D. E. Couch and A. Brenner, *J. Electrochem. Soc.*, **99**, 234 (1952).

293. J. H. Connor and A. Brenner, *J. Electrochem. Soc.*, **103**, 657 (1956).

294. W. E. Reid, Jr., J. M. Bish, and A. Brenner, *J. Electrochem. Soc.*, **104**, 21 (1957).

295. G. B. Wood and A. Brenner, *J. Electrochem. Soc.*, **104**, 29 (1957).

296. J. H. Connor, W. E. Reid, Jr., and G. B. Wood, *J. Electrochem. Soc.*, **104**, 38 (1957).

297. G. Szekely, *J. Electrochem. Soc.*, **98**, 318 (1951).

298. H. Bloom, *Rev. Pure Appl. Chem. (Aust.)*, **9**, 140 (1959).

299. B. R. Sundheim (Ed.), *Fused Salts*, McGraw-Hill Book Company, New York (1964).

300. J. D. Corbett, in *Survey of Progress in Chemistry*, A. B. Scott (Ed.), Vol. 2, pp. 91–154, Academic Press, New York (1964).

301. M. Blander (Ed.), *Molten Salt Chemistry*, Wiley-Interscience, New York (1964).

302. H. Bloom and J. W. Hastie, Ref. 10, pp. 353–390.

303. H. Bloom, *The Chemistry of Molten Salts*, W. A. Benjamin, New York (1967).

304. J. H. R. Clarke and G. J. Hills, *Chem. Br.*, **9**, 12 (1973).

304a. D. H. Kerridge, Ref. 11, Vol. 5B, pp. 269–329.

304b. B. W. Hatt and D. H. Kerridge, *Chem. Br.*, **15**, 78 (1979). Industrial applications.

305. G. J. Janz, *J. Chem. Educ.*, **39**, 59 (1962).

306. R. W. Laity, *J. Chem. Educ.*, **39**, 67 (1962).

307. H. A. Laitinen and J. W. Panky, *J. Am. Chem. Soc.*, **81**, 1053 (1959).

308. H. Lux, *Naturwissenschaften*, **28**, 92 (1940).

309. F. R. Duke, *J. Chem. Educ.*, **39**, 57 (1962).

310. J. A. Duffy and M. D. Ingram, *J. Am. Chem. Soc.*, **93**, 6448 (1971).

311. J. F. G. Hicks and W. A. Craig, *J. Phys. Chem.*, **26**, 563 (1922).

312. N. E. Norton and H. F. Johnstone, *Ind. Eng. Chem.*, **43**, 1553 (1951).

313. W. Sundermeyer and O. Glemser, *Angew. Chem.*, **70**, 625 (1958).

314. G. Jander, *Angew. Chem.*, **62**, 264 (1950); G. Jander and K. Brodersen, *Z. Anorg. Chem.*, **261**, 261 (1950); **262**, 33 (1950).

315. W. Sundermeyer, *Angew. Chem., Int. Ed. Engl.*, **4**, 222 (1965).

316. J. A. Duffy and M. D. Ingram, *J. Am. Chem. Soc.*, **93**, 6448 (1971).

317. J. Robinson, B. Gilbert, and R. A. Osteryoung, *Inorg. Chem.*, **16**, 3040 (1977).

318. K. Niki and H. A. Laitinen, *J. Inorg. Nucl. Chem.*, **37**, 91 (1975).

319. I. Uchida, K. Niki, and H. A. Laitinen, *J. Inorg. Nucl. Chem.*, **39**, 255 (1977).

319a. R. J. Gale and R. A. Osteryoung, *Inorg. Chem.*, **18**, 1603 (1979).

320. U. Anders and J. A. Plambeck, *J. Inorg. Nucl. Chem.*, **40**, 387 (1978).

321. J. D. Corbett, W. J. Burkhart, and L. F. Druding, *J. Am. Chem. Soc.*, **83**, 76 (1961).

322. J. D. Corbett, *Inorg. Chem.*, **7**, 198 (1968).

323. R. Fehrmann, N. J. Bjerrum, and F. W. Poulsen, *Inorg. Chem.*, **17**, 1195 (1978).

324. T. W. Couch, D. A. Lokken, and J. D. Corbett, *Inorg. Chem.*, **11**, 357 (1972).

325. D. J. Merryman, P. A. Edwards, J. D. Corbett, and R. E. McCarley, *Inorg. Chem.*, **13**, 1471 (1974).

326. R. Fehrmann, N. J. Bjerrum, and H. A. Audreasen, *Inorg. Chem.*, **15**, 2187 (1976).

327. R. M. Friedman and J. D. Corbett, *Inorg. Chem.*, **12**, 1134 (1973).

328. R. J. Gillespie and J. Passmore, *Chem. Br.*, **8**, 475 (1972).

329. N. J. Bjerrum, C. R. Boston, and G. P. Smith, *Inorg. Chem.*, **6**, 1162 (1967).

329a. G. Mamantov, R. Marassi, F. W. Poulsen, S. E. Springer, J. P. Wiaux, R. Huglen, and N. R. Smyrl, *J. Inorg. Nucl. Chem.*, **41**, 260 (1979).

330. R. N. Kust and F. R. Duke, *J. Am. Chem. Soc.*, **85**, 3338 (1963).

331. F. R. Duke, in *Fused Salts*, B. R. Sundheim (Ed.), Ch. 7, McGraw-Hill Book Company, New York (1964).

332. L. E. Topol, R. A. Osteryoung, and J. H. Christie, *J. Phys. Chem.*, **70**, 2857 (1966).

333. P. G. Zambonin and J. Jordan, *J. Am. Chem. Soc.*, **91**, 2225 (1969).

334. S. S. Al Omer and D. H. Kerridge, *J. Inorg. Nucl. Chem.*, **40**, 975 (1978).

335. B. J. Brough and D. H. Kerridge, *Inorg. Chem.*, **4**, 1353 (1965).

336. D. H. Kerridge and J. C. Rey, *J. Inorg. Nucl. Chem.*, **39**, 297 (1977).

337. D. H. Kerridge and J. C. Rey, *J. Inorg. Nucl. Chem.*, **39**, 405 (1977).

337a. H. Frouzanfar and D. H. Kerridge, *J. Inorg. Nucl. Chem.*, **41**, 181 (1979).

338. D. H. Kerridge, in *MTP International Review of Science, Inorganic Chemistry, Series One*, Vol. 2, Butterworth, New York (1972).

339. D. H. Kerridge, in *Advances in Molten Salt Chemistry*, J. Braunstein, G. Mamantov, and G. P. Smith (Eds.), Vol. 4, pp. 249–273, Plenum Press, New York (1975).

340. W. Giggenbach, *Inorg. Chem.*, **10**, 1308 (1971).

341. D. H. Kerridge and S. J. Walker, *J. Inorg. Nucl. Chem.*, **38**, 1795 (1976).

342. D. H. Kerridge and S. J. Walker, *J. Inorg. Nucl. Chem.*, **39**, 1579 (1977).

343. J. D. Corbett, in *Fused Salts*, B. R. Sundheim (Ed.), Ch. 6, McGraw-Hill Book Company, New York (1964).

344. L. F. Druding, J. W. Corbett, and B. N. Ramsey, *Inorg. Chem.*, **2**, 869 (1963).

345. J. D. Corbett, in *Preparative Inorganic Reactions*, W. L. Jolly (Ed.), Vol. 3, pp. 1–33, Interscience Publishers, New York (1966).

346. D. A. J. Swinkels, *Adv. Molten Salt Chem.*, **1**, 165 (1971).

347. L. F. Audrieth and T. Moeller, *J. Chem. Educ.*, **20**, 219 (1943).

348. G. Cohn, *Chem. Rev.*, **42**, 527 (1948). Reactions in general.

349. A. B. Scott, *J. Chem. Educ.*, **38**, 317 (1961). Electrochemistry.

350. G. Parravano, *Chem. Eng. News*, Mar. 19, 1962, 111. Reactions in general.

351. J. S. Anderson, *Proc. Chem. Soc. Lond.*, June 1964, p. 166. Solid state.

352. E. F. Gurnee, *J. Chem. Educ.*, **46**, 80 (1969). Semiconductors.

353. P. F. Weller, *J. Chem. Educ.*, **47**, 501 (1970). Solid-state principles.

354. K. Hardel, *Angew. Chem., Int. Ed. Engl.*, **11**, 173 (1972). Reaction mechanisms.

355. H. G. Drickamer, *Chem. Br.*, **9**, 353 (1973). Electronic effects, high pressure.

356. H. Schmalzried (translated by A. D. Pelton), *Solid State Reactions*, Academic Press, New York (1974).

357. S. M. Ariya and P. Morozova, *Zh. Obshch. Khim.*, **28**, 2617 (1958).

358. H. Schmalzried, *Angew. Chem.*, **75**, 353 (1963).

359. F. Feigl, L. J. Miranda, and H. A. Suter (translated by R. E. Oesper), *J. Chem. Educ.*, **21**, 18 (1944).

360. A. Wold and J. K. Ruff (Eds.), *Inorganic Syntheses*, Vol. 14, pp. 99–192, McGraw-Hill Book Company, New York (1973). Variety of compounds.

361. H. Schäfer (translated by H. Frankfort), *Chemical Transport Reactions*, Academic Press, New York (1964).

362. M. C. Ball, *J. Chem. Educ.*, **45**, 651 (1968).

363. K. E. Spear, *J. Chem. Educ.*, **49**, 81 (1972).

364. J. B. Adamec and D. B. Springer, in *The Encyclopedia of the Chemical Elements*, C. A. Hampel (Ed.), p. 438, Reinhold Publishing Corp., New York (1968).

365. A. E. van Arkel and J. H. de Boer, *Z. Anorg. Allg. Chem.*, **148**, 345 (1925).

366. A. E. van Arkel, *Reine Metalle*, Springer-Verlag, Berlin (1939).

367. R. F. Ralston, *Iodide Metals and Metal Iodides*, John Wiley & Sons, New York (1961).

368. J. W. van Tijen, *Phillips Tech. Rev.*, **23**, 226 (1961–1962).

369. E. S. Scott and L. F. Audrieth, *J. Chem. Educ.*, **31**, 168 (1954).

370. H. Krebs, *Angew. Chem.*, **70**, 615 (1958).

371. H. J. Eméleus, *Proc. Chem. Soc.*, **1959**, 202.

372. D. B. Sowerby and L. F. Audrieth, *J. Chem. Educ.*, **37**, 2, 86, 134 (1960).

373. *Inorganic Polymers*, Spec. Publ. 15, The Chemical Society, London (1961). Collection of papers from an international symposium.

374. F. G. A. Stone and W. A. G. Graham (Eds.), *Inorganic Polymers*, Academic Press, New York (1962).

375. K. R. Eilar and R. I. Wagner, *Chem. Eng. News*, Aug. 6, 1962, p. 138.

376. H. R. Allcock, *Sci. Am.*, **230**, 66 (1974); *Science*, **193**, 1214 (1976); *Heteroatom Ring Systems and Polymers*, Academic Press, New York (1967); *Sci. Prog., Oxf.*, **66**, 355 (1980).

377. L. G. Sillén, *Q. Rev. (Lond.)*, **13**, 146 (1959).

378. C. L. Rollinson, in *The Chemistry of the Coordination Compounds*, J. C. Bailar, Jr., and D. H. Busch (Eds.), Ch. 13, Reinhold Publishing Corp., New York (1956).

379. T. H. Whitehead, *Chem. Rev.*, **21**, 113 (1937).

380. O. Schmitz-Dumont and W. Hilger, *Z. Anorg. Allg. Chem.*, **300**, 175 (1959).

381. J. A. Seaton, F. G. Sherif, and L. F. Audrieth, *J. Inorg. Nucl. Chem.*, **9**, 222 (1959).

382. D. C. Bradley, in *Progress in Inorganic Chemistry*, F. A. Cotton (Ed.), Vol. 2, pp. 303–361, Interscience Publishers, New York (1960).

383. R. K. Iler, *The Colloid Chemistry of Silica and Silicates*, Cornell University Press, Ithaca, N.Y. (1955).

384. O. F. Hill and L. F. Audrieth, *J. Phys. Colloid Chem.*, **54**, 690 (1950).

385. L. F. Audrieth and R. N. Bell, *Inorg. Synth.*, **3**, 85 (1950).

386. L. F. Audrieth, J. R. Mills, and L. E. Netherton, *J. Phys. Chem.*, **58**, 482 (1954).

387. R. N. Bell, *Ind. Eng. Chem.*, **40**, 1464 (1948).

388. R. Klement and G. Biberacher, *Z. Anorg. Allg. Chem.*, **283**, 246 (1956).

389. A. B. Burg and F. F. Caserio, *J. Inorg. Nucl. Chem.*, **11**, 259 (1960).

390. F. G. A. Stone and A. B. Burg, *J. Am. Chem. Soc.*, **76**, 386 (1954).

391. A. B. Burg and R. I. Wagner, *J. Am. Chem. Soc.*, **76**, 3307 (1954).

392. N. Davidson and H. C. Brown, *J. Am. Chem. Soc.*, **64**, 316 (1942).

393. M. E. Kenney and A. W. Laubengayer, *J. Am. Chem. Soc.*, **76**, 4839 (1954).

394. J. Goubeau and P. Schulz, *Z. Phys., Chem.*, Neue Folge, **14**, 49 (1958).

395. R. Schenck and G. Römer, *Chem. Ber.*, **57B**, 1343 (1924).

396. R. Steinman, F. B. Schirmer, Jr., and L. F. Audrieth, *J. Am. Chem. Soc.*, **64**, 2377 (1942).

397. K. John and T. Moeller, *J. Am. Chem. Soc.*, **82**, 2647 (1960).

398. F. Fehér, *Angew. Chem.*, **67**, 337 (1955).

399. B. Meyer, in *Advances in Inorganic Chemistry and Radiochemistry*, H. J. Emeléus and A. G. Sharpe (Eds.), Vol. 18, pp. 287–318, Academic Press, New York (1976); *Chem. Rev.*, **76**, 367 (1976).

400. D. E. C. Corbridge, in *Topics in Phosphorus Chemistry*, M. Grayson and E. J. Griffith (Eds.), Vol. 3, pp. 37–394, Wiley-Interscience, New York (1966).

401. P. D. Bartlett and G. Merguerian, *J. Am. Chem. Soc.*, **78**, 3710 (1956).

402. P. D. Bartlett and R. E. Davis, *J. Am. Chem. Soc.*, **80**, 2513 (1958).

403. O. Foss, *Acta Chem. Scand.*, **4**, 404 (1950).

404. M. Schmidt and G. Talsky, *Chem. Ber.*, **90**, 1673 (1957).

405. A. Jennen and M. Hens, *C. R. Acad. Sci.*, **242**, 786 (1956).

406. P. W. Bridgman, *J. Am. Chem. Soc.*, **36**, 1344 (1914); **38**, 609 (1916).

407. H. Krebs, H. Weitz, and K. H. Worms, *Z. Anorg. Allg. Chem.*, **280**, 119 (1955).

408. T. W. Dewitt and S. Skolnik, *J. Am. Chem. Soc.*, **68**, 2305 (1946).

409. H. R. Allcock, *Heteroatom Ring Systems and Polymers*, Academic Press, New York (1967).

410. H. N. Stokes, *Am. Chem. J.*, **19**, 782 (1897).

411. J. O. Konecny and C. M. Douglas, *J. Polym. Sci.*, **36**, 195 (1959).

412. H. R. Allcock, Ref. 412, pp. 173–181; *Chem. Rev.*, **72**, 315 (1972); *Chem. Br.*, **10**, 118 (1974).

413. H. R. Allcock and R. J. Best, *Can. J. Chem.*, **42**, 447 (1964).

414. A. G. MacDiarmid, C. M. Mikulski, M. S. Saran, P. J. Russo, M. J. Cohen, A. A. Bright, A. F. Garito, and A. J. Heeger, in *Inorganic Compounds with Unusual Properties*, R. B. King (Ed.), Advances in Chemistry Series Vol. 150, pp. 63–72, American Chemical Society, Washington, D.C. (1976).

415. C. Hsu and M. M. Labes, *J. Chem. Phys.*, **61**, 4640 (1974).

416. C. K. Chiang, M. J. Cohen, A. F. Garito, A. J. Heeger, C. M. Mikulski, and A. G. MacDiarmid, *Solid State Commun.*, **18**, 1451 (1976).

417. R. L. Greene, P. M. Grant, and G. B. Street, *Phys. Rev. Lett.*, **34**, 89 (1975).

418. R. L. Greene, G. B. Street, and L. J. Suter, *Phys. Rev. Lett.*, **34**, 577 (1975).

419. L. Pintschovius, H. P. Geserich, and W. Moller, *Solid State Commun.*, **17**, 477 (1975).

420. C. M. Mikulski, A. G. MacDiarmid, A. F. Garito, and A. J. Heeger, *Inorg. Chem.*, **15**, 2943 (1976).

421. J. C. Bailar, Jr., W. C. Drinkard, Jr., and M. L. Judd, WADC Technical Report 57–391, ASTIA Document No. AD131100, Wright Air Development Center, Ohio (1957).

422. R. W. Kluiber and J. W. Lewis, *J. Am. Chem. Soc.*, **83**, 542 (1961).

423. F. S. Arimoto and A. C. Haven, Jr., *J. Am. Chem. Soc.*, **77**, 6295 (1955).

424. C. S. Marvel and N. Tarköy, *J. Am. Chem. Soc.*, **79**, 6000 (1957).

425. H. A. Goodwin and J. C. Bailar, Jr., *J. Am. Chem. Soc.*, **83**, 2466 (1961).

426. R. C. Burns, R. J. Gillespie, and W.-C. Luk, *Inorg. Chem.*, **17**, 3596 (1978).

427. R. J. Gillespie, W.-C. Luk, E. Maharajh, and D. R. Slim, *Inorg. Chem.*, **16**, 892 (1977).

428. R. C. Burns, R. J. Gillespie, W.-C. Luk, and D. R. Slim, *Inorg. Chem.*, **18**, 3086 (1979).

429. R. K. McMullan, D. J. Prince, and J. D. Corbett, *Inorg. Chem.*, **10**, 1749 (1971).

430. P. Boldrini, I. D. Brown, R. J. Gillespie, P. R. Ireland, W.-C. Luk, D. R. Slim, and J. E. Verkis, *Inorg. Chem.*, **15**, 765 (1976).

SUGGESTED SUPPLEMENTARY REFERENCES

H. R. Allcock, *Heteroatom Ring Systems and Polymers*, Academic Press, New York (1967). Detailed summaries.

L. F. Audrieth and J. Kleinberg, *Non-aqueous Solvents as Media for Chemical Reactions*, John Wiley & Sons, New York (1953). Excellent background for properties and reactions.

M. Blander (Ed.), *Molten Salt Chemistry*, Wiley-Interscience, New York (1964). Summaries of various aspects of properties and reactions.

H. Bloom, *The Chemistry of Molten Salts*, W. A. Benjamin, New York (1967). Introduction to physical and chemical behaviors.

J. Braunstein, G. Mamantov, and G. P. Smith (Eds.), *Advances in Molten Salt Chemistry*, Plenum Press, New York (1971–). Continuing series covering many areas.

E. C. Franklin, *The Nitrogen System of Compounds*, Reinhold Publishing Corp., New York (1935). Classic reference on nitrogen compounds.

V. Gutmann, *Coordination Chemistry in Non-aqueous Solutions*, Springer-Verlag, New York (1968). Donor and acceptor solvents.

W. L. Jolly (Ed.), *Metal–Ammonia Solutions*, Dowden, Hutchinson & Ross, Stroudsburg, Pa. (1972). Reprints of 63 selected papers.

J. J. Lagowski (Ed.), *The Chemistry of Non-aqueous Solvents*, Academic Press, New York (1966–). Continuing series of volumes covering aspects of specific systems, properties, reactions, theory, and concept.

O. Popovych and R. P. T. Tomkins, *Non-aqueous Solution Chemistry*, John Wiley & Sons, New York (1981). Comprehensive coverage of properties, reactions, and numerical data.

H. Schmalzried (trans. by A. D. Pelton), *Solid-State Reactions*, Academic Press, New York (1974). Overall summary.

F. G. A. Stone and W. A. G. Graham (Eds.), *Inorganic Polymers*, Academic Press, New York (1962). Detailed summaries of various polymer types.

B. R. Sundheim (Ed.), *Fused Salts*, McGraw-Hill Book Company, New York (1964). Extensive summaries, involving structures, natures of solutes, physical and chemical phenomena.

T. C. Waddington (Ed.), *Non-aqueous Solvent Systems*, Academic Press, New York (1965). Separate discussions of 10 systems.

CHAPTER ELEVEN

INORGANIC CHEMICAL REACTIONS, III REACTIONS INVOLVING COORDINATION ENTITIES

The chemical reactions of coordination entities are so numerous and so varied that it is difficult to develop a simple, but effective, system of classification. In their classic treatment of the reactions of metal complexes in solution, F. Basolo and R. G. Pearson[1] discussed substitution reactions, stereochemical changes, oxidation–reduction reactions, reactions of transition metal organometallic species, metal-ion catalysis, and photochemistry. These reactions were discussed largely in the mechanistic sense, which approach is of course directly related to reaction kinetics (Ch. 9). In a contemporary discussion, J. P. Candlin, K. A. Taylor, and D. T. Thompson[2] classified reactions of the transition metal complexes as substitution reactions, combination reactions, and redox reactions. Other less extensive approaches have appeared also.

In a very broad sense, the reactions of coordination molecules and ions are all, directly or indirectly, reactions of synthesis. Synthesis may involve the preparation of a coordination species from another species not considered to be a complex, or it may involve the conversion of one complex into another, either in a single step or via a sequence of steps. Not uncommonly, it may proceed in solution (often aqueous) and thus involve a solvated reactant. Whether or not oxidation–reduction occurs, most reactions of synthesis, and all that take place in solution, are based upon and involve ligand exchange, or ligand substitution, processes. Equilibria among complex molecules or ions in solution, which are measures of the thermodynamic stabilities of complexes under these conditions, are again exchange or substitution processes. So also are isomerization and racemization, at least in the sense that exchange of positions of ligands in coordination spheres must take place. Mechanistically, however, reactions of this type are widely variable.

In this chapter, the exchange, or substitution, concept is used as a general theme for discussion of thermodynamic stability in aqueous solution, ligand substitution in various molecular geometries, isomerism and racemization, and redox systems. Related topics already covered in Chapters 7 to 9 should be reviewed and be used to supplement the topics that follow. More comprehensive treatments can be consulted for additional details.[1-3]

1. THERMODYNAMIC STABILITY IN AQUEOUS SOLUTION

Thermodynamic stability in aqueous solution, as measured in terms of change in standard free energy (ΔG^0) or formation constant ($\log K$ or $\log \beta$, Sec. 9.2), is a measure of the displacement of coordinated water molecules in terms of the general equilibrium scheme

$$[M(H_2O)_m]^{x\pm} + L^{y\pm} \rightleftharpoons [M(H_2O)_{m-1}L]^{z\pm} + H_2O$$

$$+ L^{y\pm}$$

$$\updownarrow$$

$$[M(H_2O)_{m-2}L_2]^{z\pm} + H_2O$$

$$+ (n-2)L^{y\pm}$$

$$\updownarrow$$

$$[ML_n]^{z\pm} + (n-2)H_2O$$

where L is a unidentate ligand molecule, anion, or cation of charge $y^{\pm}$, and the charge on a product species $(z^{\pm})$ is determined by n, $x^{\pm}$, and $y^{\pm}$. Formation-constant data,* as determined by a variety of techniques,[4,5] are available at specified conditions of temperature (commonly 25°C), concentration, and ionic strength and are listed in various comprehensive summaries.[6-8]

The thermodynamic stabilities in aqueous solution of complex ions or molecules of comparable types (i.e., same metal with a series of structurally related ligands, or same ligand with a series of cations of the same charge) can be compared meaningfully in terms of their formation constants. The data summarized in Table 11-1 for selected systems are both illustrative and typical. Comparable data from a variety of sources indicate that for the dipositive $3d$-metal ions, irrespective of the donor, solution stabilities of complexes as indicated by formation-constant data vary as

$$Mn^{2+} < Fe^{2+} < Co^{2+} < Ni^{2+} < Cu^{2+} > Zn^{2+}$$

This sequence has been termed the "natural order." [9-11]

The influence of complexation on the stabilization of oxidation states has been discussed earlier (Sec. 9.5-2) in terms of electrode-potential data. In terms of the relationships

$$\Delta G_n^0 = -nE^0 \mathscr{F} = -RT\ln K_n \tag{11-1}$$

$$\Delta G_n^0 = -nE^0 \mathscr{F} = -RT\ln \beta_n \tag{11-2}$$

the stabilization of oxidation states, as effected by complexation, can be given quantitative meaning. Numerous examples are given in Tables 9-2, 9-3, and 9-4.

Factors that affect the stabilities of complexes relative to their formation in nonredox systems include the following:

1. Environmental factors such as temperature and radiant energy.
2. Concentration factors for solutions and pressure factors where one or more of the reactants are gases.
3. Nature of the metal ion.
4. Nature of ions not in the coordination sphere.
5. Chelation.
6. Relative enthalpy and entropy changes.

Complex species containing volatile ligands are uniformly decreasingly stable with respect to their components as temperature increases. This phenomenon is much more evident with solids than with solutions. A typical reaction is described by the equation

$$[Co(en)_3]Cl_3(s) \xrightarrow{210°C} [Co(en)_2Cl_2]Cl(s) + en(g)$$

Photochemical effects are less well characterized.[11a]

---

* Constants determined for equilibria as written above are formation or stability constants. Constants for equilibria written in the reverse fashion are dissociation or instability constants.

The significance of concentration factors is apparent in formation-constant data. In a number of cases, a complex ion present in a crystal lattice is destroyed upon dissolution of the crystal in water. Thus bright yellow $Cs_2[CuCl_4](s)$ yields the pale blue $[Cu(H_2O)_4]^{2+}$ ion on dissolution, as

$$[CuCl_4]^{2-} + 4H_2O \rightleftharpoons [Cu(H_2O)_4]^{2+} + 4Cl^-$$
yellow   blue

but the addition of excess chloride ion (as HCl, LiCl, etc.) effects displacement of the equilibrium toward the $[CuCl_4]^{2-}$ ion. Correspondingly, the pink hydrated cobalt(II) ion in aqueous solution is converted to blue complex anions by addition of chloride, bromide, or thiocyanate ion. Decrease in pressure results in removal of volatile ligands (e.g., ammonia from complexes such as $[Cu(NH_3)_4]^{2+}$), whereas increase in the partial pressure of a volatile ligand may enhance complexation by that ligand.

The effects of the natures and properties of the metal ions have been noted in numerous earlier discussions (Chs. 7–10) and need not be treated in detail here. The cations that serve best as acceptors are those that can interact through $(n-1)d$ orbitals, are small, and carry large ionic and/or effective nuclear charges.

Effects on stability that are dependent on the nature and type of the ligand and upon chelation need not be treated in any additional detail. Nor is it necessary in the thermodynamic sense to go beyond what has already been discussed (Chs. 7–10) in terms of enthalpy and entropy contributions to stability as indicated by the relationship

$$\Delta G^0 = -RT\ln K_n (\text{or } \beta_n) = \Delta H^0 - T\Delta S^0 \tag{11-3}$$

As noted in earlier discussions, complexes that form in aqueous solution are entropy-stabilized when coordinated water molecules are displaced by chelating ligands (the *chelate effect*, Sec. 9.2-1).*

It must be emphasized, however, that the thermodynamic stability of a complex species may or may not be related to rate at which it forms. Thermodynamically stable complexes may form rapidly or slowly, depending upon the natures of the reactions and the reactants. Kinetic studies on the formation of thermodynamically stable complex species provide useful information as to the probable mechanisms of the formation reactions. When referring to the stability of a particular species, one must, therefore, specify either thermodynamic or kinetic stability or both.

2. LIGAND SUBSTITUTION, OR EXCHANGE, REACTIONS

Reactions of this type are more numerous and have been more thoroughly studied and discussed than any other reaction type in which d-transition metal species are involved. Such a reaction may be formulated as

$$MX + Y \rightarrow MY + X$$

---

* For additional interpretations, see D. Munro, *Chem. Br.*, **13**, 100 (1977); L. E. Simmons, *J. Chem. Educ.*, **56**, 578 (1979).

Table 11-1 Typical Formation, or Stability, Constants with N or N + O Donors

Ligand L	Equilibrium Expression[a] $ML_n/M \cdot L^n$	log β_n											
		Ca^{2+}	Cr^{2+}	Cr^{3+}	Mn^{2+}	Fe^{2+}	Fe^{3+}	Co^{2+}	Co^{3+}	Ni^{2+}	Cu^{2+}	Zn^{2+}	Cd^{2+}
NH_3	$ML_2/M \cdot L^2$	−0.8[b]			1.54[c]			3.50[d]		4.89[e]	7.47[e]	4.50[e]	4.56[e]
	$ML_4/M \cdot L^4$				1.3[c]			5.07[d]		7.67[e]	11.75	8.89[e]	6.74[e]
	$ML_6/M \cdot L^6$							4.39[d]	34.36[f]	8.31[e]			
$CH_3CH_2NH_2$[g]	$ML_2/M \cdot L^2$											2.3	
	$ML_4/M \cdot L^4$										11.5		
$H_2NCH_2CH_2NH_2$[g]	$ML/M \cdot L$		5.15		2.77	4.34		5.96		7.47	10.71		
	$ML_2/M \cdot L^2$		9.19		4.87	7.66		10.8		13.82	20.04		
	$ML_3/M \cdot L^3$				5.81	9.72		14.1	48.68	18.13			
$HOCH_2CH_2NH_2$	$ML/M \cdot L$									2.98	5.7	3.7	
	$ML_2/M \cdot L^2$									5.33	9.8	6.1	
	$ML_3/M \cdot L^3$									7.33	13.0	9.4	
$CH_3CHCH_2NH_2$ $\;\;\;\vert$ $\;\;\;NH_2$	$ML/M \cdot L$									7.34	10.58	5.72	
	$ML_2/M \cdot L^2$									13.51	19.72	10.77	
	$ML_3/M \cdot L^3$									17.82			
trans-1,2-$C_6H_{10}(NH_2)_2$	$ML/M \cdot L$				2.94			6.37[h]		7.74	10.94	6.65	5.8
	$ML_2/M \cdot L^2$				5.33			11.74[h]		14.27	20.35	12.12	10.51
	$ML_3/M \cdot L^3$							15.22[h]				14.1	
$HN(CH_2CO_2)_2^{2-}$	$ML/M \cdot L$	2.59		10.94		5.8[h]	10.72[h]	6.94	29.6	8.13	10.57	7.24	5.71
	$ML_2/M \cdot L^2$			21.4		10.1		12.23		14.1	16.54	12.52	10.12
$N(CH_2CO_2)_3^-$	$ML/M \cdot L$	6.39			7.46	8.33[h]	15.9	10.38		11.50	12.94	10.66	9.78
	$ML_2/M \cdot L^2$	8.76			10.94	12.8[h]	24.3[h]	14.33		16.32	17.42	14.24	14.39

706

$(O_2CCH_2)_2N^{4-}$ $\mid$ C_2H_4	ML/M·L	10.61	13.6[h]	23.4[h]	13.81	14.27	25.1[h]	16.26	41.4	18.52	18.70	16.44	16.36
$(O_2CCH_2)_2N$ $\mid$ C_2H_4 $(O_2CCH_2)_2N^{5-}$ $\mid$ C_2H_4 $\mid$ O_2CCH_2N $\mid$ C_2H_4 $\mid$ N $\mid$ $(O_2CCH_2)_2N$	ML/M·L	10.75			15.51	16.4	28.0[h]	19.15		20.17	21.38	18.29	19.0

---

[a] Concentrations (mole liter^{-1}).
[b] At 23°C, $\mu = 2.0$ M.
[c] At 20°C, $\mu = 2.0$ M.
[d] At 20°C, $\mu = 0$ M.
[e] At 25°C, $\mu = 0$ M.
[f] At 30°C, $\mu = 1.0$ M.
[g] At 25°C, $\mu = 0.5$ M (Others at 25°C, $\mu = 0.1$ M).
[h] At 20°C, $\mu = 0.1$ M.

where X and Y are different ligands, or perhaps more realistically as

$$ML_nX + Y \rightarrow ML_nX + Y$$

where L is a ligand that completes the coordination sphere and is not displaced. Both the rates (Sec. 9.3) and mechanisms of these reactions are important.[12-17]

Complexes that undergo ligand exchange are said to be *labile*; those that undergo either no or very slow ligand exchange are said to be *inert*. In a practical sense, slow ligand exchange (half-time 1 or more minutes)[18] is measurable after mixing the reagents by changes in pH, optical density, refractive index, volume of gas liberated, and so on. Rapid ligand exchange (half-time much less than 1 minute)[18] is measurable only by the newer techniques of stop-flow or relaxation that lend themselves to instantaneous electronic recording.[19, 20]

Lability is affected by oxidation state, electronic configuration, and size of the central cation; by the natures of the ligands X, Y, and L; by the reaction medium; by the temperature; and by the relative concentrations of the reactants. Thus in the $3d$ series, all octahedral complexes except those of chromium(III) (d^3) and cobalt(III) (d^6-low spin) are labile; those of these two ions are inert. Complexes of a particular ligand with cobalt(II) are labile. Square-planar complexes of palladium(II) and platinum(II) (d^8-low spin) are also inert.

Inasmuch as the exchange of coordinated water molecules is common to the formation of many complexes in aqueous solution, it is convenient to classify cations according to the scheme of Gray and Langford:[14]

Class I cations. The exchange of water molecules is too rapid to be followed by even fast-reaction techniques. First-order rate constants are ca. 10^8 sec^{-1} or larger, and the interactions between cation and ligands are entirely electrostatic. These complexes are derived from cations of the periodic group IA and the heavier periodic group IIA elements, for which the ratio Z^2/r (Sec. 5.1-8) is below 4.0.

Class II cations. The exchange of water molecules is fast but measurable, with first-order rate constants in the range 10^4 to 10^8 sec^{-1}. Interactions are again substantially ionic but weaker than for the Class I cations. Ligand-field stabilization energies (Sec. 8.1-3) are small. Magnesium ion, bipositive d-transition metal ions, and terpositive lanthanide ions are examples. For these, values of Z^2/r are in the range 4 to 12.

Class III cations. The exchange of water molecules, although rapid (first-order rate constants 1 to 10^4 sec^{-1}), can be followed by conventional combined with flow techniques. Ligand-field stabilization energies are larger. Many of the terpositive d-transition metal ions, the Be^{2+} ion, and the Al^{3+} ion are examples. For these, values of Z^2/r are larger than 12.

Class IV cations. The exchange of water molecules is very slow (first-order rate constants 10^{-1} to 10^{-9} sec^{-1}). Ligand-field stabilization energies are large. Examples include $Cr^{3+}(d^3)$, $Co^{3+}(d^6$-low spin), and Pd^{2+} and Pt^{2+} (d^8-low spin). Values of Z^2/r are comparable with those of the Class III cations.

Earlier experimental observations were largely restricted, because of the limitations of available techniques, to Class IV cations.

Kinetic data are often in agreement with a stepwise process that can be formulated for 6-coordinate species as

$$[M(H_2O)_6]^{n+} \overset{X^-}{\rightleftharpoons} \{[M(H_2O)_6]X\}^{(n-1)+} \rightarrow \{[M(H_2O)_5]X\}^{(n-1)+} + H_2O$$

$$\downarrow \text{very fast}$$

$$[M(H_2O)_5X]^{(n-1)+}$$

wherein the ligand X^- first enters the outer coordination sphere, a water molecule is then lost from the inner coordination sphere, and the ligand X^- finally occupies the resulting vacant position in the inner coordination sphere. However, in ligand-exchange reactions where water or hydroxide ion is present, for example, in species of the type $[\overset{III}{Co}L_nX]$ for which the overall equilibrium equation already written applies, the reaction sequence is commonly

$$[\overset{III}{Co}L_nX] \underset{H_2O}{\rightleftharpoons} [\overset{III}{Co}L_n(H_2O)] \overset{Y}{\rightarrow} [\overset{III}{Co}L_nY]$$

For cobalt(III) pentaammines, the rate of hydrolysis can be formulated as

$$\text{rate} = k_A c_{[Co(NH_3)_5X]} + k_B c_{[Co(NH_3)_5X]} c_{OH^-} \tag{11-4}$$

where k_A and k_B are, respectively, acid and base hydrolysis constants and k_B is commonly ca. $10^4 k_A$.

2-1. Mechanistic Approaches

For a reaction that we have described by the general equation noted above, two broad mechanistic patterns, based largely upon the very successful treatment of formally comparable organic reactions by E. D. Hughes and C. K. Ingold,[21] were originally proposed:

1. Ligand X leaves and ligand Y then adds, as

$$[ML_nX] \underset{\text{slow}}{\longrightarrow} X + [ML_n] \underset{\text{fast}}{\overset{+Y}{\longrightarrow}} [ML_nY]$$

for which the ideal rate law is

$$\text{rate} = k c_{[ML_nX]} \tag{11-5}$$

the expression for a unimolecular reaction (Sec. 9.3-3). This mechanism is referred to as S_N1 (i.e., *substitution nucleophilic* * *unimolecular*). In the intermediate, the coordination number of M has been *decreased*.

2. Ligand Y adds and ligand X then leaves, as

---

* A *nucleophilic* reagent is an atom, molecule, or ion that *donates* electrons to another species. An atom, molecule, or ion that *accepts* electrons from another species is an *electrophilic* reagent.

$$[ML_nX] \xrightarrow[\text{slow}]{+Y} [ML_nXY] \xrightarrow[\text{fast}]{-X} [ML_nY]$$

for which the ideal rate law is

$$\text{rate} = kc_{[ML_nX]}c_Y \tag{11-6}$$

the expression for a bimolecular reaction. This mechanism is referred to as S_N2 (i.e., *substitution nucleophilic bimolecular*). Here, in the intermediate, the coordination number of M has been *increased*.*

   This classification is based upon strict adherence of the reaction in question to what Langford and Gray have referred to as *stoichiometric molecularity*, without attention to *intimate molecularity*.[13, 14] They note that when two fundamental particles collide in solution, these particles are caught in a cage of solvent molecules wherein energy exchange is sufficiently rapid that the system may remain in an activated state for a sufficiently long period of time to allow several exchanges of collision partners. The ligand may collide with a complex species with sufficient energy to promote dissociation of that complex and then enter the coordination sphere without effecting significant activation. It is only necessary that it be at the correct place in the outer sphere at the proper time. Thus a stoichiometrically bimolecular reaction may be an intimately uni-molecular reaction.[22]
   On this basis, three different simple pathways can be distinguished,[14] namely

1.   A *dissociative* pathway (D), in which the leaving ligand is given up in the initial reaction step, forming an intermediate of smaller coordination number, as

$$[MX_n] \underset{+X}{\overset{-X}{\rightleftharpoons}} [MX_{n-1}] \underset{-Y}{\overset{+Y}{\rightleftharpoons}} [MX_{n-1}Y]$$

2.   An *associative* pathway (A), in which the entering ligand is added in the initial reaction step, giving an intermediate of larger coordination number, as

$$[MX_n] \underset{-Y}{\overset{+Y}{\rightleftharpoons}} [MX_nY] \underset{+X}{\overset{-X}{\rightleftharpoons}} [MX_{n-1}Y]$$

3.   An *interchange* pathway (I), in which the leaving ligand (Y) is moved from the inner coordination sphere to the outer sphere, and the entering ligand is moved from the outer sphere to the inner one, as

$$[MX_n \cdots Y] \rightleftharpoons [MX_{n-1}Y \cdots X]$$

where the dotted lines indicate separation between the inner and outer spheres.

---

*   Much less commonly, substitution involving a second metal ion,

$$[ML_nX] + M' \rightarrow [M'L_nX] + M$$

is encountered. This type of reaction is termed an *electrophilic* substitution reaction, S_E.

In the *dissociative* mechanism, bond rupture is the important process. Reactions of the S_N1 type are thus classified as D. In the *associative* mechanism, bond formation is the important process. Reactions of the S_N2 type are thus classified as A. Both bond rupture and bond formation are characteristic of the *interchange* mechanism. If the former is the more important, the mechanism may be termed I_D; if the latter is the more important, the mechanism may be designated I_A. Gray and Langford[14] have compared these mechanistic approaches in terms of changes in potential energy in an instructive series of diagrams and plots of the energy-reaction coordinate type (Fig. 9-1). These should be consulted for additional details.

Mechanistic interpretations based upon experimentally determined rate laws are not infrequently fraught with difficulties and may be subject to ambiguity.* Very broadly, one can distinguish situations of the following types.†

1. Solvolytic reactions, as described by the general equation (S = solvent)

$$MX + S \rightarrow MS + X$$

for which the three mechanistic interpretations are consistent with rate laws as follows:

Dissociative (D): rate $= \dfrac{k_1 k_2 c_{MX} c_S}{k_3 c_S + k_2 c_X} = \dfrac{k_1 k_3' c_{MX}}{k_3' + k_2 c_X}$

$$\text{where } c_S = \text{constant)} \tag{11-7}$$

which may have the limiting forms

$$\text{rate} = k c_{MX} \qquad \text{(if } k_3' \gg k_2 c_X\text{)}$$

$$\text{rate} = \frac{k_1 k_3' c_{MX}}{k_2 c_X} \qquad \text{(if } k_3' \ll k_2 c_X\text{)} \tag{11-8}$$

Associative (A): rate $= \dfrac{k_1 k_3}{k_2 + k_3} c_{MX} c_S = k' c_{MX}$ $\qquad$ (11-9)

Interchange (I): rate $= k_1 c_{MX} c_S = k_1' c_{MX}$ $\qquad$ (11-10)

Distinction, on this basis, between I and A mechanism is not possible. Distinction between D and I or A is possible only if one of the limiting forms of the rate law for D is observed.

2. Solvent exchange reactions, as described by the general equation (S* = second solvent)

$$MS + S^* \rightarrow MS^* + S$$

---

* For a series of interesting analogies involving electrical systems, see J. P. Birk, *J. Chem. Educ.*, **47**, 805 (1970).

† Only simple skeleton chemical equations are used, and only ligands that are involved are included.

for which the three mechanistic interpretations are consistent with the rate laws as follows:

$$\text{Dissociative } (D): \quad \text{rate} = \frac{k_1 c_{MS} c_{S^*}}{c_{S(total)}} = k_1' c_{MS} c_{S^*}$$

$$\text{where } c_{S(total)} = c_S + c_{S^*} \qquad (11\text{-}11)$$

$$\text{Associative } (A): \quad \text{rate} = k_1 c_{MS} c_{S^*} \qquad (11\text{-}12)$$

$$\text{Interchange } (I): \quad \text{rate} = k_1 c_{MS} c_{S^*} \qquad (11\text{-}13)$$

Distinction, on this basis, among the three mechanisms is not possible unless it is possible to detect M or MSS* by physical or chemical means.

3. Anation, or substitution of X for S, reactions, as described by the general equation

$$MS + X \rightarrow MX + S$$

for which the three mechanistic interpretations are consistent with the rate laws as follows:

$$\text{Dissociation } (D): \quad \text{rate} = \frac{k_1 k_3 c_{MS} c_X}{k_2 c_S + k_3 c_X} = \frac{k_1 k_3 c_{MS} c_X}{k_2' + k_3 c_X} \qquad (11\text{-}14)$$

$$\text{Associative } (A): \quad \text{rate} = \left(\frac{k_1 k_3}{k_2 + k_3} \right) c_{MS} c_X = k c_{MS} c_X \qquad (11\text{-}15)$$

$$\text{Interchange } (I): \quad \text{rate} = k_1 c_{MS} c_X \qquad (11\text{-}16)$$

Distinction, on this basis, among the three mechanisms is not possible if in the equation for D, $k_2' \gg k_3 c_X$ (which is, in fact, the common situation).

4. Direct substitution reactions, as described by the general equation (X or Y $\neq$ S)

$$MX + Y \rightarrow MY + X$$

for which the three mechanistic interpretations are consistent with the rate laws as follows:

$$\text{Dissociation } (D): \quad \text{rate} = \frac{k_1 k_3 c_{MX} c_{MY}}{k_2 c_X + k_3 c_Y} \qquad (11\text{-}17)$$

$$\text{Associative } (A): \quad \text{rate} = \left(\frac{k_1 k_3}{k_2 + k_3} \right) c_{MX} c_Y$$

$$= k c_{MX} c_Y \qquad (11\text{-}18)$$

$$\text{Interchange } (I): \quad \text{rate} = k_1 c_{MX} c_Y \qquad (11\text{-}19)$$

Distinction, on this basis, between D and A or I is always possible. However, solvation followed by rapid anation can give the rate expression

$$\text{rate} = k_1 c_{MX} \qquad (11\text{-}20)$$

Table 11-2 Effects of Charge and Size upon Reaction Rates

Species (Change in Property)	Change in Rate for Mechanism Indicated		
	D	A	I
Acceptor ion			
Increase + charge	Decrease	Opposing effects	Increase
Increase size	Increase	Increase	Increase
Entering ligand			
Increase − charge	None	Increase	Increase
Increase size	None	Decrease	Decrease
Leaving ligand			
Increase − charge	Decrease	Decrease	Decrease
Increase size	Increase	Opposing effects	Decrease
Nonlabile ligand			
Increase − charge	Increase	Opposing effects	Decrease
Increase size	Increase	Opposing effects	Decrease

The effects of charge and size of the acceptor central atom and the ligand upon reaction rates are summarized in terms of mechanistic interpretation in Table 11-2.

At least one other mechanistic variation is significant, specifically for the base-promoted hydrolysis of octahedral ammine and amine complexes.[23, 24] For example, the hydrolysis of the ion $[Co(NH_3)_5Cl]^{2+}$,

$$[Co(NH_3)_5Cl]^{2+} + H_2O \rightleftharpoons [Co(NH_3)_5(OH)]^{2+} + H^+ + Cl^-$$

is both particularly rapid and proceeds by an S_N1 mechanism for all reactants. Garrick's mechanistic approach suggested that an initial rapid acid–base equilibration resulting in the formation of the conjugate base $[Co(NH_3)_4(NH_2)Cl]^+$ is significant. The reaction scheme is then

$$[Co(NH_3)_5Cl]^{2+} + OH^- \underset{\text{fast}}{\rightleftharpoons} [Co(NH_3)_4(NH_2)Cl]^+ + H_2O$$

$$[Co(NH_3)_4(NH_2)Cl]^+ \xrightarrow{\text{slow}} [Co(NH_3)_4(NH_2)]^{2+} + Cl^-$$

$$[Co(NH_3)_4(NH_2)]^{2+} + H_2O \xrightarrow{\text{fast}} [Co(NH_3)_5(OH)]^{2+}$$

This mechanism, which has been proved experimentally for a number of cases, is designated S_N1 CB, *substitution nucleophilic unimolecular conjugate base.**

Additional details on mechanistic approaches can be found in various of the excellent treatises and review articles already cited.[1–3, 12–18]

---

* For an interesting discussion of conflict of ideas relative to the mechanism of this type of reaction, in particular as they were applied to the base-catalyzed hydrolysis of the related ion *cis*-[Co(en)$_2$Cl$_2$]$^+$, see R. G. Pearson, *J. Chem. Educ.*, **55**, 720 (1978).

2-2. Ligand Substitution, or Exchange, in 4-Coordinate, Square-Planar Complexes

The regularities observed in substitution reactions of square-planar complexes of platinum(II) were known to A. Werner and were used by him in assigning *cis* and *trans* geometries.[25] By a wide margin, the bulk of the reported investigations in this area have been with platinum(II), a d^8 species.[26-30] Other less extensively studied square-planar d^8 species are palladium(II), nickel(II), gold(III), rhodium(I), and iridium(I).

Mechanistically, reactions involving this geometry are simpler than those involving octahedral geometry. The square-planar geometry is better suited to substitution by an associative (*A*) mechanism than to substitution by a dissociative (*D*) mechanism because of the availability of two open coordination positions for associative activation. Although the trigonal bipyramidal and square pyramidal geometries of 5-coordinate potential intermediates are of nearly equal energy, the former appears to be slightly the more favorable and is the one usually assumed to be important.

The associative (*A*), or S_N2, mechanism for the reaction of $[\overset{II}{Pt}(NH_3)_3Cl]^+$ ion with Cl^- ion to give *trans*-$[\overset{II}{Pt}(NH_3)_2Cl_2]$ is typical and may be described in terms of a trigonal bipyramidal intermediate, $[Pt(NH_3)_3Cl_3]$, as

wherein NH_3' is the displaced ligand. The d^8 species are diamagnetic and thus low spin in character. It appears from orbital-energy and steric considerations that the trigonal bipyramidal geometry is ideal for an associative intermediate for these d^8 species.

For a reaction of this type described by the general equation

$$ML_3X + Y \xrightarrow{\ (H_2O)\ } ML_3Y + X$$

a two-term rate law appears to apply, as

$$\text{rate} = k_1 c_{[ML_3X]} + k_2 c_{[ML_3X]} c_Y \tag{11-21}$$

where k_1 and k_2 are first- and second-order rate constants, respectively. If a large excess of Y is present, the reaction becomes pseudo first-order, and the experimentally determined first-order rate constant measured (k_{obs}) is given by the equation

$$k_{obs} = k_1 + k_2 c_Y \tag{11-22}$$

A plot of k_{obs} vs. c_Y is then linear, with slope k_2 for the reagent-independent pathway. Data for the reaction of *trans*-$[Pt(NC_5H_5)_2Cl_2]$ with a variety of nucleophiles in methanol at 30°C[31] are typical.

The two-term rate law indicated in Eq. 11-21 and 11-22 is in agreement with an overall reaction mechanism requiring two joining reaction pathways, in one of which the solvent is involved and in the other of which it is not.[26] This situation is described by the following reaction scheme (S = solvent):

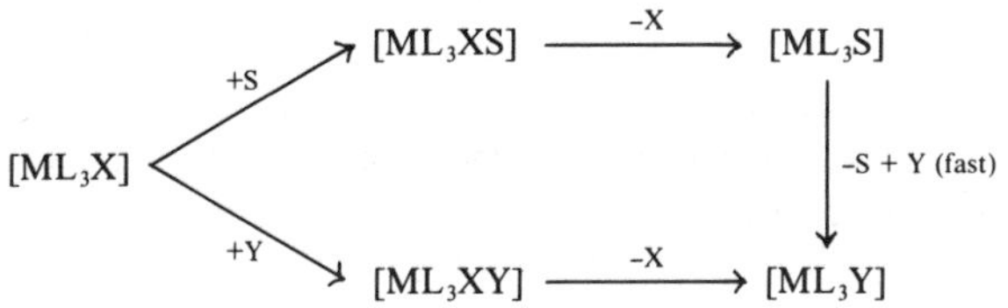

Equation 11-22 can then be rewritten as

$$k_{obs} = k_S + k_Y c_Y \tag{11-23}$$

The Trans Effect. The α and β forms of $[Pt(NH_3)_2Cl_2]$ have been mentioned earlier (Sec. 7.4-10), together with A. Werner's conclusion[25] that in a square-planar geometry the α isomer has the *cis* configuration and the β isomer the *trans* configuration. The methods of synthesis, namely

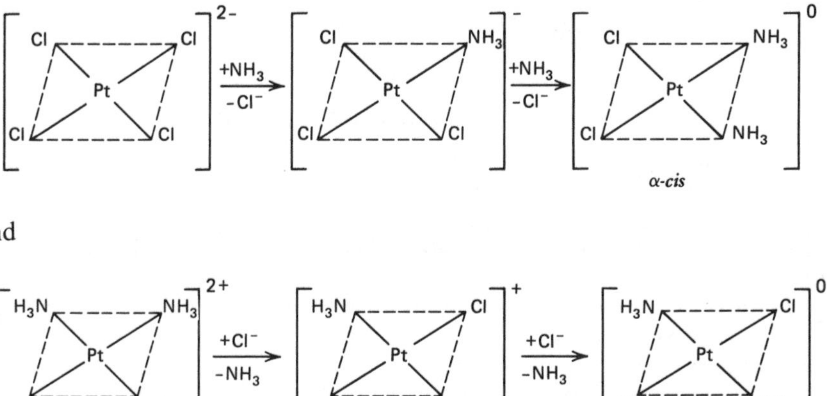

and

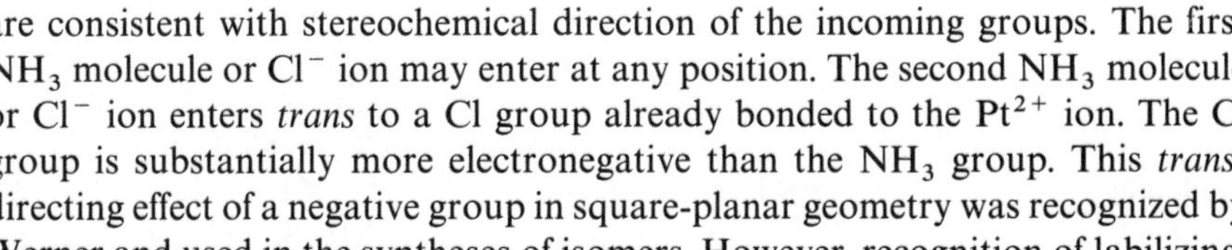

are consistent with stereochemical direction of the incoming groups. The first NH_3 molecule or Cl^- ion may enter at any position. The second NH_3 molecule or Cl^- ion enters *trans* to a Cl group already bonded to the Pt^{2+} ion. The Cl group is substantially more electronegative than the NH_3 group. This *trans*-directing effect of a negative group in square-planar geometry was recognized by Werner and used in the syntheses of isomers. However, recognition of labilizing, with respect to substitution, of groups *trans* to electronegative groups in square-planar complexes and development of this area of coordination chemistry are due to Il'ya Il'ich Chernyaev.[32-35]

The *trans* effect has been defined in various ways. The organizing committee of a conference on the *trans* effect held in 1952 offered the definition: "In compounds with square or octahedral structure with a central complex-forming cation, the rate of substitution of an atom or molecule linked to the central atom is determined by the nature of the substituent at the opposite end of the diagonal. Thus the stability of the bond between this (central) atom and any substituent is little affected by the character of neighboring atoms or molecules, but is greatly influenced by those more distant, in the *trans* position, on the diagonal of the square."[36] A somewhat more manageable definition of the *trans* effect is "the effect of a coordinated group upon the rate of substitution reactions of ligands opposite to it in a metal complex."[37]

In practice, nearly all known examples of the *trans* effect are restricted to square-planar platinum(II) species. It is evident also among other d^8 species of the same geometry, but examples involving octahedral geometry are much less well defined.

Although an order of ligands based upon relative *trans* effects is dependent also upon the nature of the initial complex, an "average order" can be derived for platinum(II), as indicated in Table 11-3[27,37] from comparisons of rate constants. The approximate order of decreasing *trans* effect relative to substitution reactions of platinum(II) compounds is, thus,

$$CO, CN^-, C_2H_4 > PR_3, H^- > CH_3^-, SC(NH_2)_2 > C_6H_5^-, NO_2^-, I^-, SCN^-$$
$$> Br^-, Cl^- > C_5H_5N, NH_3, OH^-, H_2O$$

with an overall difference in specific rate constants of ca. 10^6.

Table 11-3 Comparison of Kinetic *Trans* Effects for Various Ligands [Platinum(II) Species]

Relative Reaction Rates		Ligands
Qualitative	Approximate k	
Very large	10^5 (?)	CO, CN$^-$, $\searrow$C$=$C$\diagdown$
		P(CH$_3$)$_3$ > P(C$_2$H$_5$)$_3$ $\approx$ H$^-$
Large	10^2	SC(NH$_2$)$_2$
		NO$_2^-$, I$^-$, SCN$^-$
		CH$_3^-$ > C$_6$H$_5^-$
Moderate	$10^0 = 1$	Br$^-$, Cl$^-$
Small	10^{-1} (?)	C$_5$H$_5$N, NH$_3$
		OH$^-$

Theoretical approaches to the *trans* effect include, most usefully, the electrostatic-polarization approximation of Grinberg,[38] the π-bonding approximation of Chatt, Duncanson, and Venanzi,[39] and Orgel,[40] and a combined σ- and π-molecular orbital approximation.[26, 27] The first of these emphasizes the significance of the ground-state contribution to reaction rate; the second concerns itself with bonding in the transition state; and the third combines both σ and π contributions. Each attempts to account for a weakening of a Pt—X bond by a ligand (Y) that is placed high in the *trans*-effect series.

The electrostatic-polarization approximation assumes that the electrostatic charge of the Pt^{2+} ion induces a dipole into the ligand atom, which in turn induces a dipole into the Pt atom, as shown in Fig. 11-1. This dipole is then so oriented as to repel negative charge density in a bonded ligand X and thus, because of reduced attraction between X and Pt, lengthen and weaken the Pt—X bond. The magnitude of the *trans* effect produced by ligand L is indeed directly related to its polarizability, as so predicted. Furthermore, the *trans* effect for a given ligand is more apparent for platinum(II) species than for the less easily polarized platinum(IV) or palladium(II) species. However, an induced dipole in platinum(II) should be more dependent upon the net charge borne by ligand L and upon an induced dipole moment, suggesting that, contrary to observation, Cl^- ion should have a larger *trans* effect than I^- ion.

The latter problem is minimized if covalent interactions are considered. Of the d, s, and p^2 orbitals involved in covalent bonding in a square-planar complex (Sec. 8.1), only the p orbitals have directional properties of the *trans* type. As shown in Fig. 11-2, the leaving X ligand and a ligand L *trans* thereto share a p_σ orbital (p_x). If sigma interaction between M and L is particularly strong, the

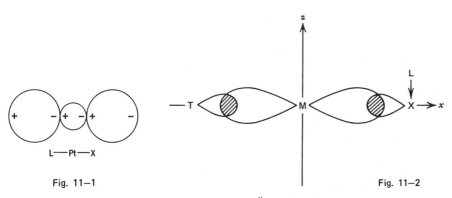

Fig. 11–1   Fig. 11–2

Figure 11-1 Induced dipoles in *trans*-[X—Pt^{II}—X]. [F. Basolo and R. G. Pearson, *Mechanisms of Inorganic Reactions. A Study of Metal Complexes in Solution*, 2nd ed., Fig. 5.5, p. 370, John Wiley & Sons, New York (1967).]

Figure 11-2 Bonding of entering L and leaving X groups involving a p orbital with a *trans*-ligand T present. [C. H. Langford and H. B. Gray, *Ligand Substitution Processes*, Fig. 2-7, p. 26, W. A. Benjamin, Menlo Park, California (1965).]

interaction between M and X will then be correspondingly weaker. As the group Y enters and X moves away, more p-orbital interaction with the *trans* ligand L is reasonable. In terms of a 5-coordinate intermediate, both entering group Y and leaving group X can be envisioned as sharing an available p_z orbital of the atom M, with the *trans* ligand L then sharing more than half of the p_x orbital of atom M (Fig. 11-3). Thus the energetics for large σ overlap with a p_σ orbital of the platinum(II) species are favorable.

Consistent with this approach is the following σ-effect series

$$H^- > PR_3 > -SCN^- > I^- > CH_3^-, CO, CN^- > Br^- > Cl^- > NH_3 > OH^-$$

The π-bonding approximation is restricted to π-bonding ligands such as C_2H_4, PR_3, CO, and CN^-. In a square-planar complex of a d-transition metal atom or ion, the d_{xy}, d_{xz}, and d_{yz} orbitals have appropriate symmetries for π interaction (Sec. 8.1). In terms of the system of axes used in Fig. 11-2, 11-3, and 11-4, only the d_{xz} and d_{yz} can interact with orbitals of a pair of *trans*-located ligands. Considered in terms of Fig. 11-4, the d_{xz} orbital is shared between the

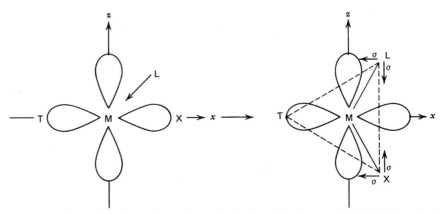

Figure 11-3. p Structures, involving substitution of L for *trans*-X. [C. H. Langford and H. B. Gray, *Ligand Substitution Processes*, Fig. 2-8, p. 26, W. A. Benjamin, Menlo Park, California (1965).]

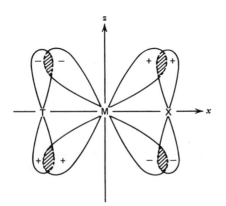

Figure 11-4 π Interaction of *trans*-group T with leaving group X. [C. H. Langford and H. B. Gray, *Ligand Substitution Processes*, Fig. 2-9, p. 28, W. A. Benjamin, Menlo Park, California (1965).]

leaving ligand X and another ligand *trans* to it. If a trigonal bipyramidal geometry is assumed for the 5-coordinate intermediate, the $d_{xz}, d_{yz}, d_{xy},$ and $d_{x^2-y^2}$ orbitals of the M atom have correct symmetries for π interaction. These orbitals share in π interaction with the three ligands in the trigonal plane, that is the leaving and remaining *trans* X ligands and the entering Y ligand. Thus the intermediate is stabilized if the *trans* group has empty orbitals of π symmetry and reasonable stability and thus helps to compensate for the extra electronic charge imposed on the central metal ion by the entering ligand. A *trans* group that is a good π acceptor thus decreases the overall energy of activation, or favors a π-*trans* effect.

Consistent with this approach is the following π-effect series

$$C_2H_4, CO > CN^- > -NO_2^- > -SCN^- > I^- > Br^- > Cl^- > NH_3 > OH^-$$

Neither of these approaches can accommodate all of the ligands that are strongly *trans* directing. Some of these ligands are strongly σ bonding (e.g., H^-, CH_3^-); others are strongly π bonding (e.g., C_2H_4, CO); and still others are moderately σ and π bonding [e.g., I^-, $SC(NH_2)_2$]. To some degree, the molecular-orbital approximation provides a better overall accommodation since it allows for both σ and π bonding and associates comparative energies with each interaction. The energy-level diagram for the $[PtCl_4]^{2-}$ ion shown in Fig. 11-5 emphasizes these energies for a typical case. Here the σ-bonding orbitals are the most stable and reside largely on the Cl^- groups. The π-bonding orbitals lie at slightly higher energies. The σ- and π-anti-bonding orbitals, at higher energies, are derived from $5d$ atomic orbitals of the Pt^{2+} ion, as well as the $6s$ and $6p$ orbitals. It is perhaps best to conclude that there are usually both *sigma* and *pi* contributions to an observed *trans* effect.

The Cis Effect. Rates of substitution are much less influenced by ligands *cis* to leaving ligands. Slight *cis* effects do exist, for example, the observation that the pyridine molecule substitutes more readily for Cl^- in the complex ion $[Pt(NC_5H_5)Cl_3]^-$ than does the ammonia molecule in the complex ion $[Pt(NH_3)Cl_3]^-$.[41] Combined with similar comparisons involving the nitrite ion, a *cis*-effect order is then

$$NC_5H_5 > NH_3 > NO_2^-$$

The *cis* effect can become more significant than the *trans* effect when ligands of nearly equal *trans* effects are compared.[42]

Other Effects. The rates of substitution reactions involving 4-coordinate square-planar complex species are also influenced by the nature of the leaving group, the charge borne by the complex species, the steric properties of other ligands in the complex, the nature of the solvent, the presence of a catalyst, and the temperature.[26,27] The cited references can be consulted for additional details.

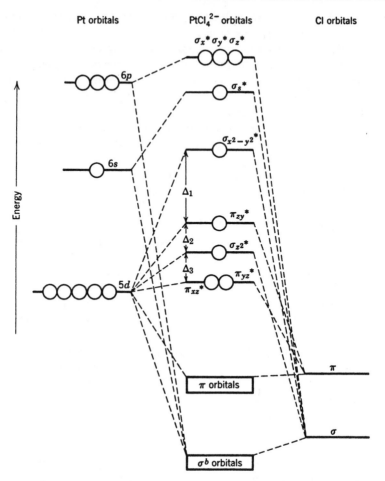

Figure 11-5 Molecular-orbital diagram for $[PtCl_4]^{2-}$ ion. [F. Basolo and R. G. Pearson, *Mechanisms of Inorganic Reactions. A Study of Metal Complexes in Solution*, 2nd ed., Fig. 5-8, p. 373, John Wiley & Sons, New York (1967).]

2-3. Ligand Substitution, or Exchange, in 6-Coordinate, Octahedral Complexes

Ligand substitution involving octahedral complexes is much more involved than that involving square-planar complexes for at least the following reasons: sheer numbers of characterized and potential examples, an additional steric complication, and lack of saturation in σ bonding.[43,44] Octahedral 6-coordinate complex ions and compounds are far more numerous than those of any other type. Unlike those of the square-planar, 4-coordinate type that are restrictive to d^8-transition metal species and almost only to platinum(II), the octahedral complexes are derived from ions in a number of oxidation states from all of the transition-element families, as well as ions from post-transition metal families. A wide variety of species with widely variable properties thus exists. Inasmuch as they differ from the square-planar species by addition of fifth and sixth ligands above and below open faces, all ligands bonded to a central atom are sterically more crowded and exhibit larger ligand–ligand repulsions. Furthermore, the p_z orbital that is available for σ bonding in the square-planar d^8 species is already involved in bonding in the octahedral complexes. On the latter two bases, both reduced availability of the *associative* (*A*) reaction pathway and probable enhanced reaction by a *dissociative* (*D* or I_d) pathway can be expected. In keeping with other studies in coordination chemistry, investigations of ligand substitution reactions in this geometry have emphasized cobalt(III) species.* The comparative inertness of cobalt(III) complexes, allowing the evaluation of reaction kinetics by classical methodology, also contributed to the number of these studies. Much less well investigated are reactions of the chromium(III), rhodium(III), iridium(III), nickel(II), platinum(IV), ruthenium(II and III), and iron(III) complexes. Discussions that follow should be supplemented by reference to typical summations and reviews.[2,43-49] These discussions are restricted to cobalt(III) species.

Rate Laws and Mechanistic Considerations. Typical of the carefully investigated reactions in acidic aqueous systems, as reported in the earlier literature,[13] are those (1) of azide and thiocyanate ions with the $[Co(en)_2(NO_2)Cl]^+$ ion;[50] (2) of thiocyanate ion with the $[Co(NH_3)_5(NO_3)]^{2+}$ and $[Co(NH_3)_5Br]^{2+}$ ions;[51] (3) in an exchange fashion, of labeled thiocyanate ion with the $[Co(en)_2(NCS)_2]^+$ ion;[52] and (4) also in an exchange fashion, of labeled chloride ion with the *cis*-$[Co(en)_2Cl_2]^+$ ion.[53] Each of these reactions proved to be zero order in the concentration of the incoming ligand and first order in the concentration of the complex ion, under the conditions of the investigation.

Reactions of these complex ions with the solvent water under acidic conditions lead to the conclusion that in each the substitution process involves two successive steps: substitution of the original anionic ligand by a water molecule (acid hydration, pH < 5) and subsequent substitution of the water molecule by an anion (anation). For a pentaammine species, $[Co(NH_3)_5X]^{2+}$ ($X^- =$ bonded anion), these reactions correspond to the equations

$$[Co(NH_3)_5X]^{2+} + H_2O \rightarrow [Co(NH_3)_5(H_2O)]^{3+} + X^-$$

and

$$[Co(NH_3)_5(H_2O)]^{3+} + Y^- \rightarrow [Co(NH_3)_5Y]^{2+} + H_2O$$

where X and Y may or may not be identical. The first reaction is rate

---

* For a detailed summary of *cis* and *trans* effects involving cobalt(III) complexes, see J. M. Pratt and R. G. Thorp, in *Advances in Inorganic Chemistry and Radiochemistry*, H. J. Eméleus and A. G. Sharpe (Eds.), Vol. 12, pp. 375–427, Academic Press, New York (1970).

determining, but the concentration of the entering anion is involved only in the rate expression for the second reaction, assuming that any outer-sphere interaction can be ignored. A third process, *base hydrolysis* at pH > 5, may be involved, as

$$[Co(NH_3)_5X]^{2+} + OH^- \rightarrow [Co(NH_3)_5(OH)]^{2+} + X^-$$

This process is often commonly second-order overall (i.e., first-order in each reactant).[43]

In these ligand substitution processes in aqueous solution, the water molecule or its anion (OH^-) is involved. For a comparable process in a different, but waterlike (Table 10-1), solvent, the solvent molecule or its anion is probably involved. It is not unreasonable to assume that the mechanisms of ligand substitution reactions involving cobalt(III) ammines in general will follow this three-equation mechanistic pattern,[43] although one cannot discount the possibility that each reaction is mechanistically unique.[43]

In terms of *intimate mechanism*, any entering ligand is no more than weakly bonded in the reaction intermediate, or transition state, and it cannot, therefore, contribute extensively in this way to the activation energy of the reaction. Loss of initial ligand is thus significant, that is, a dissociative (*d*) activation energy is the more important. Mechanistically, then, these ligand substitution reactions are classified as I_d or D, with 5-coordinate activated complexes as intermediates. The dependence of measured reaction rates upon steric patterns, charge types, isotope effects, changes in leaving groups, and electronic effects are in general agreement with dissociative processes, with an I_d mechanism the more probable.

A D mechanism does characterize certain reactions, for example, the reaction described by the equation

$$[Co(CN)_5X]^{3-} + H_2O \rightleftharpoons [Co(CN)_5(H_2O)]^{2-} + X^-$$

where $X^- = N_3^-$, SCN^-, I^-, or Br^-, and the reaction rate was studied by ^{18}O isotopic exchange methodology.[54] If the overall reaction is broken into two parts as

$$[Co(CN)_5(H_2O)]^{2-} \underset{k_{-1}}{\overset{k_1}{\rightleftharpoons}} [Co(CN)_5]^{2-} + H_2O$$

and

$$[Co(CN)_5]^{2-} + X^- \overset{k_2}{\rightarrow} [Co(CN)_5X]^{3-}$$

then the observed pseudo-first-order rate constants (*k*) can be given by the expression

$$k = \frac{k_1 c_{x^-}}{c_{x^-} + k_{-1}/k_2} \tag{11-24}$$

When evaluated at 40°C, pH = 6.4, and $\mu = 1.0\ M$, these k values for the various anions X^- gave $k_1 = 1.6 \times 10^{-3}$ sec^{-1}. The constant k_1 represents the rate of formation of the 5-coordinate intermediate $[Co(CN)_5]^{2-}$, which is also the rate of water exchange in the absence of anions as determined by oxygen-isotope experiments.

In summary, substitution reactions involving octahedral cobalt(III) complexes in acidic solution depend mechanistically primarily upon a dissociative activation process in which the bulk of the energy of activation does not reflect a significant bonding contribution by the entering ligand. The more significant evidences in support of this conclusion are:[43] accessibility to study by kinetic means of only acid hydration and anation reactions; failure to identify selective reagents for bonding to cobalt(III) even under nonaqueous conditions; and uniform interpretation of effects by leaving groups, steric arrangements, chelation, and nonlabile ligands on the basis of activation by dissociation. Rapid hydrolysis in the presence of hydroxide ion involves the conjugate base of the initial complex since (1) the presence of an acidic proton in that complex and (2) the separability of the step dependent upon hydroxide ion from the step that determines the product are essential.

Stereochemical Effects. In previous discussions, the geometry of the 5-coordinate intermediate formed by a dissociative mechanism was discussed in terms of trigonal bipyramidal, rather than a tetragonal (or square) pyramidal, geometry. Distinction between these two possibilities can sometimes be made via comparisons between the numbers and types of predicted isomers vs. the isomers detected experimentally.[55] For example, for the dissociative substitution process

$$[M(LL)_2BX] \xrightarrow{-X} ([M(LL)_2B]) \xrightarrow{+Y} [M(LL)_2BY]$$

where LL is a bidentate ligand such as ethylenediamine or oxalate ion, the predicted reaction sequences can be depicted as follows for *trans*-$[M(LL)_2BX]$ as the reactant:

1. Via a tetragonal pyramidal intermediate:

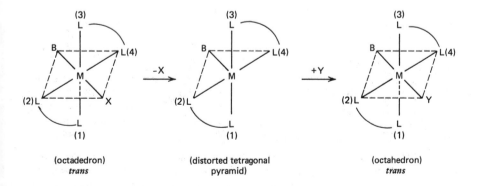

(octadedron)   (distorted tetragonal   (octahedron)
trans   pyramid)   *trans*

2. Via a trigonal bipyramidal intermediate:

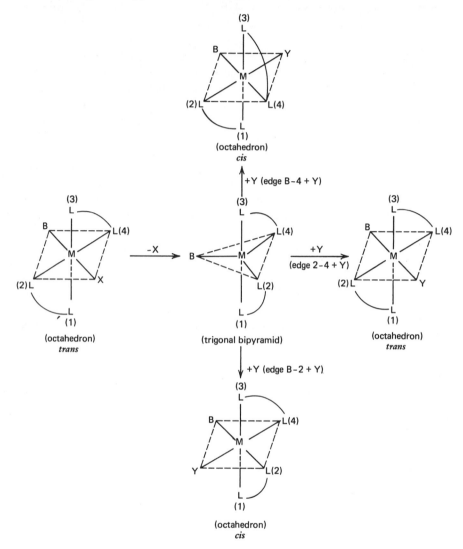

Thus if the reaction proceeds by way of a tetragonal pyramidal intermediate, only the open face left by loss of ligand X is available to an attack by ligand Y, and only *trans*-[M(LL)$_2$BY] can result. However, if the reaction proceeds by way of a trigonal bipyramidal intermediate, three essentially equivalent positions for entry of ligand Y exist: between L(2) and L(4), between L(2) and B, and between L(4) and B. The first of these yields *trans*-[M(LL)$_2$BY], and each of the other two yields *cis*-[M(LL)$_2$BY]. If only statistical factors are involved, the product via the trigonal bipyramidal intermediate should be a mixture of 66.7% *cis*-[M(LL)$_2$BY] and 33.3% *trans*-[M(LL)$_2$BY].

In a parallel fashion, the predicted reaction sequences for cis-[M(LL)$_2$BX] as the reactant can be depicted as follows:

1. Via a tetragonal pyramidal intermediate:

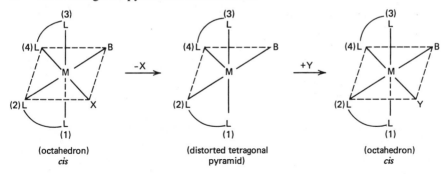

2. Via a trigonal bipyramidal intermediate:

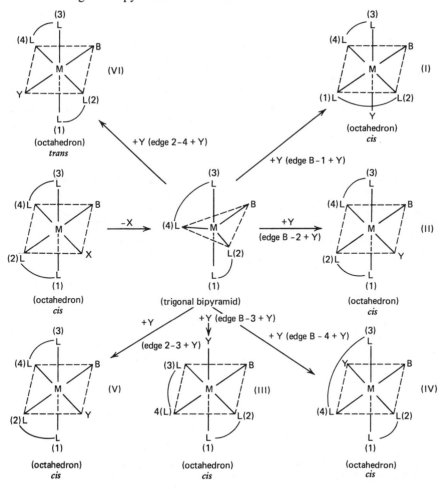

Again, if the reaction proceeds by way of a tetragonal pyramidal intermediate, only the open face left by loss of ligand X is available to an attack by ligand Y, and only cis-$[M(LL)_2BY]$ can result. If, however, the reaction proceeds by way of a trigonal bipyramidal intermediate, six positions for entry of ligand Y are potentially available: between L(2) and L(4), L(2) and L(3), B and L(1), B and L(2), B and L(3), and B and L(4). Only the first of these can yield trans-$[M(LL)_2BY]$. All of the others yield cis-$[M(LL)_2BY]$. Thus if only statistical factors are involved, the product obtained via the trigonal bipyramidal intermediate should be a mixture of 83.3% cis-$[M(LL)_2BY]$ and 16.7% trans-$[M(LL)_2BY]$. A close examination of the structures depicted for the cis isomers, particularly if three-dimensional models are used, reveals two arrangements that are mirror images of each other, for example, arrangement (IV) if rotated 90° around the L(1)–L(3) axis and then 180° around the B–L(4) axis, is the mirror image of arrangement (I). This is, of course, in keeping with the well-known asymmetry of this type of complex, as exemplified by A. Werner's initial resolution of the cis-$[Co(en)_2(NH_3)Cl]^{2+}$ ion (Sec. 7.4-10).

Proof of these considerations must come from experiments. At the outset, it must be realized that electrostatic factors, as well as steric factors, are influential in determining both the presumed geometry of the intermediate and the position where ligand Y can enter. Thus it is highly unlikely that any experiment will yield the statistical distribution of isomers. However, the facts that reactions of trans and cis complexes of the type $[M(LL)_2BX]^{n+}$ yield products better in accord with predictions based upon trigonal bipyramidal intermediates and that similar observations are recorded for other octahedral species support this geometry. Typical data for the acid hydration and base hydrolysis of cobalt(III)

Table 11-4 Distribution of Products in Acid Hydration of $[Co(en)_2BX]^{n+}$ Species $[Co(en)_2BX]^{n+} + H_2O \rightarrow [Co(en)_2B(H_2O)]^{n+ \text{ or } 1+n+} + X^{0 \text{ or } -}$

cis-$[Co(en)_2BX]^{n+}$			trans-$[Co(en)_2BX]^{n+}$		
		Percent cis Isomer[a]			Percent cis Isomer[a]
B	X	in Product	B	X	in Product
OH⁻	Cl⁻	100	OH⁻	Cl⁻	75
OH⁻	Br⁻	100	OH⁻	Br⁻	73
Br⁻	Cl⁻	100	Br⁻	Cl⁻	50
Cl⁻	Cl⁻	100	Br⁻	Br⁻	30
Cl⁻	Br⁻	100	Cl⁻	Cl⁻	35
N₃⁻	Cl⁻	100	Cl⁻	Br⁻	20
NCS⁻	Cl⁻	100	NCS⁻	Cl⁻	50–70
NCS⁻	Br⁻	100	NH₃	Cl⁻	0
NO₂⁻	Cl⁻	100	NO₂⁻	Cl⁻	0

[a] Percent trans = 100% − percent cis.

complexes of compositions $[Co(en)_2BX]^{n+}$ are summarized, respectively, in Tables 11-4 and 11-5.[56] These data are more supportive than confirmative of the approach delineated above, however.

As the data in Table 11-5 suggest, the asymmetry of *cis* isomers of the type $[M(LL)_2BX]^{n+}$ adds the complication of retention or inversion of configuration. This problem arises because of the formation, and ultimate reaction, of two distinctive trigonal bipyramidal intermediates, each of which can react in a distinctive fashion.[43] The potential reaction pattern can be illustrated for *D-cis*-$[Co(en)_2BX]^+$ ion as follows:*

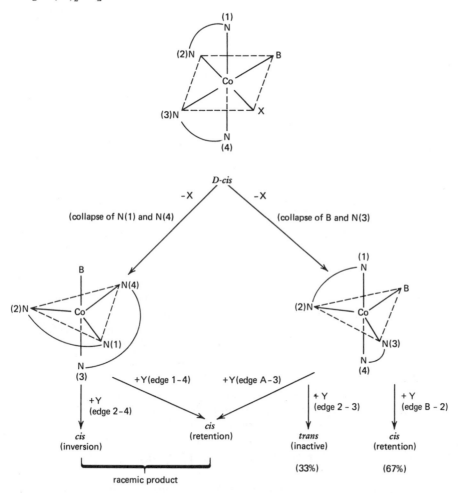

A tetragonal pyramidal intermediate should yield only a *cis* product with retention of optical activity.

---

* The assigned geometry is arbitrary, since the mirror-image form could be used just as well.

Table 11-5 Distribution of Products in Base Anation of
$[Co(en)_2BX]^{n+}$ **Species** $[Co(en)_2BX]^{n+} + OH^- \rightarrow$
$[Co(en)_2BOH]^{n+ \text{ or } -1+n+} + X^{0 \text{ or } -1}$

		D-cis-$[Co(en)_2BX]^{n+}$					trans-$[Co(en)_2BX]^{n+}$
		Percent *cis* Isomer in Product[a]					Percent *cis* Isomer in Product[a]
B	X	D-cis	Racemic[b]	L-cis	B	X	
OH^-	Cl^-	61		36	OH^-	Cl^-	94
OH^-	Br^-		96		OH^-	Br^-	90
Cl^-	Cl^-	21	16	21	Cl^-	Cl^-	5
Cl^-	Br^-		30		Cl^-	Br^-	5
Br^-	Cl^-		40		Br^-	Cl^-	0
N_3^-	Cl^-		51		N_3^-	Cl^-	13
NCS^-	Cl^-	56		24	NCS^-	Cl^-	76
NH_3	Br^-	59		26	NCS^-	Br^-	81
NH_3	Cl^-	60		24	NH_3	Cl^-	76
NO_2^-	Cl^-	46		20	NO_2^-	Cl^-	6

[a] Percent *trans* = 100% −percent *cis*.

[b] Racemic reactant; so racemic product.

Very broadly, it is often assumed that ligand Y will enter the position vacated by the loss of ligand X. It has been stated[57] that ligand-to-metal π bonding can enhance rupture of the M—X bond. The course that a given substitution reaction follows will be the one that allows maximum contribution of energy for rupture of the M—X bond. Of the ligands B given in Table 10-4 all except NH_3 and NO_2^- have electrons in p orbitals that can be donated via π bonding to empty $3d$ orbitals of the Co(III) species of proper symmetry. For a *cis* ligand, this donation into an empty d^2sp^3 hybrid orbital can occur without rearrangement, whereas for a *trans* ligand rearrangement is essential. Thus a better statement is that in a *cis* structure ligand X is replaced by ligand Y in a continuing *cis* position, whereas replacement of ligand X in a *trans* structure gives a mixture of *cis* and *trans* isomers. It should be noted also that, for a given complex, a *cis* isomer usually reacts more rapidly than a *trans* isomer.

2-4. Intramolecular Stereochemical Changes

Intramolecular stereochemical changes may involve *cis-trans* isomerization reactions, optical inversion reactions, or racemization reactions. In each type of reaction, an internal rearrangement of one or more ligands occurs. Although this rearrangement is commonly considered to be truly intramolecular, it may mechanistically involve intermolecular processes. The mechanistic aspects of these intramolecular changes have been treated in detail by J. P. Jesson.[58]

cis-trans Isomerization. Isomerization in octahedral systems was first noted by Jørgensen in 1889 in terms of displacement of the equilibrium[59]

$$trans\text{-}[Co(en)_2Cl_2]^+ \rightleftharpoons cis\text{-}[Co(en)_2Cl_2]^+$$
$$\quad\quad green \quad\quad\quad\quad\quad\quad\quad violet$$

As outlined in Table 7-13, the *trans → cis* conversion of the chloride salt is aided by concentration on a steam bath and addition of aqueous ammonia and the *cis → trans* conversion by the addition of aqueous hydrogen chloride. It is probable that these conversions have been more extensively studied than any others, although comparable changes are not uncommon among octahedral complex species (Tables 7-13, 7-14). Other species that undergo *cis → trans* conversions include $[Co(en)_2(NO_2)Cl]Cl$,[60] $[Co(en)_2(NO_2)_2]NO_3$,[61] $[Co(en)_2(H_2O)_2]^{3+}$,[62,63] $[Co(en)_2(NH_3)(OH)]^{2+}$,[64] $[Co(en)_2(NH_3)(H_2O)]^{3+}$,[64] $[Rh(C_2O_4)_2Cl_2]^{3-}$,[65] $[Ir(C_2O_4)_2Cl_2]^{3-}$,[66] and $[Cr(C_2O_4)_2(H_2O)_2]^-$.[67,68] Comparable rearrangements are much less common among species with square-planar geometry, but they do characterize, for example, compounds of the type $[Pt(ZR_3)_2X_2]$, where Z = P, As, Sb; R = CH_3 to $n\text{-}C_5H_{11}$, and X = Cl, I.[69,70]

Based upon an assumption that *trans*-$[Co(en)_2Cl_2]Cl\cdot HCl$(s) (Table 7-13) is really *trans*-$[Co(en)(enH)Cl_3]Cl$, it was suggested that isomerization to the *cis* compound proceeds by the opening of a Co-en ring, the subsequent bonding of Cl^- or OH^- ion at the vacant coordination position, and then the ultimate reclosing of the ring to yield the rearranged product.[71] However, infrared and x-ray diffraction data have indicated the parent compound (above) to be $[Co(en)_2Cl_2]\cdot HCl\cdot 2H_2O$, with the $H_5O_2^+$ and Cl^- ions in the crystal lattice together with the $[Co(en)_2Cl_2]^+$ ion.[72,73]

In the presence of radiochloride ion, isomerization results in a random distribution of labeled chloride ion with coordinated chloro ligands, and there is no direct replacement of bonded chloro groups by chloride ion.[53] Thus, isomerization may involve the equilibria

$$trans\text{-}[Co(en)_2Cl_2]^+ + H_2O \underset{2}{\overset{1}{\rightleftharpoons}} [Co(en)_2(H_2O)Cl]^{2+} + Cl^-$$

and

$$[Co(en)_2(H_2O)Cl]^{2+} + H_2O \underset{4}{\overset{3}{\rightleftharpoons}} [Co(en)_2(H_2O)_2]^{3+} + Cl^-$$

with equations 1 and 3 indicating reactions favored in the early stages of the isomerization process and equations 2 and 4 indicating reactions that occur later as water is lost by evaporation and concentration of the Cl^- ion is increased. These data[53] suggested the mechanistic sequence

$$cis\text{-}[Co(en)_2Cl_2]^+ + H_2O \rightleftharpoons cis\text{-}[Co(en)_2(H_2O)Cl]^{2+} + Cl^-$$

$$trans\text{-}[Co(en)_2Cl_2]^+ + H_2O \rightleftharpoons trans\text{-}[Co(en)_2(H_2O)Cl]^{2+} + Cl^-$$

Experimental data are in accord with the first of these equilibria, but not the second. Aquation of *trans*-$[Co(en)_2Cl_2]^+$ ion gives directly a mixture of 35% *cis*-

and 65% *trans*-[Co(en)$_2$(H$_2$O)Cl]$^{2+}$ ion. A further complication is imposed by the fact that in the absence of hydrochloric acid *cis*-[Co(en)$_2$Cl$_2$]Cl is the less soluble, whereas in the presence of hydrochloric acid *trans*-[Co(en)$_2$Cl$_2$]Cl· H$_5$O$_2$Cl is the less soluble.

Irrespective of the exact mechanism of the isomerization process, a dissociative process involving a trigonal bipyramidal intermediate.

appears to be involved.

Another interesting system involves the various isomeric [Cr(C$_2$O$_4$)$_2$(H$_2$O)$_2$]$^-$ species.[67, 68] A. Werner observed that at equilibrium in an aqueous solution of the potassium salt almost none of the *trans* isomer is present, but that on evaporation of the solution the less soluble *trans* salt crystallizes first.[74] The kinetics of both the isomerization process (*trans* → *cis*) and the racemization of *cis* optical isomers in aqueous solution have been investigated.[67, 68, 75, 76] Although racemization is considered in more detail later in this chapter, the two isomerization processes are so interrelated in this system that some mention of racemization is essential at this point. Experimentally, there is no pH dependence of either process in the pH range 3 to 7, and at 25°C both reaction rates are equal ($k = 4.3 \times 10^{-4}$ sec^{-1}). On this basis, it may be assumed that these steric changes depend upon exchange involving coordinated and outer-sphere water molecules.

In terms of the trigonal bipyramidal pathway just discussed for the isomerization of the [Co(en)$_2$Cl$_2$]$^+$ ion, the observed *trans* → *cis* process can be rationalized. The *cis* → *trans* isomerization, although not investigated because of its extreme slowness, could reasonably follow a similar pathway. The fact that racemization proceeds at the same rate as the conversion *trans* → *cis* means that racemization cannot proceed via formation of the *trans* isomer. The overall process can then be formulated as[77]

Table 11-6 Equilibrium and Thermodynamic Data for Isomerization Reactions in Benzene at 25°C[78]

Initial cis Complex[a]	Equilibrium Data			Thermodynamic Data		
	% cis Isomer	% trans Isomer	K	ΔG^0 (cal mole^{-1})	ΔH^0 (cal mole^{-1})	ΔS^0 (cal mole^{-1} deg^{-1})
[Pt{P(Et)$_3$}$_2$Cl$_2$]	7.52	92.48	12.3	−1480	2470	13.3
[Pt{P(n-Pr)$_3$}$_2$Cl$_2$]	3.28	96.72	29.5	−2005	1975	13.3
[Pt{P(n-Pr)$_3$}$_2$I$_2$]	~0.55	99.45	~180			
[Pt{P(n-Bu)$_3$}$_2$Cl$_2$]	3.77	96.23	25.5			
[Pt{As(Et)$_3$}$_2$Cl$_2$]	0.57	99.43	175	−3070	1180	14.2
[Pt{As(n-Pr)$_3$}$_2$Cl$_2$]				ca. −3840		
[Pt{Sb(Et)$_3$}$_2$Cl$_2$]	34.4	65.6	1.9	−380	2410	9.4
[Pt{Sb(n-Pr)$_3$}$_2$Cl$_2$]	20	80	4.0	−818	2200	10.1
[Pt{Sb(n-Bu)$_3$}$_2$Cl$_2$]	21	79	3.75			
[Pt{As(Me$_2$Et)}$_2$Cl$_2$]	1.8	98.2	55			
[Pd{Sb(Et)$_3$}$_2$Cl$_2$]	6.0	94.0	15.7			

^{a} Me = CH$_3$, Et = C$_2$H$_5$, n-Pr = n-C$_3$H$_7$, n-Bu = n-C$_4$H$_9$.

Cis-trans isomerization reactions have been investigated for only a limited number of square-planar, 4-coordinate complex species.[77a] Data for the equilibria

$$cis\text{-}[Pt(ZR_3)_2X_2] \rightleftharpoons trans\text{-}[Pt(ZR_3)_2X_2]$$

(where Z = P, As, Sb; R = CH_3 to $n\text{-}C_5H_{11}$; and X = Cl, I) collected for solutions in benzene at 25°C are summarized in Table 11-6.[78] Equilibration is commonly a slow process, but it can be catalyzed. Thus *cis-* and *trans-*$[Pt\{P(C_2H_5)_3\}_2Cl_2]$ appear to be kinetically stable with respect to isomerization as solutions in benzene at room temperature, but rearrangement to an equilibrium mixture occurs rapidly if a trace of triethylphosphine is added.[78]

At least two conclusions are apparent from the data in Table 11-6:

1. Contrary to the usual assumption that because of their more symmetrical molecular structures *trans* isomers should form exothermically from *cis* isomers, the reaction can be endothermic, with the enhanced stability indicated by a negative change in standard free energy being due to an increase in entropy that arises from a loss of solvating benzene molecules.

2. Inasmuch as it is assumed that metal–ligand bond strength increases with increasing degree of π bonding and phosphine or phosphinelike ligands form stronger metal-to-ligand π bonds than chloro groups, the R_3P-to-Pt bond strength is enhanced in the *cis* isomers. For these species the *trans-*R_3P—M—Cl arrangements promote competition for electron density in the same *d* orbitals of the Pt atom. In the *trans* isomers, the phosphorus ligands would compete with each other for electronic charge density.

Although the reactions being discussed are formally intramolecular, the marked influence of a catalyst upon reaction rate suggests the formation of intermediates, primarily of a trigonal bipyramidal geometry, which then by dissociation yield products, as shown on p. 733.

This tentative mechanism is consistent with isomerization data for the more labile square planar complexes of nickel(II) and palladium(II).

Optical Inversion. An often useful principle in synthesis is that the generic configuration of an optical isomer is preserved when that optical isomer is converted to another optical isomer by a ligand substitution process. For example, retention characterizes conversion of $(+)\text{-}[Co(en)_2Cl_2]^+$ ion to $(+)\text{-}[Co(en)_2(H_2O)Cl]^{2+}$, $(+)\text{-}[Co(en)_2(C_2O_4)]^+$, or $(+)\text{-}[Co(en)_2(NCS)Cl]^+$ ion. A change in generic configuration upon substitution is termed *optical inversion*. This process is an unusual one among coordination entities. It is apparently restrictive to 6-coordinate, octahedral species.

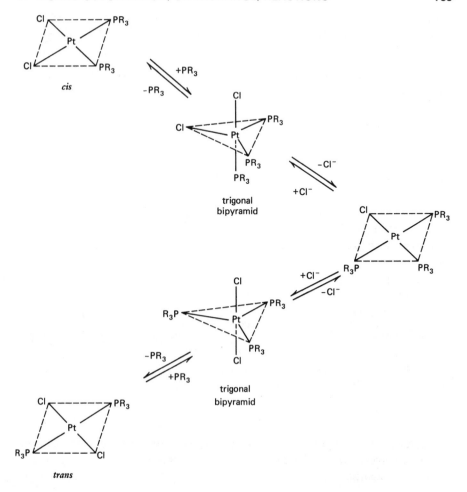

The first reported example,[79] termed a "Walden inversion" by analogy to familiar examples among organic compounds, can be summarized in terms of the reaction sequence

$$(+)\text{-}[Co(en)_2(CO_3)]^+ \xrightarrow[\text{(C}_2\text{H}_5\text{OH)}]{\text{HCl}} (-)\text{-}[Co(en)_2Cl_2]Cl$$

$$(-)\text{-}[Co(en)_2Cl_2]Cl \nearrow^{K_2CO_3\text{(aq)}} $$

$$(-)\text{-}[Co(en)_2Cl_2]Cl \searrow_{Ag_2CO_3\text{(s)}}$$

$$(-)\text{-}[Co(en)_2(CO_3)]^+ \xrightarrow[\text{(C}_2\text{H}_5\text{OH)}]{\text{HCl}} (+)\text{-}[Co(en)_2Cl_2]Cl$$

using $(+)$ for dextrorotatory and $(-)$ for levorotatory. In terms of absolute configuration relative to $D\text{-}[Co(en)_3]^{3+}$ ion, this scheme becomes

$$D\text{-}[Co(en)_2Cl_2]Cl$$

with arrows:

$$D\text{-}[Co(en)_2(CO_3)]^+ \xrightarrow[\text{(C}_2\text{H}_5\text{OH)}]{\text{HCl}} D\text{-}[Co(en)_2Cl_2]Cl$$

via K_2CO_3

$$L\text{-}[Co(en)_2(CO_3)]^+ \xrightarrow[\text{(C}_2\text{H}_5\text{OH)}]{\text{HCl}} L\text{-}[Co(en)_2Cl_2]Cl$$

via Ag_2CO_3

A more nearly complete investigation of this system[80, 81] led to the conclusion that aquation, rather than the carbonate used, is the important factor, as indicated in the reaction scheme

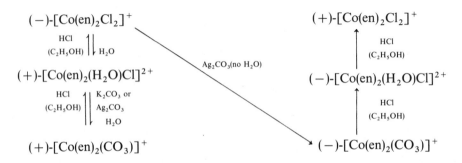

An additional investigation[82] indicated that inversions occur as a consequence of base hydrolysis of $D\text{-}[Co(en)_2Cl_2]^+$ ion in the presence of Ag^+ ion, giving the reaction sequence

$$D\text{-}[Co(en)_2Cl_2]^+ \xrightarrow{\text{H}_2\text{O}} D\text{-}[Co(en)_2(H_2O)Cl]^{2+}$$

$$\downarrow \text{OH}^-(Ag^+) \qquad\qquad \text{OH}^- \qquad \downarrow (\text{OH}^-(Ag^+))$$

$$L\text{-}[Co(en)_2(OH)Cl]^+ \qquad\qquad D\text{-}[Co(en)_2(OH)Cl]^+$$

$$\downarrow \text{OH}^-(Ag^+) \qquad\qquad\qquad \downarrow \text{OH}^-(Ag^+)$$

$$L\text{-}[Co(en)_2(OH)_2]^+ \qquad\qquad D\text{-}[Co(en)_2(OH)_2]^+$$

$$\downarrow \text{HCO}_3^- \qquad\qquad\qquad \downarrow \text{HCO}_3^-$$

$$L\text{-}[Co(en)_2(CO_3)]^+ \qquad\qquad D\text{-}[Co(en)_2(CO_3)]^+$$

Furthermore, base hydrolysis of the $D\text{-}[Co(en)_2Cl_2]^+$ ion with inversion can take place in the absence of Ag^+ ion in concentrated ($c_{OH^-} > 0.25\,M$) but not in dilute ($c_{OH^-} < 0.01\ M$) alkaline solutions, as[83]

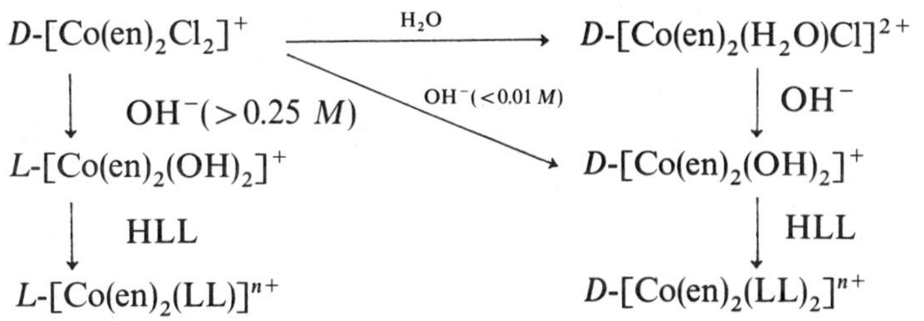

$$D\text{-}[Co(en)_2Cl_2]^+ \xrightarrow{\ H_2O\ } D\text{-}[Co(en)_2(H_2O)Cl]^{2+}$$

$$\downarrow OH^- (>0.25\ M) \qquad OH^-(<0.01\ M)\qquad \downarrow OH^-$$

$$L\text{-}[Co(en)_2(OH)_2]^+ \qquad\qquad\quad D\text{-}[Co(en)_2(OH)_2]^+$$

$$\downarrow HLL \qquad\qquad\qquad\qquad\qquad \downarrow HLL$$

$$L\text{-}[Co(en)_2(LL)]^{n+} \qquad\qquad\quad D\text{-}[Co(en)_2(LL)_2]^{n+}$$

where $HLL = HCO_3^-$ ion, acetylacetone, or trifluoroacetylacetone and is used to obtain an isolable product. Several potential mechanistic approaches to base hydrolysis in this system have been suggested.[84]

Optical inversion has been noted in a limited number of other cases, among them reaction of $D\text{-}[Co(en)_2Cl_2]^+$ ion with ammonia[85,86] and the base hydrolysis of $D\text{-}\alpha\text{-}[Co(trien)Cl_2]^+$ ion.[87,*] The reaction sequences can be summarized as

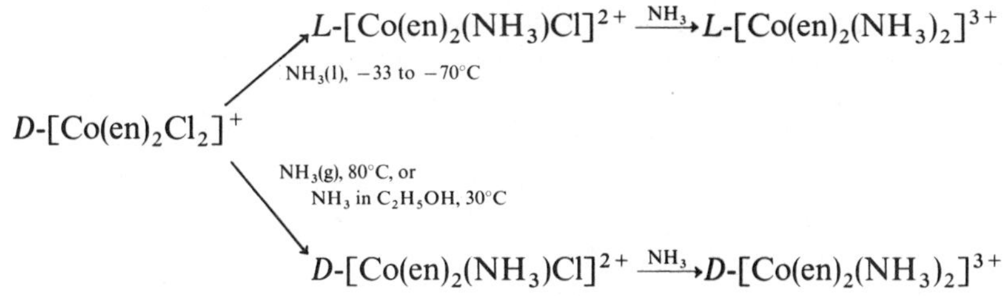

$$L\text{-}[Co(en)_2(NH_3)Cl]^{2+} \xrightarrow{NH_3} L\text{-}[Co(en)_2(NH_3)_2]^{3+}$$
$$\nearrow \quad NH_3(l),\ -33\ \text{to}\ -70°C$$
$$D\text{-}[Co(en)_2Cl_2]^+$$
$$\searrow \quad NH_3(g),\ 80°C,\ \text{or}$$
$$NH_3\ \text{in}\ C_2H_5OH,\ 30°C$$
$$D\text{-}[Co(en)_2(NH_3)Cl]^{2+} \xrightarrow{NH_3} D\text{-}[Co(en)_2(NH_3)_2]^{3+}$$

and

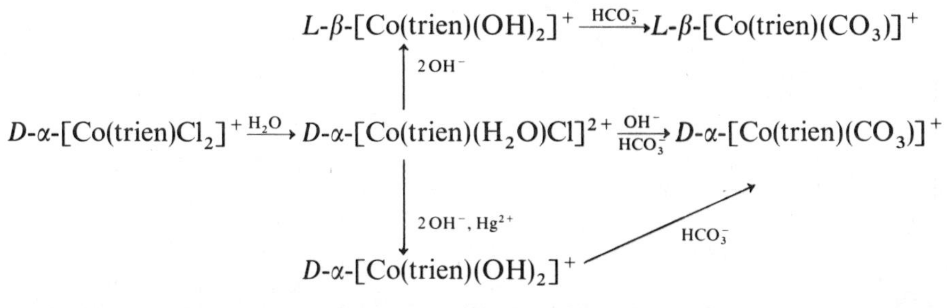

$$L\text{-}\beta\text{-}[Co(trien)(OH)_2]^+ \xrightarrow{HCO_3^-} L\text{-}\beta\text{-}[Co(trien)(CO_3)]^+$$
$$\uparrow 2OH^-$$
$$D\text{-}\alpha\text{-}[Co(trien)Cl_2]^+ \xrightarrow{H_2O} D\text{-}\alpha\text{-}[Co(trien)(H_2O)Cl]^{2+} \xrightarrow[HCO_3^-]{OH^-} D\text{-}\alpha\text{-}[Co(trien)(CO_3)]^+$$
$$\downarrow 2OH^-,\ Hg^{2+} \qquad\qquad \nearrow HCO_3^-$$
$$D\text{-}\alpha\text{-}[Co(trien)(OH)_2]^+$$

In the latter system, the α and β forms differ in orientations of the chelating rings. A possible conjugate base mechanism involving a trigonal bipyramidal intermediate can be formulated as[88]

---

* Trien = triethylenetetramine, $H_2NC_2H_4NHC_2H_4NHC_2H_4NH_2$.

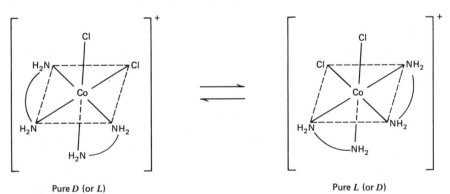

Optical inversion in 6-coordinate, octahedral species requires changes in points of spatial attachment to the metal ion by only two ligand atoms. Proper orientation of an incoming group with respect to bonded ligands may be all that is required for inversion.

Racemization. Racemization is a process in which one optical isomer (e.g., d) converts into the other (i.e., l) to such an extent that the two are present in equimolar quantities and the product is then optically inactive. Unlike many optically active organic compounds, most optically active complex species, racemize readily and often rapidly. The racemization process can be visualized for the D and L forms of the $[Co(en)_2Cl_2]^+$ ion as

with interconversions proceeding until 50 mole % of each is present. The process may be thought of as proceeding by either (1) dissociation followed by association to give the other isomer (i.e., an intermolecular process), or (2) intramolecular trading of places by ligands.

An intermolecular mechanism involving tris-chelate equilibria of the types

$$D\text{-}[M(LL)_3] \underset{+LL}{\overset{-LL}{\rightleftharpoons}} [M(LL)_2] \underset{-LL}{\overset{+LL}{\rightleftharpoons}} L\text{-}[M(LL)_3]$$

would require that ligand exchange be at least as rapid as racemization. Although there are many experimental results available that do not support intermolecular mechanisms, these mechanisms have been substantiated in a limited number of cases. Thus the rates of exchange of water molecules with the species $cis\text{-}[Co(en)_2(H_2O)_2]^{3+}$ [89] and $cis\text{-}[Co(en)_2(NH_3)(H_2O)]^{3+}$ [64] are, respectively, about 65 and 40 times the rates of racemization. Also, it has been observed that the rate of racemization in methanol, as measured by loss of optical activity, of $(+)\text{-}cis\text{-}[Co(en)_2Cl_2]^+$ ion is equal to the rate of radio-chlorine exchange with one chloro ligand.[90,91] The last of these reactions is believed to involve loss of Cl^- ion to give a 5-coordinate intermediate, which then adds Cl^- ion to give the *trans* isomer ca. 70% of the time and the racemic *cis* isomer 30% of the time.

Intramolecular racemization was first suggested by A. Werner.[92,93] Mechanistically, the necessary rearrangements can be accounted for via a bond-rupture-reforming sequence or various twist mechanisms involving octahedral geometry. The bond-rupture-reforming mechanism, which was favored by Werner,[92,93] can be visualized diagrammatically for a tris-chelate in terms of one or two bidentate ligands as

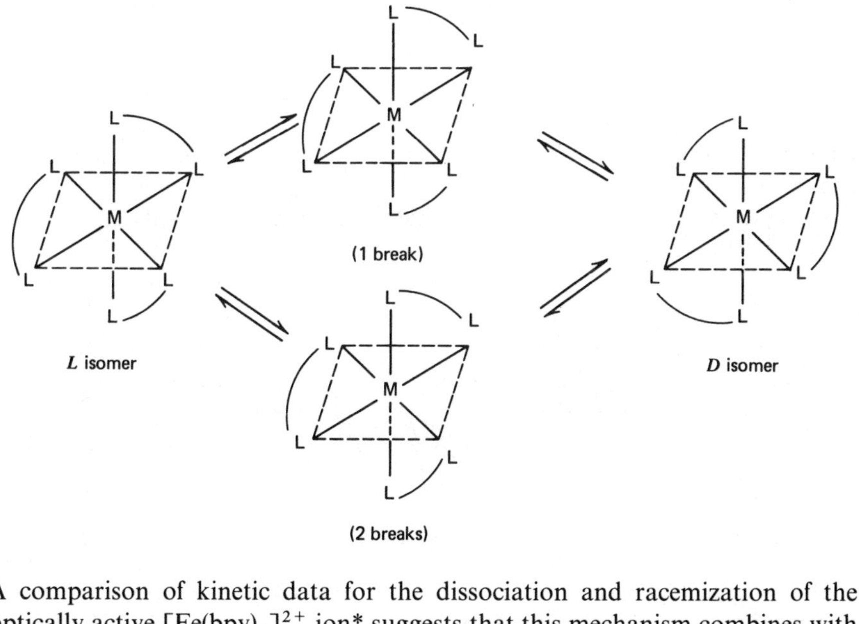

L isomer

(1 break)

(2 breaks)

D isomer

A comparison of kinetic data for the dissociation and racemization of the optically active $[Fe(bpy)_3]^{2+}$ ion* suggests that this mechanism combines with

---

* bpy = 2,2'-bipyridine.

a twist mechanism in the overall racemization process.[94] Racemization of the active $[Cr(C_2O_4)_3]^{3-}$ ion is complete in aqueous solution before radiocarbon-labeled $C_2O_4^{2-}$ ion can exchange with the complex ion.[95] Again an intra-molecular process is indicated. That the bond-rupture-reformation mechanism is operative was confirmed by exchange studies using ^{18}O.[96]

Intramolecular mechanisms involve presumed distortions, or "twists," of the octahedral arrangement which result in different geometrical arrangements of ligands. The best known of these are the *rhombic*, or *tetragonal*, twist[97] and the *trigonal* twist.[98-100]

Using a tris chelate as illustrative, the tetragonal (rhombic) twist can be illustrated diagrammatically as[101]

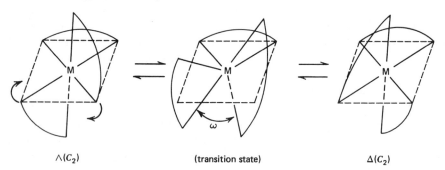

$\Lambda(C_2)$ (transition state) $\Delta(C_2)$

using the designations for absolute configurations (Sec. 7.4-10) proposed by T. S. Piper.[102] In this twist, one of the chelate rings is assumed to remain fixed in space while the other two rings rotate 90° in opposite directions about axes perpendicular to their respective planes and passing through the central metal ion. The internal angles in the chelate rings do not change, the angles (ω) between bonds in the rotating rings change, and the other nonring angles do not change.

Using the same tris chelate, the trigonal twist can be illustrated diagrammatically as[101, 103]

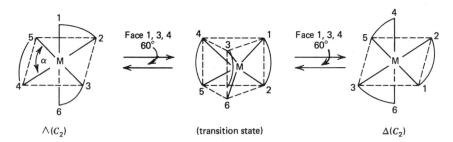

$\Lambda(C_2)$ (transition state) $\Delta(C_2)$

In this twist, a clockwise 60° rotation of the face 1, 3, 4 gives a trigonal prismatic intermediate with all three ligands parallel, and a second clockwise 60° rotation of the same face gives an octahedral product the molecular structure of which is a mirror image of that of the reactant. In other terms, a total counterclockwise rotation of three metal–ligand bonds through 120° gives the mirror-image isomer. Contraction of ring bond angles α is necessary to the formation of the transition-state species, ideally to 81°48′ if all angles are the same. It has been noted[103] that optical inversion can also result if a clockwise rotation of the face 2,3,6 through 60°, giving the transition-state configuration

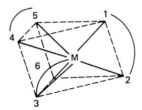

with two ligands parallel and one perpendicular thereto, is followed by a second clockwise 60° rotation of the same face.

In another proposed intramolecular twist mechanism,[101] which can be illustrated diagrammatically as

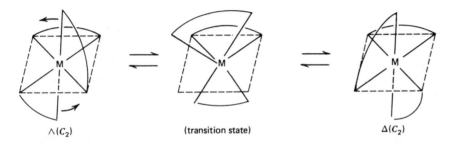

$\Lambda(C_2)$ (transition state) $\Delta(C_2)$

one ring is assumed to remain fixed in space and the other two revolve past each other while changing planes but do not rotate past each other in their own planes, as is characteristic of the "tetragonal twist." The internal ring angles remain constant, but the L—M—L angles between bonds in different rings are decreased in the transition state. Again, optical inversion results.

Both this "twist" and the "tetragonal twist" are special cases of the more general "trigonal twist."[101] Actual cases may well deviate from the rigid "trigonal twist" in L—M—L bond angles in the transition state. It must be realized also that *cis-trans* isomerization processes can occur for certain chelates along with racemization processes. Distinction among possible "twist" mechanisms is not always possible, although in specific cases such distinctions can be made. Thus the slow thermal conversion at 160°C of *cis*-α-dichlorotriethylene-tetraminecobalt(III) chloride[104] to the β isomer is explicable in terms of a trigonal-twist mechanism, but movement of any chelate ring in its own plane, as required by a tetragonal-twist mechanism cannot occur without disruption of the complex.[103]

Only a few of the many other reported studies[49, 105] need be noted here. Kinetic studies of stereochemical rearrangements of tris(trifluoroacetonato)-

metal(III) chelates (M = Al, Ga, In, Co, Cr, Rh), using ^{19}F nuclear magnetic resonance,[106] indicate that the trigonal-twist mechanism can be eliminated for M = Al, Ga, In, and cis-Cr but not for cis-Co. The activation energy for the trigonal-twist mechanism was found to be larger than that for the rhombic-twist mechanism where crystal-field stabilization is present. All stereochemical rearrangements among all compounds studied probably involve the rupture of one bond for each and the formation of a symmetrical 5-coordinate intermediate. Rates of optical inversion of various unsymmetrical β-diketone chelates of aluminum(III), as determined by ^{1}H nuclear magnetic resonance measurements,[107] did not allow clean-cut distinction among possible mechanisms but were supportive of tetragonal-twist pathways. Rate studies on optical inversion of the cis and trans isomers of tris(benzoylacetonato)cobalt(III) in various solvents[108] indicate isomerization and inversion occurring primarily by a common intramolecular mechanism; the most probable pathway is an $\sim 80\%$ rhombic-twist and $\sim 20\%$ trigonal-twist mechanism or bond rupture, giving $\sim 80\%$ axial trigonal-bipyramidal and $\sim 20\%$ square-pyramidal intermediates.

3. ELECTRON-TRANSFER, OR REDOX, REACTIONS

The general principles applicable to inorganic electron-transfer reactions, a very general treatment of mechanisms, a limited number of examples involving coordination entities, and the effects of coordination upon the stabilization of oxidation states in solution have been discussed briefly in Chapter 9. It is our purpose in this section to consider mechanistic approaches in more detail and to outline, using appropriate examples, the development of, current interest in, and importance of this broad area of the reaction chemistry of complex species. Numerous more detailed summaries[109-116a] can be consulted for additional information.

3-1. Mechanistic Approaches

Mechanistically, H. Taube's classification[117,118] of transition states in electron-transfer reactions involving complex species as "outer-sphere" and "inner-sphere" (Sec. 8.1-1) remains as the most useful approach to rationalizing and understanding how these reactions proceed. Detailed mechanistic information relating to specific reactions is available from numerous sources.[109-119]

An *outer-sphere mechanism* is one in which the inner coordination spheres of the metal ions are believed to remain unaltered in the transition state or reaction intermediate. An *inner-sphere mechanism* is one in which the two metal ions are believed to be bonded to each other in the transition state through a common bridging ligand. It is apparent that an inner-sphere mechanism requires the presence of at least one ligand that is capable of bonding simultaneously to two metal ions and the presence of at least one ligand so bonded to one of the metal ions (commonly the ion that reacts as the reductant) as to be labile to substitution.

Irrespective of which mechanism applies to a specific reaction, an initial chemical activation process is essential to provide appropriate electronic configurations for favorable overlap of orbitals. In molecular-orbital terminology, this means that both the donor and acceptor orbitals must be of the same type. Optimum conditions for outer-sphere mechanisms obtain when both of these orbitals are of the π^* type. For inner-sphere mechanisms, both must be of the π^* or the σ^* type. In addition, of course, thermodynamic properties must be favorable to lowering the activational energy barrier.

Outer-Sphere Mechanisms. If an electron-transfer reaction involving two complex species is more rapid than a substitution reaction involving these species, it is assumed that the electron-transfer process occurs via the two intact coordination spheres, assuming that neither reagent labilizes the other to substitution. Even in reactions where only one reactant is inert to substitution, for example, in species such as $[M(NH_3)_6]^{n+}$ [M = Co(III), Ru(II)], where the ligand is nonbridging, an outer-sphere mechanism is probable. An outer-sphere mechanism has been assumed to involve the successive steps (1) formation of some type of *precursor complex* which may be regarded as a cage containing the two reactants, (2) activation of this precursor complex combined with electron transfer and ultimately relaxation to a *successor complex*, and (3) dissociation of the successor complex into the final products. The precursor complex could represent a situation in which the metal ions are close enough to each other to allow electron transfer but not in a proper orientation for transfer to take place. Activation provides proper orientation, allowing electron transfer and ultimate relaxation to a different complex wherein the redox process has been completed. Orientation is then favorable to dissociation to the final products.

There is thus an outer-sphere, activated complex involved in this type of mechanism, within which electron transfer occurs via a tunneling process.[120] That an electron can "leak" through a potential energy barrier that should by classical mechanics be impenetrable is allowed as a quantum mechanical phenomenon. Thus, an electron by tunneling through such a barrier can transfer from one metal center to another over a larger distance than would be allowable by simple collision. A tunneling process is, of course, closely related to the relative extensions in space about the metal ions of the orbitals that are involved.

As noted earlier, reactions believed to proceed via outer-orbital pathways are normally slow and kinetically second order. Rate constants for a few electron-transfer *exchange reactions*, wherein two different oxidation states of the same element are involved, and *cross reactions*, wherein two different elements are involved, are summarized in Table 11-7. Although it is difficult to generalize, it appears that variations in rate constants are much smaller for outer-sphere mechanisms than for inner-sphere mechanisms.

Table 11-7 Second-Order Rate Constants for Selected Electron-Transfer Reactions[a]

k (mole^{-1} sec^{-1})[b]

Oxidant	Outer-Sphere Reductants		Inner-Sphere Reductants		
	$[Cr(bpy)_3]^{2+}$ $\mu = 0.1\ M$	$[Ru(NH_3)_6]^{2+}$ $\mu \sim 0.2\ M$	$[Cr(H_2O)_m]^{2+}$ $\mu = 1.0\ M$	$[V(H_2O)_m]^{2+}$ $\mu = 1.0\ M$	$[Co(CN)_5]^{3-}$ $\mu \sim 0.2\ M$
$[Co(NH_3)_5(NH_3)]^{3+}$	6.9×10^2	1.1×10^{-2}	8.9×10^{-5}	3.7×10^{-3}	8×10^4
$[Co(NH_3)_5(H_2O)]^{3+}$	5×10^4	3.0	5×10^{-1}	$\sim 5 \times 10^{-1}$	—
$[Co(NH_3)_5(OH)]^{2+}$	3×10^4	4×10^{-2}	1.5×10^{-6}	—	9.3×10^4
$[Co(NH_3)_5Cl]^{2+}$	8×10^5	2.6×10^2	6.0×10^5	~ 5	5×10^7
$[Co(NH_3)_5(N_3)]^{2+}$	4.1×10^4	1.2	$\sim 3 \times 10^5$	1.3×10^1	1.6×10^6
$[Co(NH_3)_5(NCS)]^{2+}$	1.0×10^4	—	1.9×10^1	3×10^{-1}	1.1×10^6
$[Co(NH_3)_5(NO_3)]^{2+}$	—	3.4×10^1	$\sim 9.0 \times 10^1$	—	2.4×10^5
$[Co(NH_3)_5(CN)]^{2+}$	—	—	6.1×10^1	—	2.9×10^2

[a] Adapted from F. Basolo and R. G. Pearson, Ref. 1, p. 481.
[b] Aqueous solutions, 25°C.

Inner-Sphere Mechanisms. An outer-sphere mechanism is said to be operative if electron transfer is from one primary bond system to another. By contrast, if the electron transfer occurs within a single primary bond system, an inner-sphere mechanism is said to be operative. If the mechanism is of the inner-sphere type, the effects on primary bonds should be sufficiently persistent as to be apparent in the reaction products.[121]

The original proof of an inner-sphere mechanism resulted from the very clever choice of the reduction of the nonlabile, low-spin $[Co(NH_3)_5Cl]^{2+}$ ion with the labile, high-spin $[Cr(H_2O)_6]^{2+}$ ion to form the nonlabile, low-spin $[Cr(H_2O)_5Cl]^{2+}$ ion and the labile, high-spin $[Co(H_2O)_6]^{2+}$ ion,[122,123]

$$[Co(NH_3)_5Cl]^{2+} + [Cr(H_2O)_6]^{2+} + 5H_3O^+ \rightarrow [Co(H_2O)_6]^{2+}$$
$$+ [Cr(H_2O)_5Cl]^{2+} + 5NH_4^+$$

as previously noted. The specific rate constant for this reduction process is larger than 10^5 mole^{-1} sec^{-1} at 25°C.[124] That in the presence of $^{36}Cl^-$ ion very little radioactivity was noted in the $[Cr(H_2O)_5Cl]^{2+}$ ion formed in the reaction proved that chlorine $(-I)$ was transferred directly from initial cobalt(III) to product chromium(III) and suggested the transition state or intermediate to be the chlorine-bridged species $[NH_3)_5Co—Cl—Cr(H_2O)_5]^{4+}$.

For a related system described by the equation

$$[Co(NH_3)_5(H_2O)]^{3+} + [Cr(H_2O)_6]^{2+} + 5H_3O^+ \rightarrow [Co(H_2O)_6]^{2+}$$
$$+ [Cr(H_2O)_6]^{3+} + 5NH_4^+$$

the experimentally determined rate law is expressed as[125,126]

$$-\frac{dc_{[Co(NH_3)_5(H_2O)]^{3+}}}{dt} = (c_{Cr^{2+}})(c_{[Co(NH_3)_5(H_2O)]^{3+}})\left(k_1 + \frac{k_2}{c_{H_3O^+}}\right) \quad (11\text{-}25)$$

where the second term can be replaced by the kinetically equivalent term

$$k_2'(c_{Cr^{2+}})(c_{[Co(NH_3)_5(OH)]^{2+}})$$

At 25°C and $\mu = 0.5\ M$, $k_1 = 0.6$ mole^{-1} sec^{-1} and $k_2' \sim 2 \times 10^6$ mole^{-1} sec^{-1}. Tracer experiments using ^{18}O proved clearly that oxygen transfer from Co(III) to Cr(II) is quantitative and indicated that an O atom in the coordination sphere of cobalt(III) is bonded in the transition state to the Cr^{2+} ion.[126,127]

It may be concluded from the examples cited and others that an inner-sphere mechanism is most probable for a reaction in which both the oxidant and the oxidized form of the initial reductant are nonlabile, or inert to substitution, and under conditions favoring simultaneous ligand and electron transfer. The reactions of $[Cr(H_2O)_6]^{2+}$ ion with various $[Co(NH_3)_5X]^{n+}$ ions all exemplify these criteria. However, ligand transfer is not itself a necessity, since the reaction described by the equation

$$[IrCl_6]^{2-} + [Cr(H_2O)_6]^{2+} \rightarrow [IrCl_6]^{3-} + [Cr(H_2O)_6]^{3+}$$

apparently proceeds by an inner-sphere mechanism.

Changes in electronic configurations and energy states are certainly of importance in the formation and destruction of the inner-sphere activated complex. These can be indicated qualitatively for the $[Cr(H_2O)_6]^{2+}$—$[Co(NH_3)_5X]^{2+}$ type of reaction as in Fig. 11-6.[111] It will be noted that the donor orbital of the Cr^{2+} ion and the acceptor orbital of the Co^{3+} ion are of the same type and symmetry (σ^*). In the activated intermediate, the single σ^* d electron is presumably shared between the two cationic species. Dissociation of this intermediate then transfers this σ^* electron to the cobalt species, converting it to cobalt(II) and forming chromium(III) with a d^3 nonbonding arrangement. In molecular-orbital terms, the highest occupied molecular orbital (HOMO) of the chromium(II) species and the lowest unoccupied molecular orbital (LUMO) of the cobalt(III) species both have the same symmetry (σ^*).

That the nature of the bridging ligand is important in inner-sphere mechanisms is suggested by the rate-constant data given in Table 11-8. Of particular

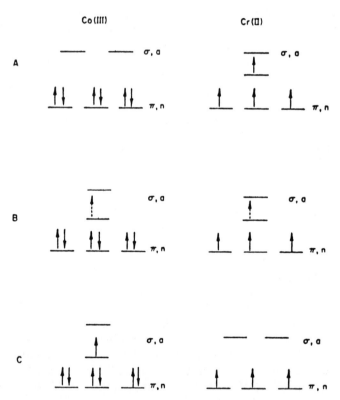

Figure 11-6 Changes in electronic configuration for the reaction of $[Cr(H_2O)_6]^{2+}$ ion with $[Co(NH_3)_5X]^{2+}$ ion. (A) before electron transfer; (B) in activated complex; (C) after electron transfer. The notations a and n refer to antibonding and nonbonding orbitals. [H. Taube, *Electron Transfer Reactions of Complex Ions in Solution*, Fig. II-1, p. 39, Academic Press, New York (1970).]

Table 11-8 Comparison of Specific Rate Constants for Reduction (at 25°C) of Azido and Isothiocyanto Complexes $[Co(NH_3)_5X]^{2+}$

Oxidant	Reductant	$k_{N_3^-}/k_{NCS^-}$	Probable Mechanism	References
$[Co(NH_3)_5X]^{2+}$	Cr^{2+}(aq)	10^4	Inner sphere	124
	V^{2+}(aq)	27	Inner sphere (?)	124
	Eu^{2+}(aq)	300	Inner sphere	124
	Fe^{2+}(aq)	$\geq 3 \times 10^3$	Inner sphere	128
	$[Cr(bpy)_3]^{2+}$	4	Outer sphere	124
	$[Co(CN)_5]^{3-}$	1	Outer sphere (?)	131
$[Cr(H_2O)_5X]^{2+}$	Cr^{2+}(aq)	4×10^4	Inner sphere	132, 133

interest are comparisons of specific rate constants for reactions of ions $[Co(NH_3)_5(N_3)]^{2+}$ and $[Co(NH_3)_5(NCS)]^{2+}$ with various reductants.[128] The N_3^- ion gives a more symmetrical Co—N bonding arrangement,

$$Co-\overset{\oplus}{\ddot{N}}=N=\overset{\ominus}{\ddot{N}}:$$

than does the NCS$^-$ ion,*

$$Co-\ddot{N}=C=\ddot{S}:$$

and should react much more rapidly with a reductant if the mechanism is of the inner-sphere type. That this is indeed the case is shown by the data in Table 11-8.

It is of interest also that when the NCS$^-$ ion is bonded through the S atom in the oxidant (e.g., in the $[Co(NH_3)_5(SCN)]^{2+}$ ion), the N atom of this ligand can then bond to the reductant ion $[Cr(H_2O)_6]^{2+}$ to give the inner-sphere bridge with a specific rate constant for the redox reaction of 1.9×10^5 mole^{-1} sec^{-1},[129] which is nearly the same as that for the azide ion (Table 11-8). This "remote" vs. "adjacent" attack on bonded NCS$^-$ groups by the $[Cr(H_2O)_6]^{2+}$ ion is also characteristic of the $[Co(CN)_5]^{3-}$ ion.[130]

Organic molecules have been particularly useful in establishing inner-sphere mechanisms because by proper choice of structure both bridging and electron transfer by tunneling, as well as inherent stability of the precursor complex, can be controlled.[134] Proper choice of organic ligand in a particular redox system allows alteration of the reaction rate by many orders of magnitude. Details can be found both in the review paper cited[134] or subsequently published papers.

Multielectron Redox Reactions. All of the redox systems described above in the mechanistic sense have been one-electron processes. Multielectron changes are rare and difficult to identify clearly. A well-characterized two-electron

---

* See Sec. 7.4.4-5 for discussion of the ambidentate characteristics of the NCS$^-$ ion.

reaction is represented by the overall equation, which does not indicate redox behavior,

$$trans\text{-}[Pt(en)_2Cl_2]^{2+} + {}^*Cl^- \xrightarrow{[Pt(en)_2]^{2+}} trans\text{-}[Pt(en)_2^* ClCl]^{2+} + Cl^-$$

for which platinum(II) is the catalyst.[135-137] The rate-law expression is

$$\text{rate} = k(c_{Pt(II)})(c_{Pt(IV)})(c_{Cl^-}) \tag{11-26}$$

To account for this kinetic relationship, it was proposed that the platinum(II) compound adds chloride ion rapidly, giving the 5-coordinate ion $[Pt(en)_2Cl]^+$, which then forms a chloride-bridged 6-coordinate intermediate with the trans-$[Pt(en)_2Cl_2]^{2+}$ ion. Both platinum atoms are now in the same environment. Transfer of two σ^* electrons and transfer in the opposite direction of two Cl^- ions then can occur readily. Establishment of the mechanism was helped by the observation that exchange of radiochloride ion occurs at the same rate as change in optical rotation in a parallel reaction involving trans-$[Pt(l\text{-}pn)_2Cl_2]^{2+}$ ion and $[Pt(en)_2]^{2+}$ ion.

Complementary vs. Noncomplementary Reactions.   A redox reaction in which the formal oxidation states of the reductant and the oxidant change by the same number of units is termed a *complementary reaction*. We have been concerned thus far with reactions of this type, wherein of necessity, molecular stoichiometries are always 1 : 1. A reaction in which the changes in formal oxidation states of the reductant and the oxidant are by different numbers of units and the molecular stoichiometries are thus not 1 : 1 is termed a *noncomplementary* reaction.

Most of the noncomplementary reactions for which the kinetics have been investigated involve reduction of chromate(VI) ion.[138] Elementary step reactions common to many of these overall processes can be formulated in general terms as

$$Cr(VI) + Red \underset{k_{56}}{\overset{k_{65}}{\rightleftharpoons}} Cr(V) + Ox$$

$$Cr(V) + Red \xrightarrow{k_{54}} Cr(IV) + Ox$$

$$Cr(IV) + Red \rightarrow Cr(III) + Ox$$

where Cr(VI) is often the $HCrO_4^-$ ion, Red is the reductant for each study, Ox is an oxidized species, and the subscripts on the k's refer to oxidation-state changes of chromium [e.g., 65, indicating $Cr(VI) \rightarrow Cr(V)$]. At a steady-state concentration of Cr(V), the general rate law becomes

$$-\frac{dc_{HCrO_4^-}}{dt} = \frac{k_{65}k_{54}c_{Cr(VI)}c_{Red}^2}{k_{54}c_{Red} + k_{56}c_{Ox}} \tag{11-27}$$

For modifications for specific cases, the literature should be consulted.[138,139]

3-2. Ligand-Field Considerations

A review of rate data for one-electron redox reactions of octahedral complexes indicates that if the added electron enters a t_{2g} orbital of the reduced product, the reaction is fast.[140] Energywise, this situation is reasonable since the alternative addition of an electron to an e_g orbital would result in increased metal ion–ligand bond distance, enhanced difference in geometry between the oxidized and the reduced forms, and additional reorganizational energy. Furthermore, t_{2g} orbitals are more sterically available to incoming electrons since e_g orbitals are shielded by ligands in this geometry. A commonly cited example[140] is a comparison between the slow oxidation of Mn^{2+} ion by cerium(IV) (e_g electron transfer, $k = 0.2$ mole^{-1} sec^{-1} at 20°C)[141] and the rapid oxidation of Fe^{2+} ion by cerium(IV) (t_{2g} electron transfer, $k = 3 \times 10^5$ mole^{-1} sec^{-1} at 0°C).[142]

The reduction of cobalt(III) complexes to cobalt(II) species has been particularly well studied (Sec. 11.3-1) and lends itself to ligand-field treatment. As noted earlier, cobalt(II) most commonly has the ground-state high-spin valency shell configuration $(t_{2g})^5(e_g)^2$, whereas cobalt (III) has the ground-state low-spin configuration $(t_{2g})^6$. Neither the simple removal nor the simple addition of a single electron can convert one of these configurations into the other. The reaction most probably involves one or more excited states, with a smaller promotional energy being required for a single such state. The magnitude of Δ, or $10\ Dq$, should affect the nature of this promotion inasmuch as promotional energy for the change $(t_{2g})^6 \to (t_{2g})^5(e_g)^1$ should increase with increasing ligand-field stabilization energy and that for the change $(t_{2g})^5(e_g)^2$ to $(t_{2g})^6(e_g)^1$ should decrease with increasing ligand-field stabilization energy.

Experimental data sometimes support and sometimes refute this reasoning. Thus for the series of ligands

$$H_2O < C_2O_4^{2-} < EDTA^{4-} < NH_3 < en < phen$$

increasing ligand-field
stabilization energy →

electron exchange between cobalt(II) and cobalt(III) is slowest at the middle of the series.[143, 144] Rapid exchange in the $[Co(H_2O)_6]^{2+}$–$[Co(H_2O)_6]^{3+}$ reaction is assumed to result from the small ligand field imposed by the water molecule that allows easy excitation of diamagnetic $[Co(H_2O)_6]^{3+}$ (t_{2g}^6) ion to paramagnetic $[Co(H_2O)_6]^{3+}$ ($t_{2g}^5 e_g^1$) ion. The more rapid exchange of the 1,10-phenanthroline complexes as compared with the ethylenediamine complexes is associated with easy conversion of low-spin cobalt(II) to high-spin cobalt(II) by the large ligand field imposed by the 1,10-phenanthroline molecule.

The rates of reduction of species $[Co(NH_3)_5X]^{(3-n)}$, where X^{n-} is an anionic ligand, appear to vary even more regularly with ligand-field stabilization energy, as indicated by the linearity of a plot of the activation energy for reduction at the dropping mercury electrode vs. ligand-field stabilization energy.[145]

Both ligand-field stabilization and geometry appear to be significant in the reduction of *cis* and *trans* isomers of the types $[Co(NH_3)_4XY]^{n+}$ and

$[Co(en)_2XY]^{n+}$. Thus the rate sequence for the reductions of *cis*- and *trans*-$[Co(en)_2ClX]^{n+}$ ions by iron(II) varies with X as

$$H_2O > SCN^- > NH_3$$

which is the order of both increasing ligand-field and bond strength.[146] For the *trans* isomers specific rate constants spread by 10^4, whereas for the *cis* isomer the spread is by 10^1. For these species and others, ligand-field effects are commonly more significant for the *trans* isomers, in keeping with the view[147] that the electron in the bridging mechanism is in an antibonding e_g orbital directed toward ligands X and Y and thus responsible for greater increase in the M—X and M—Y bond distances in the *trans* isomers.

It should be noted, however, that changes in bond strength may provide better correlations than changes in ligand-field stabilization energies. That these two changes are often parallel renders absolute decisions difficult.

3-3. Applications in Syntheses

Some synthetic routes involving redox procedures have been described in Chapter 7, and the significance of alteration of reduction potentials by complexation in indicating feasible preparations has been discussed in Chapter 9.

In some cases, redox procedures have been useful in enhancing yields of specific enantiomers over those yields that might be expected.[148] Thus treatment of a racemic solution of $[Co(en)_3]^{3+}$ ion with *d*-tartrate to precipitate selectively the *d*-tartrate salt, followed by electron-transfer equilibration with the $[Co(en)_3]^{2+}$ ions to produce more *D*-$[Co(en)_2]^{3+}$ ion and subsequent precipitation with *d*-tartrate ion, yielded ultimately 75% of the $[Co(en)_3]^{3+}$ ion in the *D* form.[149] Independent oxidation of *D*- and *L*-$[Ru(phen)_3](ClO_4)_2$ to the ruthenium(III) species with cerium(IV) with retention of optical activity allowed syntheses of all enantiomers in high states of purity.[150] Reduction of the $[Ru(phen)_3]^{3+}$ enantiomers with iron(II) sulfate yielded the dipositive species with retention of optical activity.[151] Similarly high yields of active $[Cr(phen_3]^{3+}$ and $[Co(phen)_3]^{3+}$ were obtained via selective precipitation of the dipositive enantiomers as antimonyl-*d*-tartrates.[152]

EXERCISES

11-1.   Using appropriate electrode potential or equilibrium constant data, cite a specific example to support increase of thermodynamic stability among coordination entities that is dependent upon each of the following: (1) increase in oxidation state of the acceptor atom, (2) increase in number of donor sites in the ligand, (3) increase in $+\Delta S$ for the formation reaction, (4) presence of $(n-1)d$ orbital sites in the acceptor atom. Account for each.

11-2. Account for the anomaly that exists in the position of the Cu^{2+} ion in the "natural order" of solution stabilities of complexes. Give supportive data.

11-3. Distinguish clearly between the *cis* and *trans* effects in ligand substitution reactions. Cite for each at least two well-defined specific examples.

11-4. The high-spin d^4 complex ion $[Cr(H_2O)_6]^{2+}$ is *labile*, but the low-spin d^4 complex ion $[Cr(CN)_6]^{4-}$ is *inert*. Explain.

11-5. For many substitution reactions involving coordination entities, it is observed that (1) for a given aqueous ion, the rate of displacement of a bonded water molecule is not strongly dependent upon the nature of the incoming ligand and (2) the exchange of water molecules is always more rapid than similar reactions involving other ligands. Indicate how these observations are consistent with the assumption of the initial formation of an aquo ion–ligand "outer-sphere" complex followed by the exchange of an "inner-sphere" coordinated water molecule with the "outer-sphere" coordinated ligand.

11-6. When one of the isomers of $[Pt(NH_3)_2Cl_2]$ is treated with thiourea, $(H_2N)_2CS = tu$, $[Pt(tu)_4]^{2+}$ ion is formed, but when the other isomer is similarly treated, $[Pt(NH_3)_2(tu)_2]^{2+}$ is formed. Identify each isomer and account for the stated observations.

11-7. Treatment of $CrCl_3(s)$ with $NH_3(l)$ yields very largely the complex $[Cr(NH_3)_5Cl]Cl_2(s)$, but if a small quantity of elemental potassium is first dissolved in the $NH_3(l)$, the product is largely $[Cr(NH_3)_6]Cl_3$. Explain by giving a possible mechanism.

11-8. Suggest a reason why the magnitude of Z^2/r is significant in the classification of cations in terms of exchange of coordinated water molecules in ligand substitution reactions in aqueous solution.

11-9. Cite arguments for and against applying the S_N1 and S_N2 mechanisms to ligand substitution reactions.

11-10. Distinguish clearly between *stoichiometric* and *intimate* molecularity. How can these terms be applied using the S_N1–S_N2 classification?

11-11. Define the *trans* effect and cite some specific experimental evidence for its existence. Is it significant in complexes having octahedral geometry? Justify your answer.

11-12. In general, how can one distinguish between possible triangular bipyramidal and tetragonal pyramidal intermediates in substitution reactions involving octahedral complexes.

11-13. Summarize briefly mechanistic approaches to the racemization of species of octahedral geometry.

11-14. Summarize briefly H. Taube's concept of *outer-sphere* vs. *inner-sphere* mechanisms for electron-transfer reactions involving octahedral complexes.

11-15. Summarize experimental data and its interpretation for enhancing the yield of D-$[Co(en)_3]^{3+}$ ion via a redox procedure in the resolution of racemic $[Co(en)_3]^{3+}$ ion, as delineated in Reference 149.

REFERENCES

1. F. Basolo and R. G. Pearson, *Mechanisms of Inorganic Reactions*, 2nd ed., John Wiley & Sons, New York (1967).

2. J. P. Candlin, K. A. Taylor, and D. T. Thompson, *Reactions of Transition-Metal Complexes*, Elsevier, Amsterdam (1968).

3. Various authors, in *MTP International Review of Science, Inorganic Chemistry, Series One* (1972), *Series Two* (1974), Vol. 9, M. L. Tobe (Ed.), Butterworth, London. Reaction mechanisms.

4. R. S. Tobias, *J. Chem. Educ.*, **35**, 592 (1958).

5. F. J. C. Rossotti and H. Rossotti, *The Determination of Stability Constants*, McGraw-Hill Book Company, New York (1961).

6. L. G. Sillén and A. E. Martell, *Stability Constants of Metal-Ion Complexes*, Suppl. No. 1 to Spec. Publ. No. 17, *Spec. Publ.* No. 25, The Chemical Society, London (1971).

7. R. M. Smith and A. E. Martell, *Critical Stability Constants*, Vol. 1–4, Plenum Publishing Corp., New York (1974–1977).

8. J. Kragten, *Atlas of Metal–Ligand Equilibria in Aqueous Solution*, Halsted Press, New York (1978).

9. D. P. Mellor and L. E. Maley, *Nature*, **159**, 370 (1947); **161**, 436 (1948).

10. M. Calvin and N. C. Melchior, *J. Am. Chem. Soc.*, **70**, 3270 (1948).

11. H. Irving and R. J. P. Williams, *Nature*, **162**, 746 (1948).

11a. W. L. Waltz and R. G. Sutherland, *Chem. Soc. Rev.*, **1**, 241 (1972).

12. R. F. Gould (Ed.), *Mechanisms of Inorganic Reactions*, Vol. 49, Advances in Chemistry Series, American Chemical Society, Washington, D.C. (1965).

13. C. H. Langford and H. B. Gray, *Ligand Substitution Processes*, W. A. Benjamin, New York (1965).

14. H. B. Gray and C. H. Langford, *Chem. Eng. News*, Apr. 1, 1968, p. 68.

15. Various authors, in *Progress in Inorganic Chemistry*, J. O. Edwards (Ed.), Vol. 13, (1970); Vol. 17 (1972), Interscience, New York.

16. R. G. Wilkins, *The Study of Kinetics and Mechanisms of Reactions of Transition Metal Ions*, Allyn and Bacon, Boston (1974).

17. A. E. Martell, Ed., *Coordination Chemistry II*, Vol. 174, Advances in Chemistry Series, American Chemical Society, Washington, D.C. (1978).

18. H. Taube, *Chem. Rev.*, **50**, 69 (1952).

19. Various authors, *Z. Elektrochem.*, **64**, 47–89 (1960). International colloquium on rapid reactions in solution (pp. 1–204).

20. A. A. Frost and R. G. Pearson, *Kinetics and Mechanism: A Study of Homogeneous Chemical Reactions*, 2nd ed., John Wiley & Sons, New York (1961).

21. C. K. Ingold, *Structure and Mechanism in Organic Chemistry*, 2nd ed., Cornell University Press, Ithaca, N.Y. (1969).

22. F. Basolo and R. G. Pearson, Ref. 1, Ch. 3.

23. F. J. Garrick, *Nature*, **139**, 507 (1937).

24. F. Basolo and R. G. Pearson, Ref. 1, pp. 183–193.

25. A. Werner, *Z. Anorg. Chem.*, **3**, 267 (1893).

26. F. Basolo and R. G. Pearson, Ref. 1, Ch. 5.

27. C. H. Langford and H. B. Gray, Ref. 13, Ch. 2.

28. J. S. Coe, in Ref. 3, *Series Two*, pp. 45–62.

29. F. Basolo, in Ref. 12, pp. 81–94.

30. F. R. Hartley, *Chem. Soc. Rev.*, **2**, 163 (1973).

31. U. Belluco, L. Cattalini, F. Basolo, R. G. Pearson, and A. Turco, *J. Am. Chem. Soc.*, **87**, 241 (1965).

32. I. I. Chernyaev, *Ann. Inst. Platine USSR*, **4**, 243 (1926); **5**, 118 (1927).

33. I. I. Chernyaev, *Ann. Inst. Platine USSR*, **4**, 261 (1926).

34. G. B. Kauffman, *J. Chem. Educ.*, **54**, 86 (1977). A detailed and informative account of Chernyaev's life and contributions to coordination chemistry in general and the *trans* effect in particular.

35. V. A. Golovnya, T. N. Leonova, W. Craig, and G. B. Kauffman, *Ambix*, **23**, Pt. 3, 187 (1976). Chernyaev's work and personality.

36. Quoted from F. Basolo and R. G. Pearson, Ref. 1, p. 355. See Report of Conference Organizing Committee, *Izv. Sekt. Platiny Drugikh. Blagorodn. Met. Inst. Obshch. Neorg. Khim. Akad. Nauk SSSR*, **28**, 12 (1954); I. I. Chernyaev, *Zh. Neorg. Khim.*, **2**, 475 (1957).

37. F. Basolo and R. G. Pearson, Ref. 1, p. 356; in *Progress in Inorganic Chemistry*, Vol. 4, F. A. Cotton. (Ed.), pp. 381–453, Interscience Publishers, New York (1962).

38. A. A. Grinberg, *Ann. Inst. Platine USSR*, **5**, 109 (1927); *Acta Physicochim. USSR*, **3**, 573 (1935).

39. J. Chatt, L. A. Duncanson, and L. M. Venanzi, *J. Chem. Soc.*, **1955**, 4456.

40. L. E. Orgel, *J. Inorg. Nucl. Chem.*, **2**, 137 (1956).

41. A. A. Grinberg, *Russ. J. Inorg. Chem.*, **4**, 683 (1959).

42. M. A. Tucker, C. B. Colvin, and D. S. Martin, Jr., *Inorg. Chem.*, **3**, 1373 (1964).

43. C. H. Langford and H. B. Gray, Ref. 13, Ch. 3.

44. F. Basolo and R. G. Pearson, Ref. 1, Ch. 3.

45. F. Basolo, *Chem. Rev.*, **52**, 459 (1953).

46. C. H. Langford and V. S. Sastri, in Ref. 3, *Series One*, p. 203 (1972).

47. C. H. Langford and M. Parris, in *Comprehensive Chemical Kinetics*, (C. H. Bamford and C. F. H. Tipper (Eds.), Vol. 7, Elsevier, Amsterdam (1972).

48. M. L. Tobe, *Inorganic Reaction Mechanisms*, T. Nelson, London (1972).

49. T. P. Dasgupta, in Ref. 3, *Series Two*, pp. 63–91 (1974).

50. F. Basolo, B. D. Stone, J. G. Bergmann, and R. G. Pearson, *J. Am. Chem. Soc.*, **76**, 3079 (1954).

51. R. G. Pearson and J. W. Moore, *Inorg. Chem.*, **3**, 1334 (1964).

52. A. W. Adamson and R. G. Wilkins, *J. Am. Chem. Soc.*, **76**, 3379 (1954).

53. G. W. Ettle and C. H. Johnson, *J. Chem. Soc.*, **1939**, 1490.

54. A. Haim and W. K. Wilmarth, *Inorg. Chem.*, **1**, 573, 583 (1962).

55. F. Basolo, B. D. Stone, and R. G. Pearson, *J. Am. Chem. Soc.*, **75**, 819 (1953).

56. F. Basolo and R. G. Pearson, Ref. 1, pp. 257, 262.

57. R. G. Pearson and F. Basolo, *J. Am. Chem. Soc.*, **78**, 4878 (1956).

58. J. P. Jesson, in Ref. 3, *Series Two*, pp. 227–269.

59. S. M. Jørgensen, *J. Prakt. Chem.*, **39**, 16 (1889); **41**, 449 (1890).

60. A. Werner, *Chem. Ber.*, **44**, 3272 (1911).

61. A. Werner, *Chem. Ber.*, **44**, 2445 (1911).

62. J. Bjerrum and S. E. Rasmussen, *Acta Chem. Scand.*, **6**, 1265 (1952).

63. J. Y.-P. Tong and P. E. Yankwich, *J. Am. Chem. Soc.*, **80**, 2664 (1958).

64. D. F. Martin and M. L. Tobe, *J. Chem. Soc.*, **1962**, 1388.

65. M. Delepine, *Anales Real. Soc. Esp. Fis. Quim.* (*Madrid*), **27**, 485 (1929).

66. M. Delepine, *C. R. Acad. Sci.*, **175**, 1409 (1922).

67. R. E. Hamm, *J. Am. Chem. Soc.*, **75**, 609 (1953).

68. R. E. Hamm and R. H. Perkins, *J. Am. Chem. Soc.*, **77**, 2083 (1955).

69. J. Chatt and I. Leden, *J. Chem. Soc.*, **1955**, 2936.

70. J. Chatt and G. A. Gamlen, *J. Chem. Soc.*, **1956**, 2371.

71. H. D. K. Drew and H. H. Pratt, *J. Chem. Soc.*, **1937**, 506.

72. N. F. Curtis, *Proc. Chem. Soc.*, **1960**, 410.

73. R. T. Gillard and G. Wilkinson, *J. Chem. Soc.*, **1964**, 1640.

74. A. Werner, *Ann. Chem.*, **406**, 261 (1914).

75. R. E. Hamm, R. Kollrack, G. L. Welch, and R. H. Perkins, *J. Am. Chem. Soc.*, **83**, 340 (1961).

76. D. L. Welch and R. E. Hamm, *Inorg. Chem.*, **2**, 295 (1963).

77. F. Basolo and R. G. Pearson, Ref. 1, p. 284.

77a. G. K. Anderson and R. J. Cross, *Chem. Soc. Rev.*, **9**, 185 (1980). Detailed review of isomerization mechanisms of square-planar complexes.

78. J. Chatt and R. G. Wilkins, *J. Chem. Soc.*, **1952**, 273, 4300; **1953**, 70; **1956**, 525.

79. J. C. Bailar, Jr., and R. W. Auten, *J. Am. Chem. Soc.*, **56**, 774 (1934).

80. J. C. Bailar, Jr., F. G. Jonelis, and E. H. Huffman, *J. Am. Chem. Soc.*, **58**, 2224 (1936).

81. J. C. Bailar, Jr., and D. F. Peppard, *J. Am. Chem. Soc.*, **62**, 820 (1940).

82. F. P. Dwyer, A. M. Sargeson, and I. K. Reid, *J. Am. Chem. Soc.*, **85**, 1215 (1963).

83. L. J. Boucher, E. Kyuno, and J. C. Bailar, Jr., *J. Am. Chem. Soc.*, **86**, 3656 (1964).

84. F. Basolo and R. G. Pearson, Ref. 1, pp. 268–270.

85. J. C. Bailar, Jr., J. H. Haslam, and E. M. Jones, *J. Am. Chem. Soc.*, **58**, 2226 (1936).

86. E. Kyuno and J. C. Bailar, Jr., *J. Am. Chem. Soc.*, **88**, 1125 (1966).

87. E. Kyuno, L. J. Boucher, and J. C. Bailar, Jr., *J. Am. Chem. Soc.*, **87**, 4485 (1965).

88. F. Basolo and R. G. Pearson, Ref. 1, p. 272.

89. W. Kruse and H. Taube, *J. Am. Chem. Soc.*, **83**, 1280 (1961).

90. D. D. Brown and C. K. Ingold, *J. Chem. Soc.*, **1953**, 2680.

91. D. D. Brown and R. S. Nyholm, *J. Chem. Soc.*, **1953**, 2696.

92. A. Werner, *Ann. Chem.*, **386**, 1 (1912).

93. A. Werner, *Chem. Ber.*, **45**, 1228, 3061 (1912).

94. F. Basolo, J. C. Hayes, and H. M. Neumann, *J. Am. Chem. Soc.*, **76**, 3807 (1954).

95. F. A. Long, *J. Am. Chem. Soc.*, **61**, 570 (1939); **63**, 1353 (1941).

96. S. T. Spees and A. W. Adamson, *Inorg. Chem.*, **1**, 531 (1962).

97. P. C. Rây and N. K. Dutt, *J. Indian Chem. Soc.*, **20**, 81 (1943).

98. W. G. Gehman, Doctoral dissertation, Pennsylvania State University, State College, Pa. (1954).

99. L. Seiden, Doctoral Dissertation, Northwestern University, Evanston, Ill. (1957).

100. J. C. Bailar, Jr., *J. Inorg. Nucl. Chem.*, **8**, 165 (1958).

101. C. S. Springer, Jr., and R. E. Sievers, *Inorg. Chem.*, **6**, 852 (1967).

102. T. S. Piper, *J. Am. Chem. Soc.*, **83**, 3908 (1961).

103. A. Sault, F. Fry, and J. C. Bailar, Jr., *J. Inorg. Nucl. Chem.*, **42**, 201 (1980).

104. A. M. Sargeson and G. H. Searle, *Inorg. Chem.*, **6**, 787 (1967).

105. F. Basolo and R. G. Pearson, Ref. 1, pp. 313–327.

106. R. C. Fay and T. S. Piper, *Inorg. Chem.*, **3**, 348 (1964).

107. J. J. Fortman and R. E. Sievers, *Inorg. Chem.*, **6**, 2022 (1967).

108. A. Y. Girgis and R. C. Fay, *J. Am. Chem. Soc.*, **92**, 7061 (1970).

109. F. Basolo and R. G. Pearson, Ref. 1, Ch. 6.

110. J. P. Candlin, K. A. Taylor, and D. T. Thompson, Ref. 2, Ch. 3.

111. H. Taube, *Electron Transfer Reactions of Complex Ions in Solution*, Academic Press, New York (1970).

112. J. Halpern, *Q. Rev. (Lond.)*, **15**, 207 (1961).

113. A. G. Sykes, *Chem. Br.*, **6**, 159 (1970).

114. R. G. Linck, in *MTP International Review of Science, Inorganic Chemistry, Series Two: Reaction Mechanisms in Inorganic Chemistry*, M. L. Tobe (Ed.), pp. 173–225, Butterworth, London (1974).

115. M. L. Tobe, Ref. 48, Ch. 10.

116. D. E. Pennington, in *Coordination Chemistry II*, A. E. Martell (Ed.), American Chemical Society Monograph, Vol. 174, pp. 476–590, D. Van Nostrand, New York (1978).

116a. W. L. Reynolds and R. W. Lumry, *Mechanisms of Electron Transfer*, Ronald Press, New York (1966).

117. H. Taube, *Chem. Rev.*, **50**, 69 (1952).

118. H. Taube, in *Advances in Inorganic Chemistry and Radiochemistry*, H. J. Emeléus and A. G. Sharpe (Eds.), Vol. 1, pp. 1–53, Academic Press, New York (1959).

119. D. W. Margerum, G. R. Cayley, D. C. Weatherburn, and G. K. Pagenkopf, in *Coordination Chemistry II*, A. E. Martell (Ed.), American Chemical Society Monograph, Vol. 174, pp. 1–220, D. Van Nostrand, New York (1978).

120. J. Weiss, *Proc. R. Soc. Lond. Ser. A*, **222**, 128 (1954).

121. R. J. Marcus, B. J. Zwolinski, and H. Eyring, *J. Phys. Chem.*, **58**, 432 (1954); B. J. Zwolinski, R. J. Marcus, and H. Eyring, *Chem. Rev.*, **55**, 157 (1955).

122. H. Taube, H. Myers, and R. L. Rich, *J. Am. Chem. Soc.*, **75**, 4118 (1953).

123. H. Taube and H. Myers, *J. Am. Chem. Soc.*, **76**, 2103 (1954).

124. J. P. Candlin, J. Halpern, and D. L. Trimm, *J. Am. Chem. Soc.*, **86**, 1019 (1964).

125. A. M. Zwichel and H. Taube, *J. Am. Chem. Soc.*, **81**, 1288 (1959).

126. R. K. Murmann, H. Taube, and F. A. Posey, *J. Am. Chem. Soc.*, **79**, 262 (1957).

127. W. Kruse and H. Taube, *J. Am. Chem. Soc.*, **82**, 526 (1960).

128. J. H. Espenson, *Inorg. Chem.*, **4**, 121 (1965).

129. C. J. Shea and A. Haim, *J. Am. Chem. Soc.*, **93**, 3055 (1971).

130. C. J. Shea and A. Haim, *Inorg. Chem.*, **12**, 3013 (1973).

131. J. P. Candlin, J. Halpern, and S. Nakamura, *J. Am. Chem. Soc.*, **85**, 2517 (1963).

132. D. L. Ball and E. L. King, *J. Am. Chem. Soc.*, **80**, 1091 (1958).

133. R. Snellgrove and E. L. King, *Inorg. Chem.*, **3**, 288 (1964).

134. H. Taube and E. S. Gould, *Acc. Chem. Res.*, **2**, 321 (1969).

135. F. Basolo, A. F. Messing, P. H. Wilks, R. G. Wilkins, and R. G. Pearson, *J. Inorg. Nucl. Chem.*, **6**, 161 (1958); **8**, 203 (1958).

136. R. C. Johnson and F. Basolo, *J. Inorg. Nucl. Chem.*, **13**, 36 (1960).

137. F. Basolo, M. L. Morris, and R. G. Pearson, *Disc. Faraday Soc.*, **29**, 80 (1960).

138. J. H. Espenson, *Acc. Chem. Res.*, **3**, 347 (1970). Review.

139. J. P. Birk, *J. Am. Chem. Soc.*, **91**, 3189 (1969).

140. F. Basolo and R. G. Pearson, Ref. 1, pp. 511–515.

141. M. J. Aspray, D. R. Rosseinsky, and G. B. Shaw, *Chem. Ind. (Lond.)*, **1963**, 911.

142. M. G. Adamson, F. S. Dainton, and P. Glentworth, *Trans. Faraday Soc.*, **61**, 689 (1965).

143. A. W. Adamson and K. S. Vorres, *J. Inorg. Nucl. Chem.*, **3**, 206 (1956).

144. H. L. Friedman, J. P. Hunt, R. A. Plane, and H. Taube, *J. Am. Chem. Soc.*, **73**, 4028 (1951).

145. A. A. Vlček, *Disc. Faraday Soc.*, **26**, 164 (1958).

146. R. D. Cannon and J. E. Earley, *J. Am. Chem. Soc.*, **87**, 5264 (1965).

147. L. E. Orgel, *Proc. 10th Solvay Conf.*, p. 289, Brussels (1965).

148. S. Kirschner, in *Preparative Inorganic Reactions*, W. L. Jolly (Ed.), Vol. 1, Ch. 2, John Wiley & Sons, New York (1964).

149. D. H. Busch, *J. Am. Chem. Soc.*, **77**, 2747 (1955).

150. F. P. Dwyer and E. C. Gyarfas, *J. Proc. R. Soc., N.S. Wales*, **83**, 170, 174, 263 (1949).

151. W. W. Brandt, F. P. Dwyer, and E. C. Gyarfas, *Chem. Rev.*, **54**, 997 (1954).

152. C. S. Lee, E. M. Gorton, H. M. Neumann, and H. R. Hunt, Jr., *Inorg. Chem.*, **5**, 1397 (1966).

SUGGESTED SUPPLEMENTARY REFERENCES

F. Basolo and R. G. Pearson, *Mechanisms of Inorganic Reactions*, 2nd ed., John Wiley & Sons, New York (1967). Kinetics of reactions of coordination entitities and their mechanistic interpretations.

J. P. Candlin, K. A. Taylor, and D. T. Thompson, *Reactions of Transition Metal Complexes*, Elsevier, Amsterdam (1968).

C. H. Langford and H. B. Gray, *Ligand Substitution Processes*, W. A. Benjamin, New York (1965). Mechanistic interpretations of kinetic data for transition-metal complexes.

A. E. Martell (Ed.), *Coordination Chemistry*, Vol. 2, American Chemical Society Monograph 174, American Chemical Society, Washington, D.C. (1971). Substitution reactions, coordinated-ligand reactions, electron-transfer reactions.

A. E. Martell (Ed.), *Coordination Chemistry II*, Vol. 174, Advances in Chemistry Series, American Chemical Society, Washington, D.C. (1978). Kinetics, mechanisms, and oxidation–reduction reactions.

R. G. Pearson, *Symmetry Rules for Chemical Reactions*, John Wiley & Sons, New York (1976). Application of orbital symmetry rules of Woodward and Hoffman to prediction of reaction pathways.

R. G. Pearson, *J. Chem. Educ.*, **58**, 753 (1981). Orbital topology and reaction mechanisms.

H. Taube, *Electron Transfer Reactions of Complex Ions in Solution*, Academic Press, New York (1970). Classic interpretation of kinetic data in terms of outer-sphere vs. inner-sphere mechanisms.

M. L. Tobe (Ed.), *MTP International Review of Science, Inorganic Chemistry, Series One* (1972), *Series Two* (1974), Butterworth, London. Reaction mechanisms for variety of reactions, largely of coordination entities.

Various senior reporters, *Inorganic Reaction Mechanisms*, Royal Society of Chemistry, London. Literature surveys, e.g., Vol. 6 (1980), covering July 1977–Dec. 1980.

R. G. Wilkins, *A Study of the Kinetics and Mechanisms of the Reactions of Transition Metal Complexes*, Allyn and Bacon, Boston (1974). Extensive discussions of numerous systems.

APPENDIX I

USEFUL LITERATURE SOURCES FOR INORGANIC CHEMISTRY

An acquaintance with the scientific literature is an absolute requirement for study and/or research in inorganic chemistry. The universal post World War II renaissance of inorganic chemistry has been so extensive that it is impossible to present an exhaustive listing of literature sources. Those that follow are, in the author's opinion, adequate for overall purposes, sufficiently indicative of what has been accomplished, suggestive for more comprehensive investigation of specific areas and topics, and generally available. That most of those cited are in English is not meant to demean the importance of sources in other languages but is dictated more by the very extensive use of English in chemistry on a global basis.

For purposes of presentation, inorganic literature sources are classified under the headings of comprehensive reference works, advanced-level textbooks and reference volumes, inorganic laboratory practice, summary and review publications, and journals. In addition, some notes on the organization, content, and use of *L. Gmelin's Handbuch der Anorganischen Chemie* are included.

I-1. COMPREHENSIVE REFERENCE WORKS

1. H. J. Emeléus (Consulting Ed.), *MTP International Review of Science, Inorganic Chemistry, Series One*, Vol. 1: *H and Groups I–IV*; Vol. 2: *Groups V and VI*; Vol. 3: *Group VII and Noble Gases*; Vol. 4: *Organometallic Derivatives of Main-Group Elements*; Vols. 5 and 6: *Transition Metals*; Vol. 7: *Lanthanides and Actinides*; Vol. 8: *Radiochemistry*; Vol. 9: *Reaction Mechanisms*; Vol. 10: *Solid-State Chemistry*; Butterworth, London (1972). *Series Two*, same titles (1974).

2. *L. Gmelin's Handbuch der Anorganischen Chemie*, 8th ed., Verlag Chemie Gmbh, Weinheim/-Bergstrasse (1924–1973); Springer-Verlag, Berlin-New York (1974–   ).* In German except

* See special section on the organization of the *Gmelin Handbuch* in Sec. I-6.

for a limited number of recent volumes. Newer volumes have English tables of contents and marginal notations. Future volumes will be in English.

3. C. A. Hampel (Ed.), *The Encyclopedia of the Chemical Elements*, Reinhold Publishing Corp., New York (1968).

4. J. W. Mellor's *A Comprehensive Treatise on Inorganic and Theoretical Chemistry*, Longmans, Green, London (1922–1937). Supplementary volumes, Interscience Publishers, New York (1956–   ): Vol. 2 (Suppls. 1: *Halogens*, 1956; 2, 3: *Alkali Metals*, 1961, 1963); Vol. 8 (Suppls. 1, 2: *Nitrogen*, 1964, 1967; Suppl. 3: *Phosphorus*, 1971).

5. P. Pascal (Direçtor), *Nouveau traité de chimie minérale*, Masson et Cie, Paris (1956–1970). Encyclopedic summary articles by elements.

6. A. Standen (Executive Ed.), *Kirk-Othmer's Encyclopedia of Chemical Technology*, 2nd ed., Interscience Publishers, New York (1963–   ).

7. A. F. Trotman-Dickinson (Executive Ed.), *Comprehensive Inorganic Chemistry*, Vols. 1–5, Pergamon Press, Elmsford, N.Y. (1973). Some chapters as separate volumes.

8. A. F. Wells, *Structural Inorganic Chemistry*, 4th ed., Clarendon Press, Oxford (1975).

I-2. ADVANCED-LEVEL TEXTBOOKS AND REFERENCE WORKS

1. E. de Barry Barnett and C. L. Wilson, *Inorganic Chemistry for Advanced Students*, 2nd ed., John Wiley & Sons, New York (1957). Combination of fact and theory. Limited literature citations.

2. C. F. Bell and K. A. K. Lott, *Modern Approach to Inorganic Chemistry*, 3rd ed., Butterworth, London (1972). Combination of fact and theory. Very limited literature citations.

3. B. Chiswell and D. W. James, *Fundamental Aspects of Inorganic Chemistry*, John Wiley & Sons, New York (1969). Introductory theory and concept. No literature citations.

4. F. A. Cotton and G. Wilkinson, *Advanced Inorganic Chemistry*, 4th ed., John Wiley & Sons, New York (1980). Combination of fact and theory. Moderate citations of original literature, mostly in 1970s.

5. F. A. Cotton and G. Wilkinson, *Basic Inorganic Chemistry*, John Wiley & Sons, New York (1976). More elementary than number 4, with a substantially different order of presentation.

6. M. C. Day and J. Selbin, *Theoretical Inorganic Chemistry*, 2nd ed., Reinhold Publishing Corp., New York (1969). Completely theoretical and conceptual with limited citations of original literature.

7. G. C. Demitras, C. R. Russ, J. F. Salmon, J. H. Weber, and G. S. Weiss, *Inorganic Chemistry*, Prentice-Hall, Englewood Cliffs, N.J. (1972). Largely theoretical and conceptual with some factual information. Limited number of suggested readings.

8. B. E. Douglas and D. H. McDaniel, *Concepts and Models in Inorganic Chemistry*, Blaisdell, New York (1965). Largely conceptual. General literature citations.

9. P. J. Durrant and B. Durrant, *Introduction to Advanced Inorganic Chemistry*, 2nd ed., Longmans, Green, London (1970). Combination of many facts with essential theory. Limited literature citations.

10. H. J. Eméleus and A. G. Sharpe, *Modern Aspects of Inorganic Chemistry*, 4th ed., John Wiley & Sons, New York (1973). Detailed treatment of special topics with ample literature citations.

11. E. S. Gilreath, *Fundamental Concepts of Inorganic Chemistry*, McGraw-Hill Book Company, New York (1958). Conceptual. Limited literature citations.

12. E. S. Gould, *Inorganic Reactions and Structure*, rev. ed., Holt, Rinehart and Winston, New York (1962). Combination of fact and principle. Suggested readings.

13. K. B. Harvey and G. B. Porter, *Introduction to Physical Inorganic Chemistry*, Addison-Wesley, Reading, Mass. (1963). Principles. Bibliographies.

14. R. B. Heslop and K. Jones, *Inorganic Chemistry: A Guide to Advanced Study*, Elsevier, New York (1976). Combination of fact and theory. Further readings.

15. A. F. Holleman and E. Wiberg, *Handbuch der anorganischen Chemie*, W. de Gruyter, Berlin (1971). Combination of fact, history, and theory. Some literature citations.

16. J. E. Huheey, *Inorganic Chemistry: Principles of Structure and Reactivity*, 2nd ed., Harper & Row, New York (1978). Combination of fact and principle. Numerous citations of original literature.

17. W. L. Jolly, *The Principles of Inorganic Chemistry*, McGraw-Hill Book Company, New York (1976). Primarily principles. Limited literature citations.

18. J. Kleinberg, W. J. Argersinger, Jr., and E. Griswold, *Inorganic Chemistry*, D. C. Heath, Boston (1960). Combination of fact and theory. Limited literature citations.

19. J. J. Lagowski, *Modern Inorganic Chemistry*, Marcel Dekker, New York (1973). Combination of fact and theory. Numerous literature citations.

20. K. Laidler and M. H. Ford-Smith, *The Chemical Elements*, Bogden & Quigley, Tarrytown-on-Hudson, New York (1970). Combination of fact and theory. Limited readings.

21. W. M. Latimer and J. H. Hildebrand, *Reference Book of Inorganic Chemistry*, 3rd ed., Macmillan, New York (1951). Largely factual. No literature citations.

22. K. M. Mackay and R. A. Mackay, *Introduction to Modern Inorganic Chemistry*, Intertext Books, London (1969). Combination of fact and theory. Further readings.

23. E. B. Maxted, *Modern Advances in Inorganic Chemistry*, Clarendon Press, Oxford (1947). Special topics. Useful in historical perspective. Reasonable original literature citations.

24. T. Moeller, *Inorganic Chemistry*, John Wiley & Sons, New York (1952). Combination of fact and theory. Many literature citations.

25. G. D. Parkes, *Mellor's Modern Inorganic Chemistry*, rev., Longmans, Green, London (1951). Largely factual, with some theory and history. No literature citations.

26. J. R. Partington, *General and Inorganic Chemistry*, 2nd ed., Macmillan, London (1951). Largely factual, with some theory and history. No literature citations.

27. C. S. G. Phillips and R. J. P. Williams, *Inorganic Chemistry*, Vol. 1: *Principles and Non-Metals*; Vol. 2: *Metals*, Oxford University Press, London (1965, 1966). Primarily principles, with some factual material. Bibliographies.

28. K. F. Purcell and J. C. Kotz, *Inorganic Chemistry*, W. B. Saunders, Philadelphia (1977). Combination of principles and facts, with emphasis on former. Numerous literature citations.

29. K. F. Purcell and J. C. Kotz, *An Introduction to Inorganic Chemistry*, W. B. Saunders, Philadelphia (1980). Lower-level version of number 28 with more emphasis on factual chemistry. Literature citations.

30. H. Remy, *Treatise on Inorganic Chemistry* (translated by J. S. Anderson), J. Kleinberg (Ed.), Vol. 1: *Introduction and Main Periodic Groups*; Vol. 2: *Sub Periodic Groups and General Topics*, Elsevier, Amsterdam (1956). Largely factual. Summaries of significant references.

31. R. T. Sanderson, *Inorganic Chemistry*, Reinhold Publishing Corp., New York (1967). Combination of principles and facts in unique presentation. Supplementary general references.

32. N. V. Sidgwick, *The Chemical Elements and Their Compounds*, Vols. 1 and 2, Clarendon Press, Oxford (1950). Largely factual. Many literature citations.

33. M. J. Sienko and R. A. Plane, *Physical Inorganic Chemistry*, W. A. Benjamin, New York (1963). Only theory and principles. Supplementary references.

34. R. Steudel, *Chemistry of the Non-metals* (English ed. by F. C. Nachod and J. J. Zuckerman), Walter de Gruyter, Berlin (1977). Atomic structure and bonding followed by factual chemistry. No literature citations.

35. P. C. L. Thorne and E. R. Roberts, *Ephraim's Inorganic Chemistry*, 6th ed., Interscience Publishers, New York (1954). Largely factual, with some theory and history. Limited number of literature citations.

36. S. Y. Tyree, Jr., and K. Knox, *Textbook of Inorganic Chemistry*, Macmillan, New York (1961). Largely factual. Supplementary readings.

37. C. W. Wood and A. K. Holliday, *Inorganic Chemistry*, Butterworth, London (1960). Largely factual. No literature citations.

38. D. M. Yost and H. Russell, Jr., *Systematic Inorganic Chemistry*, Prentice-Hall, Englewood Cliffs, N.J. (1944). Combination of fact and theory as restricted to fifth and sixth group nonmetals. Many literature citations plus bibliography.

39. B. J. Aylett and B. C. Smith, *Problems in Inorganic Chemistry*, American Elsevier, New York (1965). Numerical and nonnumerical based upon types of elements. With answers.

40. W. E. Hatfield and R. A. Palmer, *Problems in Structural Inorganic Chemistry*, W. A. Benjamin, New York (1971). By area. Largely numerical. With answers.

I-3. INORGANIC LABORATORY PRACTICE

1. D. M. Adams and J. B. Raynor, *Advanced Practical Inorganic Chemistry*, John Wiley & Sons, New York (1965).

2. R. J. Angelici, *Synthesis and Technique in Inorganic Chemistry*, 2nd ed., W. B. Saunders, Philadelphia (1977).

3. G. Brauer (Ed.), *Handbook of Preparative Inorganic Chemistry*, Academic Press, New York (Vol. 1, 1963; Vol. 2, 1964).

4. R. E. Dodd and P. L. Robinson, *Experimental Inorganic Chemistry*, Elsevier, New York (1954).

5. W. E. Henderson and W. C. Fernelius, *A Course in Inorganic Preparations*, McGraw-Hill Book Company, New York (1935).

6. *Inorganic Syntheses*, McGraw-Hill Book Company, New York (Vols. 1–17), John Wiley & Sons, New York (Vols. 18–). Editors: H. S. Booth: Vol. 1 (1939); W. C. Fernelius: Vol. 2 (1946); L. F. Audrieth: Vol. 3 (1950); J. C. Bailar: Vol. 4 (1953); T. Moeller: Vol. 5 (1957); E. G. Rochow: Vol. 6 (1960); J. Kleinberg: Vol. 7 (1963); H. F. Holtzclaw: Vol. 8 (1966); S. Y. Tyree: Vol. 9 (1967); E. L. Muetterties: Vol. 10 (1967); W. L. Jolly: Vol. 11 (1968); R. W. Parry: Vol. 12 (1970); F. A. Cotton: Vol. 13 (1972); A. Wold and J. K. Ruff: Vol. 14 (1973); G. Parshall: Vol. 15 (1974); F. Basolo: Vol. 16 (1976); A. G. MacDiarmid: Vol. 17 (1977); B. E. Douglas: Vol. 18 (1978); D. F. Shriver: Vol. 19 (1979); D. H. Busch: Vol. 20 (1980); J. P. Fackler, Jr.: Vol. 21 (1982); S. L. Holt: Vol. 22 (198); S. Kirschner: Vol. 23 (198).

7. W. L. Jolly, *Synthetic Inorganic Chemistry*, Prentice-Hall, Englewood Cliffs, N.J. (1960).

8. W. L. Jolly, *The Synthesis and Characterization of Inorganic Compounds*, Prentice-Hall, Englewood Cliffs, N.J. (1970).

9. W. L. Jolly (Ed.), *Preparative Inorganic Reactions*, Interscience Publishers, New York (Vol. 1, 1964; Vol. 2, 1966; Vol. 3, 1966; Vol. 4, 1968; Vol. 5, 1968; Vols. 6, 7, 1971). Specific broad topics.

10. H. B. Jonassen and A. Weissberger (Eds.), *Techniques of Inorganic Chemistry*, John Wiley & Sons, New York (Vols. 1–3, 1963; Vol. 4, 1964; Vol. 5, 1965; Vol. 6, 1966; Vol. 7, 1968).

11. R. B. King, *Organometallic Syntheses*, Academic Press, New York (1969).

12. H. Lux, *Anorganisch-Chemische Experimentier-Kunst*, 2nd ed., J. A. Barth Verlag, Leipzig (1959).

13, W. G. Palmer, *Experimental Inorganic Chemistry*, Oxford University Press, London (1954).

14. G. G. Schlessinger, *Inorganic Laboratory Preparations*, Chemical Publishing Co., New York (1962).

15. D. F. Shriver, *The Manipulation of Air-Sensitive Compounds*, McGraw-Hill Book Company, New York (1969). Very useful reference on techniques.

16. H. F. Walton, *Inorganic Preparations*, Prentice-Hall, Englewood Cliffs, N.J. (1948).

17. H. Zimmer and K. Niedenzu (Eds.), *Annual Reports in Inorganic and General Syntheses*, Academic Press, New York (1972–   ).

I-4. SUMMARY AND REVIEW PUBLICATIONS

1. The Chemical Society (London), *Annual Reports on the Progress of Chemistry* (1904–   ); *Specialist Periodical Reports, Main-Group Elements* (1971–   ), *Transition Elements* (1970–   ).

2. F. A. Cotton, S. J. Lippard (Eds.), *Progress in Inorganic Chemistry*, Interscience Publishers, New York [Vol. 1, 1959; Vol. 2, 1960; Vol. 3, 1962; Vol. 4, 1962; Vol. 5, 1963; Vol. 6, 1964; Vol. 7, 1966; Vol. 8, 1967; Vol. 9, 1968; Vol. 10, 1968; Vol. 11, 1970; Vol. 12, 1970; Vol. 13 (J. O. Edwards, Ed.), 1970; Vol. 14, 1971; Vol. 15, 1972; Vol. 16, 1972; Vol. 17 (J. O. Edwards, Ed.), 1972; Vol. 18, 1973; Vol. 19, 1975; Vol. 20, 1976; Vol. 21, 1976; Vol. 22, 1977; Vol. 23, 1977; Vol. 24, 1978; Vol. 25, 1979; Vol. 26, 1979; Vol. 27, 1980; Vol. 28, 1981; Vol. 29, 1982].

3. H. J. Emeléus and A. G. Sharpe (Eds.), *Advances in Inorganic Chemistry and Radiochemistry*, Academic Press, New York (Vol. 1, 1959; Vol. 2, 1960; Vol. 3, 1961; Vol. 4, 1962; Vol. 5, 1963; Vol. 6, 1964; Vol. 7, 1965; Vol. 8, 1966; Vol. 9, 1966; Vol. 10, 1967; Vol. 11, 1968; Vol. 12, 1970; Vol. 13, 1970; Vol. 14, 1972; Vol. 15, 1972; Vol. 16, 1973; Vol. 17, 1974; Vol. 18, 1976; Vol. 19, 1976; Vol. 20, 1977; Vol. 21, 1978; Vol. 22, 1979; Vol. 23, 1980; Vol. 24, 1981; Vol. 25, 1982).

4. A. F. Scott (Ed.), *Survey of Progress in Chemistry*, Academic Press, New York (Vol. 1, 1963; Vol. 2, 1964; Vol. 3, 1966; Vol. 4, 1968; Vol. 5, 1969; Vol. 6, 1975; Vol. 7, 1976; Vol. 8, 1978; Vol. 9, 1980).

5. E. J. Griffith and M. Grayson (Eds.), *Topics in Phosphorus Chemistry*, John Wiley & Sons, New York (Vol. 1, 1964; Vol. 2, 1965; Vol. 3, 1966; Vol. 4, 1967; Vol. 5, 1967; Vol. 6, 1969; Vol. 7, 1972; Vol. 8, 1976; Vol. 9, 1977; Vol. 10, 1980).

6. H. D. B. Jenkins and D. F. C. Morris (Eds.), *Reviews in Inorganic Chemistry*, John Wiley & Sons, New York (Vol. 1, Jan.–Dec. 1979; Vol. 2, Jan.–Dec. 1980).

7. M. Stacey, J. C. Tatlow, A. G. Sharpe, R. D. Peacock, H. Hyman (Eds.), *Advances in Fluorine Chemistry*, Butterworth, London (Vol. 1, 1960; Vol. 2, 1961; Vol. 3, 1963; Vol. 4, 1965; Vol. 6, 1970; Vol. 7, 1973).

8. F. G. A. Stone and R. West (Eds.), *Advances in Organometallic Chemistry*, Academic Press, New York (Vol. 1, 1964; Vol. 2, 1964; Vol. 3, 1965; Vol. 4, 1966; Vol. 5, 1967; Vol. 6, 1968; Vol. 7, 1968; Vol. 8, 1970; Vol. 9, 1970; Vol. 10, 1972; Vol. 11, 1973; Vol. 12, 1974; Vol. 13, 1975; Vol. 14, 1976; Vol. 15, 1977; Vol. 16, 1977; Vol. 17, 1979; Vol. 18, 1980).

9. Various editors, *Structure and Bonding*, Springer-Verlag, New York (1966–   ). Specific authorative articles on many aspects of structure and bonding.

I-5. Journals

1. Abstracting

   Chemical Abstracts (C.A.) (1907–   ).

2. Original Research

   Acta Chimica Scandinavica (Acta Chem. Scand.) (1947–   ).

   Acta Crystallographica (Acta Crystallogr.) (1947–   ).

   Angewandte Chemie (Angew. Chem.) (1888–   ).

   Angewandte Chemie, International Edition in English (Angew. Chem., Int. Ed. Engl.) (1961–   ).

   Bioinorganic Chemistry (Bioinorg. Chem.) (1971–   ).

Chemische Berichte (Chem. Ber. or Ber.) (1868–   ).

Helvetica Chimica Acta (Helv. Chim. Acta) (1918–   ).

Inorganic Chemistry (Inorg. Chem.) (1962–   ).

Inorganica Chimica Acta (Inorg. Chim. Acta) (1967–   ).

Inorganic and Nuclear Chemistry Letters (Inorg. Nucl. Chem. Lett.) (1965–1981).

Journal of the American Chemical Society (J. Am. Chem. Soc.) (1876–   ).

Journal of Chemical Physics (J. Chem. Phys.) (1933–   ).

Journal of the Chemical Society, London (J. Chem. Soc.) (1841–   ).

Journal of the Chemical Society, London, Chemical Communications (J. Chem. Soc. Chem. Commun.) (1965–   ).

Journal of Coordination Chemistry (J. Coord. Chem.) (1971–   ).

Journal of Crystal and Molecular Structure (J. Cryst. Mol. Struct.) (1973–   ).

Journal of Inorganic and Nuclear Chemistry (J. Inorg. Nucl. Chem.) (1955–1981).

Journal of the Less Common Metals (J. Less Common Met.) (1959–   ).

Journal of Organometallic Chemistry (J. Organomet. Chem.) (1963–   ).

Journal of Solid State Chemistry (J. Solid State Chem.) (1969–   ).

Organometallic Chemical Syntheses (Organomet. Chem. Synth.) (1965–   ).

Organometallics (Organomet.) (1982–   ).

Polyhedron (Polyhed.) (1982–   ).

Zeitschrift für anorganische und allgemeine Chemie (Z. Anorg. Allg. Chem. or *Z. Anorg. Chem.)* (1892–   ).

Zhurnal Neorganicheskoi Khimii (Zh. Neorg. Khim. or translated as *Russ. J. Inorg. Chem.)* (1956–   ).

3. Review

Accounts of Chemical Research (Acc. Chem. Res.) (1968–   ).

Actinides Review (Actinides Rev.) (1967–   ).

Angewandte Chemie (Angew. Chem.) (1888–   ).

Angewandte Chemie, International Edition in English (Angew. Chem., Int. Ed. Engl.) (1961–   ).

Annual Reports of the Chemical Society of London (Ann. Rep.) (1904–   ).

Chemical Reviews (Chem. Rev.) (1924–   ).

Coordination Chemistry Reviews (Coord. Chem. Rev.) (1966–   ).

Inorganic Macromolecules Review (Inorg. Macromol. Rev.) (1971–   ).

Journal of Chemical Education (J. Chem. Educ.) (1924–   ).

Organometallic Reviews (Organometallic Rev.) (1966–   ).

Quarterly Reviews of the Chemical Society [*Quart. Rev. (Lond.)*] (1947–1971); *Chemical Society Reviews (Chem. Soc. Rev.)* (1972–   ).

Revue de Chimie Minérale (Rev. Chim. Min.) (1964–   ).

Russian Chemical Reviews (Russ. Chem. Rev., in English) (1960–   ).

I-6.   L.GMELIN'S HANDBUCH DER ANORGANISCHEN CHEMIE— NOTES ON ORGANIZATION, CONTENT, USE

In 1817, Leopold Gmelin, professor of chemistry at Heidelberg University, published the first edition of his *Handbuch der theoretischen Chemie*, a three-volume compilation dealing with theoretical chemistry, "inorganic compounds of the weighable nonmetallic materials," the theory of inorganic compounds of metals, and the description of organic compounds. This handbook was designed "for the purpose of his lectures" and organized very much as is the current eighth edition. Gmelin produced four more editions, the last in 1852, the year prior to his death. Organic substances last appeared in the fourth edition, and thereafter appeared in Konrad Friedrich Beilstein's *Handbuch der organischen Chemie*. The sixth edition (1872) was prepared under the direction of Karl Kraut, and the seventh edition, compiled under the supervision of Franz Peters, was begun in 1905. In 1922, the Deutsche Chemische Gesellschaft (German Chemical Society) established a permanent group under the direction of R. J. Meyer to continue the project. The initial volume of the current, and continuing, 8th edition appeared in 1924. For a number of years, compilation of volumes of the Handbuch has been handled by the Gmelin Institut für Anorganishe Chemie der Max Planck Gesellschaft zur Förderung der Wissenschaften, located in Frankfurt (Main), West Germany.

I-6-1. Organization

To each element, or series of closely related elements such as the lanthanides and transuranium elements, and ammonium, there is assigned a System Number (System Nummer) as noted in Fig. I-1 and Table I-1. This provides a classification "of the last digit" under which the component with the highest running number determines the location of the description of a compound in question. Thus to locate NaCl, one notes that Cl is assigned System Number 6 and Na is assigned System Number 21 and thus concludes that this compound is described in System Number 21. Other examples are provided by the arrows in and the footnote to Table I-1.

I-6-2. Content

For each volume, an attempt is made to summarize and to cite all pertinent literature up through a cutoff date indicated at the beginning of the volume. Each volume provides the most nearly complete of all summations of information available about each element or compound through the indicated date. Both syntheses and properties are strongly emphasized. Each volume contains many tabulations and figures.

I-6-3. Use

As a source of information about any specific inorganic substance and as a starting point for research in any area of inorganic chemistry, the *Gmelin Handbuch* has no equal. Use is facilitated in later volumes by tables of contents and marginal indices in English.

For further information, consult available references.[1-4]

1. M. Becke-Goehring, *ACHEMA Jahrb.*, **1965/1967**, 300–305.

2. M. Becke and K.-C. Buschbeck, *CZ-Chem. Tech.*, **1**, 551 (1972).

3. W. Lippert, *Chem. Br.*, **8**, 483 (1972).

4. M. Becke-Goehring, *J. Chem. Educ.*, **50**, 406 (1973).

Figure I-1 Periodic table with Gmelin system numbers (in boldface type). Values in parentheses are mass numbers of most stable or best known isotopes.

Each cell is given as: **Gmelin system number** / atomic number / symbol / atomic weight.

IA	IIA	IIIB	IVB	VB	VIB	VIIB	VIIIB			IB	IIB	IIIA	IVA	VA	VIA	VIIA	O or VIIIA
2 1 H 1.008																	**1** 2 He 4.003
20 3 Li 6.941	**26** 4 Be 9.012											**13** 5 B 10.81	**14** 6 C 12.01	**4** 7 N 14.01	**3** 8 O 16.00	**5** 9 F 19.00	**1** 10 Ne 20.18
21 11 Na 22.99	**27** 12 Mg 24.31											**35** 13 Al 26.98	**15** 14 Si 28.09	**16** 15 P 30.97	**9** 16 S 32.06	**6** 17 Cl 35.45	**1** 18 Ar 39.95
22 19 K 39.10	**28** 20 Ca 40.08	**39** 21 Sc 44.96	**41** 22 Ti 47.90	**48** 23 V 50.94	**52** 24 Cr 52.00	**56** 25 Mn 54.94	**59** 26 Fe 55.85	**58** 27 Co 58.93	**57** 28 Ni 58.71	**60** 29 Cu 63.55	**32** 30 Zn 65.38	**36** 31 Ga 69.72	**45** 32 Ge 72.59	**17** 33 As 74.92	**10** 34 Se 78.96	**7** 35 Br 79.90	**1** 36 Kr 83.80
24 37 Rb 85.47	**29** 38 Sr 87.62	**39** 39 Y 88.91	**42** 40 Zr 91.22	**49** 41 Nb 92.91	**53** 42 Mo 95.94	**69** 43 Tc (99)	**63** 44 Ru 101.1	**64** 45 Rh 102.9	**65** 46 Pd 106.4	**61** 47 Ag 107.9	**33** 48 Cd 112.4	**37** 49 In 114.8	**46** 50 Sn 118.7	**18** 51 Sb 121.8	**11** 52 Te 127.6	**8** 53 I 126.9	**1** 54 Xe 131.3
25 55 Cs 132.9	**30** 56 Ba 137.3	**39**** 57 La 138.9	**43** 72 Hf 178.5	**50** 73 Ta 180.9	**54** 74 W 183.9	**70** 75 Re 186.2	**66** 76 Os 190.2	**67** 77 Ir 192.2	**68** 78 Pt 195.1	**62** 79 Au 197.0	**34** 80 Hg 200.6	**38** 81 Tl 204.4	**47** 82 Pb 207.2	**19** 83 Bi 209.0	**12** 84 Po (210)	**85** 85 At (210)	**1** 86 Rn (222)
87 Fr (223)	**31** 88 Ra 226.0	**40***** 89 Ac (227)															

Lanthanides (**):

39 58 Ce 140.1	**39** 59 Pr 140.9	**39** 60 Nd 144.2	**39** 61 Pm (147)	**39** 62 Sm 150.4	**39** 63 Eu 152.0	**39** 64 Gd 157.3	**39** 65 Tb 158.9	**39** 66 Dy 162.5	**39** 67 Ho 164.9	**39** 68 Er 167.3	**39** 69 Tm 168.9	**39** 70 Yb 173.0	**71** 71 Lu 175.0

Actinides (***):

44 90 Th 232.0	**51** 91 Pa 231.0	**55** 92 U 238.0	**71** 93 Np (237)	**71** 94 Pu (242)	**71** 95 Am (243)	**71** 96 Cm (247)	**71** 97 Bk (249)	**71** 98 Cf (251)	**71** 99 Es (254)	**71** 100 Fm (253)	**71** 101 Md (256)	**71** 102 No (254)	**71** 103 Lr (257)

* **23** NH₄

Table I-1   System Numbers of Volumes in Gmelin's Handbuch der Anorganischen Chemie[4]

System Number	Symbol	Element			System Number	Symbol	Element
1	He,Ne,Ar,	Noble gases			37	In	Indium
	Kr,Xe,Rn				38	Tl	Thallium
2	H	Hydrogen			39	Sc	Scandium
3	O	Oxygen				Y	Yttrium
4	N	Nitrogen				La	Lanthanum
5	F	Fluorine				Ce-Lu	Lanthanides
6	Cl	Chlorine			40	Ac	Actinium
7	Br	Bromine			41	Ti	Titanium
8	I	Iodine			42	Zr	Zirconium
	At	Astatine			43	Hf	Hafnium
9	S	Sulfur			44	Th	Thorium
10	Se	Selenium			45	Ge	Germanium
11	Te	Tellurium			46	Sn	Tin
12	Po	Polonium			47	Pb	Lead
13	B	Boron			48	V	Vanadium
14	C	Carbon			49	Nb	Niobium
15	Si	Silicon			50	Ta	Tantalum
16	P	Phosphorus			51	Pa	Protactinium
17	As	Arsenic			52	Cr	Chromium
18	Sb	Antimony			53	Mo	Molybdenum
19	Bi	Bismuth			54	W	Tungsten

HCl

$CrCl_2$

$ZnCrO_4$

Table I-1 (*Continued*)

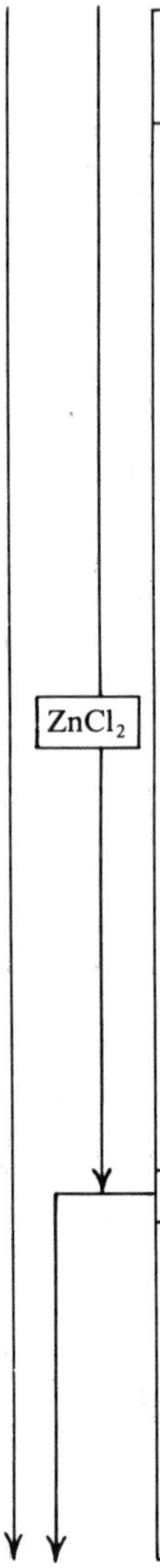

System Number	Symbol	Element		System Number	Symbol	Element
20	Li	Lithium		55	U	Uranium
21	Na	Sodium		56	Mn	Manganese
22	K	Potassium		57	Ni	Nickel
23	NH_4	Ammonium		58	Co	Cobalt
24	Rb	Rubidium		59	Fe	Iron
25	Cs	Cesium		60	Cu	Copper
	Fr	Francium		61	Ag	Silver
26	Be	Beryllium		62	Au	Gold
27	Mg	Magnesium		63	Ru	Ruthenium
28	Ca	Calcium		64	Rh	Rhodium
29	Sr	Strontium		65	Pd	Palladium
30	Ba	Barium		66	Os	Osmium
31	Ra	Radium		67	Ir	Iridium
32	Zn	Zinc		68	Pt	Platinum
33	Cd	Cadmium		69	Tc	Technetium [*a]
34	Hg	Mercury		70	Re	Rhenium
35	Al	Aluminum		71	Np,Pu,	Transuranium elements [*b]
36	Ga	Gallium				

$ZnCl_2$

"The material under each System Number contains all information on the element itself as well as on all compounds with other elements which precede this element in the Gmelin System. For example, zinc (System Number 32) as well as all zinc compounds with elements numbered from 1 through 31 are classified under number 32."

[*a] Published in 1941 under the title Masurium.

[*b] Revision to appear in the framework of supplement series to the 8th Edition.

(Reproduced, with English equivalents, from inside back cover of *Gmelin Handbuch der Anorganischen Chemie*, Springer-Verlag, Heidelberg (1980).

APPENDIX II

NOMENCLATURE OF INORGANIC SUBSTANCES

Except for applications to certain coordination compounds and ions (Sec. 7.2), inorganic nomenclature remained essentially unsystematic until publication of the so-called *1940 Rules*, a tentative guide prepared by the Commission on the Nomenclature of Inorganic Chemistry of the International Union of Pure and Applied Chemistry (IUPAC). After extensive discussion of these rules, the *Tentative Rules for Inorganic Nomenclature* were published after the 1953 meeting of IUPAC in Stockholm. Further considerations led to the first edition of definitive rules for inorganic nomenclature, the *1957 Rules*.[1] These were revised as the *1970 Rules* (the so-called *Red Book*)[2] and supplemented in 1977 by a guide to their usage.[3] Interpretations and additional suggestions for usage have appeared in a number of articles in the *Journal of Chemical Education*.[4] Of the sources cited, References 3 and 4 are particularly useful interpretations.

It is impossible to treat in detail within the space available all the definitive rules. Certain of the more important and generally useful ones are discussed, together with suitable examples. For classical coordination entities, reference should be made to the detailed discussion in Chapter 7. It must not be concluded, however, that the available definitive rules can cover all cases of nomenclature. Although there is much organization, there is still some chaos.[5,6]

II-1. GENERAL PRACTICE AND RULES

II-1-1. Oxidation State or Number

The oxidation state of a specific element in an ion or molecule may be indicated by a Roman numeral or an Arabic 0 (zero) placed in parentheses after the name of the element (Stock System). This is normally necessary only if a given element can exist in more than one oxidation state.

Examples

$FeCl_2$ iron(II) chloride
$Fe_2(SO_4)_3$ iron(III) sulfate
PF_6^- hexafluorophosphate(V) ion
$Ni(CN)_4^{4-}$ tetracyanonickelate(0) ion
MoO_4^{2-} tetraoxomolybdate(VI) ion

Alternatively, the charge carried by an ion may be indicated by an Arabic number followed by the sign of the ionic charge and placed in parentheses immediately following the name of the ion (Ewens–Bassett system).

Examples

$FeCl_2$ iron(2+) chloride
$Fe_2(SO_4)_3$ iron(3+) sulfate
PF_6^- hexafluorophosphate(1−) ion
$Ni(CN)_4^{4-}$ tetracyanonickelate(4−) ion
MoO_4^{2-} tetraoxomolybdate(2−) ion

II-1-2. Multiplying Affixes,* Enclosing Marks, Numbers, Italicized Letters

Multiplying affixes (i.e., Greek mono, di, tri, tetra, penta, hexa, hepta, octa, nona, deca, ..., poly) may indicate:†

1. Stoichiometric proportions

    Examples SO_2 sulfur *di*oxide
                         SO_3 sulfur *tri*oxide
                         S_4N_4 *tetra*sulfur *tetra*nitride

2. Extent of substitution

    Examples $SiBr_2Cl_2$ *di*bromo*di*chlorosilane
                         $PS_2O_2^{3-}$ *di*thiophosphate ion
                         CS_3^{2-} *tri*thiocarbonate ion

3. Number of identical coordinating groups

    Examples $[Co(NH_3)_3(NO_2)_3]$ *tri*ammine*tri*nitrocobalt(III)
                         $[Cr(H_2O)_4Cl_2]^+$ *tetra*qua*di*chlorochromium(III) ion

4. Number of identical central atoms in a condensed or isopoly acid or anion

    Examples H_3PO_4 *mono*phosphoric acid
                         $H_5P_3O_{10}$ *tri*phosphoric acid

5. Number of atoms of the same element in the skeleton of a molecule or ion

---

*   Important structural affixes are summarized in Table II-1.

†   Italics used in this section only for emphasis.

Examples Ge_2H_6 *di*germane
          $B_{10}H_{14}$ *deca*borane(14)
          $S_5O_6^{2-}$ *penta*thionate ion
          $Mn_2(CO)_{10}$ dimanganese *deca*carbonyl

Multiplying affixes such as bis, tris, tetrakis are normally used for complexes to indicate numbers of multidentate groups.

Examples $[Co(en)_2Cl_2]^+$ dichloro*bis*(ethylenediamine)cobalt(III) ion
          $[Cr(C_2O_4)_3]^{3-}$ *tris*(oxalato)chromate(III) ion

Table II-1 Structural Affixes or Prefixes

Affix or Prefix[a]	Meaning
antiprismo	Eight atoms bonded to central atom in rectangular antiprismatic geometry (Archimedean antiprism)
asym	Asymmetrical
catena	Chain arrangement
cis	Two groups adjacent to each other
closo	Caged, or closed, structure; in particular, a polyhedral boron cage with triangular faces
cyclo	Ring structure
dodecahedro	Eight atoms arranged at apices of a dodecahedron with triangular faces
fac	Three groups at corners of same face of an octahedron
hexahedro	Eight atoms arranged at apices of a hexahedron, or cube
hexaprismo	Twelve atoms arranged at apices of a hexagonal prism
icosahedro	Twelve atoms arranged at apices of a triangular icosahedron
mer	Meridional arrangement, as three groups on an octahedron such that one is *cis* to two others that are themselves *trans* to each other
nido	Nestlike, or open, structure; in particular, a polyhedral boron arrangement with an open side
octahedro	Six atoms arranged at apices of an octahedron
pentaprismo	Ten atoms arranged at apices of a pentagonal prism
quadro	Four atoms arranged at corners of a square
sym	Symmetrical
tetrahedro	Four atoms arranged at apices of a tetrahedron
trans	Two groups directly across from each other
triangulo	Three atoms at corners of a triangle
triprismo	Six atoms at apices of a triangular prism
η (eta or hapto)	Two or more contiguous atoms of a ligand bonded to a metal atom or ion
μ (mu)	Group so designated bridging two or more centers of coordination
σ (sigma)	One atom of a ligand bonded to a metal atom or ion

[a] Printed in italics and separated from remainder of name by a hyphen.

Enclosing marks are included in formulas to identify sets of identical groups. They are invariably used following the bis, tris, ... affixes. Square brackets are customarily used to enclose coordination entities (Ch. 7). The normal "nesting" order is $\{[(\cdot)]\}$ where more than one enclosure is necessary. Examples appear above.

Numbers, aside from those used to identify oxidation states, are uniformly Arabic and locate specific atoms in a skeletal array at which substitution, replacement, or addition occurs.

Examples $H_3Ge\text{-}GeH_2\text{-}GeHBr\text{-}GeH_4$ 2-bromotetragermane

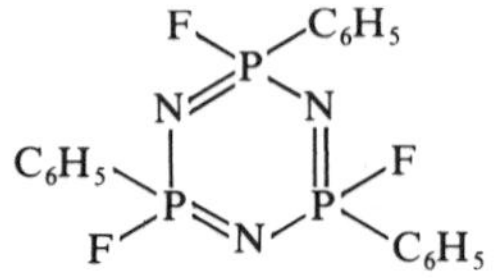

1,3,5-trifluoro-
1,3,5-triphenyl-
cyclotriphosphazene

Italicized capital letters are used to symbolize atoms of elements in formulas of compounds at which there is

1. Substitution

   Examples

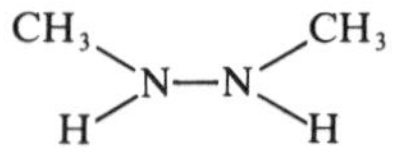

   N,N′-dimethylhydrazine

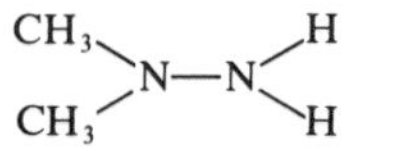

   N,N-dimethylhydrazine

2. Coordination to a metal atom or ion

   Examples $[Hg(SCN)_4]^{2-}$ *S*-tetrathiocyanatomercurate(II) ion
   $[Zn(NCS)_4]^{2-}$ *N*-tetrathiocyanatozincate(II) ion

3. Bridging between two metal centers in a coordination entity

   Example $[Be_4O(CH_3OO)_6]$ hexa-μ-acetato-(O,O')-μ_4-oxo-tetraberyllium(II)

4. Need to identify a specific isotope in an isotopically labeled species

   Example $^{15}NH_3$ ammonia$[^{15}N]$

Italicized small letters are sometimes used to identify spatial substitution positions

Example

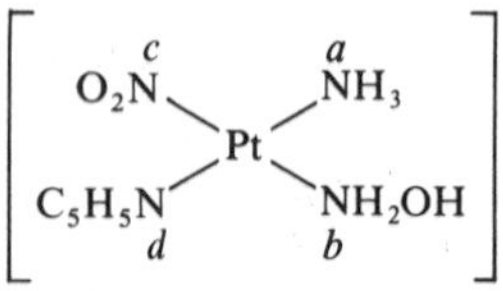

a-ammine-*b*-(hydroxylamine)-*c*-nitro-*d*(pyridine) platinum(1+) ion

II-1-3. Suffixes

Distinction between oxidation states in cations by use of the suffixes *ous* and *ic* is sufficiently nondefinitive that it is now seldom used. The *-ous, -ite* and *-ic, -ate* suffixes to distinguish among oxo acids and their salts are useful but nondefinitive as to oxidation state of the central element. These usages are exemplified by the following examples.

Examples	$\overset{I}{HOCl}$	*hypo*chlor*ous* acid	$\overset{I}{HOPH_2O}$	*hypo*phosphor*ous* acid
	ClO^-	*hypo*chlor*ite* ion	$H_2PO_2^-$	*hypo*phosph*ite* ion
	$\overset{III}{HOClO}$	chlor*ous* acid	$\overset{III}{(HO)_2PHO}$	phosphor*ous* acid
	$OClO^-$	chlor*ite* ion	HPO_3^{2-}	phosph*ite* ion
	$\overset{V}{HOClO_2}$	chlor*ic* acid	$\overset{V}{(HO)_3PO}$	phosphor*ic* acid
	$OClO_2^-$	chlor*ate* ion	PO_4^{3-}	phosph*ate* ion
	$\overset{VII}{HOClO_3}$	*per*chloric acid	$\overset{V}{(HO)_2(HOO)PO}$	*per*phosphoric acid*
	$OClO_3^-$	*per*chlorate ion	$(O_2)(OO)PO^{3-}$	*per*phosphate ion*

Except for the common acids, for which this type of nomenclature is usually adequate, more definitive names based upon IUPAC suggestions are better.

Examples	$H_3[Fe(CN)_6]$	hexacyanoferric(III) acid
	$[Fe(CN)_6]^{4-}$	hexacyanoferrate(II) ion
	$H[PF_6]$	hexafluorophosphoric(V) acid
	$[PF_6]^-$	hexafluorophosphate($1-$) ion

The suffixes *ic* and *ate* are thus used irrespective of the oxidation state of the central element, the latter being indicated by either a Stock or Ewens–Bassett number.

Preferably, the prefix *per* indicates more oxygen atoms bonded to a central atom, as with perchlorate vs. chlorate. Although it is still used to some extent to describe species containing peroxo groups [e.g., $HO(HOO)SO_2$ or H_2SO_5, $(HO)_2S(O)(O—O)S(O)(OH)_2$ or $H_2S_2O_8$], it is more definitive to use for these the peroxo terms, as, for example, H_2SO_5—peroxomonosulfuric acid, $H_2S_2O_8$—peroxodisulfuric acid.

II-2. IONS AND RADICALS

II-2-1. Cations

Monatomic cations are best named as the corresponding element without a suffix but with additions of the oxidation number in the Stock fashion.

Examples Sn^{2+} tin(II) ion
             Sn^{4+} tin(IV) ion
             Ce^{3+} cerium(III) ion
             Ce^{4+} cerium(IV) ion

Polyatomic cations may carry specific names, such as NO^+ (nitrosyl or nitrosonium ion), NO_2^+ (nitryl or nitronium ion) (see below). If they contain protons, the suffix -*onium* is commonly used with the stem name of the central element.

Examples PH_4^+ phosphonium ion
             SH_3^+ sulfonium ion

However, exceptions are common: for example, H_3O^+—hydronium, NH_3OH^+—hydroxylammonium, NH_4^+—ammonium. Protonated ions from amines are treated similarly but using amine stems.

Examples $N_2H_5^+$ hydrazinium ion
             $C_5H_6N^+$ pyridinium ion

Cationic (and sometimes neutral) radicals derived from oxygen or sulfur and another element are given provisional special names.

Examples   HO   hydroxyl
             CO   carbonyl
             PO   phosphoryl
             PS   thiophosphoryl
             ClO_2 chloryl
             ClO_3 perchloryl   (other halogens comparably)
             SO   sulfinyl (or thionyl)
             SO_2 sulfonyl (or sulfuryl)
             S_2O_5 Disulfuryl
             SeO   seleninyl
             SeO_2 selenonyl
             CrO_2 chromyl   (other members of Cr family similarly)
             UO_2 uranyl   (other actinides similarly)

II-2-2. Anions

Monoatomic anions are named by adding to the stem name of the element the suffix -*ide*.

Examples

H^-	hydride ion	O^{2-}	oxide ion
D^-	deuteride ion	Te^{2-}	telluride ion
F^-	fluoride ion	N^{3-}	nitride ion
I^-	iodide ion	C^{4-}	carbide ion
		B^{3-}	boride ion

The names of certain polyatomic anions also end in *-ide*.

Examples

HO^-	hydroxide ion	N_3^-	azide ion
HO_2^-	hydrogen peroxide ion	NH^{2-}	imide ion
O_2^{2-}	peroxide ion	NH_2^-	amide ion
O_2^-	hyperoxide ion	$NHOH^-$	hydroxylamide ion
HS^-	hydrogen sulfide ion	$N_2H_3^-$	hydrazide ion
S_2^{2-}	disulfide ion	CN^-	cyanide ion
S_4^{2-}	tetrasulfide ion	C_2^{2-}	acetylide ion
HF_2^-	hydrogendifluoride ion		
I_3^-	triiodide ion		

Other polyatomic anions are best named by adding *-ate* to the stem name of the central element and indicating both the number of bonded groups and the oxidation state of the central element as noted under "Suffixes." Oxo anions will without doubt continue to be named in the trivial system based upon the names of the parent acids and, for a given central element in more than one oxidation state, by use of suitable prefixes and suffixes as indicated under "Suffixes."

Anions derived from oxo acids by formal substitution of halide ions for hydroxyl groups or by substitution of nitrogen-system groups (Sec. 10.1-3) for oxygen-system groups are named systematically in terms of the oxo parents.

Examples

HSO_3F	fluorosulfuric acid	SO_3F^-	fluorosulfate ion
HSO_3Cl	chlorosulfuric acid	SO_3Cl^-	chlorosulfate ion
HSO_3NH_2	amidosulfuric acid	$SO_3NH_2^-$	amidosulfate ion
$(HSO_3)_2NH$	imidobis-(sulfuric) acid	$(SO_3)_2NH^{2-}$	imidobis-(sulfate) ion
$(HSO_3)_3N$	nitridotris-(sulfuric) acid	$(SO_3)_3N^{3-}$	nitridotris-(sulfate) ion

Older names, for example, fluorosulfonate ion (SO_3F^-) and sulfamate ion ($SO_3NH_2^-$), are less commonly used but will undoubtedly persist.

II-2-3. Polyanions

Isopolyanions (Sec. 8.2-4), derived formally by the condensation of molecules of the same acid and containing the central element in the oxidation state characteristic of the periodic group number [e.g., P(V), Mo(VI)], are named by indicating the numbers of atoms of the central element with Greek numerical prefixes, preferably with the Ewens–Bassett number.

Examples $S_2O_7^{2-}$ disulfate(2−)
  $P_3O_{10}^{5-}$ triphosphate(5−)
  $Cr_4O_{13}^{2-}$ tetrachromate(2−)
  $Mo_7O_{24}^{6-}$ heptamolybdate(6−)

The Ewens–Bassett number is not essential in salts.

Examples $Na_2B_4O_7 \cdot 10\,H_2O$ disodium tetraborate-10-water
  $Na_7HNb_6O_{19} \cdot 15\,H_2O$ heptasodium monohydrogen
          hexaniobate-15-water

When the central element is partially or wholly in a lower oxidation state than that corresponding to the periodic group number, a Stock number (or numbers) may be used.

Examples $S_2O_5^{2-}$ disulfate(IV)(2−) ion (disulfite ion)
  $(HO_3P\!-\!PO_3H)^{2-}$ dihydrogendiphosphate(IV)(2−) ion
          (dihydrogenhypophosphate ion)

(Common names are in parentheses.)

Cyclic and straight-chain molecular arrangements can be distinguished by use, respectively, of the prefixes *cyclo* and *catena*, although the latter is often omitted.

Examples

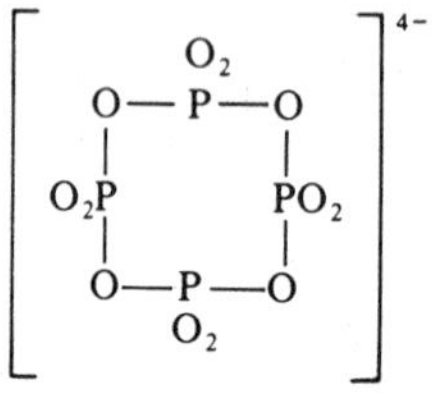

cyclo-tetraphosphate ion
(cyclic tetrametaphosphate ion)

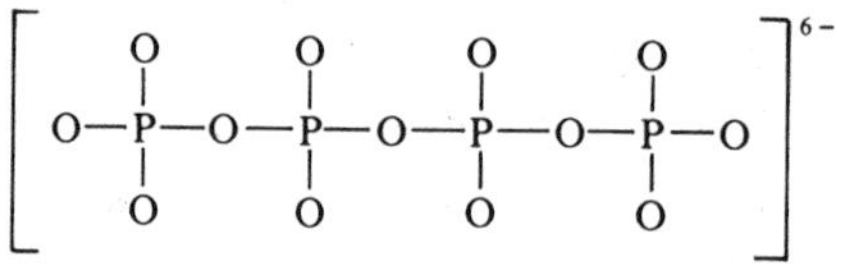

catena-tetraphosphate ion
(tetraphosphate ion)

Heteropolyanions (Sec. 8.2-4) having chain or simple ring molecular structures are commonly named by treating the parent anions in alphabetical order unless there is an obvious central atom.

Examples $[O_3S\!-\!O\!-\!CrO_3]^{2-}$ chromatosulfate(2−)
  $[O_3As\!-\!O\!-\!PO_3]^{4-}$ arsenatophosphate(4−)
  $[O_3P\!-\!O\!-\!AsO_2\!-\!O\!-$ [phosphatoarsenatochromato]-
  $CrO_2\!-\!O\!-\!SO_3]^{4-}$ sulfate(4−)

          but

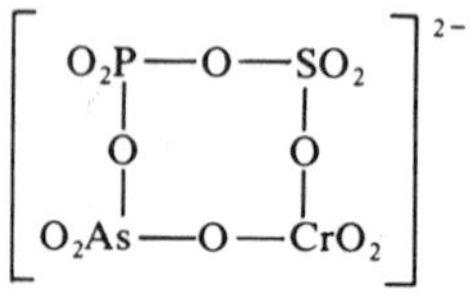

cyclo-arsenatophosphato-
sulfatochromate(2−)

Condensed *heteropolyanions* containing three-dimensional frameworks of linked MoO_6, WO_6, and so on, octahedra surrounding central AsO_4, PO_4, SiO_4, and so on, tetrahedra are named as molybdo-, tungsto-,* and so on, arsenates, phosphates, silicates, and so on, using either Greek prefixes or Arabic numbers.

Examples $[PW_{12}O_{40}]^{3-}$ dodecatungstophosphate(3−), or
                                             12-tungstophosphate(3−), ion
       $[Ce^{IV}Mo_{12}O_{42}]^{8-}$ dodecamolybdocerate(IV)(8−) ion

Salts and free heteropolyacids are named as usual.

Examples $(NH_4)_6[TeMo_6-$ hexaammoniumhexamolydotellurate-
               $O_{24}] \cdot 7 H_2O$ 7-water
       $H_4[SiW_{12}O_{40}]$ tetrahydrogen dodecatungstosilicate,
                                     or dodecatungstosilicic, acid

II-3. COORDINATION ENTITIES

Both the development of nomenclature for coordination entities and applications to many species are covered in sufficient detail in Chapter 7 and 8. It is useful here to refer to certain terms and to summarize abbreviations for a number of ligands. When the abbreviations or acronyms are used, however, it is always good practice to indicate their meanings.

II-3-1. Terms

Atoms designated A in general formulas are *nuclear* or *central* atoms; those designated B, C, ... are *ligated* atoms. A *ligating group*, or a ligand, containing more than one potential donor atom is referred to as a *multidentate* ligand. Ligands are then termed *unidentate*, *bidentate*, *quadridentate*, and so on, using Latin prefixes with the Latin-derived stem. *Chelation* refers to ring formation involving a central atom A and *two* donor atoms in a given ligand molecule or ion. A *chelating ligand* containing appropriately arranged donor atoms may simultaneously form more than a single chelate ring with a given central atom. A *bridging ligand* is one that bonds simultaneously to two or more central atoms. Coordination entities are termed *polynuclear* (e.g., *dinuclear*, *trinuclear*) if more than one central atom is present.

II-3-2. Names and Abbreviations of Ligands

The name of an *anionic* ligand is derived by adding o to its stem name. If the name of the anion ends in *-ide*, the ligand name ends in *-ido*. Anion names ending in *-ite* and *-ate* give, respectively, ligand names ending in *-ito* and *-ato*. Hydrocarbon radicals, although often bonded as anions, are given their

---

* Wolframo is preferred to tungsto by IUPAC.

common radical names (e.g., phenyl for $C_6H_5^-$). Other organic radicals formed by loss of protons are named with the *-ato* ending. Many trivial names for organic ligands, rather than systematic names, are used (e.g., cupferron, oxine, acetylacetone, dipyridyl). Neutral ligands are named as such, except for, in deference to long-standing practice, H_2O as *aqua* (formerly aquo), NH_3 as *ammine*, CO as *carbonyl*, and NO as *nitrosyl*. Cations acting as ligands are named as such, unless ambiguity results. Different donor sites for a given ligand are indicated by adding italicized symbols for the donor atoms at the end of the name of the ligand. Exceptions, again reflecting long practice are: —NO_2^-, nitro; ONO, nitrito; —SCN, thiocyanato; —NCS, isothiocyanato; —OCN, cyanato; —NCO, isocyanato; —CN, cyano; —NC, isocyano.

The use of *abbreviations*, according to IUPAC, should be governed by the following rules:

1. Each publication should explain the abbreviations used therein.
2. An abbreviation should be as short as possible, preferably not in excess of four letters.
3. An abbreviation should not cause confusion with one commonly used for an organic radical (e.g., Me for methyl).
4. Each abbreviation for a ligand, except L for a ligand in general and certain ones beginning with capital letters (Table II-2), should be in lowercase letters. M is used as an abbreviation for a metal.
5. Abbreviations should not contain hyphens (e.g., phen is preferred to o-phen for 1,10-phenanthroline).
6. A neutral compound and its ligand ion should be clearly differentiated (e.g., Hacac vs. acac for acetylacetone and acetylacetonate, respectively).
7. Abbreviations should be separated from symbols or enclosed in parentheses (e.g., $[Co(en)_3]^{3+}$ or $[Co\ en_3]^{3+}$, not $[Coen_3]^{3+}$).
8. Abbreviations combined with those commonly used for organic groups should be avoided.

Commonly used abbreviations are listed in Table II-2.

II-4. ADDITION COMPOUNDS

Addition compounds include donor–acceptor entities formed by the combination of independently existing species in definite stoichiometries as well as lattice compounds and clathrates (Sec. 5.3). For many years, the suffix -ate has been used to indicate addition compounds, for example, hydrates from water, ammoniates from ammonia, alcoholates from alcohols, and etherates from ethers. Although such usage continues, IUPAC recommends that it be allowed, but not preferred, only for hydrates, since the suffix -ate is broadly restricted to anions.

Such compounds are named by connecting the names of the individual compounds by spaced hyphens and indicating the numbers of molecules by Arabic numbers, separated by a solidus, and enclosed in parentheses.

Examples

$3\,CdSO_4 \cdot 8\,H_2O$	cadmium sulfate–water (3/8)
$Na_2CO_3 \cdot 10\,H_2O$	sodium carbonate–water (1/10), or
	sodium carbonate decahydrate
$CaCl_2 \cdot 8\,NH_3$	calcium chloride–ammonia (1/8)
$NH_3 \cdot BF_3$	ammonia–boron trifluoride (1/1)
$8\,H_2S \cdot 46\,H_2O$	hydrogen sulfide–water (8/46)
$8\,Kr \cdot 46\,H_2O$	krypton–water (8/46)
$C_6H_6 \cdot NH_3 \cdot Ni(CN)_2$	ammonia–benzene–nickel(II)
	cyanide (1/1/1)

II-5. BORANES AND THEIR DERIVATIVES

The rather specific details of the nomenclature used and limitations in space preclude more than a summary of general principles, with a few examples.*

The BH_3 group, which does not exist as a monomer is called *borane*, and both this group and isolable higher hydrides are called *boranes*. In each of these species, the number of boron atoms present per molecule is given by a Greek prefix, except for Latin nona and hendeca, and the number of hydrogen atoms is indicated by an Arabic number in parentheses after the name.

Examples

B_2H_6	diborane(6)
B_4H_{10}	tetraborane(10)
B_5H_9	pentaborane(9)
B_5H_{11}	pentaborane(11)
B_6H_{10}	hexaborane(10)
B_9H_{15}	nonaborane(15)
$B_{10}H_{14}$	decaborane(14)
$B_{20}H_{16}$	icosaborane(16)

In the polyhedral array of boron atoms characteristic of each borane, each boron atom is assigned a number that is then used to indicate positions occupied by substituent atoms or groups of atoms. Further complications result as a consequence of the existence among the higher boranes of two broad molecular structural types: (1) closed structures wherein boron atoms constitute skeletons with only triangular faces (*closo* structures), and (2) nonclosed, nestlike structures (*nido* structures). The latter, resulting from the assumed removal of one vertex boron atom from a *closo* structure, may closely approach the *closo* structure. A less common *arachno* (weblike) structure results from the assumed

*For a very detailed summary, including necessary structural diagrams, see *Inorg. Chem.*, 7, 1945–1964 (1968).

Table II-2 Some Commonly Used Ligands and Abbreviations

Chemical Name	Formula	Abbreviations Ligand and/or Ion
1. Anionic ligands (listed as parent acids)		
Acetylacetone, or 2,4-pentanedione	$CH_3COCH_2COCH_3$	Hacac, acac⁻
Benzoylacetone	$C_6H_5COCH_2COCH_3$	Hbenzac, benzac⁻
Biguanide	$H_2NC(NH)NHC(NH)NH_2$	Hbg, bg⁻
Cyclooctatetraene	C_8H_8	Cot²⁻
Cyclopentadiene	C_5H_6	HCp, Cp⁻
1,2-*trans*-Diaminocyclohexanediaminetetraacetic acid	$(HO_2CCH_2)_2NC_6H_{10}N(CH_2CO_2H)_2$	H₄dcta, dcta⁴⁻
Dibenzoylmethane	$C_6H_5COCH_2COC_6H_5$	Hdbm, dbm⁻
Diethylenetriaminepentaacetic acid	$(HO_2CCH_2)_2NC_2H_4N(CH_2CO_2H)$—$CH_2CO_2H$	H₅dtpa, dtpa⁵⁻
Diethylthiocarbamic acid	$(C_2H_5)_2NC(S)SH$	Hdtc, dtc⁻
Diethylthiophosphoric acid	$(C_2H_5O)_2P(S)OH$	Hdtp, dtp⁻
Dimethylglyoxime	$CH_3C(=NOH)C(=NOH)CH_3$	H₂dmg, dmg²⁻
Ethylenediaminetetraacetic acid	$(HO_2CCH_2)_2NC_2H_4N(CH_2CO_2H)_2$	H₄edta, edta⁴⁻
Glycine, or aminoacetic acid	$H_2C(NH_2)CO_2H$	Hgly, gly⁻
1,1,1,2,2,3,3-Heptafluoro-7,7-dimethyloctane-4,6-dione	$CF_3CF_2CF_2COCH_2COC(CH_3)_2CH_3$	Hfod, fod⁻
Hexafluoroacetylacetone	$CF_3COCH_2COCF_3$	Hfac, fac⁻
N-Hydroxyethylethylenediaminetriacetic acid	$(HO_2CCH_2)_2NC_2H_4N(CO_2H)(C_2H_4OH)$	H₃hedta, hedta³⁻
Hydrohalic acids	HX (X = F, Cl, Br, I)	HX, X⁻
Imidodiacetic acid	$HN(CH_2CO_2H)_2$	H₂imda, imda²⁻
Nitrilotriacetic acid	$N(CH_2CO_2H)_3$	H₃nta, nta³⁻
Oxalic acid	HO_2CCO_2H	H₂ox, ox²⁻
2,2,6,6-Tetramethylheptane-3,5-dione	$CH_3C(CH_3)_2COCH_2COC(CH_3)_2CH_3$	Hthd, thd⁻
Thenoyltrifluoroacetone		Htta, tta⁻

2. Neutral ligands

Name	Formula	Abbreviation
Benzene	C_6H_6	bz
2,2'-Bipyridine, or 2,2'-bipyridyl		bpy
Diethylenetriamine	$H_2NC_2H_4NHC_2H_4NH_2$	dien
Dimethylacetamide	$CH_3CON(CH_3)_2$	dma
Dimethylsulfoxide	$(CH_3)_2SO$	dmso
Ethylenebis(diphenylphosphine)	$(C_6H_5)_2PC_2H_4P(C_6H_5)_2$	diphos
Ethylenediamine	$H_2NC_2H_4NH_2$	en
1,10-Phenanthroline		phen
o-Phenylenebis(dimethylarsine)	$(CH_3)_2AsC_6H_4As(CH_3)_2$	diars
Propylenediamine, or 1,2-diaminopropane	$H_2NCH_2CH(NH_2)CH_3$	pn
Pyridine	C_5H_5N	py
2,2',2'',2'''-Tetrapyridyl		tetrpy
Thiourea	$(H_2N)_2CS$	tu
1,2,3-Triaminopropane	$CH_2\!-\!CH_2\!-\!CH_2$ $\quad NH_2 \quad NH_2 \quad NH_2$	tn
2,2',2''-Triaminotriethylamine	$N(CH_2CH_2NH_2)_3$	tren
Triethylenetetraamine	$H_2NC_2H_4NHC_2H_4NHC_2H_4NH_2$	trien
2,2',2''-Tripyridyl		tripy
Urea	$(H_2N)_2CO$	ur

removal of two vertex boron atoms from a *closo* structure. Thus a meaningful nomenclature must be based upon knowledge of molecular structures. In addition, closely related compounds result from the substitution of atoms such as carbon (i.e., carboranes) and phosphorus (i.e., phosphinoboranes) in polyhedral boron arrays. Furthermore, certain boron compounds and ions are named as derivatives of boranes.

Some randomly chosen examples of various type species are as follows:

diboranyl radical

1,2-diboranediyl radical

1,1-diboranediyl radical

B_4Cl_4 tetrachloro-*closo*-tetraborane(4)

1-chlorodiborane(6)

$[BH_4]^-$ tetrahydroborate$(1-)$ ion
$[B_{10}H_{12}]^{2-}$ dodecahydro-*nido*-decaborate$(2-)$ ion
$[B_{12}H_{12}]^{2-}$ dodecahydro-*closo*-dodecaborate$(2-)$ ion
$[B_{10}C_2H_{12}]^{2-}$ dodecahydro-1,2-dicarba-*closo*-dodecaborate$(2-)$ ion
$[BH_2(C_5H_5N)_2]^+$ dihydrobis(pyridine)boron$(1+)$ ion
$[BH_2(NH_3)_2][B_3H_8]$ diamminedihydroboron octahydrotriborate

REFERENCES

1. International Union of Pure and Applied Chemistry, *Definitive Rules for Nomenclature of Inorganic Compounds*, Butterworth, London (1959).

2. International Union of Pure and Applied Chemistry, *Pure Appl. Chem.*, **28**, 67 (1971); *Nomenclature of Inorganic Chemistry*, 2nd ed., Butterworth, London (1971).

3. International Union of Pure and Applied Chemistry, *How to Name an Inorganic Substance*, Pergamon Press, Oxford (1977).

4. W. C. Fernelius, K. L. Loening, and R. M. Adams, *J. Chem. Educ.*, **48**, 433, 594, 730 (1971); **49**, 49, 253, 333, 488, 699, 844 (1972); **50**, 123, 341 (1973); **51**, 43, 468, 603, 705 (1974); **52**, 60, 583, 793 (1975); **53**, 354, 726, 773 (1976); **54**, 30, 299, 509 (1977); **55**, 30 (1978).

5. H. Egan and E. W. Godly, *Chem. Br.*, **16**, 16 (1980).

6. R. S. Cahn and O. C. Dermer, *Introduction to Chemical Nomenclature*, 5th ed., Butterworth, Woburn, Mass. (1979).

APPENDIX III

CHRONOLOGY OF DISCOVERY OF THE ELEMENTS[1–5]

Element	Symbol	Atomic Number	Date	Discoverer(s)—Country[a]	Method
Actinium	Ac	89	1899	A. Debierne—F	Radioactivity
Aluminum	Al	13	1827	F. Wöhler—G	$Al_2Cl_6 + K$
Americium	Am	95	1944	G. T. Seaborg—USA R. A. James—USA S. G. Thompson—USA A. Ghiorso—USA	Neutron bombardment of ^{239}Pu
Antimony	Sb	51	Ancient	—	
Argon	Ar	18	1894	Lord Rayleigh—GB W. Ramsay—GB	Residue from reaction of atmospheric nitrogen with Mg
Arsenic	As	33	1250	A. Magnus—G	Heating orpiment with soap
Astatine	At	85	1940	D. R. Corson—USA K. R. MacKenzie—USA E. Segrè—USA	Neutron bombardment
Barium	Ba	56	1808	H. Davy—GB	Electrolysis of $BaCl_2$
Berkelium	Bk	97	1950	G. T. Seaborg—USA S. G. Thompson—USA A. Ghiorso—USA	Alpha particle bombardment of ^{241}Am
Beryllium	Be	4	1828	F. Wöhler—G A. A. B. Bussy—F	Anhydrous $BeCl_2 + K$
Bismuth	Bi	83	1753	C. Geoffroy—F	Distinction from Pb
Boron	B	5	1808	H. Davy—GB J. L. Gay-Lussac—F L. J. Thenard—F	Boric acid heated with K
Bromine	Br	35	1826	A. J. Balard—F	Salt spring water + Cl_2
Cadmium	Cd	48	1817	F. Stromeyer—G	Reduction of oxide from a Zn ore with C

Element	Symbol	Atomic number	Year	Discoverer	Method/Source
Calcium	Ca	20	1808	H. Davy—GB	Electrolysis of mixture with HgO and volatilizing Hg
Californium	Cf	98	1950	G. T. Seaborg—USA S. G. Thompson—USA A. Ghiorso—USA K. Street, Jr.—USA	Alpha particle bombardment of ^{242}Cm
Carbon	C	6	Ancient	—	—
Cerium	Ce	58	1803	J. J. Berzelius—S W. Hisinger—S M. H. Klaproth—G	Chemical separation from cerite
Cesium	Cs	55	1860	R. Bunsen—G G. R. Kirchhoff—G	Spectroscopic (mother liquor from a mineral water)
Chlorine	Cl	17	1774	C. W. Scheele—S	Hydrochloric acid + MnO_2
Chromium	Cr	24	1797	L. N. Vauquelin—F	Analysis of Siberian red lead ore
Cobalt	Co	27	1735	G. Brandt—G	Separation from copper ores
Copper	Cu	29	Ancient	—	—
Curium	Cm	96	1944	G. T. Seaborg—USA R. A. James—USA A. Ghiorso—USA	Alpha particle bombardment of ^{239}Pu
Dysprosium	Dy	66	1886	L. de Boisbaudran—F	Fractionation of holmia from gadolinite
Einsteinium	Es	99	1952	A. Ghiorso—USA	Debris from Nov. 1952 thermonuclear blast
Erbium	Er	68	1879	P. T. Cleve—S	Fractionation of erbia from gadolinite
Europium	Eu	63	1901	E.-A. Demarçay—F	Fractionation of impure samarium magnesium nitrate
Fermium	Fm	100	1953	A. Ghiorso—USA	Debris from thermonuclear blast
Fluorine	F	9	1886	H. Moissan—F	Electrolysis of KHF_2 in liquid HF

Appendix III (*Continued*)

Element	Symbol	Atomic Number	Discovery or Isolation			
			Date	Discoverer(s)—Country[a]	Method	
Francium	Fr	87	1939	M. Perey—F	Alpha decay of ^{227}Ac	
Gadolinium	Gd	64	1880	J. C. Marignac—F	Fractionation of terbia from gadolinite	
Gallium	Ga	31	1875	L. de Boisbaudran—F	Spectroscopic (sphalerite)	
Germanium	Ge	32	1886	C. Winkler—G	Analysis of argyrodite	
Gold	Au	79	Ancient	—		
Hafnium	Hf	72	1923	D. Coster—NL G. von Hevesy—H	X-ray spectroscopic (zircon)	
Helium	He	2	1868	P. Janssen—F	Spectroscopic (sun)	
			1895	W. Ramsay—GB	Terrestrial isolation from cleveite	
Holmium	Ho	67	1879	P. T. Cleve—S	Fractionation of impure holmia from gadolinite	
Hydrogen	H	1	1766	H. Cavendish—GB	Metals plus acids	
Indium	In	49	1863	F. Reich—G T. Richter—G	Spectroscopic (zinc blende)	
Iodine	I	53	1811	B. Courtois—F	Ash from algae leached and H_2SO_4 added	
Iridium	Ir	77	1803	S. Tennant—GB	Treatment of residue from crude platinum plus HCl–HNO_3	
Iron	Fe	26	Ancient	—		
Krypton	Kr	36	1898	W. Ramsay—GB M. W. Travers—GB	Spectrum of residue from partial evaporation of liquid air	
Lanthanum	La	57	1839	C. G. Mosander—S	Fractionation of impure ceria from cerite	

Element	Symbol	Atomic number	Year	Discoverer	Method/Notes
Lawrencium	Lr	103	1961	A. Ghiorso—USA T. Sikkeland—USA A. E. Larch—USA R. M. Latimer—USA	^{10}B or ^{11}B bombardment of $^{249-252}Cf$
Lead	Pb	82	Ancient	—	—
Lithium	Li	3	1817	A. Arfvedson—S	Analysis of petalite
Lutetium	Lu	71	1907	G. Urbain—F C. Auer von Welsbach—A	Fractionation of impure ytterbia from gadolinite
Magnesium	Mg	12	1808	H. Davy—GB	As with Ba, Ca
Manganese	Mn	25	1774	J. G. Gahn—S	Reduction of pyrolusite
Mendelevium	Md	101	1955	A. Ghiorso—USA G. R. Choppin—USA G. T. Seaborg—USA B. G. Harvey—USA S. G. Thompson—USA	Alpha particle bombardment of ^{253}Es
Mercury	Hg	80	Ancient	—	—
Molybdenum	Mo	42	1778	C. W. Scheele—S	Analysis of molybdenite
Neodymium	Nd	60	1885	C. Auer von Welsbach—A	Fractionation of didymia
Neon	Ne	10	1898	W. Ramsay—GB M. W. Travers—GB	As with krypton
Neptunium	Np	93	1940	E. M. McMillan—USA P. H. Abelson—USA	Neutron bombardment of ^{238}U
Nickel	Ni	28	1751	A. F. Cronstedt—S	Analysis of green crystals on weathered niccolite
Niobium	Nb	41	1801	C. Hatchett—GB	Analysis of columbite sample from Connecticut
Nitrogen	N	7	1772	D. Rutherford—GB	Unreactive gas remaining after removal of oxygen and carbon dioxide from air

Appendix III (*Continued*)

Element	Symbol	Atomic Number	Discovery or Isolation		
			Date	Discoverer(s)—Country[a]	Method
Nobelium	No	102	1958	A. Ghiorso—USA T. Sikkeland—USA J. R. Walton—USA G. T. Seaborg—USA	Bombardment of ^{246}Cm with ^{12}C
Osmium	Os	76	1803	S. Tennant—GB	Treatment of residue from crude platinum plus $HCl–HNO_3$
Oxygen	O	8	1774	J. Priestley—GB C. W. Scheele—S	Thermal decomposition of HgO, Ag_2CO_3, etc.
Palladium	Pd	46	1803	W. H. Wollaston—GB	Treatment of solution of crude Pt in $HCl–HNO_3$ with $Hg_2(CN)_2$
Phosphorus	P	15	1669	H. Brandt—G	Complex treatment of putrefied urine
Platinum	Pt	78	1735	A. de Ulloa—Sp	Native in Andean region
Plutonium	Pu	94	1940	G. T. Seaborg—USA E. M. McMillan—USA J. W. Kennedy—USA A. C. Wahl—USA	Deuteron bombardment of uranium
Polonium	Po	84	1898	M. Curie—F	Fractionation of residues from uranium mines
Potassium	K	19	1807	H. Davy—GB	Electrolysis of molten potash
Praseodymium	Pr	59	1885	C. Auer von Welsbach—A	Fractionation of didymia
Promethium	Pm	61	1945	J. A. Marinsky—USA L. E. Glendenin—USA C. D. Coryell—USA	Ion-exchange separation of lanthanides from neutron fission of ^{235}U

Protactinium	Pa	91	O. Hahn—G L. Meitner—G K. Fajans—G F. Soddy—GB J. A. Cranston—GB A. Fleck—GB	Fractionation of pitch-blende process residues
Radium	Ra	1898	M. Curie—F P. Curie—F	Spectroscopic examination after fractionation of mixed $BaCl_2$–$RaCl_2$
Radon	Rn	1900	F. E. Dorn—G	Emanation from decay of ^{226}Ra
Rhenium	Re	1925	W. Noddack—G I. Tacke—G O. Berg—G	X-ray spectra of concentrates from gadolinite; separation from columbite
Rhodium	Rh	1804	W. H. Wollaston—GB	Chemical treatment of solution of crude Pt in HCl–HNO_3
Rubidium	Rb	1861	R. Bunsen—G G. Kirchhoff—G	Spectroscopic in potassium separated from lepidolite
Ruthenium	Ru	1844	K. K. Klaus—R	Chemical treatment of residues from Pt refineries
Samarium	Sm	1879	L. de Boisbaudran—F	Fractionation of didymia
Scandium	Sc	1879	L. F. Nilson—S	Fractionation of ytterbia from gadolinite
Selenium	Se	1817	J. J. Berzelius—S	Residues in sulfuric acid made from pyrite
Silicon	Si	1824	J. J. Berzelius—S	Reduction of SiF_4 or K_2SiF_6 with K
Silver	Ag	Ancient	—	—
Sodium	Na	1807	H. Davy—GB	Electrolysis of molten caustic soda
Strontium	Sr	1808	H. Davy—GB	As with Ba, Ca
Sulfur	S	Ancient	—	—

Appendix III (*Continued*)

Element	Symbol	Atomic Number	Date	Discovery or Isolation	
				Discoverer(s)—Country[a]	Method
Tantalum	Ta	73	1802	A. G. Ekeberg—S	Analysis of tantalite
Technetium	Tc	43	1937	C. Perrier—I E. Segré—I	Radioactivity in products from deuteron bombardment of Mo
Tellurium	Te	52	1782	F. J. Müller—A	Extraction from bluish-white Au ore
Terbium	Tb	65	1843	C. G. Mosander—S	Fractionation of yttria from gadolinite
Thallium	Tl	81	1861	W. Crookes—GB	Spectroscopic (green line from H_2SO_4 plant residues)
Thorium	Th	90	1828	J. J. Berzelius—S	Analysis of thorite
Thulium	Tm	69	1879	P. T. Cleve—S	Fractionation of holmia from gadolinite
Tin	Sn	50	Ancient	—	
Titanium	Ti	22	1791	W. Gregor—GB	Analysis of ilmenite
Tungsten	W	74	1783	J. J. and F. de Elhuyar—Sp	Reduction of tungstic acid from wolframite with C
Uranium	U	92	1789	M. H. Klaproth—G	Yellow oxide from pitchblende and presumed U
Vanadium	V	23	1801	A. M. del Rio—Sp	New metal in "brown lead" from Mexico
			1830	N. G. Sefström—S	Analysis of cast iron from Swedish ore
Xenon	Xe	54	1898	W. Ramsay—GB M. W. Travers—GB	Spectrum from residue of distillation of krypton

Ytterbium	Yb	70	1970	G. Urbain—F	Fractionation of ytterbia from gadolinite
Yttrium	Y	39	1843	C. G. Mosander—S	Fractionation of yttria from gadolinite
Zinc	Zn	30	1746	A. S. Marggraf—G	Carbon reduction of calamine
Zirconium	Zr	40	1789	M. H. Klaproth—G	Analysis of zircon

[a] A = Austria; F = France; G = Germany; GB = Great Britain; H = Hungary; I = Italy; NL = Netherlands; R = Russia; S = Sweden; Sp = Spain; USA = United States of America.

REFERENCES

1. A. H. Ihde, *The Development of Modern Chemistry*, Harper & Row, New York (1964).

2. M. E. Weeks and H. M. Leicester, *Discovery of the Elements*, 7th ed., Journal of Chemical Education Circulation Services, Easton, Pa. (1968).

3. V. Gutmann, *Allg. Prak. Chem.*, **17**, 394 (1966). Chronology.

4. H. Goldwhite and R. C. Adams, *J. Chem. Educ.*, **47**, 808 (1970). Interesting plot of atomic number vs. date of discovery from ca. 1700 on.

5. G. P. Dinga, *Chemistry*, **41**(2), 20 (1968).

APPENDIX IV

STANDARD ELECTRODE (REDUCTION) POTENTIALS[1–4]

Couple	Equation for Half-Reaction	E^0 (V)
	ACIDIC MEDIA	
Eu(II)–Eu(0)	$Eu^{2+} + 2e^- \rightleftharpoons Eu$	−3.395
Sm(II)–Sm(0)	$Sm^{2+} + 2e^- \rightleftharpoons Sm$	−3.121
N(0)–N($-\frac{1}{3}$)	$\frac{3}{2} N_2 + H^+ + e^- \rightleftharpoons HN_3(aq)$	−3.09
Li(I)–Li(0)	$Li^+ + e^- \rightleftharpoons Li$	−3.045
Rb(I)–Rb(0)	$Rb^+ + e^- \rightleftharpoons Rb$	−2.925
K(I)–K(0)	$K^+ + e^- \rightleftharpoons K$	−2.925
Cs(I)–Cs(0)	$Cs^+ + e^- \rightleftharpoons Cs$	−2.923
Ra(II)–Ra(0)	$Ra^{2+} + 2e^- \rightleftharpoons Ra$	−2.916
Ba(II)–Ba(0)	$Ba^{2+} + 2e^- \rightleftharpoons Ba$	−2.906
Sr(II)–Sr(0)	$Sr^{2+} + 2e^- \rightleftharpoons Sr$	−2.888
Ca(II)–Ca(0)	$Ca^{2+} + 2e^- \rightleftharpoons Ca$	−2.866
Yb(II)–Yb(0)	$Yb^{2+} + 2e^- \rightleftharpoons Yb$	−2.797
Na(I)–Na(0)	$Na^+ + e^- \rightleftharpoons Na$	−2.714
Cm(III)–Cm(0)	$Cm^{3+} + 3e^- \rightleftharpoons Cm$	−2.70
Ac(III)–Ac(0)	$Ac^{3+} + 3e^- \rightleftharpoons Ac$	−2.6
La(III)–La(0)	$La^{3+} + 3e^- \rightleftharpoons La$	−2.522
No(III)–No(0)	$No^{3+} + 3e^- \rightleftharpoons No$	−2.5
Ce(III)–Ce(0)	$Ce^{3+} + 3e^- \rightleftharpoons Ce$	−2.483
Pr(III)–Pr(0)	$Pr^{3+} + 3e^- \rightleftharpoons Pr$	−2.462
Nd(III)–Nd(0)	$Nd^{3+} + 3e^- \rightleftharpoons Nd$	−2.431
Pm(III)–Pm(0)	$Pm^{3+} + 3e^- \rightleftharpoons Pm$	−2.423
Sm(III)–Sm(0)	$Sm^{3+} + 3e^- \rightleftharpoons Sm$	−2.414
Eu(III)–Eu(0)	$Eu^{3+} + 3e^- \rightleftharpoons Eu$	−2.407
Bk(III)–Bk(0)	$Bk^{3+} + 3e^- \rightleftharpoons Bk$	−2.4
Gd(III)–Gd(0)	$Gd^{3+} + 3e^- \rightleftharpoons Gd$	−2.397
Tb(III)–Tb(0)	$Tb^{3+} + 3e^- \rightleftharpoons Tb$	−2.391

Couple	Half-reaction	E
Y(III)–Y(0)	$Y^{3+} + 3e^- \rightleftharpoons Y$	−2.372
Mg(II)–Mg(0)	$Mg^{2+} + 2e^- \rightleftharpoons Mg$	−2.363
Dy(III)–Dy(0)	$Dy^{3+} + 3e^- \rightleftharpoons Dy$	−2.353
Am(III)–Am(0)	$Am^{3+} + 3e^- \rightleftharpoons Am$	−2.320
Ho(III)–Ho(0)	$Ho^{3+} + 3e^- \rightleftharpoons Ho$	−2.319
Er(III)–Er(0)	$Er^{3+} + 3e^- \rightleftharpoons Er$	−2.296
Tm(III)–Tm(0)	$Tm^{3+} + 3e^- \rightleftharpoons Tm$	−2.278
Yb(III)–Yb(0)	$Yb^{3+} + 3e^- \rightleftharpoons Yb$	−2.267
Lu(III)–Lu(0)	$Lu^{3+} + 3e^- \rightleftharpoons Lu$	−2.255
H(0)–H(−I)	$\frac{1}{2}H_2 + e^- \rightleftharpoons H^-$	−2.25
Am(IV)–Am(III)	$Am^{4+} + e^- \rightleftharpoons Am^{3+}$	−2.181
H(I)–H(0)	$H^+ + e^- \rightleftharpoons H$	−2.1065
Fm(III)–Fm(0)	$Fm^{3+} + 3e^- \rightleftharpoons Fm$	−2.1
Cf(III)–Cf(0)	$Cf^{3+} + 3e^- \rightleftharpoons Cf$	−2.1
Sc(III)–Sc(0)	$Sc^{3+} + 3e^- \rightleftharpoons Sc$	−2.077
Al(III)–Al(0)	$AlF_6^{3-} + 3e^- \rightleftharpoons Al + 6F^-$	−2.069
Pu(III)–Pu(0)	$Pu^{3+} + 3e^- \rightleftharpoons Pu$	−2.031
Pa(III)–Pa(0)	$Pa^{3+} + 3e^- \rightleftharpoons Pa$	−1.95
Th(IV)–Th(0)	$Th^{4+} + 4e^- \rightleftharpoons Th$	−1.899
Np(III)–Np(0)	$Np^{3+} + 3e^- \rightleftharpoons Np$	−1.856
Be(II)–Be(0)	$Be^{2+} + 2e^- \rightleftharpoons Be$	−1.847
U(III)–U(0)	$U^{3+} + 3e^- \rightleftharpoons U$	−1.798
Th(IV)–Th(0)	$ThO_2(s) + 4H^+ + 4e^- \rightleftharpoons Th + 2H_2O$	−1.789
Hf(IV)–Hf(0)	$Hf^{4+} + 4e^- \rightleftharpoons Hf$	−1.700
Al(III)–Al(0)	$Al^{3+} + 3e^- \rightleftharpoons Al$	−1.662
Ti(II)–Ti(0)	$Ti^{2+} + 2e^- \rightleftharpoons Ti$	−1.628
Zr(IV)–Zr(0)	$Zr^{4+} + 4e^- \rightleftharpoons Zr$	−1.529
Hf(IV)–Hf(0)	$HfO_2(s) + 4H^+ + 4e^- \rightleftharpoons Hf + 2H_2O$	−1.505
Zr(IV)–Zr(0)	$ZrO_2(s) + 4H^+ + 4e^- \rightleftharpoons Zr + 2H_2O$	−1.456
Si(IV)–Si(0)	$SiF_6^{2-} + 4e^- \rightleftharpoons Si + 6F^-$	−1.24
Yb(III)–Yb(II)	$Yb^{3+} + e^- \rightleftharpoons Yb^{2+}$	−1.205

Appendix IV (*Continued*)

Couple	Equation for Half-Reaction	E^0 (V)
Ti(IV)–Ti(0)	$TiF_6^{2-} + 4e^- \rightleftharpoons Ti + 6F^-$	-1.191
Mn(II)–Mn(0)	$Mn^{2+} + 2e^- \rightleftharpoons Mn$	-1.185
V(II)–V(0)	$V^{2+} + 2e^- \rightleftharpoons V$	-1.175
Sm(III)–Sm(II)	$Sm^{3+} + e^- \rightleftharpoons Sm^{2+}$	-1.15
Nb(III)–Nb(0)	$Nb^{3+} + 3e^- \rightleftharpoons Nb$	-1.099
Pa(V)–Pa(0)	$PaO_2^+ + 4H^+ + 5e^- \rightleftharpoons Pa + 2H_2O$	-1.0
Cr(II)–Cr(0)	$Cr^{2+} + 2e^- \rightleftharpoons Cr$	-0.913
Ti(IV)–Ti(0)	$TiO^{2+} + 2H^+ + 4e^- \rightleftharpoons Ti + H_2O$	-0.882
B(III)–B(0)	$H_3BO_3(s) + 3H^+ + 3e^- \rightleftharpoons B + 3H_2O$	-0.869
Si(IV)–Si(0)	$SiO_2(s) + 4H^+ + 4e^- \rightleftharpoons Si + 2H_2O$	-0.857
Ta(V)–Ta(0)	$Ta_2O_5(s) + 10H^+ + 10e^- \rightleftharpoons 2Ta + 5H_2O$	-0.812
Zn(II)–Zn(0)	$Zn^{2+} + 2e^- \rightleftharpoons Zn$	-0.7628
Tl(I)–Tl(0)	$TlI(s) + e^- \rightleftharpoons Tl + I^-$	-0.752
Cr(III)–Cr(0)	$Cr^{3+} + 3e^- \rightleftharpoons Cr$	-0.744
Te(0)–Te(−II)	$Te + 2H^+ + 2e^- \rightleftharpoons H_2Te(aq)$	-0.739
Tl(I)–Tl(0)	$TlBr(s) + e^- \rightleftharpoons Tl + Br^-$	-0.658
Nb(V)–Nb(0)	$Nb_2O_5(s) + 10H^+ + 10e^- \rightleftharpoons 2Nb + 5H_2O$	-0.644
U(IV)–U(III)	$U^{4+} + e^- \rightleftharpoons U^{3+}$	-0.607
As(0)–As(−III)	$As(s) + 3H^+ + 3e^- \rightleftharpoons AsH_3(g)$	-0.607
Tl(I)–Tl(0)	$TlCl(s) + e^- \rightleftharpoons Tl + Cl^-$	-0.5568
Ga(III)–Ga(0)	$Ga^{3+} + 3e^- \rightleftharpoons Ga$	-0.529
Sb(0)–Sb(−III)	$Sb + 3H^+ + 3e^- \rightleftharpoons SbH_3(g)$	-0.510
P(I)–P(0)	$H_3PO_2 + H^+ + e^- \rightleftharpoons P(white) + 2H_2O$	-0.508
P(III)–P(I)	$H_3PO_3 + 2H^+ + 2e^- \rightleftharpoons H_3PO_2 + H_2O$	-0.499
Fe(II)–Fe(0)	$Fe^{2+} + 2e^- \rightleftharpoons Fe$	-0.4402
Eu(III)–Eu(II)	$Eu^{3+} + e^- \rightleftharpoons Eu^{2+}$	-0.429

Cr(III)–Cr(II)	$Cr^{3+} + e^- \rightleftharpoons Cr^{2+}$	-0.408
Cd(II)–Cd(0)	$Cd^{2+} + 2e^- \rightleftharpoons Cd$	-0.4029
Se(0)–Se(−II)	$Se + 2H^+ + 2e^- \rightleftharpoons H_2Se(aq)$	-0.399
Ti(III)–Ti(II)	$Ti^{3+} + e^- \rightleftharpoons Ti^{2+}$	-0.369
Pb(II)–Pb(0)	$PbI_2(s) + 2e^- \rightleftharpoons Pb + 2I^-$	-0.365
Pb(II)–Pb(0)	$PbSO_4(s) + 2e^- \rightleftharpoons Pb + SO_4^{2-}$	-0.3588
In(III)–In(0)	$In^{3+} + 3e^- \rightleftharpoons In$	-0.343
Tl(I)–Tl(0)	$Tl^+ + e^- \rightleftharpoons Tl$	-0.3363
Pb(II)–Pb(0)	$PbBr_2(s) + 2e^- \rightleftharpoons Pb + 2Br^-$	-0.284
Co(II)–Co(0)	$Co^{2+} + 2e^- \rightleftharpoons Co$	-0.277
P(V)–P(III)	$H_3PO_4 + 2H^+ + 2e^- \rightleftharpoons H_3PO_3 + H_2O$	-0.276
Pb(II)–Pb(0)	$PbCl_2(s) + 2e^- \rightleftharpoons Pb + 2Cl^-$	-0.268
V(III)–V(II)	$V^{3+} + e^- \rightleftharpoons V^{2+}$	-0.256
Sn(IV)–Sn(0)	$SnF_6^{2-} + 4e^- \rightleftharpoons Sn + 6F^-$	-0.25
Ni(II)–Ni(0)	$Ni^{2+} + 2e^- \rightleftharpoons Ni$	-0.250
N(0)–N(−II)	$N_2 + 5H^+ + 4e^- \rightleftharpoons N_2H_5^+$	-0.23
Cu(I)–Cu(0)	$CuI(s) + e^- \rightleftharpoons Cu + I^-$	-0.1852
Ag(I)–Ag(0)	$AgI(s) + e^- \rightleftharpoons Ag + I^-$	-0.1518
Ge(IV)–Ge(0)	$GeO_2(s) + 4H^+ + 4e^- \rightleftharpoons Ge + 2H_2O$	-0.15
Sn(II)–Sn(0)	$Sn^{2+} + 2e^- \rightleftharpoons Sn(white)$	-0.136
Pb(II)–Pb(0)	$Pb^{2+} + 2e^- \rightleftharpoons Pb$	-0.126
W(VI)–W(0)	$WO_3(s) + 6H^+ + 6e^- \rightleftharpoons W + 3H_2O$	-0.090
P(0)–P(−III)	$P(white) + 3H^+ + 3e^- \rightleftharpoons PH_3(g)$	-0.063
Hg(I)–Hg(0)	$Hg_2I_2(s) + 2e^- \rightleftharpoons 2Hg + 2I^-$	-0.0405
Hg(II)–Hg(0)	$HgI_4^{2-} + 2e^- \rightleftharpoons Hg + 4I^-$	-0.038
Fe(III)–Fe(0)	$Fe^{3+} + 3e^- \rightleftharpoons Fe$	-0.036
D(I)–D(0)	$2D^+ + 2e^- \rightleftharpoons D_2$	-0.0034
H(I)–H(0)	$2H^+ + 2e^- \rightleftharpoons H_2$	0.0000
Ag(I)–Ag(0)	$Ag(S_2O_3)_2^{3-} + e^- \rightleftharpoons Ag + 2S_2O_3^{2-}$	$+0.017$
Cu(I)–Cu(0)	$CuBr(s) + e^- \rightleftharpoons Cu + Br^-$	$+0.033$
U(VI)–U(V)	$UO_2^{2+} + e^- \rightleftharpoons UO_2^+$	$+0.05$

Appendix IV (Continued)

Couple	Equation for Half-Reaction	E^0 (V)
Ag(I)–Ag(0)	$AgBr(s) + e^- \rightleftharpoons Ag + Br^-$	+0.0713
Si(0)–Si(−IV)	$Si + 4H^+ + 4e^- \rightleftharpoons SiH_4$	+0.102
Cu(I)–Cu(0)	$CuCl(s) + e^- \rightleftharpoons Cu + Cl^-$	+0.137
Hg(I)–Hg(0)	$Hg_2Br_2(s) + 2e^- \rightleftharpoons 2Hg + 2Br^-$	+0.1397
S(0)–S(−II)	$S(rhombic) + 2H^+ + 2e^- \rightleftharpoons H_2S(aq)$	+0.142
Np(IV)–Np(III)	$Np^{4+} + e^- \rightleftharpoons Np^{3+}$	+0.147
Sn(IV)–Sn(II)	$Sn^{4+} + 2e^- \rightleftharpoons Sn^{2+}$	+0.15
Sb(III)–Sb(0)	$Sb_2O_3(s) + 6H^+ + 6e^- \rightleftharpoons 2Sb + 3H_2O$	+0.152
Cu(II)–Cu(I)	$Cu^{2+} + e^- \rightleftharpoons Cu^+$	+0.153
S(VI)–S(IV)	$SO_4^{2-} + 4H^+ + 2e^- \rightleftharpoons H_2SO_3 + H_2O$	+0.172
Ag(I)–Ag(0)	$AgCl(s) + e^- \rightleftharpoons Ag + Cl^-$	+0.2222
Hg(II)–Hg(0)	$HgBr_4^{2-} + 2e^- \rightleftharpoons Hg + 4Br^-$	+0.223
As(III)–As(0)	$HAsO_2(aq) + 3H^+ + 3e^- \rightleftharpoons As + 2H_2O$	+0.2476
Re(IV)–Re(0)	$ReO_2(s) + 4H^+ + 4e^- \rightleftharpoons Re + 2H_2O$	+0.2513
Hg(I)–Hg(0)	$Hg_2Cl_2(s) + 2e^- \rightleftharpoons 2Hg + 2Cl^-$	+0.2676
Bi(III)–Bi(0)	$BiO^+ + 2H^+ + 3e^- \rightleftharpoons Bi + H_2O$	+0.320
U(VI)–U(IV)	$UO_2^{2+} + 4H^+ + 2e^- \rightleftharpoons U^{4+} + 2H_2O$	+0.330
Cu(II)–Cu(0)	$Cu^{2+} + 2e^- \rightleftharpoons Cu$	+0.337
S(VI)–S(0)	$SO_4^{2-} + 8H^+ + 6e^- \rightleftharpoons S + 4H_2O$	+0.3572
V(IV)–V(III)	$VO^{2+} + 2H^+ + e^- \rightleftharpoons V^{3+} + H_2O$	+0.359
Fe(III)–Fe(II)	$Fe(CN)_6^{3-} + e^- \rightleftharpoons Fe(CN)_6^{4-}$	+0.36
Re(VII)–Re(0)	$ReO_4^- + 8H^+ + 7e^- \rightleftharpoons Re + 4H_2O$	+0.362
CN(0)–CN(−I)	$C_2N_2(g) + 2H^+ + 2e^- \rightleftharpoons 2HCN(aq)$	+0.373
S(IV)–S(0)	$H_2SO_3 + 4H^+ + 4e^- \rightleftharpoons S + 3H_2O$	+0.450
Re(VII)–Re(IV)	$ReO_4^- + 4H^+ + 3e^- \rightleftharpoons ReO_2 + 2H_2O$	+0.510
Cu(I)–Cu(0)	$Cu^+ + e^- \rightleftharpoons Cu$	+0.521

Couple	Half-reaction	E°
Te(IV)–Te(0)	$TeO_2(s) + 4H^+ + 4e^- \rightleftharpoons Te + 2H_2O$	+0.529
I(0)–I(–I)	$I_2(s) + 2e^- \rightleftharpoons 2I^-$	+0.5355
	$I_3^- + 2e^- \rightleftharpoons 3I^-$	+0.536
Cu(II)–Cu(I)	$Cu^{2+} + Cl^- + e^- \rightleftharpoons CuCl(s)$	+0.538
As(V)–As(III)	$H_3AsO_4(aq) + 2H^+ + 2e^- \rightleftharpoons HAsO_2(aq) + 2H_2O$	+0.560
Mn(VII)–Mn(VI)	$MnO_4^- + e^- \rightleftharpoons MnO_4^{2-}$	+0.564
Sb(V)–Sb(III)	$Sb_2O_5(s) + 6H^+ + 4e^- \rightleftharpoons 2SbO^+ + 3H_2O$	+0.581
Tc(IV)–Tc(II)	$TcO_2(s) + 4H^+ + 2e^- \rightleftharpoons Tc^{2+} + 2H_2O$	+0.6
Hg(I)–Hg(0)	$Hg_2SO_4(s) + 2e^- \rightleftharpoons 2Hg + SO_4^{2-}$	+0.6151
U(V)–U(IV)	$UO_2^+ + 4H^+ + e^- \rightleftharpoons U^{4+} + 2H_2O$	+0.62
Cu(II)–Cu(I)	$Cu^{2+} + Br^- + e^- \rightleftharpoons CuBr$	+0.640
Po(II)–Po(0)	$Po^{2+} + 2e^- \rightleftharpoons Po$	+0.65
Pt(IV)–Pt(II)	$PtCl_6^{2-} + 2e^- \rightleftharpoons PtCl_4^{2-} + 2Cl^-$	+0.68
O(0)–O(–1)	$O_2(g) + 2H^+ + 2e^- \rightleftharpoons H_2O_2(aq)$	+0.6824
Tc(VII)–Tc(IV)	$TcO_4^- + 4H^+ + 3e^- \rightleftharpoons TcO_2(s) + 2H_2O$	+0.7
N(II)–N(I)	$2NO + 2H^+ + 2e^- \rightleftharpoons H_2N_2O_2$	+0.712
Pt(II)–Pt(0)	$PtCl_4^{2-} + 2e^- \rightleftharpoons Pt + 4Cl^-$	+0.73
Se(IV)–Se(0)	$H_2SeO_3(aq) + 4H^+ + 4e^- \rightleftharpoons Se(gray) + 3H_2O$	+0.740
Np(V)–Np(IV)	$NpO_2^+ + 4H^+ + e^- \rightleftharpoons Np^{4+} + 2H_2O$	+0.75
NCS(0)–NCS(–I)	$(NCS)_2 + 2e^- \rightleftharpoons 2NCS^-$	+0.77
Fe(III)–Fe(II)	$Fe^{3+} + e^- \rightleftharpoons Fe^{2+}$	+0.771
Hg(I)–Hg(0)	$Hg_2^{2+} + 2e^- \rightleftharpoons 2Hg$	+0.788
Ag(I)–Ag(0)	$Ag^+ + e^- \rightleftharpoons Ag$	+0.7991
Po(IV)–Po(II)	$PoO_2(s) + 4H^+ + 2e^- \rightleftharpoons Po^{2+} + 2H_2O$	+0.80
Rh(III)–Rh(0)	$Rh^{3+} + 3e^- \rightleftharpoons Rh$	+0.80
N(V)–N(IV)	$2NO_3^- + 4H^+ + 2e^- \rightleftharpoons N_2O_4(g) + 2H_2O$	+0.803
Os(VIII)–Os(0)	$OsO_4(s, yellow) + 8H^+ + 8e^- \rightleftharpoons Os + 4H_2O$	+0.85
N(III)–N(I)	$2HNO_2 + 4H^+ + 4e^- \rightleftharpoons H_2N_2O_2 + 2H_2O$	+0.86
Cu(II)–Cu(I)	$Cu^{2+} + I^- + e^- \rightleftharpoons CuI(s)$	+0.86
Au(III)–Au(0)	$AuBr_4^- + 3e^- \rightleftharpoons Au + 4Br^-$	+0.87 (60°C)
Hg(II)–Hg(I)	$2Hg^{2+} + 2e^- \rightleftharpoons Hg_2^{2+}$	+0.920

Appendix IV (*Continued*)

Couple	Equation for Half-Reaction	E^0 (V)
Pu(VI)–Pu(V)	$PuO_2^{2+} + e^- \rightleftharpoons PuO_2^+$	$+0.93$
N(V)–N(III)	$NO_3^- + 3H^+ + 2e^- \rightleftharpoons HNO_2 + H_2O$	$+0.94$
Au(I)–Au(0)	$AuBr_2^- + e^- \rightleftharpoons Au + 2Br^-$	$+0.956$
N(V)–N(II)	$NO_3^- + 4H^+ + 3e^- \rightleftharpoons NO + 2H_2O$	$+0.96$
Pd(II)–Pd(0)	$Pd^{2+} + 2e^- \rightleftharpoons Pd$	$+0.987$
N(III)–N(II)	$HNO_2 + H^+ + e^- \rightleftharpoons NO + H_2O$	$+1.00$
Au(III)–Au(0)	$AuCl_4^- + 3e^- \rightleftharpoons Au + 4Cl^-$	$+1.00$
Te(VI)–Te(IV)	$H_6TeO_6(s) + 2H^+ + 2e^- \rightleftharpoons TeO_2(s) + 4H_2O$	$+1.02$
N(IV)–N(II)	$N_2O_4 + 4H^+ + 4e^- \rightleftharpoons 2NO + 2H_2O$	$+1.03$
Pu(VI)–Pu(IV)	$PuO_2^{2+} + 4H^+ + 2e^- \rightleftharpoons Pu^{4+} + 2H_2O$	$+1.04$
Br(0)–Br(−1)	$Br_2(l) + 2e^- \rightleftharpoons 2Br^-$	$+1.0652$
N(IV)–N(III)	$N_2O_4 + 2H^+ + 2e^- \rightleftharpoons 2HNO_2$	$+1.07$
Br(0)–Br(−1)	$Br_2(aq) + 2e^- \rightleftharpoons 2Br^-$	$+1.087$
Se(VI)–Se(IV)	$SeO_4^{2-} + 4H^+ + 2e^- \rightleftharpoons H_2SeO_3 + H_2O$	$+1.15$
Np(VI)–Np(V)	$NpO_2^{2+} + e^- \rightleftharpoons NpO_2^+$	$+1.15$
O(0)–O(−II)	$O_2 + 4H^+ + 4e^- \rightleftharpoons 2H_2O(g)$	$+1.185$
Cl(VII)–Cl(V)	$ClO_4^- + 2H^+ + 2e^- \rightleftharpoons ClO_3^- + H_2O$	$+1.19$
I(V)–I(0)	$IO_3^- + 6H^+ + 5e^- \rightleftharpoons \frac{1}{2}I_2 + 3H_2O$	$+1.195$
Pt(II)–Pt(0)	$Pt^{2+} + 2e^- \rightleftharpoons Pt$	ca. 1.2
Cl(V)–Cl(III)	$ClO_3^- + 3H^+ + 2e^- \rightleftharpoons HClO_2 + H_2O$	$+1.21$
O(0)–O(−II)	$O_2 + 4H^+ + 4e^- \rightleftharpoons 2H_2O(l)$	$+1.229$
Mn(IV)–Mn(II)	$MnO_2(s) + 4H^+ + 2e^- \rightleftharpoons Mn^{2+} + 2H_2O$	$+1.23$
Tl(III)–Tl(I)	$Tl^{3+} + 2e^- \rightleftharpoons Tl^+$	$+1.25$
N(−II)–N(−III)	$N_2H_5^+ + 3H^+ + 2e^- \rightleftharpoons 2NH_4^+$	$+1.275$
Pd(IV)–Pd(II)	$PdCl_6^{2-} + 2e^- \rightleftharpoons PdCl_4^{2-} + 2Cl^-$	$+1.288$
N(III)–N(I)	$2HNO_2(aq) + 4H^+ + 4e^- \rightleftharpoons N_2O(g) + 3H_2O$	$+1.29$

796

Couple	Half-reaction	E°
Cr(VI)–Cr(III)	$Cr_2O_7^{2-} + 14H^+ + 6e^- \rightleftharpoons 2Cr^{3+} + 7H_2O$	+1.33
N(−I)–N(−III)	$NH_3OH^+ + 2H^+ + 2e^- \rightleftharpoons NH_4^+ + H_2O$	+1.35
Cl(0)–Cl(−I)	$Cl_2 + 2e^- \rightleftharpoons 2Cl^-$	+1.3595
N(−I)–N(−II)	$2NH_3OH^+ + H^+ + 2e^- \rightleftharpoons N_2H_5^+ + 2H_2O$	+1.42
Au(III)–Au(0)	$Au(OH)_3(s) + 3H^+ + 3e^- \rightleftharpoons Au + 3H_2O$	+1.45
I(I)–I(0)	$HIO + H^+ + e^- \rightleftharpoons \frac{1}{2}I_2 + H_2O$	+1.45
Pb(IV)–Pb(II)	$PbO_2(s) + 4H^+ + 2e^- \rightleftharpoons Pb^{2+} + 2H_2O$	+1.455
Au(III)–Au(0)	$Au^{3+} + 3e^- \rightleftharpoons Au$	+1.498
Mn(III)–Mn(II)	$Mn^{3+} + e^- \rightleftharpoons Mn^{2+}$	+1.51
Mn(VII)–Mn(II)	$MnO_4^- + 8H^+ + 5e^- \rightleftharpoons Mn^{2+} + 4H_2O$	+1.51
Br(V)–Br(0)	$BrO_3^- + 6H^+ + 5e^- \rightleftharpoons \frac{1}{2}Br_2(l) + 3H_2O$	+1.52
Po(VI)–Po(IV)	$PoO_3(s) + 2H^+ + 2e^- \rightleftharpoons PoO_2(s) + H_2O$	+1.52
Br(I)–Br(0)	$HBrO + H^+ + e^- \rightleftharpoons \frac{1}{2}Br_2(l) + H_2O$	+1.595
I(VII)–I(V)	$H_5IO_6 + H^+ + 2e^- \rightleftharpoons IO_3^- + 3H_2O$	+1.601
Ce(IV)–Ce(III)	$Ce^{4+} + e^- \rightleftharpoons Ce^{3+}$	+1.61
Cl(I)–Cl(0)	$HClO + H^+ + e^- \rightleftharpoons \frac{1}{2}Cl_2 + H_2O$	+1.63
Am(VI)–Am(V)	$AmO_2^{2+} + e^- \rightleftharpoons AmO_2^+$	+1.639
Cl(III)–Cl(I)	$HClO_2 + 2H^+ + 2e^- \rightleftharpoons HClO + H_2O$	+1.645
Ni(IV)–Ni(0)	$NiO_2(s) + 4H^+ + 2e^- \rightleftharpoons Ni^{2+} + 2H_2O$	+1.678
Pb(IV)–Pb(II)	$PbO_2(s) + SO_4^{2-} + 4H^+ + 2e^- \rightleftharpoons PbSO_4(s) + 2H_2O$	+1.682
Au(I)–Au(0)	$Au^+ + e^- \rightleftharpoons Au$	+1.691
Am(VI)–Am(III)	$AmO_2^{2+} + 4H^+ + 3e^- \rightleftharpoons Am^{3+} + 2H_2O$	+1.694
O(−I)–O(−II)	$H_2O_2 + 2H^+ + 2e^- \rightleftharpoons 2H_2O$	+1.776
Xe(VI)–Xe(0)	$XeO_3 + 6H^+ + 6e^- \rightleftharpoons Xe + 3H_2O$	+1.8
Co(III)–Co(II)	$Co^{3+} + e^- \rightleftharpoons Co^{2+}$	+1.808
Ag(II)–Ag(I)	$Ag^{2+} + e^- \rightleftharpoons Ag^+$	+1.980
O(−I)–O(−II)	$S_2O_8^{2-} + 2e^- \rightleftharpoons 2SO_4^{2-}$	+2.01
O(0)–O(−II)	$O_3 + 2H^+ + 2e^- \rightleftharpoons O_2 + H_2O$	+2.07
O(II)–O(−II)	$F_2O + 2H^+ + 4e^- \rightleftharpoons 2F^- + H_2O$	+2.15
Am(IV)–Am(III)	$Am^{4+} + e^- \rightleftharpoons Am^{3+}$	+2.18
Fe(VI)–Fe(III)	$FeO_4^{2-} + 8H^+ + 3e^- \rightleftharpoons Fe^{3+} + 4H_2O$	+2.20

Appendix IV (*Continued*)

Couple	Equation for Half-Reaction	E^0 (V)
O(0)–O(−II)	$O(g) + 2\,H^+ + 2\,e^- \rightleftharpoons H_2O$	+2.422
N(I)–N(0)	$H_2N_2O_2 + 2\,H^+ + 2\,e^- \rightleftharpoons N_2 + 2\,H_2O$	+2.65
Pr(IV)–Pr(III)	$Pr^{4+} + e^- \rightleftharpoons Pr^{3+}$	+2.86
F(0)–F(−I)	$F_2(g) + 2\,e^- \rightleftharpoons 2\,F^-$	+2.87
Xe(VIII)–Xe(VI)	$H_4XeO_6 + 2\,H^+ + 2\,e^- \rightleftharpoons XeO_3 + 3\,H_2O$	+3.0
F(0)–F(−I)	$F_2(g) + 2\,H^+ + 2\,e^- \rightleftharpoons 2\,HF$	+3.06

ALKALINE MEDIA

Couple	Equation for Half-Reaction	E^0 (V)
Li(I)–Li(0)	$Li^+ + e^- \rightleftharpoons Li$	−3.045
Ca(II)–Ca(0)	$Ca(OH)_2(s) + 2\,e^- \rightleftharpoons Ca + 2\,OH^-$	−3.02
Ba(II)–Ba(0)	$Ba(OH)_2 \cdot 8\,H_2O(s) + 2\,e^- \rightleftharpoons Ba + 2\,OH^- + 8\,H_2O$	−2.99
H(I)–H(0)	$H_2O + e^- \rightleftharpoons H(g) + OH^-$	−2.9345
Rb(I)–Rb(0)	$Rb^+ + e^- \rightleftharpoons Rb$	−2.925
K(I)–K(0)	$K^+ + e^- \rightleftharpoons K$	−2.925
Cs(I)–Cs(0)	$Cs^+ + e^- \rightleftharpoons Cs$	−2.923
La(III)–La(0)	$La(OH)_3(s) + 3\,e^- \rightleftharpoons La + 3\,OH^-$	−2.90
Sr(II)–Sr(0)	$Sr(OH)_2(s) + 2\,e^- \rightleftharpoons Sr + 2\,OH^-$	−2.88
Ce(III)–Ce(0)	$Ce(OH)_3(s) + 3\,e^- \rightleftharpoons Ce + 3\,OH^-$	−2.87
Pr(III)–Pr(0)	$Pr(OH)_3(s) + 3\,e^- \rightleftharpoons Pr + 3\,OH^-$	−2.85
Nd(III)–Nd(0)	$Nd(OH)_3(s) + 3\,e^- \rightleftharpoons Nd + 3\,OH^-$	−2.84
Pm(III)–Pm(0)	$Pm(OH)_3(s) + 3\,e^- \rightleftharpoons Pm + 3\,OH^-$	−2.84
Sm(III)–Sm(0)	$Sm(OH)_3(s) + 3\,e^- \rightleftharpoons Sm + 3\,OH^-$	−2.83
Eu(III)–Eu(0)	$Eu(OH)_3(s) + 3\,e^- \rightleftharpoons Eu + 3\,OH^-$	−2.83
Gd(III)–Gd(0)	$Gd(OH)_3(s) + 3\,e^- \rightleftharpoons Gd + 3\,OH^-$	−2.82
Y(III)–Y(0)	$Y(OH)_3(s) + 3\,e^- \rightleftharpoons Y + 3\,OH^-$	−2.81
Tb(III)–Tb(0)	$Tb(OH)_3(s) + 3\,e^- \rightleftharpoons Tb + 3\,OH^-$	−2.79
Dy(III)–Dy(0)	$Dy(OH)_3(s) + 3\,e^- \rightleftharpoons Dy + 3\,OH^-$	−2.78

Couple	Half-reaction	E° (V)
Ho(III)–Ho(0)	$Ho(OH)_3(s) + 3e^- \rightleftharpoons Ho + 3OH^-$	−2.77
Er(III)–Er(0)	$Er(OH)_3(s) + 3e^- \rightleftharpoons Er + 3OH^-$	−2.75
Tm(III)–Tm(0)	$Tm(OH)_3(s) + 3e^- \rightleftharpoons Tm + 3OH^-$	−2.74
Yb(III)–Yb(0)	$Yb(OH)_3(s) + 3e^- \rightleftharpoons Yb + 3OH^-$	−2.73
Lu(III)–Lu(0)	$Lu(OH)_3(s) + 3e^- \rightleftharpoons Lu + 3OH^-$	−2.72
Na(I)–Na(0)	$Na^+ + e^- \rightleftharpoons Na$	−2.714
Mg(II)–Mg(0)	$Mg(OH)_2(s) + 2e^- \rightleftharpoons Mg + 2OH^-$	−2.690
Be(II)–Be(0)	$BeO(s) + H_2O + 2e^- \rightleftharpoons Be + 2OH^-$	−2.613
Sc(III)–Sc(0)	$Sc(OH)_3(s) + 3e^- \rightleftharpoons Sc + 3OH^-$	−2.61
Hf(IV)–Hf(0)	$HfO(OH)_2(s) + H_2O + 4e^- \rightleftharpoons Hf + 4OH^-$	−2.50
Th(IV)–Th(0)	$Th(OH)_4(s) + 4e^- \rightleftharpoons Th + 4OH^-$	−2.48
Pu(III)–Pu(0)	$Pu(OH)_3(s) + 3e^- \rightleftharpoons Pu + 3OH^-$	−2.42
U(IV)–U(0)	$UO_2(s) + 2H_2O + 4e^- \rightleftharpoons U + 4OH^-$	−2.39
Zr(IV)–Zr(0)	$ZrO(OH)_2(s) + H_2O + 4e^- \rightleftharpoons Zr + 4OH^-$	−2.36
	$H_2AlO_3^- + H_2O + 3e^- \rightleftharpoons Al + 4OH^-$	−2.33
Al(III)–Al(0)	$Al(OH)_3(s) + 3e^- \rightleftharpoons Al + 3OH^-$	−2.30
U(IV)–U(III)	$U(OH)_4(s) + e^- \rightleftharpoons U(OH)_3 + OH^-$	−2.20
U(III)–U(0)	$U(OH)_3(s) + 3e^- \rightleftharpoons U + 3OH^-$	−2.17
P(I)–P(0)	$H_2PO_2^- + e^- \rightleftharpoons P + 2OH^-$	−2.05
B(III)–B(0)	$H_2BO_3^- + H_2O + 3e^- \rightleftharpoons B + 4OH^-$	−1.79
Si(IV)–Si(0)	$SiO_3^{2-} + 3H_2O + 4e^- \rightleftharpoons Si + 6OH^-$	−1.697
U(VI)–U(IV)	$Na_2UO_4(s) + 4H_2O + 2e^- \rightleftharpoons U(OH)_4(s) + 2Na^+ + 4OH^-$	−1.618
P(III)–P(I)	$HPO_3^{2-} + 2H_2O + 2e^- \rightleftharpoons H_2PO_2^- + 3OH^-$	−1.565
Mn(II)–Mn(0)	$Mn(OH)_2(s) + 2e^- \rightleftharpoons Mn + 2OH^-$	−1.55
	$MnCO_3(s) + 2e^- \rightleftharpoons Mn + CO_3^{2-}$	−1.50
Zn(II)–Zn(0)	$ZnS(s) + 2e^- \rightleftharpoons Zn + S^{2-}$	−1.405
Cr(III)–Cr(0)	$Cr(OH)_3(s) + 3e^- \rightleftharpoons Cr + 3OH^-$	−1.34
	$CrO_2^- + 2H_2O + 3e^- \rightleftharpoons Cr + 4OH^-$	−1.27
Zn(II)–Zn(0)	$Zn(OH)_2(s) + 2e^- \rightleftharpoons Zn + 2OH^-$	−1.245
Ga(III)–Ga(0)	$H_2GaO_3^- + H_2O + 3e^- \rightleftharpoons Ga + 4OH^-$	−1.219
Zn(II)–Zn(0)	$ZnO_2^{2-} + 2H_2O + 2e^- \rightleftharpoons Zn + 4OH^-$	−1.215

Appendix IV (*Continued*)

Couple	Equation for Half-Reaction	E^0 (V)
Cd(II)–Cd(0)	$CdS(s) + 2e^- \rightleftharpoons Cd + 2\,OH^-$	-1.175
Te(0)–Te($-$II)	$Te + 2e^- \rightleftharpoons Te^{2-}$	-1.143
P(V)–P(III)	$PO_4^{3-} + 2\,H_2O + 2e^- \rightleftharpoons HPO_3^{2-} + 3\,OH^-$	-1.12
Zn(II)–Zn(0)	$ZnCO_3(s) + 2e^- \rightleftharpoons Zn + CO_3^{2-}$	-1.06
W(VI)–W(0)	$WO_4^{2-} + 4\,H_2O + 6e^- \rightleftharpoons W + 8\,OH^-$	-1.05
Mo(VI)–Mo(0)	$MoO_4^{2-} + 4\,H_2O + 6e^- \rightleftharpoons Mo + 8\,OH^-$	-1.05
Ge(IV)–Ge(0)	$HGeO_3^- + 2\,H_2O + 4e^- \rightleftharpoons Ge + 5\,OH^-$	-1.03
In(III)–In(0)	$In(OH)_3(s) + 3e^- \rightleftharpoons In + 3\,OH^-$	-1.00
Pu(IV)–Pu(0)	$Pu(OH)_4(s) + 4e^- \rightleftharpoons Pu + 4\,OH^-$	-0.963
Pb(II)–Pb(0)	$PbS(s) + 2e^- \rightleftharpoons Pb + S^{2-}$	-0.93
Sn(IV)–Sn(II)	$Sn(OH)_6^{2-} + 2e^- \rightleftharpoons HSnO_2^- + H_2O + 3\,OH^-$	-0.93
S(VI)–S(IV)	$SO_4^{2-} + H_2O + 2e^- \rightleftharpoons SO_3^{2-} + 2\,OH^-$	-0.93
Se(0)–Se($-$II)	$Se + 2e^- \rightleftharpoons Se^{2-}$	-0.92
Sn(II)–Sn(0)	$HSnO_2^- + H_2O + 2e^- \rightleftharpoons Sn + 3\,OH^-$	-0.909
P(0)–P($-$III)	$P(white) + 3\,H_2O + 3e^- \rightleftharpoons PH_3 + 3\,OH^-$	-0.89
Fe(II)–Fe(0)	$Fe(OH)_2(s) + 2e^- \rightleftharpoons Fe + 2\,OH^-$	-0.877
H(I)–H(0)	$2\,H_2O + 2e^- \rightleftharpoons H_2 + 2\,OH^-$	-0.8281
Cd(II)–Cd(0)	$Cd(OH)_2(s) + 2e^- \rightleftharpoons Cd + 2\,OH^-$	-0.809
Fe(II)–Fe(0)	$FeCO_3(s) + 2e^- \rightleftharpoons Fe + CO_3^{2-}$	-0.756
Cd(II)–Cd(0)	$CdCO_3(s) + 2e^- \rightleftharpoons Cd + CO_3^{2-}$	-0.74
Co(II)–Co(0)	$Co(OH)_2(s) + 2e^- \rightleftharpoons Co + 2\,OH^-$	-0.73
Ni(II)–Ni(0)	$Ni(OH)_2(s) + 2e^- \rightleftharpoons Ni + 2\,OH^-$	-0.72
As(V)–As(III)	$AsO_4^{3-} + 2\,H_2O + 2e^- \rightleftharpoons AsO_2^- + 4\,OH^-$	-0.68
As(III)–As(0)	$AsO_2^- + 2\,H_2O + 3e^- \rightleftharpoons As + 4\,OH^-$	-0.675
Sb(III)–Sb(0)	$SbO_2^- + 2\,H_2O + 3e^- \rightleftharpoons Sb + 4\,OH^-$	-0.66
Co(II)–Co(0)	$CoCO_3(s) + 2e^- \rightleftharpoons Co + CO_3^{2-}$	-0.64

Couple	Half-reaction	E° (V)
Re(VII)–Re(IV)	$ReO_4^- + 2H_2O + 3e^- \rightleftharpoons ReO_2(s) + 4OH^-$	-0.594
Re(VII)–Re(0)	$ReO_4^- + 4H_2O + 7e^- \rightleftharpoons Re + 8OH^-$	-0.584
Pb(II)–Pb(0)	$PbO(red) + H_2O + 2e^- \rightleftharpoons Pb + 2OH^-$	-0.580
Re(IV)–Re(0)	$ReO_2(s) + 2H_2O + 4e^- \rightleftharpoons Re + 4OH^-$	-0.577
Te(IV)–Te(0)	$TeO_3^{2-} + 3H_2O + 4e^- \rightleftharpoons Te + 6OH^-$	-0.57
Fe(III)–Fe(II)	$Fe(OH)_3(s) + e^- \rightleftharpoons Fe(OH)_2(s) + OH^-$	-0.56
O(0)–O($-\tfrac{1}{2}$)	$O_2 + e^- \rightleftharpoons O_2^-$	-0.563
Pb(II)–Pb(0)	$HPbO_2^- + H_2O + 2e^- \rightleftharpoons Pb + 3OH^-$	-0.540
Pb(II)–Pb(0)	$PbCO_3(s) + 2e^- \rightleftharpoons Pb + CO_3^{2-}$	-0.509
Po(IV)–Po(0)	$PoO_3^{2-} + 3H_2O + 4e^- \rightleftharpoons Po + 6OH^-$	-0.49
Bi(III)–Bi(0)	$Bi_2O_3(s) + 3H_2O + 6e^- \rightleftharpoons 2Bi + 6OH^-$	-0.46
Ni(II)–Ni(0)	$NiCO_3(s) + 2e^- \rightleftharpoons Ni + CO_3^{2-}$	-0.45
S(0)–S($-$II)	$S(s) + 2e^- \rightleftharpoons S^{2-}$	-0.447
Se(IV)–Se(0)	$SeO_3^{2-} + 3H_2O + 4e^- \rightleftharpoons Se + 6OH^-$	-0.366
Cu(I)–Cu(0)	$Cu_2O(s) + H_2O + 2e^- \rightleftharpoons 2Cu + 2OH^-$	-0.358
Tl(I)–Tl(0)	$TlOH(s) + e^- \rightleftharpoons Tl + OH^-$	-0.343
Cr(VI)–Cr(III)	$CrO_4^{2-} + 4H_2O + 3e^- \rightleftharpoons Cr(OH)_3(s) + 5OH^-$	-0.13
Cu(II)–Cu(I)	$2Cu(OH)_2(s) + 2e^- \rightleftharpoons Cu_2O(s) + H_2O + 2OH^-$	-0.080
O(0)–O($-$I)	$O_2 + H_2O + 2e^- \rightleftharpoons HO_2^- + OH^-$	-0.076
Tl(III)–Tl(I)	$Tl(OH)_3(s) + 2e^- \rightleftharpoons TlOH + 2OH^-$	-0.05
Mn(IV)–Mn(II)	$MnO_2(s) + 2H_2O + 2e^- \rightleftharpoons Mn(OH)_2(s) + 2OH^-$	-0.05
N(V)–N(III)	$NO_3^- + H_2O + 2e^- \rightleftharpoons NO_2^- + 2OH^-$	$+0.01$
Rh(III)–Rh(0)	$Rh_2O_3(s) + 3H_2O + 6e^- \rightleftharpoons 2Rh + 6OH^-$	$+0.04$
Se(VI)–Se(IV)	$SeO_4^{2-} + H_2O + 2e^- \rightleftharpoons SeO_3^{2-} + 2OH^-$	$+0.05$
Pd(II)–Pd(0)	$Pd(OH)_2(s) + 2e^- \rightleftharpoons Pd + 2OH^-$	$+0.07$
Hg(II)–Hg(0)	$HgO(red) + H_2O + 2e^- \rightleftharpoons Hg + 2OH^-$	$+0.098$
Ir(III)–Ir(0)	$Ir_2O_3(s) + 3H_2O + 6e^- \rightleftharpoons 2Ir + 6OH^-$	$+0.098$
Pb(IV)–Pb(II)	$PbO_2(s) + H_2O + 2e^- \rightleftharpoons PbO(red) + 2OH^-$	$+0.247$
I(V)–I($-$I)	$IO_3^- + 3H_2O + 6e^- \rightleftharpoons I^- + 6OH^-$	$+0.26$
Cl(V)–Cl(III)	$ClO_3^- + H_2O + 2e^- \rightleftharpoons ClO_2^- + 2OH^-$	$+0.33$
Ag(I)–Ag(0)	$Ag_2O(s) + H_2O + 2e^- \rightleftharpoons 2Ag + 2OH^-$	$+0.345$

Appendix IV (Continued)

Couple	Equation for Half-Reaction	E^0 (V)
Cl(VII)–Cl(V)	$ClO_4^- + H_2O + 2e^- \rightleftharpoons ClO_3^- + 2OH^-$	$+0.36$
Te(VI)–Te(IV)	$TeO_4^{2-} + H_2O + 2e^- \rightleftharpoons TeO_3^{2-} + 2OH^-$	ca. $+0.4$
O(0)–O(–II)	$O_2 + 2H_2O + 4e^- \rightleftharpoons 4OH^-$	$+0.401$
Ag(I)–Ag(0)	$Ag_2CO_3(s) + 2e^- \rightleftharpoons 2Ag + CO_3^{2-}$	$+0.47$
I(I)–I(–I)	$IO^- + H_2O + 2e^- \rightleftharpoons I^- + 2OH^-$	$+0.485$
Ni(IV)–Ni(II)	$NiO_2(s) + 2H_2O + 2e^- \rightleftharpoons Ni(OH)_2(s) + 2OH^-$	$+0.490$
Mn(VII)–Mn(IV)	$MnO_4^- + 2H_2O + 3e^- \rightleftharpoons MnO_2(\text{pyrolusite}) + 4OH^-$	$+0.588$
Mn(VI)–Mn(IV)	$MnO_4^{2-} + 2H_2O + 2e^- \rightleftharpoons MnO_2(s) + 4OH^-$	$+0.60$
Ag(II)–Ag(I)	$2AgO(s) + H_2O + 2e^- \rightleftharpoons Ag_2O(s) + 2OH^-$	$+0.607$
Br(V)–Br(–I)	$BrO_3^- + 3H_2O + 6e^- \rightleftharpoons Br^- + 6OH^-$	$+0.61$
Cl(III)–Cl(I)	$ClO_2^- + H_2O + 2e^- \rightleftharpoons ClO^- + 2OH^-$	$+0.66$
I(VII)–I(V)	$H_3IO_6^{2-} + 2e^- \rightleftharpoons IO_3^- + 3OH^-$	$+0.7$
Fe(VI)–Fe(III)	$FeO_4^{2-} + 4H_2O + 3e^- \rightleftharpoons Fe(OH)_3(s) + 5OH^-$	$+0.72$
N(–I)–N(–II)	$2NH_2OH + 2e^- \rightleftharpoons N_2H_4 + 2OH^-$	$+0.73$
Ag(III)–Ag(II)	$Ag_2O_3(s) + H_2O + 2e^- \rightleftharpoons 2AgO(s) + 2OH^-$	$+0.739$
Br(I)–Br(–I)	$BrO^- + H_2O + 2e^- \rightleftharpoons Br^- + 2OH^-$	$+0.761$
O(–I)–O(–II)	$HO_2^- + H_2O + 2e^- \rightleftharpoons 3OH^-$	$+0.878$
Cl(I)–Cl(–I)	$ClO^- + H_2O + 2e^- \rightleftharpoons Cl^- + 2OH^-$	$+0.89$
Xe(VIII)–Xe(VI)	$HXeO_6^{3-} + 2H_2O + 2e^- \rightleftharpoons HXeO_4^- + 4OH^-$	$+0.9$
Xe(VI)–Xe(0)	$HXeO_4^- + 3H_2O + 6e^- \rightleftharpoons Xe + 7OH^-$	$+0.9$
Cl(IV)–Cl(III)	$ClO_2(g) + e^- \rightleftharpoons ClO_2^-$	$+1.16$
	$O_3(g) + H_2O + 2e^- \rightleftharpoons O_2 + 2OH^-$	$+1.24$
O(–I)–O(–II)	$OH(g) + e^- \rightleftharpoons OH^-$	$+2.02$

REFERENCES

1. W. M. Latimer, *The Oxidation States of the Elements and Their Potentials in Aqueous Solution*, 2nd ed., Prentice-Hall, Englewood Cliffs, N.J. (1952). A classic.

2. A. J. de Bethune and N. A. S. Loud, *Standard Aqueous Electrode Potentials and Temperature Coefficients at 25°C.*, Clifford A. Hampel, Skokie, Ill. (1964).

3. M. Pourbaix, *Atlas of Electrochemical Equilibria in Aqueous Solutions*, J. A. Franklin (transl.), Pergamon Press, Oxford (1966).

4. G. Milazzo and S. Caroli, *Tables of Standard Electrode Potentials*, John Wiley & Sons, New York (1978).

AUTHOR INDEX

SUBJECT INDEX

APR '88

DATE DUE		
SEP 28 1993		
NOV 11 1993		